Florencio Zaragoza Dörwald

Organic Synthesis on Solid Phase

Related Titles from Wiley-VCH

Nicolaou, K. C. / Hanko, Rudolf / Hartwig, Wolfgang (Eds.)
Handbook of Combinatorial Chemistry
2002. ISBN 3-527-30509-2

Bannwarth, Willi / Felder, Eduard (Eds.)
Combinatorial Chemistry
2000. ISBN 3-527-30186-0

Kobayashi, Shu / Jørgensen, Karl Anker (Eds.)
Cycloaddition Reactions in Organic Synthesis
2001. ISBN 3-527-30159-3

Yamamoto, Hisashi (Ed.)
Lewis Acids in Organic Synthesis
2000. ISBN 3-527-29579-8

Drauz, Karlheinz / Waldmann, Herbert (Eds.)
Enzyme Catalysis in Organic Synthesis
2002. ISBN 3-527-29949-1

Florencio Zaragoza Dörwald

Organic Synthesis on Solid Phase

Supports, Linkers, Reactions

Second, Completely Revised and
Enlarged Edition

WILEY-VCH

Dr. Florencio Zaragoza Dörwald
Novo Nordisk A/S MedChem Research
Novo Nordisk Park
DK-2760 Måløv
Denmark

Library of Congress Card No. applied for
A catalogue record for this book is available from the British Library

Die Deutsche Bibliothek – CIP Cataloguing-in-Publication-Data
A catalogue record for this publication is available from Die Deutsche Bibliothek

© Wiley-VCH Verlag GmbH, Weinheim, 2002

ISBN 3-527-30603-X

Printed on acid-free paper

Composition: Kühn & Weyh, Freiburg
Printing: betzdruck GmbH, Darmstadt
Bookbinding: Litges & Dopf, Heppenheim

Printed in the Federal Republic of Germany

Preface to the Second Edition

We are currently witnessing an explosive growth of solid-phase synthetic chemistry. In the years 1999 and 2000 approximately 700 articles were published, describing new synthetic sequences performed on insoluble supports. Almost all important organic reactions have now been carried out on solid phase, and the limits of solid-phase organic chemistry are becoming increasingly clear.

Encouraged by the success of the first edition of *Organic Synthesis on Solid Phase*, and by the many positive comments from the readers, the original manuscript (finished in October 1999) has been exhaustively updated. The resulting new edition covers the literature up to December 2001, which has led to a significant increase in the number of references and examples in most chapters. For instance, Chapter 3 (Linkers) has grown from 608 to 852 references, and Chapter 15 (Heterocycles) from 288 to 454 references. Despite this increased number of references, the new information could be accommodated without the need of new chapters, and it should still be easy to locate this information quickly with the aid of the index and the table of contents. I hope that this revised version of *Organic Synthesis on Solid Phase* will become an indispensable tool for all chemists interested in solid-phase organic synthesis.

Ballerup, Denmark, January 2002   Florencio Zaragoza Dörwald

Preface to the First Edition

Although the concept of performing organic synthesis on insoluble supports is almost 40 years old, there has been rapid growth of this research area in recent years. It was probably the idea of 'parallel synthesis', proposed by several groups in the late 1980s, which spurred renewed interest in solid-phase chemistry. Parallel synthesis enables the rapid production of large numbers of compounds, and is therefore of critical importance to all organizations which depend on discovering new, patentable substances. In particular the implementation of robotic high-throughput screening, which enables the testing of thousands of compounds per day, has originated the need for comparably fast compound production. Solid-phase synthesis is well suited to perform reactions in parallel, because it readily enables the automated performance of multistep synthetic sequences. Accordingly, solid-phase synthesis is becoming an increasingly important tool for the synthetic chemist, as the trend towards automation and miniaturization of synthesis continues.

My main motivation for writing this book was that my colleagues and I required a practical guide to solid-phase chemistry, in which clear-cut answers to specific questions could be readily found. Thus, the aim of this book is to give the reader a well-structured and exhaustive collection of synthetic procedures realizable on insoluble supports. The subject has been organized according to the type of product resulting from a given synthesis. All preparations or types of linker for a specific type of product should be easy to find in this book, either by using the index or with the aid of the table of contents.

In most chapters illustrative examples are listed in tables, and the precise reaction conditions are given as reported by the authors. If closely related examples have been published, these are listed under the remark 'see also'. These references need be consulted, however, only if more detailed information about a specific reaction is required.

The composition of mixtures of solvents or reagents has been specified by *volume* ratios throughout the book (e.g. TFA/DCM 1:1). Yields are given for cleavage reactions only. These yields often refer, however, not only to the cleavage reaction but to several synthetic steps before cleavage. Yields for the conversion of one support-bound intermediate into another are not given, because such yields cannot generally be accurately determined. In many of the cleavage reactions reported, products were isolated as salts (e.g. trifluoroacetates). To keep the tables more comprehensible, counter-ions have generally been omitted, and the products have been sketched in their uncharged form.

I would like to thank my colleagues and supervisors at Novo Nordisk A/S, in particular Jesper Lau and Behrend F. Lundt, for their steadfast support and motivation. Thanks are also due to Marie Grimstrup, Ulrich Sensfuß, Kilian W. Conde-Frieboes, and Bernd Peschke, who have been kind enough to read various sections of the manuscript and to give me valuable suggestions. I wish to thank also Kjeld Madsen, Nils Langeland Johansen, and Leif Christensen for helpful discussions about solid-phase peptide synthesis. Special thanks are due to Dr Anette Eckerle and to the other editors at Wiley-VCH Verlag GmbH for their assistance during the preparation of the manuscript and for the production of the finished book.

Ballerup, Denmark, October 1999 Florencio Zaragoza Dörwald

Contents

3   Linkers for Solid-Phase Organic Synthesis   39

6   Preparation of Alkyl and Aryl Halides   205

7   Preparation of Alcohols and Ethers   213

8   Preparation of Sulfur Compounds   239

9   Preparation of Organoselenium Compounds   259

10   Preparation of Nitrogen Compounds   263

Glossary and Abbreviations

Ac	acetyl, MeCO
acac	pentane-2,4-dione
Acm	acetamidomethyl, MeCONH–CH$_2$
ADDP	azodicarboxylic acid dipiperidide
Ade	adenine
Adpoc	1-(1-adamantyl)-1-methylethoxycarbonyl
agarose	*see* sepharose
AIBN	azobis(isobutyronitrile)
Ala	*S*-alanine
All	allyl
Alloc	allyloxycarbonyl
aq	aqueous
Arg	*S*-arginine
Argogel™	PEG-grafted cross-linked polystyrene
Asn	*S*-asparagine
Asp	*S*-aspartic acid
Azoc	1-(4-phenylazophenyl)-1-methylethoxycarbonyl
(B)	heterocyclic base of nucleotides
BAL	backbone amide linker
9-BBN	9-borabicyclo[3.3.1]nonane
BEMP	2-*tert*-butylimino-2-diethylamino-1,3-dimethylperhydro-1,3-diaza-2-phosphorine
BHT	2,6-di-*tert*-butyl-4-methylphenol
BINAP	2,2′-bis(diphenylphosphino)-1,1′-binaphthyl
Bn	benzyl
Bnpeoc	2,2-bis(4-nitrophenyl)ethoxycarbonyl
Boc	*tert*-butyloxycarbonyl
Bom	benzyloxymethyl
BOP	(1,2,3-benzotriazol-1-yloxy)tris(dimethylamino)phosphonium hexafluorophosphate, [(Me$_2$N)$_3$P–OBt][PF$_6$]
Bop-Cl	*N,N′*-bis(2-oxo-3-oxazolidinyl)phosphinic chloride
Bpoc	1-(4-biphenylyl)-1-methylethoxycarbonyl
BSA	*N,O*-bis(trimethylsilyl)acetimidate
Bsmoc	(1,1-dioxobenzothiophen-2-yl)methoxycarbonyl

Bt	1-benzotriazolyl
BTPP	*tert*-butylimino-tris(pyrrolidino)phosphorane
Bu	butyl
Bz	benzoyl
CAN	cerium(IV) ammonium nitrate, $(NH_4)_2Ce(NO_3)_6$
Cbz	Z, benzyloxycarbonyl, $PhCH_2OCO$
CDI	carbonyldiimidazole
CIP	2-chloro-1,3-dimethyl-2-imidazolinium hexafluorophosphate (Figure 13.6)
coll	collidine, 2,4,6-trimethylpyridine
conc	concentrated
CPG	controlled pore glass
CSA	10-camphorsulfonic acid
Cy	cyclohexyl
Cys	*R*-cysteine
Cyt	cytosine
DABCO	1,4-diazabicyclo[2.2.2]octane
DAST	(diethylamino)sulfur trifluoride
dba	1,5-diphenyl-1,4-pentadien-3-one
DBN	1,5-diazabicyclo[4.3.0]non-5-ene
DBU	1,8-diazabicyclo[5.4.0]undec-5-ene
DCC	*N*,*N*′-dicyclohexylcarbodiimide
DCE	1,2-dichloroethane
DCM	dichloromethane
DCP	1,2-dichloropropane
Dde-OH	2-acetyl-5,5-dimethyl-1,3-cyclohexanedione
DDQ	2,3-dichloro-5,6-dicyano-1,4-benzoquinone
Ddz	1-(3,5-dimethoxyphenyl)-1-methylethoxycarbonyl
de	diastereomeric excess
DEAD	diethyl azodicarboxylate, $EtO_2C–N=N–CO_2Et$
DECP	diethyl cyanophosphonate, $(EtO)_2P(O)CN$
dextran	*see* sephadex
(DHQD)$_2$PHAL	dihydroquinidine 1,4-phthalazinediyl diether
DIAD	diisopropyl azodicarboxylate, $iPrO_2C–N=N–CO_2iPr$
DIBAH	diisobutylaluminum hydride
DIC	diisopropylcarbodiimide
dipamp	1,2-bis[phenyl(2-methoxyphenyl)phosphino]ethane
DIPEA	diisopropylethylamine
DMA	*N*,*N*-dimethylacetamide
DMAD	dimethyl acetylenedicarboxylate, $MeO_2C–C≡C–CO_2Me$
DMAP	4-(dimethylamino)pyridine
DME	1,2-dimethoxyethane, glyme
DMF	*N*,*N*-dimethylformamide
DMI	1,3-dimethylimidazolidin-2-one
DMSO	dimethyl sulfoxide

DMT	4,4′-dimethoxytrityl
DNA	deoxyribonucleic acid
Dnp	2,4-dinitrophenyl
DPPA	diphenylphosphoryl azide, $(PhO)_2P(O)N_3$
dppe	1,2-bis(diphenylphosphino)ethane
dppf	1,1′-bis(diphenylphosphino)ferrocene
dppp	1,3-bis(diphenylphosphino)propane
DTBMP	2,6-di-*tert*-butyl-4-methylpyridine
DTBP	2,6-di-*tert*-butylpyridine
Dts	dithiasuccinoyl
DVB	divinylbenzene (mixture of regioisomers)
EDC	*N*-ethyl-*N*′-[3-(dimethylamino)propyl]carbodiimide hydrochloride
EDT	1,2-ethanedithiol
ee	enantiomeric excess
EE	1-ethoxyethyl
EEDQ	2-ethoxy-1-ethoxycarbonyl-1,2-dihydroquinoline
eq	equivalent
Et	ethyl
Expansin™	cross-linked polyacrylamide
Fmoc	9-fluorenylmethyloxycarbonyl
FT	Fourier transform
Gln	*S*-glutamine
Glu	*S*-glutamic acid
Gly	glycine
glycan	synonym of 'polysaccharide'
Gua	guanosine
HAL linker	hypersensitive acid-labile linker
HATU	7-aza-3-[(dimethyliminium)(dimethylamino)methyl]-1,2,3-benzo-triazol-1-ium-1-olate hexafluorophosphate (Figure 13.6)
HBTU	3-[(dimethyliminium)(dimethylamino)methyl]-1,2,3-benzotriazol-1-ium-1-olate hexafluorophosphate (Figure 13.6)
HDTU	*O*-(4-oxo-3,4-dihydro-1,2,3-benzotriazin-3-yl)-*N,N,N′,N′*-tetra-methyluronium hexafluorophosphate (Figure 13.6)
Hex	hexyl
His	*S*-histidine
Hmb	2-hydroxy-4-methoxybenzyl
HMBA	4-hydroxymethylbenzoic acid linker
HMPA	hexamethylphosphoric triamide, $(Me_2N)_3PO$
HOAt	3-hydroxy-3*H*-[1,2,3]triazolo[4,5-*b*]pyridine, 4-aza-3-hydroxy-benzotriazole
HOBt	1-hydroxybenzotriazole
HODhbt	3-hydroxy-3,4-dihydro-1,2,3-benzotriazin-4-one
HOSu	*N*-hydroxysuccinimide
HPLC	high-performance liquid chromatography
Ile	*S*-isoleucine, 2-amino-3-methylvaleric acid

IPA	2-propanol
*i*Pr	isopropyl
IR	infrared
ivDde-OH	2-(3-methylbutyryl)-5,5-dimethyl-1,3-cyclohexanedione
Kaiser oxime resin	cross-linked polystyrene with 4-nitrobenzophenone oxime linker
Lawesson's reagent	2,4-bis(4-methoxyphenyl)-1,3-dithia-2,4-diphosphetane-2,4-disulfide
LCAA	long chain alkylamine spacer
LC–MS	liquid chromatography coupled with mass spectrometry
LDA	lithium diisopropylamide
Leu	*S*-leucine
Lys	*S*-lysine
Macrosorb™	cross-linked polyacrylamide adsorbed onto kieselguhr
MALDI–TOF MS	matrix-assisted laser desorbtion/ionization time-of-flight mass spectrometry
MAS	magic-angle spinning
MBHA	4-methylbenzhydrylamine
MCPBA	3-chloroperbenzoic acid
Me	methyl
MeOPEG	poly(ethylene glycol) monomethyl ether
Merrifield resin	partially chloromethylated, cross-linked polystyrene
MES	2-(4-morpholino)ethanesulfonic acid
Met	*S*-methionine
MMT	monomethoxytrityl
Mom	methoxymethyl
Mpc	1-(4-methylphenyl)-1-methylethoxycarbonyl
Ms	methanesulfonyl
MS	molecular sieves; mass spectrometry
MSNT	1-(mesitylene-2-sulfonyl)-3-nitro-1,2,4-triazole
Mtr	4-methoxy-2,3,6-trimethyl-1-benzenesulfonyl
Mts	mesitylene-2-sulfonyl, 2,4,6-trimethylbenzene-1-sulfonyl
Multipin™	polymer crowns grafted with various supports
nbd	norbornadiene
NBS	*N*-bromosuccinimide
NCS	*N*-chlorosuccinimide
NIS	*N*-iodosuccinimide
NMM	*N*-methylmorpholine
NMO	*N*-methylmorpholine-*N*-oxide
NMP	*N*-methyl-2-pyrrolidinone
NMR	nuclear magnetic resonance
Nos	4-nitrobenzenesulfonyl
Np	4-nitrophenyl
Npeoc	2-(4-nitrophenyl)ethoxycarbonyl
Nsc	2-(4-nitrophenyl)sulfonylethoxycarbonyl
Oxone™	2 $KHSO_5 \cdot KHSO_4 \cdot K_2SO_4$, potassium peroxymonosulfate

PA	polyacrylamide, polyacrylates
(PA)	PA with linker or spacer
PAL	5-(4-aminomethyl-3,5-dimethoxyphenoxy)valeric acid linker
PAM	4-(hydroxymethyl)phenylacetic acid linker
Pbf	2,2,4,6,7-pentamethyl-2,3-dihydrobenzo[*b*]furan-5-sulfonyl
PE	polyethylene
PEG	poly(ethylene glycol)
(PEG)	PEG with linker or spacer
PEGA	polyacrylamide cross-linked with PEG
Pepsyn™	cross-linked polyacrylamide
Pepsyn K™	cross-linked polyacrylamide adsorbed onto kieselguhr
Ph	phenyl
Phe	*S*-phenylalanine
Pht	phthaloyl
Piv	pivaloyl, 2,2-dimethylpropanoyl
Pmc	2,2,5,7,8-pentamethyl-6-chromanesulfonyl
PNA	peptide nucleic acid
Pol	undefined polymeric support
Polyhipe™	polyacrylamide on macroporous polystyrene
PPTS	pyridinium tosylate
Pr	propyl
Pro	*S*-proline
PS	cross-linked polystyrene
(PS)	PS with linker or spacer
PTFE	polytetrafluoroethylene
Py	pyridine
PyAOP	(4-aza-1,2,3-benzotriazol-3-yloxy)-tris(pyrrolidino)phosphonium hexafluorophosphate (Figure 13.5)
PyBOP	(1,2,3-benzotriazol-1-yloxy)-tris(pyrrolidino)phosphonium hexafluorophosphate (Figure 13.5)
PyBrOP	bromo-tris(pyrrolidino)phosphonium hexafluorophosphate (Figure 13.5)
RAM	Rink amide linker, (2,4-dimethoxyphenyl)(4-alkoxyphenyl)-methylamine
Red-Al™	sodium bis(2-methoxyethoxy)aluminum hydride
salen	bis-imine from ethylenediamine and salicylaldehyde
Sasrin™	cross-linked polystyrene with 4-alkoxy-2-methoxybenzyl alcohol linker
satd	saturated
L-Selectride™	lithium tri(2-butyl)borohydride
sephadex	dextran; a branched glycan consisting of 1,6-α-linked glucopyranose
sepharose	agarose; an unbranched glycan consisting of D-galactose and 3,6-anhydro-L-galactose
Ser	*S*-serine
SG	silica gel

Sieber linker	XAL linker, 3-alkoxy-9*H*-9-xanthenylamine
Su	*N*-succinimidyl
TBAF	tetrabutylammonium fluoride
TBDPS	*tert*-butyldiphenylsilyl
TBS	*tert*-butyldimethylsilyl
TBTU	3-[(dimethyliminium)(dimethylamino)methyl]-1,2,3-benzotria-zol-1-ium-1-olate tetrafluoroborate (Figure 13.6)
TEMPO	2,2,6,6-tetramethyl-1-piperidinyloxy
Tentagel™	PEG-grafted cross-linked polystyrene
Teoc	2-(trimethylsilyl)ethoxycarbonyl
TES	triethylsilane
Tf	trifluoromethanesulfonyl
TFA	trifluoroacetic acid
TFFH	tetramethylfluoroformamidinium hexafluorophosphate, $[(Me_2N)_2CF][PF_6]$
TfOH	triflic acid, trifluoromethanesulfonic acid
TG	PS grafted or copolymerized with PEG (e.g. Tentagel)
(TG)	TG with linker or spacer
THF	tetrahydrofuran
THP	2-tetrahydropyranyl
Thr	*S*-threonine
Thy	thymine
TIPS	triisopropylsilyl
TMAD	*N,N,N',N'*-tetramethyl azodicarboxamide
TMEDA	*N,N,N',N'*-tetramethylethylenediamine
TMG	*N,N,N',N'*-tetramethylguanidine
TMS	trimethylsilyl
TOF–SIMS	time-of-flight secondary ion mass spectrometry
Tol	4-tolyl, 4-methylphenyl
Tosmic	tosylmethyl isocyanide, $TsCH_2NC$
TPAP	tetrapropylammonium perruthenate, $[Pr_4N][RuO_4]$
Tr	trityl
TRIS	tris(hydroxymethyl)aminomethane, $(HOCH_2)_3CNH_2$
Triton™ B	benzyltrimethylammonium hydroxide
Trp	*S*-tryptophan
Ts	tosyl, *p*-toluenesulfonyl
TSTU	*O*-(1-succinimidyl)-*N,N,N',N'*-tetramethyluronium hexafluoro-phosphate, $[(Me_2N)_2C–OSu][PF_6]$
Tyr	*S*-tyrosine
Ura	uracil
UV	ultraviolet
Val	*S*-valine
Wang resin	cross-linked polystyrene with 4-benzyloxybenzyl alcohol linker
XAL linker	Sieber linker, 3-alkoxy-9*H*-9-xanthenylamine
Z	Cbz, benzyloxycarbonyl

Experimental Procedures

1 General Techniques and Analytical Tools for Solid-Phase Organic Synthesis

In this chapter some general techniques for the performance of solid-phase reactions are presented. An overview of analytical tools suitable for the characterization of support-bound intermediates is also given, and strategies for the selection of reagents and reactions for parallel solid-phase synthesis are discussed.

1.1 General Techniques for Performing Syntheses on Insoluble Supports

Solid-phase organic synthesis seduces by its simplicity. Typically, reactions on solid phase are performed by shaking a support with a mixture of solvents and reagents for a given time, filtering the mixture, and washing the support with suitable solvents. Cleavage of the product from the support often yields products of high purity, which can either be used directly or purified further by recrystallization or chromatography.

Most supports for solid-phase synthesis are produced and sold as small spherical particles (beads, 0.04–0.15 mm; see Chapter 2). Beaded polymers can be filtered with glass or polypropylene frits, and, to keep losses of support low, reaction sequences are generally conducted in fritted reactors (Figure 1.1). When handling beaded polymers, mechanical stress on the beads should be avoided (e.g. grinding of the beads with a magnetic stirring bar or with a spatula), because powdered supports tend to clog frits irreversibly.

A typical setup for manual solid-phase synthesis at room temperature is sketched in Figure 1.1. A fritted polypropylene tube (e.g. a disposable syringe) fixed on an orbital shaker can be used as a reactor. Because polypropylene reactors are light, several reactors per shaker can be used. An additional advantage of polypropylene is that this material is almost transparent, and so enables visual inspection of the reaction mixture. Polypropylene cannot, however, be heated, and if heating is required, PTFE or glass reactors should be used. After charging with support, each reactor is closed with a septum and connected via polyethylene tubing to a filtering flask, which serves as a waste container for all the reactors.

The support does not usually need to be dried between different reactions, but only washed sufficiently (2–4 times) with a suitable solvent. Only the last wash before cleavage deserves special attention, in particular if the products are to be used without further purification. Impurities physically adsorbed by the support (DMF, NMP, DIPEA, piperidine) will usually be desorbed during cleavage and contaminate the final products. A safe washing protocol consists of 3–10 short (10–30 s) washings alternately with MeOH and DCM, followed by a long washing cycle with DCP (5–18 h). The cleavage is usually performed by treating the support with the cleavage reagent, followed by filtration and concentration of the filtrate. If gaseous cleavage reagents are used (HF, HCl, NH_3 [1,2]), or if the cleavage reagent is evaporated in the presence of the support (for instance by bubbling nitrogen through the reactor) the product has to be extracted from the support with a suitable solvent (MeOH, DCM, or DMF).

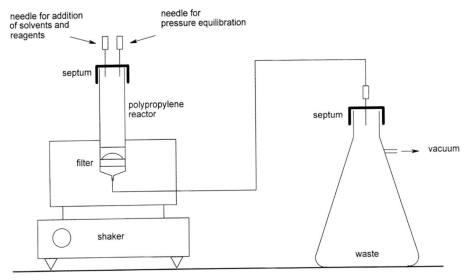

Figure 1.1. Typical setup for manual solid-phase synthesis.

Syntheses on solid phase can, in principle, be performed with standard equipment for organic synthesis. This is, however, only justifiable if large amounts of support are being handled. For small-scale preparations or for the development of solid-phase synthetic methodology, the simplicity of solid-phase synthesis should be fully exploited by conducting experiments in parallel. This requires more systematic planning of the experiments, but will be amply rewarded by the number of results obtained.

Synthetic transformations on solid phase need to be driven to completion, because the purification of intermediates is not possible. For long reaction sequences (e.g. for the synthesis of oligomeric compounds) high yields in each step are essential, but for shorter syntheses lower conversions per step might still lead to acceptably pure final products (Table 1.1). If a given transformation does not proceed smoothly, yields can often be increased by using a reagent in excess or at high concentration, by increasing the reaction temperature, or by repeating the reaction several times.

Table 1.1. Yields of final product as a function of yield per step and number of steps.

Yield per step	Number of steps								
	2	3	4	5	6	7	8	9	10
99%	98%	97%	96%	95%	94%	93%	92%	91%	90%
95%	90%	86%	82%	77%	74%	70%	66%	63%	60%
90%	81%	73%	66%	59%	53%	48%	43%	39%	35%
85%	72%	61%	52%	44%	38%	32%	27%	23%	20%
80%	64%	51%	41%	33%	26%	21%	17%	13%	11%

For the development of new reaction sequences for solid-phase synthesis, the optimum conditions for each step must be identified. This includes determination of the minimum reaction time and temperature, and the minimum amounts of reagents required to obtain sufficiently pure products with a representative selection of different reagents. If reaction times or amounts of reagents are excessive, library production will be costly and inefficient.

Because the characterization of support-bound intermediates is difficult (see below), solid-phase reactions are most conveniently monitored by cleaving the intermediates from the support and analyzing them in solution. Depending on the loading, 5–20 mg of support will usually deliver sufficient material for analysis by HPLC, LC-MS, and NMR, and enable assessment of the outcome of a reaction. Analytical tools that are particularly well suited for the rapid analysis of small samples resulting from solid-phase synthesis include MALDI-TOF MS [3–5], ion-spray MS [6–8], and tandem MS [9]. MALDI-TOF MS can even be used to analyze the product cleaved from a single bead [5], and is therefore well suited to the identification of products synthesized by the mix-and-split method (Section 1.2). The analysis and quantification of small amounts of product can be further facilitated by using supports with two linkers, which enable either release of the desired product or release of the product covalently bound to a dye [10–13], to an isotopic label [11], or to a sensitizer for mass spectrometry [6,14,15] (e.g., product–linker–dye–'analytical linker'–Pol).

The absolute amount of product resulting from solid-phase synthesis can often be readily determined by ^{1}H NMR with an internal standard or, less efficiently, by purification of the product followed by weighing and full characterization. Volatile samples can be cleaved with a calibrated mixture of hexamethyldisiloxane/TFA/CDCl$_3$, from which ^{1}H NMR spectra can be recorded directly [16]. In our laboratory, we use DMSO-d_6 as solvent and DMSO-d_5 as an internal standard for yield determination by ^{1}H NMR. The weight of crude products obtained from solid-phase synthesis is generally unsuitable for estimating the true yield, because these products are often contaminated with significant amounts of salts, which lead to overestimation of the yield.

1.2 Strategies for Parallel Synthesis

Various techniques have been developed that enable the rapid preparation and screening of large numbers of different compounds by parallel solid-phase synthesis. Reactions can, for instance, be conducted in an array of reactors such as that sketched in Figure 1.1. Synthesizers with up to approximately 600 discrete reactors are commercially available; these enable the automated preparation of compound libraries containing up to 0.1 mmol of each compound. Discrete reactors are also well suited for the development and optimization of solid-phase chemistry.

Alternatively, the so-called mix-and-split method [17–22] can be used to prepare mixtures of support particles (beads, paper disks, etc.) or of small portions of support (e.g., 'tea bags') with a well-defined quantity of one discrete compound linked to each portion of support. These compound libraries can be screened either directly on the support, or in solution after partial or total cleavage of the product from the support.

A schematic representation of the mix-and-split method is sketched in Figure 1.2. In the example shown, a support (e.g., Wang resin) is split into four portions and each portion is coupled with one of the four monomers A, B, C, and D. When coupling is complete, the support particles are combined in one flask, mixed, and again divided into four portions. Although each portion contains all four types of support particle, each particle has only a single monomer attached. Now each of the four portions is again coupled with a further pure monomer. The first portion will contain support particles with the products A–A, A–B, A–C, and A–D, but on each particle there is still only one compound. Our library at this stage consists of a total of 16 compounds dis-

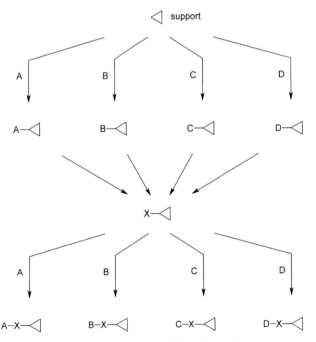

Figure 1.2. The mix-and-split method. X: A, B, C, or D.

tributed into four portions. This mix-and-split protocol can now be repeated several times, thereby, in principle, enabling the preparation of all possible oligomers of the four monomers A, B, C, and D. The amount of compound available depends on the loading of the support particles chosen.

The mix-and-split method is particularly well suited to the preparation of oligonucleotide libraries [23–25] and peptide libraries [3,19,26–30], because even small amounts of these products are sufficient for screening and their unambiguous structural elucidation [31]. Alternatively, if direct structural elucidation is not possible, each support particle might be tagged during each coupling reaction, so that the final tag of a given support particle is unique to the synthesized oligomer attached to it. If tea bags or paper disks are used, tagging can be achieved simply by marking with a pencil [23]. The tagging of single polymer beads has been accomplished by chemical means by the attachment of compounds consisting of a linker and an easily identifiable molecular fragment, such as polychloroarenes [32–35], amines [36–38], carboxylic acids [39], oligonucleotides [40], or fluoroarenes [41]. Chemical tagging generally requires a support with two types of linker: one for the product and one for the tag. Alternatively, the polystyrene support itself may be directly coupled with the tag, for instance by Friedel–Crafts alkylation [38] or by treatment of the resin with diazoketones in the presence of catalytic amounts of rhodium(II) carboxylates [33,34], which presumably leads to C,H-insertion/cyclopropanation of the polystyrene. These rather severe reaction conditions can, however, also lead to a partial derivatization of the resin-bound product. It has been shown that the tagging of polystyrene beads with diazoketones indeed leads to products contaminated with small amounts (< 1%) of tag-containing derivatives [35]. Because the identification of the tags (polyhalobenzenes) by gas chromatography coupled with electron-capture detection is much more sensitive than standard LC-MS or NMR, these tagged by-products can also be used to identify the main product [35].

Polymer beads have also been tagged by treating them after each new diversity-introducing reaction with dye-containing, colloidal silica particles, which can be irreversibly adsorbed on the surface of the beads with the aid of polyelectrolytes such as poly(diallyldimethylammonium chloride) and poly(acrylic acid) [42,43]. Larger portions of support can also be linked to a chip that enables electronic tagging with a radio emitter [44–46].

The deconvolution of compound libraries prepared by the mix-and-split method can be greatly simplified by using polymeric supports that have been labelled with various dyes prior to library synthesis. In this case, the first monomer can be identified from the color or the UV spectrum of each bead. This type of labelling can, for instance, be achieved by partial derivatization of the support with different dyes [47], or by the use of polymers prepared from monomers showing characteristic IR or Raman spectra [48].

Some types of compound are also well suited to being screened as mixtures. This is, for instance, possible for peptides, which often either bind very strongly or not at all to a given receptor. In this instance, the most potent peptide can be identified by deconvolution, and tagging of the support particles is not required. As shown in Figure 1.2, oligomers prepared by the mix-and-split method can contain positions with defined monomers and positions with an undefined monomer. If each mixture of oligomers A–X, B–X, C–X, and D–X is cleaved from the support and screened directly, and only one of the 16 oligomers X–X gives a distinct positive response in an assay, then the mixture containing this oligomer will usually also give a positive response. The active oligomer can now be identified by synthesizing and screening the four pure oligomers contained in the active mixture. Hence, 16 compounds have been screened with only eight assays. For larger oligomers, the increased efficiency achieved by this methodology is much greater than that in the example shown in Figure 1.2, and deconvolution of compound mixtures prepared by the mix-and-split method has been successfully used to screen large peptide libraries containing several million peptides [29,30,49–53]. Mixtures of less selective ligands are, however, not always easy to deconvolute, and for the identification of new, non-peptidic lead structures, parallel synthesis and screening of single compounds is generally the preferred strategy.

1.3 Analytical Methods for Support-Bound Intermediates

As alternatives to the cleavage of intermediates from a support and their characterization in solution, various methods have been developed for analyses of support-bound intermediates. The most common analytical tools include combustion analysis, colorimetric assays for specific functional groups, IR, MALDI-TOF MS, TOF–SIMS, and NMR.

1.3.1 Combustion Analysis

Combustion (elemental) analysis of polymeric supports has mainly been used to determine the amount of halogens, nitrogen, or sulfur present in samples of cross-linked polystyrene (see, e.g., [54]). This information can be used to estimate the loading of a support, or, for example, to verify that the displacement of a halide has proceeded to completion. In solid-phase peptide synthesis, nitrogen determination has been used to estimate the loading of the first amino acid [55].

1.3.2 Colorimetric Assays

One of the first assays used for monitoring the solid-phase synthesis of peptides was the reaction of ninhydrin (**1**, Figure 1.3) with primary amines to yield a blue dye ('Kaiser test', Experimental Procedure 1.1 [56]). This rather sensitive assay enables the detection of even small amounts of primary amines on a support, and can thus be used to monitor acylation reactions. Other reagents suitable for detecting amines include fluorescamine (**2**, Figure 1.3 [57]), 2,4,6-trinitrobenzenesulfonic acid (**3**), and *p*-chloranil (**4**)/RCOMe [58,59]. The latter reagent also gives blue stained beads with secondary amines, e.g. with proline, which cannot be detected with ninhydrin. Various acylating agents containing visually detectable chromophores have also been described, e.g., 4-*N*,*N'*-dimethylaminoazobenzene-4'-isothiocyanate (**5** [60]), malachite green isothiocyanate (**6** [60]), and the dye **7** [61].

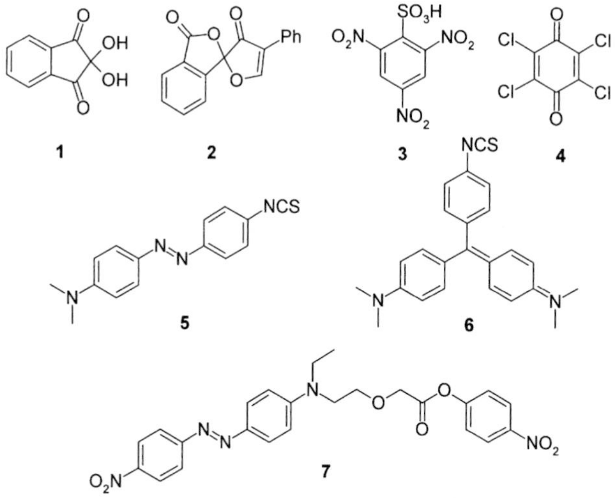

Figure 1.3. Reagents for the detection of resin-bound amines.

Experimental Procedure 1.1: Detection of primary amines by the Kaiser test [58]

Reagent A: ninhydrin (5 g) in EtOH (100 mL).
Reagent B: phenol (80 g) in EtOH (20 mL).
Reagent C: KCN (2 mL, 1 mmol/L in H$_2$O), pyridine (98 mL).
   A few beads of the support are washed with EtOH and mixed with two drops of each of the reagents A, B, and C. The resulting mixture is heated to 120 °C for 4–6 min. If primary amino groups are present on the support the beads turn blue.

Experimental Procedure 1.2: Detection of primary or secondary amines by the chloranil test [59]
   Acetone (0.2 mL; detection of secondary amines) or acetaldehyde (0.2 mL; detection of primary amines) is added to a small sample of resin (approx. 1 mg). To this suspension is added a saturated solution of chloranil (2,3,5,6-tetrachloro-1,4-benzoquinone) in toluene (0.05 mL) and the mixture is shaken at room temperature for 5 min. Blue or green beads indicate the presence of amino groups.

   There are few sensitive assays for the selective detection of resin-bound alcohols. Electrophilic derivatization with a chromophore-containing reagent (see, e.g., [62]) will usually give a positive result with resin-bound amines or thiols as well, and is therefore of limited diagnostic value.

One alcohol-specific assay has been described (Experimental Procedure 1.3), which is suitable for monitoring the esterification of support-bound alcohols [63–65]. In this assay the alcohol is first converted into a tosylate, which is then used to quaternize 4-(4-nitrobenzyl)pyridine. Upon deprotonation, the pyridinium salt absorbs visible light strongly, giving rise to a deep-blue or red color. Other alkylating agents, such as support-bound alkyl halides, as well as reactive aryl halides, Mannich bases, or reactive epoxides, can also give a positive result in this assay. Serine, phenols, carboxylic acids, and amines give a negative result. Tertiary alcohols can be detected by treatment with dichlorodiphenylsilane and then with a dye containing a carboxyl group (e.g., methyl red) [66]. This test also gives a positive result with resin-bound phenols and amines. Resin-bound phenols have been detected by treatment with iron(III) chloride and pyridine [67].

Experimental Procedure 1.3: Detection of primary or secondary aliphatic alcohols [63,64,68]

Reagent A: TsCl (5%) in pyridine/toluene (1:1).
Reagent B: 4-(4-nitrobenzyl)pyridine (2%) in acetone.
Reagent C: Na_2CO_3 (1 mol/L) in water.

A small portion of support (approximately 1 mg of dry resin) is treated with reagent A for 1.5 h and then washed with DCM and toluene. Reagent B is added and the mixture is heated to 110 °C for at least 20 min. A blue or red color upon treatment with reagent C indicates the presence of primary or secondary aliphatic alcohols on the original support.

Thiols can be detected on insoluble supports by treatment with the symmetric disulfides of 4-mercaptonitrobenzene or 5-mercapto-2-nitrobenzoic acid ('Ellman's reagent' [69,70]). These reagents are reduced by thiols to the corresponding thiophenolates, which are intensely colored.

Carboxylates and other negatively charged functional groups can be detected by treatment with a positively charged dye, such as malachite green [71]. Polystyrene-bound aldehydes can be visualized by treatment with 4-methoxybenzaldehyde under acidic conditions [72]. Quantification of resin-bound carbonyl compounds has been achieved by conversion into dansyl hydrazones [73].

The amount of acylable amino, hydroxyl, or thiol groups on a support can also be determined by derivatization of these groups with a chromophore-containing reagent, followed by release of this reagent from the support and photometric quantification. Suitable reagents for the derivatization of primary or secondary amines are 3-[(dimethoxytrityloxy)methyl]-4-nitrophenyl isothiocyanate [74], picric acid [75], and N-Fmoc amino acids. During the deprotection of DMT-protected, support-bound alcohols and thiols, the release of the DMT cation can be monitored photometrically and used to estimate the loading of alcohol. Alternatively, the amount of an acylable resin-bound functional group can be determined by acylation with a precisely quantified excess of a chromophore-containing acylating agent, followed by quantitative determination of the unreacted acylating agent. This has been realized with 9-anthracenecarbonyl cyanide and pyrenyl diazomethane for the quantification of resin-bound hydroxyl and carboxyl groups [76], respectively.

1.3.3 Infrared Spectroscopy

IR spectroscopy is a fast, simple, and cheap method for the qualitative detection of certain functional groups on insoluble supports [77–79]. Dried supports can be used directly to prepare KBr pellets for standard recording of IR spectra [54,80–82]. Newer IR-based techniques, which require much less support material than required for a KBr pellet, include single-bead FT-IR spectroscopy [16,77,83–86], single-bead Raman spectroscopy [87], near-IR multispectral imaging [88], and the simultaneous analysis of several different beads by FT-IR microscopy for analysis of combinatorial libraries [89,90].

IR spectroscopy is not a very sensitive analytical tool and is, therefore, not well suited to the detection of small amounts of material. If, however, intermediates have intense and well-resolved IR absorptions, the progress of their chemical transformation can be followed by IR spectroscopy [83,88,91–93]. Near-infrared spectroscopy, in combination with an acousto-optic tunable filter, can be sufficiently sensitive to enable the on-bead identification of polystyrene-bound di- and tripeptides, even if the peptides have very similar structures (e.g., Leu-Ala-Gly-PS and Val-Ala-Gly-PS) or differ only in their amino acid sequence (e.g., Leu-Val-Gly-PS and Val-Leu-Gly-PS) [94]. Special resins displaying an IR and Raman 'barcode' have been developed, which may facilitate the deconvolution of combinatorial compound libraries prepared by the mix-and-split method [48].

1.3.4 Mass Spectrometry

Photosensitive linkers (see Section 3.1.3) enable the direct analysis of support-bound intermediates by MALDI-TOF MS [14,95,96]. Alternatively, compounds linked to insoluble supports by non-photolabile linkers can be analyzed directly by means of TOF–SIMS [97].

In both MALDI-TOF MS and ion-spray MS, molecules must be positively charged to be detected. To facilitate detection of all types of support-bound compound, linkers incorporating a charged spacer (e.g. a quaternary ammonium salt), which is also released during MALDI, have proven particularly convenient [6,14,15].

1.3.5 Nuclear Magnetic Resonance Spectroscopy

Standard (gel-phase) NMR spectra of polymers usually show significant line broadening, mainly because of chemical shift anisotropy and dipolar coupling [98]. Only nuclei with strong chemical shift dispersion, e.g. ^{13}C [99–106], ^{15}N [107], ^{19}F [108–112], and ^{31}P [113] give sufficiently resolved gel-phase NMR spectra. The resolution of

NMR spectra is improved when the mobility of support-bound molecules increases. Hence, gel-phase NMR of PEG–polystyrene graft supports, for instance, will generally give better spectra than if normal cross-linked polystyrene is used as the support. Gel-phase ^{1}H NMR spectra, even if recorded on well-solvated and flexible supports, are, however, too poorly resolved to be of use for the characterization of support-bound intermediates.

A technique especially developed for recording NMR spectra of support-bound compounds is magic-angle spinning (MAS) NMR. This technique requires a special accessory, which keeps a sample of swollen support spinning at 1–2 kHz at the 'magic angle' relative to the magnetic field [98]. MAS NMR enables the recording of much better resolved spectra than gel-phase NMR, and ^{1}H NMR spectra of high quality can be obtained under optimized conditions [54,114–119]. Approximately 2–6 nmol of product are required for recording spectra in less than 20 min [120]. C,H-correlated and other two-dimensional NMR spectra can also be recorded with MAS NMR [96,121–124], and the direct quantification of resin-bound intermediates using MAS NMR has been reported [125,126]. Special pulse sequences also enable the selective suppression of solvent resonances and other signals resulting from non-resin-bound impurities [127]. Isotopically enriched supports and linkers (e.g. with ^{13}C [128] or ^{19}F [111,112,129,130]) or intermediates [106,131,132] have been used to facilitate the monitoring of solid-phase reactions by either standard gel-phase or MAS NMR.

1.4 Strategies for the Selection of Reactions and Reagents for Parallel Solid-Phase Synthesis

For companies depending on the identification of novel molecular entities (drugs, herbicides, pesticides, catalysts, dyes, flavors, etc.), rapid access to large numbers of different compounds is of critical importance. Large numbers of compounds are generally required when new, patentable molecules with particular properties are being sought, or when the properties of a given lead structure must be optimized. In the evaluation of compound collections purchased or prepared for this purpose, not only the size of the collection but also its quality (i.e. the diversity, purity, chemical stability, mean molecular weight, lipophilicity, etc., of the compounds) is an important issue, which must be carefully considered.

Solid-phase chemistry can readily be automated, and is therefore well suited to parallel, high-throughput compound production. Programmable synthesizers with hundreds of reactors for solid-phase synthesis have become commercially available; these enable the production of large arrays of compounds (one different compound per reactor). This is usually achieved by performing the same synthesis in each reactor, but using different reagents.

Compound libraries designed for the identification of new lead structures should not contain large numbers of closely related compounds (which are likely to have similar properties), but highly diverse compounds with widely different properties. Because compound libraries are prepared using the same reaction sequence for each member of a library, all compounds will contain a repetitive structural element. Diverse libraries will only result if the repetitive element is small and devoid of functional groups capable of strongly influencing the requisite property (e.g., a strong interaction with proteins when searching for leads for drug discovery), and if the synthesis enables the incorporation of widely different side chains (Figure 1.4). For drug discovery, repetitive structural elements containing important pharmacophores (e.g. diketopiperazines, benzodiazepines, peptides) or showing inherently poor pharmacokinetic properties (e.g., peptides, guanidines) should be avoided.

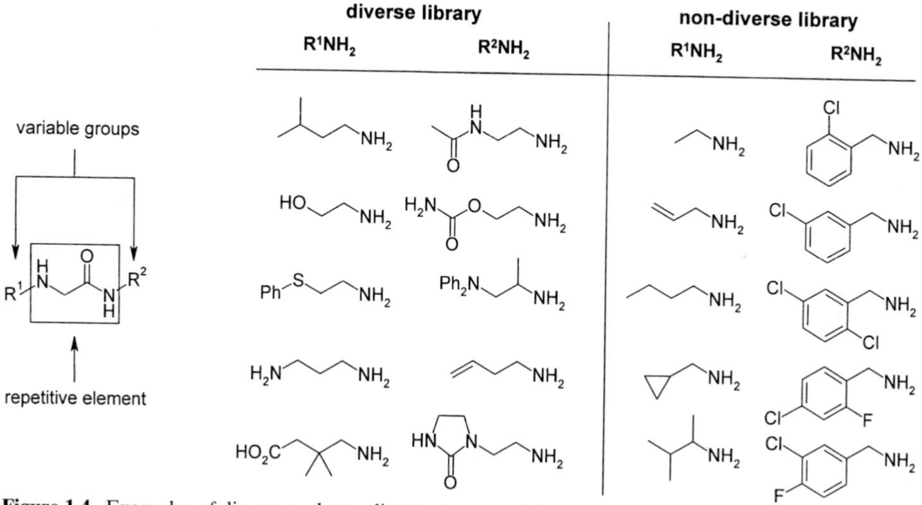

Figure 1.4. Examples of diverse and non-diverse compound libraries.

Which kind of solid-phase chemistry and which type of reagent will be most suitable for the production of high-quality, diverse libraries? As shown in Table 1.1, if the yields of the individual steps are less than 95%, the purity of the final products rapidly decreases with the total number of synthetic steps. Because reactions cannot always be optimized to such an extent, it is advisable to keep reaction sequences for library production as short as possible. Furthermore, reactions should be chosen which enable the use of unprotected, polyfunctional reagents available in large numbers [133]. Analysis of commercially available reagents reveals that few types of reagent are available which, in addition to the required reactive functionality, contain a broad selection of further functional groups. The reagent types that contain a suitable reactive group and have the highest structural diversity are amines, carboxylic acids, alcohols, and thiols.

Suitable chemistry must be chosen to allow the use of polyfunctional, unprotected reagents. In general, acylations and other reactions with electrophiles of a support-bound substrate will require the protection of functionalized side chains, both in the substrate and in the reagent. Protected, polyfunctional reagents are, however, rare and expensive, and protections/deprotections add further synthetic steps to the synthesis. Nucleophilic transformations, on the other hand, often enable the direct use of highly functionalized, unprotected reagents. Hence, nucleophilic transformations (e.g.

aliphatic or aromatic nucleophilic substitutions at support-bound electrophiles, acylations with support-bound weak acylating agents) should play a central role in reaction sequences for parallel synthesis of compound libraries [134,135].

For the preparation of large compound libraries, the cost of reagents and resins is a further issue that must be considered. Some supports, e.g. resin-bound phenols or *N*-hydroxybenzotriazole, which enable the preparation of resin-bound, reactive esters (Section 3.3.3), can be reused many times without the need to dismantle the reactor, and are therefore much more cost-efficient than supports that can only be used once [136,137]. Reactions such as the acylation of amines with resin-bound acylating agents have the additional advantage that only one equivalent of amine is needed, which again leads to a substantial reduction of costs.

If the compounds prepared by parallel solid-phase synthesis are to be tested without further purification in assays involving living cells, metal-mediated reactions should be avoided. Some metals, such as tin or silver, are highly cytotoxic, and may be detrimental to the assay even in trace amounts [138].

References for Chapter 1

[1] Kerschen, A.; Kanizsai, A.; Botros, I.; Krchnák, V. *J. Comb. Chem.* **1999**, *1*, 480–484.
[2] Lebl, M.; Pires, J.; Poncar, P.; Pokorny, V. *J. Comb. Chem.* **1999**, *1*, 474–479.
[3] Dawson, P. E.; Fitzgerald, M. C.; Muir, T. W.; Kent, S. B. H. *J. Am. Chem. Soc.* **1997**, *119*, 7917–7927.
[4] Lyttle, M. H.; Hudson, D.; Cook, R. M. *Nucleic Acids Res.* **1996**, *24*, 2793–2798.
[5] Haskins, N. J.; Hunter, D. J.; Organ, A. J.; Rahman, S. S.; Thom, C. *Rapid Commun. Mass Spectrom.* **1995**, *9*, 1437–1440.
[6] McKeown, S. C.; Watson, S. P.; Carr, R. A. E.; Marshall, P. *Tetrahedron Lett.* **1999**, *40*, 2407–2410.
[7] Bray, A. M.; Chiefari, D. S.; Valerio, R. M.; Maeji, N. J. *Tetrahedron Lett.* **1995**, *36*, 5081–5084.
[8] Gao, J. M.; Cheng, X. H.; Chen, R. D.; Sigal, G. B.; Bruce, J. E.; Schwartz, B. L.; Hofstadler, S. A.; Anderson, G. A.; Smith, R. D.; Whitesides, G. M. *J. Med. Chem.* **1996**, *39*, 1949–1955.
[9] Carlson, C. B.; Beal, P. A. *Org. Lett.* **2000**, *2*, 1465–1468.
[10] Williams, G. M.; Carr, R. A. E.; Congreve, M. S.; Kay, C.; McKeown, S. C.; Murray, P. J.; Scicinski, J. J.; Watson, S. P. *Angew. Chem. Int. Ed.* **2000**, *39*, 3293–3296.
[11] Congreve, M. S.; Ladlow, M.; Marshall, P.; Parr, N.; Scicinski, J. J.; Sheppard, T.; Vickerstaffe, E.; Carr, R. A. E. *Org. Lett.* **2001**, *3*, 507–510.
[12] Scicinski, J. J.; Congreve, M. S.; Jamieson, C.; Ley, S. V.; Newman, E. S.; Vinader, V. M.; Carr, R. A. E. *J. Comb. Chem.* **2001**, *3*, 387–396.
[13] Zaramella, A.; Conti, N.; Cin, M. D.; Paio, A.; Seneci, P.; Gehanne, S. *J. Comb. Chem.* **2001**, *3*, 410–420.
[14] Carrasco, M. R.; Fitzgerald, M. C.; Oda, Y.; Kent, S. B. H. *Tetrahedron Lett.* **1997**, *38*, 6331–6334.
[15] Lorthioir, O.; McKeown, S. C.; Parr, N. J.; Washington, M.; Watson, S. P. *Tetrahedron Lett.* **2000**, *41*, 8609–8613.
[16] Hamper, B. C.; Kolodziej, S. A.; Scates, A. M.; Smith, R. G.; Cortez, E. *J. Org. Chem.* **1998**, *63*, 708–718.
[17] Lebl, M.; Krchnák, V.; Sepetov, N. F.; Seligmann, B.; Strop, P.; Felder, S. *Biopolymers (Peptide Science)* **1995**, *37*, 177–198.
[18] Furka, A. *Drug Develop. Res.* **1995**, *36*, 1–12.
[19] Frank, R. *J. Biotech.* **1995**, *41*, 259–272.
[20] Burgess, K.; Liaw, A. I.; Wang, N. *J. Med. Chem.* **1994**, *37*, 2985–2987.
[21] Dittrich, F.; Tegge, W.; Frank, R. *Bioorg. Med. Chem. Lett.* **1998**, *8*, 2351–2356.
[22] Zhao, P. L.; Zambias, R.; Bolognese, J. A.; Boulton, D.; Chapman, K. *Proc. Natl. Acad. Sci. USA* **1995**, *92*, 10212–10216.
[23] Frank, R.; Heikens, W.; Heisterberg-Moutsis, G.; Blöcker, H. *Nucleic Acids Res.* **1983**, *11*, 4365–4377.

[24] Frank, R.; Meyerhans, A.; Schwellnus, K.; Blöcker, H. *Methods Enzymol.* **1987**, *154*, 221–249.

[25] Markiewicz, W. T.; Markiewicz, M.; Astriab, A.; Godzina, P. *Collect. Czech. Chem. Commun.* **1996**, *61*, S315–S318.

[26] Frank, R.; Döring, R. *Tetrahedron* **1988**, *44*, 6031–6040.

[27] Blankemeyer-Menge, B.; Frank, R. *Tetrahedron Lett.* **1988**, *29*, 5871–5874.

[28] Furka, A.; Sebestyén, F.; Asgedom, M.; Dibó, G. *Int. J. Pept. Prot. Res.* **1991**, *37*, 487–493.

[29] Pinilla, C.; Appel, J.; Blondelle, S. E.; Dooley, C.; Eichler, J.; Houghten, R. A. *Biopolymers (Peptide Science)* **1995**, *37*, 221–240.

[30] Houghten, R. A.; Appel, J. R.; Blondelle, S. E.; Cuervo, J. H.; Dooley, C. T.; Pinilla, C. *Pept. Res.* **1992**, *5*, 351–358.

[31] Youngquist, R. S.; Fuentes, G. R.; Lacey, M. P.; Keough, T. *J. Am. Chem. Soc.* **1995**, *117*, 3900–3906.

[32] Baldwin, J. J.; Burbaum, J. J.; Henderson, I.; Ohlmeyer, M. H. *J. Am. Chem. Soc.* **1995**, *117*, 5588–5589.

[33] Nestler, H. P.; Bartlett, P. A.; Still, W. C. *J. Org. Chem.* **1994**, *59*, 4723–4724.

[34] Ohlmeyer, M. H. J.; Swanson, R. N.; Dillard, L. W.; Reader, J. C.; Asouline, G.; Kobayashi, R.; Wigler, M.; Still, W. C. *Proc. Natl. Acad. Sci. USA* **1993**, *90*, 10922–10926.

[35] Blackwell, H. E.; Pérez, L.; Schreiber, S. L. *Angew. Chem. Int. Ed.* **2001**, *40*, 3421–3425.

[36] Boussie, T. R.; Coutard, C.; Turner, H.; Murphy, V.; Powers, T. S. *Angew. Chem. Int. Ed.* **1998**, *37*, 3272–3275.

[37] Ni, Z. J.; Maclean, D.; Holmes, C. P.; Murphy, M. M.; Ruhland, B.; Jacobs, J. W.; Gordon, E. M.; Gallop, M. A. *J. Med. Chem.* **1996**, *39*, 1601–1608.

[38] Scott, R. H.; Barnes, C.; Gerhard, U.; Balasubramanian, S. *Chem. Commun.* **1999**, 1331–1332.

[39] Hilaire, P. M. S.; Lowary, T. L.; Meldal, M.; Bock, K. *J. Am. Chem. Soc.* **1998**, *120*, 13312–13320.

[40] Nielsen, J.; Brenner, S.; Janda, K. D. *J. Am. Chem. Soc.* **1993**, *115*, 9812–9813.

[41] Pirrung, M. C.; Park, K. *Bioorg. Med. Chem. Lett.* **2000**, *10*, 2115–2118.

[42] Battersby, B. J.; Bryant, D.; Meutermans, W.; Smythe, M. L.; Trau, M. in *Innovation and Perspectives in Solid Phase Synthesis and Combinatorial Libraries*, Proceedings of the Sixth International Symposium, York, UK, **2000**, 199–202.

[43] Grøndahl, L.; Battersby, B. J.; Bryant, D.; Trau, M. *Langmuir* **2000**, *16*, 9709–9715.

[44] Moran, E. J.; Sarshar, S.; Cargill, J. F.; Shahbaz, M. M.; Lio, A.; Mjalli, A. M. M.; Armstrong, R. W. *J. Am. Chem. Soc.* **1995**, *117*, 10787–10788.

[45] Nicolaou, K. C.; Xiao, X. Y.; Parandoosh, Z.; Senyei, A.; Nova, M. P. *Angew. Chem. Int. Ed. Engl.* **1995**, *34*, 2289–2291.

[46] Xiao, X. Y.; Parandoosh, Z.; Nova, M. P. *J. Org. Chem.* **1997**, *62*, 6029–6033.

[47] Nanthakumar, A.; Pon, R. T.; Mazumder, A.; Yu, S.; Watson, A. *Bioconjugate Chem.* **2000**, *11*, 282–288.

[48] Fenniri, H.; Ding, L.; Ribbe, A. E.; Zyrianov, Y. *J. Am. Chem. Soc.* **2001**, *123*, 8151–8152.

[49] Blondelle, S. E.; Takahashi, E.; Houghten, R. A.; Pérez-Payá, E. *Biochem. J.* **1996**, *313*, 141–147.

[50] Eichler, J.; Lucka, A. W.; Pinilla, C.; Houghten, R. A. *Mol. Diversity* **1996**, *1*, 233–240.

[51] Blondelle, S. E.; Takahashi, E.; Dinh, K. T.; Houghten, R. A. *J. Appl. Bacteriol.* **1995**, *78*, 39–46.

[52] Lutzke, R. A. P.; Eppens, N. A.; Weber, P. A.; Houghten, R. A.; Plasterk, R. H. A. *Proc. Natl. Acad. Sci. USA* **1995**, *92*, 11456–11460.

[53] Pinilla, C.; Chendra, S.; Appel, J. R.; Houghten, R. A. *Pept. Res.* **1995**, *8*, 250–257.

[54] Stranix, B. R.; Gao, J. P.; Barghi, R.; Salha, J.; Darling, G. D. *J. Org. Chem.* **1997**, *62*, 8987–8993.

[55] Wang, S. *J. Org. Chem.* **1975**, *40*, 1235–1239.

[56] Kaiser, E.; Colescott, R. L.; Bossinger, C. D.; Cook, P. I. *Anal. Biochem.* **1970**, *34*, 595–598.

[57] Felix, A. M.; Jimenez, M. H. *Anal. Biochem.* **1973**, *52*, 377–381.

[58] Novabiochem Catalog & Peptide Synthesis Handbook **1999**, Läufelfingen, CH.

[59] Christensen, T. *Acta Chem. Scand. B* **1979**, *33*, 763–766.

[60] Shah, A.; Rahman, S. S.; de Biasi, V.; Camilleri, P. *Anal. Commun.* **1997**, *34*, 325–328.

[61] Madder, A.; Farcy, N.; Hosten, N. G. C.; De Muynck, H.; De Clercq, P. J.; Barry, J.; Davis, A. P. *Eur. J. Org. Chem.* **1999**, 2787–2791.

[62] Attardi, M. E.; Falchi, A.; Taddei, M. *Tetrahedron Lett.* **2000**, *41*, 7395–7399.

[63] Kuisle, O.; Quiñoá, E.; Riguera, R. *Tetrahedron Lett.* **1999**, *40*, 1203–1206.

[64] Kuisle, O.; Lolo, M.; Quiñoá, E.; Riguera, R. *Tetrahedron* **1999**, *55*, 14807–14812.

[65] Kuisle, O.; Quiñoá, E.; Riguera, R. *J. Org. Chem.* **1999**, *64*, 8063–8075.

[66] Burkett, B. A.; Brown, R. C. D.; Meloni, M. M. *Tetrahedron Lett.* **2001**, *42*, 5773–5775.

[67] Breitenbucher, J. G.; Johnson, C. R.; Haight, M.; Phelan, J. C. *Tetrahedron Lett.* **1998**, *39*, 1295–1298.

[68] Pomonis, J. G.; Severson, R. F.; Freeman, P. J. *J. Chromatogr.* **1969**, *40*, 78–84.

[69] Ellman, G. L. *Arch. Biochem. Biophys.* **1959**, *82*, 70–77.
[70] Badyal, J. P.; Cameron, A. M.; Cameron, N. R.; Coe, D. M.; Cox, R.; Davis, B. G.; Oates, L. J.; Oye, G.; Steel, P. G. *Tetrahedron Lett.* **2001**, *42*, 8531–8533.
[71] Attardi, M. E.; Porcu, G.; Taddei, M. *Tetrahedron Lett.* **2000**, *41*, 7391–7394.
[72] Vázquez, J.; Albericio, F. *Tetrahedron Lett.* **2001**, *42*, 6691–6693.
[73] Yan, B.; Li, W. *J. Org. Chem.* **1997**, *62*, 9354–9357.
[74] Chu, S. S.; Reich, S. H. *Bioorg. Med. Chem. Lett.* **1995**, *5*, 1053–1058.
[75] Gisin, B. F. *Anal. Chim. Acta* **1972**, *58*, 248–249.
[76] Yan, B.; Liu, L.; Astor, C. A.; Tang, Q. *Anal. Chem.* **1999**, *71*, 4564–4571.
[77] Luo, Y.; Ouyang, X. H.; Armstrong, R. W.; Murphy, M. M. *J. Org. Chem.* **1998**, *63*, 8719–8722.
[78] Bing, Y. *Acc. Chem. Res.* **1998**, *31*, 621–630.
[79] de Miguel, Y. R.; Shearer, A. S. *Biotechnol. Bioeng. (Comb. Chem.)* **2000**, *71*, 119–129.
[80] Fréchet, J. M.; Schuerch, C. *J. Am. Chem. Soc.* **1971**, *93*, 492–496.
[81] Chen, C.; Randall, L. A. A.; Miller, R. B.; Jones, A. D.; Kurth, M. J. *J. Am. Chem. Soc.* **1994**, *116*, 2661–2662.
[82] Gendre, F.; Diaz, P. *Tetrahedron Lett.* **2000**, *41*, 5193–5197.
[83] Yan, B.; Fell, J. B.; Kumaravel, G. *J. Org. Chem.* **1996**, *61*, 7467–7472.
[84] Yan, B.; Sun, Q.; Wareing, J. R.; Jewell, C. F. *J. Org. Chem.* **1996**, *61*, 8765–8770.
[85] Grice, P.; Leach, A. G.; Ley, S. V.; Massi, A.; Mynett, D. M. *J. Comb. Chem.* **2000**, *2*, 491–495.
[86] Yan, B.; Yan, H. *J. Comb. Chem.* **2001**, *3*, 78–84.
[87] Rahman, S. S.; Busby, D. J.; Lee, D. C. *J. Org. Chem.* **1998**, *63*, 6196–6199.
[88] Fischer, M.; Tran, C. D. *Anal. Chem.* **1999**, *71*, 2255–2261.
[89] Haap, W. J.; Walk, T. B.; Jung, G. *Angew. Chem. Int. Ed.* **1998**, *37*, 3311–3314.
[90] Snively, C. M.; Oskarsdottir, G.; Lauterbach, J. *J. Comb. Chem.* **2000**, *2*, 243–245.
[91] Li, W.; Yan, B. *J. Org. Chem.* **1998**, *63*, 4092–4097.
[92] Marti, R. E.; Yan, B.; Jarosinski, M. A. *J. Org. Chem.* **1997**, *62*, 5615–5618.
[93] Haap, W. J.; Kaiser, D.; Walk, T. B.; Jung, G. *Tetrahedron* **1998**, *54*, 3705–3724.
[94] Alexander, T.; Tran, C. D. *Anal. Chem.* **2001**, *73*, 1062–1067.
[95] Fitzgerald, M. C.; Harris, K.; Shevlin, C. G.; Siuzdak, G. *Bioorg. Med. Chem. Lett.* **1996**, *6*, 979–982.
[96] Halkes, K. M.; Gotfredsen, C. H.; Grøtli, M.; Miranda, L. P.; Duus, J. Ø.; Meldal, M. *Chem. Eur. J.* **2001**, *7*, 3584–3591.
[97] Enjalbal, C.; Maux, D.; Subra, G.; Martinez, J.; Combarieu, R.; Aubagnac, J.-L. *Tetrahedron Lett.* **1999**, *40*, 6217–6220.
[98] Keifer, P. A.; Baltusis, L.; Rice, D. M.; Tymiak, A. A.; Shoolery, J. N. *J. Magn. Reson. Ser. A* **1996**, *119*, 65–75.
[99] Gordeev, M. F.; Patel, D. V.; Wu, J.; Gordon, E. M. *Tetrahedron Lett.* **1996**, *37*, 4643–4646.
[100] Barn, D. R.; Morphy, J. R.; Rees, D. C. *Tetrahedron Lett.* **1996**, *37*, 3213–3216.
[101] Vidal-Ferran, A.; Bampos, N.; Moyano, A.; Pericàs, M. A.; Riera, A.; Sanders, J. K. M. *J. Org. Chem.* **1998**, *63*, 6309–6318.
[102] Lee, H. B.; Balasubramanian, S. *J. Org. Chem.* **1999**, *64*, 3454–3460.
[103] Epton, R.; Wellings, D. A.; Williams, A. *Reactive Polymers* **1987**, *6*, 143–157.
[104] Look, G. C.; Murphy, M. M.; Campbell, D. A.; Gallop, M. A. *Tetrahedron Lett.* **1995**, *36*, 2937–2940.
[105] Gordeev, M. F.; Patel, D. V.; Gordon, E. M. *J. Org. Chem.* **1996**, *61*, 924–928.
[106] Look, G. C.; Holmes, C. P.; Chinn, J. P.; Gallop, M. A. *J. Org. Chem.* **1994**, *59*, 7588–7590.
[107] Swayze, E. E. *Tetrahedron Lett.* **1997**, *38*, 8643–8646.
[108] Shapiro, M. J.; Kumaravel, G.; Petter, R. C.; Beveridge, R. *Tetrahedron Lett.* **1996**, *37*, 4671–4674.
[109] Svensson, A.; Bergquist, K. E.; Fex, T.; Kihlberg, J. *Tetrahedron Lett.* **1998**, *39*, 7193–7196.
[110] Albericio, F.; Pons, M.; Pedroso, E.; Giralt, E. *J. Org. Chem.* **1989**, *54*, 360–366.
[111] Svensson, A.; Fex, T.; Kihlberg, J. *J. Comb. Chem.* **2000**, *2*, 736–748.
[112] Drew, M.; Orton, E.; Krolikowski, P.; Salvino, J. M.; Kumar, N. V. *J. Comb. Chem.* **2000**, *2*, 8–9.
[113] Johnson, C. R.; Zhang, B. R. *Tetrahedron Lett.* **1995**, *36*, 9253–9256.
[114] Wehler, T.; Westman, J. *Tetrahedron Lett.* **1996**, *37*, 4771–4774.
[115] Pop, I. E.; Dhalluin, C. F.; Déprez, B. P.; Melnyk, P. C.; Lippens, G. M.; Tartar, A. L. *Tetrahedron* **1996**, *52*, 12209–12222.
[116] Chin, J.; Fell, B.; Shapiro, M. J.; Tomesch, J.; Wareing, J. R.; Bray, A. M. *J. Org. Chem.* **1997**, *62*, 538–539.
[117] Riedl, R.; Tappe, R.; Berkessel, A. *J. Am. Chem. Soc.* **1998**, *120*, 8994–9000.
[118] Rademann, J.; Meldal, M.; Bock, K. *Chem. Eur. J.* **1999**, *5*, 1218–1225.

[119] Rousselot-Pailley, P.; Maux, D.; Wieruszeski, J.-M.; Aubagnac, J.-L.; Martinez, J.; Lippens, G. *Tetrahedron* **2000**, *56*, 5163–5167.

[120] Gotfredsen, C. H.; Grøtli, M.; Willert, M.; Meldal, M.; Duus, J. Ø. *J. Chem. Soc., Perkin Trans. 1* **1999**, 1167–1171.

[121] Anderson, R. C.; Jarema, M. A.; Shapiro, M. J.; Stokes, J. P.; Ziliox, M. *J. Org. Chem.* **1995**, *60*, 2650–2651.

[122] Anderson, R. C.; Stokes, J. P.; Shapiro, M. J. *Tetrahedron Lett.* **1995**, *36*, 5311–5314.

[123] Ruhland, T.; Andersen, K.; Pedersen, H. *J. Org. Chem.* **1998**, *63*, 9204–9211.

[124] Bianco, A.; Furrer, J.; Limal, D.; Guichard, G.; Elbayed, K.; Raya, J.; Piotto, M.; Briand, J.-P. *J. Comb. Chem.* **2000**, *2*, 681–690.

[125] Warrass, R.; Lippens, G. *J. Org. Chem.* **2000**, *65*, 2946–2950.

[126] Hany, R.; Rentsch, D.; Dhanapal, B.; Obrecht, D. *J. Comb. Chem.* **2001**, *3*, 85–89.

[127] Warrass, R.; Wieruszeski, J.-M.; Lippens, G. *J. Am. Chem. Soc.* **1999**, *121*, 3787–3788.

[128] Jamieson, C.; Congreve, M. S.; Hewitt, P. R.; Scicinski, J. J.; Ley, S. V. *J. Comb. Chem.* **2001**, *3*, 397–399.

[129] Mogemark, M.; Elofsson, M.; Kihlberg, J. *Org. Lett.* **2001**, *3*, 1463–1466.

[130] Salvino, J. M.; Patel, S.; Drew, M.; Krowlikowski, P.; Orton, E.; Kumar, N. V.; Caulfield, T.; Labaudiniere, R. *J. Comb. Chem.* **2001**, *3*, 177–180.

[131] Sarkar, S. K.; Garigipati, R. S.; Adams, J. L.; Keifer, P. A. *J. Am. Chem. Soc.* **1996**, *118*, 2305–2306.

[132] Kanemitsu, T.; Kanie, O.; Wong, C.-H. *Angew. Chem. Int. Ed.* **1998**, *37*, 3415–3418.

[133] Teague, S. J.; Davis, A. M.; Leeson, P. D.; Oprea, T. *Angew. Chem. Int. Ed.* **1999**, *38*, 3743–3748.

[134] Grimstrup, M.; Zaragoza, F. *Eur. J. Org. Chem.* **2001**, 3233–3246.

[135] Zaragoza, F.; Stephensen, H. *Angew. Chem. Int. Ed.* **2000**, *39*, 554–556.

[136] Irving, M. M.; Kshirsagar, T.; Figliozzi, G. M.; Yan, B. *J. Comb. Chem.* **2001**, *3*, 407–409.

[137] Pon, R. T.; Yu, S.; Guo, Z.; Deshmukh, R.; Sanghvi, Y. S. *J. Chem. Soc., Perkin Trans. 1* **2001**, 2638–2643.

[138] Scheuerman, R. A.; Tumelty, D. *Tetrahedron Lett.* **2000**, *41*, 6531–6535.

2 Supports for Solid-Phase Organic Synthesis

Solid-phase organic synthesis refers to syntheses in which the starting material and synthetic intermediates are linked to an insoluble material (support), which enables the facile mechanical separation of the intermediates from reactants and solvents (Figure 2.1).

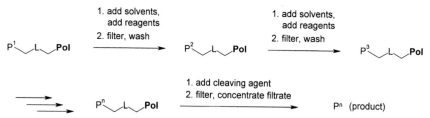

Figure 2.1. Schematic representation of a synthesis on solid phase. Pol: support, L: linker, P: synthetic intermediate.

Supports of different macroscopic shape have been used for solid-phase synthesis. The most common are spherical particles (beads, 0.04–0.15 mm), which are readily weighed, filtered, and dried, and are well-suited for most applications. Other forms of insoluble support include sheets [1], crown-shaped 'pins' [2], or small discs [3].

The general requirements for a support are mechanical stability and chemical inertness under the reaction conditions to be used. Mechanical stability is required to avoid the breaking down of the polymer into smaller particles, which could lead to the clogging of filters. Supports also need to be chemically functionalized, so that the synthetic intermediates can be covalently attached to the support via a suitable linker. Moreover, if the intermediates are located *within* the support (and not only on the surface), diffusion of reagents into the support particles will be necessary, and materials with sufficient permeability or swelling capacity need to be chosen.

Soluble polymers, such as non-cross-linked polystyrene or poly(ethylene glycol), can also be used as supports for organic synthesis [4]. These polymers can generally be precipitated from certain solvents or purified by membrane filtration or recrystallization [5,6], and in this way enable the separation of the desired intermediate from the reagents. Unlike insoluble polymers, soluble supports enable the use of insoluble reagents or catalysts and also the characterization of intermediates by NMR in a homogeneous phase. However, synthesis on soluble supports has a number of drawbacks, which make this technology less attractive for high-throughput synthesis than

the use of insoluble supports. Most importantly, synthesis on soluble supports is difficult to automate [7]. Furthermore, the precipitation of a soluble polymer is not always
quantitative when lipophilic intermediates are attached to it [8], and it may well be
the case that a reagent or solvent shows strong physical binding to the polymer, which
then precipitates as a sticky mass that can no longer be filtered [9]. It has been shown
that reactions on soluble polymers do not proceed significantly faster than those on
insoluble, cross-linked polymers [10,11].

Several different support materials have proven useful for solid-phase organic synthesis, but not all materials are compatible with all types of solvents and reagents.
Therefore, for each application the proper type of support has to be selected. Some
review articles on supports for solid-phase synthesis have recently appeared [1,9,
12–15].

2.1 Polystyrene

In this section the use of polystyrene and copolymers of styrene with various cross-
linking agents as supports for solid-phase organic synthesis is discussed. Copolymers
of styrene with divinylbenzene are the most common supports for solid-phase synthesis. Depending on the kind of additives used during the polymerization and on the
styrene/divinylbenzene ratio, various different types of polystyrene can be prepared.
However, non-cross-linked polystyrene has also been used as a support for organic
synthesis [10,16–22]. Linear, non-cross-linked polystyrene is soluble in organic solvents such as toluene, pyridine, ethyl acetate, THF, chloroform, or DCM, even at low
temperatures, but can be selectively precipitated by the addition of methanol or
water.

2.1.1 Microporous Styrene–Divinylbenzene Copolymers

One of the support types most frequently used for solid-phase organic synthesis are
styrene–divinylbenzene copolymers (cross-linked polystyrene, Figure 2.2).

styrene   polystyrene

divinylbenzene   cross-linked polystyrene

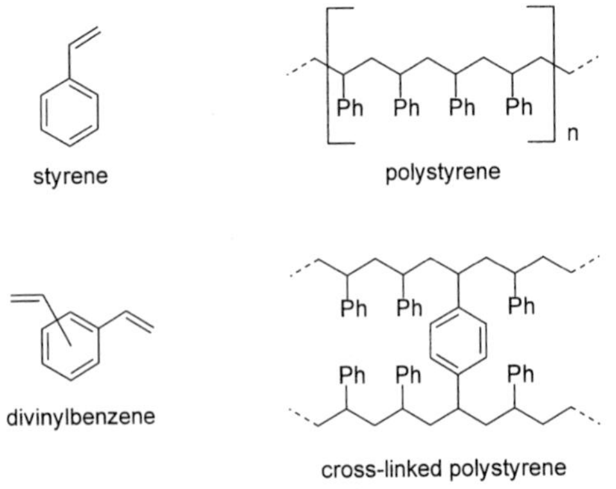

Figure 2.2.

Copolymers of styrene and divinylbenzene were initially developed for the production of ion-exchange resins [23], and are still being used for this purpose [9]. These polymers are essentially insoluble if cross-linking exceeds 0.2%, but can swell to a variable extent in organic solvents (Table 2.1). The capacity of polystyrene to swell generally decreases with increasing cross-linking [24]. Reaction kinetics are also influenced by the degree of cross-linking, with reactions normally proceeding more rapidly on sparsely cross-linked polymers [25].

Cross-linked polystyrenes are usually prepared by radical polymerization of suspensions of styrene and divinylbenzene in such a way that polymer beads (0.04–0.15 mm) are directly obtained. The size of the beads can be controlled by the addition of surfactants and by adjustment of the stirring speed [26]. Depending on the precise polymerization conditions, polymers of different porosities might result. If the polymerization is conducted in water or other solvents that neither dissolve the monomers nor swell the polymer, microporous, gel-type resins with low internal surface area are formed. Microporous polystyrene cross-linked with 1–2% divinylbenzene is the most common polymer for solid-phase synthesis, and was the support initially chosen by Merrifield for solid-phase peptide synthesis [27]. Resins with less cross-linking (e.g. 0.5% divinylbenzene [28]) have also been used, but resins with 1–2% cross-linking remain the most common type of polystyrene support. These polymers are commercially available as beads with a broad choice of functional groups and linkers. These groups are not only located on the surface of the beads, but are uniformly distributed throughout the polymer [29–31]. Highly cross-linked, non-porous polystyrene has been used for the solid-phase synthesis of oligonucleotides [32].

2.1.1.1 Swelling Behavior

Solvents can penetrate into cross-linked polystyrene to varying extents, causing the size of the beads to increase (Table 2.1). Swelling is most pronounced in solvents that can bind to the polymer in a non-covalent manner. Because polystyrene is a hydrophobic, polarizable material, swelling is generally strong in dipolar aprotic solvents but poor in alkanes, protic solvents, or water. The volumes of swollen, polystyrene-based supports in a selection of solvents are given in Table 2.1. Because swelling will vary between different batches of polymer, the values given in Table 2.1 should only be considered as an approximate guideline.

The size of the beads is also affected by the amount of synthetic intermediate attached to them. For instance, when a peptide with a molecular weight of 5957 g/mol is prepared on polystyrene (1% cross-linked; loading: 0.95 mmol/g), the volume of the beads (which consist of 81% peptide) becomes five times as large as that of the unloaded support [29,34]. When this resin is suspended in DMF, the beads swell further, attaining a volume 26 times greater than that of the original unloaded, non-swollen polystyrene beads. This strong swelling makes it difficult to use gel-type cross-linked polystyrene for continuous-flow synthesis (in which reagents are pumped through a column containing the support).

Table 2.1. Swelling of polystyrene and Tentagel in different solvents [33]. Volume of swollen, drained resin (1.0 g) in mL.

Solvent	Wang resin (PS, 2% DVB, 0.6 mmol/g)	Tentagel S RAM [a] (0.3 mmol/g)
NMP	6.4	4.4
pyridine	6.0	4.6
THF	6.0	4.0
N,N-dimethylacetamide	5.8	4.0
quinoline [b]	5.7	5.2
DMPU [b]	5.7	4.5
CHCl$_3$	5.6	5.6
2,6-dimethylpyridine [b]	5.6	5.1
NMM [b]	5.6	4.4
dioxane	5.6	4.2
CH$_2$Cl$_2$	5.4	5.6
DMF	5.2	4.4
1,2-dichlorobenzene	4.8	5.2
MeOCH$_2$CH$_2$OMe	4.8	2.0
1,2-dichloropropane [b]	4.5	4.5
ClCH$_2$CH$_2$Cl	4.4	5.4
benzene	4.4	4.4
2-butanone	4.4	2.0
nitrobenzene [b]	4.3	4.8
DMSO	4.2	3.8
AcOEt	4.2	2.0
toluene	4.0	3.6
MeOCH$_2$CH$_2$OH	4.0	2.4
acetone	3.6	2.8
EtOCH$_2$CH$_2$OH	3.6	2.0
xylene	3.0	2.0
THF/H$_2$O 1:1	2.8	5.2
AcOH	2.8	5.2
Et$_2$O	2.8	2.0
CCl$_4$	2.4	2.8
tert-butyl methyl ether	2.4	1.6
TFA	2.0	6.4
CF$_3$CH$_2$OH [b]	2.0	5.8
MeNO$_2$	2.0	4.4
MeCN	2.0	4.0
MeCN/H$_2$O 1:1	2.0	4.0
DMF/H$_2$O 1:1	2.0	4.0
1-butanol	2.0	2.0
1-propanol	2.0	2.0
2-propanol	2.0	2.0
ethanol	2.0	1.8
DIPEA [b]	1.8	1.6
methanol	1.6	3.6
water	1.6	3.6
HOCH$_2$CH$_2$OH [b]	1.6	1.7
tert-butanol	1.6	1.6
heptane	1.6	1.6

[a] Tentagel with Rink amide linker.   [b] determined by the author.

Diffusion of lipophilic reagents into swollen, microporous polystyrene is generally fast [24,35]. For instance, N-acylation of polystyrene-bound alanine (0.05 mm diameter beads) with the symmetric anhydride of *N*-Bpoc leucine reaches 99% completion after only 14 s [34]. On the other hand, reactions involving ionic reagents are often sluggish in polystyrene; this might be because of rate-limiting diffusion of charged particles into the hydrophobic support.

2.1.1.2 Chemical Stability

Cross-linked polystyrene tolerates well a broad range of reaction conditions, including treatment with weak oxidants (ozone, DDQ), strong bases (LDA), and acids (HBr, TfOH). Saponifications with 40% aqueous sodium hydroxide at 180 °C for 10 h can be realized on 5% cross-linked macroporous polystyrene without deterioration of the polymer [36]. Not surprisingly, however, strong oxidants at high temperatures, and other reagents which lead to a chemical modification of alkylbenzenes, will attack polystyrene.

2.1.1.3 Functionalization

Polystyrene does not enable the covalent, reversible attachment of synthetic intermediates unless the support is derivatized with suitable functional groups. The most common groups for the reversible attachment of intermediates are chloromethyl and hydroxymethyl, whereas aminomethyl groups are mainly used as non-cleavable points of attachment for linkers (see Chapter 3).

Functionalized polymers can be prepared either by chemical transformation of the unfunctionalized polymer or by copolymerization of functionalized monomers. Neither of these strategies provides for perfectly homogeneous distribution of the functional groups: during chemical modification of polymer beads, their exterior will be more exposed to reagents than the core, and this can lead to an uneven distribution of functionality. On the other hand, different monomers (e.g. styrene and 4-(chloromethyl)styrene or divinylbenzene) do not polymerize at the same rate and will be incorporated into the polymer to variable extents as the polymerization proceeds [30,37]. For this reason, the degree of cross-linking of non-functionalized cross-linked polystyrene will also be different in the core from that at the exterior of the beads. Accordingly, the quality of a functionalized support depends strongly on the monomers used and on the precise conditions used for its preparation.

Cross-linked polystyrene can be functionalized in many ways [38–41]. Those functionalized resins that are frequently used as supports for solid-phase synthesis are commercially available, and their preparation will be mentioned only in brief here.

Almost all electrophilic substitutions known to proceed in solution with isopropylbenzene can also be performed with polystyrene, using solvents such as nitrobenzene, carbon disulfide, or carbon tetrachloride. These substitutions include bromination [42], nitration [43,44], sulfonylation, Friedel–Crafts acylations [45–49], and alkylations

[50–52]. Chloromethyl polystyrene (Merrifield resin) has been prepared by chloro-methylation of polystyrene [23,27,53,54], by copolymerization of 4-chloromethylsty-rene with styrene [20,26,55,56], and by chlorination of poly(4-methylstyrene) [57,58]. Aminomethyl polystyrene is most conveniently prepared by direct amidomethylation of polystyrene with (hydroxymethyl)amides or (halomethyl)amides under acidic con-ditions followed by hydrolysis [59–62], but it has also been prepared directly from chloromethyl polystyrene ([63,64]; see also Section 10.1.1.1).

Metallated polystyrenes are versatile intermediates for the preparation of a number of polystyrene derivatives. Metallated polystyrene has been prepared from haloge-nated polystyrenes by halogen–metal exchange [41,42,65,66] and by direct metallation of polystyrene [67–69] (see Chapter 4). Electrophiles suitable for the derivatization of metallated polystyrene include carbon dioxide, carbonyl compounds, sulfur, trimethyl borate, isocyanates, chlorosilanes, alkyl bromides, chlorodiphenylphosphine, DMF, oxirane, selenium [70], dimethyldiselenide [71], organotin halides [69], oxygen [72], etc. [41,42,65–67].

The loading of a support with functional groups is usually expressed in mmol/g. This means that the loading will decrease with increasing weight of the compound attached to the support, because the number of attachment sites remains constant but the weight of the resin increases. Poly(4-chloromethyl)styrene would have a chloride loading of 6.55 mmol/g. Such high loadings cannot generally be attained by direct chloromethylation of polystyrene, because interstrand cross-linking by Friedel–Crafts alkylation efficiently competes with chloromethylation in the later stages of the reac-tion and leads to a highly cross-linked polymer with poor swelling properties [23,55,73]. A similar effect is also observed in the amidomethylation of polystyrene [60].

Commercially available functionalized polystyrenes generally have loadings of 0.5–1.5 mmol/g; this corresponds roughly to 20% derivatization of all available phenyl groups. Higher loadings can also be realized, e.g. by polymerization of functionalized styrenes [74] or by multiplying the number of attachment sites through the grafting on of dendrimer-like structures [14,75–78], but these supports have not yet found broad application. The reason for this might be that for the solid-phase synthesis of large peptides high loading leads to less predictable swelling properties of the support in later stages of the synthesis. However, despite scepticism among peptide chemists, it has been shown that there is sufficient 'space' in 1–2% cross-linked polystyrene to accommodate large peptides [29]. Coupling difficulties in peptide synthesis are gener-ally not related to excessive loading of a support but to folding of the peptide onto itself (β-sheet formation), or to phase transitions from a polystyrene-like resin to a polyamide-like resin [79]. Peptides have, in fact, been successfully prepared on poly-acrylamide supports with initial loadings as high as 5 mmol/g [80]. High loading will not usually be a problem for the solid-phase synthesis of small molecules, but rather an advantage, because smaller reactors and less solvent will be required for the syn-thesis of a given amount of product.

High loading can, however, be an inconvenience if resin-bound substrates are able to react with themselves (e.g. during olefin cross-metathesis or macrocyclizations), or if bi- or polyfunctional, unprotected reagents are to be used in solid-phase synthesis.

It is, for example, possible to mono-derivatize symmetrical, bifunctional reagents with resin-bound acylating or alkylating agents [81,82]. However, the higher the loading of the support, the higher will be the degree of bis-derivatization of the bifunctional reagent (i.e. the degree of cross-linking). This unwanted reaction can be suppressed by choosing a support with lower loading, or by partial capping (e.g. by silylation) of the attachment sites on the support [83].

2.1.2 Macroporous Styrene–Divinylbenzene Copolymers

Polymerization of a suspension of undissolved styrene and divinylbenzene in water leads to the formation of microporous, gel-type supports with low internal surface area. Macroporous resins with high internal surface area result when a solution of *dissolved* monomers is polymerized [84,85]. As solvents, for instance, toluene, xylene, or diphenylmethane [86] may be used, and the polymerization can also be conducted in the presence of higher alcohols or linear polystyrene ('porogens'). During the polymerization, porogens are trapped within the cross-linked polymer, generating large pores. When polymerization is complete, the porogens are removed either by washing or by evaporation under reduced pressure. The porous structure of the resulting polymers (Figure 2.3) remains stable even in the absence of solvents, leading to a large internal surface area [9,87]. To provide for sufficient mechanical stability these polymers are generally highly cross-linked (> 10% divinylbenzene); this implies limited swelling properties.

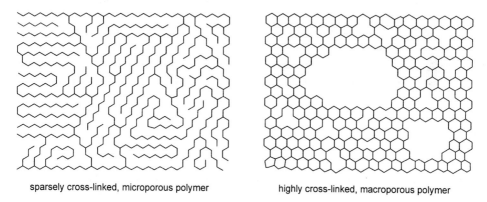

sparsely cross-linked, microporous polymer   highly cross-linked, macroporous polymer

Figure 2.3. Schematic representation of polymer chains in microporous and macroporous polymers.

Macroporous, highly cross-linked styrene–divinylbenzene copolymers are mainly used as ion-exchange resins (after sulfonylation or aminomethylation). However, macroporous (or 'macroreticular') polystyrene [84,85] is increasingly receiving attention as a support for solid-phase synthesis [12,39,88–90]. Although these polymers do not swell significantly, a large internal surface area is available for functionalization and remains accessible in a variety of solvents, including alcohols and water. Hence, the choice of solvent is less critical when using macroporous polystyrene than in the case of microporous polystyrene. The capacity of macroporous polystyrene can reach up to 0.8–1.0 mmol/g [12,88].

The fact that macroporous, highly cross-linked polystyrene does not swell makes this support particularly interesting for continuous-flow synthesis in columns. This support has also been successfully used as an alternative to CPG for the solid-phase synthesis of oligonucleotides [90,91]. Furthermore, because reagents do not need to penetrate into the polystyrene network, enzyme-mediated reactions should also proceed smoothly on macroporous polystyrene [85].

2.1.3 Miscellaneous Polystyrenes

With the aim of fine-tuning the physicochemical properties of polystyrene-based supports and thereby improving their suitablity for solid-phase synthesis, cross-linking agents other than divinylbenzene have been investigated. These include ethylene glycol dimethacrylate [92], 1,4-butanediol dimethacrylate [93,94], hexane-1,6-diol diacrylate [95–97], 1,4-(4-vinylphenoxy)butane [98–100], 1,3-(4-vinylphenoxy)propan-2-ol [101], and tetraethylene glycol diacrylate [102–105], which lead to supports that swell more than Merrifield resin. Supports with higher ethylene glycol content have been prepared by polymerization of styrene with α,ω-bis(4-vinylphenoxy)oligo- or poly-(ethylene glycol) [106,107]. Unlike divinylbenzene-cross-linked polystyrene, these supports swell both in dipolar aprotic solvents and in protic solvents. Supports of this type devoid of benzyl ether groups have also been prepared [108], and proved to be more stable towards acids than conventional PEG-polystyrene composite supports [108]. Copolymers of styrene, α,ω-bis(3-aminopropyl) PEG monomethacrylate, and 5% divinylbenzene have been described and used for solid-phase peptide synthesis [109].

Linear polystyrene can be generated on insoluble polymers by γ-irradiation of the latter in a solution of styrene [110]. Polystyrene grafted onto polytetrafluoroethylene [111–116], polyethylene [2,110,117], or polypropylene [15] can be functionalized in the same way as cross-linked polystyrene, and loadings of up to 1.0 mmol/g can be attained. These supports, which are also available as crown-shaped pins (Multipin, 2–3 mm diameter, 8–10 µmol per crown), have been used for the synthesis of peptides [2,110,111,118], oligonucleotides [112–115,117,119], and small molecules [120–122].

Chloromethyl polystyrene can be converted to a free-radical initiator by reaction with 2,2,6,6-tetramethylpiperidine-*N*-oxyl (TEMPO). Radical polymerization of various substituted alkenes on this resin has been used to prepare new types of polystyrene-based supports [123]. Alternatively, cross-linked vinyl polystyrene can be copolymerized with functionalized norbornene derivatives by ruthenium-mediated ring-opening metathesis polymerization [124].

Beads of cross-linked polystyrene containing magnetite crystals (Fe_3O_4, a ferromagnetic material) have been prepared and successfully used for solid-phase synthesis [125–127]. The magnetic beads could be readily separated from the reaction mixture with a magnet.

2.2 Poly(ethylene glycol)–Polystyrene Graft Polymers

PEG grafted onto cross-linked polystyrene can be prepared either by linking PEG to suitably functionalized polystyrene [128,129] or by polymerization of oxirane on a hydroxylated polystyrene support [5,51,130] (Figure 2.4). The former strategy generally gives inferior results, because PEG or symmetrical derivatives thereof are bifunctional and can lead to extensive cross-linking [5,7]. An additional strategy for the production of PEG-grafted polystyrene supports is the partial derivatization of aminomethyl polystyrene with PEG monomethyl ether [62] (Figure 2.4).

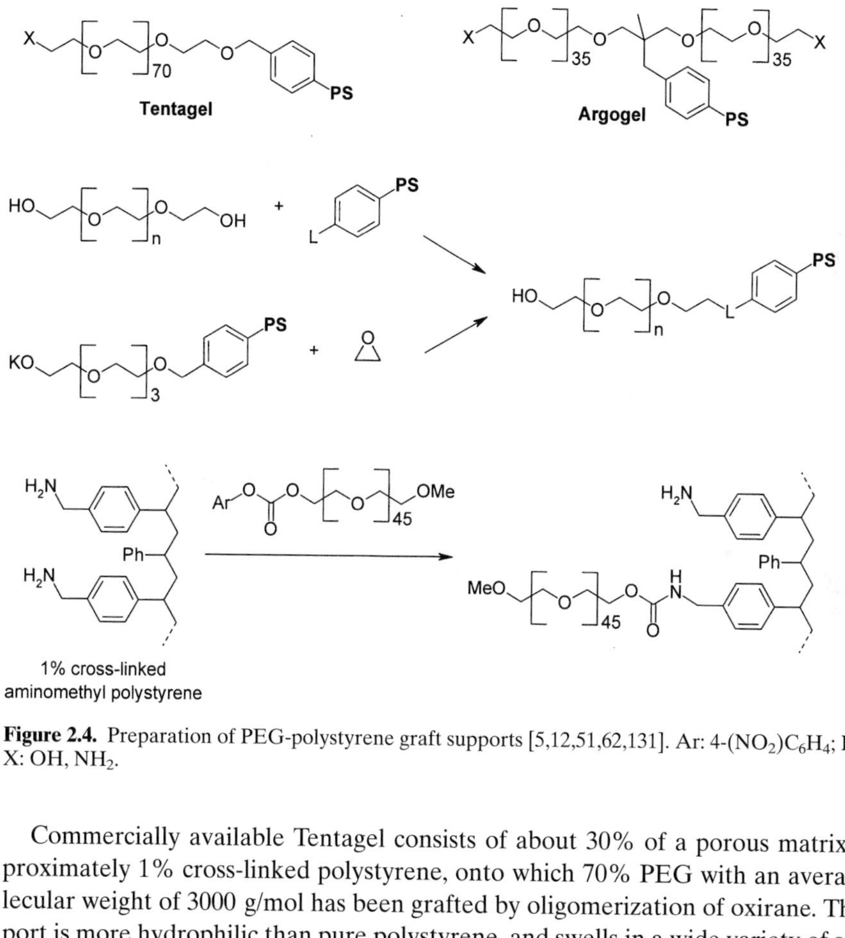

Figure 2.4. Preparation of PEG-polystyrene graft supports [5,12,51,62,131]. Ar: 4-$(NO_2)C_6H_4$; L: linker; X: OH, NH_2.

Commercially available Tentagel consists of about 30% of a porous matrix of approximately 1% cross-linked polystyrene, onto which 70% PEG with an average molecular weight of 3000 g/mol has been grafted by oligomerization of oxirane. This support is more hydrophilic than pure polystyrene, and swells in a wide variety of solvents (Table 2.1). Loadings of commercially available Tentagel are in the range 0.15–0.30 mmol/g. Supports closely related to Tentagel are available commercially under the tradename Argogel [131,132] (Figure 2.4).

In swollen Tentagel and related supports, the PEG chains are more mobile than the cross-linked polystyrene matrix. This enables the recording of well-resolved NMR spectra of samples linked to Tentagel [5,7,133–135]. Polystyrene and other less mobile supports generally give broad NMR signals, unless special techniques are used (e.g. magic-angle spinning NMR [136–138]).

It has been proposed that the increased mobility of intermediates linked to Tentagel also provides for a more 'solution-like' environment and higher reaction rates [5,11,139]. Recent investigations have shown, however, that reaction rates on polymeric supports depend on the precise type of reaction, reagents, and solvents used [25], and can be higher on cross-linked polystyrene than on Tentagel [35,140]. Polymeric supports generally need to be swollen if dissolved reagents are to reach the interior of the polymer particles [84]. Only reactions involving small ions and/or protic solvents will therefore usually proceed faster on Tentagel than on cross-linked polystyrene.

Unfortunately, PEG-grafted polystyrene supports have some disadvantages, the most critical being the low loading and the release of PEG upon treatment with TFA [9,62,141] or upon heating [142]. Further, because of their high PEG content, these supports can sometimes become adherent and are difficult to dry [9]. Increased stability towards acid has been achieved with PEG-grafted polystyrene devoid of benzylic C–O bonds [12,51,62,132].

2.3 Poly(ethylene glycol)

The most common PEG used as a support is poly(ethylene glycol) monomethyl ether with a molecular weight of about 5000 g/mol (MeOPEG$_{5000}$). Occasionally non-etherified PEG is used and both hydroxyl groups of each oligomer (e.g. PEG$_{3400}$) are coupled with the starting material [143]. These polymers are soluble in water and most organic solvents, but can be precipitated with hexane, diethyl ether, or *tert*-butyl methyl ether. Applications of this class of support in organic synthesis have been reviewed [4,144].

The synthesis of peptides on soluble poly(ethylene glycol) (PEG) was described as early as 1971 [145]. The main motivation for seeking new supports at that time was that some peptides could not be prepared on cross-linked polystyrene. It was later found that this was not because of the support, but mainly due to folding of the growing peptide onto itself or because of β-sheet formation [146,147]. Such problems are solved today by introducing *N*-Hmb-amino acids at certain positions of the peptide (see Section 16.1.5), or by extending the coupling times, but generally not by choosing a more hydrophilic or a soluble support.

Despite the drawbacks of soluble supports mentioned in the introduction to this chapter, soluble PEG has often been used as a support for the synthesis of both biopolymers [6,148–153] and small organic molecules [18,154–158]. Related supports with higher loadings have been described [159], in which 3,5-bis(chloromethyl)phenoxy groups were linked to both ends of PEG$_{5000}$.

Cross-linked, insoluble PEG can be prepared by copolymerizing PEG with epichlorohydrin [107,160] or with the tosylate of 3-methyl-3-(hydroxymethyl)oxetane [161,162]. These supports are highly permeable and hydrophilic, and enable the use of enzymes such as subtilisin (27 kDa) to catalyze transformations of support-bound substrates [161], as well as the recording of well-resolved MAS NMR spectra.

2.4 Polyacrylamides

With the aim of finding supports suitable for solid-phase peptide synthesis that are more hydrophilic than polystyrene, polyacrylamides have been extensively investigated [163,164]. As with polystyrene, non-cross-linked polyacrylamides are soluble and hence are not well suited for solid-phase synthesis; indeed, they have only occasionally been used for this purpose [4,165,166]. Cross-linked polyacrylamides, however, do not dissolve but only gelate in various solvents. Because polyacrylamide is very hydrophilic and therefore incompatible with most organic solvents, derivatives of poly-*N,N*-dimethylacrylamide (e.g. Pepsyn), poly(*N*-acryloylpyrrolidine) [167–169] (e.g. Expansin [170]), or poly(*N*-acryloylmorpholine) [166,171] have usually been chosen as supports. A selection of typical monomers used for the preparation of polyacrylamide-based supports is given in Figure 2.5. Unlike polystyrene, polyacrylamides swell strongly in both aprotic (DMF, pyridine, DCM) and protic (methanol, water) solvents [172].

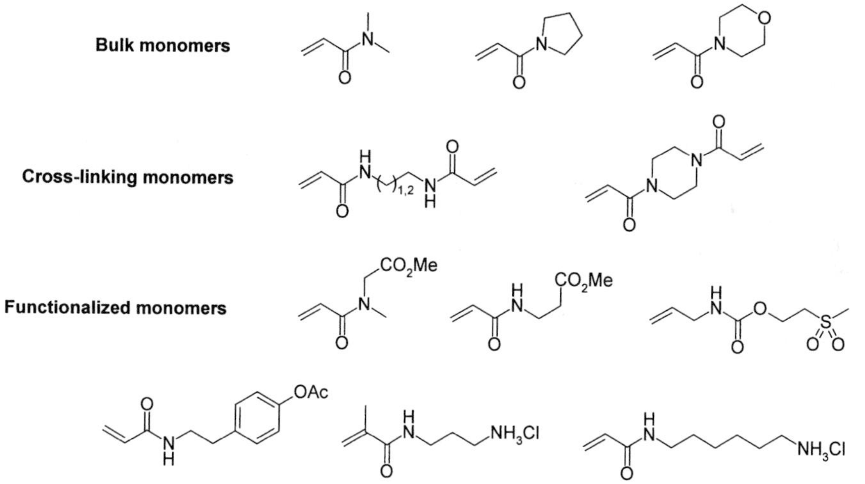

Figure 2.5. Typical components of cross-linked polyacrylamide supports for solid-phase synthesis.

As an illustrative example, the monomers used for the preparation of a typical polyacrylamide support (Pepsyn) and its schematic representation are sketched in Figure 2.6. The ester functionality, obtained by copolymerization with *N*-acryloylsarcosine methyl ester, can be used as the attachment point for a suitable linker. This can be achieved by aminolysis with ethylenediamine, followed by acylation of the resulting primary amine with 3-(4-hydroxymethyl)phenylpropionic acid [163] or other linkers. Similar supports have been prepared by copolymerization of protected allylamine [37] or *N*-(acryloyl)-1,3-diaminopropane [173] with *N,N*-dimethylacrylamide and

N,N′-bis(acryloyl)-1,3-diaminopropane. Amine deprotection after polymerization leads directly to the functionalized support.

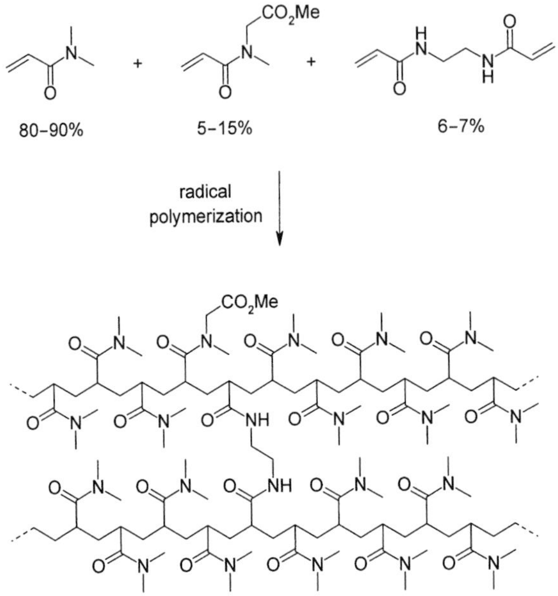

Figure 2.6. Preparation of a functionalized, cross-linked poly-*N,N*-dimethylacrylamide (Pepsyn).

Polyacrylamides in which all bulk monomers bore a site of attachment have also been prepared [80]. *N*-Acryloyl-*N*-2-(4-acetoxyphenyl)ethylamine (Figure 2.5) has been polymerized in the presence of 1,4-bis(acryloyl)piperazine to yield a support with an initial loading of 5.0 mmol/g. The successful synthesis of a decapeptide on this support demonstrated that the use of highly loaded supports is feasible and offers a cost-efficient alternative to standard supports [80].

Cross-linked polyacrylamides, such as that sketched in Figure 2.6, are not sufficiently mechanically stable to allow their packing into columns for continuous-flow synthesis. To overcome this problem, polyacrylamides can be generated on kieselguhr (macroporous silicon dioxide); this results in more rigid particles, which are suitable for continuous-flow peptide synthesis [174,175]. This material is called Pepsyn K or Macrosorb SPR, and has been used for the solid-phase synthesis of peptides [176,177], oligonucleotides [178], and glycopeptides [179]. Macroporous polystyrene (Polyhipe [180]) has also been used as a mechanical support for polyacrylamides.

Polyacrylamides are chemically stable towards acids (TFA, hydrogen fluoride [163]), bases, and weak oxidants or reducing agents. Problems, which might be related to cleavage of the amide bonds in poly(*N,N*-dialkylacrylamides), were encountered upon treatment of such supports with sodium in liquid ammonia [169].

2.5 Polyacrylamide–PEG Copolymers

Meldal and coworkers developed polyacrylamides cross-linked with poly(ethylene glycol), referred to as PEGA, as supports for solid-phase synthesis and on-bead enzymatic assays [181–183]. Functionalization of the polymer was performed in a similar fashion as in the case of other polyacrylamides, i.e. either by copolymerization with *N*-acryloylsarcosine ethyl ester followed by aminolysis with ethylenediamine, or by copolymerization with an amino group containing monomer. The monomers used for a high-capacity (0.4–0.8 mmol/g [182]) and a low-capacity (0.2–0.4 mmol/g [181]) PEGA support are sketched in Figure 2.7.

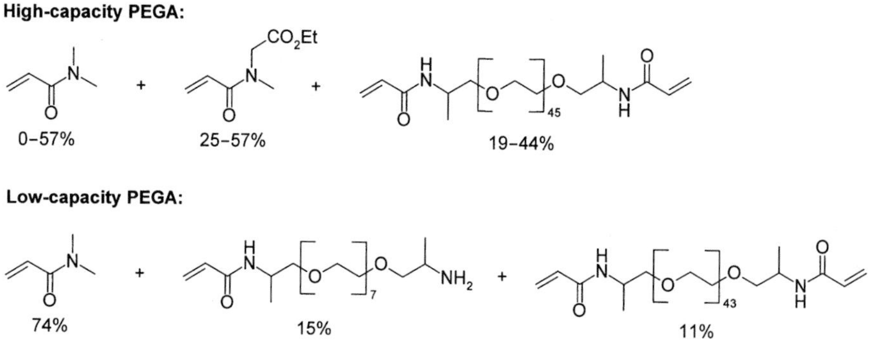

High-capacity PEGA:

0–57%   25–57%   19–44%

Low-capacity PEGA:

74%   15%   11%

Figure 2.7. Monomers and approximate composition of two types of PEGA support [181,182].

PEGA supports swell in a wide range of solvents. Most notably, enzymes can efficiently diffuse into these polymers, making PEGA the support of choice for on-bead screening [184–186] and enzyme-mediated solid-phase chemistry [187]. Furthermore, PEGA is stable towards hydrogen fluoride and is therefore suitable for the solid-phase synthesis of peptides using the Boc methodology [188]. The main disadvantage of PEGA-based supports is their limited mechanical stability, which occasionally leads to problems during filtration of the polymer.

2.6 Silica

Different forms of silicon dioxide have been used as supports for solid-phase organic synthesis. Silica gel is a rigid, insoluble material, which does not swell in organic solvents. Commercially available silica gel differs in particle size, pore size (typically 2–10 nm), and surface area (typically 200–800 m^2/g). Like macroporous, highly cross-linked polystyrene, silica gel enables efficient and rapid transfer of solvents and reagents to its entire surface. Because the synthetic intermediates are only located on the surface of the support, enzyme-mediated reactions can be realized on silica [189,190]. Silica gel is particularly well suited for continuous-flow synthesis because its volume stays constant and diffusion rates are high.

In 1970, Bayer and coworkers investigated the synthesis of peptides on silica gel [191]. The first amino acid was attached to the support as the ester of 1,4-bis(hydroxy-methyl)benzene (Figure 2.8); loadings of 0.006–0.06 mmol/g were attained. Although higher coupling rates than those on polystyrene were reported [191], silica gel never became an established support for solid-phase peptide synthesis.

For the automated solid-phase synthesis of oligonucleotides, however, silica was found to be the support of choice [192–197]. Silica with large pore size (25–300 nm), so-called 'controlled pore glass' (CPG), is generally used for this purpose. The main advantages of CPG, as compared with silica gel, are its more regular particle size and shape, and greater mechanical stability.

Attachment of suitable linkers to the surface of silica can be achieved by transester-ification with (3-aminopropyl)triethoxysilane, which leads to the support **2** (Figure 2.8) [198–200]. Alternatively, silica can be functionalized by reaction with alkyltri-chlorosilanes [201]. For the solid-phase synthesis of oligonucleotides, supports with a longer spacer, such as that in **3**, have proven more convenient than **2** [202–206]. Supports **3**, so-called LCAA-CPG (long chain alkylamine CPG [194,195]), are commer-cially available (typical loading: 0.1 mmol/g) and are currently the most commonly used supports for the synthesis of oligonucleotides. For this purpose, protected nucleo-sides are converted into succinic acid monoesters, and then coupled to LCAA-CPG. CPG functionalized with a 3-mercaptopropyl linker has been used for the solid-phase synthesis of oligosaccharides [207].

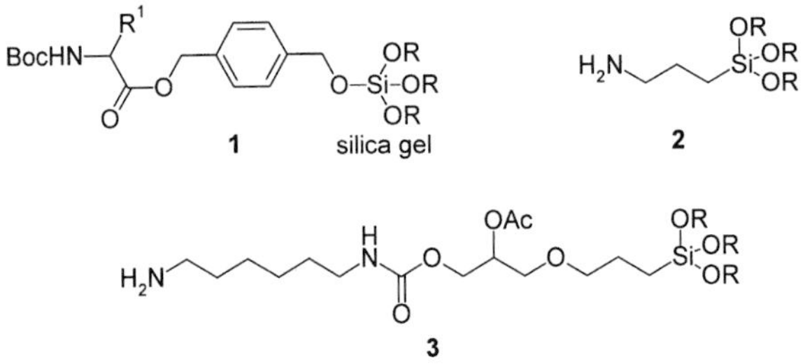

Figure 2.8. Linkers used to attach synthetic intermediates to silica. **1** [191], **2** [198], **3** [202].

The main disadvantages of CPG are its low loading, its high cost, and its high hydrophilicity. Water is difficult to remove quantitatively from this support, which can reduce the yields of acylation reactions [90].

2.7 Polysaccharides

Cellulose is an unbranched polysaccharide consisting of 1,4-linked β-D-glucose and can attain lengths of up to 15000 pyranose units. It neither dissolves nor swells in most solvents, but can be hydrolyzed upon prolonged treatment with acids. Cellulose pow-der was the first type of support to be used (with limited success) for solid-phase pep-tide synthesis [208]. Limitations were mainly due to the low loading (0.1 mmol/g) attained in these initial experiments, and the high reactivity of cellulose. Despite these problems, cellulose has occasionally been used for solid-phase synthesis, e.g. in the form of cotton [209], paper [210–216], and Perloza (beaded cellulose) [217].

Cellulose contains one primary and two secondary hydroxyl groups per pyranose, which form intra- and interstrand hydrogen bonds and thereby stabilize the compact structure of this polysaccharide. The acylation of all three hydroxyl groups therefore requires harsh reaction conditions [209]. The usual strategy for derivatization of a polysaccharide involves partial O-acylation or O-alkylation with a suitable linker or intermediate, followed by acetylation of the remaining hydroxyl groups. Loadings of up to 0.27 mmol/g have been attained by acylation of cotton with Fmoc glycine (DIC, HOBt, *N*-methylimidazole, DMF, 2 × 4 h [209]). Mixed [211] and symmetric anhydrides [218] can also be used to acylate cellulose. Some strategies for the functionalization of polysaccharides are outlined in Figure 2.9.

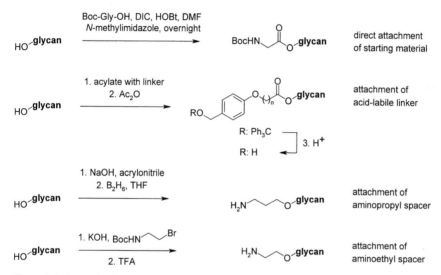

Figure 2.9. Strategies for the derivatization of polysaccharides [209,211,217–219].

After capping (i.e. acetylation of the remaining hydroxyl groups), the resulting functionalized cellulose can be used for the synthesis of peptides using both the Boc and Fmoc methodologies. Because cellulose is hydrolyzed by hydrogen fluoride (but not by TFA), nucleophilic saponification must be used to cleave the peptides from the support [209]. Alternatively, the support can be derivatized with a TFA-labile linker [211].

Another polysaccharide that has proven suitable as a support for the synthesis of peptides [219] and oligonucleotides [220] is sephadex (dextran, a branched glycan consisting of 1,6-α-linked glucopyranose). Sepharose (agarose, an unbranched glycan consisting of D-galactose and 3,6-anhydro-L-galactose) has been successfully used as a support for the enzyme-mediated synthesis of oligosaccharides [221].

Polysaccharides are chemically less stable than most other supports presented in this chapter. Chemical modification of polysaccharides can increase the solubility or change the mechanical properties of these materials. This can lead to a deterioration and partial loss of the support during longer synthetic sequences.

2.8 Miscellaneous Supports

Polyethylene has been grafted with acrylic acid, 2-hydroxyethyl methacrylate/*N*,*N*-dimethylacrylamide, or methacrylic acid/*N*,*N*-dimethylacrylamide to yield supports (e.g. Figure 2.10) suitable for the synthesis of peptides [2,222–224] and other compounds [216,225,226]. A similar support, also suitable for the synthesis of peptides, is polypropylene grafted with hydroxypropyl acrylate [227]. These supports can be used as membranes or as crown-shaped pinheads (Multipin; 2–4 mm diameter) with loadings of 1.2–2.2 µmol per crown.

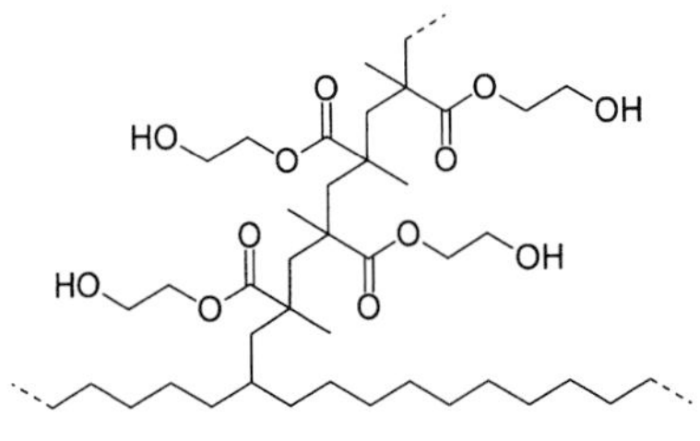

Figure 2.10. Hydroxyethyl methacrylate grafted onto polyethylene [2,224].

Treatment of polyethylene with $Cr_2O_3/H_2SO_4/H_2O$ leads to partial oxidation of its surface and the generation of carboxyl groups [228]. These have been used as attachment points for a functionalized polysaccharide (carboxymethyldextran) to yield a hydrophilic support suitable for the preparation of peptide libraries [228].

Copolymers of methacrylates with vinyl alcohol (Toyopearl, Fractogel) have, moreover, been successfully used as supports for the synthesis of peptides [229], oligonucleotides [230], and PNA [231]. Cross-linked poly(methyl methacrylates) [12,92,232] have also been investigated and used for solid-phase peptide synthesis.

With the discovery of ruthenium carbene complexes as highly effective catalysts for olefin metathesis under mild reaction conditions [233,234], the scope of ring-opening metathesis polymerization could be extended to include functionalized and sensitive monomers. The resulting (soluble) polymers have been used as supports for simple synthetic transformations [235–237]. Insoluble polymers have been prepared by ring-opening metathesis copolymerization of norbornene with 1,4,4a,5,8,8a-hexahydro-1,4,5,8-*exo-endo*-dimethanonaphthalene. These polymers have been used as supports for ruthenium carbene complexes [238].

Further materials that have been evaluated as supports for solid-phase synthesis include phenol–formaldehyde polymers [239,240], platinum electrodes coated with polythiophenes [241], proteins (bovine serum albumin) [242], polylysine [243], soluble poly(vinyl alcohol) [244], various copolymers of vinyl alcohol [4,245,246], and soluble dendrimers [14,247].

References for Chapter 2

[1] Lebl, M. *Biopolymers* **1998**, *47*, 397–404.
[2] Maeji, N. J.; Valerio, R. M.; Bray, A. M.; Campbell, R. A.; Geysen, H. M. *Reactive Polymers* **1994**, *22*, 203–212.
[3] Hird, N.; Hughes, I.; Hunter, D.; Morrison, M. G. J. T.; Sherrington, D. C.; Stevenson, L. *Tetrahedron* **1999**, *55*, 9575–9584.
[4] Gravert, D. J.; Janda, K. D. *Chem. Rev.* **1997**, *97*, 489–509.
[5] Bayer, E. *Angew. Chem. Int. Ed. Engl.* **1991**, *30*, 113–129.
[6] Mutter, M.; Uhmann, R.; Bayer, E. *Liebigs Ann. Chem.* **1975**, 901–915.
[7] Bayer, E.; Dengler, M.; Hemmasi, B. *Int. J. Pept. Prot. Res.* **1985**, *25*, 178–186.
[8] Jesberger, M.; Jaunzems, J.; Jung, A.; Jas, G.; Schönberger, A.; Kirschning, A. *Synlett* **2000**, 1289–1293.
[9] Sherrington, D. C. *Chem. Commun.* **1998**, 2275–2286.
[10] Andreatta, R. H.; Rink, H. *Helv. Chim. Acta* **1973**, *56*, 1205–1218.
[11] Gerritz, S. W.; Trump, R. P.; Zuercher, W. J. *J. Am. Chem. Soc.* **2000**, *122*, 6357–6363.
[12] Labadie, J. W. *Curr. Opinion Chem. Biol.* **1998**, *2*, 346–352.
[13] Blackburn, C. *Biopolymers* **1998**, *47*, 311–351.
[14] Haag, R. *Chem. Eur. J.* **2001**, *7*, 327–335.
[15] Rasoul, F.; Ercole, F.; Pham, Y.; Bui, C. T.; Wu, Z.; James, S. N.; Trainor, R. W.; Wickham, G.; Maeji, N. J. *Biopolymers (Peptide Science)* **2000**, *55*, 207–216.
[16] Shemyakin, M. M.; Ovchinnikov, Y. A.; Kinyushkin, A. A.; Kozhevnikova, I. V. *Tetrahedron Lett.* **1965**, 2323–2327.
[17] Chiu, S. H. L.; Anderson, L. *Carbohyd. Res.* **1976**, *50*, 227–238.
[18] Chen, S.; Janda, K. D. *J. Am. Chem. Soc.* **1997**, *119*, 8724–8725.
[19] Hayatsu, H.; Khorana, H. G. *J. Am. Chem. Soc.* **1967**, *89*, 3880–3887.
[20] Narita, M. *Bull. Chem. Soc. Jpn.* **1978**, *51*, 1477–1480.
[21] Cramer, F.; Helbig, R.; Hettler, H.; Scheit, K. H.; Seliger, H. *Angew. Chem.* **1966**, *78*, 640–641; *Angew. Chem. Int. Ed. Engl.* **1966**, *5*, 601–602.
[22] Enholm, E. J.; Gallagher, M. E.; Jiang, S.; Batson, W. A. *Org. Lett.* **2000**, *2*, 3355–3357.
[23] Pepper, K. W.; Paisley, H. M.; Young, M. A. *J. Chem. Soc.* **1953**, 4097–4105.
[24] Pickup, S.; Blum, F. D.; Ford, W. T.; Periyasami, M. *J. Am. Chem. Soc.* **1986**, *108*, 3987–3990.
[25] Rana, S.; White, P.; Bradley, M. *J. Comb. Chem.* **2001**, *3*, 9–15.
[26] Balakrishnan, T.; Ford, W. T. *J. Appl. Polym. Sci.* **1982**, *27*, 133–138.
[27] Merrifield, R. B. *J. Am. Chem. Soc.* **1963**, *85*, 2149–2154.
[28] Letsinger, R. L.; Kornet, M. J. *J. Am. Chem. Soc.* **1963**, *85*, 3045–3046.
[29] Sarin, V. K.; Kent, S. B. H.; Merrifield, R. B. *J. Am. Chem. Soc.* **1980**, *102*, 5463–5470.
[30] Shea, K. J.; Stoddard, G. J. *Macromolecules* **1991**, *24*, 1207–1209.
[31] Kress, J.; Rose, A.; Frey, J. G.; Brocklesby, W. S.; Ladlow, M.; Mellor, G. W.; Bradley, M. *Chem. Eur. J.* **2001**, *7*, 3880–3883.
[32] Köster, H.; Geussenhainer, S. *Angew. Chem. Int. Ed. Engl.* **1972**, *11*, 713–714.
[33] Santini, R.; Griffith, M. C.; Qi, M. *Tetrahedron Lett.* **1998**, *39*, 8951–8954.
[34] Merrifield, R. B. *Brit. Polym. J.* **1984**, *16*, 173–178.
[35] Yan, B.; Fell, J. B.; Kumaravel, G. *J. Org. Chem.* **1996**, *61*, 7467–7472.
[36] Naumann, G. in *Ullmanns Encyklopädie der technischen Chemie*; Verlag Chemie: Weinheim, New York, **1977**; p. 303.
[37] Kanda, P.; Kennedy, R. C.; Sparrow, J. T. *Int. J. Pept. Prot. Res.* **1991**, *38*, 385–391.
[38] Akelah, A.; Sherrington, D. C. *Chem. Rev.* **1981**, *81*, 557–587.
[39] Stranix, B. R.; Gao, J. P.; Barghi, R.; Salha, J.; Darling, G. D. *J. Org. Chem.* **1997**, *62*, 8987–8993.
[40] Stranix, B. R.; Darling, G. D. *J. Org. Chem.* **1997**, *62*, 9001–9004.
[41] Darling, G. D.; Fréchet, J. M. J. *J. Org. Chem.* **1986**, *51*, 2270–2276.
[42] Farrall, M. J.; Fréchet, J. M. J. *J. Org. Chem.* **1976**, *41*, 3877–3882.
[43] Seliger, H. *Makromol. Chem.* **1973**, *169*, 83–93.
[44] Dowling, L. M.; Stark, G. R. *Biochemistry* **1969**, *8*, 4728–4734.
[45] Mizoguchi, T.; Shigezane, K.; Takamura, N. *Chem. Pharm. Bull.* **1970**, *18*, 1465–1474.
[46] Matsueda, G. R.; Stewart, J. M. *Peptides* **1981**, *2*, 45–50.
[47] Ajayaghosh, A.; Pillai, V. N. R. *Tetrahedron* **1988**, *44*, 6661–6666.
[48] Kobayashi, S.; Moriwaki, M. *Tetrahedron Lett.* **1997**, *38*, 4251–4254.
[49] Scarr, R. B.; Findeis, M. F. *Pept. Res.* **1990**, *3*, 238–241.
[50] Wang, S.; Merrifield, R. B. *J. Am. Chem. Soc.* **1969**, *91*, 6488–6491.

[51] Park, B. D.; Lee, H. I.; Ryoo, S. J.; Lee, Y. S. *Tetrahedron Lett.* **1997**, *38*, 591–594.
[52] Tomoi, M.; Kori, N.; Kakiuchi, H. *Reactive Polymers* **1985**, *5*, 341–349.
[53] Pinnell, R. P.; Khune, G. D.; Khatri, N. A.; Manatt, S. L. *Tetrahedron Lett.* **1984**, *25*, 3511–3514.
[54] Gerigk, U.; Gerlach, M.; Neumann, W. P.; Vieler, R.; Weintritt, V. *Synthesis* **1990**, 448–452.
[55] Ford, W. T.; Yacoub, S. A. *J. Org. Chem.* **1981**, *46*, 819–821.
[56] Arshady, R.; Kenner, G. W.; Ledwith, A. *Makromol. Chem.* **1976**, *177*, 2911–2918.
[57] Sheng, Q.; Stöver, H. D. H. *Macromolecules* **1997**, *30*, 6712–6714.
[58] Mohanraj, S.; Ford, W. T. *Macromolecules* **1986**, *19*, 2470–2472.
[59] Mitchell, A. R.; Kent, S. B. H.; Erickson, B. W.; Merrifield, R. B. *Tetrahedron Lett.* **1976**, 3795–3798.
[60] Zikos, C. C.; Ferderigos, N. G. *Tetrahedron Lett.* **1995**, *36*, 3741–3744.
[61] Mitchell, A. R.; Kent, S. B. H.; Engelhard, M.; Merrifield, R. B. *J. Org. Chem.* **1978**, *43*, 2845–2852.
[62] Adams, J. H.; Cook, R. M.; Hudson, D.; Jammalamadaka, V.; Lyttle, M. H.; Songster, M. F. *J. Org. Chem.* **1998**, *63*, 3706–3716.
[63] Sparrow, J. T. *J. Org. Chem.* **1976**, *41*, 1350–1353.
[64] Sampson, D. F. J.; Simmonds, R. G.; Bradley, M. *Tetrahedron Lett.* **2001**, *42*, 5517–5519.
[65] Bernard, M.; Ford, W. T. *J. Org. Chem.* **1983**, *48*, 326–332.
[66] O'Brien, R. A.; Chen, T.; Rieke, R. D. *J. Org. Chem.* **1992**, *57*, 2667–2677.
[67] Fyles, T. M.; Leznoff, C. C. *Can. J. Chem.* **1976**, *54*, 935–942.
[68] Lochmann, L.; Fréchet, J. M. J. *Macromolecules* **1996**, *29*, 1767–1771.
[69] Ruel, G.; The, N. K.; Dumartin, G.; Delmond, B.; Pereyre, M. *J. Organomet. Chem.* **1993**, *444*, C18–C20.
[70] Ruhland, T.; Andersen, K.; Pedersen, H. *J. Org. Chem.* **1998**, *63*, 9204–9211.
[71] Nicolaou, K. C.; Pastor, J.; Barluenga, S.; Winssinger, N. *Chem. Commun.* **1998**, 1947–1948.
[72] Nicolaou, K. C.; Winssinger, N.; Pastor, J.; DeRoose, F. *J. Am. Chem. Soc.* **1997**, *119*, 449–450.
[73] Neumann, W. P.; Peterseim, M. *Reactive Polymers* **1993**, *20*, 189–205.
[74] Deleuze, H.; Sherrington, D. C. *J. Chem. Soc., Perkin Trans. 2* **1995**, 2217–2221.
[75] Mahajan, A.; Chhabra, S. R.; Chan, W. C. *Tetrahedron Lett.* **1999**, *40*, 4909–4912.
[76] Wells, N. J.; Basso, A.; Bradley, M. *Biopolymers* **1998**, *47*, 381–396.
[77] Fromont, C.; Bradley, M. *Chem. Commun.* **2000**, 283–284.
[78] Basso, A.; Evans, B.; Pegg, N.; Bradley, M. *Chem. Commun.* **2001**, 697–698.
[79] Tam, J. P.; Lu, Y. A. *J. Am. Chem. Soc.* **1995**, *117*, 12058–12063.
[80] Epton, R.; Wellings, D. A.; Williams, A. *Reactive Polymers* **1987**, *6*, 143–157.
[81] Dixit, D. M.; Leznoff, C. C. *J. Chem. Soc., Chem. Commun.* **1977**, 798–799.
[82] Farrall, M. J.; Fréchet, J. M. J. *J. Am. Chem. Soc.* **1978**, *100*, 7998–7999.
[83] Biagini, S. C. G.; Gibson, S. E.; Keen, S. P. *J. Chem. Soc., Perkin Trans. 1* **1998**, 2485–2499.
[84] Guyot, A. *Pure Appl. Chem.* **1988**, *60*, 365–376.
[85] Svec, F.; Fréchet, J. M. J. *Science* **1996**, *273*, 205–211.
[86] Millar, J. R.; Smith, D. G.; Marr, W. E.; Kressman, T. R. E. *J. Chem. Soc.* **1963**, 218–224.
[87] Kunin, R.; Meitzner, E.; Bortnick, N. *J. Am. Chem. Soc.* **1962**, *84*, 305–306.
[88] Hori, M.; Gravert, D. J.; Wentworth, P.; Janda, K. D. *Bioorg. Med. Chem. Lett.* **1998**, *8*, 2363–2368.
[89] Malenfant, P. R. L.; Fréchet, J. M. J. *Chem. Commun.* **1998**, 2657–2658.
[90] McCollum, C.; Andrus, A. *Tetrahedron Lett.* **1991**, *32*, 4069–4072.
[91] Köster, H.; Cramer, F. *Liebigs Ann. Chem.* **1974**, 946–958.
[92] Sophiamma, P. N.; Sreekumar, K. *Indian J. Chem., Sect. B - Org. Chem.* **1997**, *36*, 995–999.
[93] Roice, M.; Kumar, K. S.; Pillai, V. N. R. *Tetrahedron* **2000**, *56*, 3725–3734.
[94] Roice, M.; Kumar, K. S.; Pillai, V. N. R. *Macromolecules* **1999**, *32*, 8807–8815.
[95] Varkey, J. T.; Pillai, V. N. R. *J. Pept. Res.* **1998**, *51*, 49–54.
[96] Varkey, J. T.; Pillai, V. N. R. *J. Appl. Polym. Sci.* **1999**, *71*, 1933–1939.
[97] Varkey, J. T.; Pillai, V. N. R. *J. Pept. Sci.* **1999**, *5*, 577–581.
[98] Toy, P. H.; Janda, K. D. *Tetrahedron Lett.* **1999**, *40*, 6329–6332.
[99] Toy, P. H.; Reger, T. S.; Garibay, P.; Garno, J. C.; Malikayil, J. A.; Liu, G.-Y.; Janda, K. D. *J. Comb. Chem.* **2001**, *3*, 117–124.
[100] Garibay, P.; Toy, P. H.; Hoeg-Jensen, T.; Janda, K. D. *Synlett* **1999**, 1438–1440.
[101] Dickerson, T. J.; Reed, N. N.; Janda, K. D. *Bioorg. Med. Chem. Lett.* **2001**, *11*, 1507–1509.
[102] Renil, M.; Pillai, V. N. R. *J. Appl. Polym. Sci.* **1996**, *61*, 1585–1594.
[103] Renil, M.; Nagaraj, R.; Pillai, V. N. R. *Tetrahedron* **1994**, *50*, 6681–6688.
[104] Renil, M.; Pillai, V. N. R. *Tetrahedron Lett.* **1994**, *35*, 3809–3812.
[105] Kumar, K. S.; Pillai, V. N. R. *Tetrahedron* **1999**, *55*, 10437–10446.
[106] Wilson, M. E.; Paech, K.; Zhou, W.; Kurth, M. J. *J. Org. Chem.* **1998**, *63*, 5094–5099.
[107] Renil, M.; Meldal, M. *Tetrahedron Lett.* **1996**, *37*, 6185–6188.

[108] Buchardt, J.; Meldal, M. *Tetrahedron Lett.* **1998**, *39*, 8695–8698.
[109] Cho, J. K.; Park, B.-D.; Lee, Y.-S. *Tetrahedron Lett.* **2000**, *41*, 7481–7485.
[110] Berg, R. H.; Almdal, K.; Pedersen, W. B.; Holm, A.; Tam, J. P.; Merrifield, R. B. *J. Am. Chem. Soc.* **1989**, *111*, 8024–8026.
[111] Kent, S. B. H.; Merrifield, R. B. *Israel J. Chem.* **1978**, *17*, 243–247.
[112] Birch-Hirschfeld, E.; Foldes-Papp, Z.; Gührs, K. H.; Seliger, H. *Nucleic Acids Res.* **1994**, *22*, 1760–1761.
[113] Birch-Hirschfeld, E.; Földes-Papp, Z.; Gührs, K. H.; Seliger, H. *Helv. Chim. Acta* **1996**, *79*, 137–150.
[114] Witkowski, W.; Birch-Hirschfeld, E.; Weiss, R.; Zarytova, V. F.; Gorn, V. V. *J. Prakt. Chem.* **1984**, *326*, 320–328.
[115] Weiss, R.; Birch-Hirschfeld, E.; Witkowski, W.; Friese, K. *Z. Chem.* **1986**, *26*, 127–130.
[116] Zhao, C.; Shi, S.; Mir, D.; Hurst, D.; Li, R.; Xiao, X. Y.; Lillig, J.; Czarnik, A. W. *J. Comb. Chem.* **1999**, *1*, 91–95.
[117] Devivar, R. V.; Koontz, S. L.; Peltier, W. J.; Pearson, J. E.; Guillory, T. A.; Fabricant, J. D. *Bioorg. Med. Chem. Lett.* **1999**, *9*, 1239–1242.
[118] Bray, A. M.; Jhingran, A. G.; Valerio, R. M.; Maeji, N. J. *J. Org. Chem.* **1994**, *59*, 2197–2203.
[119] Juby, C. D.; Richardson, C. D.; Brousseau, R. *Tetrahedron Lett.* **1991**, *32*, 879–882.
[120] Takahashi, T.; Tomida, S.; Inoue, H.; Doi, T. *Synlett* **1998**, 1261–1263.
[121] Tommasi, R. A.; Nantermet, P. G.; Shapiro, M. J.; Chin, J.; Brill, W. K. D.; Ang, K. *Tetrahedron Lett.* **1998**, *39*, 5477–5480.
[122] Takahashi, T.; Ebata, S.; Doi, T. *Tetrahedron Lett.* **1998**, *39*, 1369–1372.
[123] Hodges, J. C.; Harikrishnan, L. S.; Ault-Justus, S. *J. Comb. Chem.* **2000**, *2*, 80–88.
[124] Barrett, A. G. M.; Cramp, S. M.; Roberts, R. S. *Org. Lett.* **1999**, *1*, 1083–1086.
[125] Sucholeiki, I.; Perez, J. M. *Tetrahedron Lett.* **1999**, *40*, 3531–3534.
[126] Szymonifka, M. J.; Chapman, K. T. *Tetrahedron Lett.* **1995**, *36*, 1597–1600.
[127] Sucholeiki, I.; Perez, J. M.; Owens, P. D. *Tetrahedron Lett.* **2001**, *42*, 3279–3282.
[128] Hellermann, H.; Lucas, H. W.; Maul, J.; Pillai, V. N. R.; Mutter, M. *Makromol. Chem.* **1983**, *184*, 2603–2617.
[129] Kates, S. A.; McGuinness, B. F.; Blackburn, C.; Griffin, G. W.; Solé, N. A.; Barany, G.; Albericio, F. *Biopolymers* **1998**, *47*, 365–380.
[130] Wright, P.; Lloyd, D.; Rapp, W.; Andrus, A. *Tetrahedron Lett.* **1993**, *34*, 3373–3376.
[131] Gooding, O. W.; Baudart, S.; Deegan, T. L.; Heisler, K.; Labadie, J. W.; Newcomb, W. S.; Porco, J. A.; van Eikeren, P. *J. Comb. Chem.* **1999**, 113–122.
[132] Porco, J. A.; Deegan, T.; Devonport, W.; Gooding, O. W.; Heisler, K.; Labadie, J. W.; Newcomb, B.; Nguyen, C.; van Eikeren, P.; Wong, J.; Wright, P. *Mol. Diversity* **1997**, *2*, 197–206.
[133] Keifer, P. A. *J. Org. Chem.* **1996**, *61*, 1558–1559.
[134] Pursch, M.; Schlotterbeck, G.; Tseng, L. H.; Albert, K. *Angew. Chem. Int. Ed. Engl.* **1997**, *35*, 2867–2869.
[135] Bettinger, T.; Remy, J. S.; Erbacher, P.; Behr, J. P. *Bioconjugate Chem.* **1998**, *9*, 842–846.
[136] Keifer, P. A.; Baltusis, L.; Rice, D. M.; Tymiak, A. A.; Shoolery, J. N. *J. Magn. Reson. Ser. A* **1996**, *119*, 65–75.
[137] Wehler, T.; Westman, J. *Tetrahedron Lett.* **1996**, *37*, 4771–4774.
[138] Anderson, R. C.; Jarema, M. A.; Shapiro, M. J.; Stokes, J. P.; Ziliox, M. *J. Org. Chem.* **1995**, *60*, 2650–2651.
[139] Li, W.; Xiao, X.; Czarnik, A. W. *J. Comb. Chem.* **1999**, *1*, 127–129.
[140] Li, W.; Yan, B. *J. Org. Chem.* **1998**, *63*, 4092–4097.
[141] Swali, V.; Wells, N. J.; Langley, G. J.; Bradley, M. *J. Org. Chem.* **1997**, *62*, 4902–4903.
[142] Hutchins, S. M.; Chapman, K. T. *Tetrahedron Lett.* **1996**, *37*, 4869–4872.
[143] Nouvet, A.; Binard, M.; Lamaty, F.; Martinez, J.; Lazaro, R. *Tetrahedron* **1999**, *55*, 4685–4698.
[144] Harris, J. M. *J. Macromol. Sci. C., Rev. Macromol. Chem. Phys.* **1985**, *C25*, 325–373.
[145] Mutter, M.; Hagenmaier, H.; Bayer, E. *Angew. Chem.* **1971**, *83*, 883–884; *Angew. Chem. Int. Ed. Engl.* **1971**, *10*, 811–812.
[146] Novabiochem Catalog & Peptide Synthesis Handbook **1999**, Läufelfingen, CH.
[147] Simmonds, R. G. *Int. J. Pept. Prot. Res.* **1996**, *47*, 36–41.
[148] Bonora, G. M.; Biancotto, G.; Maffini, M.; Scremin, C. L. *Nucleic Acids Res.* **1993**, *21*, 1213–1217.
[149] Wang, Y.; Zhang, H.; Voelter, W. *Chem. Lett.* **1995**, 273–274.
[150] Brandstetter, F.; Schott, H.; Bayer, E. *Tetrahedron Lett.* **1973**, 2997–3000.
[151] Ito, Y.; Kanie, O.; Ogawa, T. *Angew. Chem. Int. Ed. Engl.* **1996**, *35*, 2510–2512.
[152] Douglas, S. P.; Whitfield, D. M.; Krepinsky, J. J. *J. Am. Chem. Soc.* **1995**, *117*, 2116–2117.
[153] Bayer, E.; Mutter, M.; Uhmann, R.; Polster, J.; Mauser, H. *J. Am. Chem. Soc.* **1974**, *96*, 7333–7336.

[154] Molteni, V.; Annunziata, R.; Cinquini, M.; Cozzi, F.; Benaglia, M. *Tetrahedron Lett.* **1998**, *39*, 1257–1260.
[155] Park, W. K. C.; Auer, M.; Jaksche, H.; Wong, C. H. *J. Am. Chem. Soc.* **1996**, *118*, 10150–10155.
[156] Far, A. R.; Tidwell, T. T. *J. Org. Chem.* **1998**, *63*, 8636–8637.
[157] Yoon, J.; Cho, C. W.; Han, H.; Janda, K. D. *Chem. Commun.* **1998**, 2703–2704.
[158] Shey, J. Y.; Sun, C. M. *Synlett* **1998**, 1423–1425.
[159] Benaglia, M.; Annunziata, R.; Cinquini, M.; Cozzi, F.; Ressel, S. *J. Org. Chem.* **1998**, *63*, 8628–8629.
[160] Rademann, J.; Meldal, M.; Bock, K. *Chem. Eur. J.* **1999**, *5*, 1218–1225.
[161] Rademann, J.; Grøtli, M.; Meldal, M.; Bock, K. *J. Am. Chem. Soc.* **1999**, *121*, 5459–5466.
[162] Grøtli, M.; Gotfredsen, C. H.; Rademann, J.; Buchardt, J.; Clark, A. J.; Duus, J. Ø.; Meldal, M. *J. Comb. Chem.* **2000**, *2*, 108–119.
[163] Arshady, R.; Atherton, E.; Clive, D. L. J.; Sheppard, R. C. *J. Chem. Soc., Perkin Trans. 1* **1981**, 529–537.
[164] McMurray, J. S. *Biopolymers* **1998**, *47*, 405–411.
[165] Bergbreiter, D. E.; Case, B. L.; Liu, Y. S.; Caraway, J. W. *Macromolecules* **1998**, *31*, 6053–6062.
[166] Bonora, G. M.; Baldan, A.; Schiavon, O.; Ferruti, P.; Veronese, F. M. *Tetrahedron Lett.* **1996**, *37*, 4761–4764.
[167] Calas, B.; Méry, J.; Parello, J.; Cave, A. *Tetrahedron* **1985**, *41*, 5331–5339.
[168] Smith, C. W.; Stahl, G. L.; Walter, R. *Int. J. Pept. Prot. Res.* **1979**, *13*, 109–112.
[169] Stahl, G. L.; Walter, R.; Smith, C. W. *J. Am. Chem. Soc.* **1979**, *101*, 5383–5394.
[170] Mendre, C.; Sarrade, V.; Calas, B. *Int. J. Pept. Prot. Res.* **1992**, *39*, 278–284.
[171] Epton, R.; Goddard, P.; Marr, G.; McLaren, J. V.; Morgan, G. J. *Polymer* **1979**, *20*, 1444–1446.
[172] Atherton, E.; Clive, D. L. J.; Sheppard, R. C. *J. Am. Chem. Soc.* **1975**, *97*, 6584–6585.
[173] Sparrow, J. T.; Knieb-Cordonier, N. G.; Obeyseskere, N. U.; McMurray, J. S. *Pept. Res.* **1996**, *9*, 297–304.
[174] Dryland, A.; Sheppard, R. C. *J. Chem. Soc., Perkin Trans. 1* **1986**, 125–137.
[175] Atherton, E.; Brown, E.; Sheppard, R. C.; Rosevear, A. *J. Chem. Soc., Chem. Commun.* **1981**, 1151–1152.
[176] Dryland, A.; Sheppard, R. C. *Tetrahedron* **1988**, *44*, 859–876.
[177] Atherton, E.; Sheppard, R. C. *Solid Phase Peptide Synthesis; A Practical Approach*; Oxford University Press: Oxford, **1989**.
[178] Gait, M. J.; Matthes, H. W. D.; Singh, M.; Sproat, B. S.; Titmas, R. C. *Nucleic Acids Res.* **1982**, *10*, 6243–6254.
[179] Chadwick, R. J.; Thompson, J. S.; Tomalin, G. *Biochem. Soc. Trans.* **1991**, *19*, 406S.
[180] Small, P. W.; Sherrington, D. C. *J. Chem. Soc., Chem. Commun.* **1989**, 1589–1591.
[181] Meldal, M. *Tetrahedron Lett.* **1992**, *33*, 3077–3080.
[182] Renil, M.; Meldal, M. *Tetrahedron Lett.* **1995**, *36*, 4647–4650.
[183] Renil, M.; Ferreras, M.; Delaisse, J. M.; Foged, N. T.; Meldal, M. *J. Pept. Sci.* **1998**, *4*, 195–210.
[184] Hilaire, P. M. S.; Lowary, T. L.; Meldal, M.; Bock, K. *J. Am. Chem. Soc.* **1998**, *120*, 13312–13320.
[185] Meldal, M.; Svendsen, I. *J. Chem. Soc., Perkin Trans. 1* **1995**, 1591–1596.
[186] Meldal, M.; Svendsen, I.; Breddam, K.; Auzanneau, F. I. *Proc. Natl. Acad. Sci. USA* **1994**, *91*, 3314–3318.
[187] Meldal, M.; Auzanneau, F. I.; Hindsgaul, O.; Palcic, M. M. *J. Chem. Soc., Chem. Commun.* **1994**, 1849–1850.
[188] Camarero, J. A.; Cotton, G. J.; Adeva, A.; Muir, T. W. *J. Pept. Res.* **1998**, *51*, 303–316.
[189] Singh, J.; Allen, M. P.; Ator, M. A.; Gainor, J. A.; Whipple, D. A. *J. Med. Chem.* **1995**, *38*, 217–219.
[190] Schuster, M.; Wang, P.; Paulson, J. C.; Wong, C. H. *J. Am. Chem. Soc.* **1994**, *116*, 1135–1136.
[191] Bayer, E.; Jung, G.; Halász, I.; Sebastian, I. *Tetrahedron Lett.* **1970**, 4503–4505.
[192] Köster, H. *Tetrahedron Lett.* **1972**, 1527–1530.
[193] Pon, R. T.; Ogilvie, K. K. *Tetrahedron Lett.* **1984**, *25*, 713–716.
[194] Engels, J. W.; Uhlmann, E. *Angew. Chem. Int. Ed. Engl.* **1989**, *28*, 716–734.
[195] Ghosh, P. K.; Kumar, P.; Gupta, K. C. *J. Indian Chem. Soc.* **1998**, *75*, 206–218.
[196] Ogilvie, K. K.; Nemer, M. J. *Tetrahedron Lett.* **1980**, *21*, 4159–4162.
[197] Köster, H.; Biernat, J.; McManus, J.; Wolter, A.; Stumpe, A.; Narang, C. K.; Sinha, N. D. *Tetrahedron* **1984**, *40*, 103–112.
[198] Matteucci, M. D.; Caruthers, M. H. *J. Am. Chem. Soc.* **1981**, *103*, 3185–3191.
[199] Matteucci, M. D.; Caruthers, M. H. *Tetrahedron Lett.* **1980**, *21*, 719–722.
[200] Keana, J. F. W.; Shimizu, M.; Jernsted, K. K. *J. Org. Chem.* **1986**, *51*, 1641–1644.
[201] Parr, W.; Grohmann, K. *Angew. Chem. Int. Ed. Engl.* **1972**, *11*, 314–315.

[202] Adams, S. P.; Kavka, K. S.; Wykes, E. J.; Holder, S. B.; Galluppi, G. R. *J. Am. Chem. Soc.* **1983**, *105*, 661–663.

[203] Gough, G. R.; Brunden, M. J.; Gilham, P. T. *Tetrahedron Lett.* **1981**, *22*, 4177–4180.

[204] Usman, N.; Ogilvie, K. K.; Jiang, M. Y.; Cedergren, R. J. *J. Am. Chem. Soc.* **1987**, *109*, 7845–7854.

[205] Katzhendler, J.; Cohen, S.; Rahamim, E.; Weisz, M.; Ringel, I.; Deutsch, J. *Tetrahedron* **1989**, *45*, 2777–2792.

[206] Van Aerschot, A.; Herdewijn, P.; van der Haeghe, H. *Nucleosides and Nucleotides* **1988**, *7*, 75–90.

[207] Heckel, A.; Mross, E.; Jung, K. H.; Rademann, J.; Schmidt, R. R. *Synlett* **1998**, 171–173.

[208] Merrifield, B. in *Peptides; Synthesis, Structures, and Applications*; Gutte, B., Ed.; Academic Press: London, **1995**.

[209] Eichler, J.; Bienert, M.; Stierandova, A.; Lebl, M. *Pept. Res.* **1991**, *5*, 296–307.

[210] Blankemeyer-Menge, B.; Frank, R. *Tetrahedron Lett.* **1988**, *29*, 5871–5874.

[211] Frank, R.; Döring, R. *Tetrahedron* **1988**, *44*, 6031–6040.

[212] Ott, J.; Eckstein, F. *Nucleic Acids Res.* **1984**, *12*, 9137–9142.

[213] Frank, R.; Heikens, W.; Heisterberg-Moutsis, G.; Blöcker, H. *Nucleic Acids Res.* **1983**, *11*, 4365–4377.

[214] Frank, R.; Meyerhans, A.; Schwellnus, K.; Blöcker, H. *Methods Enzymol.* **1987**, *154*, 221–249.

[215] Kramer, A.; Schuster, A.; Reineke, U.; Malin, R.; Volkmer-Engert, R.; Landgraf, C.; Schneider-Mergener, J. *Methods: A Companion to Methods in Enzymology* **1994**, *6*, 388–395.

[216] Scharn, D.; Wenschuh, H.; Reineke, U.; Schneider-Mergener, J.; Germeroth, L. *J. Comb. Chem.* **2000**, *2*, 361–369.

[217] Englebretsen, D. R.; Harding, D. R. K. *Int. J. Pept. Prot. Res.* **1992**, *40*, 487–496.

[218] Dittrich, F.; Tegge, W.; Frank, R. *Bioorg. Med. Chem. Lett.* **1998**, *8*, 2351–2356.

[219] Annis, I.; Chen, L.; Barany, G. *J. Am. Chem. Soc.* **1998**, *120*, 7226–7238.

[220] Köster, H.; Heyns, K. *Tetrahedron Lett.* **1972**, 1531–1534.

[221] Blixt, O.; Norberg, T. *J. Org. Chem.* **1998**, *63*, 2705–2710.

[222] Geysen, H. M.; Rodda, S. J.; Mason, T. J.; Tribbick, G.; Schoofs, P. G. *J. Immunol. Meth.* **1987**, *102*, 259–274.

[223] Valerio, R. M.; Bray, A. M.; Maeji, N. J. *Int. J. Pept. Prot. Res.* **1994**, *44*, 158–165.

[224] Valerio, R. M.; Bray, A. M.; Campbell, R. A.; Dipasquale, A.; Margellis, C.; Rodda, S. J.; Geysen, H. M.; Maeji, N. J. *Int. J. Pept. Prot. Res.* **1993**, *42*, 1–9.

[225] Valerio, R. M.; Bray, A. M.; Patsiouras, H. *Tetrahedron Lett.* **1996**, *37*, 3019–3022.

[226] Bray, A. M.; Chiefari, D. S.; Valerio, R. M.; Maeji, N. J. *Tetrahedron Lett.* **1995**, *36*, 5081–5084.

[227] Daniels, S. B.; Bernatowicz, M. S.; Coull, J. M.; Köster, H. *Tetrahedron Lett.* **1989**, *30*, 4345–4348.

[228] Luo, K. X.; Zhou, P.; Lodish, H. F. *Proc. Natl. Acad. Sci. USA* **1995**, *92*, 11761–11765.

[229] Buettner, J. A.; Dadd, C. A.; Baumbach, G. A.; Masecar, B. L.; Hammond, D. J. *Int. J. Pept. Prot. Res.* **1996**, *47*, 70–83.

[230] Reddy, M. P.; Michael, M. A.; Farooqui, F.; Girgis, S. *Tetrahedron Lett.* **1994**, *35*, 5771–5774.

[231] Stetsenko, D. A.; Lubyako, E. N.; Potapov, V. K.; Azhikina, T. L.; Sverdlov, E. D. *Tetrahedron Lett.* **1996**, *37*, 3571–3574.

[232] Kempe, M.; Barany, G. *J. Am. Chem. Soc.* **1996**, *118*, 7083–7093.

[233] Zaragoza, F. *Metal Carbenes in Organic Synthesis*; Wiley-VCH: Weinheim, New York, **1999**.

[234] Ivin, K. J.; Mol, J. C. *Olefin Metathesis and Metathesis Polymerization*; Academic Press: London, **1997**.

[235] Barrett, A. G. M.; Cramp, S. M.; Roberts, R. S.; Zecri, F. *J. Org. Lett.* **1999**, *1*, 579–582.

[236] Ball, C. P.; Barrett, A. G. M.; Poitout, L. F.; Smith, M. L.; Thorn, Z. E. *Chem. Commun.* **1998**, 2453–2454.

[237] Enholm, E. J.; Gallagher, M. E. *Org. Lett.* **2001**, *3*, 3397–3399.

[238] Mayr, M.; Mayr, B.; Buchmeiser, M. R. *Angew. Chem. Int. Ed.* **2001**, *40*, 3839–3842.

[239] Inukai, N.; Nakano, K.; Murakami, M. *Bull. Chem. Soc. Jpn.* **1968**, *41*, 182–186.

[240] Flanigan, E.; Marshall, G. R. *Tetrahedron Lett.* **1970**, 2403–2406.

[241] Marchand, G.; Pilard, J. F.; Simonet, J. *Tetrahedron Lett.* **2000**, *41*, 883–885.

[242] Hansen, P. R.; Holm, A.; Houen, G. *Int. J. Pept. Prot. Res.* **1993**, *41*, 237–245.

[243] Chapman, T. M.; Kleid, D. G. *J. Chem. Soc., Chem. Commun.* **1973**, 193–194.

[244] Schott, H.; Brandstetter, F.; Bayer, E. *Makromol. Chem.* **1973**, *173*, 247–251.

[245] Eynde, J. J. V.; Rutot, D. *Tetrahedron* **1999**, *55*, 2687–2694.

[246] Seliger, H.; Aumann, G. *Tetrahedron Lett.* **1973**, 2911–2914.

[247] Zhang, J.; Aszodi, J.; Chartier, C.; L'hermite, N.; Weston, J. *Tetrahedron Lett.* **2001**, *42*, 6683–6686.

3 Linkers for Solid-Phase Organic Synthesis

Linkers are molecules which keep the intermediates in solid-phase synthesis bound to the support. Linkers should enable the easy attachment of the starting material to the support, be stable under a broad variety of reaction conditions, and yet enable selective cleavage at the end of a synthesis without damage to the product. Several types of linker have been developed, which meet these conflicting requirements to different extents [1]. Newer developments include linkers containing fluorine to facilitate the monitoring of solid-phase chemistry by NMR (see Section 1.3.5), and enantiomerically pure linkers that enable the synthesis on solid phase of enantiomerically enriched products [2].

Linkers enable the attachment of a variety of functional groups to a solid support and upon cleavage either the originally attached or a new functional group may be generated. In this chapter, the known linkers are organized according to the functional group *that results upon cleavage* from the support.

In many of the resins used for solid-phase synthesis, the linkers are attached to the support by means of a spacer (Figure 3.1). It has been proposed [3] that flexible spacers facilitate the diffusion of reagents to the resin-bound substrate by increasing its distance from the support. Although this might be of importance for the synthesis of certain 'difficult' peptides, there is little evidence that spacers improve the synthesis of small molecules on insoluble supports [4]. For the characterization of compounds attached to a polymeric support by NMR spectroscopy, however, long spacers are an advantage, because they increase the mobility of the substrate and reduce the line-broadening usually observed in the NMR spectra of polymers.

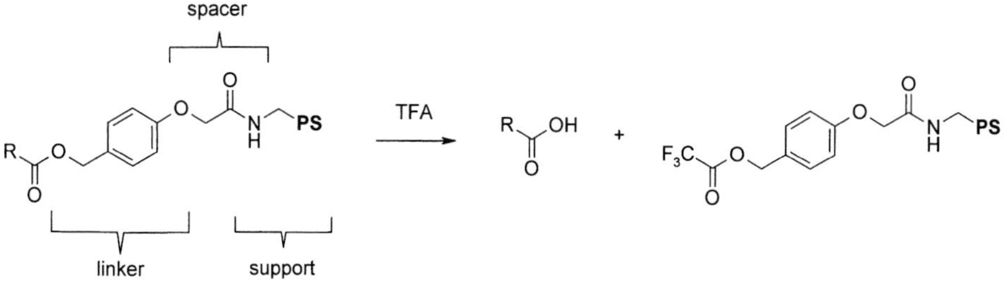

Figure 3.1. Acidolysis of a typical, acid-labile linker.

Linkers must be attached to the support in such a way that no release of the linker occurs either during a synthetic sequence or upon cleavage of the linker–product bond. For this purpose, linkers often contain a carboxyl group, which enables irreversible attachment of the linker to amino group containing supports through amide formation. In the case of polystyrene, aminomethyl polystyrene or amino(4-methylphenyl)methyl polystyrene ('methylbenzhydrylamine', MBHA resin, Figure 3.2) are generally used to attach other, more elaborate linkers. Aminomethyl polystyrene can be prepared directly from polystyrene by acid-catalyzed aminomethylation with *N*-(hydroxymethyl)phthalimide or *N*-(chloromethyl)phthalimide [5,6], or by ammonolysis of chloromethyl polystyrene [7] (see Section 10.1). MBHA resin is prepared by Friedel–Crafts acylation of polystyrene with 4-methylbenzoyl chloride, followed by reductive amination of the resulting benzophenone [8], or by amidoalkylation of polystyrene with *N*-[1-chloro-1(4-tolyl)methyl]phthalimide [6]. Unlike N-acylated aminomethyl polystyrene, N-acylated MBHA resin is not stable towards strong acids, and can, in fact, be used as a support for amides cleavable by hydrogen fluoride (Section 3.3).

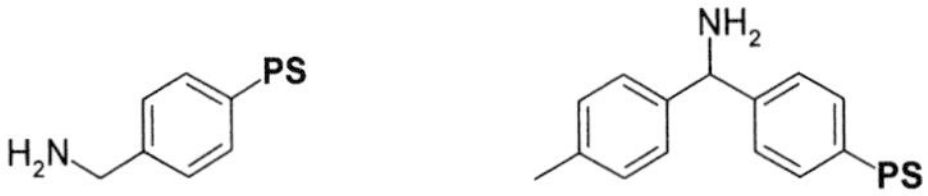

Figure 3.2. Polystyrene derivatives suitable for the irreversible attachment of linkers.

Amino group bearing derivatives of most other commonly used supports for solid-phase synthesis (PEG-grafted polystyrene, PEG, polyacrylamides, CPG) are commercially available; these enable the irreversible attachment of different linkers.

For some applications it might be desirable to cleave the product from a support in two or more portions. This can be realized by derivatizing a functionalized support with a mixture of different linkers that enable a sequential cleavage [9]. The resulting support can, for instance, be used to prepare and screen combinatorial peptide libraries by the mix-and-split method ([10–12]; one different peptide on each bead). The first portion of peptide released would be tested for biological activity, and, once an active peptide had been identified, the remaining peptide on the support could be used for structure elucidation.

Before cleaving products from a support, either extensive washing of the resin with volatile solvents or drying of the resin is required, because certain solvents (e.g. NMP, DMF) and reagents (e.g. DIPEA, DBU) tend to adhere strongly to some supports. These reagents can inhibit the cleavage reaction or lead to impure products. Extensive washing is particularly relevant if the products are to be analyzed or screened without prior purification.

Most acid-labile linkers and protective groups used in solid-phase synthesis lead to the formation of carbocations during acidolytic cleavage. Products containing electron-rich structural elements, such as pyrroles, indoles, thiols, phenols, etc. (but not ammonium salts) can be irreversibly alkylated by these carbocations, thereby leading to low yields or to mixtures of products. This can be avoided by adding scavengers to the cleavage reagent. Typical carbocation scavengers include thiols (EDT, PhSH), thioethers (Me$_2$S, PhSMe), phenols (cresol, PhOH), arylethers (PhOMe), and silanes (Et$_3$SiH, iPr$_3$SiH). If non-volatile scavengers are used, or when non-volatile by-products are formed during cleavage and deprotection (e.g. during detritylation), the crude products cannot usually be used directly but require further purification.

3.1 Linkers for Carboxylic Acids

Carboxylic acids are generally attached to polymeric supports as esters or amides. Depending on the type of linker and on the cleavage conditions used, cleavage can lead either to the regeneration of a carboxylic acid, or to the formation of a new product, such as an ester, amide, ketone, or alcohol. This section covers only linkers which lead, upon cleavage, to the release of carboxylic acids.

3.1.1 Linkers for Acids Cleavable by Acids or Other Electrophiles

3.1.1.1 Benzyl Alcohol Linkers

Acid-labile linkers are the oldest and still the most commonly used linkers for carboxylic acids. Most are based on the acidolysis of benzylic C–O bonds. Benzyl esters cleavable under acidic conditions were the first type of linker to be investigated in detail. The reason for this was probably the initial choice of polystyrene as an insoluble support for solid-phase synthesis [13]. Polystyrene-derived benzyl esters were initially prepared by the treatment of partially chloromethylated polystyrene with salts of carboxylic acids (Figure 3.3).

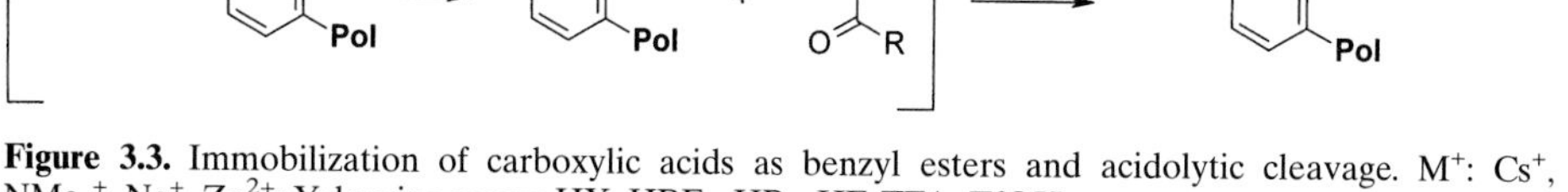

Figure 3.3. Immobilization of carboxylic acids as benzyl esters and acidolytic cleavage. M$^+$: Cs$^+$, NMe$_4^+$, Na$^+$, Zn^{2+}; Y: leaving group; HX: HBF$_4$, HBr, HF, TFA, TfOH.

Acidolysis of benzylic C–O bonds becomes easier with increasing stability of the corresponding benzylic cation. The sensitivity of benzylic linkers towards acids can therefore be fine-tuned by varying the substitution pattern on the arene. Electron-donating substituents will increase the sensitivity towards acids, whereas electron-acceptors will diminish acid-sensitivity. This order of reactivity towards electrophiles is the opposite to that towards nucleophiles: the more electron-rich the aromatic, the more resistant are the corresponding benzyl esters towards nucleophilic attack.

Table 3.1 lists the most commonly used types of acid-sensitive benzyl alcohol linker. All these can be attached to various supports by the use of different types of spacer. Because resins bearing these linkers are commercially available, their preparation will not be discussed here.

The 4-alkylbenzyl alcohol linker was the first type of linker to be used in solid-phase peptide synthesis [13]. Esters of this alcohol can readily be prepared from chloromethyl polystyrene (Merrifield resin) and carboxylates [13] or by acylation of hydroxymethyl polystyrene (see Section 13.4). Acidolysis of this linker requires acids with high ionizing power, such as hydrogen fluoride [14–16], trifluoromethanesulfonic (triflic) acid [17], trimethylsilyl triflate/TFA [18,19], trimethylsilyl bromide/TFA [20], hydrogen bromide/AcOH [21,22], or aluminum chloride/DCM/MeNO$_2$ [23,24]. The stability of 4-alkylbenzyl esters towards weak acids enables the deprotection of Boc-protected amino groups with TFA without cleavage of the product from the resin.

Esters of the PAM linker are slightly more resistant towards acids than the corresponding 4-alkylbenzyl esters [5,25–27] (Table 3.1). The PAM linker is particularly well suited for solid-phase peptide synthesis using *N*-Boc amino acids because less than 0.02% cleavage of the peptide from the support occurs during the acidolytic deprotection steps [27]. Esters of both the 4-alkylbenzyl alcohol and PAM linkers can also be cleaved by nucleophiles (see Sections 3.1.2 and 3.3.3).

4-Alkoxybenzyl alcohol was first used for solid-phase synthesis by Wang in 1973 [28], and has become one of the most widely used linkers (Wang linker). Cross-linked polystyrene bearing the Wang linker is often referred to as Wang resin. Esters of the Wang linker are difficult to cleave from the support by treatment with nucleophiles or with weak acids (< 10% cleavage occurs upon treatment with neat acetic acid for 4 h

Table 3.1. Acid-labile benzyl alcohol linkers. See text for additional cleavage reagents.

Name of linker/resin	Structure	Cleavage reagent for benzyl esters	Literature
4-alkylbenzyl alcohol (e.g., hydroxymethyl polystyrene)		HF/Me$_2$S/*p*-cresol 25:65:10, 0 °C, 2 h	[16]
4-(hydroxymethyl)phenylacetamide (PAM linker)		HF/*p*-cresol 9:1, 0 °C, 1 h	[5,16,25,34]
4-alkoxybenzyl alcohol (Wang linker)		TFA/DCM 1:1, 20 °C, 0.5 h	[28,35]
4-alkoxy-2-methoxybenzyl alcohol [e.g. Sasrin (Pol = PS); *super acid-sensitive resin*]		1% TFA in DCM, 5 min	[33,35,36]
4-alkoxy-2,6-dimethoxybenzyl alcohol (HAL linker)		0.1% TFA in DCM, 25 °C, 1 h	[37]

at 100 °C [29]), but are readily hydrolyzed by treatment with 50% TFA in DCM or with Lewis acids such as Et_2AlCl [30]. The Wang linker is well suited for the preparation of peptides using the Fmoc methodology (Section 16.1.3). A fluorinated version of the Wang linker has been described (4-alkoxy-2-fluorobenzyl alcohol [31]), which enables the monitoring of solid-phase reactions by gel-phase ^{19}F NMR spectroscopy without the need for a MAS probe. Esters of this linker are more resistant towards acids than those of the Wang linker, but can be cleaved with warm TFA (60 °C, 2 h). Slightly more acid-labile than the Wang linker is the corresponding α-methyl Wang linker ($Me(OH)CH–C_6H_4–O–Pol$ [32]).

Dialkoxy- and trialkoxybenzyl esters are even more acid-labile than the Wang linker, and can, for example, be cleaved with dilute TFA, acetic acid, or hexafluoroisopropanol without simultaneous acidolysis of Boc groups. These linkers thus enable the solid-phase synthesis of protected peptide fragments or other acid-sensitive products [33].

Three different strategies are generally used for the attachment of carboxylic acids to resins as benzyl esters: (a) acylation of resin-bound benzyl alcohols [38–40], (b) O-alkylation of carboxylates by resin-bound benzylic halides [41–43], or (c) O-alkylation of carboxylic acids under Mitsunobu conditions [44,45] (Figure 3.3). These reactions are treated in detail in Section 13.4.

3.1.1.2 Diarylmethanol (Benzhydrol) Linkers

The trialkoxy benzhydrol linker, developed by Rink in 1987 [46] ('Rink acid resin', Figure 3.4) is a further acid-labile linker for carboxylic acids. Esters of this linker can, like trityl esters, be cleaved with acids as weak as acetic acid or HOBt [47], and care must be taken to avoid loss of the product during synthetic operations.

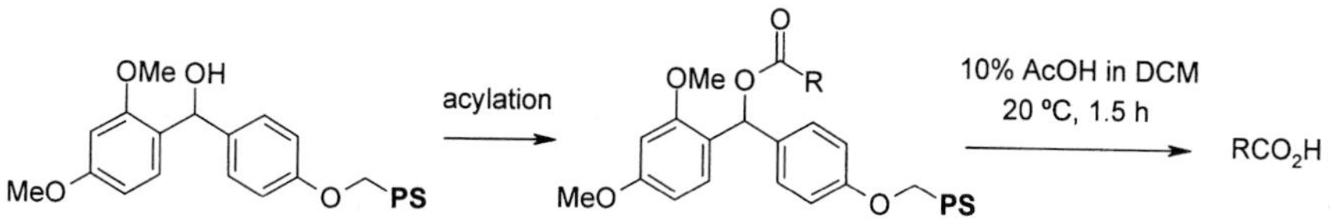

Figure 3.4. The Rink acid linker [46].

Neither the trialkoxybenzhydryl alcohol linker nor other types of benzhydryl alcohols [44,45,48] have found widespread use as linkers for carboxylic acids. These linkers do not seem to offer special advantages compared with benzyl alcohol or trityl linkers.

3.1.1.3 Trityl Alcohol Linkers

Esters of unsubstituted, polystyrene-bound trityl alcohol are too acid-sensitive to be useful for solid-phase synthesis [49]. Even treatment with alcohols or HOBt can lead to significant product release from the support. More stable towards solvolysis is the 2-chlorotrityl alcohol linker [47], and numerous examples of the use of this linker for the attachment of carboxylic acids to polymeric supports have been reported [50–57]. Cleavage can be effected by treatment with dilute TFA, acetic acid, or hexafluoro-isopropanol [58]. Trityl alcohol linkers substituted with electron-withdrawing groups have also been described (e.g., resin-bound 4-carboxytrityl alcohol [59,60], 9-hydroxy-9-(4-carboxyphenyl)fluorene [61], or 9-hydroxy-9-(4-alkoxy)fluorene [62]); these are sufficiently stable to enable the attachment of carboxylic acids as triarylmethyl esters.

The most valuable property of 2-chlorotrityl esters is their high stability towards nucleophiles and bases. 2-Chlorotrityl esters have been successfully used at elevated temperatures (e.g. 80 °C, 5 h, toluene [63]), although prolonged heating can lead to cleavage [63].

Attachment of carboxylic acids to supports as trityl esters is achieved by treatment of the corresponding trityl chloride resin with the acid in the presence of an excess of a tertiary amine (Figure 3.5; see also Section 13.4.2). This esterification usually proceeds more quickly than the acylation of benzyl alcohol linkers. Less racemization is generally observed during the esterification of *N*-protected α-amino acids with trityl linkers than with benzyl alcohol linkers [47]. If valuable acids are to be linked to insoluble supports, quantitative esterification can be accomplished by using excess 2-chlorotrityl chloride resin, followed by displacement of the remaining chloride with methanol [64].

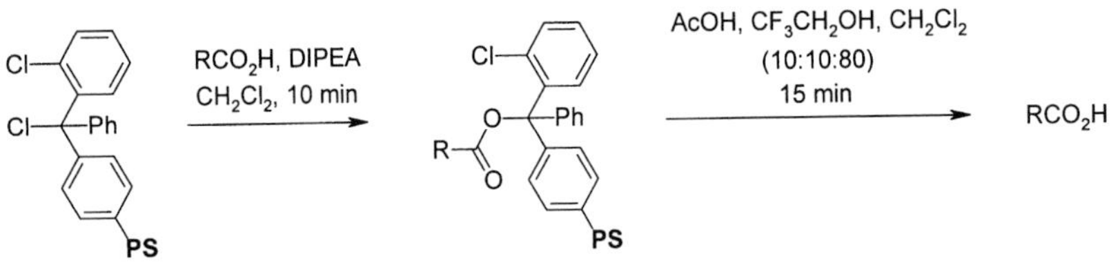

Figure 3.5. Immobilization of carboxylic acids as 2-chlorotrityl esters [47].

3.1.1.4 Non-Benzylic Alcohol Linkers

Few examples have been reported of the use of resin-bound tertiary alcohols as linkers for carboxylic acids. Some examples are sketched in Figure 3.6. Esters of the tertiary alcohol linkers **1**, **2**, and **4** can be cleaved with 50% TFA, and can therefore be used for peptide synthesis using the Fmoc method but not using the Boc method. In linker **3**, only the primary hydroxyl group is utilized for the attachment of carboxylic acids. Cleavage requires previous dehydration with TFA, leading to a vinyl ester that is readily hydrolyzed [65] (no examples given). The secondary alcohol linker **5** is stable towards TFA, but can be cleaved by hydrogen fluoride, thereby enabling the synthesis of peptides by means of the Boc methodology [66].

The main advantage of *tert*-alkyl esters as linkers is their stability towards nucleophiles. For instance, no diketopiperazine formation is observed during the preparation of peptides containing carboxy-terminal proline, an otherwise common side reaction when using benzyl alcohol linkers (Section 15.22.1).

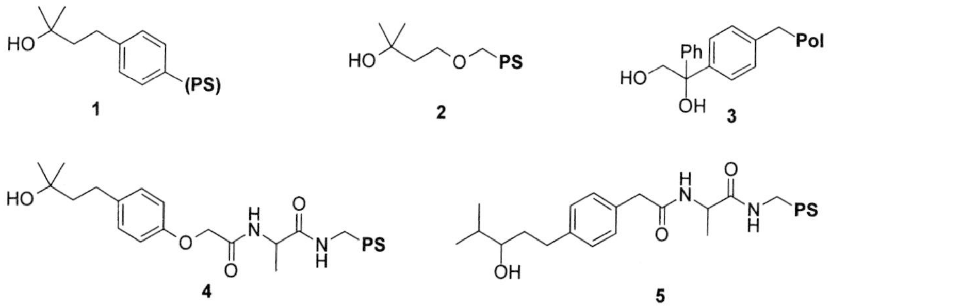

Figure 3.6. Non-benzylic alcohols as acid-labile linkers for carboxylic acids (**1**: [67–70]; **2**: [71]; **3**: [65]; **4**: [72]; **5**: [66]).

Tertiary aliphatic alcohol linkers have only occasionally been used in solid-phase organic synthesis [73]. This might be because of the vigorous conditions required for their acylation. Esterification of resin-bound linker **4** with *N*-Fmoc-proline [72,74] could not be achieved with the symmetric anhydride in the presence of DMAP (20 h), but required the use of *N*-Fmoc-prolyl chloride (10–40% pyridine in DCM, 25 °C, 10–20 h [72]). A further problem with these linkers is that they can undergo elimination, a side reaction that cannot occur with benzyl or trityl linkers. Hence, for most applications in which a nucleophile-resistant linker for carboxylic acids is needed, 2-chlorotrityl- or 4-acyltrityl esters will probably be a better choice than *tert*-alkyl esters.

Resin-bound (4-acyloxy-2-buten-1-yl)silanes, which can be prepared from resin-bound allylsilanes and allyl esters by cross-metathesis, react with dilute TFA to yield free carboxylic acids (Figure 3.7 [75]). However, the scope of this strategy remains to be explored. Similarly, esters of polystyrene-bound (2-hydroxyethyl)silanes readily undergo acidolysis and have been used as acid-labile linkers (Figure 3.7 [76]).

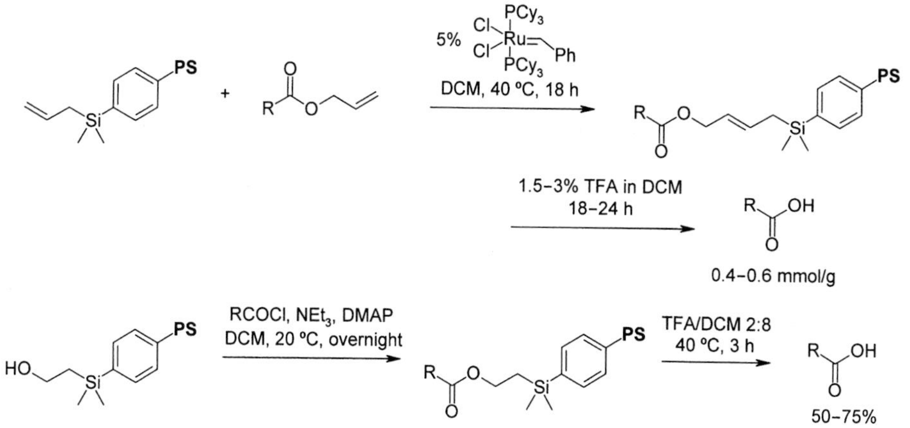

Figure 3.7. (4-Acyloxy-2-buten-1-yl)silanes and 2-acyloxyethylsilanes as acid-labile linkers for carboxylic acids.

3.1.2 Linkers for Acids Cleavable by Bases or Nucleophiles

Linkers for carboxylic acids cleavable by nucleophiles were mainly developed for the preparation of peptide amides or cyclic peptides. These linkers are, however, increasingly being used for the preparation of non-peptides. Several different carboxylic acid derivatives are available by nucleophilic cleavage, the most common being carboxylates, esters, and amides. In addition to these, ketones or primary alcohols can be obtained by the use of organometallic reagents or reducing agents to effect nucleophilic cleavage. These applications will be discussed in sections dealing with linkers for the resulting functional groups.

Most of the reagents required for the saponification of resin-bound esters are, unfortunately, non-volatile, and purification of the products is generally necessary. This is probably the main reason for the scarcity of examples of cleavage by saponification as compared with other methods of cleavage.

The type of support chosen can have an impact on the facility with which nucleophilic cleavage takes place. Polystyrene, a very hydrophobic support, is difficult to perfuse with small ions such as hydroxide, and for this reason the saponification of polystyrene-bound esters usually proceeds more slowly than the corresponding solution reaction. Tentagel, polyacrylamides, or other more hydrophilic supports are generally a better choice if saponifications or other reactions involving small ions are to be performed.

3.1.2.1 Benzyl Alcohol Linkers

Polystyrene-supported benzyl esters can be cleaved by treatment with various types of nucleophile. Benzyl esters become more sensitive towards nucleophiles with decreasing electron density of the benzene ring [77]. On the other hand, acid-sensitive benzyl alcohol linkers, in particular those with electron-donating alkoxy groups (e.g. the Wang linker), are rather resistant towards nucleophilic attack. 4-(Hydroxymethyl)benzoic acid (HMBA linker, Table 3.1) was especially designed for nucleophilic cleavage and enables, for example, the preparation of peptides using Boc methodology and cleavage of the peptides from the support by treatment with various nucleophiles. Similarly sensitive towards nucleophilic attack are esters of 4-hydroxymethyl-3-nitrobenzamides (a photocleavable linker, Entry **5**, Table 3.2). Esters of 4-hydroxybenzyl alcohol are also highly sensitive towards bases, and undergo cleavage through a 1,6-elimination. Entry **6** in Table 3.2 is an example of a so-called 'safety-catch' linker, which requires activation by Boc-group removal to mediate an intramolecular aminolysis of the aryl acetate and the generation of an unprotected 4-hydroxybenzyl ester. Fluoride-sensitive benzyl esters have been developed, which enable nucleophilic cleavage under mild conditions (Entries **7** and **8**, Table 3.2 [78,79]). Entry **9** in Table 3.2 is an example of a Dde-based linker, which is resistant towards both the conditions of Fmoc-group removal (e.g. 20% piperidine or 2% DBU in DMF, 3 h) and the conditions of Boc-group removal (e.g. 90% TFA in DMF, 4 h [80]). However, cleavage occurs upon treatment with hydrazine, to yield, along with the product, an aniline derivative and a resin-bound pyrazole.

Table 3.2. Saponification of resin-bound benzyl esters.

Entry	Loaded resin	Cleavage conditions	Product, yield (purity)	Ref.
1		Bu$_4$NOH, THF, MeOH, 70 °C, 24 h	100%	[81] see also [82–84]
2		THF/satd aq K$_2$CO$_3$ 10:1, [Bu$_4$N][HSO$_4$], 25 °C, 2–24 h	70–100% (85–90%)	[85]
3	(PAM linker)	DBU (2 eq), LiBr (5 eq), THF/H$_2$O 20:1, 25 °C, 4 h	81%	[86] see also [87]
4		TBAF·3 H$_2$O (0.05 mol/L), DMF, 0.5 h	48%	[77] see also [88,89]
5		TBAF (8 eq), MeCN, 25 °C, 40 min; or LiOH (15 eq), H$_2$O, 25 °C, 10 min	54%	[90]
6		1. TFA/DCM 1:1, 20 °C, 1 h 2. K$_2$HPO$_4$ (0.05 mol/L, pH 8), 4 h	100%	[91]
7		TBAF (3 eq), DMF, 65 °C, 1 h	78%	[92] see also [93]
8		TBAF·3 H$_2$O (1 eq), PhSH (1.2 eq), DMF, 5 min; or TFA/DCM/Me$_2$S 5:4:1, PhSH (1.2 eq), 0.5 h	100%	[94]
9		N$_2$H$_4$, H$_2$O, DMF, 2 × 20 min		[80]

3.1.2.2 Non-Benzylic Alcohol Linkers

Support-bound non-benzylic alcohols can also be used to immobilize carboxylic acids as esters (Table 3.3). The advantage of this type of linker is its stability towards electrophiles. Attachment of carboxylic acids is usually realized by acylation of the resin-bound alcohol with a reactive acid derivative.

PEG-grafted polystyrene resins (e.g. Tentagel) enable the direct attachment of carboxylic acids as 2-alkoxyethyl esters. PEG esters are usually cleaved with nucleophiles, to yield carboxylates, esters, amides, or hydrazides. Entries 4 and 5 in Table 3.3 are interesting examples of linkers that are sensitive towards reducing agents. Treatment of the azide-containing linker (Entry 5) with tributylphosphine in the presence of water leads to reduction of the azido group to an amino group, which cleaves the benzamide by intramolecular transamidation. The disulfide-based linker (Entry 4) also undergoes reductive cleavage upon treatment with phosphines to yield a 2-mer-

Table 3.3. Saponification of non-benzylic, resin-bound esters.

Entry	Loaded resin	Cleavage conditions	Product, yield (purity)	Ref.
1		TBAF (0.1 mol/L), EDT (1 eq), DMF, 25 °C, 0.5 h	35%	[88] see also [96]
2		NaOH (0.01 mol/L), H_2O/iPrOH 3:7 or H_2O, 3 h		[97] see also [98]
3		aq NaOH (1 mol/L)/ iPrOH 5:2, 50 °C, 8 h	71% (88%)	[99]
4		1. PR_3, pH 4.5 2. pH 8–9		[95]
5		PBu_3 (3 eq), buffer (pH 7), DMF, 10 h	60%	[100]
6		1. TFA/DCM 1:1 2. phosphate buffer (0.01 mol/L, pH 7.5), 50 °C, 25 min	33–80%	[101] see also [102]
7		1. $NaBH_4$, THF, MeOH, 20 °C, 0.5 h 2. TBAF, THF, 20 h	35%	[103]

captoethyl ester. This ester is hydrolyzed under mildly basic conditions as a result of intramolecular nucleophilic catalysis by the mercapto group [95].

Entries **6** and **7** in Table 3.3 are additional linkers based on intramolecular nucleophilic cleavage. In Entry **6**, it is an imidazole that efficiently catalyzes the saponification of the ester linkage, whereas in Entry **7** a hydroquinone is O-alkylated intramolecularly, with the carboxylate acting as a leaving group. Esters of resin-bound (2-hydroxyethyl)silanes have also been used as linkers, which can be cleaved by treatment with either TFA or fluoride ions (TBAF in THF [76]).

A special group of base-labile linkers for carboxylic acids rely on cleavage by β-elimination. Here, the resin-bound alcohol must bear an electron-withdrawing group in the β position (Figure 3.8), which facilitates elimination by acidifying this position. Mechanistically, these linkers are closely related to the Fmoc protective group, and

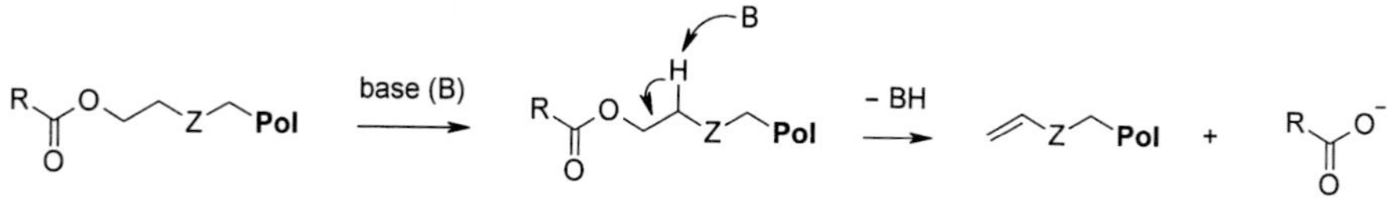

Z = electron-withdrawing group

Figure 3.8. Cleavage of linkers by β-elimination.

Table 3.4. Linkers for carboxylic acids cleavable by base-induced β-elimination.

Entry	Loaded resin	Cleavage conditions	Product, yield (purity)	Ref.
1	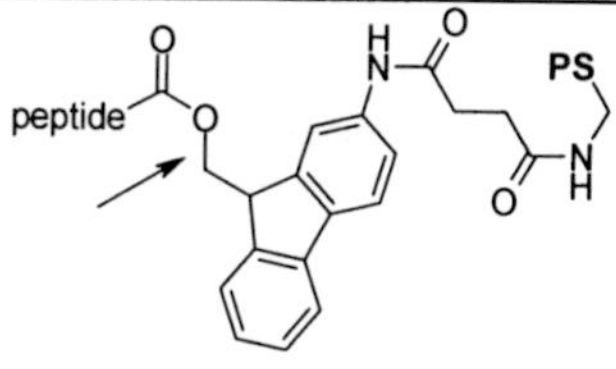	morpholine/DMF 1:4, 25 °C, 0.5 h	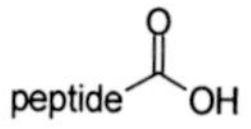 100% (> 95%)	[104], see also [105–107]
2	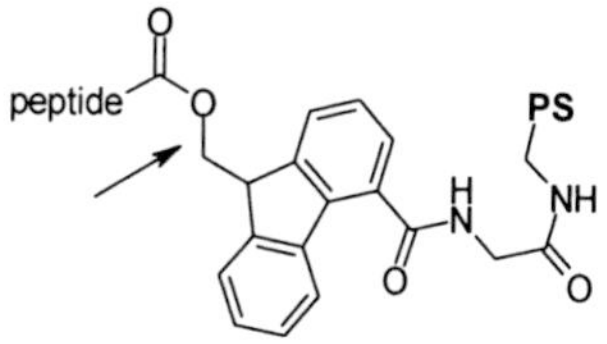	piperidine/DMF 15:85, 3 × 5 min	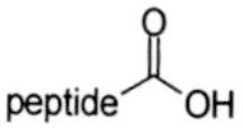 72% (90%)	[108]
3	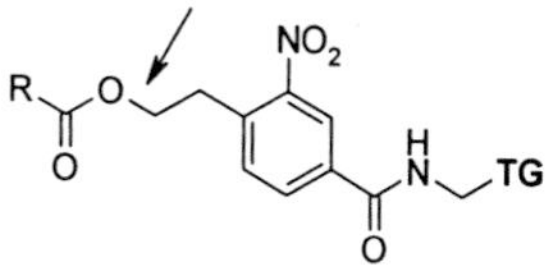	satd aq NH₃/dioxane 9:1, 50 °C, overnight	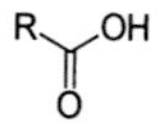	[109]
4	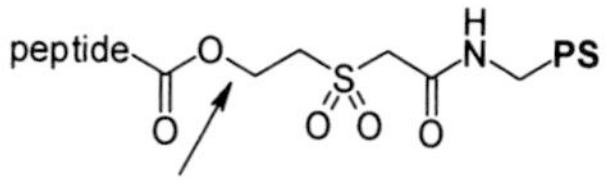	NaOH (4 mol/L)/ MeOH/dioxane 1:9:30, 0.5 h	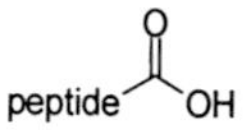 54–60%	[110] see also [111]
5	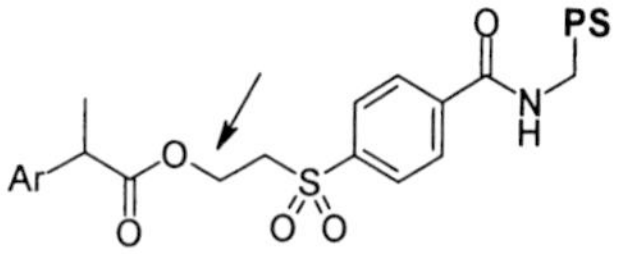	Me₂NH (20%) in THF, 4 h	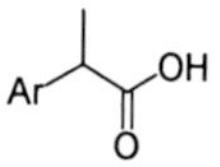 100%	[112]

some of them are indeed based on 9-fluorenylmethanol (Entries **1** and **2**, Table 3.4). The stability of these linkers towards acids permits their use for the preparation of peptides using Boc methodology.

Support-bound phenols, oximes, and related compounds yield, upon acylation, esters that are highly susceptible to nucleophilic cleavage. These esters are often used as insoluble acylating agents for the preparation of amides or esters, but only occasionally as linkers for carboxylic acids [113]. These linkers are considered in Sections 3.3.3 and 3.5.1.

3.1.2.3 Miscellaneous Linkers Cleavable by Bases or Nucleophiles

Carboxylic acids can also be attached to solid supports as amides, imides, and thiol esters. Illustrative examples of the saponification of such linkers are listed in Table 3.5. Thiol esters are more sensitive towards nucleophilic attack than the corresponding esters, and can be readily saponified. Resin-bound thiol esters have, however, mainly been used for the preparation of amides by nucleophilic cleavage with amines (see Section 3.3.3).

Resin-bound amides generally need to be activated to make them susceptible to saponification under acceptably mild reaction conditions [114] (Table 3.5). Particularly elegant are those linkers that allow this activation to be realized as the final synthetic step before cleavage (safety-catch linkers [115–117]). The activation of some amide-based safety-catch linkers is outlined in Figure 3.9.

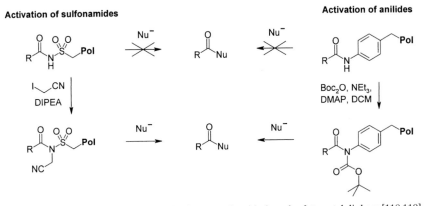

Figure 3.9. Activation and nucleophilic cleavage of amide-based safety-catch linkers [118,119].

Table 3.5. Saponification of resin-bound thiol esters, imides, and related compounds.

Entry	Loaded resin	Cleavage conditions	Product, yield (purity)	Ref.
1		LiOH, THF/H$_2$O 3:1, 12 h		[120] see also [121–123]
2	(Fmoc)peptide (PA)	NaOH (1 eq), CaCl$_2$, iPrOH/H$_2$O 7:3, 2 × 1.5 h	(Fmoc)peptide OH 72%	[124] see also [125]
3	(PS)	LiOH, 5% H$_2$O$_2$, H$_2$O, THF		[119]
4		OH$^-$, H$_2$O, 20 °C	93%	[126] see also [115]
5		1. PPTS, PhMe, 50 °C, 16 h (formation of *N*-acylindole) 2. aq NaOH (1 mol/L)/ MeOH/dioxane 1:1:3, 16 h	91% (> 98%)	[117]
6	(PS)	CAN (2 eq), THF, H$_2$O, 10 min	57%	[127]
7	(PS)	aq NaOH (1 mol/L)/ dioxane 1:4, 100 °C, 6 h	59%	[128]

3.1.3 Photocleavable Linkers for Carboxylic Acids

The development of light-sensitive protective groups began in the early 1960s [129–131] and led to the identification of several functionalities that could be selectively cleaved under UV irradiation (for a review, see [132]). Some of these protective groups, such as 2-nitrobenzyl esters, carbonates, or carbamates [131,133–135], benzoin [136–139], and other phenacyl esters [140] were also found to be useful as photocleavable linkers.

Photocleavable linkers can be stable towards both bases and acids, thereby overcoming some of the inherent limitations of other types of linker. Cleavage by photolysis is, however, complicated by other problems. UV-absorbing by-products, which are, unfortunately, often formed during synthetic transformations, can completely obstruct the passage of light through the polymer and thereby inhibit photolytic cleavage. For this reason, the use of photocleavable linkers is restricted to synthetic sequences in which no, or only insignificant amounts of, UV-absorbing by-products are formed. Moreover, the products must not contain chromophores that absorb light of the same wavelength as the photocleavable linker, because this would also lead to an inhibition of the cleavage reaction.

Most photocleavable linkers for carboxylic acids used today are based on the photoisomerization of 2-nitrobenzyl esters and on the light-induced cleavage of phenacyl esters (Figure 3.10). Several possible mechanisms have been proposed for the photolytic cleavage of benzoin esters. One of the most recent is the dissociation of the excited phenacyl ester into a carboxylate and a phenacyl cation ('photosolvolysis', Figure 3.10 [136]).

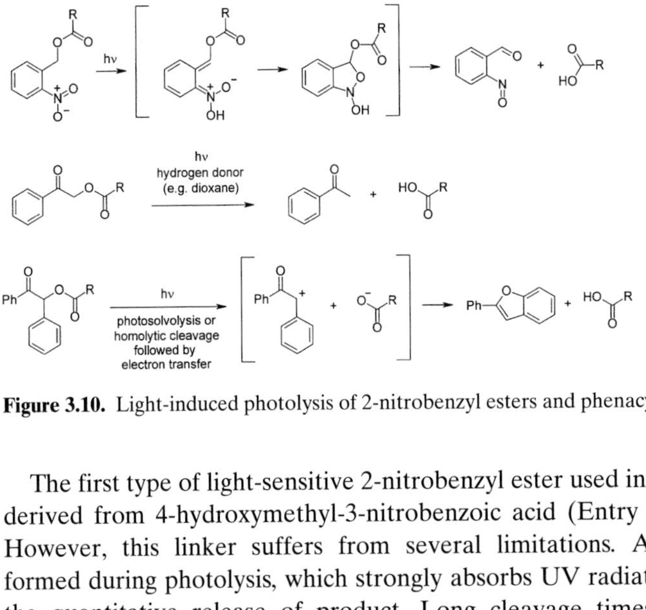

Figure 3.10. Light-induced photolysis of 2-nitrobenzyl esters and phenacyl esters [136,138,140].

The first type of light-sensitive 2-nitrobenzyl ester used in solid-phase synthesis was derived from 4-hydroxymethyl-3-nitrobenzoic acid (Entry **1**, Table 3.6 [7,141,142]). However, this linker suffers from several limitations. A nitrosobenzaldehyde is formed during photolysis, which strongly absorbs UV radiation and thereby prevents the quantitative release of product. Long cleavage times (> 10 h) are generally required, and this can lead to other undesired photochemical reactions of the products. For example, photodimerization of thymine has been observed on using photolabile linkers for the solid-phase synthesis of oligonucleotides [143,144]. Moreover, thioethers, such as methionine-containing peptides, might be oxidized to the corresponding sulfoxides upon prolonged irradiation. These problems were partly overcome by the development of 4,5-dialkoxy-2-nitrobenzyl alcohol linkers (Entry **4**, Table 3.6 [145–148]), which can be cleaved photolytically within 2 h to yield crude products of high purity. These linkers are stable towards both TFA and piperidine, thereby enabling the preparation of peptides using both the Boc and Fmoc methodologies. It has been observed that the addition of hydrazine or ethanolamine as scavengers of resin-bound nitrosoarenes formed during photolysis often improves the yield of released product [149]. Phenacyl and benzoin esters can also serve as photolabile linkers (Entries **5** and **6**, Table 3.6), but have not received as much attention as nitrobenzyl derivatives. Entry **6** in Table 3.6 is an example of a 'safety-catch' photolinker, in which the ketone, which enables the photolytic cleavage, is masked as a 1,3-dithiane.

In 1998, a new type of light-sensitive linker based on the photolysis of *tert*-butyl ketones was developed by Peukert and Giese (Entry **7**, Table 3.6). The proposed

Table 3.6. Photolytic cleavage of resin-bound esters.

Entry	Loaded resin	Cleavage conditions	Product, yield (purity)	Ref.
1		hv, CF₃CH₂OH/ DCM 1:4, 15 h	peptide–OH 82%	[150]
2		hv (350 nm), CuSO₄, EtOH/DCM 1:1, 24 h	peptide–OH 40–48%	[151] see also [152, 153]
3		hv (350 nm), CuSO₄, EtOH, 24 h	peptide–OH 60%	[154]
4		hv (354 nm)	R–OH 92%	[155]
5		hv (350 nm), DMF, 72 h	peptide–OH 77%	[156] see also [140]
6		1. PhI(O₂CCF₃)₂ or Hg(ClO₄)₂ or MeOTf or HIO₄; THF/H₂O, 20 °C, 18 h 2. hv (350 nm), 2 h, THF/MeOH, 28 °C	Fmoc-N–OH 65–75%	[157, 158]
7		hv (320–340 nm), THF/H₂O, 12 min	peptide–OH 85% (93%)	[159]

mechanism of cleavage is outlined in Figure 3.11. A photolabile linker for alcohols based on the same mechanism has also been developed [160].

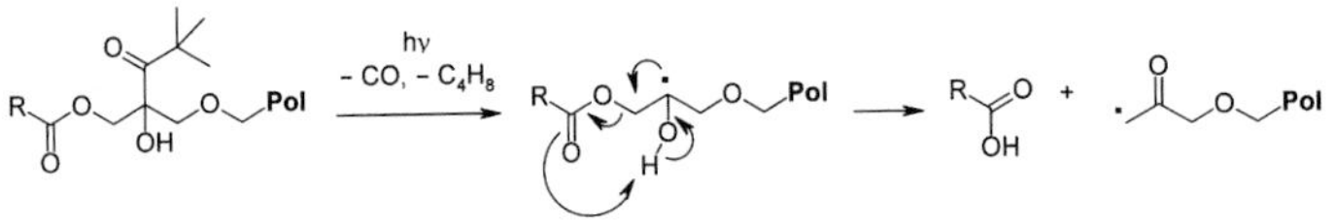

Figure 3.11. Mechanism of photolytic cleavage of 2-acyloxyethyl *tert*-butyl ketones [159].

This linker proved to be stable under a variety of reaction conditions (e.g. 50% TFA or 5% $BF_3 \cdot OEt_2$ in DCM, 20 °C, 2 h; 5% DBU in toluene, 80 °C, 2 h). An additional advantage of this linker, as compared with 2-nitrobenzyl alcohol derivatives, is that no UV-absorbing products are formed during photolysis. This enables fast and complete photolytic detachment of the resin-bound product. Relatively short wavelengths (< 340 nm) are, however, required to effect cleavage.

3.1.4 Linkers for Acids Cleavable by Transition Metal Catalysis

Benzyl alcohol linkers, such as those described in Section 3.1.1.1, can also be cleaved by palladium-catalyzed hydrogenolysis. Carboxylic acids have, for example, been obtained by hydrogenolysis of insoluble benzyl esters with $Pd(OAc)_2/DMF/H_2$ [89,161]. Resin-bound benzylic carbamates [162,163] and amides [164] can also be released by treatment with $Pd(OAc)_2$ in DMF in the presence of a hydrogen source, such as 1,4-cyclohexadiene or ammonium formate. These reactions are quite surprising, because they require the formation of metallic palladium within the gelated beads.

Allyl esters, carbonates, and carbamates readily undergo C–O bond cleavage upon reaction with palladium(0) to yield allyl palladium(II) complexes. These complexes are electrophilic and can react with nucleophiles to form products of allylic nucleophilic substitution. Linkers based on this reaction have been designed, which are cleavable by treatment with catalytic amounts of palladium complexes [165,166]. For the immobilization of carboxylic acids, support-bound allyl alcohols have proven suitable (Figure 3.12, Table 3.7).

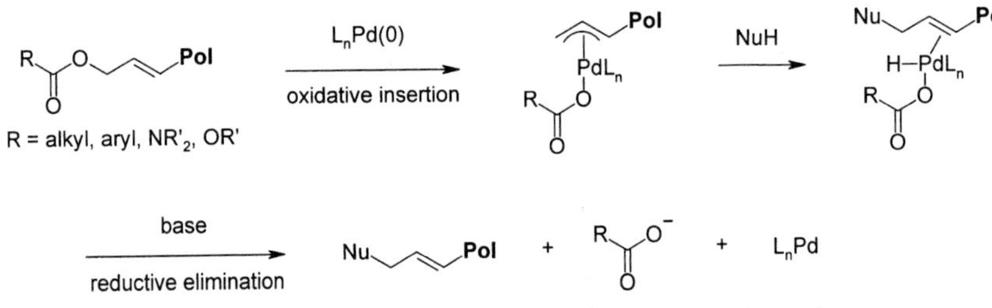

Figure 3.12. Cleavage of support-bound allyl esters by allyl complex formation.

The two most commonly used types of allyl alcohol linker are 4-hydroxycrotonic acid derivatives (Entry **1**, Table 3.7) and (*Z*)- or (*E*)-2-butene-1,4-diol derivatives (Entries **2** and **3**, Table 3.7). The former are well suited for solid-phase peptide synthesis using Boc methodology, but give poor results when using the Fmoc technique, probably because of Michael addition of piperidine to the α,β-unsaturated carbonyl compound [167]. Butene-1,4-diol derivatives, however, are tolerant to acids, bases, and weak nucleophiles, and are therefore suitable linkers for a broad range of solid-phase chemistry.

Table 3.7. Linkers cleavable by palladium(0)-catalyzed allylic nucleophilic substitution.

Entry	Loaded resin	Cleavage conditions	Product, yield (purity)	Ref.
1		Pd(PPh$_3$)$_4$ (1.4 mmol/L), THF/ morpholine 10:1, 20 °C, 2 h	98%	[168]
2		PdCl$_2$(PPh$_3$)$_2$ (0.7 mmol/L), Bu$_3$SnH (0.05 mol/L), DCM, 20 °C, 20 min; then AcOEt/HCl	80% (> 95%)	[169] see also [170]
3		Pd(PPh$_3$)$_4$ (2.9 mmol/L), PhNHMe (0.34 mol/L), DCM/DMSO 1:1, 15 h	62%	[167] see also [171]

Hydrazides have been used as oxidant-labile linkers for carboxylic acids (Figure 3.13 [172–174]). Cleavage is effected by copper(II)-catalyzed or NBS-mediated oxidation of the hydrazide to an azo compound, which is highly electrophilic and decomposes to yield the free carboxylic acid or amide, depending on the nucleophile present during cleavage. The same type of cleavage methodology has been used for the generation of arenes ('traceless linker', see Section 3.16).

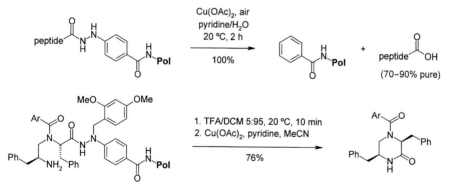

Figure 3.13. Oxidative cleavage of hydrazides [175,176].

3.1.5 Linkers for Acids Cleavable by Enzymes

The efficiency of enzyme catalysis on solid supports depends strongly on the type of support and the size of the enzyme chosen. Enzymes can generally only be functional in water or in solvent mixtures with high water content, and can therefore only be used with water-compatible supports. Microporous polystyrene, being a highly hydrophobic support, is not well suited for enzyme-catalyzed reactions [177,178]. The suitability of PEG-grafted polystyrene supports, such as Tentagel or Argogel, for reactions involving enzymes seems to depend on the size of the enzyme used [179], and both positive (lipase RB 001-05 [180], penicillin amidase [181], porcine pancreas lipase VI-S [178]) and negative results (papain [182], chymotrypsin, elastase, pepsin [183]) have been reported. Authors who examined the suitability of different types of support for the realization of enzyme-catalyzed reactions concluded that supports such as Tentagel or Argogel are not generally suitable for these reactions. Selective and quantitative hydrolysis of peptides with papain could, for example, only be achieved on PEGA and not on PEG-grafted polystyrene [182]. Similar results were obtained by Meldal and co-workers, who used PEGA as a support for combinatorial

Table 3.8. Linkers for carboxylic acids cleavable by enzymes.

Entry	Loaded resin	Cleavage conditions	Product, yield (purity)	Ref.
1	(racemic)	lipase VI-S (porcine pancreas), *t*BuOH/THF 5:1, phosphate buffer (pH 7.8), 30 °C, 13 h	16% (88% ee)	[178]
2	(racemic)	lipase VI-S (porcine pancreas), *t*BuOH/THF 5:1, phosphate buffer (pH 7.8), 30 °C, 6 h	30% (90% ee)	[178]
3		α-chymotrypsin, H$_2$O (pH 7.0)	> 95%	[188]
4		penicillin amidase, phosphate buffer (pH 7.5), 25 °C, 16 h (also cleavable by TFA, HCl, or NaOH)	50% (25% on PEGA)	[181]
5		lipase RB 001-05, MES buffer (0.05 mol/L, pH 5.8), 30 °C	70–80%	[180]

peptide and glycopeptide libraries. On-bead screening of these libraries with enzymes enabled the rapid identification of new enzyme substrates [184–187]. Further supports that have been claimed to be compatible with enzymatic catalysis are CPG [177,188], polyacrylamides [186,189,190], and sepharose [191]. Macroporous polystyrene or other supports with the attachment sites located exclusively on the surface should also be suitable for enzyme-mediated reactions.

There have been a few reports of enzymatic cleavage of linkers for carboxylic acids from PEG-grafted polystyrene supports, such as Tentagel (Table 3.8). Some of these linkers are also sensitive towards acids or bases, and will, therefore, only remain uncleaved under a narrow range of reaction conditions.

3.2 Linkers for Thiocarboxylic, Boronic, Phosphonic, Phosphoric, and Sulfonic Acids

Support-bound thiols can be suitable linkers for thiocarboxylic acids. Peptide thiocarboxylic acids are interesting synthetic intermediates that react smoothly with various alkylating agents, including bromoacetyl peptides, to yield the corresponding thiol esters; these have been used as peptide mimetics [192,193]. Furthermore, thiocarboxylic acids react with diaryl disulfides to yield acyl disulfides, which can be used as acylating agents [194]. Amines are directly acylated by thiocarboxylic acids in the presence of silver nitrate to yield amides [195]. As linkers for thiocarboxylic acids, 4-alkoxybenzhydrylthiols (Figure 3.14) have most often been used. Cleavage of the benzylic C–S bond of these linkers requires strongly ionizing acids, such as HF.

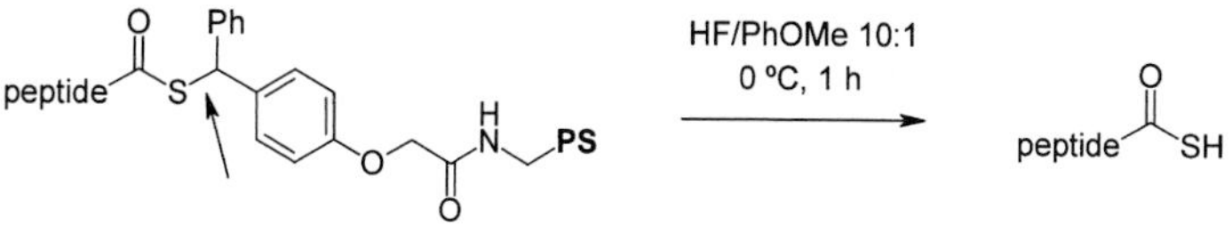

Figure 3.14. Benzhydrylthiol-based linker for thiocarboxylic acids [194,196].

Boronic acids can be reversibly esterified with resin-bound diols (Figure 3.15). The resulting boronic esters are stable under the standard conditions of amide bond formation, but can be cleaved by treatment with water under acidic or neutral conditions to yield boronic acids. Treatment of the resin-bound boronic esters with alcohols yields the corresponding boronic esters [197]. Resin-bound boronic esters are suitable intermediates for the Suzuki reaction [198]. Treatment with H_2O_2 leads to the formation of alcohols (Entry **8**, Table 3.36), while treatment of resin-bound aryl boronates with silver ammonium nitrate leads to the conversion of the C–B bond into a C–H bond (Entry **14**, Table 3.46).

Only a few examples of solid-phase syntheses of phosphonic, phosphoric, and sulfonic acids have been reported (Figure 3.16). Benzyl esters of these strong acids can act as alkylating agents, and may therefore be too labile to serve as linkers for long synthetic sequences on solid phase. However, if cross-linked polystyrene is used as the support, the reactivity of, for example, benzyl sulfonates is strongly reduced, and even

Sasrin-derived sulfonates have been prepared in the presence of DMAP, without significant loss of sulfonate through displacement by DMAP [201].

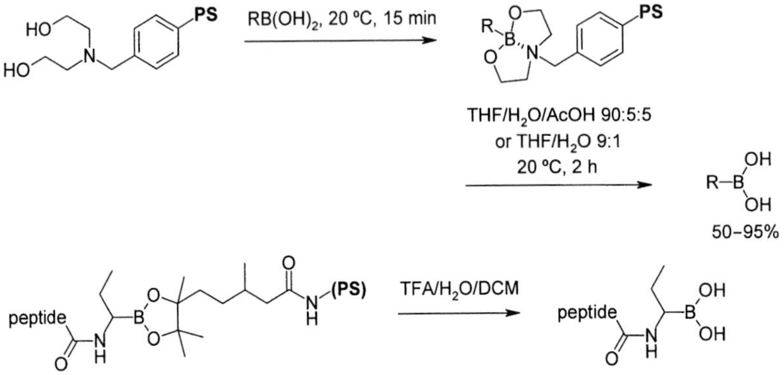

Figure 3.15. Diols as linkers for boronic acids [199,200].

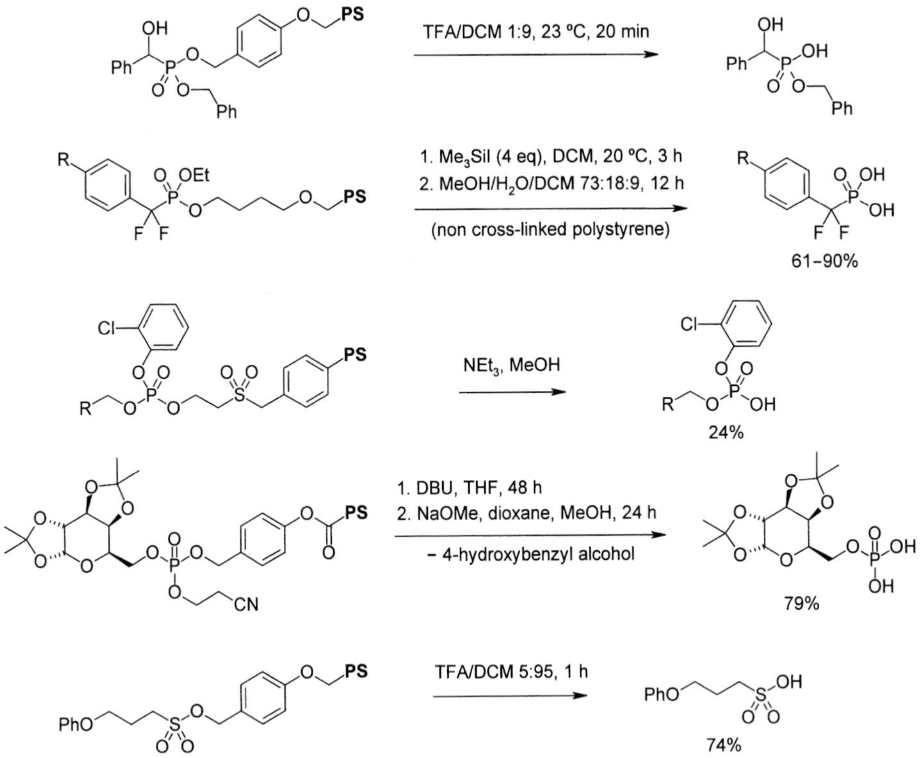

Figure 3.16. Linker strategies for phosphonic acids [202,203], phosphoric acids [204,205], and sulfonic acids [201].

3.3 Linkers for Amides, Sulfonamides, Carbamates, and Ureas

Several different types of linker have been developed that yield amides upon cleavage. These linkers can often also be used to prepare sulfonamides, carbamates, or ureas. There are essentially three different strategies for the release of amides from insoluble supports: (a) cleavage of the benzylic C–N bond of resin-bound *N*-alkyl-*N*-benzylamides (backbone amide linkers, BAL linkers), (b) nucleophilic cleavage of resin-bound acylating agents with amines, and (c) acylation/debenzylation of resin-bound *N*-benzyl-*N,N*-dialkylamines.

3.3.1 Benzylamine Linkers

A series of linkers has been developed for the solid-phase preparation of peptide amides (peptide–CONH$_2$). Most of these are based on the scission of benzylic C–N bonds (Figure 3.17). Because of the lower polarization of C–N bonds compared with C–O bonds, benzylic amines or amides are generally more difficult to cleave with electrophiles than benzylic ethers or esters. The reactivity of the *N*-benzylamides towards electrophiles, however, depends to a great extent on the electronic properties of all the substituents, and is not always easy to predict. Anilides (R: Ar, Figure 3.17) are usually more difficult to cleave from a support by acidolysis than amino acid amides (R: CHR^1–CONHR^2), and these, in turn, are more difficult to cleave than simple *N*-alkylamides (e.g., R: Et [206]). Strongly acidic amides, such as sulfonamides, are more readily cleaved from an acid-labile linker than normal amides [207].

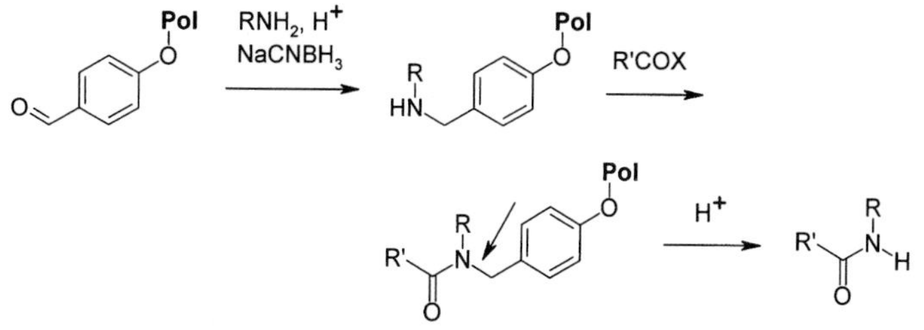

Figure 3.17. Backbone amide linking.

As illustrated by the examples in Table 3.9, resin-bound 4-alkoxybenzylamides often require higher concentrations of TFA and longer reaction times than carboxylic acids esterified to Wang resin. For this reason, the more acid-sensitive di- or (trialkoxybenzyl)amines [208] are generally preferred as backbone amide linkers. The required resin-bound, secondary benzylamines can readily be prepared by reductive amination of resin-bound benzaldehydes (Section 10.1.4 and Figure 3.17 [209]) or by *N*-alkylation of primary amines with resin-bound benzyl halides or sulfonates (Section 10.1.1.1). Sufficiently acidic amides can also be *N*-alkylated by resin-bound benzyl alcohols under Mitsunobu conditions (see, e.g., [210]; attachment to Sasrin of Fmoc cycloserine, an *O*-alkyl hydroxamic acid).

4-Alkoxy-2-hydroxybenzylamides (Entry **13**, Table 3.9) can also be cleaved by treatment with TFA (see also Section 16.1.5). If the phenolic hydroxyl group is acylated, however, acidolysis proceeds more slowly. Hence, 4-alkoxy-2-(acyloxy)benzylamides can serve as linkers that are stable towards both acids and bases, but which can be activated towards acidolysis by saponification to the corresponding 4-alkoxy-2-hydroxybenzylamide [211].

One problem occasionally encountered with BAL linkers attached to polystyrene as aryl benzyl ethers (Table 3.9, Figure 3.18) is that cleavage of the whole linker from the support can compete with C–N bond cleavage. Surprisingly, cleavage of polystyrene-bound aryl ethers can even occur under mildly acidic conditions, under which such cleavage would not have been expected. For example, the cleavage sketched in Figure 3.18 would only have been expected to occur upon treatment with hydrogen fluoride (acidolysis of a 4-alkylbenzyl ester; see Table 3.1). Similar effects have been observed with the Rink amide linker (Section 3.3.2). Because of this problem, amide backbone linkers are often attached to aminomethyl polystyrene or similar supports not directly as benzyl ethers but with a spacer, such as an ω-aryloxyalkanoic acid (see, e.g., Entry **14**, Table 3.9).

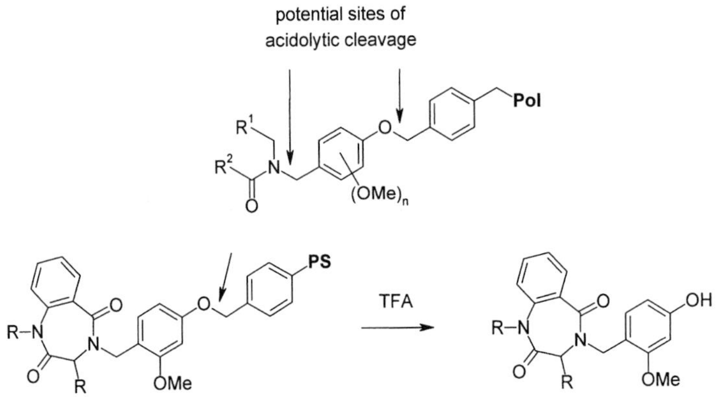

Figure 3.18. Acidolytic cleavage of aryl benzyl ethers as competing reaction during the acidolysis of *N*-benzylamides [212].

A further problematic class of substrates are amides containing a basic functional group near the attachment site of the backbone linker (Figure 3.19). These may, for instance, be non-acylated α-amino acid amides or other amides containing a tertiary amine or a pyridine ring close to the amido group (Entries **5**, **8**, **9**, and **14**, Table 3.9). Protonation of the basic functionality reduces the basicity of the amide, which needs to be protonated for cleavage to occur. Because the concentration of protonated amide is reduced, the cleavage of amides containing additional basic groups proceeds more slowly than that of similar, non-basic amides (Figure 3.19). As illustrated by Entries **8** and **9** (Table 3.9), this inhibition of the cleavage reaction can be rather strong. In such cases, the use of neat TFA as the cleavage reagent and long cleavage times may be required [213].

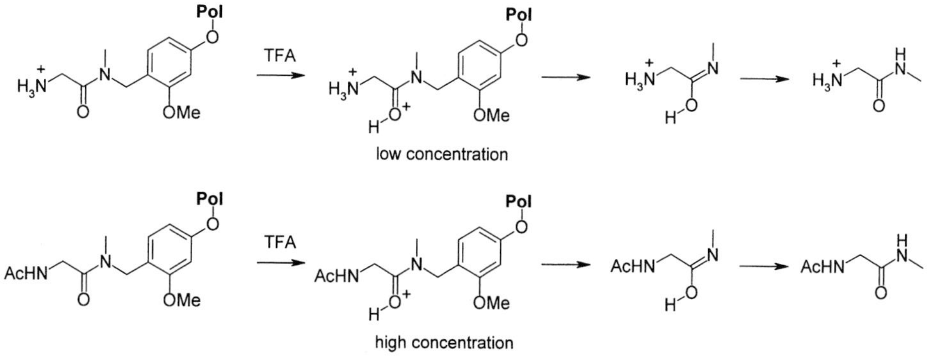

Figure 3.19. Inhibition of the cleavage of benzylic C–N bonds by basic functional groups in the support-bound amide.

Treatment of tertiary benzylamines with acylating agents can lead to debenzylation. If the benzyl group is linked to an insoluble polymer, acylation and debenzylation will lead to the release of an acylated amine into solution (Entry **6**, Table 3.9). These cleavage reactions generally yield products that are contaminated with acylating agent and so require further purification.

Table 3.9. Acidolytic cleavage of resin-bound *N*-benzylamine derivatives.

Entry	Loaded resin	Cleavage conditions	Product, yield (purity)	Ref.
1	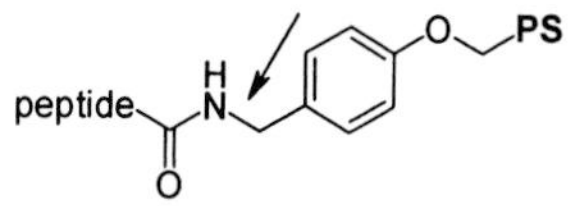	TFA/DCM 1:1, 0.5 h	no cleavage	[214]
2	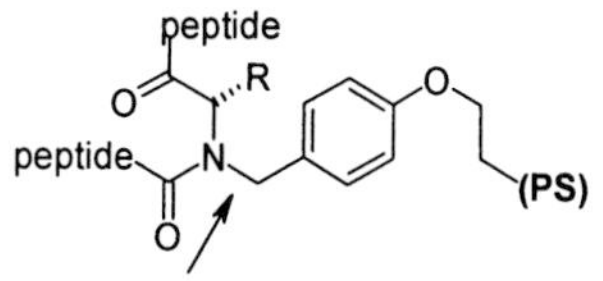	HF/*p*-cresol 10:1, −5 °C, 1 h (no cleavage with neat TFA, 25 °C, 2 h)	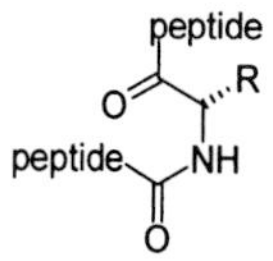	[215] see also [216, 217]
3	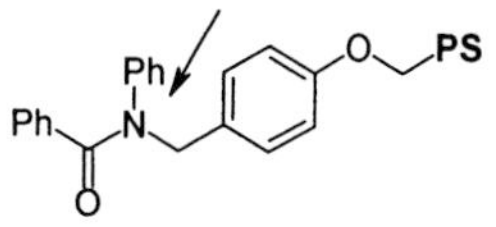	TFA/H$_2$O 95:5, 20 °C, 24 h	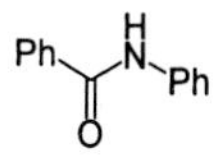 53% (95%)	[218] see also [213]
4	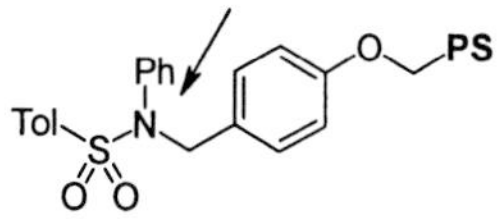	TFA/H$_2$O 95:5, 20 °C, 24 h	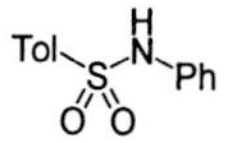 56% (96%)	[218]
5	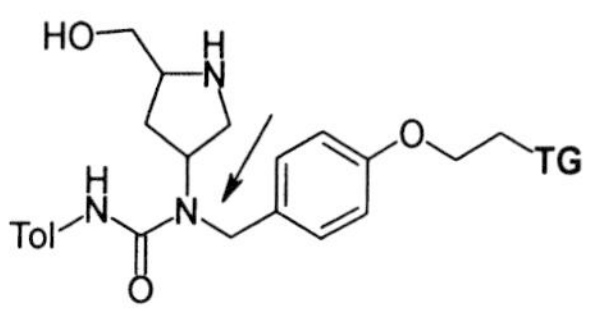	TFA/Et$_3$SiH 95:5	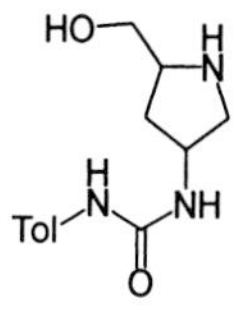 100% (> 90%)	[219]

Table 3.9. continued.

Entry	Loaded resin	Cleavage conditions	Product, yield (purity)	Ref.
6		PhCOCl (0.4 mol/L, 7 eq), DCM, 26 h	69% (> 95%)	[220] see also [221]
7		TFA/DCM > 2:8		[222] see also [223]
8		TFA/DCM 7:3, 25 °C, 2 h	24% (incomplete cleavage)	[224]
9		2.5% Et₃SiH in TFA, overnight (incomplete cleavage after 4 h)	99% (> 90%)	[213]
10		TFA/DCM 5:95, 20 °C, 10 min Ar: 4-(MeO)C₆H₄	79% (90%)	[207] see also [210,212, 225,226]
11		TFA/DCM 5:95, 20 °C, 10 min	67% (95%)	[207] see also [227]
12		TFA/DCM 5:95, 20 °C, 10 min	63% (90%)	[207] see also [228]
13		1. piperidine/DMF 2:8, 0.5 h 2. TFA/H₂O 95:5, 2 h (no cleavage if step 1 is omitted)	97%	[211]
14		TFA/DCM 7:3, 25 °C, 2 h	91%	[224] see also [229, 230]

Table 3.9. continued.

Entry	Loaded resin	Cleavage conditions	Product, yield (purity)	Ref.
15		TFA/H$_2$O 95:5, 0.5 h	91%	[231]
16		TFA/Me$_2$S/H$_2$O 90:5:5, 36 h	75%	[212]
17		TFA/H$_2$O/Me$_2$S 100:5:5, 85 °C, 12 h; or TFA/H$_2$O/*i*Pr$_3$SiH 100:6:6, 85 °C, 12 h		[232]
18		TFA/EDT/PhOH/PhSMe 90:5:3:2, 20 °C, 1 h (no cleavage with TBAF (0.2 mol/L), DMF, 2 h)	peptide—NH$_2$ 43% (> 90%)	[233]
19		TFA/DCM 1:1, 4 h	88%	[234]

Photolabile linkers for amides are most often based on 2-nitrobenzyl derivatives (Table 3.10; for the preparation of these and similar linkers, see also [147,153,235]). The mechanism of photolysis is the same as for the related 2-nitrobenzyl alcohol linkers (Section 3.1.3).

Table 3.10. Photolytic release of amides from supports.

Entry	Loaded resin	Cleavage conditions	Product, yield (purity)	Ref.
1		hν (354 nm), MeOH, 40 h	50%	[236]
2		hν (365 nm), 5% DMSO in phosphate-buffered saline (pH 7.4), 3 h	> 90% (95%)	[145,149]
3		hν (365 nm), MeOH/H$_2$O 1:4, 4 h	86%	[237]
4		hν (350 nm), MeOH, 18 h	70%	[238] see also [239,240]

3.3.2 (Diarylmethyl)amine and Tritylamine Linkers

One of the oldest linkers for amides is the (4-methylbenzhydryl)amine linker (MBHA; Entry **1**, Table 3.11). In contrast to the corresponding benzhydrol linker (which is cleavable by 5% TFA in DCM, 5 min [45]), acidolysis of the benzylic C–N bond of the MBHA linker requires treatment with hydrogen fluoride or a similar acid. As for *N*-benzylamides, the acid-lability of *N*-(diarylmethyl)amides increases with the number of electron-donating substituents on the aryl groups.

Several different alkoxy-substituted (diarylmethyl)amine linkers have been described (Table 3.11), but it is the 'Rink' amide linker (Entry **9**, Table 3.11) that is most frequently used for the synthesis of amides RCONH$_2$. Because this linker is attached to polystyrene as a phenyl ether, strong acids can also lead to cleavage of the linker from the support. With some types of product this undesirable cleavage of the support–linker bond can even occur upon treatment with TFA. This happens particularly readily with amides containing free amino groups, thioethers, or other basic functional groups close to the attachment site. The Rink linker has also been attached to amine-functionalized polystyrene (e.g. MBHA or aminomethyl polystyrene) as an aryloxy-acetamide, resulting in more robust linkage to the support. A further side reaction that can occur during the cleavage of amides RCONH$_2$ from the Rink or related linkers is dehydration to nitriles. This reaction readily occurs if TFA containing trifluoroacetic anhydride is used for the cleavage reaction [241].

The attachment of amides to supports as *N*-(diarylmethyl)amides can be achieved either by acylation of resin-bound (diarylmethyl)amines, or by acid-catalyzed N-alkylation of amides with resin-bound benzhydryl alcohols [46]. The former strategy is by far the more general. For the preparation of secondary amides RNHCOR by backbone amide linking, benzylamine linkers (Table 3.9) are more appropriate than (diarylmethyl)amine linkers, because *N*-alkyl-*N*-(diarylmethyl)amines are often difficult to acylate because of steric hindrance [242,243]. A few examples of the preparation of secondary amides by backbone amide linking to (diarylmethyl)amine linkers have been reported (Table 3.11).

Tritylamine linkers have not been extensively used for the attachment of amides, probably because *N*-tritylamides are difficult to prepare. This is not the case for strongly acidic amides and cyclic imides, which are readily N-tritylated with trityl chlo-

Table 3.11. Acidolytic cleavage of resin-bound (diarylmethyl)amine and tritylamine derivatives.

Entry	Loaded resin	Cleavage conditions	Product, yield (purity)	Ref.
1	(MBHA linker)	HF/PhOMe 93:7		[247] see also [248,249]
2		TFA/DCM/H₂O 9:90:1, 20 °C, 1 h	49% (90%)	[250]
3		TFA/H₂O/PhSMe/ EtSMe/EDT/PhSH 82:5:5:3:3:2, 20 °C, 6 h	54%	[251] see also [252]
4		TFA/PhSMe/EDT 90:5:5	43%	[253,254]
5		TFA/PhSMe 4:1, 3 h	100%	[214]

Table 3.11. continued.

Entry	Loaded resin	Cleavage conditions	Product, yield (purity)	Ref.
6	peptide–NH linker (PS), MeO, MeO	PhSMe (1 mol/L), TFA, 28 °C, 1 h	peptide–NH$_2$ 41%	[255]
7	peptide–NH linker, Pol (XAL or Sieber linker)	TFA/DCM 1:99, 0.5 h	peptide–NH$_2$ 84% (> 90%)	[256]
8	Fmoc–NH linker, MeO, MeO, OMe, PS	TFA/DCM/H$_2$O 57:40:3, 20 °C, 1 h	Fmoc-NH$_2$ 100%	[257]
9	peptide–NH linker, PS, MeO, OMe (Rink amide linker)	TFA/DCM 1:1, 20 °C, 15 min	peptide–NH$_2$ 80%	[46]
10	R–O–NH linker, PS, MeO, OMe	TFA/H$_2$O 9:1	R–O–NH$_2$ 70–90%	[258] ureas: [259]
11	Ph, FmocHN, Ph, N linker, PS	TFA/iPr$_3$SiH//H$_2$O 90:2:8, 20 °C, 0.5 h	FmocHN, Ph, Ph 55% (98%)	[242]
12	N, N linker, PS, MeO, OMe	TFA/DCM 20:80	71%	[260] see also [261]
13	N, cyclohexyl, linker, PS, MeO, OMe	TFA/DCM/H$_2$O 95:5:1, 23 °C, 40 min	99% (> 90%)	[262]

Table 3.11. continued.

Entry	Loaded resin	Cleavage conditions	Product, yield (purity)	Ref.
14	H$_2$N structure on dimethoxy benzhydryl PS resin	TFA/DCM 5:95, 0.5 h	H$_2$N naphthyl amide, 75%	[263] see also [264]
15	Tol-SO$_2$-N(CH$_2$Ph) dimethoxy benzhydryl PS resin	TFA/DCM 2:8, 15 min	Tol-SO$_2$-NH-CH$_2$Ph, (73%)	[265]
16	peptide-C(O)-NH benzhydryl, MeO, MeO, sulfoxide (PS)	TFA/DCM/PhOMe/ PhSMe/EDT/SiCl$_4$ 150:15:3:8:6:15, 25 °C, 3 h	peptide-C(O)-NH$_2$, 62%	[266] see also [267,268]
17	FmocHN isoxazolidinone, N-CH(Ph)(2-Cl-C$_6$H$_4$) PS	1% TFA in DCM, 2 h	FmocHN isoxazolidinone NH, 94% (91%)	[210]
18	RO-C$_6$H$_4$-CH$_2$ thiazolidinedione, N-CH(Ph)(Cl-C$_6$H$_4$) PS	10% TFA in DCM, 0.5 h	RO-C$_6$H$_4$-CH$_2$ thiazolidinedione NH, 80–90%	[269]

ride resins [244] to yield sufficiently stable, resin-bound amides (Entries **17** and **18**, Table 3.11). Sulfamates RO–SO$_2$–NH$_2$ react with polystyrene-based trityl chloride resin to yield the *N*-tritylated sulfamates. Treatment with TFA leads to release of the unchanged sulfamate into solution [245]. Thioureas and thiosemicarbazones have been prepared from polystyrene-bound trityl isothiocyanate [246]. Cleavage from the support required treatment of the *N*-trityl derivatives with TIPS/TFA/DCM (5:15:80) for 2–3 min.

Primary or secondary amines bound to benzhydryl- or trityl linkers can be cleaved from the support with simultaneous conversion to amides by treatment with acyl halides [243]. This strategy, however, generally leads to products contaminated with acylating agent and various salts, which require further purification. Entry **16** in Table 3.11 is an example of a safety-catch linker: cleavage of the benzylic C–N bond requires a reducing agent (e.g. EDT + silyl halide) to convert the electron-withdrawing sulfoxide into an electron-donating thioether.

3.3.3 Linkers for Amides Cleavable by Nucleophiles

Amides, carbamates, and ureas can be generated by nucleophilic cleavage of resin-bound esters, carbonates, and carbamates, respectively, with amines (Figure 3.20). These reactions only proceed well if sufficiently reactive resin-bound derivatives are used.

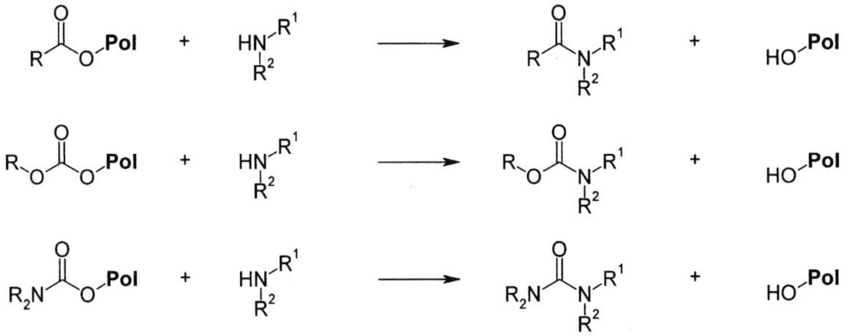

Figure 3.20. Formation of amides, carbamates, and ureas by nucleophilic cleavage with amines.

3.3.3.1 Nucleophilic Cleavage of Alkyl Esters

The examples of nucleophilic cleavage of resin-bound benzyl-, PEG-, or other alkyl esters by amines (Table 3.12) show that this reaction generally requires forcing conditions to proceed. For this reason, it is often possible to perform reactions of amines or other nucleophiles with compounds linked to benzyl alcohol resins as esters without significant loss of product (e.g., removal of Fmoc protective groups from Wang resin bound peptides with piperidine). If peptides are cleaved from benzyl alcohol linkers by aminolysis, partial racemization of the C-terminal amino acid is often observed [270]. Thiol esters are more susceptible to nucleophilic cleavage (Entry **8**, Table 3.12 [271–273]). Thus, for example, thiol ester attachment cannot be used for standard solid-phase peptide synthesis using Fmoc methodology [274].

A particularly interesting variant of nucleophilic cleavage is the intramolecular attack of the linking ester by a substrate-bound nucleophile. In this way, simultaneous cleavage from the support and formation of a cyclic acid derivative is realized (Figure 3.21).

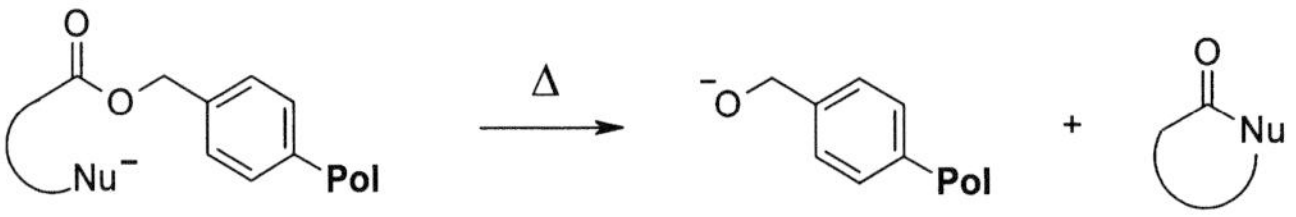

Figure 3.21. Intramolecular nucleophilic cleavage of esters.

Numerous examples of different variants of this cyclization/cleavage protocol have been reported. Diketopiperazine formation (Section 15.22.1), an unwanted side reaction in solid-phase peptide synthesis, is also an example of this type of compound release. Because intramolecular processes generally take place more readily than the corresponding intermolecular reactions, cyclization/cleavage can occur with alkyl

Table 3.12. Aminolysis of resin-bound alkyl esters.

Entry	Loaded resin	Cleavage conditions	Product, yield (purity)	Ref.
1		neat BnNH$_2$, 20 °C, 48 h; or neat PrNH$_2$, 20 °C, 22 h	benzylamide, 43% (80%) propylamide, 96% (96%)	[270]
2		1-methylpiperazine/ AlCl$_3$ 4:1, DCM, 20 °C, 18 h	45% (84%)	[275]
3		70% EtNH$_2$ in H$_2$O/THF 1:1, 20 °C, overnight	98% (94%)	[276]
4		(neat), 55 °C, 18 h	61%	[277] see also [278,279]
5		HexNH$_2$ (0.5 mol/L), DMF, 25 °C, 16 h; or excess Me$_2$NH in DCM, 25 °C, 15 h	hexylamide, 25% dimethylamide, 2%	[90]
6		NH$_3$ (satd in CF$_3$CH$_2$OH), 20 °C, 17 h	peptide—NH$_2$ > 98% (> 95%)	[98] see also [280]
7		BnNH$_2$, AlMe$_3$, DCM, PhMe, 22 °C, 20 h		[281] see also [282]
8		AgNO$_3$, piperidine (each 0.07 mol/L), DMF, 1 h	97%	[283]

esters under conditions which would not lead to intermolecular nucleophilic cleavage. In particular, five- or six-membered rings are readily formed. This cleavage strategy is discussed in the sections dedicated to the type of heterocycle formed upon cleavage.

3.3.3.2 Nucleophilic Cleavage of Aryl Esters

Aryl esters are generally more readily cleaved by nucleophiles than alkyl esters. This sensitivity towards nucleophiles becomes more pronounced with increasing acidity of the corresponding phenol. Resin-bound phenyl esters are convenient reagents for the acylation of amines (Table 3.13). A particularly interesting type of phenol linker is the 4-(alkylthio)phenol linker, which allows for further activation of the corresponding phenyl esters through oxidation of the thioether to a sulfoxide or sulfone. This type of linker is also suitable for the preparation of cyclic peptides by intramolecular, nucleophilic cleavage [284]. However, as shown by the examples in Table 3.13, additional activation by conversion into a sulfone is not always required as even the unoxidized 4-(alkylthio)phenol is a good leaving group.

Immobilized, highly reactive phenyl esters can be prepared by acylating resin-bound 4-acyl-2-nitrophenol (Entry **4**, Table 3.13 [285–288]) or 4-(aminocarbonyl)-2,3,5,6-tetrafluorophenol (Entries **7** and **8**, Table 3.13). These esters are similar to oxime esters (see Section 3.3.3.3), and even react with weak nucleophiles such as anilines or alcohols. This type of linker is not, therefore, well suited for long synthetic sequences on insoluble supports, but only for the preparation of simple acid derivatives. Because cleavage yields the unchanged phenol, these resins can be reused several times, which renders this strategy of preparing acid derivatives quite cost-effective.

Table 3.13. Nucleophilic cleavage of resin-bound aryl esters.

Entry	Loaded resin	Cleavage conditions	Product, yield (purity)	Ref.
1		BuNH$_2$, pyridine, DIPEA, 20 °C, 24–36 h	80%	[291] see also [292]
2		NEt$_3$/DCM 1:1, 27 °C, 4 h	66%	[293]
3		ethyl glycinate, DMF, 24 h	38%	[294]
4		3-aminobenzophenone, NEt$_3$, MeCN, 70 °C, 24 h		[295] see also [296,297]

Table 3.13. continued.

Entry	Loaded resin	Cleavage conditions	Product, yield (purity)	Ref.
5		DCM, 20 °C, 40 h	95%	[286]
6		PhCH₂NH₂ (1.2 eq), NMM (1 eq), THF, 3 h	86%	[298] see also [299]
7		4-cyclohexylaniline (0.07 mol/L, 0.8 eq), DMF, 16 h	84% (99%)	[300]
8		1-(4-fluorophenyl)-piperazine (0.09 mol/L, 0.8 eq), DMF, 16 h	96% (97%)	[300] see also [301]
9		1. TFA/TIPS/DCM 9:2:89, 0.5 h 2. morpholine (0.77 mol/L, 9 eq), THF, 18 h Ar: 2-chlorophenyl	88% (90%)	[289]

The preparation of cyclic peptides by intramolecular nucleophilic cleavage of polystyrene-bound 2-nitrophenyl esters has also been reported [287].

Entry **9** in Table 3.13 is an example of a safety-catch linker, which requires activation by TFA-mediated cleavage of a *tert*-butyl ether. The unactivated 2-(*tert*-butoxy)-phenyl esters are cleaved by amines 700 times more slowly than the corresponding 2-hydroxyphenyl esters [289]. A similar linker has been described [290], in which a benzyl ether is used instead of a *tert*-butyl ether. Activation of this linker by debenzylation was achieved by treatment with HF or HBr/TFA [290].

3.3.3.3 Nucleophilic Cleavage of Oxime and Related Esters

A further type of linker, especially designed for facile nucleophilic cleavage, is the oxime linker. Oximes of resin-bound ketones or aldehydes can be acylated to yield *O*-acyl oximes, which are readily cleaved by a variety of nucleophiles (Table 3.14). A widely used support of this type is *p*-nitrobenzophenone oxime resin (Kaiser oxime resin; Entries **1** and **2**, Tabel 3.14; for its preparation, see [302]), which has been successfully used for the synthesis of protected peptide fragments [303–306], diketopiperazines [307], macrocyclic peptides [308–310], and macrocyclic lactams [311].

Resin-bound 1-hydroxybenzotriazole (HOBt) has also been prepared [312–314], and yields, upon O-acylation, highly reactive esters [312,315,316]. These are even cleaved by weak nucleophiles, such as anilines or even *N*-hydroxysuccinimide (Entry **7**, Table 3.14). The reactivity of resin-bound HOBt esters is greater than that of resin-bound 4-acyl-2-nitrophenyl esters [296]. Similarly, resin-bound *N*-hydroxysuccinimide has also been used for the preparation of polymeric acylating agents (Entry **6**, Table 3.14 [317–319]).

Because of their sensitivity towards nucleophiles, O-acylated, resin-bound hydroxybenzotriazoles and related compounds are not suitable for multistep synthetic

Table 3.14. Nucleophilic cleavage of resin-bound oxime, HOBt, and related esters.

Entry	Loaded resin	Cleavage conditions	Product, yield (purity)	Ref.
1		satd NH$_3$ in MeOH/ DCM 1:1, 2 h	96%	[322]
2		BnNH$_2$ (0.5 mol/L), CHCl$_3$, 20 °C, 4 h	peptide... 87%	[306]
3		PhNH$_2$ (0.03 mol/L), CHCl$_3$, 20 °C, 3 h	50%	[323]
4		PhNH$_2$ (0.1 mol/L, 1 eq), DMF, 20 °C, 20 h	64%	[320]
5	BocHN...	ethyl glycinate (0.05 mol/L, 0.5 eq), CHCl$_3$, 30 °C, 14 h	BocHN...OEt 53%	[324]
6	BnO...	*tert*-butyl glycinate (7 mmol/L, 0.16 eq), DMF/CHCl$_3$ 1:1, overnight	BnO...CO$_2$tBu 98%	[325]
7	Ph...	*N*-hydroxy-succinimide (0.8 eq), DCM, 20 °C, 7 h	Ph... 86%	[326]

sequences on solid phase. These intermediates can, however, be used as insoluble acylating agents, which offer a practical alternative to soluble reagents for the derivatization of amines or alcohols [320,321], and are well suited to parallel automated synthesis.

3.3.3.4 Nucleophilic Cleavage of Amides and Carbamates

Amides are generally very resistant towards nucleophilic cleavage. 'Safety-catch' linkers, such as those described in Section 3.1.2.3, can, however, be cleaved by amines to yield amides (Entries **1** and **2**, Table 3.15). Entry **4** in Table 3.15 is an example of a

Table 3.15. Nucleophilic cleavage of resin-bound amides and carbamates.

Entry	Loaded resin	Cleavage conditions	Product, yield (purity)	Ref.
1		RCH$_2$NH$_2$, DMF or THF or NMP, 55 °C, 10 h R: alkyl	80–97% (87–99%)	[327] see also [328–330]
2		PhNH$_2$ (1 mol/L, 14 eq), dioxane, 100 °C, 15 h (no reaction with 4-nitroaniline)	84%	[331] see also [118]
3		1. PPTS, PhMe, 50 °C, 16 h (formation of N-acylindole) 2. pyrrolidine (15 eq), THF, 20 °C, 72 h	96% (> 95%)	[117]
4		morpholine (0.04 mol/L, 10 eq), hv (> 290 nm), THF, 25 °C, 6–12 h	70% (90%)	[332]
5		NH$_3$ (satd in iPrOH), 24 h	60–80%	[125]
6		4-(MeO)C$_6$H$_4$NH$_2$ (0.13 mol/L, 4 eq), DCM/PhMe, 80 °C, overnight	88% (87%)	[333] see also [334]

Table 3.15. continued.

Entry	Loaded resin	Cleavage conditions	Product, yield (purity)	Ref.
7		morpholine (0.13 mol/L, 4 eq), DCM/PhMe, 80 °C, overnight	77% (92%)	[333,335]
8		BnNH₂ (1.3 eq), NEt₃ (4 eq), MeCN, 60 °C, 24 h	89% (> 98%)	[336]
9		BnNH₂ (0.74 eq), THF, 66 °C, overnight	80% (47%)	[337]
10		leucinol (0.7 mol/L), PhMe, 60 °C, 12 h	69%	[338]
11		LiOH, H₂O, THF, 20 °C, 54 h	63% (96%)	[339]
12		NaOMe (1.5 mol/L), MeOH/THF 1:1, 20 °C, 16 h	88% (98%)	[339] see also [340]
13		NH₃ (2 mol/L), MeOH, 20 °C, 2 h	74%	[340]
14		LiN(SiMe₃)₂ (1 eq), THF, −78 °C, 2 h, 20 °C, 0.5 h	86% (96%)	[341]

linker that requires photolytic activation in order to undergo nucleophilic cleavage. Photolysis probably leads to an intramolecular O-acylation of the nitro group, yielding an ester that is susceptible to intermolecular attack by nucleophiles. Carbamates are generally more resistant towards nucleophilic cleavage than amides or carboxylic esters. Some phenol- or oxime-derived carbamates, however, react readily with amines, and resin-bound carbamates of this type have been successfully used for the conversion of amines into ureas (Table 3.15).

Support-bound *N*-sulfonyl carbamates, which can be prepared by N-sulfonylation of resin-bound carbamates, are susceptible to nucleophilic cleavage. These intermediates enable the solid-phase preparation of *N*-aryl- or *N*-alkylsulfonamides using inexpensive hydroxymethyl polystyrene (Entries **8** and **9**, Table 3.15). Polystyrene-bound carbamates can also be cleaved by treatment with acyl halides in the presence of Lewis acids (Entry **4**, Table 3.16).

3.3.4 Miscellaneous Linkers for Amides and Ureas

Resin-bound triazenes with a free NH group can be acylated by treatment with acyl halides, or carbamoylated by treatment with isocyanates [342]. The resulting triazene derivatives are stable towards strong bases, but undergo acidolysis when treated with TFA or TMSCl, yielding amides and ureas, respectively (Entries **1** and **2**, Table 3.16). Polystyrene-bound triazenes devoid of a free NH group or carbamates can be cleaved from the support by treatment with acyl halides to yield amides (Entries **3** and **4**, Table 3.16).

O-Alkylhydroxamic acids undergo reductive N–O bond cleavage when treated with samarium(II) iodide. This reaction can be used to cleave amides from insoluble supports (Entry **5**, Table 3.16). Further reactions leading to the formation of amides and ureas include the oxidation of resin-bound *N*-(4-alkoxyphenyl)amides with CAN, which can also be applied to the release of β-lactams, and the acidolysis of isothioureas (Entries **6** and **7**, Table 3.16). Resin-bound hydrazides undergo facile copper-mediated oxidation to yield *N*-acylazo compounds, which are strong acylating agents and can be used to acylate amines intramolecularly with simultaneous release of the cyclic amide from the support. This strategy has been used to prepare 2-oxopiperazines (Figure 3.13 [176]) and cyclic peptides [343].

Table 3.16. Miscellaneous linkers for amides and ureas.

Entry	Loaded resin	Cleavage conditions	Product, yield (purity)	Ref.
1		TFA/DCM 5:95, 20 °C, 3 × 5 min		[342]
2		TMSCl/DCM 1:9, 20 °C, 3 × 5 min		[342]
3		AcCl, THF, 20 °C, 12 h	(> 90%)	[344]
4		PhCOCl (0.6 mmol/L, 3 eq), ZnBr$_2$ (0.5 eq), 20 °C, 24 h, then NEt$_3$ (1.5 eq)	86%	[345]
5		SmI$_2$ (0.1 mol/L, 2 eq), THF, 25 °C, 3 h Ar: 4-MeOC$_6$H$_4$	54% (99%)	[346] see also [347]
6		CAN (5 eq), MeCN/H$_2$O 2:1, 20 °C, 0.5 h,	70% (95%)	[348]
7		AcOH/H$_2$O/dioxane/ EtOH 8:8:42:42, 78 °C, 16 h	71%	[349]
8		TFA/H$_2$O/Et$_3$SiH 94:3:3, 20 °C, 6 h Ar: 3-pyridyl	73% (97%)	[350]

3.4 Linkers for Hydroxamic Acids and Hydrazides

Several strategies enable the generation of hydroxamic acids or hydrazides upon cleavage of a carboxylic acid derivative from a support (Figure 3.22). N-Protected hydroxylamines can be linked to supports via the oxygen atom as the benzyl ether [351–354], as the benzhydryl ether [48,355], or as the trityl ether [356–358]. After deprotection, these intermediates can be acylated at nitrogen and, upon cleavage from the support, hydroxamic acids result (Table 3.17). Resin-bound hydroxylamine can be prepared by treatment of an appropriate resin-bound benzyl halide or sulfonate [352] with N-hydroxyphthalimide in the presence of a base, or by Mitsunobu reaction of N-hydroxyphthalimide with resin-bound benzyl alcohols [351]. The phthaloyl protective group of the resulting intermediate is removed by treatment with hydrazine. Alternatively, resin-bound hydroxylamine can be prepared from 2-chlorotrityl chloride resin and Fmoc- or Dde-protected hydroxylamine [356]. Hydroxylamine can also be attached to a backbone amide linker via the nitrogen atom (Entry 6, Table 3.17).

Hydrazine can readily be linked to acid-labile benzyl alcohol resins either as N-benzylhydrazine (Entry 7, Table 3.17) or as carbamate (Entries 8 and 9, Table 3.17). The

Table 3.17. Linkers for hydroxamic acids and hydrazides.

Entry	Loaded resin	Cleavage conditions	Product, yield (purity)	Ref.
1		TFA/*i*Pr₃SiH/DCM 50:5:45		[352] see also [351,361]
2		TFA/DCM 5:95, ultrasound, 15 min		[362]
3		TFA/*i*Pr₃SiH/EDT/H₂O 90:1:4:5, 30 °C, 4 h	peptide 80–90%	[355] see also [48]
4		HCO₂H/THF 1:3, 1 h	50–78% (> 84%)	[357]
5		TFA/DCM 5:95, 25 °C, 1 h	89% (> 95%)	[358] see also [356]
6		1. TFA/DCM/H₂O 3:96:1, 1 h 2. TFA/DCM/H₂O 50:49:1, 1 h	88%	[363]

Table 3.17. continued.

Entry	Loaded resin	Cleavage conditions	Product, yield (purity)	Ref.
7		1. piperidine/DMF 2:8, 0.5 h 2. HF, anisole, 0 °C, 0.5 h (no cleavage with TFA)	64%	[211]
8		TFA/DCM 1:1, 20 °C, 0.5 h	76%	[67] see also [68]
9		TFA/DCM 1:1, 20 °C, 0.5 h	42%	[28]
10		aq NH$_2$OH (50%, 25 eq), H$_2$O, THF, 2 d	81% (87%)	[364] see also [365]
11		1. NH$_2$OTMS (30 eq), PhMe or THF, 25 °C, 16–24 h 2. TFA/DCM 5:95, 10 min	(80–100%)	[359]
12		N$_2$H$_4$/DMF 1:9, 24 h; or N$_2$H$_4$·H$_2$O/EtOH 25:100, 6 h	30–78%	[366,367] see also [368]
13		2.5% N$_2$H$_4$·H$_2$O in MeOH, 20 °C, 3 h	88%	[96]

unacylated amino group can then be acylated, resulting in resin-bound hydrazides. These are stable towards nucleophilic attack, but can be cleaved by acids (depending on the type of benzyl alcohol resin chosen) to yield hydrazides. Resin-bound hydrazides can also be cleaved by oxidants (see Section 3.1.4) to yield carboxylic acids, amides, or esters [175]. Entry **7** in Table 3.17 is an example of a linker that requires activation prior to cleavage. The (2-acetoxybenzyl)hydrazine is stable towards acid treatment, but upon saponification of the aryl ester, the benzylic C–N bond becomes more prone to acidolysis [211]. The strongly acidic conditions required in this instance are probably due to the deactivating effect of the free, protonated hydrazinium group (see Figure 3.19).

Hydroxamic acids and hydrazides can also be prepared by nucleophilic cleavage of resin-bound esters, activated amides, or thiol esters [359] with hydroxylamine and hydrazine, respectively. The preparation of hydroxamic acids by nucleophilic cleavage of simple alkyl esters with hydroxylamine proceeds only slowly (Entry **10**, Table 3.17), and more reactive esters (e.g. oxime esters [360] or thiol esters, Entry **11**, Table 3.17) have to be used if higher reaction rates are required.

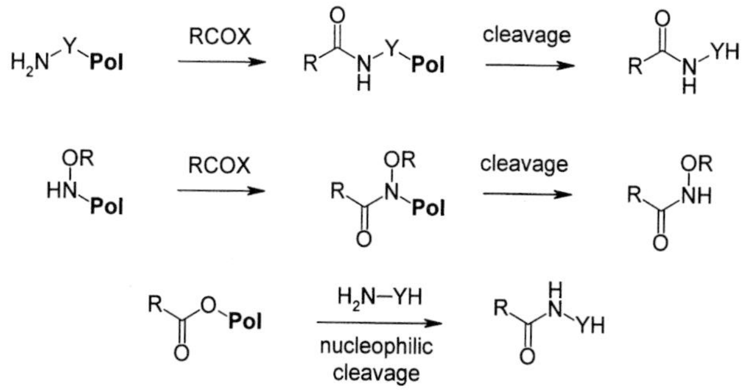

Figure 3.22. Strategies for the solid-phase preparation of hydroxamic acids and hydrazides. X: leaving group, Y: NH, O.

3.5 Linkers for Carboxylic Esters and Thiol Esters

Esters, which have no possible site of attachment, cannot be directly linked to supports, but may be generated upon cleavage from a support. This cleavage can be mediated by electrophiles, nucleophiles, or oxidants. Only a few examples have been reported of the preparation of esters by O-alkylation of carboxylates by resin-bound alkylating agents, such as sulfonic esters [369–372] or diazonium salts [373] (see also Section 3.13).

3.5.1 Linkers for Esters Cleavable by Nucleophiles

Resin-bound carboxylic acid derivatives can be susceptible to nucleophilic attack by alcohols. The preparation of esters by nucleophilic cleavage with alcohols is generally only practical when using low-boiling alcohols, which can be readily removed by evaporation after the cleavage. The free acid is often observed as a by-product in this type of cleavage. The nucleophilic cleavage of acid derivatives by alcohols can be catalyzed by tertiary amines [374,375], alcoholates, alkali metal cyanides, or acids (Table 3.18). As in the saponification or aminolysis of resin-bound carboxylic acid derivatives, nucleophilic cleavage with alcohols under basic conditions proceeds more readily with increasing electrophilicity of the resin-bound acid derivative (hydroxybenzotriazole esters > aryl esters > alkyl esters > alkoxy-substituted benzyl esters > trityl esters). Attempts to prepare macrocyclic lactones by intramolecular alcoholysis of ω-hydroxyalkanoic acids bound to polystyrene as 4-sulfonylphenyl esters [376] or thiol esters [271] led to mixtures of cyclic oligoesters. For the preparation of esters by treatment of carboxylates with resin-bound alkylating agents, see Section 3.13 and Entry **7**, Table 3.19.

Thiol esters RC(O)SR have been prepared by nucleophilic cleavage of polystyrene-bound *N*-acylsulfonamides with mixtures of a thiol and sodium thiophenolate [377] or LiBr [378], or by treatment of Wang or PAM resin bound carboxylic esters with ethanethiol in the presence of a Lewis acid (AlMe$_2$Cl or AlMe$_3$, DCM, 20 °C, 3 h [379,380]). The latter method can also give rise to the formation of orthoesters RC(SEt)$_3$ and ketene thioacetals, which can, however, be hydrolyzed to the desired thiol esters by treatment with TFA [379].

Table 3.18. Generation of esters by alcoholysis of resin-bound carboxylic acid derivatives.

Entry	Loaded resin	Cleavage conditions	Product, yield (purity)	Ref.
1		NaOMe (0.25 mol/L), MeOH/THF 1:4, 20 °C, 10 h	79%	[381] see also [86]
2		MeOH/NEt$_3$/DMF 9:1:1, 60 °C	95%	[382] see also [383]
3		NaOMe (0.02 mol/L), MeOH/THF 1:4, reflux overnight	> 95% (> 90%)	[384] see also [385]
4		NEt$_3$/MeOH 1:4, 20 °C, 20 h	58%	[386] see also [387]
5		Ti(OEt)$_4$ (5 eq), neat allyl alcohol, 120 °C, 2 h	42%	[388]
6		*i*PrOH/MeOH/ Me$_3$SiCl 7:3:1	(> 95%)	[389]
7	(HMBA linker)	NEt$_3$/MeOH 1:9, 50 °C, 20 h	98% (> 96%)	[390] see also [391]
8		KCN (0.3 mol/L), MeOH, 25 °C, 6 h	49%	[90]

Table 3.18. continued.

Entry	Loaded resin	Cleavage conditions	Product, yield (purity)	Ref.
9		NEt$_3$/MeOH 2:8, 20 °C, 23 h	63%	[96]
10		NEt$_3$/MeOH 2:8, 20 °C, 24 h	> 98% (> 95%)	[98]
11		NaCN, MeOH, 20 °C, 16 h	12–29% (83–95%)	[392] see also [393]
12		NaOMe, MeOH/THF 1:1, 20 h		[119] see also [394]
13		1. PPTS, PhMe, 50 °C, 16 h (formation of *N*-acylindole) 2. MeOH/THF 5:95, NaNH$_2$, 0.5 h	89% (> 98%)	[117]
14		LiOCH$_2$Ph	26%	[30]
15		EtOH/THF/H$_2$SO$_4$/ H$_2$O 90:100:4:5, reflux, 7 d	68%	[395]
16		TBAF, AcOH, THF, 40 °C, 6 h	61%	[396]
17		TFA/DCM 3:97 (Wang resin)	70%	[397]
18		TBAF, AcOH, THF, 40 °C, 14 h	41% 14%	[398]

3.5.2 Linkers for Esters Cleavable by Electrophiles or Oxidants

An interesting variant of electrophilic cleavage is intramolecular attack at the linking carboxylic acid derivative by a substrate-bound electrophile (Figure 3.23). This cleavage strategy generally produces lactones.

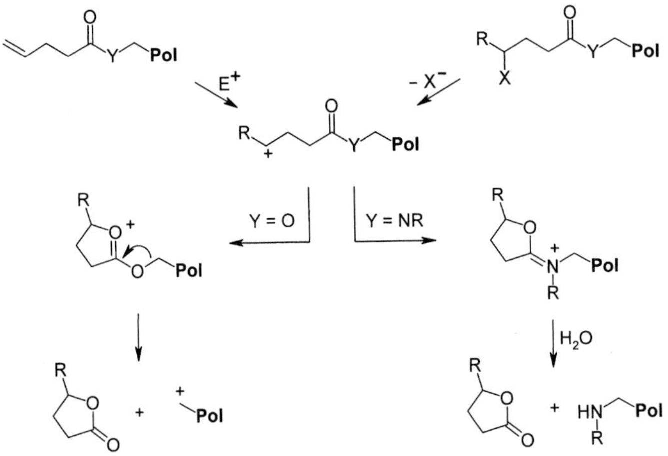

Figure 3.23. Intramolecular electrophilic cleavage of carboxylic acid derivatives from polymeric supports. X: leaving group; Y: NR, O.

Because of the special structural requirements of the resin-bound substrate, this type of cleavage reaction lacks general applicability. Some of the few examples that have been reported are listed in Table 3.19. Lactones have also been obtained by acid-catalyzed lactonization of resin-bound 4-hydroxy or 3-oxiranyl carboxylic acids [399]. Treatment of polystyrene-bound cyclic acetals with Jones' reagent also leads to the release of lactones into solution (Entry 5, Table 3.19). Resin-bound benzylic aryl or alkyl carbonates have been converted into esters by treatment with acyl halides and Lewis acids (Entry 6, Table 3.19). Similarly, alcohols bound to insoluble supports as benzyl ethers can be cleaved from the support and simultaneously converted into esters by treatment with acyl halides [400]. Esters have also been prepared by treatment of carboxylic acids with an excess of polystyrene-bound triazenes; here, diazonium salts are released into solution, which serve to O-alkylate the acid (Entry 7, Table 3.19). This strategy can also be used to prepare sulfonates [401].

Table 3.19. Formation of esters during cleavage with electrophiles and oxidants.

Entry	Loaded resin	Cleavage conditions	Product, yield (purity)	Ref.
1		I_2, THF/H_2O, 20 °C, 3 d	40%; 30% ee	[402] see also [403]
2		BrCN, CHCl$_3$/H$_2$O, TFA, 20 °C, 24 h		[404]
3		TFA/DCM 1:1, 20 °C, 2 h	60% (90%)	[399,405]
4		TFA/DCM 1:1, 20 °C, 2 h	57% (90%)	[399]
5		Jones reagent (CrO$_3$/H$_2$SO$_4$; 2 eq), acetone, 20 °C, 3 h		[406]
6		PhCOCl (0.6 mmol/L, 3 eq), ZnBr$_2$ (0.5 eq), 20 °C, 24 h Ar: 4-(MeO$_2$C)C$_6$H$_4$	97%	[345]
7		benzilic acid (0.2 eq), DCM or THF or DCM/MeOH 9:1 or dioxane, 20 °C, 6 h	(94%)	[373]

3.6 Linkers for Primary and Secondary Aliphatic Amines

Primary and secondary aliphatic amines can be linked to polymeric supports by acid-labile linkers or by linkers sensitive to nucleophiles. Linkers cleavable by light or by transition metal catalysis have also been described. The main types of linker for amines are sketched in Figure 3.24.

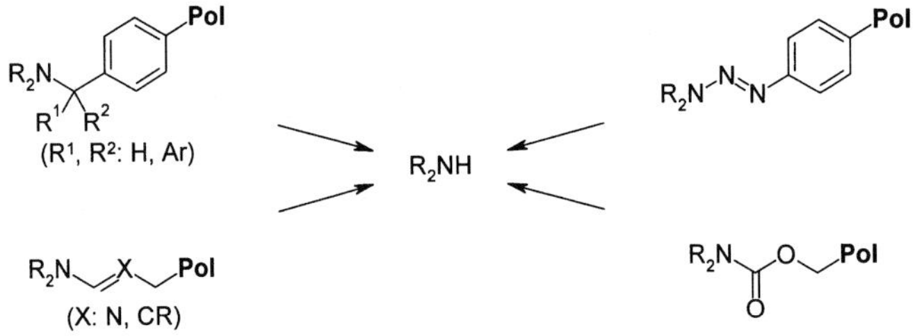

Figure 3.24. Strategies for the attachment of amines to insoluble supports.

3.6.1 Benzylamine, (Diarylmethyl)amine, and Tritylamine Linkers

Aliphatic primary and secondary amines can be linked to insoluble supports as benzylamines by reductive alkylation with support-bound benzaldehydes or by N-alkylation with support-bound benzyl halides or sulfonates (Figure 3.25; see also Section 10.1). Benzhydrylamines and tritylamines are usually prepared by N-alkylation with the corresponding halides.

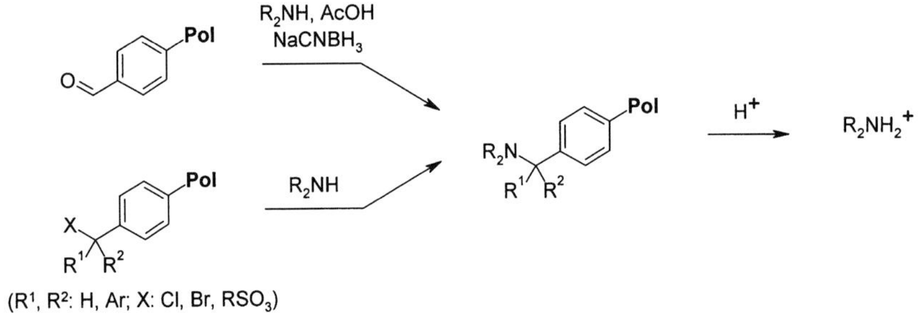

$(R^1, R^2: H, Ar; X: Cl, Br, RSO_3)$

Figure 3.25. Immobilization of amines as benzylamines.

Benzylic C–N bonds in amines are generally more difficult to cleave with electrophiles than those in *N*-benzylamides. Therefore, as illustrated by the examples in Table 3.20, most of the linkers suitable for backbone amide linking (Section 3.3) cannot be used for amines, unless more vigorous cleavage conditions are applied. As an alternative to cleavage with acids, tertiary benzylamines can be debenzylated by treatment with alkyl chloroformates. If 1-chloroethyl chloroformate is used as the debenzylating agent, free amines can easily be obtained because the 1-chloroethylcarbamates formed during cleavage from the support undergo facile hydrolysis in methanol (Entries **3** and **4**, Table 3.20). Electron-rich benzylamines can also be cleaved by treatment with DDQ (Entry **5**, Table 3.20). Support-bound 2-nitrobenzylamines can be cleaved photolytically in the presence of small amounts of TFA (J. C. Tomesch, private communication). As with the related 2-nitrobenzyl alcohol linkers for acids, photolytic cleavage of 2-nitrobenzylamines proceeds by intramolecular disproportionation (Figure 3.10).

Table 3.20. Benzylamine linkers for aliphatic amines.

Entry	Loaded resin	Cleavage conditions	Product, yield (purity)	Ref.
1		TFA/H$_2$O 95:5	no cleavage	[207]
2		TFA/DCM 95:5, 4 h	no cleavage	[234]
3		(10 eq), DCM, 20 °C, 3 h; then MeOH, 65 °C, 3 h	95% (95%)	[407]
4		(0.93 mol/L, 10 eq), 23 °C, 3 h; then MeOH, 65 °C, 3 h	56%	[408]
5		DDQ (0.1 mol/L), C$_6$H$_6$, 20 °C, 3 h	84%	[409]

Benzhydrylamines are better suited than benzylamines as acid-labile linkers for amines. The MBHA linker ('methylbenzhydrylamine'), which is usually used to prepare peptide amides (see Section 3.3), can also be used as a linker for amines (Entry 1, Table 3.21). Hydrogen fluoride is, however, required as the cleavage reagent. Easier to cleave are alkoxy-substituted benzhydrylamines (Entries 2–5, Table 3.21), which can be prepared from the corresponding benzhydryl chlorides [263] or by reductive alkylation [410] or solvolysis [411] of the Rink amide linker. In the case of benzhydrylamines linked to polystyrene as benzyl ethers, treatment with TFA can lead to the release of the linker into solution (acidolysis of the benzylic C–O bond, see Figure 3.18).

Tritylamines can serve as both linkers and protective groups for aliphatic amines because, unlike benzhydrylamines, they do not usually undergo acylation when treated with activated acid derivatives. Tritylation of aliphatic amines is readily accomplished by adding excess amine to a support-bound trityl chloride. Illustrative cleavage reactions are listed in Table 3.21.

Table 3.21. Benzhydrylamine and tritylamine linkers for aliphatic amines.

Entry	Loaded resin	Cleavage conditions	Product, yield (purity)	Ref.
1	(MBHA linker)	HF/PhOMe, 0 °C, 9 h	> 70% (> 95%)	[412] see also [413]
2		TFA/DCM 5:95, 0.5 h	95% (96%)	[263] see also [411,414]
3	Ar: 4-FC₆H₄	TFA/DCM 1:1, 3 h Ar: 4-FC$_6$H$_4$	45% (96%)	[410] see also [415]
4		TFA/H$_2$O/DCM 5:5:90, 20 °C, 5 h Ar: 4-pyridyl	82% (78%)	[416]
5		TFA/iPr$_3$SiH/ DCM 50:2:50, 0.5 h	81%	[417]
6		TFA/DCM 1:3, 1 min	(> 90%)	[418] see also [419–422]
7		TFA/DCM 2:98, 0 °C, 5 min	82%	[423]
8		95% TFA, 16 h	66%	[424] see also [62,425]
9		TFA/DCM 1:99, 20 °C, 3 × 5 min		[426]

3.6.2 Carbamate Attachment

Amines can be linked to polymeric alcohols as carbamates. Carbamate attachment of amines can be achieved by reaction of isocyanates with alcohol linkers, or by treatment of alcohol linkers with phosgene [339,427,428] or a synthetic equivalent thereof, followed by exposure to the amine (Figure 3.26). The reagents most commonly used for the 'activation' of alcohol linkers are 4-nitrophenyl chloroformate [69,429–436] and carbonyl diimidazole [427,437–440]. The preparation of support-bound carbamates is discussed in Section 14.6.

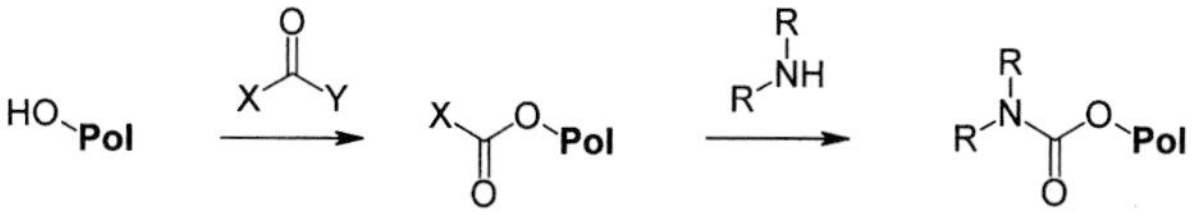

Figure 3.26. Carbamate attachment of amines to polymeric alcohols. X, Y: leaving groups.

Carbamate attachment allows the wide variety of cleavage strategies developed for carboxylic acids (for the synthesis of peptides on solid phase; acidolysis, nucleophilic cleavage, β-elimination, photolysis, and transition metal catalysis) to be extended to the solid-phase synthesis of amines (Table 3.22). Benzyl carbamates are usually more stable than amides towards nucleophilic attack, and are therefore suitable linkers if reactions with nucleophiles or bases are to be performed. The cleavage conditions for carbamates prepared from resin-bound benzyl alcohols are usually the same as those for the corresponding carboxylic esters (Table 3.22). For instance, carbamates prepared from Wang resin (Entries **2** and **3**, Table 3.22) can be cleaved by treatment with TFA/DCM (1:1) for 20 min at room temperature, to yield the amine as its trifluoroacetate salt. Carbamate attachment of amines to Wang resin is a widely used and convenient means of immobilizing amines.

An alternative method for cleaving carbamates is exhaustive reduction with LiAlH$_4$ to yield methylamines (Entry **3**, Table 3.22). Entry **9** in Table 3.22 is an example of the nucleophilic cleavage of a carbamate with sodium methoxide. The mild reaction conditions required in the case are attributable to the structure of the amine (a vinylogous amide); these conditions are unlikely to lead to the cleavage of simple *N*-alkyl- or *N,N*-dialkylcarbamates, although *N*-arylcarbamates are also susceptible to nucleophilic cleavage (Entry **6**, Table 3.26).

Table 3.22. Carbamate attachment of aliphatic amines.

Entry	Loaded resin	Cleavage conditions	Product, yield (purity)	Ref.
1		TFA/Me$_2$S 4:1, 20 °C, overnight	R–NH$_2$ 60% (95%)	[441]
2		TFA/DCM 1:1, 20 °C, 1 h	R–NH$_2$ 80% (85%)	[442]
3		LiAlH$_4$ (10 eq), THF, 60 °C, 14 h	 84% (> 95%)	[434] see also [443,444]
4		Et$_3$SiH (1 eq), TFA/DCM 2:1, 48 h	 21% (88%)	[445]
5		3% TFA in DCM, 25 °C, 69 h	 83%	[446]
6		TFA/DCM 1:9, 20 °C, 4.5 h	R–NH$_2$	[73] see also [69]
7		hv (350 nm), 3 h, MeCN/H$_2$O 9:1	R–NH$_2$ up to 98%	[447] see also [448,449]
8		(Ph$_3$P)$_2$PdCl$_2$, Bu$_3$SnH, AcOH, DMSO/DCM 1:1, 1 h	R–NH$_2$ 92%	[450] see also [444]
9		NaOMe (0.05 mol/L), THF, 1 h	 65% (98%)	[451]
10		NaOH, phosphate buffer (pH 12), 3 min		[452] see also [112,443, 444]
11		HF (0.4 mol/L, 3 eq), MeCN, 20 °C, 5 h	 > 75%	[453]

3.6.3 Miscellaneous Linkers for Aliphatic Amines

A series of special linkers and cleavage strategies has been developed for the release of amines from insoluble supports (Table 3.23). These include the attachment of amines as triazenes, enamines, aminals, amidines, sulfonamides, sulfinamides, hydrazines, or amides.

Triazenes have been prepared by the treatment of resin-bound aromatic diazonium salts with secondary amines (Figure 3.27). Regeneration of the amine can be effected by mild acidolysis (Entry **1**, Table 3.23). Triazenes have been shown to be stable towards bases such as TBAF, potassium hydroxide, or potassium *tert*-butoxide [454], and under the conditions of the Heck reaction [455]. Primary amines cannot be linked to supports as triazenes because treatment of triazenes such as R–HN–N=N–Ar–Pol with acid leads to the release of aliphatic diazonium salts into solution [373]. Triazenes derived from primary amines can, however, be used for the preparation of amides and ureas (see Section 3.3.4).

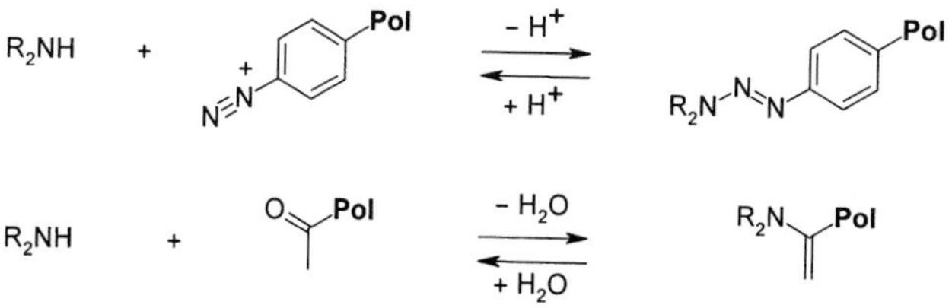

Figure 3.27. Immobilization of amines as triazenes and as enamines.

Table 3.23. Special linkers for aliphatic amines.

Entry	Loaded resin	Cleavage conditions	Product, yield (purity)	Ref.
1	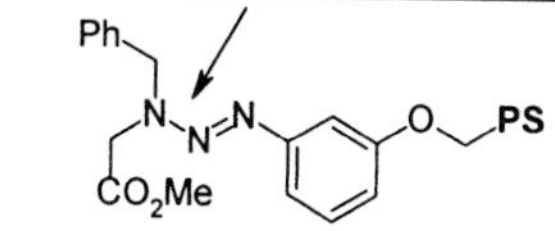	TFA/DCM 1:9, 20 °C, 5 min	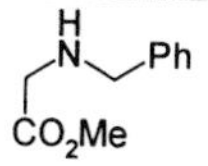 (> 90%)	[344] see also [454]
2	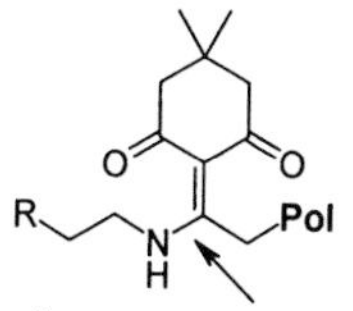	5% N$_2$H$_4$•H$_2$O or 10% PrNH$_2$ in THF/H$_2$O 1:1 (no cleavage with piperidine/DMF 2:8 or TFA/DCM 1:1, 24 h)	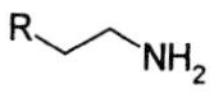 (> 90%)	[458] see also [459]
3	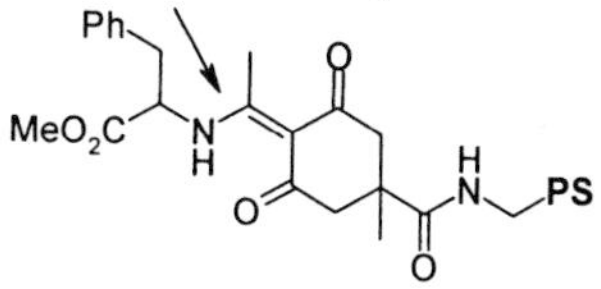	2% N$_2$H$_4$•H$_2$O, DMF, 5 min	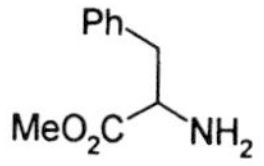 100%	[460]
4	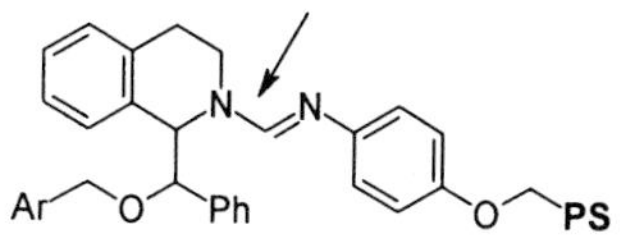	N$_2$H$_4$/AcOH/EtOH/THF 1:0.7:40:40, 60 °C, overnight Ar: 3,5-dimethoxyphenyl	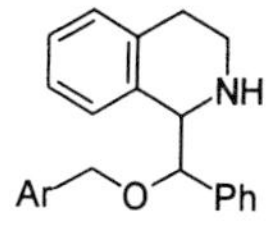 40%	[461] see also [462]

Table 3.23. continued.

Entry	Loaded resin	Cleavage conditions	Product, yield (purity)	Ref.
5		PhSK (0.08 mol/L), MeCN, 20 h, or 2-mercaptoethanol, DBU, MeCN (6:21:73), 20 °C	65%	[463,464]
6		HCl (0.7 mol/L, 25 eq), DCM/BuOH 1:1, 1 h	90%	[465]
7		TfOH (0.1 mol/L), DCM, 1 h	45%	[465]
8		α-chymotrypsin, TRIS-HCl buffer (pH 7.8), 40 °C, 24 h	72%	[189]
9		1. BH$_3$ (20 eq), THF, 67 °C, 4 h 2. aq HCl (3 mol/L), 2 h		[457]
10		LiAlH$_4$ (0.5 mol/L, 10 eq), THF, 20 °C, 18 h	41%	[466]
11		TFA/DCM 1:9 Ar: 4-methoxyphenyl		[467]

Support-bound triacylmethanes (e.g. 2-acetyldimedone) readily react with primary aliphatic amines to yield enamines. These are stable towards weak acids and bases, and can be used as linkers for solid-phase peptide synthesis using either the Boc or Fmoc methodologies, as well as for the solid-phase synthesis of oligosaccharides [456]. Cleavage of these enamines can be achieved by treatment with primary amines or hydrazine (Entries **2** and **3**, Table 3.23; see also Section 10.1.10.4).

Amidines and sulfonamides have also been used as linkers for primary or secondary aliphatic amines (Entries **4**, **5**, and **7**, Table 3.23). These derivatives are stable under basic and acidic reaction conditions and can only be cleaved by strong nucleophiles. Phenylalanine amides can be hydrolyzed by treatment with certain enzymes (Entry **8**, Table 3.23), and can therefore be used for linking amines to supports compatible with enzyme-mediated reactions (CPG, some polyacrylamides, macroporous polystyrene, etc.).

The N–N bond of polystyrene-bound hydrazines, which are prepared by reaction of organolithium compounds with resin-bound hydrazones [457], can be cleaved by treatment with borane to yield α-branched, primary amines (Entry **9**, Table 3.23). An additional example of reductive cleavage to yield amines is shown in Entry **10** (Table 3.23), in which a resin-bound α,α-disubstituted nitroacetic ester undergoes decarboxylation and reduction to the primary amine upon treatment with lithium aluminum hydride.

3.7 Linkers for Tertiary Amines

The preparation of tertiary amines with the aid of insoluble supports has mostly been performed by β-elimination of support-bound quaternary ammonium salts or by N-alkylation of secondary amines with support-bound alkylating agents (Figure 3.28).

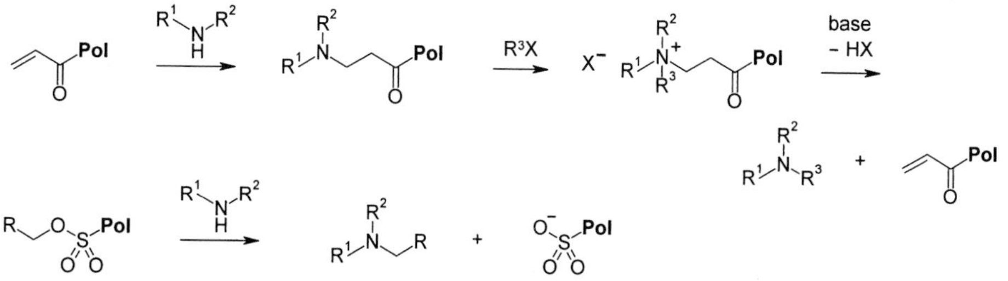

Figure 3.28. Solid-phase preparation of tertiary amines.

Base-induced β-elimination of quaternized amines can be mediated by amines (including ammonia [468]) or by basic ion-exchange resins [469,470]. Because non-quaternized amines, which are potential by-products in this synthetic sequence (Figure 3.28), do not undergo β-elimination as readily as the quaternized ammonium salts, pure tertiary amines are generally obtained by this technique.

Tertiary amines have also been prepared by N-alkylating primary or secondary amines with resin-bound alkylating agents, such as sulfonates, allyl esters (Entries **3–5**, Table 3.24), or Michael acceptors (Table 15.23 [471]). Furthermore, mixed aminals of support-bound benzotriazole and secondary amines can be cleaved with carbon nucleophiles or reducing agents to yield tertiary amines (Entries **6** and **7**, Table 3.24). Quaternization of polystyrene-bound *O*-benzyl-*N,N*-dialkylhydroxylamines with methyl triflate yields intermediates that undergo base-induced fragmentation to yield a tertiary amine and a resin-bound benzaldehyde (Entry **8**, Table 3.24). A similar strategy is based on the debenzylation of resin-bound benzylammonium salts by treatment with morpholine (Entry **9**, Table 3.24). Tertiary amines can also be obtained by reduction of resin-bound carbamates (Entry **3**, Table 3.22).

The only drawback of strategies for the synthesis of tertiary amines that are based on the quaternization of polystyrene-bound *N*-nucleophiles is that quaternizations on polystyrene only proceed sluggishly and generally require the use of highly reactive alkylating agents (methyl, allyl, or benzyl halides or sulfonates; see Section 10.2).

Under these conditions, other nucleophilic functional groups may also be alkylated, which limits the range of compounds accessible by this methodology. Moreover, volatile alkylating agents are often highly toxic and therefore problematic to handle.

Table 3.24. Linkers for tertiary amines.

Entry	Loaded resin	Cleavage conditions	Product, yield (purity)	Ref.
1		DIPEA/DCM 4:96, 20 °C, 2.5 h	75% (99%)	[472] see also [469,473]
2		DIPEA/DCM 3:10, 20 °C, 0.5 h	(0.25 mmol/g)	[244] see also [112,474, 475]
3		piperidine (0.5 mol/L), MeCN, 60 °C, 18 h (anilines were not alkylated by this reagent)	68% (84%)	[476] see also [477]
4		morpholine (5 eq), DCM, 25 °C, 12 h	95%	[371]
5		morpholine (2–3 eq), 7% Pd(PPh₃)₄, THF, 50 °C, 8 h	86%	[478] see also [479]
6		BuMgBr, THF, 67 °C, 4 h	89%	[480] see also [314,481]
7		NaBH₄ (10 eq), THF, 60 °C, 15 h	41% (79%)	[481]
8		NEt₃ (5 eq), DCM, 20 °C, 16 h	57%	[482,483] see also [484]
9		morpholine (neat), 110 °C, 10 h	85% (90%)	[485]

3.8 Linkers for Aryl- and Heteroarylamines

Aromatic and heteroaromatic amines can be linked to insoluble supports using strategies similar to those used for aliphatic amines. Because of the lower basicities of aromatic amines, however, their *N*-benzyl derivatives will usually be more susceptible to acidolytic cleavage than aliphatic *N*-benzylamines. For the same reason, *N*-acyl derivatives of aromatic amines will generally be more sensitive towards nucleophiles than the corresponding derivatives of aliphatic amines.

Illustrative examples of cleavage reactions of *N*-arylbenzylamine derivatives are listed in Table 3.25. Aromatic amines can be immobilized as *N*-benzylanilines by reductive amination of resin-bound aldehydes or by nucleophilic substitution of resin-bound benzyl halides (Chapter 10). The attachment of the amino group of 5-aminoindoles to 2-chlorotrityl chloride resin has been reported [486]. Anilines have also been linked to resin-bound dihydropyran as aminals [487].

The attachment of anilines to benzyl alcohol linkers as carbamates can be achieved either by reaction of aryl isocyanates with a resin-bound alcohol [498–500] or by treat-

Table 3.25. Benzylamine and benzhydrylamine linkers for aromatic amines.

Entry	Loaded resin	Cleavage conditions	Product, yield (purity)	Ref.
1		TFA/H$_2$O/Me$_2$S 95:5:5, 1 h	51–85%	[488]
2		TFA/DCM 7:3, 1 h	90%	[411] see also [263]
3		3% TFA in DCM, 45 min Ar: 4-(NO$_2$)C$_6$H$_4$	75%	[489] see also [490]
4		TFA/DCM 1:1, 4 h	> 99%	[234] see also [491]
5		TFA, 20 °C, 2 h	76% (87%)	[492]

Table 3.25. continued.

Entry	Loaded resin	Cleavage conditions	Product, yield (purity)	Ref.
6		TFA/H$_2$O 95:5, 50 °C, 4 h Ar: 4-ClC$_6$H$_4$	76% (94%)	[493] see also [494,495]
7		DCE/TFA/H$_2$O 20:19:1, 1 h HNR$_2$: 4-(ethoxy-carbonyl)piperidine	(95%)	[496]
8		TFA/H$_2$O 95:5, 2 h	87% (99%)	[493]
9		TFA/DCM 3:6, 0.3 h	45%	[497]

ment of alcohol-functionalized supports with phosgene [428,501] or a synthetic equivalent thereof, followed by exposure to the aromatic amine (Section 14.6). The latter strategy generally requires the use of more reactive activating agents than those used with aliphatic amines. For instance, resin-bound benzyloxycarbonyl imidazoles, obtained by the reaction of resin-bound benzyl alcohols with carbonyldiimidazole, must be activated by N-alkylation (e.g. with MeOTf [73]) in order to yield an imidazolium salt that is sufficiently electrophilic to undergo reaction with anilines.

Table 3.26 lists illustrative examples of cleavage reactions of support-bound N-aryl-carbamates, anilides, and N-arylsulfonamides. N-Arylcarbamates are more susceptible to attack by nucleophiles than N-alkylcarbamates, and, if strong bases or nucleophiles are to be used in a reaction sequence, it might be a better choice to link the aniline to the support as an N-benzyl derivative. Entry 7 (Table 3.26) is an example of a safety-catch linker for anilines, in which activation is achieved by enzymatic hydrolysis of a phenylacetamide to liberate a primary amine, which then cleaves the anilide.

Some heteroarylamines have been prepared by aromatic nucleophilic substitution of suitable support-bound arylating agents with amines (Table 3.27). This technique has been successfully employed in the synthesis of 2-(alkylamino)pyrimidines [507,508], 2-(arylamino)pyrimidines [509], aminopurines [510–512], and 1,3,5-triazines [513]. When the heteroarene is bound to the support as a thioether, nucleophilic clea-

Table 3.26. Carbamate and amide attachment of aromatic and heteroaromatic amines.

Entry	Loaded resin	Cleavage conditions	Product, yield (purity)	Ref.
1		HF/PhOMe 9:1, 0 °C, 1 h		[501]
2		TFA/DCM 1:1, 5 min	> 95% (90%)	[500]
3		TFA/DCM 1:1, 5 min	> 95% (90%)	[500]
4		TFA/DCM 1:9, 20 °C, 4.5 h		[73] see also [502]
5		10% NH$_4$OH in CF$_3$CH$_2$OH, 40 °C, 4 h	43–74%	[499]
6		NaOH (0.5 mol/L), 90 °C, 0.5 h	95–97%	[503]
7		penicillin G acylase, H$_2$O/MeOH 9:1, pH 7, 20 °C, 48 h, then 60 °C, 4 h,	60–67%	[504,505]
8		Na$_2$S, DMF, 20 °C, 20 h	45–82% (72–100%)	[506]

vage is facilitated by oxidation to a sulfone (e.g. with MCPBA or *N*-benzenesulfonyl-3-phenyloxaziridine). Aryl thiolate (e.g. Entry **2**, Table 3.27) is a significantly better leaving group than alkyl thiolate in this reaction [514]. Alternatively, polystyrene-bound *N*-hydroxybenzotriazole can be used as a leaving group for aromatic nucleophilic displacement (Entry **6**, Table 3.27). If these cleavage reactions are to be performed with non-volatile amines, then less than one equivalent of amine should be used. This will generally result in complete conversion of the amine, so that crude products of high purity can be obtained.

Secondary heteroarylamines have also been prepared by reductive cleavage of resin-bound benzotriazole-derived aminals (Entry **7**, Table 3.27). Few successful examples of this strategy have been reported, however, which may be due to the difficulty in preparing the required aminals [481].

Table 3.27. Miscellaneous linkers for aryl and heteroaryl amines.

Entry	Loaded resin	Cleavage conditions	Product, yield (purity)	Ref.
1		1. MCPBA (1.2 eq), DCM, 20 °C, 18 h 2. pyrrolidine (0.3 mol/L), dioxane, 20 °C, 3 h	32%	[507] see also [511,515]
2		butylamine (5 eq), cat HCl, DMF/PhMe 1:1, 90 °C, 48 h		[514]
3		pyrrolidine (1.5 eq), dioxane, 20 °C, 6 h	90% (98%)	[508]
4		pyrrolidine (0.2 mol/L), dioxane, 60 °C, 6 h	85%	[513]
5		4-methoxyaniline (0.3 mol/L), dioxane, 80 °C, 15 h		[513]
6		4-iodoaniline (0.43 mol/L, 10 eq), DCM, 25 °C, 20 h	49% (> 95%)	[516]
7		NaBH$_4$ (10 eq), THF, 60 °C, 15 h	55% (86%)	[481]

3.9 Linkers for Guanidines and Amidines

Several strategies enable the attachment of guanidines and amidines to insoluble supports (Figure 3.29, Table 3.28). These include attachment as *N*-benzyl, *N*-acyl, and *N*-alkoxycarbonyl derivatives. Furthermore, support-bound isothioureas can be used to convert amines into guanidines. The synthesis of support-bound guandines is considered in Section 14.3.

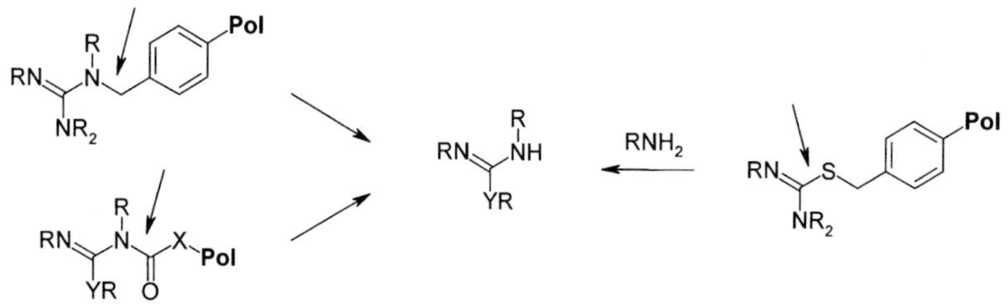

Figure 3.29. Attachment of guanidines and amidines to insoluble support. Y: CR$_2$, NR; X: CR$_2$, O.

Few examples of benzylamine-type linkers for guanidines have been described. It has been reported that *N*-benzylguanidines undergo acidolysis more easily than the corresponding *N*-benzylamines, but more slowly than comparable *N*-benzylamides [412]. The acidic desulfonylation of *N*-(arylsulfonyl)guanidines generally requires prolonged treatment with strong acids [517], unless the arene bears electron-donating substituents. The N–S bond of (4-alkoxybenzene)sulfonylguanidines can be cleaved by treatment with hydrogen fluoride, and such derivatives have proven useful as linkers for guanidines (Entry **5**, Table 3.28). Resin-bound (alkoxyarene)sulfonylguanidines have been prepared by reaction of support-bound sulfonyl chlorides with guanidines in the presence of a base. This reaction is slow, however, and requires several days to reach completion (e.g. attachment of *N*-Boc arginine as sulfonamide: KOH (0.8 mol/L) in dioxane, 75 °C, 2 d [518]).

Carbamate-bound guanidines have been prepared by the condensation of amines with thioureas linked to insoluble supports as carbamates [519]. The direct reaction of guanidines with resin-bound carbonates or other alkoxycarbonylating agents requires the use of chloroformates or other reactive carbonic acid derivatives [520,521]. Carbamate-bound amidines (Entries **7** and **8**, Table 3.28) have been prepared by the reaction of amidines with resin-bound 4-nitrophenyl carbonates (0.9 mol/L amidine in DMF/DIPEA, > 4 h [522–524]).

An additional strategy for generating guanidines from insoluble supports is the nucleophilic cleavage of resin-bound isothioureas with amines (Entries **9–12**, Table 3.28). This reaction is closely related to the preparation of 2-aminopyrimidines by nucleophilic cleavage (Table 3.27) and is generally limited to the use of volatile, low-molecular-weight amines if the aim is to obtain crude products of high purity. In the case of polystyrene-bound *N,N*′-disubstituted isothioureas (e.g., Entry **11**, Table 3.28), only primary aliphatic amines led to the expected guanidines; with secondary amines the reaction did not proceed, not even under forcing conditions (140 °C [525]). Triazene-bound guanidines (Entry **13**, Table 3.28) have been prepared from triazene-bound thioureas, and these are cleaved under mildly acidic conditions [526].

Table 3.28. Linkers for guanidines and amidines.

Entry	Loaded resin	Cleavage conditions	Product, yield (purity)	Ref.
1		TFA/DCM 1:1, 4 h Ar: 4-(MeO)C$_6$H$_4$	 96%	[234]
2		HF, PhOMe, 0 °C, 9 h; (100% TFA, 20 °C, 0.5 h yielded 30–50% guanidine)	 > 70%	[412,527]
3		TFA/DCM 25:75, 20 °C, 1 h	 70% (56%)	[528]
4		TFA/CHCl$_3$/MeOH 1:1:1, 60 °C, 24–72 h	 48% (90%)	[529]
5		HF/PhOMe 8:2, 0 °C, 4 h		[518] see also [530,531]
6		TFA/DCM/iPr$_3$SiH 49:49:2	 > 85% (> 90%)	[519] see also [520,521, 532]
7		TFA/H$_2$O 95:5, 40 min	 > 90% (> 70%)	[522] see also [523,524]

Table 3.28. continued.

Entry	Loaded resin	Cleavage conditions	Product, yield (purity)	Ref.
8		hν (350 nm), dioxane	20% (> 65%)	[523]
9		NH₃/MeOH/DMF, 15 h	95% (100%)	[533]
10		BnNH₂ (0.1 mol/L), DMF, 50 °C, 16 h (no reaction with secondary or aromatic amines)	90%	[533]
11		3-(aminomethyl)pyridine (0.4 mol/L, 3 eq), PhMe, 100 °C, 60 h	50%	[525]
12		NH₄OAc (0.6 mol/L, 10 eq), dioxane/MeCN 1:1, 81 °C, 8 h	62%	[349]
13		TFA/DCM 1:9, 20 °C, 5 min	84%	[526]

3.10 Linkers for Azoles

Heterocycles containing an NH group, such as pyrroles, indoles, imidazoles, triazoles, etc., can be linked to insoluble supports as *N*-alkyl, *N*-aryl, or *N*-acyl derivatives (Table 3.29). The optimal choice depends mainly on the NH acidity of the heterocycle in question. Increasing acidity will facilitate the acidolytic cleavage of *N*-benzyl groups and the nucleophilic cleavage of *N*-acyl groups from these heterocycles.

Histidine and histamine derivatives, as well as other imidazoles, have been successfully immobilized by N-tritylation of the imidazole ring with trityl chloride resin [534] (Entry **2**, Table 3.29; see also Section 15.8) or with 2-chlorotrityl chloride resin [535–537]. Histidine can also be linked to insoluble supports as the *N*-dinitrophenyl derivative (Entry **3**, Table 3.29), which is prepared by sequential nucleophilic substitutions at 1,5-difluoro-2,4-dinitrobenzene [89,538].

Some heterocycles can be linked to supports as tetrahydropyranyl derivatives. Attachment of indoles, purines, or tetrazoles (Table 3.29) has been achieved by treatment of a support-bound dihydropyran with the heterocycle in the presence of catalytic amounts of pyridinium tosylate [487], camphorsulfonic acid [539], or TFA [540] in DCE at 60–80 °C for 16–24 h. Indole-derived orthoesters, such as that in Entry **7** (Table 3.29), can be prepared by heating the indole with triethyl orthoformate (160 °C, 24 h) followed by acid-catalyzed reaction of the resulting orthoester with a resin-bound diol [541,542]. As illustrated by Entry **8** (Table 3.29), indoles can also be linked to the Wang resin or related supports as carbamates. Cleavage by TFA is, how-

Table 3.29. Linkers for azoles.

Entry	Loaded resin	Cleavage conditions	Product, yield (purity)	Ref.
1		AcOH, 100 °C, 2 h (no cleavage with TFA/H$_2$O 9:1, 1 h)	72% (94%)	[544]
2		TFA/H$_2$O 95:5		[545] see also [534]
3		2-mercaptoethanol/ NEt$_3$/DMF 9:0.6:90, 12 h	82%	[89,538]
4		NEt$_3$/DCM 5:95, 20 °C, 18 h Ar: 2-cyanophenyl	29% (96%)	[546]
5		TFA/DCM 1:9, 2 × 15 min	63%	[487]
6		TFA/DCM 1:9, 2 × 15 min		[487]
7		1. HCl (2 mol/L)/ dioxane 1:1, 40 °C, 3 h 2. NaOH (2 mol/L), 20 °C, 0.5 h	66% (98%)	[541,542]

Table 3.29. continued.

Entry	Loaded resin	Cleavage conditions	Product, yield (purity)	Ref.
8		pyrrolidine/DMF 5:95, 90 °C, 4 h, or AcOH, 110 °C, 4 h (PS): Wang resin	65% (94%)	[543]
9		TBAF (0.1 mol/L, 5 eq), THF, 70 °C, 5 h; or KO*t*Bu (10 eq), THF, 20 °C, 5 h	100% (95%)	[547,548]
10		TFA/DCE 2:8, 7 × 0.5 min R: isobutyl	97%	[549]
11		TFA/DCM 1:8, 20 °C, 10 min	95% (97%)	[539] see also [550]
12		3% HCl in MeOH, 24 h	58%	[540]
13		TFA/DCM 5:95, 20 °C, 18 h Ar: 4-bromophenyl	78%	[551]

ever, problematic [543] because indoles are easily C-alkylated by benzylic carbocations. Better results have been obtained in this case by nucleophilic cleavage or by hydrolysis with acetic acid. Imidazole derivatives have also been linked to supports by N-alkylation of the imidazole moiety with support-bound alkylating agents [476].

3.11 Linkers for Alcohols and Phenols

The attachment of alcohols to insoluble supports has been extensively investigated, in particular with regard to the solid-phase synthesis of oligosaccharides and oligonucleotides. The linking strategies and cleavage methods most commonly used are outlined in Figure 3.30.

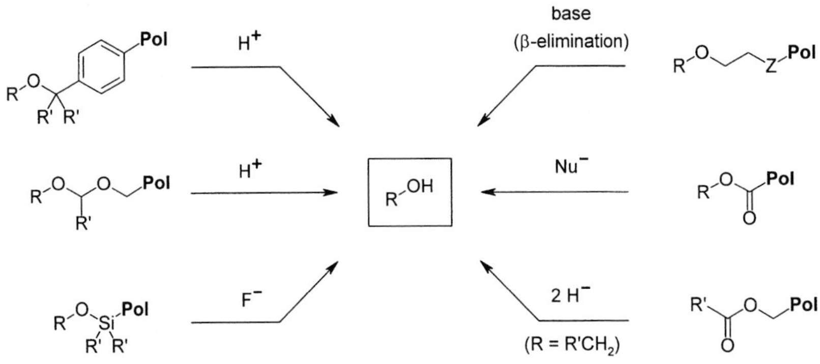

Figure 3.30. Strategies for the release of alcohols from insoluble supports. R, R′: H, alkyl, aryl; Z: electron-withdrawing group.

3.11.1 Attachment as Ethers

Most acid-labile benzyl alcohol linkers suitable for the attachment of carboxylic acids to insoluble supports can also be used to attach aliphatic or aromatic alcohols as ethers. The attachment of alcohols as ethers is less easily accomplished than esterification, and might require the use of strong bases (Williamson ether synthesis [395,552,553]) or acids. These harsh reaction conditions limit the range of additional functional groups that may be present in the alcohol. Some suitable etherification strategies are outlined in Figure 3.31. Etherifications are treated in detail in Section 7.2.

Phenols can be etherified with resin-bound benzyl alcohols by the Mitsunobu reaction [554,555], or, alternatively, by nucleophilic substitution of resin-bound benzyl halides or sulfonates [556,557]. Both reactions proceed smoothly under mild conditions. Aliphatic alcohols have been etherified with Wang resin by conversion of the latter into a trichloroacetimidate (Cl₃CCN/DCM/DBU (15:100:1), 0 °C, 40 min), fol-

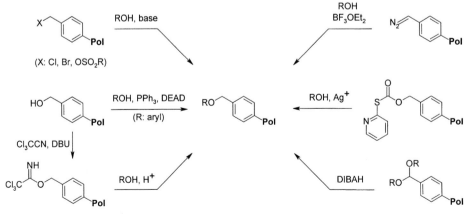

Figure 3.31. Strategies for the etherification of alcohols with insoluble supports.

lowed by nucleophilic substitution with the alcohol under slightly acidic conditions (0.07 mol/L ROH in DCM/C_6H_{12} (1:1), 0.17% (*v/v*) $BF_3 \cdot OEt_2$, 10 min) [558–561]. A similar protocol can also be performed on hydroxymethyl polystyrene [562]. Support-bound aryl diazomethanes, which can be prepared by oxidation of hydrazones or by thermolysis of the sodium salt of sulfonyl hydrazones (Section 10.5), react with alcohols in the presence of Lewis acids to yield ethers. Resin-bound benzylic thiocarbonates react with aliphatic alcohols in the presence of silver(I) salts to yield resin-bound benzyl ethers [397,563]. The nucleophilic substitution of resin-bound benzyl

Table 3.30. Cleavage of support-bound ethers to yield aliphatic alcohols.

Entry	Loaded resin	Cleavage conditions	Product, yield (purity)	Ref.
1		HCl, dioxane		[395]
2		1% TFA in DCM, 2–4 h; or 10% TFA in DCM, 0.5 h	98%	[559] see also [558]
3		TFA/DCM 1:1, 0.5 h	45–64%	[552]
4		BF_3OEt_2 (0.075 mol/L), DCM, 20 °C, 3 h	26–30%, 67% de	[560]
5		TFA/DCM 7:3, 1 h	59%	[411] see also [48,263, 575]
6		TFA vapor, 20 °C, overnight	73–84% (76–96%)	[576]
7		HCl (0.35 mol/L), dioxane, 20 °C, 48 h; or TsOH (0.2 mol/L), THF/MeOH 1:1, 22 h; or DIBAH, C_6H_6, 80 °C, 16 h	38–60%	[81,577] see also [568,569]
8		0.1% Cl_2CHCO_2H in $CHCl_3$, 25 °C, 1 h; or AcOH/$CHCl_3$ 2:8, 25 °C, 72 h	100%	[570]

Table 3.30. continued.

Entry	Loaded resin	Cleavage conditions	Product, yield (purity)	Ref.
9	(structure: pivaloylated sugar linked via fluorenylmethyl to PEG)	NEt₃/DCM 2:8	(structure: OPiv sugar, OH) (90%)	[578]
10	(structure: O₂N-nitrobenzyl linked sugar to Pol)	hν, THF, 25 °C	(structure: HO, BzO, OBz sugar) 95%	[579] see also [580]
11	(structure: R-O- pivaloyl linker to PS)	hν (300 nm), DCM, 20 °C, 2 h	R⌒OH 61–78%	[160]
12	(structure: AcO sugar linked to aryl amide PS)	CAN (5 eq), MeCN/H₂O 10:1	(structure: AcO sugar, OAc, OH) 70%	[581]
13	(structure: HO, BnO, OBn sugar-OMe linked benzyl amide PS)	DDQ (0.03 mol/L, 1.2 eq), DCM/H₂O 20:1, 20 °C, 2 × 4 h	(structure: HO, BnO, OBn sugar-OMe) 91%	[582] see also [583]
14	(structure: R-propyl-O aryl ketone OMe linked PS)	CAN, MeCN/H₂O 1:1, hexane, ultrasound	R⌒⌒OH	[584]

halides with aliphatic alcoholates requires strong bases, such as sodium hydride [552,553], and might therefore be difficult to automate.

Only a few examples have been reported of the etherification of alcohols with resin-bound diarylmethyl alcohols (Entry 5, Table 3.30; Entry 5, Table 3.31 [564]). Diarylmethyl ethers do not seem to offer advantages over the more readily accessible trityl ethers, which are widely used as linkers for both phenols and aliphatic alcohols. Attachment of alcohols to trityl linkers is usually effected by treating trityl chloride resin or 2-chlorotrityl chloride resin with the alcohol in the presence of a base (phenols: pyridine/THF, 50 °C [565] or DIPEA/DCM [566]; aliphatic alcohols: pyridine, 20–70 °C, 3 h–5 d [567–572] or collidine, Bu₄NI, DCM, 20 °C, 65 h [81]). Aliphatic or aromatic alcohols can be attached as ethers to the same type of light-sensitive linker as used for carboxylic acids (Section 3.1.3).

Ethers are generally inert towards nucleophilic attack and are therefore suitable linkers for solid-phase chemistry involving strong nucleophiles.

Cleavage conditions for alkyl benzyl ethers prepared from acid-labile benzyl alcohols are similar to those for the corresponding benzyl esters (Table 3.30). Aryl benzyl ethers, however, are generally cleaved more easily by acidolysis than esters or alkyl ethers. Phenols etherified with hydroxymethyl polystyrene, for instance, can even be released by treatment with TFA (Entry **1**, Table 3.31). It has also been shown that Wang resin derived phenyl ethers are less stable than Wang resin derived esters towards refluxing acetic acid [29]. Alternatively, boron tribromide may be used to cleave aryl ethers from hydroxymethyl polystyrene [573].

Illustrative examples of the cleavage of support-bound ethers are listed in Tables 3.30 and 3.31. Acidolytic cleavage is the most commonly used strategy, but base-mediated, photolytic, and oxidative cleavage have also been reported. Wang linker

Table 3.31. Cleavage of support-bound ethers to yield phenols.

Entry	Loaded resin	Cleavage conditions	Product, yield (purity)	Ref.
1		TFA/DCM 65:35, 20 °C, 3 h R: alkyl, acyl	51–70%	[585] see also [586]
2		TMSOTf (0.3 mol/L), DCM, 3 h	98%	[409]
3		1% TFA in DCM, 0.5 h	98%	[559] see also [555, 587–590]
4		TFA/DCM/Me₂S 45:50:5, 1.5 h	72%	[591]
5		TFA/DCM 5:95, 0.5 h	96% (91%)	[263]
6		TFA/DCM 1:99; or TFA/DCM/MeOH 2:7:1	60–95%	[565,566]
7		hv (350 nm)		[592] see also [593,594]

derived ethers can be oxidized to acetals under mild conditions (0.06 mol/L DDQ in DCM, 20 °C, 3 h [574]), and these are easy to hydrolyze. TFA-mediated cleavage of alcohols from supports occasionally leads to the formation of TFA esters of the released alcohol. This esterification can sometimes be avoided by using wet TFA (e.g. containing 5% water) instead of anhydrous TFA.

3.11.2 Attachment as Silyl Ethers

Both aliphatic alcohols and phenols [595,596] can be linked to insoluble supports as silyl ethers (Figure 3.32). This form of attachment can be realized by treatment of support-bound silyl chlorides [597–602], silyl triflates [603], or silyl trifluoroacetates [604] with alcohols in the presence of a base (imidazole, DIPEA, lutidine, or DMAP in DCM [81,598,605,606]). Alternatively, silyl ethers can be prepared by heating resin-bound silanes R$_3$SiH [598,599] with aromatic or aliphatic alcohols, ketones, or aldehydes in the presence of catalytic amounts of a rhodium complex ([(PPh$_3$)$_3$RhCl] [607], rhodium(II) perfluorobutyrate [598]), or by reaction of a resin-bound silyl nitrile with aldehydes or ketones to yield resin-bound cyanohydrins [608]. Dialkoxysilanes (e.g. Entry **3**, Table 3.32) have been prepared by first treating an alcohol in solution with a dialkyldihalosilane, and then coupling the resulting monohalosilane with a suitable polymeric alcohol [609].

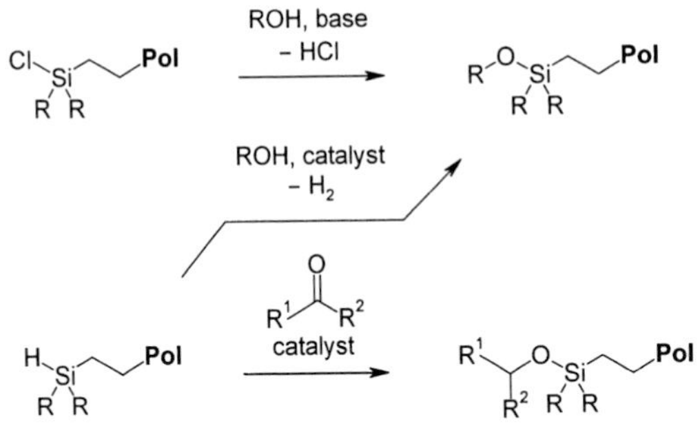

Figure 3.32. Preparation of support-bound silyl ethers.

Silyl ethers of aliphatic alcohols are inert towards strong bases, oxidants (ozone [81], Dess–Martin periodinane [605], iodonium salts [610,611], sulfur trioxide–pyridine complex [398]), and weak acids (e.g., 1 mol/L HCO$_2$H in DCM [605]), but can be selectively cleaved by treatment with HF in pyridine or with TBAF (Table 3.32). Phenols can also be linked to insoluble supports as silyl ethers, but these are less stable than alkyl silyl ethers and can even be cleaved by treatment with acyl halides under basic reaction conditions [595]. Silyl ether attachment has been successfully used for the solid-phase synthesis of oligosaccharides [600,601,612,613] and peptides [614].

Table 3.32. Silyl ethers as linkers for alcohols.

Entry	Loaded resin	Cleavage conditions	Product, yield (purity)	Ref.
1		HF/pyridine (0.4 mol/L), THF, 2 h	50%	[607] see also [599,605]
2		TBAF (0.25 mol/L), AcOH (0.13 mol/L), THF, 40 °C, 18 h	0.61 mmol/g	[600] see also [613,615]
3		HF/pyridine (0.57 mol/L), PhOMe (0.15 mol/L), DCM, −10 °C, 4 h, 20 °C, 24 h		[609] see also [616]
4		3% TFA in DCM, 18 h	0.41 mmol/g	[75]
5		HCl (g), EtOH, DCM, 0 °C to 20 °C	63%	[608]
6		CsF (0.04 mol/L, 1 eq), AcOH (5 eq), DMF, overnight	88%	[614]

3.11.3 Attachment as Acetals

Acetals constitute a further functional group suitable for linking alcohols to insoluble supports (Table 3.33). A frequently used linker of this type is resin-bound dihydropyran [617–622], which forms mixed acetals (tetrahydropyranyl ethers) with aliphatic [623] or aromatic alcohols [624] upon acid catalysis (e.g. PPTS, DCE, 80 °C, 16 h). The resulting acetals are stable towards strongly basic or nucleophilic reagents, such as organolithium compounds [624], organocuprates [617], or Grignard reagents [618,620]. Inexpensive tri-*O*-acetyl-D-glucal has been irreversibly linked to Merrifield resin and used as a linker for alcohols [625].

Mixed acetals of a support-bound and a non-support-bound alcohol with acetaldehyde have also been used as linkers (Entry 3, Table 3.33). Such acyclic mixed acetals are, however, not easy to prepare on solid supports and are more conveniently synthesized by conventional solution-phase chemistry and then loaded onto the support

Table 3.33. Acetals as linkers for alcohols and phenols.

Entry	Loaded resin	Cleavage conditions	Product, yield (purity)	Ref.
1		BuOH/DCE 1:1, PPTS (2 eq), 60 °C, 16 h	95%	[623]
2		TFA/DCM/MeOH 1:5:1	32–50%	[624]
3		TFA/DCM 3:7, 3 h	(30–70%)	[626] see also [627]
4		TFA/DCM 5:95, 20 °C, 1.5 h	73–98% (70–95%)	[629] see also [626]
5		TFA/DCM/MeOH 10:90:1, 20 °C, 1 h	88%	[630] see also [631]
6		TFA/H₂O 90:2.5, scavengers	(octreotide) 74%	[632] see also [633,634]
7		MeOTf, MeSSMe, DTBMP, DCE, 40 °C, 21 h (ROH: 2,3,6-tri-*O*-benzyl-*β*-D-glucose)	50%	[574]
8		HCl (2 mol/L); or NaOH (2 mol/L); or penicillin amidase, 25 °C, 16 h; or TFA/DCM/H₂O 9:10:1, 3 h	(HO,H₂N) + R—OH	[181]
9		TBAF (0.1 mol/L), tetramethylurea, ultrasound, 15 min ROH: subst. cyclohexanol	R-OH 71%	[628]

[626]. Strategies for preparing mixed, acyclic acetals on insoluble supports include the oxidative haloalkoxylation of support-bound enol ethers (Entry **6**, Table 6.1) and the acid-catalyzed reaction of alcohols with resin-bound enol ethers [627]. Alternatively, resin-bound α-chloro ethers can be converted to mixed acetals by reaction with alcohols or phenols in the presence of strong bases (KO*t*Bu, HO*t*Bu, DMF, 5 h) [550,628]. Polystyrene-bound α-(phenylseleno)ethers react with aliphatic alcohols under slightly acidic conditions (NIS, TfOH, DCM/dioxane (1:1), 0 °C to 20 °C, 1 h) to yield mixed, acyclic acetals [628].

Resin-bound aldehydes and ketones have been used as linkers for 1,2- and 1,3-diols (Entries **4–6**, Table 3.33). Cleavage of acetal-based linkers is usually effected by acid-catalyzed transacetalization or by hydrolysis.

3.11.4 Attachment as Esters

Both aliphatic alcohols and phenols have been immobilized as esters of support-bound carboxylic acids. The esterification can be achieved by treatment of resin-bound acids with alcohols and a carbodiimide, under Mitsunobu conditions, or by acylation of alcohols with support-bound acyl halides (see Section 13.4).

Table 3.34. Ester attachment of alcohols and phenols cleavable by nucleophiles.

Entry	Loaded resin	Cleavage conditions	Product, yield (purity)	Ref.
1		NaOMe (satd in MeOH)/THF 1:4		[643] see also [644]
2		NaOH (0.5 mol/L), dioxane/H_2O 1:1, 20 °C, 20 h or 60 °C, 3 h	74%	[645] see also [81]
3		NEt_3/MeOH/Me_2S 15:75:10, 20 °C, 6 × 3 h	92%	[591] see also [646]
4		NaOMe (0.3 mol/L), THF/MeOH 60:3, 2–4 h		[647,648] see also [649]
5		30% NH_3 in H_2O, 20 °C, 1 h		[650] see also [651]
6		NaOH/MeCN/H_2O, 3 h		[652]

Table 3.34. continued.

Entry	Loaded resin	Cleavage conditions	Product, yield (purity)	Ref.
7		NaBH$_4$ (0.01 mol/L, 2 eq), hot EtOH, 3 h	76% (major triazole regioisomer shown)	[653]
8		NaHSO$_3$, H$_2$O/THF 5:8, 20 °C, 2.5 h	R$\sim$OH 70–89%	[642]
9		penicillin G acylase, H$_2$O/MeOH 9:1, pH 7, 37 °C, 48 h	R$\sim$OH 73–94%	[504,505]
10		1. Na$_2$S$_2$O$_4$ (1 mol/L), THF/H$_2$O/HCO$_2$H 6:6:5, 20 °C, 2 h 2. NEt$_3$ (0.5 mol/L), C$_6$H$_6$, 20 °C, 1 h	R$\sim$OH (9-fluorenyl)methanol 83%	[654]
11		K$_2$CO$_3$ (1 eq), H$_2$O/ THF 1:1, 65 °C, 0.5 h	95%	[371]
12		Et$_2$NH/DCM 2:8, 55 °C, 12 h	80% (99%)	[245]

Ester attachment of alcohols is particularly useful when acidic reaction conditions are to be employed in a synthetic sequence. The solid-phase synthesis of oligosaccharides [635,636] and oligonucleotides (Chapter 16) is often performed with ester linkage to the support, because glycosylations, for instance, also generally require acid catalysis. Valuable alcohols can be quantitatively esterified by using excess support-bound acylating agent [637]. The excess acylating agent can then be capped by treatment of the support with methanol.

Cleavage of support-bound esters can be effected by a variety of reagents (Table 3.34). Saponifications with alkali metal hydroxides or transesterifications with alcoholates [638,639] proceed efficiently, but yield products contaminated with the (non-volatile) cleavage reagent. If crude products of high purity are to be obtained and used without further purification (as, for example, in arrays of compounds prepared by parallel solid-phase synthesis), volatile nucleophiles should be used for ester cleavage. These include ammonia, which is the common cleavage agent for resin-bound oligonucleotide hemisuccinates (Entry 5, Table 3.34), other low-molecular-weight amines (e.g. MeNH$_2$ [636]), mixtures of methanol and triethylamine [640], hydrazine [641], or hydrogen peroxide in the presence of triethylamine [203]. Aryl esters are generally more easily cleaved than esters of aliphatic alcohols.

In 1999, a linker for alcohols was described that can be cleaved by reducing agents (Entry **8**, Table 3.34). This linker is based on a quinone, which, after reduction to the corresponding hydroquinone, undergoes intramolecular nucleophilic ester cleavage, releasing the alcohol and leaving a resin-bound lactone [642]. Entry **9** in Table 3.34 is also based on the intramolecular nucleophilic cleavage of an ester. In this case, the attacking nucleophile is a primary amine, which is generated by enzymatic hydrolysis of a phenylacetamide.

Polystyrene-bound aliphatic sulfonic esters and *O*-aryl sulfamates can be cleaved by treatment with potassium carbonate or other nucleophiles, whereby the corresponding alcohols are released into solution (Entries **11** and **12**, Table 3.34). *O*-Alkyl sulfamates, on the other hand, do not react with nucleophiles and cannot be used as linkers for alcohols [245].

Alcohols can also be generated by hydride reduction of esters of support-bound alcohols or thiols [396,655,656], or by reduction of resin-bound imides (Table 3.35). Similarly, the reaction of esters of support-bound alcohols or thiols with Grignard or

Table 3.35. Linkers for the preparation of alcohols by reductive cleavage.

Entry	Loaded resin	Cleavage conditions	Product, yield (purity)	Ref.
1		LiBH$_4$, 25 °C		[659] see also [660]
2		DIBAH (10 eq), PhMe, 0 °C, 12 h	51%	[661]
3		NaBH$_4$/LiBr (3 eq) in THF/EtOH 6:1, 20 °C, 24 h (*N*-Fmoc groups are partially cleaved)	93% (91%)	[662]
4		DIBAH, PhMe, 0 °C	26%	[663]
5		DIBAH, PhMe, −78 °C	85%	[664]

related reagents leads to the release of tertiary alcohols [657,658]. These cleavage strategies generally require aqueous work-up and are not well-suited for the parallel synthesis of large numbers of compounds.

3.11.5 Miscellaneous Linkers for Alcohols and Phenols

Alcohols and phenols can be attached to support-bound alcohol linkers as carbonates [467,665,666], although few examples of this have been reported. For the preparation of carbonates, the support-bound alcohol needs to be converted into a reactive carbonic acid derivative by reaction with phosgene or a synthetic equivalent thereof, e.g. disuccinimidyl carbonate [665], carbonyl diimidazole [157], or 4-nitrophenyl chloroformate [467] (see Section 14.7). The best results are usually obtained with support-bound chloroformates. The resulting intermediate is then treated with an alcohol and a base (DIPEA, DMAP, or DBU), which furnishes the unsymmetrical carbonate. Carbonates are generally more resistant towards nucleophilic cleavage than esters, but are less stable than carbamates. Aryl carbonates are easily cleaved by nucleophiles and are therefore of limited utility as linkers for phenols.

Examples of the cleavage of support-bound carbonates are given in Table 3.36. Depending on the structure of the carbonate, acidolytic, base-induced, nucleophilic, or photolytic cleavage can be used to release the alcohol. Acidolysis of the benzylic C–O bond of resin-bound benzyl carbonates leads to the release of an unstable carbonic acid ester, which undergoes decarboxylation to yield the alcohol.

Few examples have been described of nucleophilic cleavage of carbonate- or carbamate-linked alcohols from insoluble supports. A serine-based linker for phenols releases the phenol upon fluoride-induced intramolecular nucleophilic cleavage of an aryl carbamate (Entry 2, Table 3.36). A linker for oligonucleotides has been described, in which the carbohydrate is bound as a carbonate to resin-bound 2-(2-nitrophenyl)ethanol, and which is cleaved by base-induced β-elimination (Entry 3, Table 3.36). Trichloroethyl carbonates, which are susceptible to cleavage by reducing agents such as zinc or phosphines, have been successfully used to link aliphatic alcohols to silica gel (Entry 4, Table 3.36). These carbonates can also be cleaved by acidolysis (Table 3.22).

Entry 7 in Table 3.36 is a rare example of the use of a phosphodiester as a linker for alcohols. This linker, when used in combination with an enzyme-compatible support, can be selectively cleaved with a phosphodiesterase. To obtain the free alcohol, the released phosphate must be subjected to an additional enzymatic dephosphorylation.

Polystyrene-derived phenylboronic acids have been used for the attachment of diols (carbohydrates) as boronic esters [667]. Cleavage was effected by treatment with acetone/water or THF/water. This high lability towards water and alcohols severely limits the range of reactions that can be performed without premature cleavage of this linker. Arylboronic acids esterified with resin-bound diols can be oxidatively cleaved to yield phenols (Entry 8, Table 3.36). Alcohols have also been prepared by nucleophilic allylation of aldehydes with polystyrene-bound, enantiomerically enriched allylsilanes [668], as well as by Pummerer reaction followed by reduction of resin-bound sulfoxides [669].

Table 3.36. Miscellaneous linkers for alcohols and phenols.

Entry	Loaded resin	Cleavage conditions	Product, yield (purity)	Ref.
1		HF/PhOMe 9:1, 0 °C, 1 h	> 95%	[670]
2		TBAF (1 mol/L), THF, 1 h	78% (92%)	[671]
3		DBU (0.5 mol/L), dioxane; or NH₃, 55 °C, 5 h; or piperidine/ DMF 2:8, 20 °C, 3 h		[672] see also [673]
4	3-cholesteryl (SG)	PBu₃/NEt₃/DMF 2:1:4, 80 °C, 5 h	cholesterol (0.14 mmol/g)	[446]
5		hv (365 nm), MeCN/H₂O 9:1, 2 h R: oligonucleotide	83%	[674] see also [670]
6		hv (350 nm), THF/MeOH 3:1, 28 °C, 3 h (ROH: cholesterol)	cholesterol 72% (> 95%)	[157]
7	peptide (PA)	1. bovine spleen phosphodiesterase, pH 5.7 2. alkaline phospho- monoesterase	peptide > 83%	[190]
8		H₂O₂ (1.5 eq), NaOH (1 eq), THF, H₂O, 0 °C, 1 h, then 20 °C, 2 h	63% (99%)	[197]

3.12 Linkers for Thiols

Thiols have been linked to insoluble supports as acid-labile benzyl thioethers, as aryl thioethers, as *S*-carbamoyl derivatives, and as unsymmetrical disulfides (Table 3.37). Because thiols often undergo oxidative dimerization in air to yield symmetric

Table 3.37. Linkers for thiols.

Entry	Loaded resin	Cleavage conditions	Product, yield (purity)	Ref.
1		HF/*p*-cresol 9:1, −5 °C, 1 h (no cleavage with neat TFA, 60 h)	> 30%	[682]
2		NCS (0.1 mol/L, 4 eq) Me$_2$S (5 eq), DCM, 0 °C, 4 h (PS): Wang resin	13%; E: CO$_2$Me	[683] see also [684]
3		TFA/DCM 5:95, 0.5 h	92% (95%)	[263]
4		TFA/H$_2$O 95:5, 3 h	100%	[575] see also [263]
5		TFA/DCM/TES 50:47:3, 3 × 5 min	90%	[675]
6		NaOH (0.1 mol/L), MeOH/H$_2$O 9:1, 15 min; or liquid ammonia	59%	[89]
7		2-mercaptoethanol, NMM, AcOH, DMF, 2 × 24 h	58%	[685]
8		P(CH$_2$CH$_2$CO$_2$H)$_3$ (4 mmol/L), dioxane/H$_2$O 9:1, 8 h	(not isolated)	[686]
9		dithiothreitol (0.01 mol/L), TRIS buffer (pH 7.5), 20 °C, 3 × 1 h		[687]

disulfides, the latter might be the only product isolated if cleavage is not conducted under an inert atmosphere (e.g. Entry **4**, Table 3.37).

Benzylic thioethers can be significantly more stable towards acidolytic solvolysis than the corresponding benzylic ethers. As illustrated by Entry **1** in Table 3.37, Wang linker derived thioethers are not cleaved by TFA, but only by acids with high ionizing power, such as hydrogen fluoride. Thioethers derived from the Wang resin can, however, be cleaved by oxidants. The example shown in Entry **2** in Table 3.37 illustrates the release of cyclic disulfides by treatment of Wang resin bound dithiols with a mild oxidant. It is advisable to use scavengers (e.g. silanes) during the acidolysis of polymeric thioethers, because the process is reversible and high yields will only be attainable if the resin-bound carbocations are efficiently scavenged [675]. The mercapto group of cysteine has been attached to insoluble supports as both the thiocarbamate and the aryl thioether (Entries **6** and **7**, Table 3.37). Both types of linker are susceptible to nucleophilic cleavage.

Thiols can be linked to insoluble supports as disulfides by disulfide interchange. Mixed disulfides can be prepared on insoluble supports by treating support-bound thiols with excess 'activated' disulfide (e.g. 2-benzothiazolyl, 2-nitrophenyl, 3-nitro-2-pyridyl disulfides [60,676] or a methanethiosulfonate MeSO$_2$–SR [677]; Figure 3.33), or by treating a support-bound disulfide (e.g. a 2-pyridyl disulfide [191]) with a thiol. Resin-bound disulfides are stable under the conditions of standard Fmoc peptide synthesis, but can be cleaved by reducing agents (Entries **8** and **9**, Table 3.37 [191,676,678–681]).

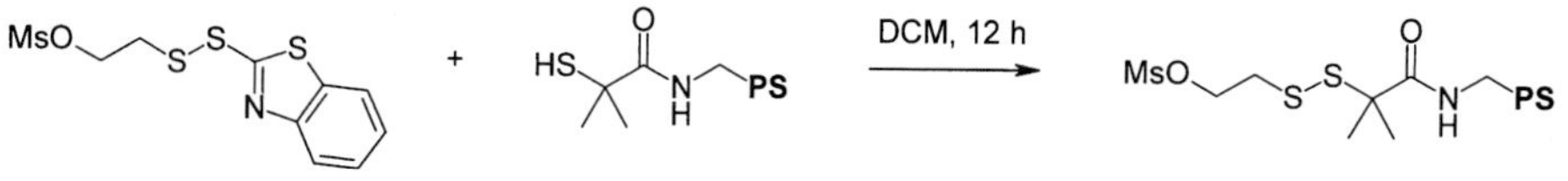

Figure 3.33. Preparation of support-bound unsymmetric disulfides [676].

3.13 Linkers for Alkyl and Aryl Halides, Azides, Diazonium Salts, and Nitriles

Alkyl or aryl halides can be generated upon cleavage from a polymeric support by treatment of silanes, organogermanium compounds, or stannanes with halogens (Entries **1–4**, Table 3.38), by solvolysis of resin-bound alcohols with hydrogen halides (Entry **5**, Table 3.38), or by treatment of immobilized alkylating agents (e.g. resin-bound sulfonates) with halides. Nucleophilic cleavage with iodide of resin-bound sulfonic esters and dialkyl aryl sulfonium salts (prepared in situ from alkyl aryl sulfides and methyl iodide) has, for instance, been used to prepare iodinated carbohydrates and other alkyl iodides (Entries **6** and **7**, Table 3.38). This strategy can also be used for the preparation of azides and esters (Entry **6**, Table 3.38), and should also enable the preparation of nitriles.

Aryl iodides have been prepared by thermolysis of resin-bound triazenes with methyl iodide (Entry **8**, Table 3.38 [688]). This reaction probably involves methylation

Table 3.38. Generation of halides and nitriles upon cleavage from insoluble supports.

Entry	Loaded resin	Cleavage conditions	Product, yield (purity)	Ref.
1		ICl (3 eq), DCM, 10 min	> 90%	[690]
2		Br$_2$ (6 eq), pyridine (3 eq), DCM, 0 °C, 2 h	97%	[690] see also [691,692]
3		Br$_2$ (4 eq), DCM, 2 × 5 min Ar: 3-(MeO)C$_6$H$_4$	59%	[693] see also [694,695]
4		I$_2$ (1.1 eq), THF, 23 °C, 2 h	100%	[696]
5		30% HBr in AcOH, CCl$_4$, 10 min	30%	[567]
6		NaI, 2-butanone, 65 °C; cleavage also succeeded with NaN$_3$ (0.1 mol/L), DMF, 60 °C, 12 h or with CsOAc (0.1 mol/L), 18-crown-6, DMF, 60 °C, 12 h	85–91%	[370,372] see also [369]
7		MeI, NaI, DMF, 75 °C, 20 h	82%	[697]
8		MeI, 110 °C, 12 h	94% (97%)	[455]
9		ICl, DCM, −78 °C, 1.5 h	0.26 mmol/g	[698,699]
10		(CF$_3$CO)$_2$O (5 eq), pyridine (10 eq), DCM, 20 °C, 16 h	93% (95%)	[241]

of the trisubstituted nitrogen atom of the triazene, followed by release of an aryldiazonium iodide into solution. Thermal decomposition of this diazonium salt yields the observed aryl iodide. Diazonium salts can also be generated by acidolysis of resin-bound triazenes R–NH–N=N–Ar–Pol, prepared from resin-bound aryl diazonium salts and primary aliphatic amines (Entry **7**, Table 3.19 [373]). The resulting aliphatic diazonium salts are highly reactive intermediates, which quickly decompose to yield nitrogen and carbocations, and these then either undergo rearrangement [689] or alkylate preferentially hard nucleophiles, such as carboxylic acids to yield esters [373], sulfonic acids to yield sulfonates [401], or halides to yield alkyl halides [689].

In addition to the procedures listed in Table 3.38, further reactions have been used to generate halides upon cleavage. In Section 3.5.2, iodolactonization is presented as a method for the preparation of iodomethyl lactones from resin-bound pentenoic or hexenoic acid derivatives. Closely related to the iodolactonization is the iodine-mediated formation of 2-(iodomethyl)tetrahydrofurans from resin-bound isoxazolidines (Entry **9**, Table 3.38; for the mechanism, see Figure 15.5). Nitriles can also be prepared by cleavage and simultaneous dehydration of amides $RCONH_2$ from the Rink or Sieber linkers with TFA anhydride (Entry **10**, Table 3.38).

3.14 Linkers for Aldehydes and Ketones

Interest in linkers for carbonyl compounds has only slowly emerged in recent years. The main driving force for the development of such linkers was the need for methods to prepare peptide aldehydes and related compounds (e.g. peptide trifluoromethyl ketones), which can be highly specific and valuable enzyme inhibitors [700,701], and are potentially useful for the treatment of various diseases.

The main strategies for the release of aldehydes or ketones from insoluble supports are sketched in Figure 3.34. These include the hydrolysis of acetals and related derivatives, the treatment of support-bound carboxylic acid derivatives with carbon nucleophiles, and the ozonolysis of resin-bound alkenes.

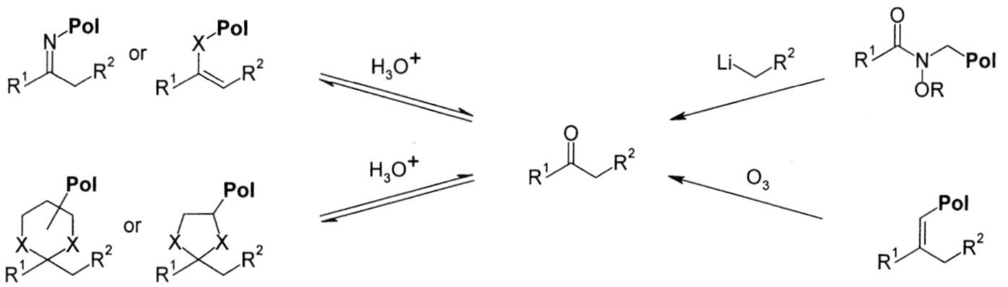

Figure 3.34. Strategies for the preparation of carbonyl compounds during cleavage from insoluble supports. X: NR, O, S.

Table 3.39. Attachment of aldehydes and ketones as enol ethers, enamines, and semicarbazides.

Entry	Loaded resin	Cleavage conditions	Product, yield (purity)	Ref.
1		TFA/DCM 3:97, 20 min	80% (> 95%)	[702]
2		TFA/acetone 5:95, 0.5 h	42% (> 95%)	[702]
3		TFA/CDCl₃ 1:99, 25 °C	30% (> 90%; endo/exo 97:3)	[704] see also [709,717]
4		TFA/DCM 1:99, 25 °C	74% (> 90%)	[704] see also [706]
5		TFA/DCM 1:9	77% (> 90%)	[708]
6		TFA/DCM 3:97, 10 min	78% (99%)	[712]
7		TFA/DCM 3:97, 10 min Ar: 4-BrC₆H₄	63% (90%)	[715]
8		aq HCl (1 mol/L)/ THF/AcOH/H₂O 1:75:2:6, 65 °C, 4 h	40%	[701] see also [718]
9		dilute aqueous acid, HCHO		[700]

3.14.1 Attachment as Enol Ethers, Enamines, Imines, and Hydrazones

Insoluble supports bearing hydroxyl groups can be used to immobilize aldehydes and ketones, either as acetals or as enol ethers (Table 3.39). 1,3-Dicarbonyl compounds react smoothly with resin-bound alcohols to yield enol ethers upon azeotropic removal of water [702,703]. Enol ethers can also be prepared by carbonyl methylenation of resin-bound esters with the Tebbe reagent [704,705] or other titanium-derived alkylidene complexes [706]. The preparation of support-bound silyl enol ethers from resin-bound silyl triflates and silyl esters [707] and ketones or aldehydes [708] has also been reported. Most of these enol ethers can be cleaved by mild acidic hydrolysis, whereby the corresponding carbonyl compounds are released into solution (Table 3.39). Cleavage of enol ethers can also be effected by treatment with oxidizing agents, whereby α-functionalized ketones are formed and released into solution. For instance, treatment of resin-bound enol ethers with dimethyl dioxirane gives α-hydroxy ketones, whereas cleavage with 1-(chloromethyl)-4-fluoro-1,4-diazoniabicyclo[2.2.2]octane bis(tetrafluoroborate) (Selectfluor) gives α-fluoro ketones [709].

Resin-bound amines can be converted into imines [710,711] or enamines by reaction with carbonyl compounds (Entries **6** and **7**, Table 3.39). Resin-bound enamines have also been prepared by Michael addition of resin-bound secondary amines to acceptor-substituted alkynes [712], by Hg(II)-catalyzed addition of resin-bound secondary amines to unactivated alkynes [713], by addition of C-nucleophiles to resin-bound imino ethers [714], and by chemical modification of other resin-bound enamines [712,713,715]. Acceptor-substituted enamines ('push–pull' alkenes) are not always susceptible to hydrolytic cleavage by TFA alone and might require aqueous acids to undergo hydrolysis [716].

3.14.2 Attachment of Carbonyl Compounds as Acetals

Support-bound alcohols and thiols can be used to immobilize aldehydes and ketones as acetals. Mixed acetals of carbonyl compounds with support-bound alcohols can be prepared by transacetalization of a symmetric acetal under acidic conditions [719]. The formation of mixed acetals on solid phase is, however, not always easy to perform and control, and so prior preparation of a mixed acetal in solution followed by loading onto a support is often the preferred protocol [626,637]. Carbohydrates can be linked to resin-bound alcohols or thiols as glycosides (Table 3.40).

Resin-bound diols, amino alcohols, and dithiols, which reversibly form cyclic acetals with aldehydes and ketones, have been successfully used as linkers for carbonyl compounds (Entries **5–11**, Table 3.40). Acetal formation on insoluble supports can be achieved by azeotropic removal of water (C_6H_6, TsOH, reflux [720]), whereas dithioacetals can be prepared by acid-catalysis alone ($BF_3 \cdot OEt_2$ or TMSCl; $CHCl_3$, 0 °C, 2 h [721]). N-Acylaminals such as R–CH(OMe)NH–CO–Pol have been prepared by treatment of resin-bound amides H_2NCO–Pol with aldehydes in the presence of $HC(OMe)_3$ and TFA [722].

Table 3.40. Attachment of aldehydes and ketones as acetals or related functional groups.

Entry	Loaded resin	Cleavage conditions	Product, yield (purity)	Ref.
1		NBS (4 eq), DTBP, THF/MeOH 10:1, 1.5 h		[725] see also [726,727]
2		Hg(O₂CCF₃)₂, DCM, H₂O, 20 °C, 5 h	64%	[728]
3		CSA (3 eq), DCM/H₂O 2:1, 25 °C, 40 h	up to 75%	[637]
4		TFA/DCM/H₂O 6:3:1	17–83% (95–98%)	[729] see also [730,731]
5		aq HCl (3 mol/L)/dioxane 1:1, 80 °C, 48 h	> 95%	[732] see also [733– 737]
6		TFA/H₂O 95:5, 15 min	96%	[738]
7		PPTS, dioxane/H₂O 8:2, 95 °C, 9 h	0.68 mmol/g (92%)	[720] see also [739]
8		AcOH/H₂O 5:95, 60 °C, 0.5 h (no cleavage with 95% TFA)	(94%)	[740] see also [741]

Table 3.40. continued.

Entry	Loaded resin	Cleavage conditions	Product, yield (purity)	Ref.
9		AcOH/H$_2$O 5:95, 60 °C, 0.5 h (no cleavage with 95% TFA)	10%	[740] see also [742]
10		H$_5$IO$_6$ (2.8 eq), THF, 5 h	92%	[721,724]
11		Hg(ClO$_4$)$_2$·3 H$_2$O (3 eq)	76%	[721]
12		DMSO, THF, 20 °C, 1 h		[743]

Acetals are usually easy to cleave by acid-catalyzed transacetalization or hydrolysis (Table 3.40). Dithioacetals, on the other hand, tend to be more resistant to hydrolysis, but cleavage can be achieved by treatment with mercury(II) salts or by oxidation with either [bis(trifluoroacetoxy)iodo]benzene [723] or periodic acid [724]. Use of the latter reagent can, however, also lead to the conversion of methyl ketones into iodomethyl ketones [721].

3.14.3 Miscellaneous Linkers for Aldehydes and Ketones

Aldehydes and ketones have also been prepared by nucleophilic cleavage of resin-bound *O*-alkyl hydroxamic acids (Weinreb amides [744]) with lithium aluminum hydride [745] or Grignard reagents (Entries 1 and 2, Table 3.41). Similarly, support-bound thiol esters can be cleaved with Grignard reagents to yield ketones [272], or with reducing agents to yield aldehydes (Entry 3, Table 3.41). Polystyrene-bound selenol esters (RCO–Se–Pol) react with alkynyl cuprates to yield alkynyl ketones [746].

Intramolecular Dieckmann cyclization of polystyrene-bound pimelates has been used to prepare β-keto esters (Entry 4, Table 3.41). Oxidative cleavage reactions leading to the formation of aldehydes include the ozonolysis of resin-bound alkenes, the periodate-mediated cleavage of 1,2-diols, and the oxidation of Wang resin derived ethers (Entries 5–7, Table 3.41).

Table 3.41. Formation of aldehydes and ketones by cleavage of carboxylic acid derivatives, alkenes, diols, and ethers.

Entry	Loaded resin	Cleavage conditions	Product, yield (purity)	Ref.
1		LiAlH₄, THF, 0 °C, 0.5 h	21% (80%)	[353] see also [747]
2		PhMgCl (15 eq), THF, 60 °C, 15 h	33%	[748] see also [749]
3		DIBAH, DCM, −78 °C, 19 h	73%	[128]
4		KOCEt₃, PhMe, 110 °C, 2 min	46%	[750]
5		O₃, DCM, −78 °C, 5 min, then Me₂S, DCM, 3 h	45%	[751] see also [752]
6		1. TFA/H₂O/*i*Pr₃SiH 95:2.5:2.5, 20 °C, 3 h 2. NaIO₄ (6 eq), H₂O/AcOH 5:1, 2 min R: peptide	38%	[753] see also [754,755]
7		DDQ (0.07 mol/L, 3 eq), DCM/H₂O 10:1, 40 min	36%	[583]

3.15 Linkers for Alkenes

The main strategies used for the preparation of alkenes by cleavage from insoluble supports are β-elimination and olefin metathesis (Figure 3.35). Because some of these strategies enable the preparation of pure alkenes, devoid of additional functional groups, the linkers are sometimes also called 'traceless' linkers, although the C=C double bond reveals the original point of attachment to the support.

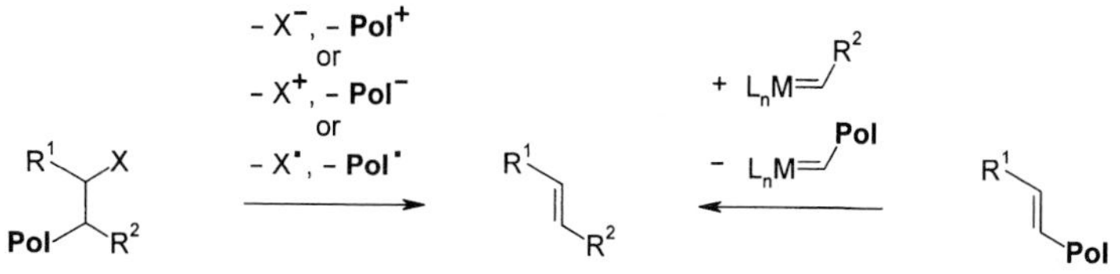

Figure 3.35. Strategies for the generation of alkenes upon cleavage from a support.

3.15.1 Linkers for Alkenes Cleavable by β-Elimination

During the release of alkenes by β-elimination, the polymeric support might act as a leaving group for nucleophilic displacement (e.g. Pol–SO$_2^-$, Pol–P(O)R$_2$, Pol–O$^-$), as a group capable of yielding a stable cation (e.g. Pol–SiR$_2^+$), or as a group with a weak covalent bond to carbon that is prone to homolytic cleavage with the generation of a radical (e.g. Pol–SnR$_2$). Most examples of these cleavage strategies reported to date have been Wittig reactions, in which phosphorus is irreversibly bound to the support (Table 3.42 [756,757]). Treatment of the immobilized ylide precursor with a base

Table 3.42. Generation of alkenes from support-bound phosphorus ylides.

Entry	Loaded resin	Cleavage conditions	Product, yield (purity)	Ref.
1	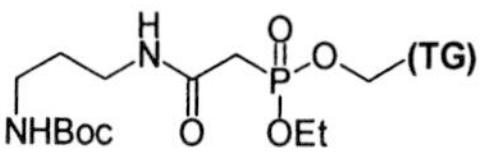	Me$_2$CHCHO (10 eq), LiBr, NEt$_3$, MeCN, 24 h	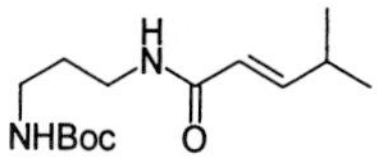 72%	[758] see also [761]
2	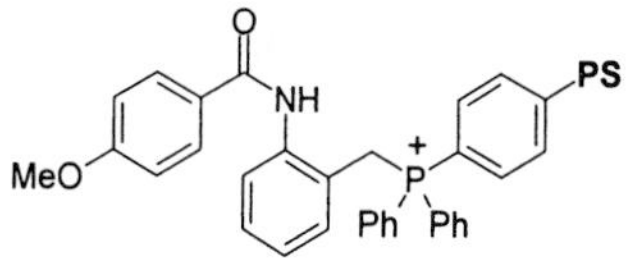	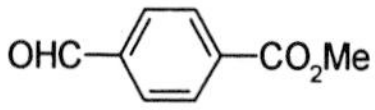 OHC—⬡—CO$_2$Me NaOMe, MeOH, 65 °C, 2 h Ar: 4-(MeO$_2$C)C$_6$H$_4$	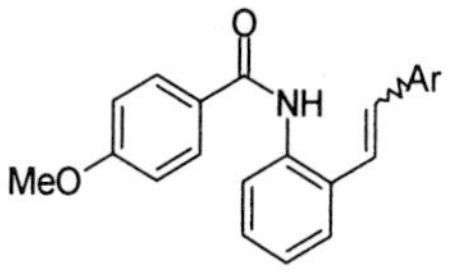 82%	[759]
3	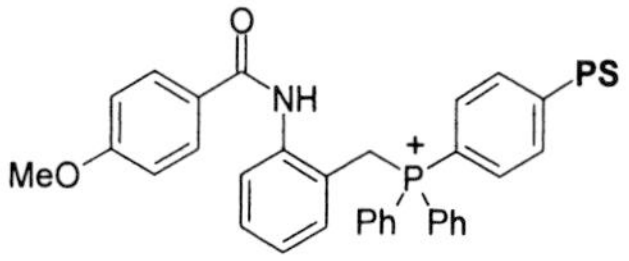	KO*t*Bu, DMF, PhMe, 110 °C, 45 min	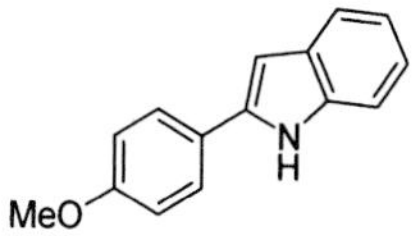 78%	[759]
4	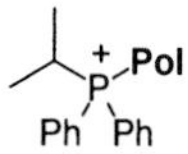	1. NaN(SiMe$_3$)$_2$, THF, 20 °C, 0.5 h 2. ArCHO (0.5 eq), THF, 20 °C, 20 min	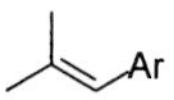 70–95%	[760]
5	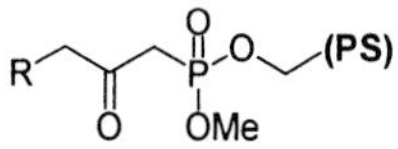	PhCHO (10 eq), K$_2$CO$_3$, 18-crown-6, PhMe, 65 °C, 3 h R: alkyl	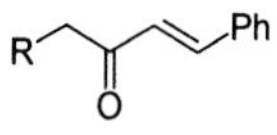 87%	[762]
6	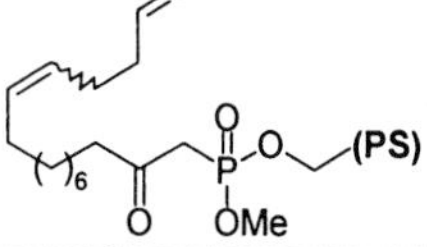	K$_2$CO$_3$, 18-crown-6, PhH, 65 °C, 12 h	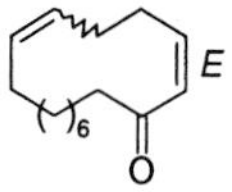 35–65%	[762]

and a carbonyl compound leads to carbonyl olefination, with simultaneous release of the alkene into solution.

If excess carbonyl compound is used in the product-releasing Wittig reaction, the product will be contaminated by the carbonyl compound. Removal of the excess carbonyl compound can be accomplished by extraction with aqueous bisulfite [758], by imine formation with an aminomethyl resin [759], or by formation of a water-soluble hydrazone with Girard's Reagent T [(carboxymethyl)trimethylammonium chloride hydrazide] [759]. A more elegant strategy is, however, the use of excess resin-bound Wittig reagent [760], whereby pure alkenes can be obtained directly.

Few examples have been reported of cleavage by β-elimination in which the support acts an as anionic leaving group (Table 3.43; see also Table 15.23). Elimination of an anionic, resin-bound leaving group has been used for the preparation of various heterocycles (see also the relevant chapters). For instance, polystyrene-bound vinyl sulfones react with tosylmethyl isonitrile (Tosmic) to yield 2-tosylpyrroles by elimination of sulfinate [763]. 3-Arylbenzofurans have also been prepared by elimination of polystyrene-bound sulfinate (Entry **12**, Table 15.9 [764]). 2,4-Diaminothiazoles can be prepared by elimination of a resin-bound thiolate (Entry **4**, Table 15.18 [765]). Alkenes can also be generated by reductive cleavage of resin-bound allyl sulfones or allyl esters with hydride or with carbon nucleophiles (see, e.g., Table 3.47). Resin-bound benzocyclobutane has been used as a precursor to *o*-quinodimethanes, which readily undergo Diels–Alder reactions with a variety of dienophiles (Figure 3.36; Entries **2** and **3**, Table 3.43).

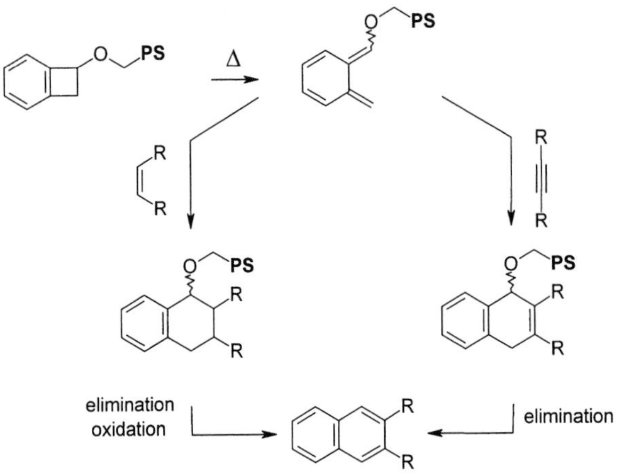

Figure 3.36. Preparation of substituted naphthalenes from resin-bound *o*-quinodimethanes [562].

The palladium-catalyzed coupling of aryl iodides with vinylstannanes (Stille coupling) leads to the formation of styrenes. With resin-bound vinylstannanes, this reaction can be conducted in such a way that simultaneous detachment from the support of the newly formed styrenes occurs. This has been realized intramolecularly in the preparation of macrocyclic lactones (Entry **4**, Table 3.43). The required resin-bound vinylstannanes were prepared either by hydrostannylation of alkynes with a resin-bound stannane HR$_2$Sn–Pol or by treatment of a resin-bound trialkyltin chloride with vinyl lithium compounds (Section 4.3). In a similar approach, the Suzuki reaction has been used to prepare macrocyclic biphenyl derivatives ([766], see Section 3.16.2).

Alkyl sulfoxides undergo thermal β-elimination to yield alkenes. This strategy for the preparation of alkenes has also been applied to solid-phase synthesis, but only substrates with a high tendency to undergo elimination (e.g., γ-oxo sulfoxides) could be thermally released from the support (Entry 5, Table 3.43). Unactivated sulfoxides could not be cleaved, not even under forcing conditions (199°C [767]).

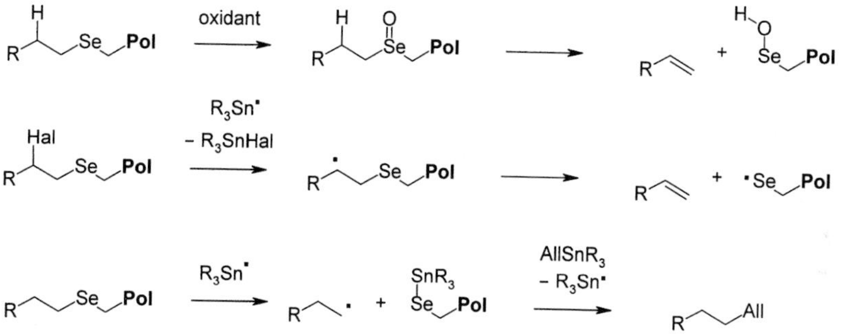

Figure 3.37. Selenides as linkers for alkenes.

Table 3.43. Generation of alkenes upon cleavage from supports by β-elimination and by vinylic substitution.

Entry	Loaded resin	Cleavage conditions	Product, yield (purity)	Ref.
1	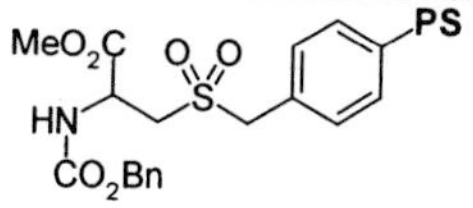	DBU (1 eq), DCM, 25 °C, 5 min	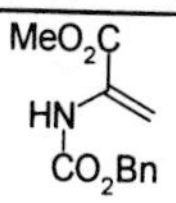 86% (96%)	[775]
2	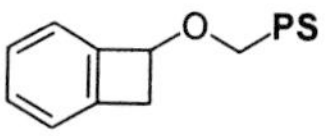	DMAD (1 eq), PhMe, 105 °C, 14 h	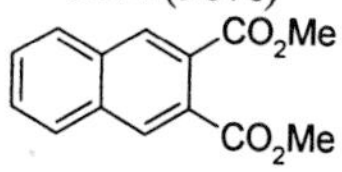 41%	[562]
3	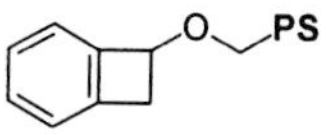	benzoquinone (1 eq), PhMe, 105 °C, 14 h	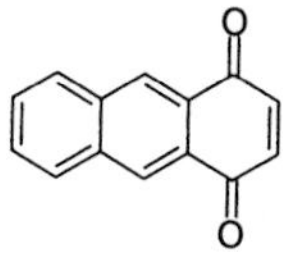 39%	[562]
4	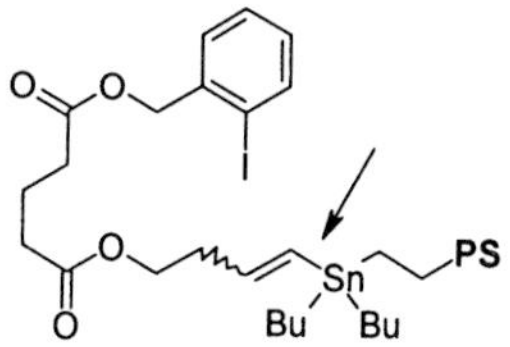	Pd(PPh$_3$)$_4$ (0.1 eq), PhMe, 100 °C, 48 h	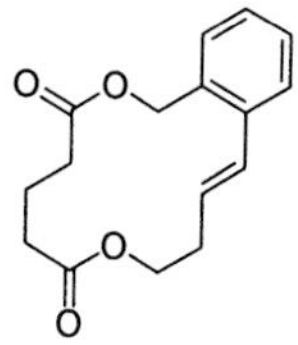 51%	[696]

Table 3.43. continued.

Entry	Loaded resin	Cleavage conditions	Product, yield (purity)	Ref.
5		dioxane, 100 °C	93:7; 45% (95%)	[767]
6		30% H₂O₂/THF 1:10, 30 °C, 2 h	91% (95%)	[776]
7		30% H₂O₂ (1 eq), THF, 23 °C, 12 h	78%	[777]
8		*tert*-BuOOH, DCM, CF₃CH₂OH, 20 °C, 7 h		[778]
9		Bu₃SnH (4 eq), AIBN (0.01 eq), PhMe, 100 °C, 6 h	92%	[777]
10		Bu₃SnCH₂CHCH₂ (5 eq), AIBN (0.1 eq), PhH, 80 °C, 2 h	37%	[768]
11		1. MCPBA (3 eq), DCM, −78 °C, 10 min 2. vinyl acetate/DIPEA/ PhMe 2:1:2, 140 °C, 12 h	79%	[779]
12		HO⌒OH, Na, 198 °C, 2 h	77%	[780]

Resin-bound selenium has been used as a linker for alkenes in two ways: (a) as an oxidant-sensitive linker (selenoxides readily undergo β-elimination at room temperature; Entries 6–8, Table 3.43 [767–773]), or (b) as a linker cleavable by tin radicals (Figure 3.37; Entries 9 and 10, Table 3.43). The main advantages of selenides as linkers are their stability under a broad variety of (non-oxidizing) reaction conditions, including high temperatures and treatment with acids or bases, and the mild conditions required for their cleavage.

Support-bound sulfonylhydrazones can also be used as linkers for alkenes. Cleavage is effected by heating in the presence of an alcoholate, whereby diazoalkanes are initially formed; these then undergo thermal fragmentation into the alkene and nitrogen (Entry 12, Table 3.43; Bamford–Stevens reaction). Polystyrene-bound alkynyl iodonium salts react with nucleophiles (sulfinates, benzotriazoles) according to an addition/elimination mechanism, thereby yielding the alkynylated nucleophile and the resin-bound aryl iodide [774].

3.15.2 Linkers for Alkenes Cleavable by Olefin Metathesis and Miscellaneous Linkers for Alkenes

Since the discovery of ruthenium and molybdenum carbene complexes that efficiently catalyze olefin metathesis under mild reaction conditions and that are compatible with a broad range of functional groups, olefin metathesis has increasingly been used for the preparation of alkenes on insoluble supports. In particular, the ruthenium complexes $Cl_2(PCy_3)_2Ru=CHR$, developed by Grubbs, show sufficient catalytic activity even in the presence of air and water [781] and are well suited for solid-phase synthesis.

For the cleavage of alkenes from a support by metathesis, several strategies can be envisaged. In most of the examples reported to date, ring-closing metathesis of resin-bound dienes has been used to release either a cycloalkene or an acyclic alkene into solution (Figure 3.38, Table 3.44). Further metathesis of the products in solution occurs only to a small extent when the initially released products are internal alkenes, because these normally react more slowly with the catalytically active carbene complex than terminal alkenes. If, however, terminal alkenes are to be prepared, self-metathesis of the product (to yield ethene and a symmetrically disubstituted ethene) is likely to become a serious side reaction. This side reaction can be suppressed by conducting the metathesis reaction in the presence of ethene [782,783].

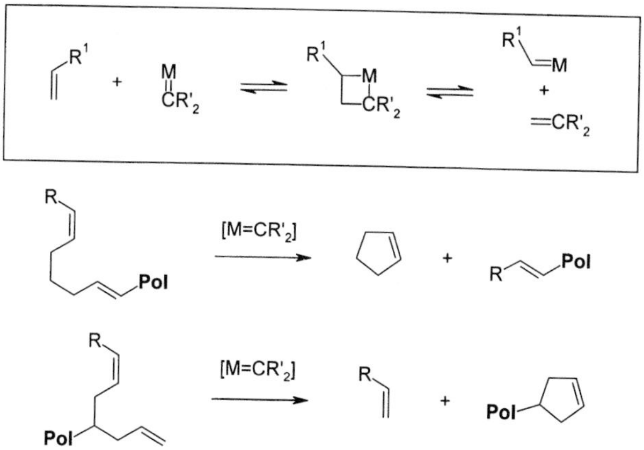

Figure 3.38. Mechanism of olefin metathesis and strategies for the cleavage of alkenes from polymeric supports by olefin metathesis.

Although five- and six-membered carbo- or heterocycles are most easily formed by ring-closing metathesis, macrocyclizations with simultaneous cleavage from the support have also been successfully performed [784]. Illustrative examples are listed in Table 3.44.

Table 3.44. Generation of alkenes upon cleavage from insoluble supports by olefin metathesis.

Entry	Loaded resin	Cleavage conditions	Product, yield (purity)	Ref.
1		Cl$_2$(PCy$_3$)$_2$Ru=CHPh (0.03–0.23 eq), DCM, 20 °C, 12 h	24–55%	[785]
2		Cl$_2$(PCy$_3$)$_2$Ru=CHPh (0.12 eq; 0.7 mM), DCM, 20 °C, 4 h	13%	[786,787] see also [782]
3		Cl$_2$(PCy$_3$)$_2$Ru=CHPh (1 eq), PhMe, 50 °C, 16 h	54%	[788] see also [789–791]
4		Cl$_2$(PCy$_3$)$_2$Ru=CHPh (0.05 eq), DCM, 20 °C, 16 h Ar: 2,4-dinitrophenyl	62%	[792] see also [793]
5		Cl$_2$(PCy$_3$)$_2$Ru=CHPh (0.2 eq), DCM, ethylene, 36 h	9% (9 steps)	[783] see also [794]

For the preparation of cycloalkenes (and heterocycles; see the relevant sections) by ring-closing metathesis with simultaneous cleavage from the support, the addition of a terminal alkene (e.g. styrene or ethene) to the reaction mixture can lead to increased yields of the cycloalkene [793]. This effect is probably related to the regeneration of the catalyst by reaction of support-bound carbene complexes with the alkene (Figure 3.39). As the rate of catalyst regeneration depends directly on the concentration of alkene RCH=CH$_2$ (Figure 3.39), high concentrations of this alkene will lead to fast catalyst regeneration. Excessively high concentrations of an additional alkene can, however, be detrimental, because cross-metathesis of the support-bound diene with the added alkene might compete with ring-closing metathesis. Cross-metathesis will become a serious side reaction when 'difficult' ring-formations are to be performed (e.g., the synthesis of seven- or eight-membered or larger rings), or when large amounts of catalyst are used.

Alkenes have also been prepared by retro-Diels–Alder reaction of resin-bound cyclohexenes. Figure 3.40 depicts an interesting example of this strategy, in which the retro-Diels–Alder reaction is induced by conversion of a thermally stable, resin-bound 7-oxanorbornadiene into a thermally unstable 7-oxanorbornene [795].

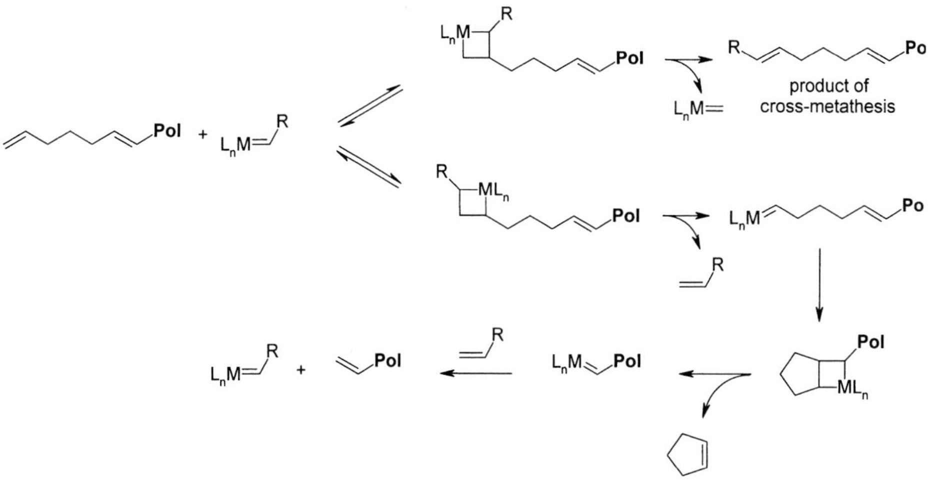

Figure 3.39. Ring-closing metathesis with simultaneous cleavage from the support, and the mechanism of catalyst regeneration.

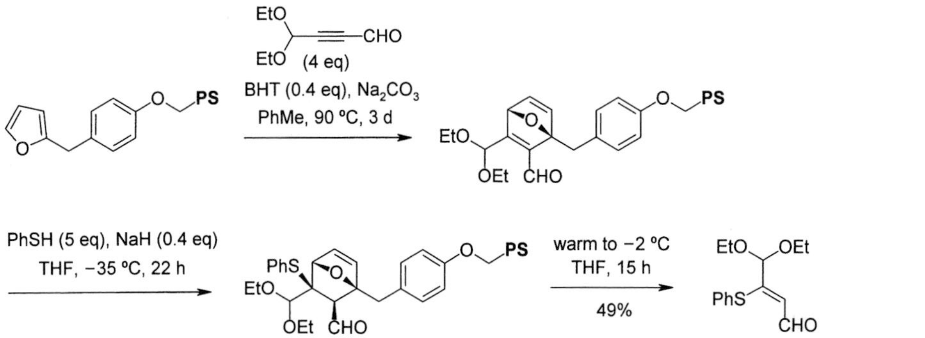

Figure 3.40. Retro-Diels–Alder reaction for the release of alkenes from cross-linked polystyrene [795].

3.16 Linkers for Alkanes and Arenes

To expand the range of products available by solid-phase synthesis, a series of strategies have been developed that enable the generation of C–H and C–C bonds upon cleavage from a support, and in this way enable the preparation of unfunctionalized hydrocarbons. These linkers are sometimes called 'traceless' linkers, because in some types of product the attachment point to the support can no longer be located.

The strategies described to date for the generation of C–H and C–C bonds during cleavage include decarboxylative cleavage, acidolysis of silanes, reductive cleavage of acetals, thioethers, selenides, sulfones, sulfonates, triazenes, sulfonylhydrazones, or organometallic compounds (Figure 3.41), the nucleophilic cleavage of resin-bound alkylating agents by carbon nucleophiles, and the oxidative cleavage of hydrazides.

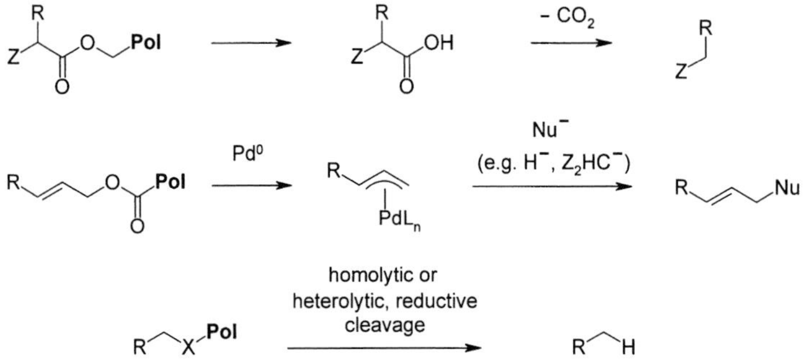

Figure 3.41. Generation of C–C and C–H bonds upon cleavage from supports. Z: electron-withdrawing group; X: metal, N_2NR, PR_2^+, O, SO_n, Se, etc.

3.16.1 Cleavage Followed by Decarboxylation

Some types of acceptor-substituted carboxylic acid readily undergo thermal decarboxylation. If such acids are released from a support under acidic conditions, decarboxylation can ensue either spontaneously or upon heating to yield a compound lacking an obvious 'attachment point'. This cleavage strategy has mainly been used for the preparation of ketones and nitriles (Table 3.45). Nitroacetic acid derivatives of polymeric alcohols can be cleaved from the polymer by treatment with a reducing agent, thereby yielding primary amines after decarboxylation (Entry **10**, Table 3.23). Carboxyl radicals can also undergo spontaneous decarboxylation to yield a carbon-centered radical and carbon dioxide. In Entry **7** (Table 3.45), a carboxyl radical is generated by photolysis, and the resulting C-radical is reduced to an alkane using a thiol.

Table 3.45. Generation of C–H bonds upon decarboxylative cleavage from supports.

Entry	Loaded resin	Cleavage conditions	Product, yield (purity)	Ref.
1		AcOH (neat), 8 h	(95%)	[52]
2		TFA/DCM/Et₃SiH 7:2:1, 1 h	80%	[796]
3		TFA/DCM/Et₃SiH 7:2:1, 1 h	42%	[796] see also [797]
4		TFA/DCM 1:1, 20 °C, 35 min	71% (71%)	[798]
5		(Me₃Si)₂NH, TFA/CDCl₃ 1:1	17%	[799]
6		NaI (0.17 mol/L), Me₃SiCl (0.4 mol/L), dioxane/MeCN 1:1, 75 °C, 72 h	69% (95%)	[800]
7		hv (350 nm), *tert*-BuSH/THF 1:40, 0.5 h		[801]

3.16.2 Cleavage of Silanes, Organogermanium, and Organoboron Compounds

The C–Si bonds of aryl-, heteroaryl-, vinyl-, and allylsilanes are stable towards alcoholates or weak reducing agents, but can be cleaved under mild conditions by treatment with acids or fluoride to yield a hydrocarbon and a silyl ester or silyl fluoride. Several linkers of this type have been tested and have proven useful for the preparation of unfunctionalized arenes and alkenes by cleavage from insoluble supports. Typical loading procedures for these linkers are sketched in Figure 3.42.

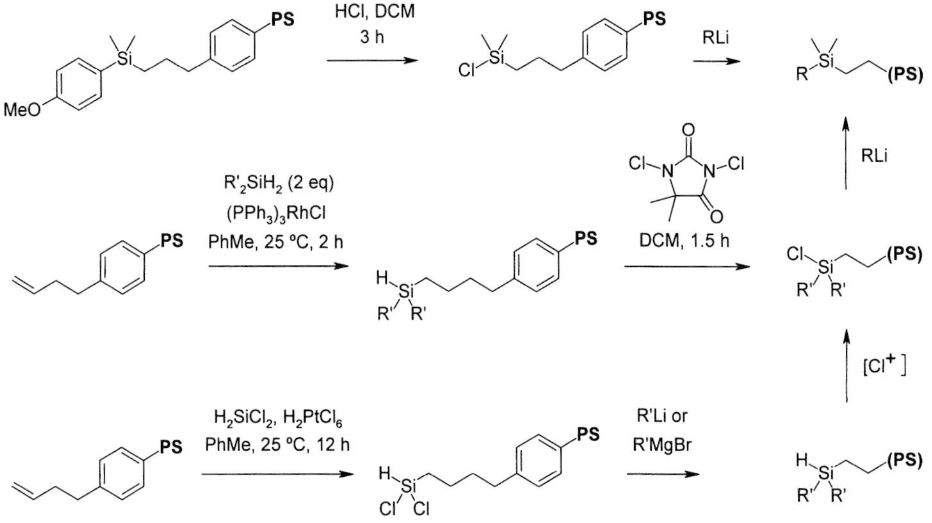

Figure 3.42. Preparation and loading of silane-based linkers for hydrocarbons [598,802,803].

The optimal conditions for the cleavage of resin-bound arylsilanes depend on the substitution pattern of the arene. Some donor-substituted arenes can even be cleaved from silyl linkers by treatment with TFA [804]. Particularly acid-sensitive are resin-bound 3-(dialkylarylsilyl)propionamides (Entry **9**, Table 3.46). As an alternative, resin-bound arylsilanes may also be cleaved by treatment with 1,2-dihydroxybenzene (5 equiv., MeCN, 50 °C, 20 h) or with glycolic acid, whereby bis(diolato)silicates are formed [805]. Arylsilanes bearing electron-withdrawing groups on the arene are more difficult to desilylate with weak acids (compare, e.g., Entries **1** and **6**, Table 3.46). When TFA fails to promote protiodesilylation, hydrogen fluoride, cesium fluoride, or TBAF might bring about the cleavage (Table 3.46). Support-bound allylsilanes can be cleaved by treatment with carbon electrophiles. Entry **12** in Table 3.46 is an example of such a cleavage, in which the electrophile is an α-alkoxy carbocation generated from an acetal and TiCl$_4$.

Arylboronic acids esterified with support-bound 1,2-diols undergo Suzuki reaction with aryl iodides, whereby biaryls are released into solution (Entry **13**, Table 3.46). This technique has also been used to prepare β-turn mimetics by simultaneous macrocyclization and cleavage from the support [766]. Alternatively, the C–B bond of a resin-bound boronate may be converted to a C–H bond by treatment with aqueous silver ammonium nitrate (Entry **14**, Table 3.46).

Table 3.46. Generation of C–H and C–C bonds upon cleavage of support-bound silanes, organogermanium compounds, and boronic esters.

Entry	Loaded resin	Cleavage conditions	Product, yield (purity)	Ref.
1		TFA/DCM 1:1, 25 °C, 3 h	80%	[598]
2		TFA/DCM 1:1, 20 °C, 24 h	93%	[806] see also [691]
3		TBAF (1 mol/L), THF, 12 h	58%	[598] see also [802,807]
4		HF, 12 h (no cleavage by TFA/Me₂S/H₂O 85:10:5) Ar: 4-(MeO)C₆H₄	68%	[693]
5		TFA, 60 °C, 24 h Ar: 4-(MeO)C₆H₄	58%	[693] see also [695]
6		CsF, DMF/H₂O 4:1, 110 °C (no cleavage by neat TFA, 25 °C) Ar: 4-formylphenyl	66%	[808]
7		TFA/DCM 1:1, 5% Me₂S, 24 h Ar: 4-(MeO)C₆H₄	60%	[690]

Table 3.46. continued.

Entry	Loaded resin	Cleavage conditions	Product, yield (purity)	Ref.
8		TBAF, DMF, 65 °C, 1 h Ar: 4-(MeO)C$_6$H$_4$		[809] see also [810]
9		TFA/DCM 1:1, 20 °C, 2 h Ar: 1-naphthyl	100%	[811]
10		1.5% TFA in DCM	0.35–0.52 mmol/g	[812]
11		3% TFA in DCM, 18 h	0.5 mmol/g	[75]
12		MeCH(OEt)$_2$, TiCl$_4$ (both 0.07 mol/L), DCM, −78 °C, 22 h	0.5 mmol/g	[75]
13		4-iodoanisole (5 eq), aq K$_3$PO$_4$ (2 mol/L, 3 eq), PdCl$_2$BINAP (0.05 eq), DMF, 60 °C, 24 h	85% (> 95%)	[766] see also [198]
14		Ag(NH$_3$)$_2$NO$_3$ (0.25 mol/L, 10 eq), H$_2$O/THF 1:1, 67 °C, 8 h R$_2$N: (Bn)N-Val-NHBn	57% (> 90%)	[813]

3.16.3 Reductive Cleavage of Carbon–Oxygen and Carbon–Nitrogen Bonds

The direct homolytic or heterolytic reductive cleavage of carbon–heteroatom bonds can be used to release products from polymeric supports. Whereas C–O bonds are too strong to undergo homolytic cleavage under acceptably mild reaction conditions, acetals and allyl esters can be smoothly cleaved heterolytically. When resin-bound acetals are treated with Lewis acids in the presence of a reducing agent or a carbon nucleophile, reductive cleavage from the support can occur. Ethers and sulfonamides have been prepared using this cleavage strategy (Entries **1** and **2**, Table 3.47).

Esters of allylic alcohols with resin-bound carboxylic acids can be converted into allyl palladium complexes, which react with carbon nucleophiles and with hydride sources to yield the formally reduced allyl derivatives (Entries **3** and **4**, Table 3.47). Alkyl sulfonates have been reduced to alkanes with NaBH$_4$ (Entry **5**, Table 3.47). Aryl sulfonates (Entry **6**, Table 3.47) and aryl perfluoroalkylsulfonates [814] can be reduced to alkanes by treatment with catalytic amounts of Pd(II) and formic acid as a hydride source.

Table 3.47. Formation of C–H and C–C bonds upon reductive cleavage of C–O and C–N bonds.

Entry	Loaded resin	Cleavage conditions	Product, yield (purity)	Ref.
1		TFA (5 eq), Et$_3$SiH (10 eq), DCM, 20 °C, 16–24 h	41%	[562]
2		SnCl$_4$ (1.1 eq), allyltrimethylsilane (2.5 eq), DCM, 20 °C, 16–24 h	47%	[562]
3		triethyl ammonium formate (5 eq), 7% Pd(PPh$_3$)$_4$, THF, 70 °C	69%	[478]
4		dimethyl malonate sodium salt (3 eq), 7% Pd(PPh$_3$)$_4$, THF, 50 °C, 8 h	78%	[478]
5		NaBH$_4$ (0.1 mol/L), DMSO, 60 °C, 12 h	(43–90%)	[372]

Table 3.47. continued..

Entry	Loaded resin	Cleavage conditions	Product, yield (purity)	Ref.
6	R–O–C(O)–C₆H₄–O–SO₂–C₆H₄–PS	HCO₂H (7.5 eq), NEt₃ (8 eq), Pd(OAc)₂ (0.2 eq), dppp, 110 °C, 12 h	R–O–C(O)–Ph, 36–74%	[815] see also [814]
7	(steroid-type structure with –O–SO₂–(PS))	OSiMe₃/Br structure (15 eq), Mg (15 eq), CuBr·SMe₂ (1 eq), THF, 20 °C, 3 h, then CSA, MeOH, H₂O	(steroid with OH), 47%	[816]
8	tBu ester diol aryl triazene N=N–N(CH₂Ph)–PS	THF/conc HCl 10:1, 50 °C, ultrasound, 5 min	tBu ester diol phenyl, 53%	[817]
9	aryl triazene N=N–N(CH₂Ph)–PS with =CH–CO₂tBu	HCl/THF or H₃PO₂/Cl₂HCCO₂H or HSiCl₃, DCM, 32 °C, 15 min	Ph–CH=CH–CO₂tBu, 81%	[817,818] see also [819]
10	aryl triazene N=N–N(piperazine)–PS with CH₂OH	≡–C(CH₃)₃ (alkyne); Pd(OAc)₂ (5%), TFA, MeOH, 40 °C, 2–12 h	aryl alkyne with CH₂OH, 53% (85%)	[820]
11	Ph–C≡C–C₆H₄–NH–NH–C(O)–(PS)	Cu(OAc)₂ (0.5 eq), pyridine (10 eq), air, MeOH, 20 °C, 2 h	Ph–C≡C–Ph, 93% (> 90%)	[174]
12	Ph–CH(CH₃)–N=N–NH–SO₂–C₆H₄–PS	NaBH₄ (1.1 mol/L, 8 eq), THF, 67 °C, 8 h	Ph–CH₂–CH₃, 27%	[780]

Support-bound triazenes, which can be prepared from resin-bound secondary aliphatic amines and aromatic diazonium salts [455], undergo cleavage upon treatment with acids, leading to regeneration of the aromatic diazonium salts. In cross-linked polystyrene, these decompose to yield nitrogen and, preferentially, radical-derived products. If the acidolysis of polystyrene-bound triazenes is conducted in the presence of hydrogen-atom donors (e.g. THF), unsubstituted arenes can be obtained (Entries **8** and **9**, Table 3.47). In the presence of alkenes or alkynes and Pd(OAc)₂, the initially formed diazonium salts undergo Heck reaction to yield vinylated or alkynylated arenes (Entry **10**, Table 3.47). Similarly, unsubstituted arenes can be obtained by oxidative cleavage of support-bound *N*-aryl-*N'*-acylhydrazines (Entry **11**, Table 3.47). Oxidation leads to the formation of *N*-aryl-*N'*-acyldiazenes, which, in the presence of nucleophiles, undergo deacylation to yield acid derivatives and aryldiazenes. The latter are unstable and decompose to give arenes and nitrogen. Air, in the presence of catalytic amounts of Cu(OAc)$_2$, or NBS [174,175] can be used as oxidants for hydrazides. Support-bound sulfonylhydrazones can be reduced to alkanes using sodium borohydride (Entry **12**, Table 3.47). This reaction, which has not yet been fully optimized for solid-phase synthesis, should enable the conversion of ketones into alkanes under mild reaction conditions.

3.16.4 Reductive Cleavage of Carbon–Phosphorus, Carbon–Sulfur, and Carbon–Selenium Bonds

Phosphonium salts can be dealkylated by treatment with alkoxides to yield alkanes. Although the hydrolytic cleavage of phosphonium salts in solution has been well investigated, the solid-phase variant of this reaction has not yet found broad application. One example, in which traceless linking was based on the alkoxide-induced dealkylation of a resin-bound phosphonium salt, is given in Table 3.48 (Entry **1**).

Hydrocarbons can be generated by nucleophilic cleavage of resin-bound allyl sulfones with carbon nucleophiles (e.g. Entry **3**, Table 3.48), whereby the resin-bound sulfinate acts as the leaving group. Thioethers, sulfoxides, and sulfones can also undergo C–S bond cleavage upon photolysis or upon treatment with reducing agents

Table 3.48. Formation of C–H and C–C bonds by reductive cleavage of C–P, C–S, and C–Se bonds.

Entry	Loaded resin	Cleavage conditions	Product, yield (purity)	Ref.
1		NaOMe (0.11 mol/L), MeOH, 65 °C, 4.5 h	81%	[759]
2		(0.61 mol/L, 0.9 eq), NaH (0.9 eq), DMSO, 100 °C, 3.5 h	62%	[824]
3		PhLi, THF, CuI, 0 °C, 4 h	20%	[825] see also [826]
4		Bu$_3$SnH, AIBN, PhH, 80 °C, 18 h; or H$_2$, Raney Ni, MeOH/EtOH 1.5:1, 20 °C, 3 h	40% (Bu$_3$SnH) 94% (H$_2$)	[827] see also [828]

Table 3.48. continued..

Entry	Loaded resin	Cleavage conditions	Product, yield (purity)	Ref.
5		hν (350 nm), MeCN, 5 h Ar: 4-PhC$_6$H$_4$	58%	[679] see also [829]
6		5% Na/Hg, Na$_2$HPO$_4$, MeOH, −40 °C to 0 °C, 2 h	98%	[830]
7		1. Et$_3$OBF$_4$ (5 eq), DCM, 20 °C, 8 h 2. PhB(OH)$_2$, (2 eq), Pd(dppf)Cl$_2$ (0.2 eq), K$_2$CO$_3$ (3 eq), THF, 60 °C, 14 h	98%	[831]
8		Bu$_3$SnH (2 eq), AIBN (0.005 eq), PhMe, 110 °C, 6 h	89%	[777]
9		Bu$_3$SnH (2 eq), AIBN (0.005 eq), PhMe, 110 °C, 8 h		[777]
10		Bu$_3$SnH (4 eq), AIBN (1.3 eq), PhMe, 90 °C, 4 h	13%	[823]

such as tin hydrides, sodium amalgam, or Raney nickel (Entries **4–6**, Table 3.48). These reducing agents are, unfortunately, non-volatile, and so further purification of the crude products will be necessary in most instances, making this cleavage strategy unsuitable for parallel synthesis.

Resin-bound benzylic thioethers can be converted into sulfonium salts by S-alkylation with triethyloxonium tetrafluoroborate. These sulfonium salts react with palladium(0) complexes to yield benzylpalladium complexes, which undergo Suzuki coupling with arylboronic acids (Entry **7**, Table 3.48).

Selenides are more readily cleaved by tin radicals than are thioethers. Two examples of the tin radical mediated cleavage of selenides are listed in Table 3.48 (Entries **8** and **9**); more examples have been reported [768,773,821–823]. The carbon-centered radicals initially formed by homolytic C–Se bond cleavage can either be directly reduced to the alkane by treatment with a tin hydride, or may add to multiple bonds before reduction. Entry **10** in Table 3.48 is an example of the formation of a polycyclic indoline through radical cyclization. Radical-mediated cleavage proceeds under mild, essentially neutral reaction conditions and is well suited for the release of sensitive organic compounds. Purification of the resulting products will, however, generally be required.

3.17 Non-Covalent Linkers

3.17.1 Ion-Exchange Resins

Charged organic compounds can be immobilized on ion-exchange resins. Release is achieved by displacement with salts, acids, or bases. The facility with which displacement of ionic products from ion-exchange resins occurs severely limits the choice of reactions that can be performed on these supports without premature product release. For this reason, ion-exchange resins are generally only used as either a convenient means of purifying charged [832] or neutral [833,834] organic products, or for the immobilization of reagents (e.g. alcoholates [835], thiolates [836], carboxylates [837], phosphorus ylides [838], diazonium arenes [839], or thiocyanate [840]).

3.17.2 Transition Metal Complexes

Kinetically stable complexes can be used as linkers for solid-phase synthesis. The cobalt(III) complex shown in Entry **1** in Table 3.49 was prepared in solution and then loaded onto polystyrene. Less than 5% cleavage occurred upon treatment of this support-bound complex with TFA/DCM (1:1) for 12 h or with 20% piperidine in DMF

Table 3.49. Transition metal complexes as linkers.

Entry	Loaded resin	Cleavage conditions	Product, yield (purity)	Ref.
1		DMF, dithiothreitol (0.5 mol/L), DIPEA (0.5 mol/L), 0.5 h	74–97%	[843,844]
2		hv, DCM, O₂, 72 h	70%	[845]
3		pyridine, heat, 2 h, or I₂, THF, 20 °C, 20 h	90%	[841,846] see also [847,848]
4		I₂, DCM, 20 °C, 1 h or hv, O₂, 48 h	80%	[842]

for 6 h, and it appears, therefore, to be suitable for the solid-phase preparation of small peptides.

The cobalt(0) complex shown in Entry **2** (Table 3.49) could be prepared either by heating a mixture of an alkyne cobalt carbonyl complex with polystyrene-bound triphenylphosphine, or by pretreating resin-bound triphenylphosphine with dicobalt octacarbonyl and then treating the resulting support with the alkyne.

Chromium(0) arene complexes can be used as linkers for arenes (Entries **3** and **4**, Table 3.49). Attachment is achieved by photolyzing a chromium(0) arene tricarbonyl complex in the presence of polystyrene-bound triphenylphosphine or isonitrile, whereby one carbonyl ligand is replaced by the resin-bound ligand. These linkers are stable towards amines, reducing agents (LiAlH$_4$), and acylating agents (acetyl chloride), but the arene can be selectively cleaved from the support by oxidation, by ligand exchange with pyridine, or by photolysis in the presence of air [841,842].

3.17.3 Miscellaneous Non-Covalent Linkers

Tetrabenzo[*a,c,g,i*]fluorene has been used to selectively link synthetic intermediates to charcoal, for the purpose of their purification. In polar solvents, the tetrabenzofluorene is strongly adsorbed by charcoal; this enables efficient separation of the intermediate from reagents. After centrifugation and washing, the intermediate is displaced from the charcoal and released into solution by addition of a non-polar solvent, and a new synthetic operation in solution can be conducted (Figure 3.43). Tetrabenzofluorene has also been used for the purification of peptides [849] and oligonucleotides [850], and for the synthesis of quinolones [851].

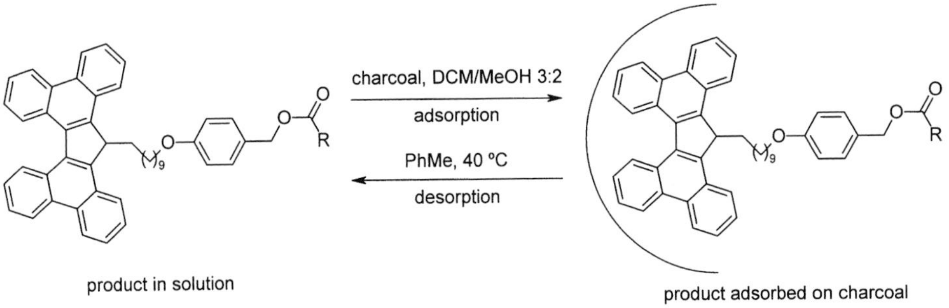

Figure 3.43. Use of tetrabenzofluorene derivatives for the reversible adsorption of compounds on charcoal [852].

References for Chapter 3

[1] Blackburn, C. *Biopolymers* **1998**, *47*, 311–351.
[2] Enholm, E. J.; Gallagher, M. E.; Jiang, S.; Batson, W. A. *Org. Lett.* **2000**, *2*, 3355–3357.
[3] Sparrow, J. T. *J. Org. Chem.* **1976**, *41*, 1350–1353.
[4] Sarin, V. K.; Kent, S. B. H.; Mitchell, A. R.; Merrifield, R. B. *J. Am. Chem. Soc.* **1984**, *106*, 7845–7850.
[5] Mitchell, A. R.; Kent, S. B. H.; Engelhard, M.; Merrifield, R. B. *J. Org. Chem.* **1978**, *43*, 2845–2852.
[6] Adams, J. H.; Cook, R. M.; Hudson, D.; Jammalamadaka, V.; Lyttle, M. H.; Songster, M. F. *J. Org. Chem.* **1998**, *63*, 3706–3716.
[7] Rich, D. H.; Gurwara, S. K. *J. Am. Chem. Soc.* **1975**, *97*, 1575–1579.
[8] Matsueda, G. R.; Stewart, J. M. *Peptides* **1981**, *2*, 45–50.
[9] Cardno, M.; Bradley, M. *Tetrahedron Lett.* **1996**, *37*, 135–138.
[10] Furka, A.; Sebestyén, F.; Asgedom, M.; Dibó, G. *Int. J. Pept. Prot. Res.* **1991**, *37*, 487–493.
[11] Zhao, P. L.; Zambias, R.; Bolognese, J. A.; Boulton, D.; Chapman, K. *Proc. Natl. Acad. Sci. USA* **1995**, *92*, 10212–10216.
[12] Burgess, K.; Liaw, A. I.; Wang, N. *J. Med. Chem.* **1994**, *37*, 2985–2987.
[13] Merrifield, R. B. *J. Am. Chem. Soc.* **1963**, *85*, 2149–2154.
[14] Meutermans, W. D. F.; Alewood, P. F. *Tetrahedron Lett.* **1995**, *36*, 7709–7712.
[15] Merrifield, R. B.; Vizioli, L. D.; Boman, H. G. *Biochemistry* **1982**, *21*, 5020–5031.
[16] Tam, J. P.; Heath, W. F.; Merrifield, R. B. *J. Am. Chem. Soc.* **1983**, *105*, 6442–6455.
[17] Yajima, H.; Fujii, N.; Ogawa, H.; Kawatani, H. *J. Chem. Soc., Chem. Commun.* **1974**, 107–108.
[18] Yajima, H.; Fujii, N.; Funakoshi, S.; Watanabe, T.; Murayama, E.; Otaka, A. *Tetrahedron* **1988**, *44*, 805–819.
[19] Fujii, N.; Otaka, A.; Ikemura, O.; Hatano, M.; Okamachi, A.; Funakoshi, S.; Sakurai, M.; Shioiri, T.; Yajima, H. *Chem. Pharm. Bull.* **1987**, *35*, 3447–3452.
[20] Nomizu, M.; Inagaki, Y.; Yamashita, T.; Ohkubo, A.; Otaka, A.; Fujii, N.; Roller, P. P.; Yajima, H. *Int. J. Pept. Prot. Res.* **1991**, *37*, 145–152.
[21] Yan, B.; Gstach, H. *Tetrahedron Lett.* **1996**, *37*, 8325–8328.
[22] Blake, J.; Li, C. H. *J. Am. Chem. Soc.* **1968**, *90*, 5882–5884.
[23] Mata, E. G. *Tetrahedron Lett.* **1997**, *38*, 6335–6338.
[24] Delpiccolo, C. M. L.; Mata, E. G. *Tetrahedron: Asymmetry* **1999**, *10*, 3893–3897.
[25] Mitchell, A. R.; Erickson, B. W.; Ryabtsev, M. N.; Hodges, R. S.; Merrifield, R. B. *J. Am. Chem. Soc.* **1976**, *98*, 7357–7362.
[26] Mitchell, A. R.; Erickson, B. W.; Ryabtsev, M. N.; Hodges, R. S.; Merrifield, R. B. *J. Am. Chem. Soc.* **1976**, *98*, 7357–7362.
[27] Kent, S. B. H.; Merrifield, R. B. *Israel J. Chem.* **1978**, *17*, 243–247.
[28] Wang, S. *J. Am. Chem. Soc.* **1973**, *95*, 1328–1333.
[29] Sarshar, S.; Siev, D.; Mjalli, A. M. M. *Tetrahedron Lett.* **1996**, *37*, 835–838.
[30] Winkler, J. D.; McCoull, W. *Tetrahedron Lett.* **1998**, *39*, 4935–4936.
[31] Svensson, A.; Fex, T.; Kihlberg, J. *J. Comb. Chem.* **2000**, *2*, 736–748.
[32] Bui, C. T.; Bray, A. M.; Nguyen, T.; Ercole, F.; Rasoul, F.; Sampson, W.; Maeji, N. J. *J. Pept. Sci.* **2000**, *6*, 49–56.
[33] Mergler, M.; Gosteli, J.; Grogg, P.; Nyfeler, R.; Tanner, R. *Chimia* **1999**, *53*, 29–34.
[34] Tam, J. P.; Kent, S. B. H.; Wong, T. W.; Merrifield, R. B. *Synthesis* **1979**, 955–957.
[35] Sheppard, R. C.; Williams, B. J. *Int. J. Pept. Prot. Res.* **1982**, *20*, 451–454.
[36] Mergler, M.; Tanner, R.; Gosteli, J.; Grogg, P. *Tetrahedron Lett.* **1988**, *29*, 4005–4008.
[37] Albericio, F.; Barany, G. *Tetrahedron Lett.* **1991**, *32*, 1015–1018.
[38] Blankemeyer-Menge, B.; Nimtz, M.; Frank, R. *Tetrahedron Lett.* **1990**, *31*, 1701–1704.
[39] Sieber, P. *Tetrahedron Lett.* **1987**, *28*, 6147–6150.
[40] Albericio, F.; Barany, G. *Int. J. Pept. Prot. Res.* **1998**, *26*, 92–97.
[41] Gisin, B. F. *Helv. Chim. Acta* **1973**, *56*, 1476–1482.
[42] Collini, M. D.; Ellingboe, J. W. *Tetrahedron Lett.* **1997**, *38*, 7963–7966.
[43] Nugiel, D. A.; Wacker, D. A.; Nemeth, G. A. *Tetrahedron Lett.* **1997**, *38*, 5789–5790.
[44] Barlos, K.; Gatos, D.; Kallitsis, J.; Papaioannou, D.; Sotiriu, P.; Schäfer, W. *Liebigs Ann. Chem.* **1987**, 1031–1035.
[45] Barlos, K.; Gatos, D.; Hondrelis, J.; Matsoukas, J.; Moore, G. J.; Schäfer, W.; Sotiriu, P. *Liebigs Ann. Chem.* **1989**, 951–955.
[46] Rink, H. *Tetrahedron Lett.* **1987**, *28*, 3787–3790.
[47] Barlos, K.; Chatzi, O.; Gatos, D.; Stavropoulos, G. *Int. J. Pept. Prot. Res.* **1991**, *37*, 513–520.

[48] Atkinson, G. E.; Fischer, P. M.; Chan, W. C. *J. Org. Chem.* **2000**, *65*, 5048–5056.
[49] Barlos, K.; Gatos, D.; Kallitsis, J.; Papaphotiou, G.; Sotiriu, P.; Wenqing, Y.; Schäfer, W. *Tetrahedron Lett.* **1989**, *30*, 3943–3946.
[50] Yang, L.; Chiu, K. *Tetrahedron Lett.* **1997**, *38*, 7307–7310.
[51] Richter, H.; Jung, G. *Tetrahedron Lett.* **1998**, *39*, 2729–2730.
[52] Garibay, P.; Nielsen, J.; Høeg-Jensen, T. *Tetrahedron Lett.* **1998**, *39*, 2207–2210.
[53] Ede, N. J.; Ang, K. H.; James, I. W.; Bray, A. M. *Tetrahedron Lett.* **1996**, *37*, 9097–9100.
[54] Barlos, K.; Gatos, D.; Kapolos, S.; Poulos, C.; Schäfer, W.; Wenqing, Y. *Int. J. Pept. Prot. Res.* **1991**, *38*, 555–561.
[55] Barlos, K.; Gatos, D.; Kutsogianni, S.; Papaphotiou, G.; Poulos, C.; Tsegenidis, T. *Int. J. Pept. Prot. Res.* **1991**, *38*, 562–568.
[56] Barlos, K.; Gatos, D.; Kapolos, S.; Papaphotiou, G.; Schäfer, W.; Wenqing, Y. *Tetrahedron Lett.* **1989**, *30*, 3947–3950.
[57] Barlos, K.; Gatos, D.; Papaphotiou, G.; Schäfer, W. *Liebigs Ann. Chem.* **1993**, 215–220.
[58] Bollhagen, R.; Schmiedberger, M.; Barlos, K.; Grell, E. *J. Chem. Soc., Chem. Commun.* **1994**, 2559–2560.
[59] Zikos, C. C.; Ferderigos, N. G. *Tetrahedron Lett.* **1994**, *35*, 1767–1768.
[60] Novabiochem Catalog & Peptide Synthesis Handbook **1999**, Läufelfingen, CH.
[61] Henkel, B.; Bayer, E. *Tetrahedron Lett.* **1998**, *39*, 9401–9402.
[62] Bleicher, K. H.; Lutz, C.; Wüthrich, Y. *Helv. Chim. Acta* **2000**, *41*, 9037–9042.
[63] Gordeev, M. F. *Biotechnology and Bioengineering* **1998**, *61*, 13–16.
[64] Xiao, X. Y.; Parandoosh, Z.; Nova, M. P. *J. Org. Chem.* **1997**, *62*, 6029–6033.
[65] Wieland, T.; Birr, C.; Fleckenstein, P. *Liebigs Ann. Chem.* **1972**, *756*, 14–19.
[66] Rosenthal, K.; Erlandsson, M.; Undén, A. *Tetrahedron Lett.* **1999**, *40*, 377–380.
[67] Wang, S.; Merrifield, R. B. *J. Am. Chem. Soc.* **1969**, *91*, 6488–6491.
[68] Wang, S. *J. Org. Chem.* **1975**, *40*, 1235–1239.
[69] Léger, R.; Yen, R.; She, M. W.; Lee, V. J.; Hecker, S. J. *Tetrahedron Lett.* **1998**, *39*, 4171–4174.
[70] Wolters, E. T. M.; Tesser, G. I.; Nivard, R. J. F. *J. Org. Chem.* **1974**, *39*, 3388–3392.
[71] Hernández, A. S.; Hodges, J. C. *J. Org. Chem.* **1997**, *62*, 3153–3157.
[72] Akaji, K.; Kiso, Y.; Carpino, L. A. *J. Chem. Soc., Chem. Commun.* **1990**, 584–586.
[73] Wilson, M. W.; Hernández, A. S.; Calvet, A. P.; Hodges, J. C. *Mol. Diversity* **1998**, *3*, 95–112.
[74] Blackburn, C.; Pingali, A.; Kehoe, T.; Herman, L. W.; Wang, H. Q.; Kates, S. A. *Bioorg. Med. Chem. Lett.* **1997**, *7*, 823–826.
[75] Schuster, M.; Lucas, N.; Blechert, S. *Chem. Commun.* **1997**, 823–824.
[76] Alonso, C.; Nantz, M. H.; Kurth, M. J. *Tetrahedron Lett.* **2000**, *41*, 5617–5622.
[77] Ueki, M.; Kai, K.; Amemiya, M.; Horino, H.; Oyamada, H. *J. Chem. Soc., Chem. Commun.* **1988**, 414–415.
[78] Chao, H. G.; Bernatowicz, M. S.; Reiss, P. D.; Klimas, C. E.; Matsueda, G. R. *J. Am. Chem. Soc.* **1994**, *116*, 1746–1752.
[79] Ramage, R.; Barron, C. A.; Bielecki, S.; Holden, R.; Thomas, D. W. *Tetrahedron* **1992**, *48*, 499–514.
[80] Chhabra, S. R.; Parekh, H.; Khan, A. N.; Bycroft, B. W.; Kellam, B. *Tetrahedron Lett.* **2001**, *42*, 2189–2192.
[81] Gennari, C.; Ceccarelli, S.; Piarulli, U.; Aboutayab, K.; Donghi, M.; Paterson, I. *Tetrahedron* **1998**, *54*, 14999–15016.
[82] Yedidia, V.; Leznoff, C. C. *Can. J. Chem.* **1980**, *58*, 1144–1150.
[83] ApSimon, J. W.; Dixit, D. M. *Can. J. Chem.* **1982**, *60*, 368–370.
[84] Kirschbaum, T.; Briehn, C. A.; Bäuerle, P. *J. Chem. Soc., Perkin Trans. 1* **2000**, 1211–1216.
[85] Anwer, M. K.; Spatola, A. F. *Tetrahedron Lett.* **1992**, *33*, 3121–3124.
[86] Seebach, D.; Thaler, A.; Blaser, D.; Ko, S. Y. *Helv. Chim. Acta* **1991**, *74*, 1102–1118.
[87] Whitney, D. B.; Tam, J. P.; Merrifield, R. B. *Tetrahedron* **1984**, *40*, 4237–4244.
[88] Kiso, Y.; Kimura, T.; Fujiwara, Y.; Shimokura, M.; Nishitani, A. *Chem. Pharm. Bull.* **1988**, *36*, 5024–5027.
[89] Stahl, G. L.; Walter, R.; Smith, C. W. *J. Am. Chem. Soc.* **1979**, *101*, 5383–5394.
[90] Nicolás, E.; Clemente, J.; Ferrer, T.; Albericio, F.; Giralt, E. *Tetrahedron* **1997**, *53*, 3179–3194.
[91] Chitkul, B.; Atrash, B.; Bradley, M. *Tetrahedron Lett.* **2001**, *42*, 6211–6214.
[92] Routledge, A.; Stock, H. T.; Flitsch, S. L.; Turner, N. J. *Tetrahedron Lett.* **1997**, *38*, 8287–8290.
[93] Ramage, R.; Barron, C. A.; Bielecki, S.; Thomas, D. W. *Tetrahedron Lett.* **1987**, *28*, 4105–4108.
[94] Mullen, D. G.; Barany, G. *J. Org. Chem.* **1988**, *53*, 5240–5248.
[95] Aldrian-Herrada, G.; Rabié, A.; Wintersteiger, R.; Brugidou, J. *J. Pept. Sci.* **1998**, *4*, 266–281.
[96] Mizoguchi, T.; Shigezane, K.; Takamura, N. *Chem. Pharm. Bull.* **1970**, *18*, 1465–1474.
[97] Baleux, F.; Calas, B.; Méry, J. *Int. J. Pept. Prot. Res.* **1986**, *28*, 22–28.

[98] Baleux, F.; Daunis, J.; Jacquier, R.; Calas, B. *Tetrahedron Lett.* **1984**, *25*, 5893–5896.

[99] Fancelli, D.; Fagnola, M. C.; Severino, D.; Bedeschi, A. *Tetrahedron Lett.* **1997**, *38*, 2311–2314.

[100] Osborn, N. J.; Robinson, J. A. *Tetrahedron* **1993**, *49*, 2873–2884.

[101] Hoffmann, S.; Frank, R. *Tetrahedron Lett.* **1994**, *35*, 7763–7766.

[102] Panke, G.; Frank, R. *Tetrahedron Lett.* **1998**, *39*, 17–18.

[103] Zheng, A.; Shan, D.; Shi, X.; Wang, B. *J. Org. Chem.* **1999**, *64*, 7459–7466.

[104] Rabanal, F.; Giralt, E.; Albericio, F. *Tetrahedron* **1995**, *51*, 1449–1458.

[105] Rabanal, F.; Giralt, E.; Albericio, F. *Tetrahedron Lett.* **1992**, *33*, 1775–1778.

[106] Nishiuchi, Y.; Nishio, H.; Inui, T.; Bódi, J.; Kimura, T. *J. Pept. Sci.* **2000**, *6*, 84–93.

[107] Rabanal, F.; Pastor, J. J.; Nicolás, E.; Albericio, F.; Giralt, E. *Tetrahedron Lett.* **2000**, *41*, 8093–8096.

[108] Mutter, M.; Bellof, D. *Helv. Chim. Acta* **1984**, *67*, 2009–2016.

[109] de la Torre, B. G.; Aviñó, A.; Tarrason, G.; Piulats, J.; Albericio, F.; Eritja, R. *Tetrahedron Lett.* **1994**, *35*, 2733–2736.

[110] Katti, S. B.; Misra, P. K.; Haq, W.; Mathur, K. B. *J. Chem. Soc., Chem. Commun.* **1992**, 843–844.

[111] Tesser, G. I.; Buis, J. T. W. A. R. M.; Wolters, E. T. M.; Bothé-Helmes, E. G. A. M. *Tetrahedron* **1976**, *32*, 1069–1072.

[112] Wade, W. S.; Yang, F.; Sowin, T. J. *J. Comb. Chem.* **2000**, *2*, 266–275.

[113] Dolence, E. K.; Dolence, J. M.; Poulter, C. D. *J. Comb. Chem.* **2000**, *2*, 522–536.

[114] Flynn, D. L.; Zelle, R. E.; Grieco, P. A. *J. Org. Chem.* **1983**, *48*, 2424–2426.

[115] Kenner, G. W.; McDermott, J. R.; Sheppard, R. C. *J. Chem. Soc., Chem. Commun.* **1971**, 636–637.

[116] Pátek, M.; Lebl, M. *Biopolymers* **1998**, *47*, 353–363.

[117] Todd, M. H.; Oliver, S. F.; Abell, C. *Org. Lett.* **1999**, *1*, 1149–1151.

[118] Backes, B. J.; Virgilio, A. A.; Ellman, J. A. *J. Am. Chem. Soc.* **1996**, *118*, 3055–3056.

[119] Hulme, C.; Peng, J.; Morton, G.; Salvino, J. M.; Herpin, T.; Labaudiniere, R. *Tetrahedron Lett.* **1998**, *39*, 7227–7230.

[120] Allin, S. M.; Shuttleworth, S. J. *Tetrahedron Lett.* **1996**, *37*, 8023–8026.

[121] Purandare, A. V.; Natarajan, S. *Tetrahedron Lett.* **1997**, *38*, 8777–8780.

[122] Sola, R.; Méry, J.; Pascal, R. *Tetrahedron Lett.* **1996**, *37*, 9195–9198.

[123] Phoon, C. W.; Abell, C. *Tetrahedron Lett.* **1998**, *39*, 2655–2658.

[124] Pascal, R.; Sola, R. *Tetrahedron Lett.* **1998**, *39*, 5031–5034.

[125] Sola, R.; Saguer, P.; David, M. L.; Pascal, R. *J. Chem. Soc., Chem. Commun.* **1993**, 1786–1788.

[126] Backes, B. J.; Ellman, J. A. *J. Am. Chem. Soc.* **1994**, *116*, 11171–11172.

[127] Li, W.-R.; Hsu, N.-M.; Chou, H.-H.; Lin, S. T.; Lin, Y.-S. *Chem. Commun.* **2000**, 401–402.

[128] Kobayashi, S.; Hachiya, I.; Yasuda, M. *Tetrahedron Lett.* **1996**, *37*, 5569–5572.

[129] Birr, C.; Lochinger, W.; Stahnke, G.; Lang, P. *Liebigs Ann. Chem.* **1972**, *763*, 162–172.

[130] Chamberlin, J. W. *J. Org. Chem.* **1966**, *31*, 1658–1660.

[131] Barltrop, J. A.; Plant, P. J.; Schofield, P. *J. Chem. Soc., Chem. Commun.* **1966**, 822–823.

[132] Pillai, V. N. R. *Synthesis* **1980**, 1–26.

[133] Ramesh, D.; Wieboldt, R.; Billington, A. P.; Carpenter, B. K.; Hess, G. P. *J. Org. Chem.* **1993**, *58*, 4599–4605.

[134] Amit, B.; Zehavi, U.; Patchornik, A. *J. Org. Chem.* **1974**, *39*, 192–196.

[135] Patchornik, A.; Amit, B.; Woodward, R. B. *J. Am. Chem. Soc.* **1970**, *92*, 6333–6335.

[136] Pirrung, M. C.; Shuey, S. W. *J. Org. Chem.* **1994**, *59*, 3890–3897.

[137] Corrie, J. E. T.; Trentham, D. R. *J. Chem. Soc., Perkin Trans. 1* **1992**, 2409–2417.

[138] Givens, R. S.; Athey, P. S.; Kueper, L. W.; Matuszewski, B.; Xue, J. Y. *J. Am. Chem. Soc.* **1992**, *114*, 8708–8710.

[139] Sheehan, J. C.; Wilson, R. M.; Oxford, A. W. *J. Am. Chem. Soc.* **1971**, *93*, 7222–7228.

[140] Sheehan, J. C.; Umezawa, K. *J. Org. Chem.* **1973**, *38*, 3771–3774.

[141] Baldwin, J. J.; Burbaum, J. J.; Henderson, I.; Ohlmeyer, M. H. J. *J. Am. Chem. Soc.* **1995**, *117*, 5588–5589.

[142] Lloyd-Williams, P.; Gairí, M.; Albericio, F.; Giralt, E. *Tetrahedron* **1991**, *47*, 9867–9880.

[143] Venkatesan, H.; Greenberg, M. M. *J. Org. Chem.* **1996**, *61*, 525–529.

[144] Greenberg, M. M.; Gilmore, J. L. *J. Org. Chem.* **1994**, *59*, 746–753.

[145] Holmes, C. P.; Jones, D. G. *J. Org. Chem.* **1995**, *60*, 2318–2319.

[146] Teague, S. J. *Tetrahedron Lett.* **1996**, *37*, 5751–5754.

[147] Åkerblom, E. B.; Nygren, A. S.; Agback, K. H. *Mol. Diversity* **1998**, *3*, 137–148.

[148] Yoo, D. J.; Greenberg, M. M. *J. Org. Chem.* **1995**, *60*, 3358–3364.

[149] Holmes, C. P.; Jones, D. G.; Frederick, B. T.; Dong, L. C. *Peptides: Chemistry, Structure and Biology; Proceedings of the 14th American Peptide Symposium, 1995; Mayflower Scientific Ltd.* **1996**, 44–45.

[150] Lloyd-Williams, P.; Gairí, M.; Albericio, F.; Giralt, E. *Tetrahedron* **1993**, *49*, 10069–10078.
[151] Ajayaghosh, A.; Pillai, V. N. R. *Tetrahedron* **1988**, *44*, 6661–6666.
[152] Rich, D. H.; Gurwara, S. K. *J. Chem. Soc., Chem. Commun.* **1973**, 610–611.
[153] Ryba, T. D.; Harran, P. G. *Org. Lett.* **2000**, *2*, 851–853.
[154] Ajayaghosh, A.; Pillai, V. N. R. *J. Org. Chem.* **1987**, *52*, 5714–5717.
[155] Whitehouse, D. L.; Savinov, S. N.; Austin, D. J. *Tetrahedron Lett.* **1997**, *38*, 7851–7852.
[156] Wang, S. *J. Org. Chem.* **1976**, *41*, 3258–3261.
[157] Routledge, A.; Abell, C.; Balasubramanian, S. *Tetrahedron Lett.* **1997**, *38*, 1227–1230.
[158] Lee, H. B.; Balasubramanian, S. *J. Org. Chem.* **1999**, *64*, 3454–3460.
[159] Peukert, S.; Giese, B. *J. Org. Chem.* **1998**, *63*, 9045–9051.
[160] Glatthar, R.; Giese, B. *Org. Lett.* **2000**, *2*, 2315–2317.
[161] Schlatter, J. M.; Mazur, R. H. *Tetrahedron Lett.* **1977**, 2851–2852.
[162] Pande, C. S.; Gupta, N.; Bhardwaj, A. *J. Appl. Polym. Sci.* **1995**, *56*, 1127–1131.
[163] Botta, M.; Corelli, F.; Maga, G.; Manetti, F.; Renzulli, M.; Spadari, S. *Tetrahedron* **2001**, *57*, 8357–8367.
[164] Colombo, R. *J. Chem. Soc., Chem. Commun.* **1981**, 1012–1013.
[165] Seitz, O.; Kunz, H. *Angew. Chem. Int. Ed. Engl.* **1995**, *34*, 803–805.
[166] Zhang, X. H.; Jones, R. A. *Tetrahedron Lett.* **1996**, *37*, 3789–3790.
[167] Seitz, O.; Kunz, H. *J. Org. Chem.* **1997**, *62*, 813–826.
[168] Kunz, H.; Dombo, B. *Angew. Chem. Int. Ed. Engl.* **1988**, *27*, 711–713.
[169] Guibé, F.; Dangles, O.; Balavoine, G.; Loffet, A. *Tetrahedron Lett.* **1989**, *30*, 2641–2644.
[170] Lloyd-Williams, P.; Jou, G.; Albericio, F.; Giralt, E. *Tetrahedron Lett.* **1991**, *32*, 4207–4210.
[171] Seitz, O. *Tetrahedron Lett.* **1999**, *40*, 4161–4164.
[172] Semenov, A. N.; Gordeev, K. Y. *Int. J. Pept. Prot. Res.* **1995**, *45*, 303–304.
[173] Wieland, T.; Lewalter, J.; Birr, C. *Liebigs Ann. Chem.* **1970**, *740*, 31–47.
[174] Stieber, F.; Grether, U.; Waldmann, H. *Angew. Chem. Int. Ed.* **1999**, *38*, 1073–1077.
[175] Millington, C. R.; Quarrell, R.; Lowe, G. *Tetrahedron Lett.* **1998**, *39*, 7201–7204.
[176] Berst, F.; Holmes, A. B.; Ladlow, M.; Murray, P. J. *Tetrahedron Lett.* **2000**, *41*, 6649–6653.
[177] Singh, J.; Allen, M. P.; Ator, M. A.; Gainor, J. A.; Whipple, D. A. *J. Med. Chem.* **1995**, *38*, 217–219.
[178] Nanda, S.; Rao, A. B.; Yadav, J. S. *Tetrahedron Lett.* **1999**, *40*, 5905–5908.
[179] Quarrell, R.; Claridge, T. D. W.; Weaver, G. W.; Lowe, G. *Mol. Diversity* **1996**, *1*, 223–232.
[180] Sauerbrei, B.; Jungmann, V.; Waldmann, H. *Angew. Chem. Int. Ed.* **1998**, *37*, 1143–1146.
[181] Böhm, G.; Dowden, J.; Rice, D. C.; Burgess, I.; Pilard, J. F.; Guilbert, B.; Haxton, A.; Hunter, R. C.; Turner, N. J.; Flitsch, S. L. *Tetrahedron Lett.* **1998**, *39*, 3819–3822.
[182] Leon, S.; Quarrell, R.; Lowe, G. *Bioorg. Med. Chem. Lett.* **1998**, *8*, 2997–3002.
[183] Vagner, J.; Barany, G.; Lam, K. S.; Krchnák, V.; Sepetov, N. F.; Ostrem, J. A.; Strop, P.; Lebl, M. *Proc. Natl. Acad. Sci. USA* **1996**, *93*, 8194–8199.
[184] Hilaire, P. M. S.; Lowary, T. L.; Meldal, M.; Bock, K. *J. Am. Chem. Soc.* **1998**, *120*, 13312–13320.
[185] Renil, M.; Ferreras, M.; Delaisse, J. M.; Foged, N. T.; Meldal, M. *J. Pept. Sci.* **1998**, *4*, 195–210.
[186] Meldal, M.; Svendsen, I.; Breddam, K.; Auzanneau, F. I. *Proc. Natl. Acad. Sci. USA* **1994**, *91*, 3314–3318.
[187] Meldal, M.; Svendsen, I. *J. Chem. Soc., Perkin Trans. 1* **1995**, 1591–1596.
[188] Schuster, M.; Wang, P.; Paulson, J. C.; Wong, C. H. *J. Am. Chem. Soc.* **1994**, *116*, 1135–1136.
[189] Yamada, K.; Nishimura, S. I. *Tetrahedron Lett.* **1995**, *36*, 9493–9496.
[190] Elmore, D. T.; Guthrie, D. J. S.; Wallace, A. D.; Bates, S. R. E. *J. Chem. Soc., Chem. Commun.* **1992**, 1033–1034.
[191] Blixt, O.; Norberg, T. *J. Org. Chem.* **1998**, *63*, 2705–2710.
[192] Schnölzer, M.; Kent, S. B. H. *Science* **1992**, *256*, 221–225.
[193] Dawson, P. E.; Kent, S. B. H. *J. Am. Chem. Soc.* **1993**, *115*, 7263–7266.
[194] Yamashiro, D.; Li, C. H. *Int. J. Pept. Prot. Res.* **1988**, *31*, 322–334.
[195] Blake, J. *Int. J. Pept. Prot. Res.* **1981**, *17*, 273–274.
[196] Canne, L. E.; Walker, S. M.; Kent, S. B. H. *Tetrahedron Lett.* **1995**, *36*, 1217–1220.
[197] Carboni, B.; Pourbaix, C.; Carreaux, F.; Deleuze, H.; Maillard, B. *Tetrahedron Lett.* **1999**, *40*, 7979–7983.
[198] Gravel, M.; Bérubé, C. D.; Hall, D. G. *J. Comb. Chem.* **2000**, *2*, 228–231.
[199] Hall, D. G.; Tailor, J.; Gravel, M. *Angew. Chem. Int. Ed.* **1999**, *38*, 3064–3067.
[200] Dunsdon, R. M.; Greening, J. R.; Jones, P. S.; Jordan, S.; Wilson, F. X. *Bioorg. Med. Chem. Lett.* **2000**, *10*, 1577–1579.
[201] Hari, A.; Miller, B. L. *Org. Lett.* **1999**, *1*, 2109–2111.
[202] Cao, X. D.; Mjalli, A. M. M. *Tetrahedron Lett.* **1996**, *37*, 6073–6076.
[203] Zhu, T.; Boons, G.-J. *J. Am. Chem. Soc.* **2000**, *122*, 10222–10223.

[204] Sasaki, S.; Ehara, T.; Alam, M. R.; Fujino, Y.; Harada, N.; Kimura, J.; Nakamura, H.; Maeda, M. *Bioorg. Med. Chem. Lett.* **2001**, *11*, 2581–2584.
[205] Parang, K.; Fournier, E. J.-L.; Hindsgaul, O. *Org. Lett.* **2001**, *3*, 307–309.
[206] Mergler, M.; Dick, F.; Gosteli, J.; Nyfeler, R.; Sax, B. *Innovation and Perspectives in Solid Phase Synthesis and Combinatorial Libraries*, Proceedings of the Sixth International Symposium, 2000, York, UK, **2000**, 313–316.
[207] Fivush, A. M.; Willson, T. M. *Tetrahedron Lett.* **1997**, *38*, 7151–7154.
[208] Alsina, J.; Jensen, K. J.; Albericio, F.; Barany, G. *Chem. Eur. J.* **1999**, *5*, 2787–2795.
[209] Bui, C. T.; Rasoul, F. A.; Ercole, F.; Pham, Y.; Maeji, N. J. *Tetrahedron Lett.* **1998**, *39*, 9279–9282.
[210] Gordeev, M. F.; Luehr, G. W.; Hui, H. C.; Gordon, E. M.; Patel, D. V. *Tetrahedron* **1998**, *54*, 15879–15890.
[211] Okayama, T.; Burritt, A.; Hruby, V. J. *Org. Lett.* **2000**, *2*, 1787–1790.
[212] Boojamra, C. G.; Burow, K. M.; Thompson, L. A.; Ellman, J. A. *J. Org. Chem.* **1997**, *62*, 1240–1256.
[213] Swayze, E. E. *Tetrahedron Lett.* **1997**, *38*, 8465–8468.
[214] Penke, B.; Rivier, J. *J. Org. Chem.* **1987**, *52*, 1197–1200.
[215] Bourne, G. T.; Meutermans, W. D. F.; Alewood, P. F.; McGeary, R. P.; Scanlon, M.; Watson, A. A.; Smythe, M. L. *J. Org. Chem.* **1999**, *64*, 3095–3101.
[216] Bourne, G. T.; Meutermans, W. D. F.; Smythe, M. L. *Tetrahedron Lett.* **1999**, *40*, 7271–7274.
[217] Aoki, Y.; Kobayashi, S. *J. Comb. Chem.* **1999**, *1*, 371–372.
[218] Raju, B.; Kogan, T. P. *Tetrahedron Lett.* **1997**, *38*, 4965–4968.
[219] Swayze, E. E. *Tetrahedron Lett.* **1997**, *38*, 8643–8646.
[220] Miller, M. W.; Vice, S. F.; McCombie, S. W. *Tetrahedron Lett.* **1998**, *39*, 3429–3432.
[221] Lee, H. B.; Balasubramanian, S. *Org. Lett.* **2000**, *2*, 323–326.
[222] Bui, C. T.; Bray, A. M.; Ercole, F.; Pham, Y.; Rasoul, F. A.; Maeji, N. J. *Tetrahedron Lett.* **1999**, *40*, 3471–3474.
[223] Bui, C. T.; Bray, A. M.; Nguyen, T.; Ercole, F.; Maeji, N. J. *J. Pept. Sci.* **2000**, *6*, 243–250.
[224] Albericio, F.; Barany, G. *Int. J. Pept. Prot. Res.* **1987**, *30*, 206–216.
[225] Sarantakis, D.; Bicksler, J. J. *Tetrahedron Lett.* **1997**, *38*, 7325–7328.
[226] Yu, K. L.; Civiello, R.; Roberts, D. G. M.; Seiler, S. M.; Meanwell, N. A. *Bioorg. Med. Chem. Lett.* **1999**, *9*, 663–666.
[227] Migawa, M. T.; Swayze, E. E. *Org. Lett.* **2000**, *2*, 3309–3311.
[228] Ngu, K.; Patel, D. V. *Tetrahedron Lett.* **1997**, *38*, 973–976.
[229] Albericio, F.; Kneib-Cordonier, N.; Biancalana, S.; Gera, L.; Masada, R. I.; Hudson, D.; Barany, G. *J. Org. Chem.* **1990**, *55*, 3730–3743.
[230] Olsen, J. A.; Jensen, K. J.; Nielsen, J. *J. Comb. Chem.* **2000**, *2*, 143–150.
[231] Tolborg, J. F.; Jensen, K. J. *Chem. Commun.* **2000**, 147–148.
[232] Brummond, K. M.; Lu, J. *J. Org. Chem.* **1999**, *64*, 1723–1726.
[233] Chao, H. G.; Bernatowicz, M. S.; Matsueda, G. R. *J. Org. Chem.* **1993**, *58*, 2640–2644.
[234] Estep, K. G.; Neipp, C. E.; Stramiello, L. M. S.; Adam, M. D.; Allen, M. P.; Robinson, S.; Roskamp, E. J. *J. Org. Chem.* **1998**, *63*, 5300–5301.
[235] Hammer, R. P.; Albericio, F.; Gera, L.; Barany, G. *Int. J. Pept. Prot. Res.* **1990**, *36*, 31–45.
[236] Gennari, C.; Longari, C.; Ressel, S.; Salom, B.; Piarulli, U.; Ceccarelli, S.; Mielgo, A. *Eur. J. Org. Chem.* **1998**, 2437–2449.
[237] Sternson, S. M.; Schreiber, S. L. *Tetrahedron Lett.* **1998**, *39*, 7451–7454.
[238] Ajayaghosh, A.; Pillai, V. N. R. *J. Org. Chem.* **1990**, *55*, 2826–2829.
[239] Renil, M.; Pillai, V. N. R. *Tetrahedron Lett.* **1994**, *35*, 3809–3812.
[240] Kumar, K. S.; Pillai, V. N. R. *Tetrahedron* **1999**, *55*, 10437–10446.
[241] Hone, N. D.; Payne, L. J.; Tice, C. M. *Tetrahedron Lett.* **2001**, *42*, 1115–1118.
[242] Chan, W. C.; Mellor, S. L. *J. Chem. Soc., Chem. Commun.* **1995**, 1475–1477.
[243] Chen, J. J.; Golebiowski, A.; McClenaghan, J.; Klopfenstein, S. R.; West, L. *Tetrahedron Lett.* **2001**, *42*, 2269–2271.
[244] Heinonen, P.; Lönnberg, H. *Tetrahedron Lett.* **1997**, *38*, 8569–8572.
[245] Ciobanu, L. C.; Maltais, R.; Poirier, D. *Org. Lett.* **2000**, *2*, 445–448.
[246] Pirrung, M. C.; Pansare, S. V. *J. Comb. Chem.* **2001**, *3*, 90–96.
[247] Dörner, B.; Husar, G. M.; Ostresh, J. M.; Houghten, R. A. *Bioorg. Med. Chem.* **1996**, *4*, 709–715.
[248] Nefzi, A.; Ostresh, J. M.; Meyer, J. P.; Houghten, R. A. *Tetrahedron Lett.* **1997**, *38*, 931–934.
[249] Wang, S. S.; Wang, B. S. H.; Hughes, J. L.; Leopold, E. J.; Wu, C. R.; Tam, J. P. *Int. J. Pept. Prot. Res.* **1992**, *40*, 344–349.
[250] Brown, D. S.; Revill, J. M.; Shute, R. E. *Tetrahedron Lett.* **1998**, *39*, 8533–8536.
[251] Noda, M.; Yamaguchi, M.; Ando, E.; Takeda, K.; Nokihara, K. *J. Org. Chem.* **1994**, *59*, 7968–7975.

[252] Patterson, J. A.; Ramage, R. *Tetrahedron Lett.* **1999**, *40*, 6121–6124.
[253] Breipohl, G.; Knolle, J.; Stüber, W. *Tetrahedron Lett.* **1987**, *28*, 5651–5654.
[254] Stüber, W.; Knolle, J.; Breipohl, G. *Int. J. Pept. Prot. Res.* **1989**, *34*, 215–221.
[255] Funakoshi, S.; Murayama, E.; Guo, L.; Fujii, N.; Yajima, H. *J. Chem. Soc., Chem. Commun.* **1988**, 382–384.
[256] Han, Y. X.; Bontems, S. L.; Hegyes, P.; Munson, M. C.; Minor, C. A.; Kates, S. A.; Albericio, F.; Barany, G. *J. Org. Chem.* **1996**, *61*, 6326–6339.
[257] Breipohl, G.; Knolle, J.; Stüber, W. *Int. J. Pept. Prot. Res.* **1989**, *34*, 262–267.
[258] Paikoff, S. J.; Wilson, T. E.; Cho, C. Y.; Schultz, P. G. *Tetrahedron Lett.* **1996**, *37*, 5653–5656.
[259] Kim, J. M.; Bi, Y. Z.; Paikoff, S. J.; Schultz, P. G. *Tetrahedron Lett.* **1996**, *37*, 5305–5308.
[260] Kim, S. W.; Bauer, S. M.; Armstrong, R. W. *Tetrahedron Lett.* **1998**, *39*, 6993–6996.
[261] Tempest, P. A.; Brown, S. D.; Armstrong, R. W. *Angew. Chem. Int. Ed. Engl.* **1996**, *35*, 640–642.
[262] Brown, E. G.; Nuss, J. M. *Tetrahedron Lett.* **1997**, *38*, 8457–8460.
[263] Garigipati, R. S. *Tetrahedron Lett.* **1997**, *38*, 6807–6810.
[264] Edvinsson, K. M.; Herslöf, M.; Holm, P.; Kann, N.; Keeling, D. J.; Mattson, J. P.; Nordén, B.; Shcherbukhin, V. *Bioorg. Med. Chem. Lett.* **2000**, *10*, 503–507.
[265] Beaver, K. A.; Siegmund, A. C.; Spear, K. L. *Tetrahedron Lett.* **1996**, *37*, 1145–1148.
[266] Kimura, T.; Fukui, T.; Tanaka, S.; Akaji, K.; Kiso, Y. *Chem. Pharm. Bull.* **1997**, *45*, 18–26.
[267] Pátek, M.; Lebl, M. *Tetrahedron Lett.* **1991**, *32*, 3891–3894.
[268] Brik, A.; Keinan, E.; Dawson, P. E. *J. Org. Chem.* **2000**, *65*, 3829–3835.
[269] Tomkinson, N. C. O.; Sefler, A. M.; Plunket, K. D.; Blanchard, S. G.; Parks, D. J.; Willson, T. M. *Bioorg. Med. Chem. Lett.* **1997**, *7*, 2491–2496.
[270] Mergler, M.; Nyfeler, R. In *Solid Phase Synthesis*; Epton, R. Ed.; Intercept: Andover, **1992**; pp 429–432.
[271] Mohanraj, S.; Ford, W. T. *J. Org. Chem.* **1985**, *50*, 1616–1620.
[272] Vlattas, I.; Dellureficio, J.; Dunn, R.; Sytwu, I. I.; Stanton, J. *Tetrahedron Lett.* **1997**, *38*, 7321–7324.
[273] Camarero, J. A.; Cotton, G. J.; Adeva, A.; Muir, T. W. *J. Pept. Res.* **1998**, *51*, 303–316.
[274] Li, X.; Kawakami, T.; Aimoto, S. *Tetrahedron Lett.* **1998**, *39*, 8669–8672.
[275] Barn, D. R.; Morphy, J. R.; Rees, D. C. *Tetrahedron Lett.* **1996**, *37*, 3213–3216.
[276] Yang, L. H.; Guo, L. Q. *Tetrahedron Lett.* **1996**, *37*, 5041–5044.
[277] Baird, E. E.; Dervan, P. B. *J. Am. Chem. Soc.* **1996**, *118*, 6141–6146.
[278] Lelièvre, D.; Chabane, H.; Delmas, A. *Tetrahedron Lett.* **1998**, *39*, 9675–9678.
[279] Fathi, R.; Patel, R.; Cook, A. F. *Mol. Diversity* **1997**, *2*, 125–134.
[280] Bray, A. M.; Jhingran, A. G.; Valerio, R. M.; Maeji, N. J. *J. Org. Chem.* **1994**, *59*, 2197–2203.
[281] Rölfing, K.; Thiel, M.; Künzer, H. *Synlett* **1996**, 1036–1038.
[282] Prien, O.; Rölfing, K.; Thiel, M.; Künzer, H. *Synlett* **1997**, 325–326.
[283] Kaljuste, K.; Tam, J. P. *Tetrahedron Lett.* **1998**, *39*, 9327–9330.
[284] Flanigan, E.; Marshall, G. R. *Tetrahedron Lett.* **1970**, 2403–2406.
[285] Fridkin, M.; Patchornik, A.; Katchalski, E. *J. Am. Chem. Soc.* **1966**, *88*, 3164–3165.
[286] Kalir, R.; Fridkin, M.; Patchornik, A. *Eur. J. Biochem.* **1974**, *42*, 151–156.
[287] Fridkin, M.; Patchornik, A.; Katchalski, E. *J. Am. Chem. Soc.* **1965**, *87*, 4646–4648.
[288] Fridkin, M.; Hazum, E.; Kalir, R.; Rotman, M.; Koch, Y. *J. Solid-Phase Biochem.* **1977**, *2*, 175–182.
[289] Beech, C. L.; Coope, J. F.; Fairley, G.; Gilbert, P. S.; Main, B. G.; Plé, K. *J. Org. Chem.* **2001**, *66*, 2240–2245.
[290] Bourne, G. T.; Golding, S. W.; McGeary, R. P.; Meutermans, W. D. F.; Jones, A.; Marshall, G. R.; Alewood, P. F.; Smythe, M. L. *J. Org. Chem.* **2001**, *66*, 7706–7713.
[291] Breitenbucher, J. G.; Johnson, C. R.; Haight, M.; Phelan, J. C. *Tetrahedron Lett.* **1998**, *39*, 1295–1298.
[292] Boldi, A. M.; Dener, J. M.; Hopkins, T. P. *J. Comb. Chem.* **2001**, *3*, 367–373.
[293] Fantauzzi, P. P.; Yager, K. M. *Tetrahedron Lett.* **1998**, *39*, 1291–1294.
[294] Marshall, D. L.; Liener, I. E. *J. Org. Chem.* **1970**, *35*, 867–868.
[295] Parlow, J. J.; Normansell, J. E. *Mol. Diversity* **1996**, *1*, 266–269.
[296] Cohen, B. J.; Karoly-Hafeli, H.; Patchornik, A. *J. Org. Chem.* **1984**, *49*, 922–924.
[297] Hahn, H. G.; Chang, K. H.; Nam, K. D.; Bae, S. Y.; Mah, H. *Heterocycles* **1998**, *48*, 2253–2261.
[298] Masala, S.; Taddei, M. *Org. Lett.* **1999**, *1*, 1355–1357.
[299] Díaz, D. D.; Yao, S.; Finn, M. G. *Tetrahedron Lett.* **2001**, *42*, 2617–2619.
[300] Salvino, J. M.; Kumar, N. V.; Orton, E.; Airey, J.; Kiesow, T.; Crawford, K.; Mathew, R.; Krolikowski, P.; Drew, M.; Engers, D.; Krolikowski, D.; Herpin, T.; Gardyan, M.; McGeehan, G.; Labaudiniere, R. *J. Comb. Chem.* **2000**, *2*, 691–697.
[301] Gong, Y.; Becker, M.; Choi-Sledeski, Y. M.; Davis, R. S.; Salvino, J. M.; Chu, V.; Brown, K. D.; Pauls, H. W. *Bioorg. Med. Chem. Lett.* **2000**, *10*, 1033–1036.

[302] Scarr, R. B.; Findeis, M. A. *Pept. Res.* **1990**, *3*, 238–241.
[303] Sasaki, T.; Findeis, M. A.; Kaiser, E. T. *J. Org. Chem.* **1991**, *56*, 3159–3168.
[304] DeGrado, W. F.; Kaiser, E. T. *J. Org. Chem.* **1980**, *45*, 1295–1300.
[305] DeGrado, W. F.; Kaiser, E. T. *J. Org. Chem.* **1982**, *47*, 3258–3261.
[306] Voyer, N.; Lavoie, A.; Pinette, M.; Bernier, J. *Tetrahedron Lett.* **1994**, *35*, 355–358.
[307] Smith, R. A.; Bobko, M. A.; Lee, W. *Bioorg. Med. Chem. Lett.* **1998**, *8*, 2369–2374.
[308] Ösapay, G.; Taylor, J. W. *J. Am. Chem. Soc.* **1990**, *112*, 6046–6051.
[309] Ösapay, G.; Profit, A.; Taylor, J. W. *Tetrahedron Lett.* **1990**, *31*, 6121–6124.
[310] Nishino, N.; Xu, M.; Mihara, H.; Fujimoto, T.; Ohba, M.; Ueno, Y.; Kumagai, H. *J. Chem. Soc., Chem. Commun.* **1992**, 180–181.
[311] Hodge, P.; Peng, P. P. *Polymer* **1999**, *40*, 1871–1879.
[312] Kalir, R.; Warshawsky, A.; Fridkin, M.; Patchornik, A. *Eur. J. Biochem.* **1975**, *59*, 55–61.
[313] Cohen, B. J.; Kraus, M. A.; Patchornik, A. *J. Am. Chem. Soc.* **1981**, *103*, 7620–7629.
[314] Schiemann, K.; Showalter, H. D. H. *J. Org. Chem.* **1999**, *64*, 4972–4975.
[315] Mokotoff, M.; Zhao, M.; Roth, S. M.; Shelley, J. A.; Slavosky, J. N.; Kouttab, N. M. *J. Med. Chem.* **1990**, *33*, 354–360.
[316] Mokotoff, M.; Patchornik, A. *Int. J. Pept. Prot. Res.* **1983**, *21*, 145–154.
[317] Adamczyk, M.; Fishpaugh, J. R.; Mattingly, P. G. *Bioorg. Med. Chem. Lett.* **1999**, *9*, 217–220.
[318] Katoh, M.; Sodeoka, M. *Bioorg. Med. Chem. Lett.* **1999**, *9*, 881–884.
[319] Shao, H.; Zhang, Q.; Goodnow, R.; Chen, L.; Tam, S. *Tetrahedron Lett.* **2000**, *41*, 4257–4260.
[320] Pop, I. E.; Déprez, B. P.; Tartar, A. L. *J. Org. Chem.* **1997**, *62*, 2594–2603.
[321] Salmon-Chemin, L.; Lemaire, A.; De Freitas, S.; Deprez, B.; Sergheraert, C.; Davioud-Charvet, E. *Bioorg. Med. Chem. Lett.* **2000**, *10*, 631–635.
[322] Mohan, R.; Chou, Y. L.; Morrissey, M. M. *Tetrahedron Lett.* **1996**, *37*, 3963–3966.
[323] Kumari, K. A.; Sreekumar, K. *Polymer* **1996**, *37*, 171–176.
[324] Sophiamma, P. N.; Sreekumar, K. *Indian J. Chem. Sect. B - Org. Chem.* **1997**, *36*, 995–999.
[325] Adamczyk, M.; Fishpaugh, J. R.; Mattingly, P. G. *Tetrahedron Lett.* **1999**, *40*, 463–466.
[326] Dendrinos, K. G.; Kalivretenos, A. G. *Tetrahedron Lett.* **1998**, *39*, 1321–1324.
[327] Golisade, A.; Herforth, C.; Wieking, K.; Kunick, C.; Link, A. *Bioorg. Med. Chem. Lett.* **2001**, *11*, 1783–1786.
[328] Link, A.; van Calenbergh, S.; Herdewijn, P. *Tetrahedron Lett.* **1998**, *39*, 5175–5176.
[329] Overkleeft, H. S.; Bos, P. R.; Hekking, B. G.; Gordon, E. J.; Ploegh, H. L.; Kessler, B. M. *Tetrahedron Lett.* **2000**, *41*, 6005–6009.
[330] Shen, D.-M.; Shu, M.; Chapman, K. T. *Org. Lett.* **2000**, *2*, 2789–2792.
[331] Backes, B. J.; Ellman, J. A. *J. Org. Chem.* **1999**, *64*, 2322–2330.
[332] Nicolaou, K. C.; Safina, B. S.; Winssinger, N. *Synlett* **2001**, 900–903.
[333] Scialdone, M. A.; Shuey, S. W.; Soper, P.; Hamuro, Y.; Burns, D. M. *J. Org. Chem.* **1998**, *63*, 4802–4807.
[334] Hamuro, Y.; Marshall, W. J.; Scialdone, M. A. *J. Comb. Chem.* **1999**, *1*, 163–172.
[335] Scialdone, M. A. *Tetrahedron Lett.* **1996**, *37*, 8141–8144.
[336] Dressman, B. A.; Singh, U.; Kaldor, S. W. *Tetrahedron Lett.* **1998**, *39*, 3631–3634.
[337] Fitzpatrick, L. J.; Rivero, R. A. *Tetrahedron Lett.* **1997**, *38*, 7479–7482.
[338] Gomez, L.; Gellibert, F.; Wagner, A.; Mioskowski, C. *J. Comb. Chem.* **2000**, *2*, 75–79.
[339] Raju, B.; Kogan, T. P. *Tetrahedron Lett.* **1997**, *38*, 3373–3376.
[340] Maclean, D.; Hale, R.; Chen, M. *Org. Lett.* **2001**, *3*, 2977–2980.
[341] Schunk, S.; Enders, D. *Org. Lett.* **2001**, *3*, 3177–3180.
[342] Bräse, S.; Dahmen, S.; Pfefferkorn, M. *J. Comb. Chem.* **2000**, *2*, 710–715.
[343] Rosenbaum, C.; Waldmann, H. *Tetrahedron Lett.* **2001**, *42*, 5677–5680.
[344] Bräse, S.; Köbberling, J.; Enders, D.; Lazny, R.; Wang, M.; Brandtner, S. *Tetrahedron Lett.* **1999**, *40*, 2105–2108.
[345] Li, W.-R.; Lin, Y.-S.; Yo, Y.-C. *Tetrahedron Lett.* **2000**, *41*, 6619–6622.
[346] Myers, R. M.; Langston, S. P.; Conway, S. P.; Abell, C. *Org. Lett.* **2000**, *2*, 1349–1352.
[347] Meloni, M. M.; Taddei, M. *Org. Lett.* **2001**, *3*, 337–340.
[348] Gordon, K. H.; Balasubramanian, S. *Org. Lett.* **2001**, *3*, 53–56.
[349] Kappe, C. O. *Bioorg. Med. Chem. Lett.* **2000**, *10*, 49–51.
[350] Doi, T.; Hijikuro, I.; Takahashi, T. *Synlett* **1999**, 1751–1753.
[351] Floyd, C. D.; Lewis, C. N.; Patel, S. R.; Whittaker, M. *Tetrahedron Lett.* **1996**, *37*, 8045–8048.
[352] Richter, L. S.; Desai, M. C. *Tetrahedron Lett.* **1997**, *38*, 321–322.
[353] Salvino, J. M.; Mervic, M.; Mason, H. J.; Kiesow, T.; Teager, D.; Airey, J.; Labaudiniere, R. *J. Org. Chem.* **1999**, *64*, 1823–1830.
[354] Robinson, D. E.; Holladay, M. W. *Org. Lett.* **2000**, *2*, 2777–2779.

[355] Mellor, S. L.; Chan, W. C. *Chem. Commun.* **1997**, 2005–2006.
[356] Mellor, S. L.; McGuire, C.; Chan, W. C. *Tetrahedron Lett.* **1997**, *38*, 3311–3314.
[357] Bauer, U.; Ho, W. B.; Koskinen, A. M. P. *Tetrahedron Lett.* **1997**, *38*, 7233–7236.
[358] Khan, S. I.; Grinstaff, M. W. *Tetrahedron Lett.* **1998**, *39*, 8031–8034.
[359] Zhang, W.; Zhang, L.; Li, X.; Weigel, J. A.; Hall, S. E.; Mayer, J. P. *J. Comb. Chem.* **2001**, *3*, 151–153.
[360] Thouin, E.; Lubell, W. D. *Tetrahedron Lett.* **2000**, *41*, 457–460.
[361] Bui, C. T.; Maeji, N. J.; Bray, A. M. *Biotechnol. Bioeng. (Comb. Chem.)* **2000**, *71*, 91–93.
[362] Barlaam, B.; Koza, P.; Berriot, J. *Tetrahedron* **1999**, *55*, 7221–7232.
[363] Ngu, K.; Patel, D. V. *J. Org. Chem.* **1997**, *62*, 7088–7089.
[364] Dankwardt, S. M. *Synlett* **1998**, 761.
[365] Dankwardt, S. M.; Billedeau, R. J.; Lawley, L. K.; Abbot, S. C.; Martin, R. L.; Chan, C. S.; Van Wart, H. E.; Walker, K. A. M. *Bioorg. Med. Chem. Lett.* **2000**, *10*, 2513–2516.
[366] Chang, J. K.; Shimizu, M.; Wang, S. S. *J. Org. Chem.* **1976**, *41*, 3255–3258.
[367] Kessler, W.; Iselin, B. *Helv. Chim. Acta* **1966**, *49*, 1330–1344.
[368] Ohno, M.; Anfinsen, C. B. *J. Am. Chem. Soc.* **1967**, *89*, 5994–5995.
[369] Takahashi, T.; Tomida, S.; Inoue, H.; Doi, T. *Synlett* **1998**, 1261–1263.
[370] Hunt, J. A.; Roush, W. R. *J. Am. Chem. Soc.* **1996**, *118*, 9998–9999.
[371] Nicolaou, K. C.; Baran, P. S.; Zhong, Y.-L. *J. Am. Chem. Soc.* **2000**, *122*, 10246–10248.
[372] Takahashi, T.; Inoue, H.; Yamamura, Y.; Doi, T. *Angew. Chem. Int. Ed.* **2001**, *40*, 3230–3233.
[373] Rademann, J.; Smerdka, J.; Jung, G.; Grosche, P.; Schmid, D. *Angew. Chem. Int. Ed.* **2001**, *40*, 381–385.
[374] Beyerman, H. C.; Hindriks, H.; De Leer, E. W. B. *J. Chem. Soc., Chem. Commun.* **1968**, 1668.
[375] de Bont, D. B. A.; Moree, W. J.; Liskamp, R. M. J. *Bioorg. Med. Chem.* **1996**, *4*, 667–672.
[376] Rothe, M.; Zieger, M. *Tetrahedron Lett.* **1994**, *35*, 9011–9012.
[377] Ingenito, R.; Bianchi, E.; Fattori, D.; Pessi, A. *J. Am. Chem. Soc.* **1999**, *121*, 11369–11374.
[378] Quaderer, R.; Hilvert, D. *Org. Lett.* **2001**, *3*, 3181–3184.
[379] Swinnen, D.; Hilvert, D. *Org. Lett.* **2000**, *2*, 2439–2442.
[380] Sewing, A.; Hilvert, D. *Angew. Chem. Int. Ed.* **2001**, *40*, 3395–3396.
[381] Bhalay, G.; Blaney, P.; Palmer, V. H.; Baxter, A. D. *Tetrahedron Lett.* **1997**, *38*, 8375–8378.
[382] Annis, D. A.; Helluin, O.; Jacobsen, E. N. *Angew. Chem. Int. Ed.* **1998**, *37*, 1907–1909.
[383] Bryan, W. M.; Huffman, W. F.; Bhatnagar, P. K. *Tetrahedron Lett.* **2000**, *41*, 6997–7000.
[384] Frenette, R.; Friesen, R. W. *Tetrahedron Lett.* **1994**, *35*, 9177–9180.
[385] Shimizu, H.; Ito, Y.; Kanie, O.; Ogawa, T. *Bioorg. Med. Chem. Lett.* **1996**, *6*, 2841–2846.
[386] Beyerman, H. C.; Kranenburg, P.; Syrier, J. L. M. *Recl. Trav. Chim. Pays-Bas* **1971**, *90*, 791–800.
[387] Barton, M. A.; Lemieux, R. U.; Savoie, J. Y. *J. Am. Chem. Soc.* **1973**, *95*, 4501–4506.
[388] O'Donnell, M. J.; Zhou, C. Y.; Scott, W. L. *J. Am. Chem. Soc.* **1996**, *118*, 6070–6071.
[389] Sylvain, C.; Wagner, A.; Mioskowski, C. *Tetrahedron Lett.* **1997**, *38*, 1043–1044.
[390] Hutchins, S. M.; Chapman, K. T. *Tetrahedron Lett.* **1996**, *37*, 4869–4872.
[391] Cheng, Y.; Chapman, K. T. *Tetrahedron Lett.* **1997**, *38*, 1497–1500.
[392] Reichwein, J. F.; Liskamp, R. M. J. *Tetrahedron Lett.* **1998**, *39*, 1243–1246.
[393] Far, A. R.; Tidwell, T. T. *J. Org. Chem.* **1998**, *63*, 8636–8637.
[394] Hulme, C.; Ma, L.; Cherrier, M.-P.; Romano, J. J.; Morton, G.; Duquenne, C.; Salvino, J.; Labaudiniere, R. *Tetrahedron Lett.* **2000**, *41*, 1883–1887.
[395] Colwell, A. R.; Duckwall, L. R.; Brooks, R.; McManus, S. P. *J. Org. Chem.* **1981**, *46*, 3097–3102.
[396] Kobayashi, S.; Wakabayashi, T.; Yasuda, M. *J. Org. Chem.* **1998**, *63*, 4868–4869.
[397] Hanessian, S.; Ma, J.; Wang, W. *Tetrahedron Lett.* **1999**, *40*, 4631–4634.
[398] Reggelin, M.; Brenig, V.; Welcker, R. *Tetrahedron Lett.* **1998**, *39*, 4801–4804.
[399] Le Hetet, C.; David, M.; Carreaux, F.; Carboni, B.; Sauleau, A. *Tetrahedron Lett.* **1997**, *38*, 5153–5156.
[400] Ma, S.; Duan, D.; Shi, Z. *Org. Lett.* **2000**, *2*, 1419–1422.
[401] Vignola, N.; Dahmen, S.; Enders, D.; Bräse, S. *Tetrahedron Lett.* **2001**, *42*, 7833–7836.
[402] Moon, H.; Schore, N. E.; Kurth, M. J. *J. Org. Chem.* **1992**, *57*, 6088–6089.
[403] Moon, H.; Schore, N. E.; Kurth, M. J. *Tetrahedron Lett.* **1994**, *35*, 8915–8918.
[404] Ko, D. H.; Kim, D. J.; Lyu, C. S.; Min, I. K.; Moon, H. S. *Tetrahedron Lett.* **1998**, *39*, 297–300.
[405] Gouault, N.; Cupif, J.-F.; Sauleau, A.; David, M. *Tetrahedron Lett.* **2000**, *41*, 7293–7297.
[406] Watanabe, Y.; Ishikawa, S.; Takao, G.; Toru, T. *Tetrahedron Lett.* **1999**, *40*, 3411–3414.
[407] Conti, P.; Demont, D.; Cals, J.; Ottenheijm, H. C. J.; Leysen, D. *Tetrahedron Lett.* **1997**, *38*, 2915–2918.
[408] Manov, N.; Bienz, S. *Tetrahedron* **2001**, *57*, 7893–7898.
[409] Kobayashi, S.; Aoki, Y. *Tetrahedron Lett.* **1998**, *39*, 7345–7348.

[410] Purandare, A. V.; Poss, M. A. *Tetrahedron Lett.* **1998**, *39*, 935–938.
[411] Tommasi, R. A.; Nantermet, P. G.; Shapiro, M. J.; Chin, J.; Brill, W. K. D.; Ang, K. *Tetrahedron Lett.* **1998**, *39*, 5477–5480.
[412] Ostresh, J. M.; Schoner, C. C.; Hamashin, V. T.; Nefzi, A.; Meyer, J. P.; Houghten, R. A. *J. Org. Chem.* **1998**, *63*, 8622–8623.
[413] Nefzi, A.; Ostresh, J. M.; Houghten, R. A. *Tetrahedron* **1999**, *55*, 335–344.
[414] Manku, S.; Laplante, C.; Kopac, D.; Chan, T.; Hall, D. G. *J. Org. Chem.* **2001**, *66*, 874–885.
[415] Gauzy, L.; Le Merrer, Y.; Depezay, J. C.; Clerc, F.; Mignani, S. *Tetrahedron Lett.* **1999**, *40*, 6005–6008.
[416] Katritzky, A. R.; Xie, L.; Zhang, G.; Griffith, M.; Watson, K.; Kiely, J. S. *Tetrahedron Lett.* **1997**, *38*, 7011–7014.
[417] Boyd, E. A.; Chan, W. C.; Loh, V. M. *Tetrahedron Lett.* **1996**, *37*, 1647–1650.
[418] Youngman, M. A.; Dax, S. L. *Tetrahedron Lett.* **1997**, *38*, 6347–6350.
[419] McNally, J. J.; Youngman, M. A.; Dax, S. L. *Tetrahedron Lett.* **1998**, *39*, 967–970.
[420] Hoekstra, W. J.; Maryanoff, B. E.; Andrade-Gordon, P.; Cohen, J. H.; Costanzo, M. J.; Damiano, B. P.; Haertlein, B. J.; Harris, B. D.; Kauffman, J. A.; Keane, P. M.; McComsey, D. F.; Villani, F. J.; Yabut, S. C. *Bioorg. Med. Chem. Lett.* **1996**, *6*, 2371–2376.
[421] Hoekstra, W. J.; Greco, M. N.; Yabut, S. C.; Hulshizer, B. L.; Maryanoff, B. E. *Tetrahedron Lett.* **1997**, *38*, 2629–2632.
[422] Strømgaard, K.; Brier, T. J.; Andersen, K.; Mellor, I. R.; Saghyan, A.; Tikhonov, D.; Usherwood, P. N. R.; Krogsgaard-Larsen, P.; Jaroszewski, J. W. *J. Med. Chem.* **2000**, *43*, 4526–4533.
[423] Barlos, K.; Gatos, D.; Kallitsis, I.; Papaioannou, D.; Sotiriu, P. *Liebigs Ann. Chem.* **1988**, 1079–1081.
[424] Bleicher, K. H.; Wareing, J. R. *Tetrahedron Lett.* **1998**, *39*, 4591–4594.
[425] Gosselin, F.; Betsbrugge, J. V.; Hatam, M.; Lubell, W. D. *J. Org. Chem.* **1999**, *64*, 2486–2493.
[426] Hidai, Y.; Kan, T.; Fukuyama, T. *Tetrahedron Lett.* **1999**, *40*, 4711–4714.
[427] Hauske, J. R.; Dorff, P. *Tetrahedron Lett.* **1995**, *36*, 1589–1592.
[428] Smith, A. L.; Thomson, C. G.; Leeson, P. D. *Bioorg. Med. Chem. Lett.* **1996**, *6*, 1483–1486.
[429] Marsh, I. R.; Smith, H. K.; Leblanc, C.; Bradley, M. *Mol. Diversity* **1997**, *2*, 165–170.
[430] Dixit, D. M.; Leznoff, C. C. *Israel J. Chem.* **1978**, *17*, 248–252.
[431] Dressman, B. A.; Spangle, L. A.; Kaldor, S. W. *Tetrahedron Lett.* **1996**, *37*, 937–940.
[432] Kaljuste, K.; Undén, A. *Tetrahedron Lett.* **1995**, *36*, 9211–9214.
[433] Dixit, D. M.; Leznoff, C. C. *J. Chem. Soc., Chem. Commun.* **1977**, 798–799.
[434] Ho, C. Y.; Kukla, M. J. *Tetrahedron Lett.* **1997**, *38*, 2799–2802.
[435] Zaragoza, F.; Petersen, S. V. *Tetrahedron* **1996**, *52*, 5999–6002.
[436] Meester, W. J. N.; Rutjes, F. P. J. T.; Hermkens, P. H. H.; Hiemstra, H. *Tetrahedron Lett.* **1999**, *40*, 1601–1604.
[437] Wang, F.; Hauske, J. R. *Tetrahedron Lett.* **1997**, *38*, 6529–6532.
[438] Cuny, G. D.; Cao, J.; Hauske, J. R. *Tetrahedron Lett.* **1997**, *38*, 5237–5240.
[439] Rotella, D. P. *J. Am. Chem. Soc.* **1996**, *118*, 12246–12247.
[440] Hiroshige, M.; Hauske, J. R.; Zhou, P. *J. Am. Chem. Soc.* **1995**, *117*, 11590–11591.
[441] Yang, L. *Tetrahedron Lett.* **2000**, *41*, 6981–6984.
[442] Stephensen, H.; Zaragoza, F. *J. Org. Chem.* **1997**, *62*, 6096–6097.
[443] Veerman, J. J. N.; Rutjes, F. P. J. T.; van Maarseveen, J. H.; Hiemstra, H. *Tetrahedron Lett.* **1999**, *40*, 6079–6082.
[444] van Maarseveen, J. H.; Meester, W. J. N.; Veerman, J. J. N.; Kruse, C. G.; Hermkens, P. H. H.; Rutjes, F. P. J. T.; Hiemstra, H. *J. Chem. Soc., Perkin Trans. 1* **2001**, 994–1001.
[445] Chen, C.; Munoz, B. *Tetrahedron Lett.* **1998**, *39*, 3401–3404.
[446] Keana, J. F. W.; Shimizu, M.; Jernsted, K. K. *J. Org. Chem.* **1986**, *51*, 1641–1644.
[447] McMinn, D. L.; Greenberg, M. M. *Tetrahedron* **1996**, *52*, 3827–3840.
[448] McKeown, S. C.; Watson, S. P.; Carr, R. A. E.; Marshall, P. *Tetrahedron Lett.* **1999**, *40*, 2407–2410.
[449] Johansson, A.; Åkerblom, E.; Ersmark, K.; Lindeberg, G.; Hallberg, A. *J. Comb. Chem.* **2000**, *2*, 496–507.
[450] Kaljuste, K.; Undén, A. *Tetrahedron Lett.* **1996**, *37*, 3031–3034.
[451] Chen, C.; McDonald, I. A.; Munoz, B. *Tetrahedron Lett.* **1998**, *39*, 217–220.
[452] Canne, L. E.; Winston, R. L.; Kent, S. B. H. *Tetrahedron Lett.* **1997**, *38*, 3361–3364.
[453] Lipshutz, B. H.; Shin, Y.-J. *Tetrahedron Lett.* **2001**, *42*, 5629–5633.
[454] Bursavich, M. G.; Rich, D. H. *Org. Lett.* **2001**, *3*, 2625–2628.
[455] Nelson, J. C.; Young, J. K.; Moore, J. S. *J. Org. Chem.* **1996**, *61*, 8160–8168.
[456] Drinnan, N.; West, M. L.; Broadhurst, M.; Kellam, B.; Toth, I. *Tetrahedron Lett.* **2001**, *42*, 1159–1162.

[457] Kirchhoff, J. H.; Bräse, S.; Enders, D. *J. Comb. Chem.* **2001**, *3*, 71–77.
[458] Chhabra, S. R.; Khan, A. N.; Bycroft, B. W. *Tetrahedron Lett.* **1998**, *39*, 3585–3588.
[459] Chhabra, S. R.; Khan, A. N.; Bycroft, B. W. *Tetrahedron Lett.* **2000**, *41*, 1099–1102.
[460] Bannwarth, W.; Huebscher, J.; Barner, R. *Bioorg. Med. Chem. Lett.* **1996**, *6*, 1525–1528.
[461] Furth, P. S.; Reitman, M. S.; Gentles, R.; Cook, A. F. *Tetrahedron Lett.* **1997**, *38*, 6643–6646.
[462] Furth, P. S.; Reitman, M. S.; Cook, A. F. *Tetrahedron Lett.* **1997**, *38*, 5403–5406.
[463] Kay, C.; Murray, P. J.; Sandow, L.; Holmes, A. B. *Tetrahedron Lett.* **1997**, *38*, 6941–6944.
[464] Williams, G. M.; Carr, R. A. E.; Congreve, M. S.; Kay, C.; McKeown, S. C.; Murray, P. J.; Scicinski, J. J.; Watson, S. P. *Angew. Chem. Int. Ed.* **2000**, *39*, 3293–3296.
[465] Dragoli, D. R.; Burdett, M. T.; Ellman, J. A. *J. Am. Chem. Soc.* **2001**, *123*, 10127–10128.
[466] Kuster, G. J.; Scheeren, H. W. *Tetrahedron Lett.* **2000**, *41*, 515–519.
[467] Gordon, K.; Bolger, M.; Khan, N.; Balasubramanian, S. *Tetrahedron Lett.* **2000**, *41*, 8621–8625.
[468] Brown, A. R. *J. Comb. Chem.* **1999**, *1*, 283–285.
[469] Yamamoto, Y.; Tanabe, K.; Okonogi, T. *Chem. Lett.* **1999**, 103–104.
[470] Alhambra, C.; Castro, J.; Chiara, J. L.; Fernández, E.; Fernández-Mayoralas, A.; Fiandor, J. M.; García-Ochoa, S.; Martín-Ortega, M. D. *Tetrahedron Lett.* **2001**, *42*, 6675–6678.
[471] Barco, A.; Benetti, S.; De Risi, C.; Marchetti, P.; Pollini, G. P.; Zanirato, V. *Tetrahedron Lett.* **1998**, *39*, 7591–7594.
[472] Brown, A. R.; Rees, D. C.; Rankovic, Z.; Morphy, J. R. *J. Am. Chem. Soc.* **1997**, *119*, 3288–3295.
[473] Ouyang, X. H.; Armstrong, R. W.; Murphy, M. M. *J. Org. Chem.* **1998**, *63*, 1027–1032.
[474] Kroll, F. E. K.; Morphy, R.; Rees, D.; Gani, D. *Tetrahedron Lett.* **1997**, *38*, 8573–8576.
[475] Heinonen, P.; Virta, P.; Lönnberg, H. *Tetrahedron* **1999**, *55*, 7613–7624.
[476] Rueter, J. K.; Nortey, S. O.; Baxter, E. W.; Leo, G. C.; Reitz, A. B. *Tetrahedron Lett.* **1998**, *39*, 975–978.
[477] Bicak, N.; Senkal, B. F. *Reactive and Functional Polymers* **1996**, *29*, 123–128.
[478] Schürer, S. C.; Blechert, S. *Synlett* **1998**, 166–168.
[479] Fisher, M.; Brown, R. C. D. *Tetrahedron Lett.* **2001**, *42*, 8227–8230.
[480] Katritzky, A. R.; Belyakov, S. A.; Tymoshenko, D. O. *J. Comb. Chem.* **1999**, *1*, 173–176.
[481] Paio, A.; Zaramella, A.; Ferritto, R.; Conti, N.; Marchioro, C.; Seneci, P. *J. Comb. Chem.* **1999**, *1*, 317–325.
[482] Blaney, P.; Grigg, R.; Rankovic, Z.; Thoroughgood, M. *Tetrahedron Lett.* **2000**, *41*, 6639–6642.
[483] Blaney, P.; Grigg, R.; Rankovic, Z.; Thoroughgood, M. *Tetrahedron Lett.* **2000**, *41*, 6635–6638.
[484] Gustafsson, M.; Olsson, R.; Andersson, C.-M. *Tetrahedron Lett.* **2001**, *42*, 133–136.
[485] Cai, J.; Wathey, B. *Tetrahedron Lett.* **2001**, *42*, 1383–1385.
[486] Heinelt, U.; Herok, S.; Matter, H.; Wildgoose, P. *Bioorg. Med. Chem. Lett.* **2001**, *11*, 227–230.
[487] Smith, A. L.; Stevenson, G. I.; Swain, C. J.; Castro, J. L. *Tetrahedron Lett.* **1998**, *39*, 8317–8320.
[488] Gray, N. S.; Kwon, S.; Schultz, P. G. *Tetrahedron Lett.* **1997**, *38*, 1161–1164.
[489] Gordeev, M. F.; Patel, D. V.; Gordon, E. M. *J. Org. Chem.* **1996**, *61*, 924–928.
[490] Gordeev, M. F.; Patel, D. V.; England, B. P.; Jonnalagadda, S.; Combs, J. D.; Gordon, E. M. *Bioorg. Med. Chem.* **1998**, *6*, 883–889.
[491] Dorff, P. H.; Garigipati, R. S. *Tetrahedron Lett.* **2001**, *42*, 2771–2773.
[492] Krchňák, V.; Smith, J.; Vágner, J. *Tetrahedron Lett.* **2001**, *42*, 2443–2446.
[493] Kearney, P. C.; Fernandez, M.; Flygare, J. A. *J. Org. Chem.* **1998**, *63*, 196–200.
[494] Krchňák, V.; Szabo, L.; Vágner, J. *Tetrahedron Lett.* **2000**, *41*, 2835–2838.
[495] Mazurov, A. *Bioorg. Med. Chem. Lett.* **2000**, *10*, 67–70.
[496] Wu, Z.; Kim, J.; Soll, R. M.; Dhanoa, D. S. *Biotechnol. Bioeng. (Comb. Chem.)* **2000**, *71*, 87–90.
[497] Di Lucrezia, R.; Gilbert, I. H.; Floyd, C. D. *J. Comb. Chem.* **2000**, *2*, 249–253.
[498] Buchstaller, H. P. *Tetrahedron* **1998**, *54*, 3465–3470.
[499] García Echeverría, C. *Tetrahedron Lett.* **1997**, *38*, 8933–8934.
[500] Sunami, S.; Sagara, T.; Ohkubo, M.; Morishima, H. *Tetrahedron Lett.* **1999**, *40*, 1721–1724.
[501] Burdick, D. J.; Struble, M. E.; Burnier, J. P. *Tetrahedron Lett.* **1993**, *34*, 2589–2592.
[502] Zucca, C.; Bravo, P.; Volonterio, A.; Zanda, M.; Wagner, A.; Mioskowski, C. *Tetrahedron Lett.* **2001**, *42*, 1033–1035.
[503] Han, H. S.; Wolfe, M. M.; Brenner, S.; Janda, K. D. *Proc. Natl. Acad. Sci. USA* **1995**, *92*, 6419–6423.
[504] Grether, U.; Waldmann, H. *Angew. Chem. Int. Ed.* **2000**, *39*, 1629–1632.
[505] Grether, U.; Waldmann, H. *Chem. Eur. J.* **2001**, *7*, 959–971.
[506] Pattarawarapan, M.; Chen, J.; Steffensen, M.; Burgess, K. *J. Comb. Chem.* **2001**, *3*, 102–116.
[507] Masquelin, T.; Sprenger, D.; Baer, R.; Gerber, F.; Mercadal, Y. *Helv. Chim. Acta* **1998**, *81*, 646–660.
[508] Obrecht, D.; Abrecht, C.; Grieder, A.; Villalgordo, J. M. *Helv. Chim. Acta* **1997**, *80*, 65–72.

[509] Gayo, L. M.; Suto, M. J. *Tetrahedron Lett.* **1997**, *38*, 211–214.
[510] Brun, V.; Legraverend, M.; Grierson, D. S. *Tetrahedron Lett.* **2001**, *42*, 8161–8164.
[511] Brun, V.; Legraverend, M.; Grierson, D. S. *Tetrahedron Lett.* **2001**, *42*, 8165–8167.
[512] Brun, V.; Legraverend, M.; Grierson, D. S. *Tetrahedron Lett.* **2001**, *42*, 8169–8171.
[513] Masquelin, T.; Meunier, N.; Gerber, F.; Rossé, G. *Heterocycles* **1998**, *48*, 2489–2505.
[514] Parrot, I.; Wermuth, C.-G.; Hibert, M. *Tetrahedron Lett.* **1999**, *40*, 7975–7978.
[515] Lorthioir, O.; McKeown, S. C.; Parr, N. J.; Washington, M.; Watson, S. P. *Tetrahedron Lett.* **2000**, *41*, 8609–8613.
[516] Scicinski, J. J.; Congreve, M. S.; Jamieson, C.; Ley, S. V.; Newman, E. S.; Vinader, V. M.; Carr, R. A. E. *J. Comb. Chem.* **2001**, *3*, 387–396.
[517] Fields, G. B.; Noble, R. L. *Int. J. Pept. Prot. Res.* **1990**, *35*, 161–214.
[518] Zhong, H. M.; Greco, M. N.; Maryanoff, B. E. *J. Org. Chem.* **1997**, *62*, 9326–9330.
[519] Josey, J. A.; Tarlton, C. A.; Payne, C. E. *Tetrahedron Lett.* **1998**, *39*, 5899–5902.
[520] Pátek, M.; Smrcina, M.; Nakanishi, E.; Izawa, H. *J. Comb. Chem.* **2000**, *2*, 370–377.
[521] Ghosh, A. K.; Hol, W. G. J.; Fan, E. *J. Org. Chem.* **2001**, *66*, 2161–2164.
[522] Mohan, R.; Yun, W. Y.; Buckman, B. O.; Liang, A.; Trinh, L.; Morrissey, M. M. *Bioorg. Med. Chem. Lett.* **1998**, *8*, 1877–1882.
[523] Roussel, P.; Bradley, M.; Matthews, I.; Kane, P. *Tetrahedron Lett.* **1997**, *38*, 4861–4864.
[524] Kim, S. W.; Hong, C. Y.; Koh, J. S.; Lee, E. J.; Lee, K. *Mol. Diversity* **1998**, *3*, 133–136.
[525] Gomez, L.; Gellibert, F.; Wagner, A.; Mioskowski, C. *Chem. Eur. J.* **2000**, *6*, 4016–4020.
[526] Dahmen, S.; Bräse, S. *Org. Lett.* **2000**, *2*, 3563–3565.
[527] Acharya, A. N.; Nefzi, A.; Ostresh, J. M.; Houghten, R. A. *J. Comb. Chem.* **2001**, *3*, 189–195.
[528] Li, M.; Wilson, L. J.; Portlock, D. E. *Tetrahedron Lett.* **2001**, *42*, 2273–2275.
[529] Wilson, L. J.; Klopfenstein, S. R.; Li, M. *Tetrahedron Lett.* **1999**, *40*, 3999–4002.
[530] Bonnat, M.; Bradley, M.; Kilburn, J. D. *Tetrahedron Lett.* **1996**, *37*, 5409–5412.
[531] Davies, M.; Bonnat, M.; Guillier, F.; Kilburn, J. D.; Bradley, M. *J. Org. Chem.* **1998**, *63*, 8696–8703.
[532] Zapf, C. W.; Creighton, C. J.; Tomioka, M.; Goodman, M. *Org. Lett.* **2001**, *3*, 1133–1136.
[533] Dodd, D. S.; Wallace, O. B. *Tetrahedron Lett.* **1998**, *39*, 5701–5704.
[534] Eleftheriou, S.; Gatos, D.; Panagopoulos, A.; Stathopoulos, S.; Barlos, K. *Tetrahedron Lett.* **1999**, *40*, 2825–2828.
[535] Saha, A. K.; Liu, L.; Marichal, P.; Odds, F. *Bioorg. Med. Chem. Lett.* **2000**, *10*, 2735–2739.
[536] Saha, A. K.; Liu, L.; Simoneaux, R. L.; Kukla, M. J.; Marichal, P.; Odds, F. *Bioorg. Med. Chem. Lett.* **2000**, *10*, 2175–2178.
[537] Gelens, E.; Koot, W. J.; Menge, W. M. P. B.; Ottenheijm, H. C. J.; Timmerman, H. *Bioorg. Med. Chem. Lett.* **2000**, *10*, 1935–1938.
[538] Glass, J. D.; Schwartz, I. L.; Walter, R. *J. Am. Chem. Soc.* **1972**, *94*, 6209–6211.
[539] Nugiel, D. A.; Cornelius, L. A. M.; Corbett, J. W. *J. Org. Chem.* **1997**, *62*, 201–203.
[540] Yoo, S. E.; Seo, J. S.; Yi, K. Y.; Gong, Y. D. *Tetrahedron Lett.* **1997**, *38*, 1203–1206.
[541] Hübner, H.; Kraxner, J.; Gmeiner, P. *J. Med. Chem.* **2000**, *43*, 4563–4569.
[542] Kraxner, J.; Arlt, M.; Gmeiner, P. *Synlett* **2000**, 125–127.
[543] Smith, A. L.; Stevenson, G. I.; Lewis, S.; Patel, S.; Castro, J. L. *Bioorg. Med. Chem. Lett.* **2000**, *10*, 2693–2696.
[544] Bilodeau, M. T.; Cunningham, A. M. *J. Org. Chem.* **1998**, *63*, 2800–2801.
[545] Sabatino, G.; Chelli, M.; Mazzucco, S.; Ginanneschi, M.; Papini, A. M. *Tetrahedron Lett.* **1999**, *40*, 809–812.
[546] Tumelty, D.; Cao, K.; Holmes, C. P. *Org. Lett.* **2001**, *3*, 83–86.
[547] Zhang, H.-C.; Ye, H.; Moretto, A. F.; Brumfield, K. K.; Maryanoff, B. E. *Org. Lett.* **2000**, *2*, 89–92.
[548] Wu, T. Y. H.; Ding, S.; Gray, N. S.; Schultz, P. G. *Org. Lett.* **2001**, *3*, 3827–3830.
[549] Brill, W. K.-D.; Riva-Toniolo, C. *Tetrahedron Lett.* **2001**, *42*, 6515–6518.
[550] Kim, K.; Wang, B. *Chem. Commun.* **2001**, 2268–2269.
[551] Matthews, D. P.; Green, J. E.; Shuker, A. J. *J. Comb. Chem.* **2000**, *2*, 19–23.
[552] Lee, C. E.; Kick, E. K.; Ellman, J. A. *J. Am. Chem. Soc.* **1998**, *120*, 9735–9747.
[553] Phoon, C. W.; Abell, C. *J. Comb. Chem.* **1999**, *1*, 485–492.
[554] Richter, L. S.; Gadek, T. R. *Tetrahedron Lett.* **1994**, *35*, 4705–4706.
[555] Hamper, B. C.; Dukesherer, D. R.; South, M. S. *Tetrahedron Lett.* **1996**, *37*, 3671–3674.
[556] Hollinshead, S. P. *Tetrahedron Lett.* **1996**, *37*, 9157–9160.
[557] Garigipati, R. S.; Adams, B.; Adams, J. L.; Sarkar, S. K. *J. Org. Chem.* **1996**, *61*, 2911–2914.
[558] Hanessian, S.; Xie, F. *Tetrahedron Lett.* **1998**, *39*, 737–740.
[559] Hanessian, S.; Xie, F. *Tetrahedron Lett.* **1998**, *39*, 733–736.
[560] Furman, B.; Thürmer, R.; Kaluza, Z.; Lysek, R.; Voelter, W.; Chmielewski, M. *Angew. Chem. Int. Ed.* **1999**, *38*, 1121–1123.

[561] Rottländer, M.; Knochel, P. *J. Comb. Chem.* **1999**, *1*, 181–183.
[562] Craig, D.; Robson, M. J.; Shaw, S. J. *Synlett* **1998**, 1381–1383.
[563] Hanessian, S.; Huynh, H. K. *Tetrahedron Lett.* **1999**, *40*, 671–674.
[564] Mergler, M.; Dick, F.; Gosteli, J.; Nyfeler, R. *Tetrahedron Lett.* **1999**, *40*, 4663–4664.
[565] Shankar, B. B.; Yang, D. Y.; Girton, S.; Ganguly, A. K. *Tetrahedron Lett.* **1998**, *39*, 2447–2448.
[566] Zhu, Z. M.; McKittrick, B. *Tetrahedron Lett.* **1998**, *39*, 7479–7482.
[567] Fréchet, J. M. J.; Nuyens, L. J. *Can. J. Chem.* **1976**, *54*, 926–934.
[568] Fréchet, J. M. J.; Haque, K. E. *Tetrahedron Lett.* **1975**, 3055–3056.
[569] Hayatsu, H.; Khorana, H. G. *J. Am. Chem. Soc.* **1966**, *88*, 3182–3183.
[570] Hayatsu, H.; Khorana, H. G. *J. Am. Chem. Soc.* **1967**, *89*, 3880–3887.
[571] Chen, C.; Randall, L. A. A.; Miller, R. B.; Jones, A. D.; Kurth, M. J. *J. Am. Chem. Soc.* **1994**, *116*, 2661–2662.
[572] Pernerstorfer, J.; Schuster, M.; Blechert, S. *Synthesis* **1999**, 138–144.
[573] Stauffer, S. R.; Katzenellenbogen, J. A. *J. Comb. Chem.* **2000**, *2*, 318–329.
[574] Ito, Y.; Ogawa, T. *J. Am. Chem. Soc.* **1997**, *119*, 5562–5566.
[575] Brill, W. K. D.; Schmidt, E.; Tommasi, R. A. *Synlett* **1998**, 906–908.
[576] Krchnák, V.; Weichsel, A. S. *Tetrahedron Lett.* **1997**, *38*, 7299–7302.
[577] Fyles, T. M.; Leznoff, C. C.; Weatherston, J. *Can. J. Chem.* **1977**, *55*, 4135–4143.
[578] Wang, Y.; Zhang, H.; Voelter, W. *Chem. Lett.* **1995**, 273–274.
[579] Nicolaou, K. C.; Winssinger, N.; Pastor, J.; DeRoose, F. *J. Am. Chem. Soc.* **1997**, *119*, 449–450.
[580] Zehavi, U.; Patchornik, A. *J. Am. Chem. Soc.* **1973**, *95*, 5673–5677.
[581] Fukase, K.; Egusa, K.; Nakai, Y.; Kusumoto, S. *Mol. Diversity* **1997**, *2*, 182–188.
[582] Fukase, K.; Nakai, Y.; Egusa, K.; Porco, J. A.; Kusumoto, S. *Synlett* **1999**, 1074–1078.
[583] Reiser, U.; Jauch, J. *Synlett* **2001**, 90–92.
[584] Nestler, H. P.; Bartlett, P. A.; Still, W. C. *J. Org. Chem.* **1994**, *59*, 4723–4724.
[585] Yamamoto, Y.; Ajito, K.; Ohtsuka, Y. *Chem. Lett.* **1998**, 379–380.
[586] Cabrele, C.; Langer, M.; Beck-Sickinger, A. G. *J. Org. Chem.* **1999**, *64*, 4353–4361.
[587] Tempest, P. A.; Armstrong, R. W. *J. Am. Chem. Soc.* **1997**, *119*, 7607–7608.
[588] Yun, W. Y.; Mohan, R. *Tetrahedron Lett.* **1996**, *37*, 7189–7192.
[589] Phillips, G. B.; Wei, G. P. *Tetrahedron Lett.* **1996**, *37*, 4887–4890.
[590] Plunkett, M. J.; Ellman, J. A. *J. Am. Chem. Soc.* **1995**, *117*, 3306–3307.
[591] Devraj, R.; Cushman, M. *J. Org. Chem.* **1996**, *61*, 9368–9373.
[592] Burbaum, J. J.; Ohlmeyer, M. H. J.; Reader, J. C.; Henderson, I.; Dillard, L. W.; Li, G.; Randle, T. L.; Sigal, N. H.; Chelsky, D.; Baldwin, J. J. *Proc. Natl. Acad. Sci. USA* **1995**, *92*, 6027–6031.
[593] Tremblay, M. R.; Poirier, D. *Tetrahedron Lett.* **1999**, *40*, 1277–1280.
[594] Tremblay, M. R.; Poirier, D. *J. Comb. Chem.* **2000**, *2*, 48–65.
[595] Maltais, R.; Tremblay, M. R.; Poirier, D. *J. Comb. Chem.* **2000**, *2*, 604–614.
[596] Heinze, K. *Chem. Eur. J.* **2001**, *7*, 2922–2932.
[597] Farrall, M. J.; Fréchet, J. M. J. *J. Org. Chem.* **1976**, *41*, 3877–3882.
[598] Hu, Y.; Porco, J. A.; Labadie, J. W.; Gooding, O. W.; Trost, B. M. *J. Org. Chem.* **1998**, *63*, 4518–4521.
[599] Stranix, B. R.; Liu, H. Q.; Darling, G. D. *J. Org. Chem.* **1997**, *62*, 6183–6186.
[600] Zheng, C.; Seeberger, P. H.; Danishefsky, S. J. *J. Org. Chem.* **1998**, *63*, 1126–1130.
[601] Doi, T.; Sugiki, M.; Yamada, H.; Takahashi, T.; Porco, J. A. *Tetrahedron Lett.* **1999**, *40*, 2141–2144.
[602] Tremblay, M. R.; Wentworth, P.; Lee, G. E.; Janda, K. D. *J. Comb. Chem.* **2000**, *2*, 698–709.
[603] Lee, D.; Sello, J. K.; Schreiber, S. L. *Org. Lett.* **2000**, *2*, 709–712.
[604] Fréchet, J. M. J.; Darling, G. D.; Itsuno, S.; Lu, P. Z.; de Meftahi, M. V.; Rolls, W. A. *Pure Appl. Chem.* **1988**, *60*, 353–364.
[605] Thompson, L. A.; Moore, F. L.; Moon, Y. C.; Ellman, J. A. *J. Org. Chem.* **1998**, *63*, 2066–2067.
[606] Chan, T.; Huang, W. *J. Chem. Soc., Chem. Commun.* **1985**, 909–911.
[607] Hu, Y. H.; Porco, J. A. *Tetrahedron Lett.* **1998**, *39*, 2711–2714.
[608] Missio, A.; Marchioro, C.; Rossi, T.; Panunzio, M.; Selva, S.; Seneci, P. *Biotech. Bioeng. (Comb. Chem.)* **2000**, *71*, 38–43.
[609] Savin, K. A.; Woo, J. C. G.; Danishefsky, S. J. *J. Org. Chem.* **1999**, *64*, 4183–4186.
[610] Zheng, C. S.; Seeberger, P. H.; Danishefsky, S. J. *Angew. Chem. Int. Ed.* **1998**, *37*, 786–789.
[611] Pelish, H. E.; Westwood, N. J.; Feng, Y.; Kirchhausen, T.; Shair, M. D. *J. Am. Chem. Soc.* **2001**, *123*, 6740–6741.
[612] Danishefsky, S. J.; McClure, K. F.; Randolph, J. T.; Ruggeri, R. B. *Science* **1993**, *260*, 1307–1309.
[613] Nakamura, K.; Hanai, N.; Kanno, M.; Kobayashi, A.; Ohnishi, Y.; Ito, Y.; Nakahara, Y. *Tetrahedron Lett.* **1999**, *40*, 515–518.

[614] Ishii, A.; Hojo, H.; Kobayashi, A.; Nakamura, K.; Nakahara, Y.; Ito, Y. *Tetrahedron* **2000**, *56*, 6235–6243.
[615] Nakamura, K.; Ishii, A.; Ito, Y.; Nakahara, Y. *Tetrahedron* **1999**, *55*, 11253–11266.
[616] Routledge, A.; Wallis, M. P.; Ross, K. C.; Fraser, W. *Bioorg. Med. Chem. Lett.* **1995**, *5*, 2059–2064.
[617] Chen, S.; Janda, K. D. *Tetrahedron Lett.* **1998**, *39*, 3943–3946.
[618] Wallace, O. B. *Tetrahedron Lett.* **1997**, *38*, 4939–4942.
[619] Koh, J. S.; Ellman, J. A. *J. Org. Chem.* **1996**, *61*, 4494–4495.
[620] Liu, G. C.; Ellman, J. A. *J. Org. Chem.* **1995**, *60*, 7712–7713.
[621] Kick, E. K.; Ellman, J. A. *J. Med. Chem.* **1995**, *38*, 1427–1430.
[622] Basso, A.; Ernst, B. *Tetrahedron Lett.* **2001**, *42*, 6687–6690.
[623] Ellman, J. A.; Thompson, L. A. *Tetrahedron Lett.* **1994**, *35*, 9333–9336.
[624] Pearson, W. H.; Clark, R. B. *Tetrahedron Lett.* **1997**, *38*, 7669–7672.
[625] Dahl, R. S.; Finney, N. S. *J. Comb. Chem.* **2001**, *3*, 329–331.
[626] Wang, G. T.; Li, S.; Wideburg, N.; Krafft, G. A.; Kempf, D. J. *J. Med. Chem.* **1995**, *38*, 2995–3002.
[627] Yoo, S. E.; Gong, Y.-D.; Choi, M.-Y.; Seo, J. S.; Yi, K. Y. *Tetrahedron Lett.* **2000**, *41*, 6415–6418.
[628] Koot, W.-J. *J. Comb. Chem.* **1999**, *1*, 467–473.
[629] Wendeborn, S.; De Mesmaeker, A.; Brill, W. K. D. *Synlett* **1998**, 865–868.
[630] Hanessian, S.; Huynh, H. K. *Synlett* **1999**, 102–104.
[631] Fréchet, J. M. J.; Pellé, G. *J. Chem. Soc., Chem. Commun.* **1975**, 225–226.
[632] Wu, Y. T.; Hsieh, H. P.; Wu, C. Y.; Yu, H. M.; Chen, S. T.; Wang, K. T. *Tetrahedron Lett.* **1998**, *39*, 1783–1784.
[633] Hsieh, H. P.; Wu, Y. T.; Chen, S. T.; Wang, K. T. *Bioorg. Med. Chem.* **1999**, *7*, 1797–1803.
[634] Bozzoli, A.; Kazmierski, W.; Kennedy, G.; Pasquarello, A.; Pecunioso, A. *Bioorg. Med. Chem. Lett.* **2000**, *10*, 2759–2763.
[635] Zhu, T.; Boons, G. J. *Angew. Chem. Int. Ed.* **1998**, *37*, 1898–1900.
[636] Wu, X.; Grathwohl, M.; Schmidt, R. R. *Org. Lett.* **2001**, *3*, 747–750.
[637] Nicolaou, K. C.; Winssinger, N.; Vourloumis, D.; Ohshima, T.; Kim, S.; Pfefferkorn, J.; Xu, J. Y.; Li, T. *J. Am. Chem. Soc.* **1998**, *120*, 10814–10826.
[638] Berteina, S.; Wendeborn, S.; De Mesmaeker, A. *Synlett* **1998**, 1231–1233.
[639] Nizi, E.; Botta, M.; Corelli, F.; Manetti, F.; Messina, F.; Maga, G. *Tetrahedron Lett.* **1998**, *39*, 3307–3310.
[640] Rano, T. A.; Cheng, Y.; Huening, T. T.; Zhang, F.; Schleif, W. A.; Gabryelski, L.; Olsen, D. B.; Kuo, L. C.; Lin, J. H.; Xu, X.; Olah, T. V.; McLoughlin, D. A.; King, R.; Chapman, K. T.; Tata, J. R. *Bioorg. Med. Chem. Lett.* **2000**, *10*, 1527–1530.
[641] Swistok, J.; Tilley, J. W.; Danho, W.; Wagner, R.; Mulkerins, K. *Tetrahedron Lett.* **1989**, *30*, 5045–5048.
[642] Zheng, A.; Shan, D.; Wang, B. *J. Org. Chem.* **1999**, *64*, 156–161.
[643] Meyers, H. V.; Dilley, G. J.; Durgin, T. L.; Powers, T. S.; Winssinger, N. A.; Zhu, H.; Pavia, M. R. *Mol. Diversity* **1995**, *1*, 13–20.
[644] Barber, A. M.; Hardcastle, I. R.; Rowlands, M. G.; Nutley, B. P.; Marriott, J. H.; Jarman, M. *Bioorg. Med. Chem. Lett.* **1999**, *9*, 623–626.
[645] Leznoff, C. C.; Dixit, D. M. *Can. J. Chem.* **1977**, *55*, 3351–3355.
[646] Molteni, V.; Annunziata, R.; Cinquini, M.; Cozzi, F.; Benaglia, M. *Tetrahedron Lett.* **1998**, *39*, 1257–1260.
[647] Kurth, M. J.; Randall, L. A. A.; Takenouchi, K. *J. Org. Chem.* **1996**, *61*, 8755–8761.
[648] Kantorowski, E. J.; Kurth, M. J. *J. Org. Chem.* **1997**, *62*, 6797–6803.
[649] Pavia, M. R.; Cohen, M. P.; Dilley, G. J.; Dubuc, G. R.; Durgin, T. L.; Forman, F. W.; Hediger, M. E.; Milot, G.; Powers, T. S.; Sucholeiki, I.; Zhou, S.; Hangauer, D. G. *Bioorg. Med. Chem.* **1996**, *4*, 659–666.
[650] Adams, S. P.; Kavka, K. S.; Wykes, E. J.; Holder, S. B.; Galluppi, G. R. *J. Am. Chem. Soc.* **1983**, *105*, 661–663.
[651] Eckstein, F. *Oligonucleotides and Analogues; A Practical Approach*; Oxford University Press: Oxford, UK, **1991**.
[652] Krchnák, V.; Weichsel, A. S.; Lebl, M.; Felder, S. *Bioorg. Med. Chem. Lett.* **1997**, *7*, 1013–1016.
[653] Moore, M.; Norris, P. *Tetrahedron Lett.* **1998**, *39*, 7027–7030.
[654] Xiao, X.-Y.; Nova, M. P.; Czarnik, A. W. *J. Comb. Chem.* **1999**, *1*, 379–382.
[655] Kobayashi, S.; Hachiya, I.; Suzuki, S.; Moriwaki, M. *Tetrahedron Lett.* **1996**, *37*, 2809–2812.
[656] May, P. J.; Bradley, M.; Harrowven, D. C.; Pallin, D. *Tetrahedron Lett.* **2000**, *41*, 1627–1630.
[657] Chandrasekhar, S.; Padmaja, M. B.; Raza, A. *J. Comb. Chem.* **2000**, *2*, 246–248.
[658] Han, Y.; Giroux, A.; Lépine, C.; Laliberté, F.; Huang, Z.; Perrier, H.; Bayly, C. I.; Young, R. N. *Tetrahedron* **1999**, *55*, 11669–11685.

[659] Burgess, K.; Lim, D. *Chem. Commun.* **1997**, 785–786.
[660] Faita, G.; Paio, A.; Quadrelli, P.; Rancati, F.; Seneci, P. *Tetrahedron Lett.* **2000**, *41*, 1265–1269.
[661] Tietze, L. F.; Hippe, T.; Steinmetz, A. *Chem. Commun.* **1998**, 793–794.
[662] Mergler, M.; Nyfeler, R. Rapid Synthesis of fully Protected Peptide Alcohols, Communication from Bachem, Bubendorf, CH, **1990**.
[663] Kurth, M. L.; Randall, L. A. A.; Chen, C.; Melander, C.; Miller, R. B.; McAlister, K.; Reitz, G.; Kang, R.; Nakatsu, T.; Green, C. *J. Org. Chem.* **1994**, *59*, 5862–5864.
[664] Ley, S. V.; Mynett, D. M.; Koot, W. J. *Synlett* **1995**, 1017–1020.
[665] Alsina, J.; Rabanal, F.; Chiva, C.; Giralt, E.; Albericio, F. *Tetrahedron* **1998**, *54*, 10125–10152.
[666] Choo, H.; Chong, Y.; Chu, C. K. *Org. Lett.* **2001**, *3*, 1471–1473.
[667] Fréchet, J. M. J.; Nuyens, L. J.; Seymour, E. *J. Am. Chem. Soc.* **1979**, *101*, 432–436.
[668] Suginome, M.; Iwanami, T.; Ito, Y. *J. Am. Chem. Soc.* **2001**, *123*, 4356–4357.
[669] Rolland, C.; Hanquet, G.; Ducep, J.-B.; Solladié, G. *Tetrahedron Lett.* **2001**, *42*, 7563–7566.
[670] Alsina, J.; Chiva, C.; Ortiz, M.; Rabanal, F.; Giralt, E.; Albericio, F. *Tetrahedron Lett.* **1997**, *38*, 883–886.
[671] Chou, Y. L.; Morrissey, M. M.; Mohan, R. *Tetrahedron Lett.* **1998**, *39*, 757–760.
[672] Eritja, R.; Robles, J.; Fernández Forner, D.; Albericio, F.; Giralt, E.; Pedroso, E. *Tetrahedron Lett.* **1991**, *32*, 1511–1514.
[673] Eritja, R.; Robles, J.; Aviñó, A.; Albericio, F.; Pedroso, E. *Tetrahedron* **1992**, *48*, 4171–4182.
[674] Venkatesan, H.; Greenberg, M. M. *J. Org. Chem.* **1996**, *61*, 525–529.
[675] Mourtas, S.; Gatos, D.; Barlos, K. *Tetrahedron Lett.* **2001**, *42*, 2201–2204.
[676] Souers, A. J.; Virgilio, A. A.; Rosenquist, A.; Fenuik, W.; Ellman, J. A. *J. Am. Chem. Soc.* **1999**, *121*, 1817–1825.
[677] Davis, B. G.; Ward, S. J.; Rendle, P. M. *J. Chem. Soc., Chem. Commun.* **2001**, 189–190.
[678] Annis, I.; Chen, L.; Barany, G. *J. Am. Chem. Soc.* **1998**, *120*, 7226–7238.
[679] Sucholeiki, I. *Tetrahedron Lett.* **1994**, *35*, 7307–7310.
[680] Souers, A. J.; Virgilio, A. A.; Schürer, S. S.; Ellman, J. A.; Kogan, T. P.; West, H. E.; Ankener, W.; Vanderslice, P. *Bioorg. Med. Chem. Lett.* **1998**, *8*, 2297–2302.
[681] Kurokawa, K.; Kumihara, H.; Kondo, H. *Bioorg. Med. Chem. Lett.* **2000**, *10*, 1827–1830.
[682] Englebretsen, D. R.; Garnham, B. G.; Bergman, D. A.; Alewood, P. F. *Tetrahedron Lett.* **1995**, *36*, 8871–8874.
[683] Zoller, T.; Ducep, J.-B.; Tahtaoui, C.; Hibert, M. *Tetrahedron Lett.* **2000**, *41*, 9989–9992.
[684] Rietman, B. H.; Smulders, R. H. P. H.; Eggen, L. F.; van Vliet, A.; van de Werken, G.; Tesser, G. I. *Int. J. Pept. Prot. Res.* **1994**, *44*, 199–206.
[685] Glass, J. D.; Talansky, A.; Grzonka, Z.; Schwartz, I. L.; Walter, R. *J. Am. Chem. Soc.* **1974**, *96*, 6476–6480.
[686] Virgilio, A. A.; Schürer, S. C.; Ellman, J. A. *Tetrahedron Lett.* **1996**, *37*, 6961–6964.
[687] Lee, T. R.; Lawrence, D. S. *J. Med. Chem.* **1999**, *42*, 784–787.
[688] Moore, J. S.; Weinstein, E. J.; Wu, Z. *Tetrahedron Lett.* **1991**, *32*, 2465–2466.
[689] Pilot, C.; Dahmen, S.; Lauterwasser, F.; Bräse, S. *Tetrahedron Lett.* **2001**, *42*, 9179–9181.
[690] Han, Y. X.; Walker, S. D.; Young, R. N. *Tetrahedron Lett.* **1996**, *37*, 2703–2706.
[691] Lee, Y.; Silverman, R. B. *Org. Lett.* **2000**, *2*, 303–306.
[692] Lee, Y.; Silverman, R. B. *Tetrahedron* **2001**, *57*, 5339–5352.
[693] Plunkett, M. J.; Ellman, J. A. *J. Org. Chem.* **1997**, *62*, 2885–2893.
[694] Spivey, A. C.; Diaper, C. M.; Rudge, A. J. *Chem. Commun.* **1999**, 835–836.
[695] Spivey, A. C.; Diaper, C. M.; Adams, H.; Rudge, A. J. *J. Org. Chem.* **2000**, *65*, 5253–5263.
[696] Nicolaou, K. C.; Winssinger, N.; Pastor, J.; Murphy, F. *Angew. Chem. Int. Ed.* **1998**, *37*, 2534–2537.
[697] Crosby, G. A.; Kato, M. *J. Am. Chem. Soc.* **1977**, *99*, 278–280.
[698] Beebe, X.; Chiappari, C. L.; Kurth, M. J.; Schore, N. E. *J. Org. Chem.* **1993**, *58*, 7320–7321.
[699] Beebe, X.; Schore, N. E.; Kurth, M. J. *J. Org. Chem.* **1995**, *60*, 4196–4203.
[700] Murphy, A. M.; Dagnino, R.; Vallar, P. L.; Trippe, A. J.; Sherman, S. L.; Lumpkin, R. H.; Tamura, S. Y.; Webb, T. R. *J. Am. Chem. Soc.* **1992**, *114*, 3156–3157.
[701] Poupart, M. A.; Fazal, G.; Goulet, S.; Mar, L. T. *J. Org. Chem.* **1999**, *64*, 1356–1361.
[702] Fraley, M. E.; Rubino, R. S. *Tetrahedron Lett.* **1997**, *38*, 3365–3368.
[703] Gutke, H.-J.; Spitzner, D. *Tetrahedron* **1999**, *55*, 3931–3936.
[704] Ball, C. P.; Barrett, A. G. M.; Commercon, A.; Compère, D.; Kuhn, C.; Roberts, R. S.; Smith, M. L.; Venier, O. *Chem. Commun.* **1998**, 2019–2020.
[705] Barrett, A. G. M.; Procopiou, P. A.; Voigtmann, U. *Org. Lett.* **2001**, *3*, 3165–3168.
[706] Guthrie, E. J.; Macritchie, J.; Hartley, R. C. *Tetrahedron Lett.* **2000**, *41*, 4987–4990.
[707] Hu, Y.; Porco, J. A. *Tetrahedron Lett.* **1999**, *40*, 3289–3292.
[708] Smith, E. M. *Tetrahedron Lett.* **1999**, *40*, 3285–3288.

[709] Wendeborn, S. *Synlett* **2000**, 45–48.
[710] Worster, P. M.; McArthur, C. R.; Leznoff, C. C. *Angew. Chem. Int. Ed. Engl.* **1979**, *18*, 221–222.
[711] McArthur, C. R.; Worster, P. M.; Jiang, J.; Leznoff, C. C. *Can. J. Chem.* **1982**, *60*, 1836–1841.
[712] Hird, N. W.; Irie, K.; Nagai, K. *Tetrahedron Lett.* **1997**, *38*, 7111–7114.
[713] Aznar, F.; Valdés, C.; Cabal, M.-P. *Tetrahedron Lett.* **2000**, *41*, 5683–5687.
[714] Hong, B.-C.; Chen, Z.-Y.; Chen, W.-H. *Org. Lett.* **2000**, *2*, 2647–2649.
[715] Crawshaw, M.; Hird, N. W.; Irie, K.; Nagai, K. *Tetrahedron Lett.* **1997**, *38*, 7115–7118.
[716] Wilson, R. D.; Watson, S. P.; Richards, S. A. *Tetrahedron Lett.* **1998**, *39*, 2827–2830.
[717] Chen, C.; Munoz, B. *Tetrahedron Lett.* **1999**, *40*, 3491–3494.
[718] Lee, A.; Huang, L.; Ellman, J. A. *J. Am. Chem. Soc.* **1999**, *121*, 9907–9914.
[719] Vojkovsky, T.; Weichsel, A.; Pátek, M. *J. Org. Chem.* **1998**, *63*, 3162–3163.
[720] Chandrasekhar, S.; Padmaja, M. B. *Synth. Commun.* **1998**, *28*, 3715–3720.
[721] Bertini, V.; Lucchesini, F.; Pocci, M.; De Munno, A. *Tetrahedron Lett.* **1998**, *39*, 9263–9266.
[722] Vanier, C.; Wagner, A.; Mioskowski, C. *Chem. Eur. J.* **2001**, *7*, 2318–2323.
[723] Huwe, C. M.; Künzer, H. *Tetrahedron Lett.* **1999**, *40*, 683–686.
[724] Bertini, V.; Lucchesini, F.; Pocci, M.; De Munno, A. *J. Org. Chem.* **2000**, *65*, 4839–4842.
[725] Rademann, J.; Schmidt, R. R. *J. Org. Chem.* **1997**, *62*, 3650–3653.
[726] Rademann, J.; Schmidt, R. R. *Tetrahedron Lett.* **1996**, *37*, 3989–3990.
[727] Kallus, C.; Opatz, T.; Wunberg, T.; Schmidt, W.; Henke, S.; Kunz, H. *Tetrahedron Lett.* **1999**, *40*, 7783–7786.
[728] Yan, L.; Taylor, C. M.; Goodnow, R.; Kahne, D. *J. Am. Chem. Soc.* **1994**, *116*, 6953–6954.
[729] Siev, D. V.; Gaudette, J. A.; Semple, J. E. *Tetrahedron Lett.* **1999**, *40*, 5123–5127.
[730] Ho, J. Z.; Levy, O. E.; Gibson, T. S.; Nguyen, K.; Semple, J. E. *Bioorg. Med. Chem. Lett.* **1999**, *9*, 3459–3464.
[731] Tamura, S. Y.; Weinhouse, M. I.; Roberts, C. A.; Goldman, E. A.; Masukawa, K.; Anderson, S. M.; Cohen, C. R.; Bradbury, A. E.; Bernardino, V. T.; Dixon, S. A.; Ma, M. G.; Nolan, T. G.; Brunck, T. K. *Bioorg. Med. Chem. Lett.* **2000**, *10*, 983–987.
[732] Chamoin, S.; Houldsworth, S.; Kruse, C. G.; Bakker, W. I.; Snieckus, V. *Tetrahedron Lett.* **1998**, *39*, 4179–4182.
[733] Leznoff, C. C.; Sywanyk, W. *J. Org. Chem.* **1977**, *42*, 3203–3205.
[734] Wong, J. Y.; Manning, C.; Leznoff, C. C. *Angew. Chem. Int. Ed. Engl.* **1974**, *13*, 666–667.
[735] Xu, Z. H.; McArthur, C. R.; Leznoff, C. C. *Can. J. Chem.* **1983**, *61*, 1405–1409.
[736] Leznoff, C. C.; Greenberg, S. *Can. J. Chem.* **1976**, *54*, 3824–3829.
[737] Maltais, R.; Bérubé, M.; Marion, O.; Labrecque, R.; Poirier, D. *Tetrahedron Lett.* **2000**, *41*, 1691–1694.
[738] Metz, W. A.; Jones, W. D.; Ciske, F. L.; Peet, N. P. *Bioorg. Med. Chem. Lett.* **1998**, *8*, 2399–2402.
[739] Ren, Q.; Huang, W.; Ho, P. *Reactive Polymers* **1989**, *11*, 237–244.
[740] Ede, N. J.; Bray, A. M. *Tetrahedron Lett.* **1997**, *38*, 7119–7122.
[741] Yao, W.; Xu, Y. *Tetrahedron Lett.* **2001**, *42*, 2549–2552.
[742] Ede, N. J.; Eagle, S. N.; Wickham, G.; Bray, A. M.; Warne, B.; Shoemaker, K.; Rosenberg, S. *J. Pept. Sci.* **2000**, *6*, 11–18.
[743] Huang, X.; Sheng, S.-R. *Tetrahedron Lett.* **2001**, *42*, 9035–9037.
[744] Nahm, S.; Weinreb, S. M. *Tetrahedron Lett.* **1981**, *22*, 3815–3818.
[745] Fehrentz, J. A.; Paris, M.; Heitz, A.; Velek, J.; Liu, C. F.; Winternitz, F.; Martinez, J. *Tetrahedron Lett.* **1995**, *36*, 7871–7874.
[746] Qian, H.; Shao, L.-X.; Huang, X. *Synlett* **2001**, 1571–1572.
[747] Caulfield, T. J.; Patel, S.; Salvino, J. M.; Liester, L.; Labaudiniere, R. *J. Comb. Chem.* **2000**, *2*, 600–603.
[748] Dinh, T. Q.; Armstrong, R. W. *Tetrahedron Lett.* **1996**, *37*, 1161–1164.
[749] O'Donnell, M. J.; Drew, M. D.; Pottorf, R. S.; Scott, W. L. *J. Comb. Chem.* **2000**, *2*, 172–181.
[750] Crowley, J. I.; Rapoport, H. *J. Am. Chem. Soc.* **1970**, *92*, 6363–6365.
[751] Hall, B. J.; Sutherland, J. D. *Tetrahedron Lett.* **1998**, *39*, 6593–6596.
[752] Fréchet, J. M.; Schuerch, C. *J. Am. Chem. Soc.* **1971**, *93*, 492–496.
[753] Fruchart, J.-S.; Gras-Masse, H.; Melnyk, O. *Tetrahedron Lett.* **1999**, *40*, 6225–6228.
[754] Melnyk, O.; Fruchart, J.-S.; Grandjean, C.; Gras-Masse, H. *J. Org. Chem.* **2001**, *66*, 4153–4160.
[755] Schlienger, N.; Bryce, M. R.; Hansen, T. K. *Tetrahedron Lett.* **2000**, *41*, 5147–5150.
[756] Bernard, M.; Ford, W. T. *J. Org. Chem.* **1983**, *48*, 326–332.
[757] Claffey, D. J.; Ruth, J. A. *Tetrahedron: Asymmetry* **1997**, *8*, 3715–3716.
[758] Johnson, C. R.; Zhang, B. R. *Tetrahedron Lett.* **1995**, *36*, 9253–9256.
[759] Hughes, I. *Tetrahedron Lett.* **1996**, *37*, 7595–7598.
[760] Bolli, M. H.; Ley, S. V. *J. Chem. Soc., Perkin Trans. 1* **1998**, 2243–2246.

[761] Barrett, A. G. M.; Cramp, S. M.; Roberts, R. S.; Zecri, F. J. *Org. Lett.* **1999**, *1*, 579–582.
[762] Nicolaou, K. C.; Pastor, J.; Winssinger, N.; Murphy, F. *J. Am. Chem. Soc.* **1998**, *120*, 5132–5133.
[763] Cheng, W.-C.; Olmstead, M. M.; Kurth, M. J. *J. Org. Chem.* **2001**, *66*, 5528–5533.
[764] Nicolaou, K. C.; Snyder, S. A.; Bigot, A.; Pfefferkorn, J. A. *Angew. Chem. Int. Ed.* **2000**, *39*, 1093–1096.
[765] Baer, R.; Masquelin, T. *J. Comb. Chem.* **2001**, *3*, 16–19.
[766] Li, W.; Burgess, K. *Tetrahedron Lett.* **1999**, *40*, 6527–6530.
[767] Russell, H. E.; Luke, R. W. A.; Bradley, M. *Tetrahedron Lett.* **2000**, *41*, 5287–5290.
[768] Nicolaou, K. C.; Pfefferkorn, J. A.; Cao, G.-Q.; Kim, S.; Kessabi, J. *Org. Lett.* **1999**, *1*, 807–810.
[769] Nicolaou, K. C.; Cao, G.-Q.; Pfefferkorn, J. A. *Angew. Chem. Int. Ed.* **2000**, *39*, 739–743.
[770] Nicolaou, K. C.; Winssinger, N.; Hughes, R.; Smethurst, C.; Cho, S. Y. *Angew. Chem. Int. Ed.* **2000**, *39*, 1084–1088.
[771] Fujita, K.-I.; Watanabe, K.; Oishi, A.; Ikeda, Y.; Taguchi, Y. *Synlett* **1999**, 1760–1762.
[772] Nicolaou, K. C.; Pfefferkorn, J. A.; Roecker, A. J.; Cao, G.-Q.; Barluenga, S.; Mitchell, H. J. *J. Am. Chem. Soc.* **2000**, *122*, 9939–9953.
[773] Li, Z.; Kulkarni, B. A.; Ganesan, A. *Biotechnol. Bioeng. (Comb. Chem.)* **2000**, *71*, 104–106.
[774] Huang, X.; Zhu, Q. *Tetrahedron Lett.* **2001**, *42*, 6373–6375.
[775] Yamada, M.; Miyajima, T.; Horikawa, H. *Tetrahedron Lett.* **1998**, *39*, 289–292.
[776] Michels, R.; Kato, M.; Heitz, W. *Makromol. Chem.* **1976**, *177*, 2311–2320.
[777] Nicolaou, K. C.; Pastor, J.; Barluenga, S.; Winssinger, N. *Chem. Commun.* **1998**, 1947–1948.
[778] Horikawa, E.; Kodaka, M.; Nakahara, Y.; Okuno, H.; Nakamura, K. *Tetrahedron Lett.* **2001**, *42*, 8337–8339.
[779] Nicolaou, K. C.; Mitchell, H. J.; Fylaktakidou, K. C.; Suzuki, H.; Rodríguez, R. M. *Angew. Chem. Int. Ed.* **2000**, *39*, 1089–1093.
[780] Kamogawa, H.; Kanzawa, A.; Kadoya, M.; Naito, T.; Nanasawa, M. *Bull. Chem. Soc. Jpn.* **1983**, *56*, 762–765.
[781] Zaragoza, F. *Metal Carbenes in Organic Synthesis*; Wiley-VCH: Weinheim, New York, **1999**.
[782] Andrade, R. B.; Plante, O. J.; Melean, L. G.; Seeberger, P. H. *Org. Lett.* **1999**, *1*, 1811–1814.
[783] Melean, L. G.; Haase, W.-C.; Seeberger, P. H. *Tetrahedron Lett.* **2000**, *41*, 4329–4333.
[784] Nicolaou, K. C.; Winssinger, N.; Pastor, J.; Ninkovic, S.; Sarabia, F.; He, Y.; Vourloumis, D.; Yang, Z.; Li, T.; Giannakakou, P.; Hamel, E. *Nature* **1997**, *387*, 268–272.
[785] Peters, J. U.; Blechert, S. *Synlett* **1997**, 348–350.
[786] Knerr, L.; Schmidt, R. R. *Eur. J. Org. Chem.* **2000**, 2803–2808.
[787] Knerr, L.; Schmidt, R. R. *Synlett* **1999**, 1802–1804.
[788] van Maarseveen, J. H.; den Hartog, J. A. J.; Engelen, V.; Finner, E.; Visser, G.; Kruse, C. G. *Tetrahedron Lett.* **1996**, *37*, 8249–8252.
[789] Piscopio, A. D.; Miller, J. F.; Koch, K. *Tetrahedron Lett.* **1998**, *39*, 2667–2670.
[790] Piscopio, A. D.; Miller, J. F.; Koch, K. *Tetrahedron* **1999**, *55*, 8189–8198.
[791] Brown, R. C. D.; Castro, J. L.; Moriggi, J.-D. *Tetrahedron Lett.* **2000**, *41*, 3681–3685.
[792] Piscopio, A. D.; Miller, J. F.; Koch, K. *Tetrahedron Lett.* **1997**, *38*, 7143–7146.
[793] Veerman, J. J. N.; van Maarseveen, J. H.; Visser, G. M.; Kruse, C. G.; Schoemaker, H. E.; Hiemstra, H.; Rutjes, F. P. J. T. *Eur. J. Org. Chem.* **1998**, 2583–2589.
[794] Hewitt, M. C.; Seeberger, P. H. *Org. Lett.* **2001**, *3*, 3699–3702.
[795] Blanco, L.; Bloch, R.; Bugnet, E.; Deloisy, S. *Tetrahedron Lett.* **2000**, *41*, 7875–7878.
[796] Sim, M. M.; Lee, C. L.; Ganesan, A. *Tetrahedron Lett.* **1998**, *39*, 2195–2198.
[797] Sim, M. M.; Lee, C. L.; Ganesan, A. *Tetrahedron Lett.* **1998**, *39*, 6399–6402.
[798] Zaragoza, F. *Tetrahedron Lett.* **1997**, *38*, 7291–7294.
[799] Hamper, B. C.; Gan, K. Z.; Owen, T. J. *Tetrahedron Lett.* **1999**, *40*, 4973–4976.
[800] Cobb, J. M.; Fiorini, M. T.; Goddard, C. R.; Theoclitou, M. E.; Abell, C. *Tetrahedron Lett.* **1999**, *40*, 1045–1048.
[801] Horton, J. R.; Stamp, L. M.; Routledge, A. *Tetrahedron Lett.* **2000**, *41*, 9181–9184.
[802] Woolard, F. X.; Paetsch, J.; Ellman, J. A. *J. Org. Chem.* **1997**, *62*, 6102–6103.
[803] Whitfield, D. M.; Ogawa, T. *Glycoconjugate J.* **1998**, *15*, 75–78.
[804] Curtet, S.; Langlois, M. *Tetrahedron Lett.* **1999**, *40*, 8563–8566.
[805] Tacke, R.; Ulmer, B.; Wagner, B.; Arlt, M. *Organometallics* **2000**, *19*, 5297–5309.
[806] Lee, Y.; Silverman, R. B. *J. Am. Chem. Soc.* **1999**, *121*, 8407–8408.
[807] Briehn, C. A.; Kirschbaum, T.; Bäuerle, P. *J. Org. Chem.* **2000**, *65*, 352–359.
[808] Chenera, B.; Finkelstein, J. A.; Veber, D. F. *J. Am. Chem. Soc.* **1995**, *117*, 11999–12000.
[809] Boehm, T. L.; Showalter, H. D. H. *J. Org. Chem.* **1996**, *61*, 6498–6499.
[810] Harikrishnan, L. S.; Showalter, H. D. H. *Tetrahedron* **2000**, *56*, 515–519.

[811] Hone, N. D.; Davies, S. G.; Devereux, N. J.; Taylor, S. L.; Baxter, A. D. *Tetrahedron Lett.* **1998**, *39*, 897–900.
[812] Schuster, M.; Blechert, S. *Tetrahedron Lett.* **1998**, *39*, 2295–2298.
[813] Pourbaix, C.; Carreaux, F.; Carboni, B.; Deleuze, H. *J. Chem. Soc., Chem. Commun.* **2000**, 1275–1276.
[814] Pan, Y.; Holmes, C. P. *Org. Lett.* **2001**, *3*, 2769–2771.
[815] Jin, S.; Holub, D. P.; Wustrow, D. J. *Tetrahedron Lett.* **1998**, *39*, 3651–3654.
[816] Hijikuro, I.; Doi, T.; Takahashi, T. *J. Am. Chem. Soc.* **2001**, *123*, 3716–3722.
[817] Bräse, S.; Enders, D.; Köbberling, J.; Avemaria, F. *Angew. Chem. Int. Ed.* **1998**, *37*, 3413–3415.
[818] Lormann, M.; Dahmen, S.; Bräse, S. *Tetrahedron Lett.* **2000**, *41*, 3813–3816.
[819] Schunk, S.; Enders, D. *Org. Lett.* **2000**, *2*, 907–910.
[820] Bräse, S.; Schroen, M. *Angew. Chem. Int. Ed.* **1999**, *38*, 1071–1073.
[821] Uehlin, L.; Wirth, T. *Org. Lett.* **2001**, *3*, 2931–2933.
[822] Ruhland, T.; Andersen, K.; Pedersen, H. *J. Org. Chem.* **1998**, *63*, 9204–9211.
[823] Nicolaou, K. C.; Roecker, A. J.; Pfefferkorn, J. A.; Cao, G.-Q. *J. Am. Chem. Soc.* **2000**, *122*, 2966–2967.
[824] Hennequin, L. F.; Piva-Le Blanc, S. *Tetrahedron Lett.* **1999**, *40*, 3881–3884.
[825] Halm, C.; Evarts, J.; Kurth, M. J. *Tetrahedron Lett.* **1997**, *38*, 7709–7712.
[826] Cheng, W.-C.; Halm, C.; Evarts, J. B.; Olmstead, M. M.; Kurth, M. J. *J. Org. Chem.* **1999**, *64*, 8557–8562.
[827] Jung, K. W.; Zhao, X. Y.; Janda, K. D. *Tetrahedron Lett.* **1996**, *37*, 6491–6494.
[828] Jung, K. W.; Zhao, X. Y.; Janda, K. D. *Tetrahedron* **1997**, *53*, 6645–6652.
[829] Forman, F. W.; Sucholeiki, I. *J. Org. Chem.* **1995**, *60*, 523–528.
[830] Zhao, X. Y.; Jung, K. W.; Janda, K. D. *Tetrahedron Lett.* **1997**, *38*, 977–980.
[831] Vanier, C.; Lorgé, F.; Wagner, A.; Mioskowski, C. *Angew. Chem. Int. Ed.* **2000**, *39*, 1679–1683.
[832] Kulkarni, B. A.; Ganesan, A. *Angew. Chem. Int. Ed. Engl.* **1997**, *36*, 2454–2455.
[833] Gayo, L. M.; Suto, M. J. *Tetrahedron Lett.* **1997**, *38*, 513–516.
[834] Siegel, M. G.; Hahn, P. J.; Dressman, B. A.; Fritz, J. E.; Grunwell, J. R.; Kaldor, S. W. *Tetrahedron Lett.* **1997**, *38*, 3357–3360.
[835] Parlow, J. J. *Tetrahedron Lett.* **1996**, *37*, 5257–5260.
[836] Bandgar, B. P.; Ghorpade, P. K.; Shrotri, N. S.; Patil, S. V. *Indian J. Chem. Sect. B - Org. Chem.* **1995**, *34*, 153–155.
[837] Regen, S. L.; Kimura, Y. *J. Am. Chem. Soc.* **1982**, *104*, 2064–2065.
[838] Cainelli, G.; Contento, M.; Manescalchi, F.; Regnoli, R. *J. Chem. Soc., Perkin Trans. 1* **1980**, 2516–2519.
[839] Khound, S.; Das, P. J. *Tetrahedron* **1997**, *53*, 9749–9754.
[840] Tamami, B.; Kiasat, A. R. *Synth. Commun.* **1996**, *26*, 3953–3958.
[841] Gibson, S. E.; Hales, N. J.; Peplow, M. A. *Tetrahedron Lett.* **1999**, *40*, 1417–1418.
[842] Maiorana, S.; Baldoli, C.; Licandro, E.; Casiraghi, L.; de Magistris, E.; Paio, A.; Provera, S.; Seneci, P. *Tetrahedron Lett.* **2000**, *41*, 7271–7275.
[843] Arbo, B. E.; Isied, S. S. *Int. J. Pept. Prot. Res.* **1993**, *42*, 138–154.
[844] Mensi, N.; Isied, S. S. *J. Am. Chem. Soc.* **1987**, *109*, 7882–7884.
[845] Comely, A. C.; Gibson, S. E.; Hales, N. J. *J. Chem. Soc., Chem. Commun.* **1999**, 2075–2076.
[846] Rigby, J. H.; Kondratenko, M. A. *Org. Lett.* **2001**, *3*, 3683–3686.
[847] Semmelhack, M. F.; Hilt, E.; Colley, J. H. *Tetrahedron Lett.* **1998**, *39*, 7683–7686.
[848] Comely, A. C.; Gibson, S. E.; Hales, N. J.; Peplow, M. A. *J. Chem. Soc., Perkin Trans. 1* **2001**, 2526–2531.
[849] Ramage, R.; Raphy, G. *Tetrahedron Lett.* **1992**, *33*, 385–388.
[850] Ramage, R.; Wahl, F. O. *Tetrahedron Lett.* **1993**, *34*, 7133–7136.
[851] Hay, A. M.; Hobbs-DeWitt, S.; MacDonald, A. A.; Ramage, R. *Synthesis* **1999**, 1979–1985.
[852] Hay, A. M.; Hobbs-DeWitt, S.; MacDonald, A. A.; Ramage, R. *Tetrahedron Lett.* **1998**, *39*, 8721–8724.

4 Preparation of Organometallic Compounds

Organometallic compounds are usually generated on insoluble supports only as synthetic intermediates, and not as target molecules. Organometallic compounds cover a broad range of reactivities, and are highly versatile reagents. In this chapter, the stoichiometric generation of main-group and transition metal derived organometallic compounds on supports is discussed.

4.1 Group I and II Organometallic Compounds

A flexible means of access to functionalized supports for solid-phase synthesis is based on metallated, cross-linked polystyrene, which reacts smoothly with a wide range of electrophiles. Cross-linked polystyrene can be lithiated directly by treatment with *n*-butyllithium and TMEDA in cyclohexane at 60–70 °C [1–3] to yield a product containing mainly *meta*- and *para*-lithiated phenyl groups [4]. Metallation of non-cross-linked polystyrene with potassium *tert*-amylate/3-(lithiomethyl)heptane has also been reported [5]. The latter type of base can, unlike butyllithium/TMEDA [6], also lead to benzylic metallation [7]. The C-lithiation of more acidic arenes or heteroarenes, such as imidazoles [8], thiophenes [9], and furans [9], has also been performed on insoluble supports (Figure 4.1). These reactions proceed, like those in solution, with high regioselectivity.

Alternatively, lithiated polystyrene or lithiated resin-bound aryl ethers can be prepared from the corresponding bromo- or iodoarenes by halogen–lithium exchange [1,11–15] (Figure 4.2, Experimental Procedure 4.1) under conditions similar to those used in solution. Activated calcium, which can be generated by reduction of calcium (II) iodide with lithium biphenylide, reacts with fluorinated, chlorinated, or brominated polystyrene to yield polymeric arylcalcium compounds, which react with a number of different electrophiles [16]. Resin-bound aryl-, heteroaryl-, or vinylmagnesium reagents have been prepared by treatment of the corresponding iodides or bromides with ethylmagnesium bromide [17] or isopropylmagnesium bromide [18,19] (Figure 4.2), as well as by transmetallation of lithiated polystyrene with magnesium bromide [14]. 4-Iodobenzoic acid esterified with cross-linked hydroxymethyl polystyrene can be converted into the corresponding *tert*-butyl zincate by halogen–metal exchange with lithium tri-*tert*-butylzincate (THF, 0 °C, 4 h [20]). This zincate can be successfully

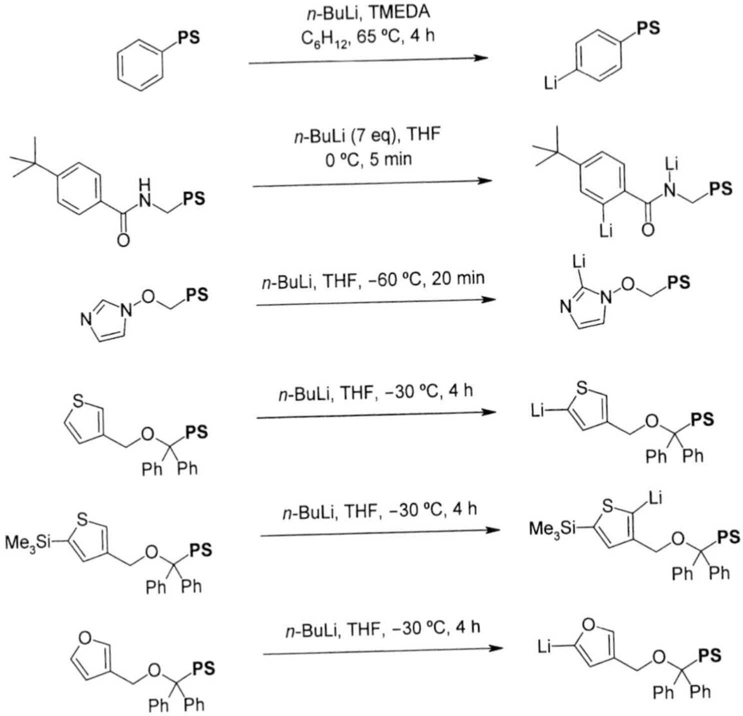

Figure 4.1. Lithiation of polystyrene-bound arenes and heteroarenes [1,8–10].

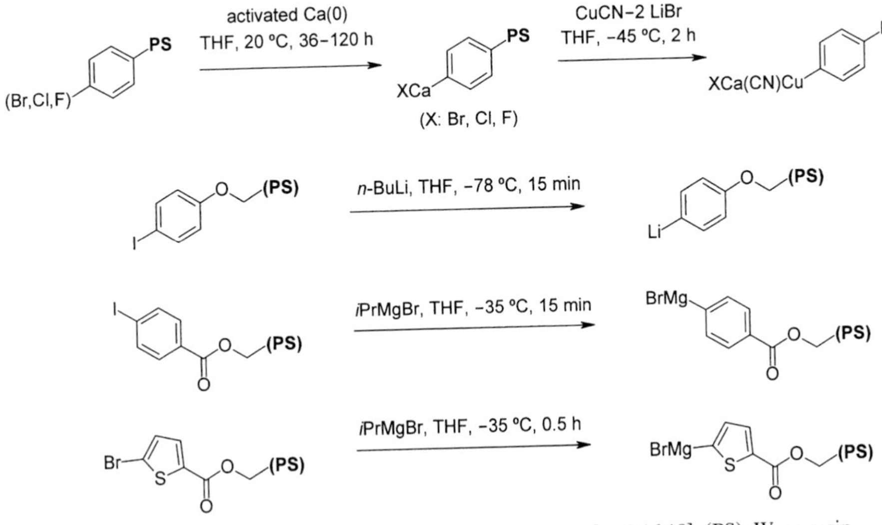

Figure 4.2. Halogen–metal exchange on cross-linked polystyrene [1,13,16,18]. (PS): Wang resin.

transmetallated with lithium cyano(2-thienyl)cuprate [20] (Entry **7**, Table 12.4). Metallated polystyrene-bound esters cannot usually be warmed to room temperature because nucleophilic attack at the ester will lead to rapid decomposition of these reagents. These metallations must therefore be carefully optimized to identify the most suitable temperature and reaction time.

> **Experimental Procedure 4.1: Lithiation of brominated, cross-linked polystyrene [11]** 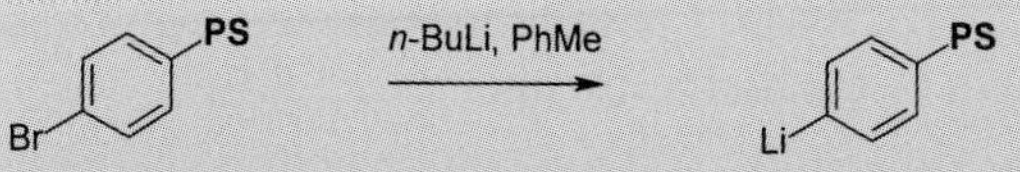
>
>
> *n*-Butyllithium (100 mL, 1.6 mol/L in hexane, 160 mmol) was added to a suspension of brominated, cross-linked polystyrene (24.0 g, 70.6 mmol bromide; for preparation, see Experimental Procedure 6.2) in toluene (200 mL). The mixture was stirred at room temperature for 2 h, and then the resin was allowed to settle. After decantation, further butyllithium (200 mL, 1.6 mol/L in hexane, 320 mmol) and toluene (200 mL) were added and the mixture was heated to 60 °C for 3 h. After cooling to room temperature and decantation, the lithiated polystyrene was used directly without further washing.

Cross-linked chloromethyl polystyrene cannot be lithiated directly by treatment with lithium or methyllithium, because lithium is insoluble in polystyrene-compatible solvents and the use of methyllithium leads to Wurtz coupling [21]. Lithiomethyl and potassiomethyl polystyrene are, however, available by transmetallation of tributyl-stannylmethyl polystyrene, which is prepared from Merrifield resin and (tributylstan-nyl)lithium [21]. Lithiomethyl polystyrene has also been prepared by ether cleavage of ethoxymethyl polystyrenes with lithium biphenylide in THF [22]. Direct halogen–metal exchange in partially chloromethylated polystyrene can also be achieved with activated calcium [16] (Figure 4.3) and with magnesium/anthracene [23,24].

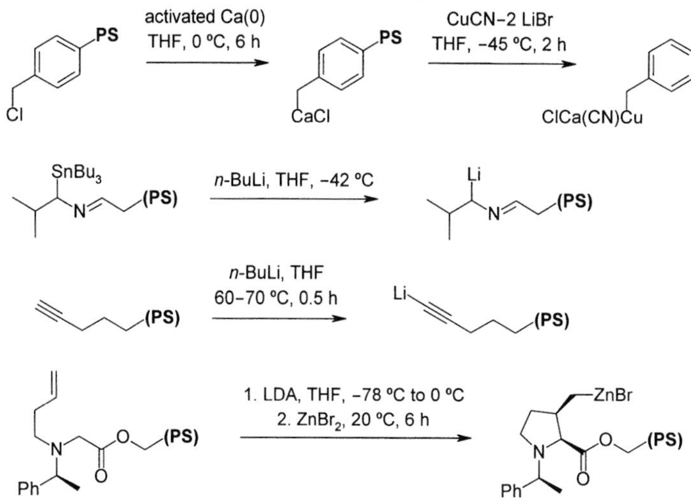

Figure 4.3. Preparation of polystyrene-bound organometallic compounds by transmetallation and direct lithiation [16,25,26,29].

Non-aromatic organolithium compounds can be prepared by transmetallation of resin-bound stannanes [25] or by deprotonation of alkynes [26], triphenylmethane [27], or other resin-bound C–H acidic compounds with lithium amides or similar bases (Figure 4.3). The reaction of polystyrene-bound trialkylboranes with diethylzinc yields resin-bound alkylzinc derivatives [28].

Polystyrene-bound alkenes react with alcohols or amines in the presence of mercury(II) trifluoroacetate to yield 2-alkoxy- or 2-aminoethylmercury compounds [30]. The C–Hg bond can be reduced to a C–H bond by treatment with LiBH$_4$, or converted into a C–I bond by treatment with iodine [30]. Organomercury compounds have been immobilized with polystyrene-bound carboxylates [31]. The resulting product was used as starting material for the preparation of radiolabelled 6-iodo DOPA (Figure 4.4).

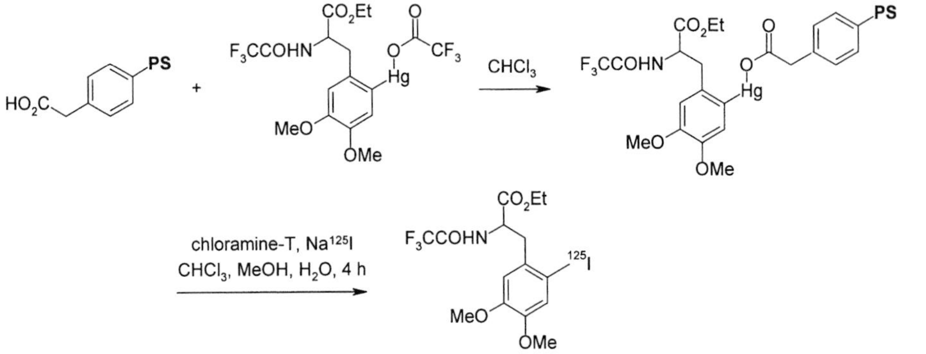

Figure 4.4. Preparation of radioiodinated DOPA from a polystyrene-bound arylmercury precursor [31].

4.2 Group III Organometallic Compounds

Support-bound boranes have been prepared by hydroboration of vinyl polystyrene with 9-BBN, and used as intermediates for the preparation of hydroxyethyl polystyrene (Figure 4.5 [32]) and alkylated polystyrenes [33]. Hydroboration of vinyl poly-

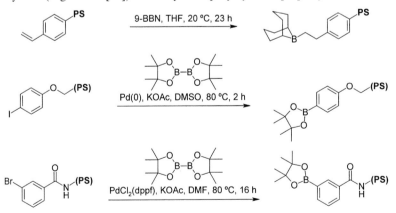

Figure 4.5. Preparation of boron derivatives on insoluble supports [13,34,35]. (PS): Wang resin.

styrene with diborane, on the other hand, proceeds with low regioselectivity [34]. Boronic esters and acids, which are useful intermediates for C–C bond-forming reactions, can be obtained from the corresponding aryllithium compounds by treatment with trimethyl borate [1] followed by hydrolysis. Alternatively, support-bound aryl bromides or iodides can be converted into boronates by treatment with diboronates in the presence of catalytic amounts of palladium(0) complexes (Figure 4.5).

4.3 Group IV Organometallic Compounds

Support-bound silanes are useful linkers and intermediates, and can be prepared by several routes. The most versatile approaches include the reaction of resin-bound organolithium compounds with chlorosilanes [36,37] and the hydrosilylation of resin-bound alkenes (Figure 4.6). Further transformations of resin-bound silanes are discussed in Section 3.11.2.

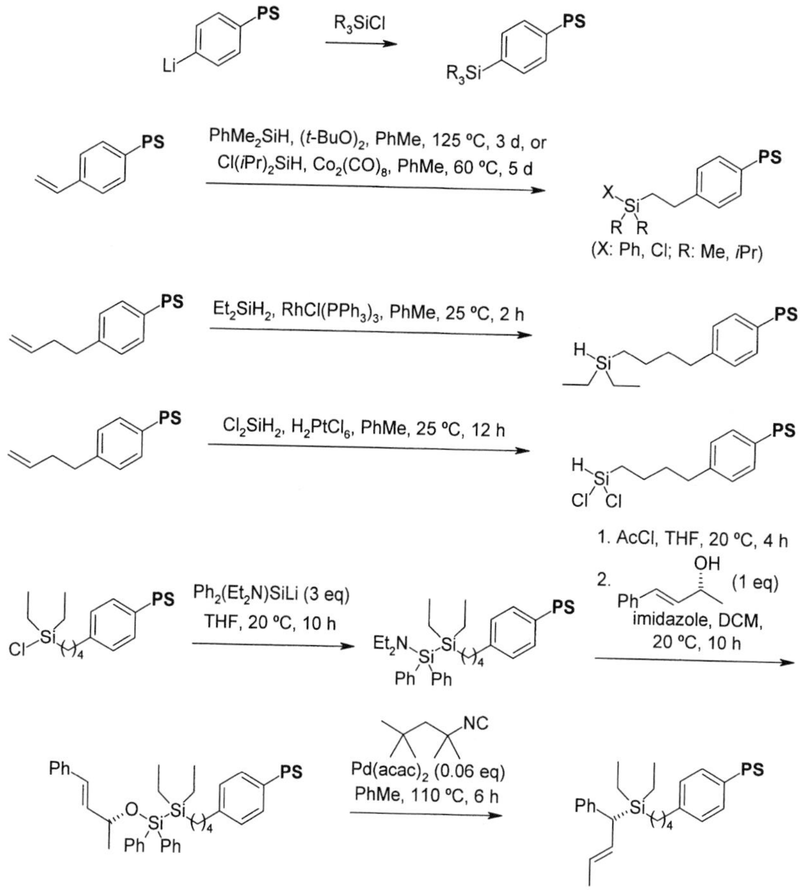

Figure 4.6. Generation of silanes on cross-linked polystyrene [1,34,38–40].

For the preparation of support-bound stannanes, similar strategies to those for silanes can be used (Figure 4.7). The preparation of stannanes on solid phase has not yet received much attention, but with the continuing development of solid-phase protocols for the preparation of elaborate carbon frameworks increased use of immobilized stannanes is to be expected.

Support-bound stannanes have been prepared from phenyllithium bound to macroporous polystyrene and chlorostannanes [14,41], by treatment of support-bound alkyl chlorides with lithiated stannanes [21,41], and by radical or palladium-mediated addition of stannanes to alkenes and alkynes (Figure 4.7 [42–47]). The chloride of polystyrene-bound chlorostannanes can be displaced by treatment with arylzinc reagents, thereby yielding resin-bound arylstannanes [46]. Polystyrene-bound stannanes have also been prepared by copolymerization of 4-[2-(dibutylchlorostannyl)ethyl]styrene with styrene and divinylstyrene [48].

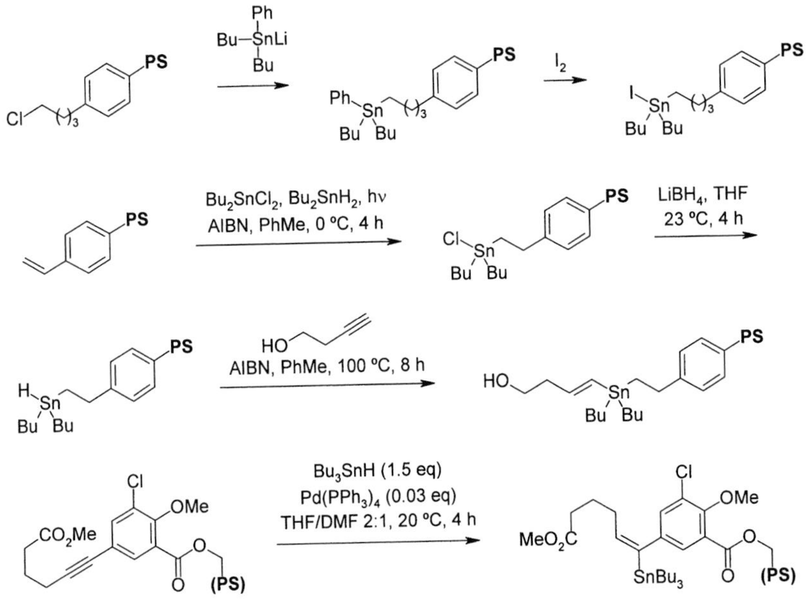

Figure 4.7. Preparation of polystyrene-bound stannanes [41,47,49,50].

4.4 Transition Metal Complexes

Support-bound transition metal complexes have mainly been prepared as insoluble catalysts. Table 4.1 lists representative examples of such polymer-bound complexes. Polystyrene-bound molybdenum carbonyl complexes have been prepared for the study of ligand substitution reactions and oxidative eliminations [51]. Moreover, well-defined molybdenum, rhodium, and iridium phosphine complexes have been prepared on copolymers of PEG and silica [52]. Several reviews have covered the preparation and application of support-bound reagents, including transition metal complexes [53–59]. Examples of the preparation and uses of organomercury and organozinc compounds are discussed in Section 4.1.

Table 4.1. Support-bound transition metal complexes as heterogeneous catalysts.

Metal	Support	Support-bound Ligand	Application	Ref.
Sc	PE	sulfonamide	Lewis acid	[60,61]
Ti	PS-micro[a]	cyclopentadiene	ethylene polymerization	[62]
Ti	PS-macro[a]	cyclopentadiene	hydrogenation	[63]
Ti	TG, PS-micro	sulfonamide	asymmetric addition of Et_2Zn to aldehydes	[64]
Zr	PS-micro	phenolate	asymmetric aza-Diels–Alder reaction	[65, 66]
Cr	PS, PE	pyridine, ammonium	oxidation of alcohols	[53]
Cr	PS, TG	phosphine	preparation of resin-bound chromium(II) carbene complexes	[67]
Mn	PS, PA	salen	asymmetric epoxidation	[68,69]
Mn	TG, PS-macro	glycine amides	oxidation of alkanes and alkenes	[70]
Ru, Ir	PS-micro	aminosulfonamide	asymmetric reduction of ketones	[71]
Ru	silica	2-aminoethanol	asymmetric reduction of ketones	[72]
Ru	PS-micro	porphyrin	epoxidation	[73]
Ru	PS-micro	carboxylate	olefin metathesis	[74]
Ru	polyester	phosphine	asymmetric hydrogenation	[75]
RuO_4^-	PS-macro	ammonium	oxidation of alcohols	[76]
OsO_4	PS, PE	pyridine, amine	dihydroxylation of alkenes	[53,77,78]
Co	PS-micro	phosphine	Pauson-Khand reaction	[79]
Co	PS, silica	salen	asymmetric ring-opening of epoxides	[80,81]
Co	polyaniline		Mannich reaction	[82]

Table 4.1. continued.

Metal	Support	Support-bound Ligand	Application	Ref.
Rh	PS-macro	cyclopentadiene	hydroformylation, hydrogenation	[83]
Rh	PS-micro	carboxylate	hydroformylation, hydrogenation	[84]
Rh	PS-micro	2-aminophosphine	asymmetric hydrogenation	[85]
Rh	PS-micro	phosphine	hydroformylation	[86]
Rh	PE	phosphine	hydrogenation	[87]
Rh	PS	bis(diphenyl)phosphine	asymmetric hydroformylation, carbonylation of RLi	[88]
Rh	PS-macro	phosphine	hydrogenation, hydroboration	[89]
Ni, Pd	PS-micro	1,2-diimines	olefin polymerization	[90]
Pd	TG	1,1'-binaphthyl-2-phosphine	asymmetric allylic substitution	[91]
Pd	TG	phosphine	Suzuki coupling	[92]
Pd	TG	phosphine	carboxylation of aryl halides	[93]
Pd	PS-macro	phosphine	allylic substitution	[94]
Pd	macroporous polysiloxane	thiourea	Suzuki coupling	[95]
Pd	poly(benz-imidazole)	nitrile	oxidation of alkenes to ketones (Wacker reaction)	[96]
Cu	silica	imine of 2-formylpyridine	radical cyclization	[97]
Cu	PS-macro	bis-oxazoline	asymmetric (Mukaiyama) aldol addition	[98]
Cu	PS	iminodiacetic acid	resolution of amino acids	[88]
Zn	PS-micro	sulfonamide	asymmetric cyclopropanation	[99]
various	PS-micro	imines	asymmetric epoxidation	[100]

[a] PS-micro: microporous, cross-linked polystyrene; PS-macro: macroporous, cross-linked polystyrene.

References for Chapter 4

[1] Farrall, M. J.; Fréchet, J. M. J. *J. Org. Chem.* **1976**, *41*, 3877–3882.
[2] Kobayashi, S.; Furuta, T.; Sugita, K.; Okitsu, O.; Oyamada, H. *Tetrahedron Lett.* **1999**, *40*, 1341–1344.
[3] Halm, C.; Evarts, J.; Kurth, M. J. *Tetrahedron Lett.* **1997**, *38*, 7709–7712.
[4] Broaddus, C. D. *J. Org. Chem.* **1970**, *35*, 10–15.
[5] Lochmann, L.; Fréchet, J. M. J. *Macromolecules* **1996**, *29*, 1767–1771.
[6] Chalk, A. J. *Polymer Lett.* **1968**, *6*, 649–651.
[7] Schlosser, M.; Strunk, S. *Tetrahedron Lett.* **1984**, *25*, 741–744.
[8] Havez, S.; Begtrup, M.; Vedsø, P.; Andersen, K.; Ruhland, T. *J. Org. Chem.* **1998**, *63*, 7418–7420.
[9] Li, Z.; Ganesan, A. *Synlett* **1998**, 405–406.
[10] Garibay, P.; Vedsø, P.; Begtrup, M.; Hoeg-Jensen, T. *J. Comb. Chem.* **2001**, *3*, 332–340.
[11] Ruhland, T.; Andersen, K.; Pedersen, H. *J. Org. Chem.* **1998**, *63*, 9204–9211.
[12] Bernard, M.; Ford, W. T. *J. Org. Chem.* **1983**, *48*, 326–332.
[13] Tempest, P. A.; Armstrong, R. W. *J. Am. Chem. Soc.* **1997**, *119*, 7607–7608.
[14] Weinshenker, N. M.; Crosby, G. A.; Wong, J. Y. *J. Org. Chem.* **1975**, *40*, 1966–1971.
[15] Crosby, G. A.; Weinshenker, N. M.; Uh, H. S. *J. Am. Chem. Soc.* **1975**, *97*, 2232–2235.
[16] O'Brien, R. A.; Chen, T.; Rieke, R. D. *J. Org. Chem.* **1992**, *57*, 2667–2677.
[17] Gelens, E.; Koot, W. J.; Menge, W. M. P. B.; Ottenheijm, H. C. J.; Timmerman, H. *Bioorg. Med. Chem. Lett.* **2000**, *10*, 1935–1938.
[18] Boymond, L.; Rottländer, M.; Cahiez, G.; Knochel, P. *Angew. Chem. Int. Ed.* **1998**, *37*, 1701–1703.
[19] Rottländer, M.; Knochel, P. *J. Comb. Chem.* **1999**, *1*, 181–183.
[20] Kondo, Y.; Komine, T.; Fujinami, M.; Uchiyama, M.; Sakamoto, T. *J. Comb. Chem.* **1999**, *1*, 123–126.
[21] Brix, B.; Clark, T. *J. Org. Chem.* **1988**, *53*, 3365–3366.

[22] Mix, H. *Z. Chem.* **1979**, *19*, 148–149.
[23] Itsuno, S.; Darling, G. D.; Stöver, H. D. H.; Fréchet, J. M. J. *J. Org. Chem.* **1987**, *52*, 4644–4645.
[24] Harvey, S.; Raston, C. L. *J. Chem. Soc., Chem. Commun.* **1988**, 652–653.
[25] Pearson, W. H.; Clark, R. B. *Tetrahedron Lett.* **1997**, *38*, 7669–7672.
[26] Fyles, T. M.; Leznoff, C. C.; Weatherston, J. *Can. J. Chem.* **1977**, *55*, 4135–4143.
[27] Cohen, B. J.; Kraus, M. A.; Patchornik, A. *J. Am. Chem. Soc.* **1981**, *103*, 7620–7629.
[28] Jackson, R. F. W.; Oates, L. J.; Block, M. H. *Chem. Commun.* **2000**, 1401–1402.
[29] Karoyan, P.; Triolo, A.; Nannicini, R.; Giannotti, D.; Altamura, M.; Chassaing, G.; Perrotta, E. *Tetrahedron Lett.* **1999**, *40*, 71–74.
[30] Raghavan, S.; Tony, K. A.; Reddy, S. R. *Tetrahedron Lett.* **2001**, *42*, 8383–8386.
[31] Kawai, K.; Ohta, H.; Channing, M. A.; Kubodera, A.; Eckelman, W. C. *Appl. Radiat. Isot.* **1996**, *47*, 37–44.
[32] Sylvain, C.; Wagner, A.; Mioskowski, C. *Tetrahedron Lett.* **1998**, *39*, 9679–9680.
[33] Vanier, C.; Wagner, A.; Mioskowski, C. *Tetrahedron Lett.* **1999**, *40*, 4335–4338.
[34] Stranix, B. R.; Gao, J. P.; Barghi, R.; Salha, J.; Darling, G. D. *J. Org. Chem.* **1997**, *62*, 8987–8993.
[35] Piettre, S. R.; Baltzer, S. *Tetrahedron Lett.* **1997**, *38*, 1197–1200.
[36] Chan, T.; Huang, W. *J. Chem. Soc., Chem. Commun.* **1985**, 909–911.
[37] Schuster, M.; Lucas, N.; Blechert, S. *Chem. Commun.* **1997**, 823–824.
[38] Stranix, B. R.; Liu, H. Q.; Darling, G. D. *J. Org. Chem.* **1997**, *62*, 6183–6186.
[39] Hu, Y.; Porco, J. A.; Labadie, J. W.; Gooding, O. W.; Trost, B. M. *J. Org. Chem.* **1998**, *63*, 4518–4521.
[40] Suginome, M.; Iwanami, T.; Ito, Y. *J. Am. Chem. Soc.* **2001**, *123*, 4356–4357.
[41] Ruel, G.; The, N. K.; Dumartin, G.; Delmond, B.; Pereyre, M. *J. Organomet. Chem.* **1993**, *444*, C18–C20.
[42] Neumann, W. P.; Peterseim, M. *Reactive Polymers* **1993**, *20*, 189–205.
[43] Whitfield, D. M.; Ogawa, T. *Glycoconjugate J.* **1998**, *15*, 75–78.
[44] Enholm, E. J.; Schulte II, J. P. *Org. Lett.* **1999**, *1*, 1275–1277.
[45] Lee, C. Y.; Hanson, R. N. *Tetrahedron* **2000**, *56*, 1623–1629.
[46] Zhu, X.; Blough, B. E.; Carroll, F. I. *Tetrahedron Lett.* **2000**, *41*, 9219–9222.
[47] Xu, G.; Loftus, T. L.; Wargo, H.; Turpin, J. A.; Buckheit, R. W.; Cushman, M. *J. Org. Chem.* **2001**, *66*, 5958–5964.
[48] Hunter, D. H.; McRoberts, C. *Organometallics* **1999**, *18*, 5577–5583.
[49] Nicolaou, K. C.; Winssinger, N.; Pastor, J.; Murphy, F. *Angew. Chem. Int. Ed.* **1998**, *37*, 2534–2537.
[50] Gerigk, U.; Gerlach, M.; Neumann, W. P.; Vieler, R.; Weintritt, V. *Synthesis* **1990**, 448–452.
[51] Heinze, K. *Chem. Eur. J.* **2001**, *7*, 2922–2932.
[52] Büchele, J.; Mayer, H. A. *Chem. Commun.* **1999**, 2165–2166.
[53] Maud, J. M. in *Traditional Organic Reactions Aided by Solid Supports or Catalysts*, **1995**, pp 171–192.
[54] Fréchet, J. M. J.; Darling, G. D.; Itsuno, S.; Lu, P. Z.; de Meftahi, M. V.; Rolls, W. A. *Pure Appl. Chem.* **1988**, *60*, 353–364.
[55] Bailey, D. C.; Langer, S. H. *Chem. Rev.* **1981**, *81*, 109–148.
[56] Shuttleworth, S. J.; Allin, S. M.; Sharma, P. K. *Synthesis* **1997**, 1217–1239.
[57] Burgess, K.; Porte, A. M. *Advances in Catalytic Processes* **1997**, *2*, 69–82.
[58] Bergbreiter, D. E. *Macromol. Symposia* **1996**, *105*, 9–16.
[59] Drewry, D. H.; Coe, D. M.; Poon, S. *Med. Res. Rev.* **1999**, *19*, 97–148.
[60] Kobayashi, S.; Nagayama, S. *J. Am. Chem. Soc.* **1996**, *118*, 8977–8978.
[61] Kobayashi, S.; Nagayama, S.; Busujima, T. *Tetrahedron Lett.* **1996**, *37*, 9221–9224.
[62] Barrett, A. G. M.; de Miguel, Y. R. *Chem. Commun.* **1998**, 2079–2080.
[63] Bonds, W. D.; Brubaker, C. H.; Chandrasekaran, E. S.; Gibbons, C.; Grubbs, R. H.; Kroll, L. C. *J. Am. Chem. Soc.* **1975**, *97*, 2128–2132.
[64] Brouwer, A. J.; Van der Linden, H. J.; Liskamp, R. M. J. *J. Org. Chem.* **2000**, *65*, 1750–1757.
[65] Kobayashi, S.; Kusakabe, K.; Ishitani, H. *Org. Lett.* **2000**, *2*, 1225–1227.
[66] Matsunaga, S.; Ohshima, T.; Shibasaki, M. *Tetrahedron Lett.* **2000**, *41*, 8473–8478.
[67] Maiorana, S.; Seneci, P.; Rossi, T.; Baldoli, C.; Ciraco, M.; de Magistris, E.; Licandro, E.; Papagni, A.; Provera, S. *Tetrahedron Lett.* **1999**, *40*, 3635–3638.
[68] Canali, L.; Cowan, E.; Deleuze, H.; Gibson, C. L.; Sherrington, D. C. *Chem. Commun.* **1998**, 2561–2562.
[69] Angelino, M. D.; Laibinis, P. E. *Macromolecules* **1998**, *31*, 7581–7587.
[70] Havranek, M.; Singh, A.; Sames, D. *J. Am. Chem. Soc.* **1999**, *121*, 8965–8966.
[71] ter Halle, R.; Schulz, E.; Lemaire, M. *Synlett* **1997**, 1257–1258.

[72] Sandee, A. J.; Petra, D. G. I.; Reek, J. N. H.; Kamer, P. C. J.; Leeuwen, P. W. N. M. *Chem. Eur. J.* **2001**, *7*, 1202–1208.

[73] Yu, X.-Q.; Huang, J.-S.; Yu, W.-Y.; Che, C.-M. *J. Am. Chem. Soc.* **2000**, *122*, 5337–5342.

[74] Nieczypor, P.; Buchowicz, W.; Meester, W. J. N.; Rutjes, F. P. J. T.; Mol, J. C. *Tetrahedron Lett.* **2001**, *42*, 7103–7105.

[75] Fan, Q.-H.; Ren, C.-Y.; Yeung, C.-H.; Hu, W.-H.; Chan, A. S. C. *J. Am. Chem. Soc.* **1999**, *121*, 7407–7408.

[76] Ley, S. V.; Bolli, M. H.; Hinzen, B.; Gervois, A. G.; Hall, B. J. *J. Chem. Soc., Perkin Trans. 1* **1998**, 2239–2241.

[77] Nagayama, S.; Endo, M.; Kobayashi, S. *J. Org. Chem.* **1998**, *63*, 6094–6095.

[78] Kobayashi, S.; Endo, M.; Nagayama, S. *J. Am. Chem. Soc.* **1999**, *121*, 11229–11230.

[79] Comely, A. C.; Gibson, S. E.; Hales, N. J. *Chem. Commun.* **2000**, 305–306.

[80] Annis, D. A.; Jacobsen, E. N. *J. Am. Chem. Soc.* **1999**, *121*, 4147–4154.

[81] Peukert, S.; Jacobsen, E. N. *Org. Lett.* **1999**, *1*, 1245–1248.

[82] Prabhakaran, E. N.; Iqbal, J. *J. Org. Chem.* **1999**, *64*, 3339–3341.

[83] Dygutsch, D. P.; Eilbracht, P. *Tetrahedron* **1996**, *52*, 5461–5468.

[84] Andersen, J. A. M.; Karodia, N.; Miller, D. J.; Stones, D.; Gani, D. *Tetrahedron Lett.* **1998**, *39*, 7815–7818.

[85] Gilbertson, S. R.; Wang, X. F. *Tetrahedron Lett.* **1996**, *37*, 6475–6478.

[86] Arya, P.; Rao, N. V.; Singkhonrat, J.; Alper, H.; Bourque, S. C.; Manzer, L. E. *J. Org. Chem.* **2000**, *65*, 1881–1885.

[87] Bergbreiter, D. E.; Chandran, R. *J. Am. Chem. Soc.* **1987**, *109*, 174–179.

[88] Akelah, A.; Sherrington, D. C. *Chem. Rev.* **1981**, *81*, 557–587.

[89] Taylor, R. A.; Santora, B. P.; Gagné, M. R. *Org. Lett.* **2000**, *2*, 1781–1783.

[90] Boussie, T. R.; Murphy, V.; Hall, K. A.; Coutard, C.; Dales, C.; Petro, M.; Carlson, E.; Turner, H. W.; Powers, T. S. *Tetrahedron* **1999**, *55*, 11699–11710.

[91] Uozumi, Y.; Danjo, H.; Hayashi, T. *Tetrahedron Lett.* **1998**, *39*, 8303–8306.

[92] Uozumi, Y.; Danjo, H.; Hayashi, T. *J. Org. Chem.* **1999**, *64*, 3384–3388.

[93] Uozumi, Y.; Watanabe, T. *J. Org. Chem.* **1999**, *64*, 6921–6923.

[94] Trost, B. M.; Warner, R. W. *J. Am. Chem. Soc.* **1983**, *105*, 5940–5942.

[95] Zhang, T. Y.; Allen, M. J. *Tetrahedron Lett.* **1999**, *40*, 5813–5816.

[96] Tang, H. G.; Sherrington, D. C. *J. Mol. Catal.* **1994**, *94*, 7–17.

[97] Clark, A. J.; Filik, R. P.; Haddleton, D. M.; Radigue, A.; Sanders, C. J.; Thomas, G. H.; Smith, M. E. *J. Org. Chem.* **1999**, *64*, 8954–8975.

[98] Orlandi, S.; Mandoli, A.; Pini, D.; Salvadori, P. *Angew. Chem. Int. Ed.* **2001**, *40*, 2519–2521.

[99] Halm, C.; Kurth, M. J. *Angew. Chem. Int. Ed.* **1998**, *37*, 510–512.

[100] Francis, M. B.; Jacobsen, E. N. *Angew. Chem. Int. Ed.* **1999**, *38*, 937–941.

5 Preparation of Hydrocarbons

5.1 Preparation of Alkanes

The preparation of unfunctionalized alkanes on insoluble supports has only recently received attention. With the aim of preparing ever more elaborate molecules on solid phase, chemists are currently searching for robust methods to assemble complex carbon frameworks on insoluble supports. The synthesis of alkanes has also been investigated in this context.

Most of the reported preparations of alkanes on insoluble supports can be categorized as either hydrolyses of organometallic compounds, reductions, hydrogenations, or coupling reactions.

5.1.1 Preparation of Alkanes by Hydrolysis of Organometallic Compounds

This reaction is only of limited synthetic utility, and has mainly been used to verify that metallations had indeed taken place [1–3], or for the regioselective introduction of deuterium or tritium into a molecule. The solvolysis of silanes, organogermanium compounds, and phosphonium salts to yield alkanes with simultaneous cleavage from the support is discussed in Section 3.16.

5.1.2 Preparation of Alkanes by Hydrogenation and Reduction

Heterogeneous catalysts are generally incompatible with insoluble supports. Although the hydrogenolytic cleavage of benzylic C–N [4] and C–O bonds [5] on cross-linked polystyrene with palladium(II) acetate and hydrogen or cyclohexadiene has been reported, catalysis with elemental palladium has not been widely used to hydrogenate alkenes on solid supports. The hydrogenation of C–C double bonds on solid phase is generally accomplished using soluble reagents or catalysts. One reagent that has been successfully used to hydrogenate alkenes and alkynes on cross-linked polystyrene is diimide (Figure 5.1). This reagent can be generated by copper-catalyzed oxidation of hydrazine or by thermolysis of sulfonyl hydrazides. As illustrated by the examples in Figure 5.1, hydrogenations with diimide leave esters and nitro groups intact.

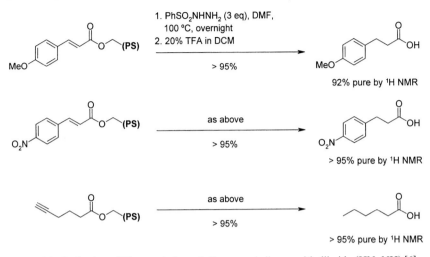

Figure 5.1. Reduction of Wang resin bound alkenes and alkynes with diimide (HN=NH) [6].

Further reagents suitable for the hydrogenation of alkenes on insoluble supports are silanes in the presence of TFA [7] and the copper(I) hydride complex [CuH(PPh$_3$)]$_6$ [8] (Figure 5.2). Catalytic hydrogenations can be performed with soluble catalysts such as RhCl(PPh$_3$)$_3$ (Wilkinson's catalyst) or related complexes. An example [9] of the use of a chiral rhodium(I) complex as a catalyst for the asymmetric hydrogenation of a 1-aminocinnamic acid derivative on polystyrene is shown in Figure 5.2. The yield and stereoselectivity of the hydrogenation proved to be highly depen-

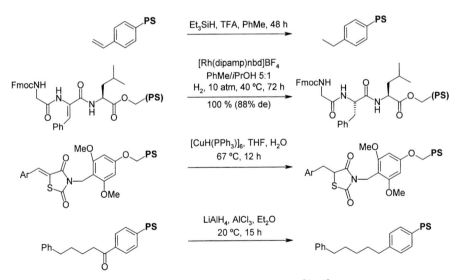

Figure 5.2. Reduction of polystyrene-bound alkenes and ketones [7–10].

dent on the solvent, but, under optimized conditions (Figure 5.2), attained values similar to those achievable in solution. THF, dioxane, and DCM proved unsuitable as solvents for the stereoselective hydrogenation [9].

Polystyrene-bound acetophenones have been directly reduced to the corresponding alkanes using lithium aluminum hydride in the presence of aluminum trichloride (Figure 5.2 [10]).

5.1.3 Preparation of Alkanes by Carbon–Carbon Bond Formation

C-Alkylations have been performed with both support-bound carbon nucleophiles and support-bound carbon electrophiles. Benzyl, allyl, and aryl halides or triflates have generally been used as the carbon electrophiles. Suitable carbon nucleophiles are boranes, organozinc and organomagnesium compounds. C-Alkylations have also been accomplished by the addition of radicals to alkenes. Polystyrene can also be alkylated under harsh conditions, e.g. by Friedel–Crafts alkylation [11–16] in the presence of strong acids. This type of reaction is incompatible with most linkers and is generally only suitable for the preparation of functionalized supports. Few examples have been reported of the preparation of alkanes by C–C bond formation on solid phase, and general methodologies for such preparations are still scarce.

5.1.3.1 Coupling Reactions with Group I Organometallic Compounds

Lithium-, sodium-, and potassium-derived organometallic reagents are highly reactive and can be handled only under strict exclusion of water and oxygen. As discussed in Section 4.1, organolithium and related reagents can be prepared on cross-linked polystyrene by direct lithiation or halogen–metal exchange. The treatment of these immobilized organometallic compounds with alkyl halides has been used as a means of preparing hydrocarbons on insoluble supports. Some of the few examples reported are listed in Table 5.1. Support-bound cuprates have been prepared by transmetallation of organocalcium compounds [1], and these also react with alkyl halides to yield alkanes (Entry **4**, Table 5.1). Alternatively, support-bound carbon electrophiles can be alkylated by treatment with organometallic reagents. This strategy has, for example, been used to prepare polystyrene-bound cyclopentadienes (Entry **5**, Table 5.1), which can be used as ligands for transition metals and for the preparation of 2,2-dialkyl-malonic acid derivatives [17–19].

Table 5.1. Alkylations with organolithium, organosodium, and organocopper compounds.

Entry	Starting resin	Conditions	Product	Ref.
1	Li—(aryl)—PS	R$\diagdown$Br PhMe, 65 °C, 24 h	R$\diagdown$(aryl)—PS	[20]
2	Li—(aryl)—PS	Cl$\diagdown$Br (10 eq), THF, 20 °C, 4 h (macroporous PS)	Cl$\diagdown$(aryl)—PS	[21]
3	Li—(thiophene)—CH$_2$O—C(Ph)$_2$—PS	MeI, THF, −30 °C to 20 °C, 2 h	(methyl-thiophene)—CH$_2$O—C(Ph)$_2$—PS	[22]
4	ClCa(CN)Cu—CH$_2$—(aryl)—PS	EtO$_2$C$\diagdown$Br THF, −45 °C, 2 h, 20 °C, 18 h	EtO$_2$C$\diagdown$(aryl)—PS	[1]
5	Cl—CH$_2$—(aryl)—PS	cyclopentadiene, NaNH$_2$, THF, 67 °C, 150 h (macroporous PS)	(cyclopentadienyl)—CH$_2$—(aryl)—PS (mixture of isomers)	[23] see also [20,24]
6	Li—(dithiane with Ph, S)—CH$_2$—PS	C$_6$H$_{13}$Br, THF, −20 °C, 1 h, −15 °C, 15 h	(Ph, dithiane)—CH$_2$—PS	[25]

5.1.3.2 Coupling Reactions with Group II Organometallic Compounds

Support-bound arylmagnesium and arylzinc compounds can be prepared from the corresponding haloarenes (Section 4.1) and alkylated with alkyl halides (Entries **1** and **2**, Table 5.2). Alternatively, support-bound carbon electrophiles, such as benzyl halides or aryl triflates, can be coupled with organomagnesium and organozinc compounds (Entries **3–6**, Table 5.2). Organozinc reagents are less reactive than organomagnesium or organolithium compounds, and are therefore compatible with several functional groups. However, alkylations with organozinc reagents often require catalysis by palladium or nickel complexes [26]. Unfortunately, organozinc compounds are not always easy to prepare. In view of the potential of these reagents, further developments in this field are to be expected.

Table 5.2. Alkylations with organomagnesium and organozinc compounds.

Entry	Starting resin	Conditions	Product	Ref.
1	BrMg (PS) resin	allyl bromide (50 eq), LiCl, CuCN, THF, −35 °C, 40 min	(PS) product	[27]
2	Li(tBu)₂Zn PS resin	MeI, THF, 0–20 °C	PS product	[28]
3	Cl–PS resin	allyl MgCl (2.5 eq), PhMe, THF, 60 °C, 12 h	PS product	[29]
4	TsO ... Si-PS resin	OSiMe₃ MgBr (30 eq), CuBr•Me₂S (3 eq), THF, −10 °C, 6 h	OSiMe₃ ... Si-PS product	[30] see also [31]
5	TfO ... (PS) resin	ZnBr with Cl groups, Pd(dba)₂, dppf, 65 °C, 16 h	(PS) product	[32]
6	I-benzamide-(PS) resin	NC-benzyl-ZnBr, Pd(dba)₂, PR₃, THF, 25 °C, 20 h	NC ... (PS) product	[32] see also [26]

5.1.3.3 Coupling Reactions with Boranes

Polystyrene-bound trialkylboranes, which can be prepared by hydroboration of support-bound alkenes with 9-BBN, undergo palladium-mediated coupling with alkyl, vinyl, and aryl iodides (Suzuki coupling; Entries **1** and **2**, Table 5.3; for vinylations, see Section 5.2.4). Because boranes are compatible with many functional groups and do not react with water, these coupling reactions could become a powerful tool for solid-phase synthesis. To date, however, few examples have been reported.

Alternatively, boranes can be prepared in solution and then coupled with support-bound carbon electrophiles. The Suzuki coupling of alkylboranes, generated in situ from 9-BBN and alkenes, with brominated cross-linked polystyrene has been used to link substituted alkyl chains directly to the polymer (Entry **4**, Table 5.3). Alkylboranes have also been used to alkylate polystyrene-bound aryl iodides (Entries **3** and **5**, Table 5.3).

Table 5.3. Alkylations with boranes.

Entry	Starting resin	Conditions	Product	Ref.
1	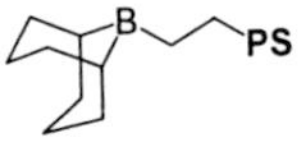	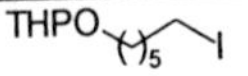 THPO—(⌇)₅—I (4 eq), PdCl₂(dppf) (0.3 eq), K₂CO₃ (3 eq), Triton B (1.5 eq), DMF, 85 °C, 14 h	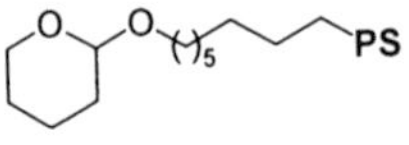	[33]
2	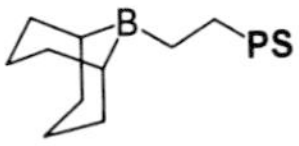	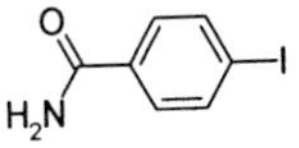 (4 eq), Pd(OAc)₂ (0.3 eq), PPh₃ (0.9 eq), NaOH (3 eq), Triton B (1.5 eq), DMF, 85 °C, 14 h	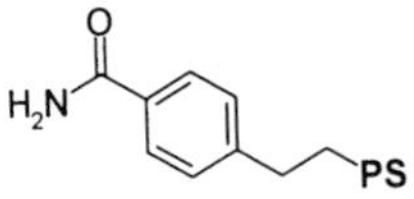	[33]
3	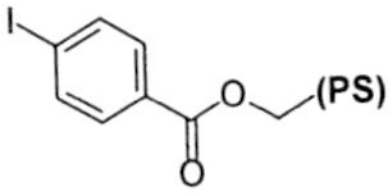	BBu₃, Pd(PPh₃)₄ or PdCl₂(dppf), K₂CO₃, DMF, 20 °C, 20 h	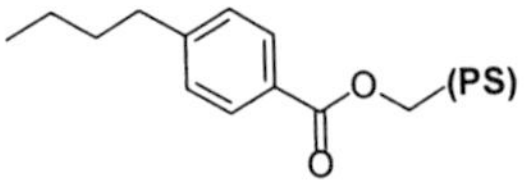	[34]
4	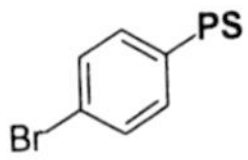	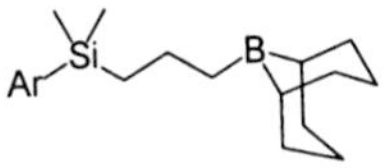 Pd(PPh₃)₄, Na₂CO₃, THF Ar: 4-(MeO)C₆H₄	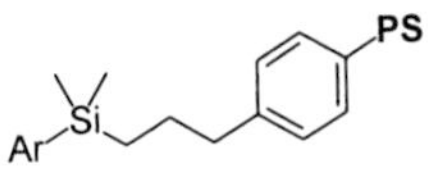	[35] see also [36]
5	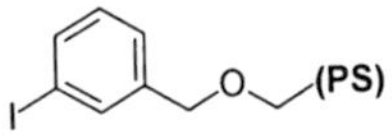	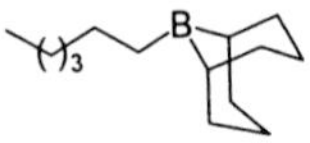 PdCl₂(MeCN)₂ (3 eq), dppf (0.2 eq), K₃PO₄ (6 eq), dioxane/DMF 1:6, 50 °C, 30 h	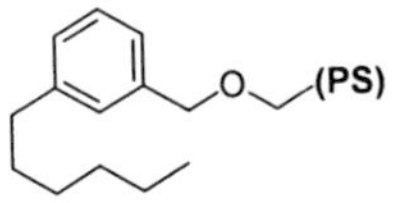	[37]

5.1.3.4 Coupling Reactions with Arylpalladium Compounds

The arylation shown in Figure 5.3 is a rare example of a palladium-mediated hydro-arylation of an alkene. Because of the polycyclic structure of the alkene, the inter-mediate formed by insertion of the alkene into the Pd–Ar bond does not undergo β-elimination (to yield the product of a normal Heck reaction), but remains unchanged. Reduction of this stable alkylpalladium intermediate with formic acid furnishes the formally hydrogenated Heck product [38].

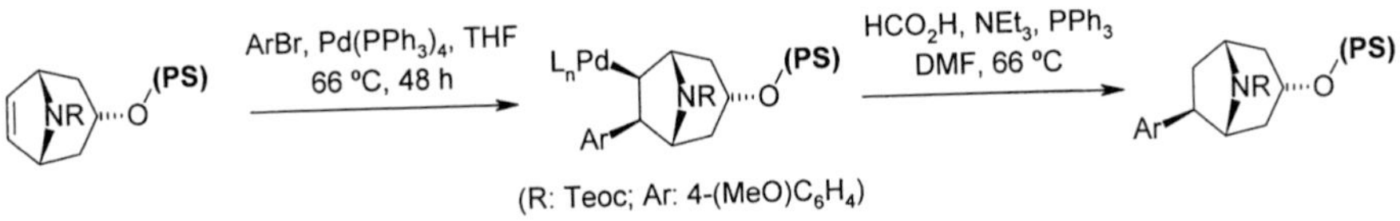

(R: Teoc; Ar: 4-(MeO)C₆H₄)

Figure 5.3. Palladium-mediated hydroarylation of polystyrene-bound alkenes [38].

5.1.3.5 Alkylations with Alkyl Radicals

Alkanes can be prepared by the addition of carbon radicals to C=C double bonds (Figure 5.4). The highest yields are usually obtained when electron-rich radicals (e.g. alkyl radicals or heteroatom-substituted radicals) add to acceptor-substituted alkenes, or when electron-poor radicals add to electron-rich double bonds. These reactions have also been performed on solid phase, and polystyrene-based supports seem to be particularly well suited for radical-mediated processes [39,40].

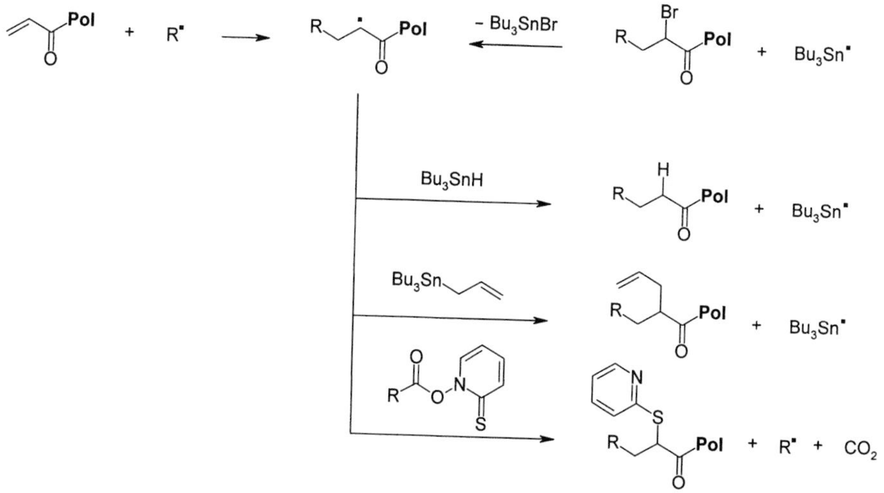

Figure 5.4. Generation, and illustrative transformations of support-bound, electron-poor radicals.

Table 5.4. Alkylations with carbon-centered radicals.

Entry	Starting resin	Conditions	Product	Ref.
1	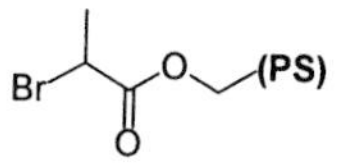 Br—⟍⟋O⟍(PS)	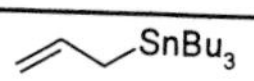 ⟍⟍SnBu₃ / AIBN, C₆H₆, 80 °C, 14 h	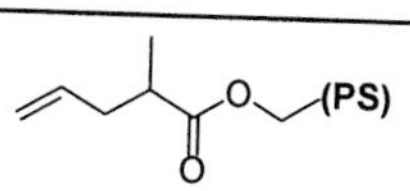 ⟍⟍⟍O⟍(PS)	[42] see also [41]
2	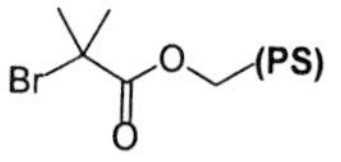 Br—⟍O⟍(PS)	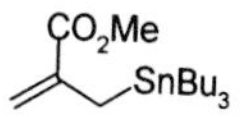 CO₂Me / ⟍SnBu₃ / AIBN, C₆H₆, 80 °C, 14 h	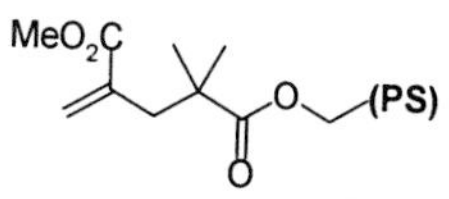 MeO₂C⟍⟍O⟍(PS)	[42]
3	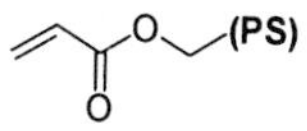 ⟍⟍O⟍(PS)	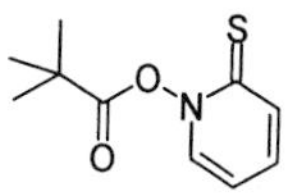 (10 eq), hv, DCM, 0 °C, 4 h	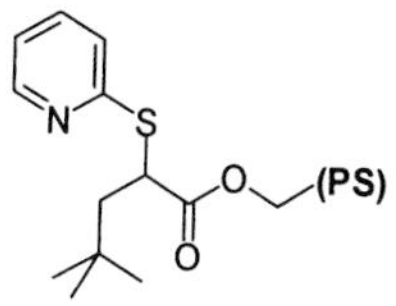 (pyridyl-S)⟍O⟍(PS)	[39]
4	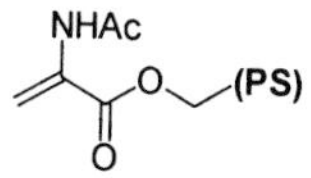 NHAc / ⟍O⟍(PS)	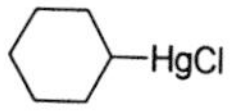 ⟨cyclohexyl⟩—HgCl (3 eq), NaBH₄ (8 eq), DCM, H₂O, 2 h	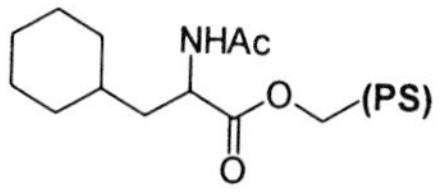 NHAc / ⟨cyclohexyl⟩⟍O⟍(PS)	[43]

Examples of radical-mediated C-alkylations are listed in Table 5.4. In these examples, radicals are formed by halogen abstraction with tin radicals (Entries **1** and **2**), by photolysis of Barton esters (Entry **3**), and by the reduction of organomercury compounds (Entry **4**). Carbohydrate-derived, polystyrene-bound α-haloesters undergo radical allylation with allyltributyltin with high diastereoselectivity (97% *de* [41]). Cleavage from supports by homolytic bond fission with simultaneous formation of C–H or C–C bonds is considered in Section 3.16.

5.1.3.6 Preparation of Cycloalkanes

In addition to the alkylations discussed above, some special reactions have been reported that enable the solid-phase synthesis of cycloalkanes. These include the intramolecular ene reaction and the cyclopropanation of alkenes (Figure 5.5; see also [44]). Cyclobutanes have been prepared by the reaction of polystyrene-bound carbanions with epichlorohydrin, and by [2 + 2] cycloadditions of ketenes to resin-bound alkenes.

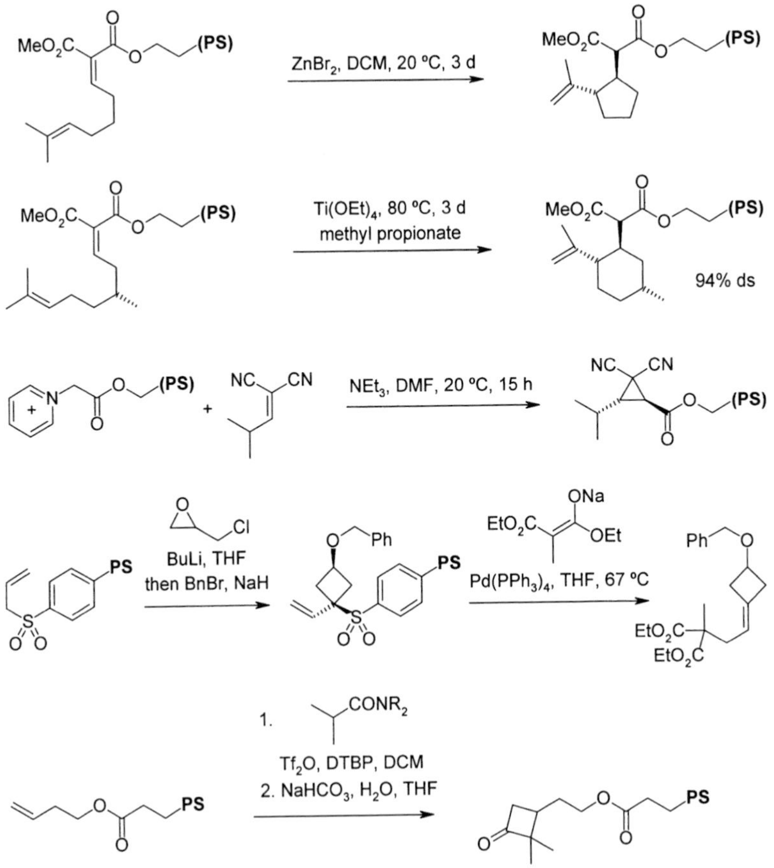

Figure 5.5. Preparation of cycloalkanes on cross-linked polystyrene [45–48].

Large cycloalkanes have been prepared by C-alkylation of resin-bound 2-alkylmalonic acid esters [18].

5.2 Preparation of Alkenes

5.2.1 Preparation of Alkenes by β-Elimination and Reduction

Linkers have been developed that enable the release of amines or other compounds by base-induced β-elimination; these linkers are thereby converted into alkenes (Chapter 3). Similarly, the β-elimination of resin-bound leaving groups has been used as a cleavage strategy for the release of alkenes from supports (Section 3.15.1).

Examples of the generation of alkenes by β-elimination are sketched in Figure 5.6. These reactions generally proceed under conditions similar to those used in solution. Further examples have been reported [49,50]. 1,3-Dienes have been prepared by thermolysis of polystyrene-bound 2,5-dihydrothiophene-1,1-diones [51]. Internal alkynes attached to cross-linked polystyrene can be stereoselectively reduced to Z-olefins by

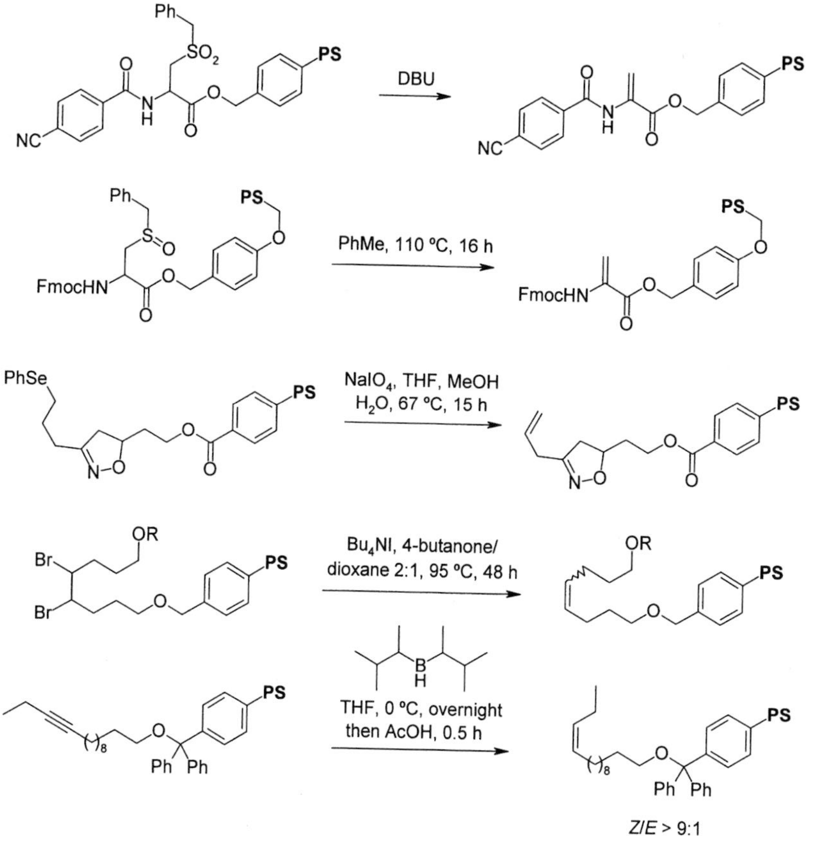

Figure 5.6. Preparation of resin-bound alkenes by β-elimination and reduction [52–56].

treatment with bis(1,2-dimethylpropyl)borane (Figure 5.6 [52]). Other reducing agents, such as 9-BBN, catechol borane, or diisobutylaluminum hydride, require higher reaction temperatures and lead to poor stereoselectivity.

Vinylpolystyrene, a useful intermediate for the preparation of various functionalized supports for solid-phase synthesis [7,57–59], has been prepared by the polymerization of divinylbenzene [7], by Wittig reaction of a Merrifield resin derived phosphonium salt with formaldehyde [59–62], or, most conveniently, by treatment of Merrifield resin with trimethylsulfonium iodide and a base [63] (Figure 5.7).

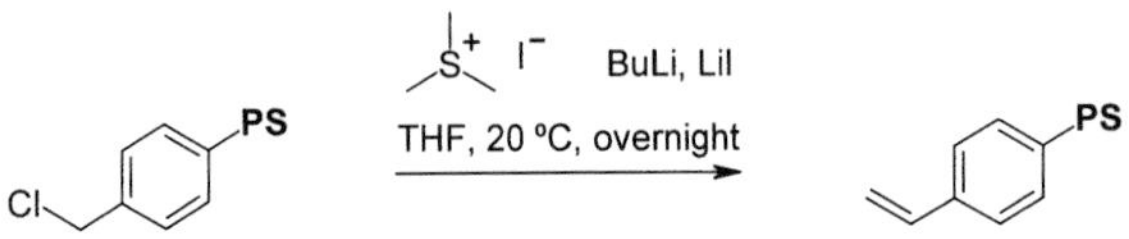

Figure 5.7. One-step preparation of vinylpolystyrene from Merrifield resin [63].

5.2.2 Preparation of Alkenes by Carbonyl Olefination

Most carbonyl olefinations known to proceed in solution have also been successfully accomplished on insoluble supports. Those Wittig reactions that lead to release of the alkene into solution (because phosphorus is irreversibly bound to the support) are considered in Section 3.15.

5.2.2.1 By Wittig Reaction

Support-bound aldehydes or ketones can be readily olefinated with various types of phosphorus ylide (Table 5.5). Stabilized (acceptor-substituted) ylides, which can even be generated with weak bases, generally react more slowly than non-stabilized ylides, which, accordingly, are only formed upon treatment with strong bases. For Wittig reactions requiring strong bases, hydrophobic supports, such as cross-linked polystyrene, are generally more convenient than hydrophilic supports, because the latter are more difficult to dry.

Polystyrene-bound benzaldehydes can be smoothly olefinated with benzyl- or cinnamylphosphonium salts in DMF or THF using sodium methoxide as a base (Entry **1**, Table 5.5 [64–67]). Alkylphosphonium salts, however, only react with resin-bound aldehydes upon deprotonation with stronger bases, such as butyllithium [30,68–70]. The more acidic acceptor-substituted phosphonium salts, on the other hand, even react with resin-bound aldehydes and ketones upon treatment with tertiary amines, DBU, sodium ethoxide, or lithium hydroxide [71–75], but stronger bases are also used occasionally [76].

Wittig reactions can also be performed with support-bound phosphorus ylides. Polystyrene-bound alkylphosphonium salts have been prepared from the corresponding alkyl mesylates or halides and trialkyl- or triarylphosphines (Figure 5.8 [60,80]). Because polystyrene is a hydrophobic support, salt formation does not proceed smoothly and quaternization of phosphines generally requires forcing conditions. The

Table 5.5. Wittig olefination of support-bound aldehydes and ketones.

Entry	Starting resin	Conditions	Product	Ref.
1		BnPPh₃Br (0.15 mol/L), NaOMe (0.77 mol/L), DMF, 20 °C, 24 h		[77] see also [30]
2		MeO₂C⌒PPh₃ THF, 65 °C, 2 h		[71] see also [78]
3		EtO₂C⌒PO(OEt)₂ DBU, LiBr, MeCN, 50 °C, 10 h		[71]
4		Ph₃PMeBr (5 eq), NaN(SiMe₃)₂, THF, 20 °C, 18 h		[79]

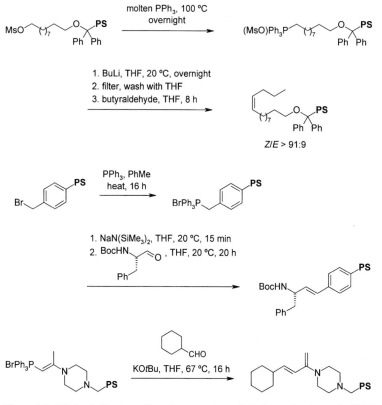

Figure 5.8. Wittig olefination with polystyrene-bound alkylphosphonium salts [68,83,84].

resulting phosphonium salts generally require the presence of a strong base such as butyllithium [68,81,82] or sodium bis(trimethylsilyl)amide [80,83] in order to react with carbonyl compounds (Figure 5.8).

Highly reactive, polystyrene-bound alkylating agents, such as halo ketones [85], haloacetic acid derivatives [86,87], or benzyl halides [62,81,83] can be converted into phosphonium salts under milder conditions than those used with alkyl halides (Figure 5.9). Resin-bound, acceptor-substituted phosphonium salts or phosphonates can undergo Wittig olefination in the presence of weak bases such as tertiary amines [87–90], DBU [86,91], or sodium hydroxide [85]. However, stronger bases (BuLi [86]; LiN(SiMe$_3$)$_2$ [92,93]; KO*t*Bu [94]) are often used to achieve quantitative ylide formation and thereby accelerate the reaction (Figure 5.9).

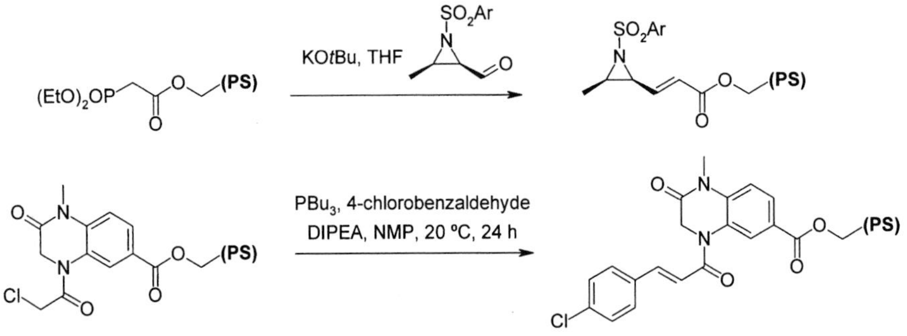

Figure 5.9. Wittig olefination with acceptor-substituted, polystyrene-bound phosphonates and phosphonium salts [87,94]. Ar: 2-nitrophenyl.

5.2.2.2 By Aldol and Related Condensations

Most C,H-acidic compounds can be condensed with aldehydes or ketones to yield alkenes. Some of these reactions have also been realized on insoluble supports, with either the C,H-acidic (nucleophilic) reactant or the electrophilic reactant linked to the support. Some illustrative examples are listed in Table 5.6. Polystyrene-bound malonic esters or amides, cyanoacetamides, nitroacetic ester [95], and 3-oxo esters undergo Knoevenagel condensation with aromatic or aliphatic aldehydes. Catalytic amounts of piperidine and heating are generally required, although reactive substrates can react at room temperature.

One problem that may be encountered with this type of reaction is related to the fact that in solid-phase synthesis reagents generally have to be used in excess. This can occasionally lead to side reactions. If, for instance, a resin-bound aldehyde is condensed under basic conditions with excess ketone, the resulting enone might undergo Michael addition with an additional equivalent of ketone (Figure 5.10). For some of these enones, the Wittig reaction may represent a superior synthetic method [75].

Table 5.6. Aldol-type condensation reactions on insoluble supports.

Entry	Starting resin	Conditions	Product	Ref.
1		(3 eq), piperidinium acetate (0.2 eq), DCM, 20 °C, 3 h		[45] see also [96–98]
2		piperidine, iPrOH/C$_6$H$_6$, 60 °C		[99] see also [100]
3		1-methylisatin (10 eq), DMF/piperidine 6:1, 20 °C, 20 h		[101]
4		ArCHO (0.22 mol/L, 10 eq), piperidinium acetate (1 eq), C$_6$H$_6$, 80 °C, 24 h Ar: 4-hydroxyphenyl		[8] see also [102]
5		ArCHO, NaOMe (0.25 mol/L), MeOH/ THF 1:1, 20 °C, 4 d Ar: 4-methoxyphenyl		[103] see also [104–106]
6		malonic acid (0.22 mol/L), piperidine/pyridine 1:60, 115 °C, 4 h		[81]
7		MeNO$_2$/AcOH 1:1, NH$_4$OAc (0.87 mol/L), 116 °C, 15 h		[107]
8		PhCOMe (0.4 mol/L), DME, LiOH, 20 °C, 16 h		[75] see also [77]

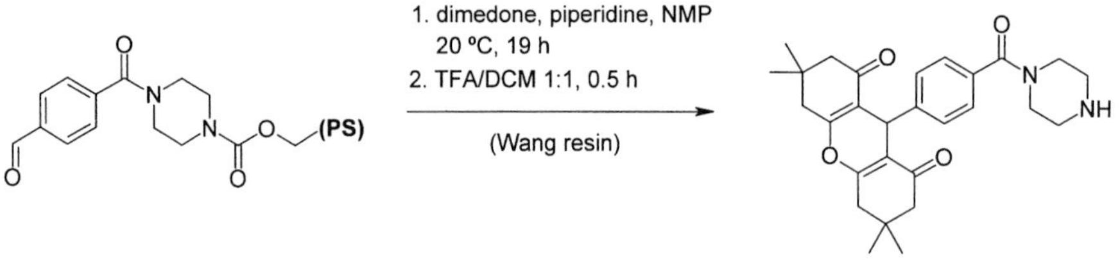

1. dimedone, piperidine, NMP
   20 °C, 19 h
2. TFA/DCM 1:1, 0.5 h

(Wang resin)

Figure 5.10. Double condensation of resin-bound aldehydes with 1,3-diketones [108].

5.2.2.3 By Other Carbonyl Olefinations

Phosphorus ylides are generally not suitable for the olefination of esters. This reaction can, however, be accomplished with titanium carbene complexes. The Tebbe reagent, a precursor of $Cp_2Ti=CH_2$, as well as other titanium alkylidenes, have been used to olefinate polystyrene-bound esters [109,110] (Figure 5.11). Substituted titanium carbene complexes $Cp_2Ti=CR_2$ can be prepared by mixing dithioacetals with $Cp_2Ti[P(OR)_3]_2$, which is obtained by reducing Cp_2TiCl_2 with magnesium in the presence of a phosphite [111]. The resulting enol ethers can either be cleaved from the support by hydrolysis, thereby yielding methyl ketones, or be used as starting materials for Diels–Alder reactions or the synthesis of thiazoles [109] (Section 15.13) or benzofurans [110].

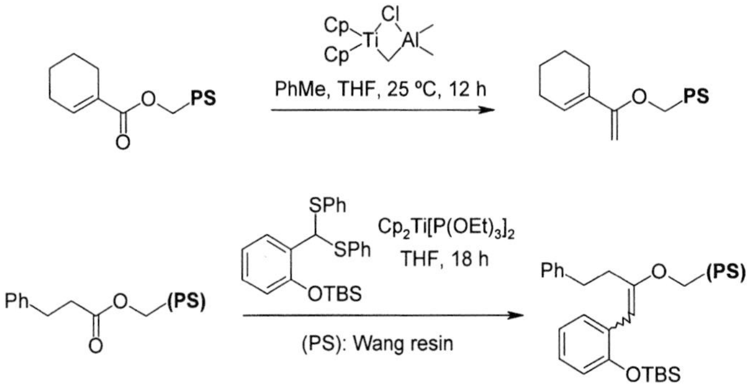

PhMe, THF, 25 °C, 12 h

$Cp_2Ti[P(OEt)_3]_2$

THF, 18 h

(PS): Wang resin

Figure 5.11. Olefination of esters using the Tebbe reagent [109,110].

5.2.3 Preparation of Alkenes by Olefin Metathesis

With the discovery by Grubbs of ruthenium carbene complexes such as $Cl_2(PCy_3)_2Ru=CHR$, which mediate olefin metathesis under mild reaction conditions and which are compatible with a broad range of functional groups [111], the application of olefin metathesis to solid-phase synthesis became a realistic approach for the preparation of alkenes. Both ring-closing metathesis and cross-metathesis of alkenes and alkynes bound to insoluble supports have been realized (Figure 5.12).

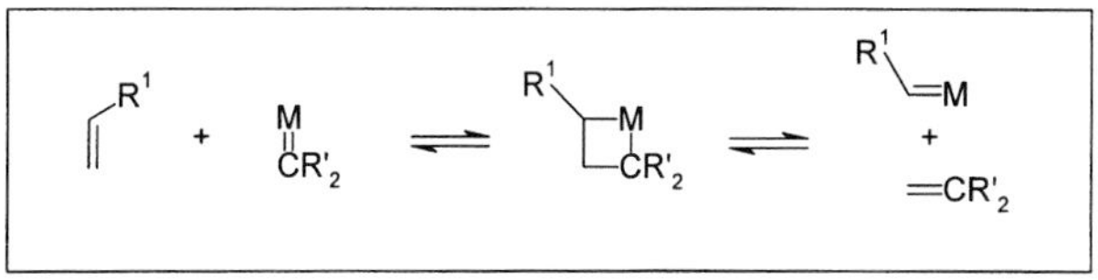

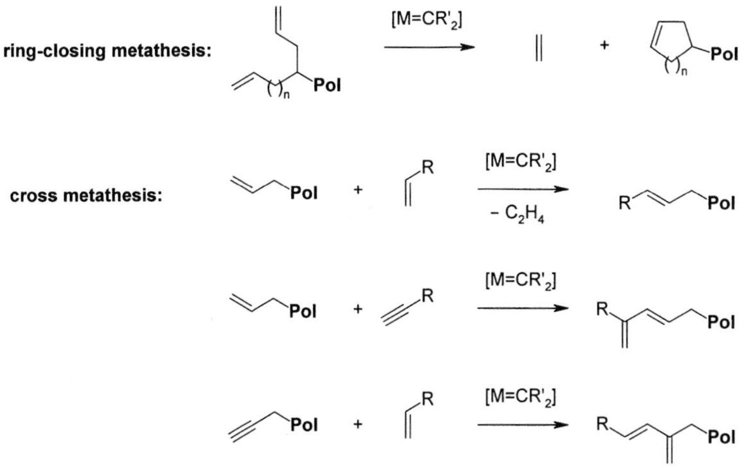

ring-closing metathesis:

cross metathesis:

Figure 5.12. Main strategies for the application of olefin metathesis to solid-phase synthesis.

Carbene complex catalyzed olefin metathesis proceeds most efficiently in solvents of low nucleophilicity, such as DCM or toluene [111,112]. Tertiary amines, pyridine, or other nucleophiles can poison the (electrophilic) catalyst, and should be carefully removed (e.g. by washing with acetic acid) before addition of the carbene complex. Alternatively, Lewis acids (e.g. Ti(OiPr)$_4$ [113]) can be added to the reaction mixture to mask nucleophilic sites and thereby improve the performance of the catalyst. Electron-rich alkenes, such as enol ethers or enamines, do not generally undergo metathesis, because the intermediate donor-substituted carbene complexes are no longer sufficiently electrophilic to catalyze olefin metathesis. The rate of metathesis sharply decreases in the series terminal alkene > internal disubstituted alkene > trisubstituted alkene. Because olefin metathesis is reversible, ring-closing metathesis is not usually well suited to the preparation of strained rings, but works best if five- or six-membered rings are formed. Five- to eight-membered nitrogen-containing heterocycles [114–118] and macrocyclic peptoids [119,120] have been prepared by ring-closing metathesis on cross-linked polystyrene and polystyrene–PEG graft polymers. Some illustrative examples of such cyclizations are shown in Figure 5.13. Ring-closing metathesis has also been used as a cleavage reaction, to release acyclic alkenes or cycloalkenes into solution. These reactions are discussed in Section 3.15.2.

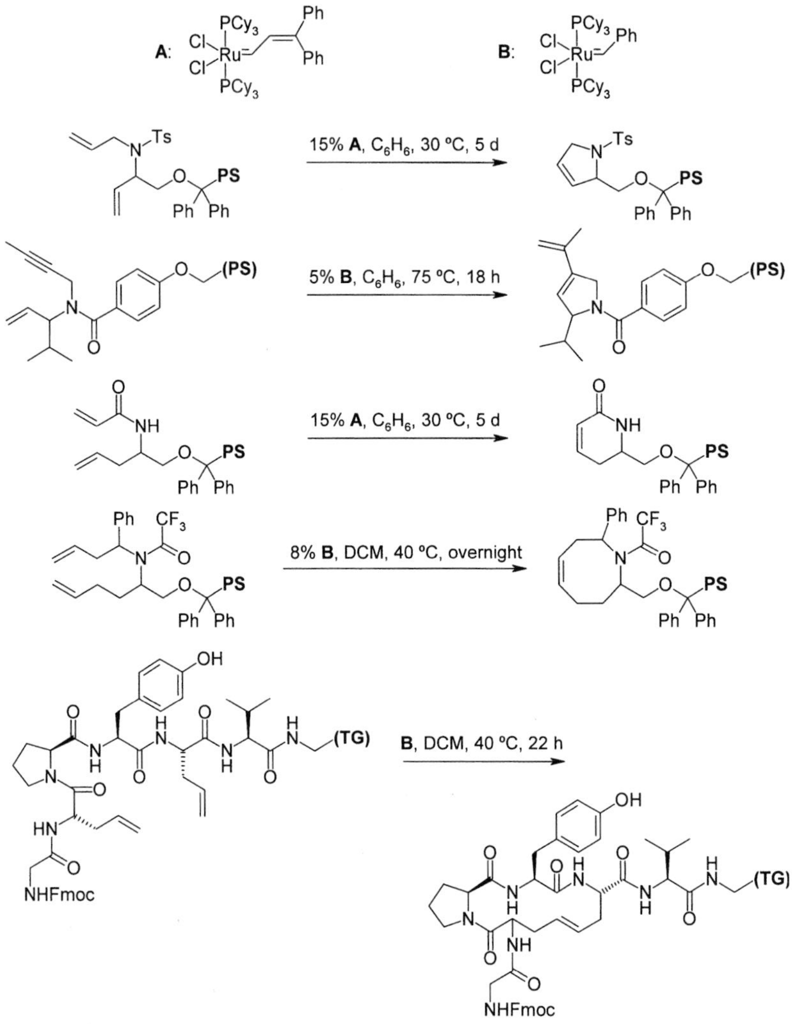

Figure 5.13. Ring-closing metathesis on insoluble supports [114–116,119].

Polycyclic structures have been prepared on solid phase by cascade reactions involving ring-opening and ring-closing metathesis. An impressive example, reported by Schreiber et al. [121], is outlined in Figure 5.14. The precursor for this ring-opening, ring-closing metathesis cascade was prepared by an Ugi reaction followed by intramolecular Diels–Alder reaction and N-allylation of the resulting bis-amide.

Various research groups have investigated the preparation of cyclic peptides by ring-closing metathesis on solid phase. Liskamp et al. have investigated the cyclization of *N*-alkenylated peptides by ring-closing metathesis (Figure 5.15 [113,122,123]). The peptides were prepared by solid-phase synthesis, and the *N*-alkenylation was effected during the assembly of the peptide by *N*-sulfonylation with 2-nitrobenzenesulfonyl chloride followed by *N*-alkenylation under Mitsunobu conditions and sulfonamide

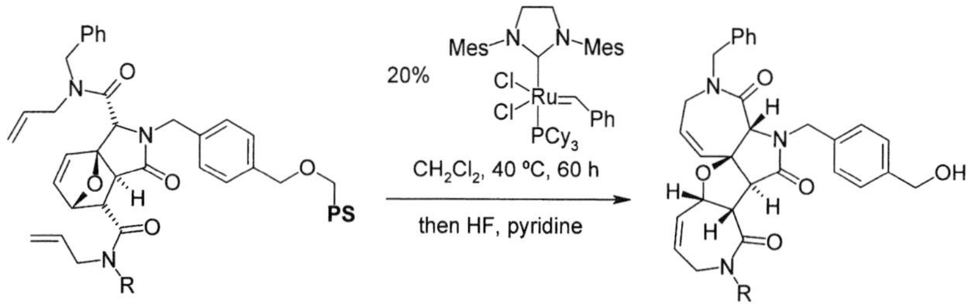

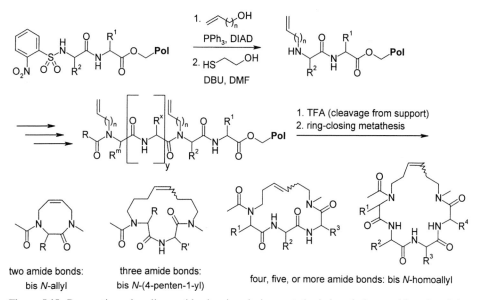

Figure 5.14. Synthesis of polycyclic structures using a ring-opening, ring-closing metathesis cascade [121].

cleavage by treatment with mercaptoethanol/DBU. Ring-closing metathesis could be performed either in solution or on solid phase, although higher yields were usually obtained in solution [113]. The main side reactions were the formation of dimers by cross-metathesis and isomerization of the double bonds. This isomerization appeared to be catalyzed by degradation products of the catalyst, and should therefore be avoidable by using more efficient and stable catalysts than $Cl_2(PCy_3)_2Ru=CHR$, such as ruthenium carbene complexes with heterocyclic carbene ligands (see, e.g., Figure 5.14 and [124–126]).

Experiments aimed at the cyclization of the peptide-derived dienes showed that the length of the *N*-alkenyl group was crucial for ring-closure. *N*-Allyl peptides could only be cyclized to yield eight-membered rings; larger ring sizes required the use of *N*-homoallyl or *N*-(4-penten-1-yl) peptides. The cyclization of tripeptides (to form a 15-membered ring) was particularly difficult, and only proceeded satisfactorily with *N*-(4-penten-1-yl) substitution (Figure 5.15).

Figure 5.15. Preparation of cyclic peptides by ring-closing metathesis in solution, and lengths of the *N*-alkenyl substituent required for ring formation [123].

Cross-metathesis enables the efficient preparation of acyclic alkenes and 1,3-dienes on insoluble supports (Figure 5.16). Unfortunately, some types of substrate show a high tendency to yield products of self-metathesis, i.e. symmetrical internal alkenes produced by dimerization of the resin-bound alkene. This is the case, for instance, with allylglycine and homoallylglycine derivatives. Dimerization of the resin-bound alkene can, however, be effectively suppressed by reducing the loading of the support [127].

An interesting variant of cross-metathesis is the so-called ring-opening cross-metathesis [128]. When strained cyclic alkenes (e.g. cyclobutene, norbornene) are treated with metathesis catalysts, rapid and irreversible ring-opening occurs. The newly formed carbene complexes can then react with a second alkene to yield the products of cross-metathesis. One example of ring-opening cross-metathesis on cross-linked polystyrene [128] is shown in Figure 5.16.

Polystyrene cross-linked with 1–2% DVB is sufficiently flexible to allow intermediates attached to it to react with each other. Terminal alkenes linked to this support can therefore undergo self-metathesis to yield symmetrical, internal alkenes when treated with a suitable catalyst [133,134]. Self-metathesis of support-bound *N*-alkenoylated peptides has been used by Conde-Frieboes et al. [133] for the preparation of symmetrical peptidomimetics (Figure 5.17). Various peptides were prepared on cross-

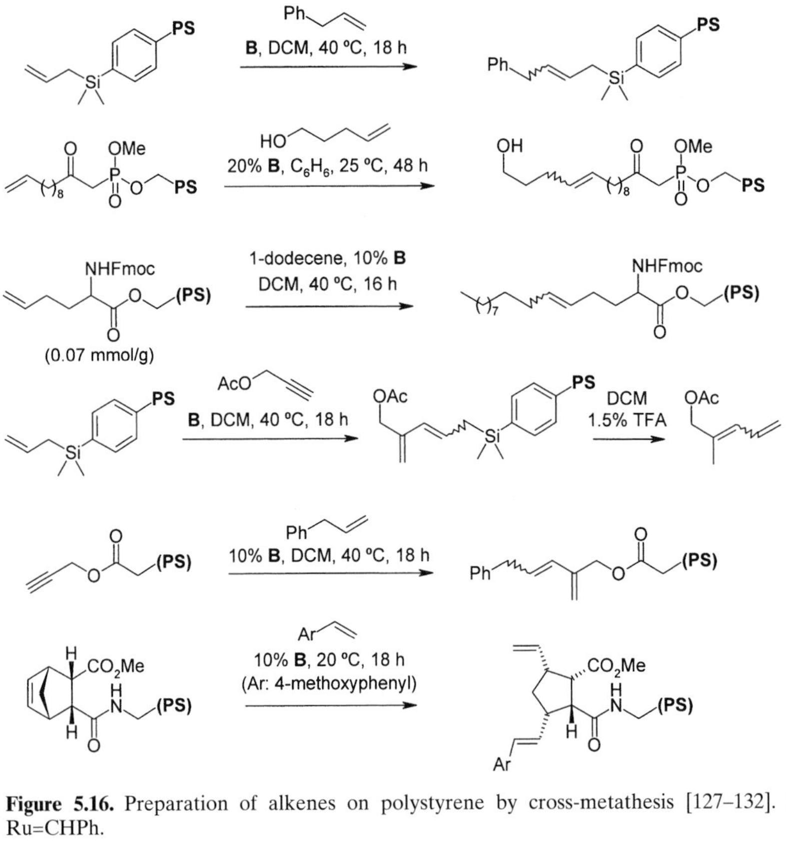

Figure 5.16. Preparation of alkenes on polystyrene by cross-metathesis [127–132]. B: $Cl_2(Cy_3P)_2$-Ru=CHPh.

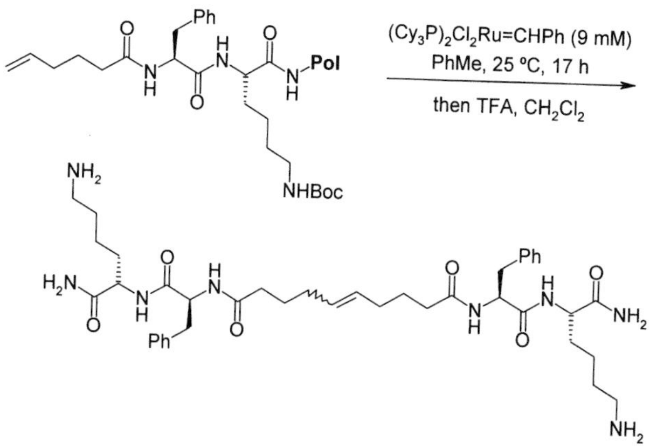

Figure 5.17. Self-metathesis of support-bound *N*-alkenoyl peptides [133].

linked polystyrene using Fmoc methodology and then acylated with an ω-alkenoic acid. The authors found that neither 3-butenamides nor 4-pentenamides could undergo efficient self-metathesis, which may be due to the formation of stable chelate complexes. With longer alkenoic acids, however, the peptide dimers were formed in high yield and with high purity as mixtures of *E/Z* isomers.

5.2.4 Preparation of Alkenes by C-Vinylation

Some types of alkene can be conveniently prepared by the alkylation or arylation of vinyl groups. The Heck, Stille, and Suzuki couplings all belong to this group of reactions. In the Heck reaction, an arene or heteroarene with a good leaving group (iodide, bromide, triflate, phenyliodonium, diazonium) is coupled with an alkene (usually electron-poor) in the presence of palladium(0) and a base. This useful reaction has also been realized on solid phase, with either the aryl halide or the alkene linked to the support (Table 5.7). Further examples have been reported [40,135–140]. Aryl iodides, substituted with either electron-withdrawing or electron-donating groups, have usually been used as the resin-bound aryl halide. Typical alkenes are acrylic esters or amides, acrylonitrile, methacrylates, crotonates, vinylsulfones, α,β-unsaturated aldehydes or ketones, styrene, cinnamates, and 4-vinylbenzoates. The Heck reaction has also been used for macrocyclizations on insoluble supports [136]. Heck reactions that lead to heterocycles are considered in Chapter 15.

Table 5.7. C-Vinylation by Heck reaction on insoluble supports.

Entry	Starting resin	Conditions	Product	Ref.
1		PhI, Pd(OAc)$_2$, NEt$_3$, DMF, Bu$_4$NCl, 90 °C, 16 h		[141]
2		2-bromonaphthalene, Pd$_2$(dba)$_3$, NEt$_3$, DMF, P(*o*-Tol)$_3$, 100 °C, 20 h		[141]
3		[Ph$_2$I][BF$_4$] (0.05 mol/L, 1.5 eq), Pd$_2$(dba)$_3$ (0.05 eq), NaHCO$_3$ (2 eq), DMF, P(*o*-Tol)$_3$ (0.1 eq), 40 °C, 20 h		[142]
4		4-Tol–I (0.2 mol/L), Pd(OAc)$_2$ (0.01 mol/L), NaHCO$_3$ (0.01 mol/L), DMF, 145 °C, 20 h		[143]
5		Pd(OAc)$_2$, NEt$_3$, DMF, Bu$_4$NCl, 90 °C, 16 h	(E: CO$_2$Me)	[141]

Table 5.7. continued.

Entry	Starting resin	Conditions	Product	Ref.
6		methacrylamide (0.3 mol/L), Pd(OAc)$_2$ (0.05 mol/L), PPh$_3$ (0.1 mol/L), Bu$_4$NCl (0.1 mol/L), NEt$_3$/DMF/H$_2$O 1:9:1, 37 °C, 4 h		[144]
7		styrene (8 eq), Pd(OAc)$_2$ (0.25 eq), NaOAc (3 eq), Bu$_4$NCl (2 eq), DMA, 100 °C, 24 h		[145]
8		(15 eq), Pd(OAc)$_2$ (0.5 eq), NaOAc (5 eq), Bu$_4$NCl (4 eq), DMA, 100 °C, 2 × 24 h		[145]
9		cyclohexene, Pd(OAc)$_2$, PPh$_3$, NEt$_3$, DMF, 80 °C, ultrasound, 24 h		[40]
10		(4 eq), Pd(PPh$_3$)$_4$ (0.15 eq), DIPEA (8 eq), MeCN, 70 °C, 3 d		[146]

Experimental Procedure 5.1: Heck reaction with a polystyrene-bound aryl iodide [145]

To Rink amide resin acylated with 3-iodobenzyloxyacetic acid (0.2 g, 0.094 mmol) were added DMA (4.7 mL), sodium acetate (23 mg, 0.28 mmol, 3 equiv.), Bu$_4$NCl (56 mg, 0.20 mmol, 2 equiv.), and dimethyl itaconate (119 mg, 0.75 mmol, 8 equiv.), and the mixture was degassed for 20 min. Palladium(II) acetate (5 mg, 0.022 mmol, 0.2 equiv.) was then added and the mixture was shaken at 100 °C for 24 h. The support was subsequently washed with dioxane, water, ethanol, and further dioxane, and the coupling reaction was repeated once more. After extensive washing, the support was dried under reduced pressure and treated with TFA/DCM (2:8; 5 mL, 4 × 1 min). The combined filtrates were concentrated. The residue was dried by coevaporation of the solvents with toluene (4 ×) and dried under reduced pressure to yield 33 mg of the coupling product, which was slightly contaminated with tetrabutylammonium salts. The yield after purification was 67%.

In the Suzuki reaction, an aryl iodide or synthetic equivalent thereof is coupled with an arylboronic acid or a borane, again using palladium(0) as the catalyst. This reaction is usually used to prepare biaryls, and few examples have been reported of the solid-phase synthesis of alkenes by means of a Suzuki coupling (Table 5.8).

Vinylstannanes have also been successfully used on solid phase for transition metal mediated coupling reactions (Stille coupling). As for the Suzuki reaction, most of the reported examples have been biaryl formations. Table 5.9 lists illustrative examples of Stille couplings leading to the formation of alkenes on solid phase. Further examples include the arylation of polystyrene-bound, 17α-ethynyl estradiol-derived vinylstannanes [151], and the formation of 1,1-diarylethenes from resin-bound vinylstannanes [152]. From the results reported to date, it might be concluded that for Stille couplings catalytic systems consisting of Pd$_2$(dba)$_3$ and tris(2-furyl)phosphine [153], triphenylarsine [37,152], or tris(2-tolyl)phosphine are superior to Pd(PPh$_3$)$_4$.

Table 5.8. C-Vinylation by Suzuki coupling on insoluble supports.

Entry	Starting resin	Conditions	Product	Ref.
1		Ph—, B(OR)$_2$ Pd(PPh$_3$)$_2$Cl$_2$, KOH (3 mol/L), H$_2$O, DME, 80 °C, 18 h		[147,148]
2		9-hexyl-9-BBN, Pd(PPh$_3$)$_4$, Na$_2$CO$_3$ (2 mol/L), H$_2$O, THF, 65 °C Ar: 4-(MeO)C$_6$H$_4$		[149,150]
3		9-hexyl-9-BBN, Pd(PPh$_3$)$_4$, Na$_2$CO$_3$ (2 mol/L), H$_2$O, THF, 65 °C Ar: 4-(MeO)C$_6$H$_4$		[149]
4		TMSO, B (1.3 eq), Pd(PPh$_3$)$_4$ or PdCl$_2$(dppf) (0.1 eq), K$_2$CO$_3$ (2 eq), DMF, 20 °C, 20 h		[34]
5		THPO, I (4 eq), PdCl$_2$(dppf) (0.3 eq), K$_2$CO$_3$ (3 eq), Triton B (1.5 eq), DMF, 85 °C, 14 h		[33]

Table 5.9. C-Vinylation with stannanes on insoluble supports.

Entry	Starting resin	Conditions	Product	Ref.
1		EtO‿SnBu$_3$ (5 eq), Pd$_2$(dba)$_3$ (0.05 eq), AsPh$_3$ (0.2 eq), NMP, 70 °C, 15 h		[135]
2		(3 eq), Pd$_2$(dba)$_3$ (0.1 eq), AsPh$_3$ (0.4 eq), dioxane, 50 °C, 24 h		[37] see also [154]
3		Ph‿SnBu$_3$ CuI (0.1 eq), NaCl (2 eq), NMP, 100 °C, 20 h		[155]
4		‿SnBu$_3$ (3 eq), Pd(PPh$_3$)$_4$ (0.05 eq), DMF, 60 °C, 24 h		[153]
5		HO‿SnBu$_3$ (2 eq), Pd$_2$(dba)$_3$ (0.1 eq), AsPh$_3$ (0.4 eq), NMP, 20 °C, overnight		[89]

5.2.5 Preparation of Cycloalkenes by Cycloaddition

Diels–Alder reactions have been conducted on solid phase, with either the dienophile or the diene linked to the support [156]. The reaction conditions and the regio- and stereoselectivities observed are similar to those in solution [58,157,158]. Illustrative examples of Diels–Alder reactions leading to support-bound cyclohexenes are listed in Table 5.10. Further examples include the cycloaddition of polystyrene-bound 2-sulfonyl-1,3-butadiene and N-phenylmaleimide [51], the high-pressure cycloaddition of 1,3-butadienes to resin-bound 1-nitroacrylates [95], and the intramolecular Diels–Alder reaction of styrenes with acrylates [159].

Further cycloadditions used to prepare cycloalkenes on insoluble supports include the cyclopropanation of resin-bound alkynes and of polystyrene [165] (Figure 5.18). The latter reaction has been used to introduce 'tags' onto polystyrene beads, which enable the recognition of a certain bead in compound libraries produced using the mix-and-split method (Section 1.2 [165–167]). The structure of polystyrene tagged in this way has not, however, been rigorously determined.

Table 5.10. Preparation of cyclohexenes by Diels–Alder reaction on insoluble supports.

Entry	Starting resin	Conditions	Product	Ref.
1		(0.52 mol/L), xylene, 140 °C, 18 h	94:6	[157] see also [160]
2		PhMe, 110 °C, 24 h	77:23	[157]
3		cyclopentadiene, DCM, Et₂AlCl, −78 °C to −40 °C (without catalyst no reaction occurred)	86% ee, *endo/exo* 21:1	[158] see also [50]
4		Me₃SiO (10 eq), PhMe, 100 °C, 72 h		[53]
5		cyclopentadiene (neat), 100 °C, 24 h	mixture of *endo/exo*	[40]

Table 5.10. continued.

Entry	Starting resin	Conditions	Product	Ref.
6	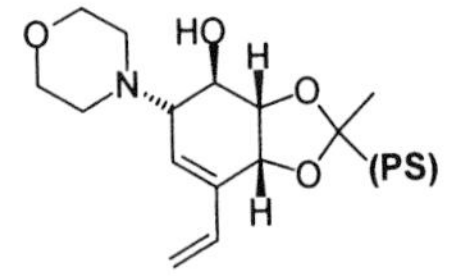	*N*-ethylmaleimide (8 eq), PhMe, 50 °C, 50 h R$_2$NH: morpholine	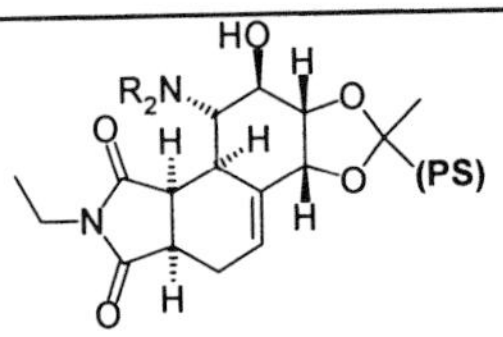	[154]
7	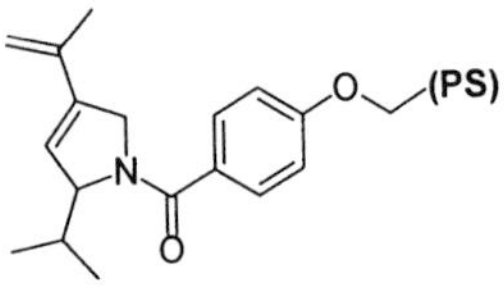	maleimide, PhMe, 105 °C, 18 h	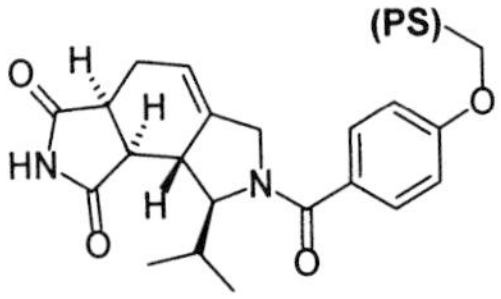	[116] see also [161]
8	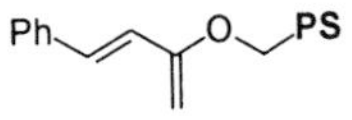	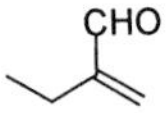 80–100 °C, PhMe	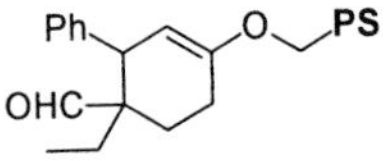 mixture of diastereomers 5:3	[109] see also [162,163]
9	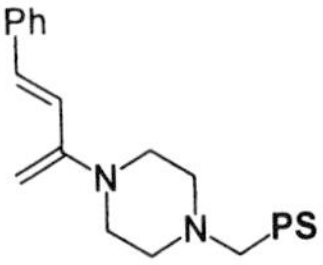	*N*-methylmaleimide (0.16 mol/L, 10 eq), THF, 4–8 h	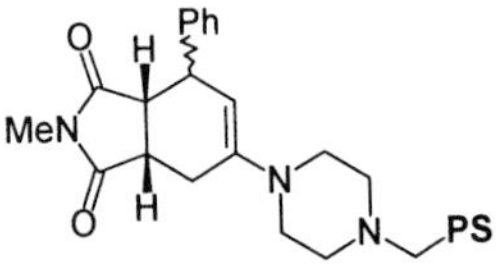	[164]
10	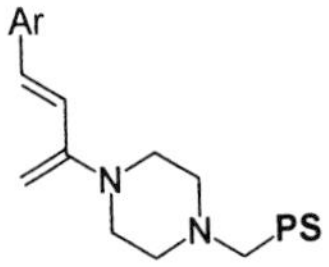	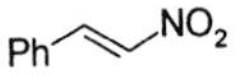 THF, 2 h	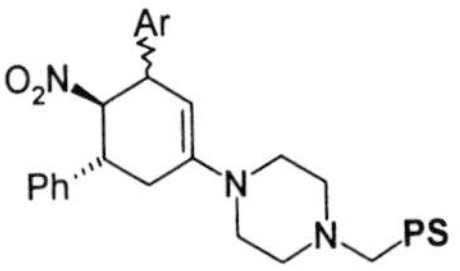	[164]

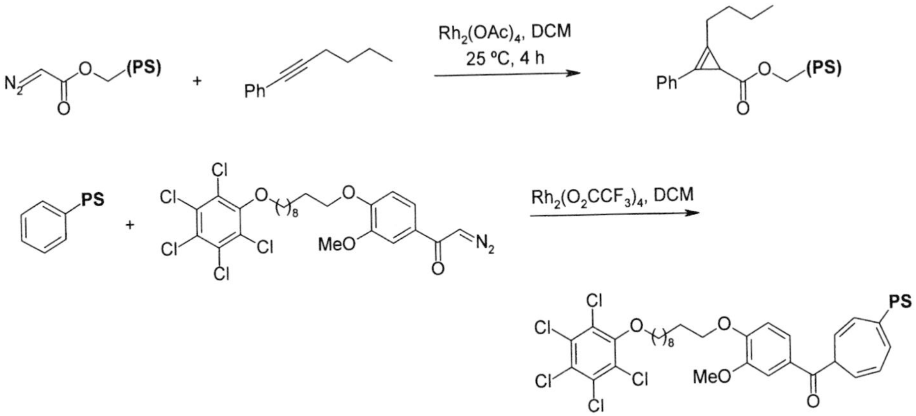

Figure 5.18. Preparation of cycloalkenes from rhodium carbene complexes [165,168].

5.3 Preparation of Alkynes

The preparations of alkynes on insoluble supports include alkylations, vinylations, arylations, and alkynylations of other alkynes. Most of these reactions can be realized with either the alkyne or the alkylating agent linked to the support.

Table 5.11. Alkylation and arylation of alkynes on insoluble supports.

Entry	Starting resin	Conditions	Product	Ref.
1		THF/HMPA 5:3, 20 °C, overnight		[68]
2		$Me_3SiC \equiv CH$ (7 eq), CuI (0.1 eq), $Pd(PPh_3)_4$ (0.1 eq), THF/NEt_3 1:1, 20 °C, 40 h		[172]
3		PhCCH, $Pd_2(dba)_3$, P(o-Tol)$_3$, NEt_3, DMF, 100 °C, 20 h (does not proceed with $HC \equiv C-CO_2R$)		[141]
4		Bu——$\equiv$——$B(OiPr)_2$ $PdCl_2(dppf)$ (0.1 eq), K_2CO_3 (2 eq), DMF, 20 °C, 20 h		[34]
5		2-propyn-1-ol (4 eq), CuI (0.2 eq), $PdCl_2(PPh_3)_2$ (0.1 eq), NEt_3/dioxane 1:2, 20 °C, 24 h		[145] see also [173]
6		Ph——$\equiv$——$SnBu_3$ CuI (0.1 eq), NaCl (2 eq), NMP, 100 °C, 24 h		[155]
7		butyl bromide, THF/HMPA 1:1, 60–70 °C, 4 h		[52]
8		ArCHO (4 eq), CuCl (0.1 eq), piperazine (4 eq), dioxane, 90 °C, 36 h Ar: 4-MeC$_6$H$_4$		[172] see also [174,175]
9		5-iodouridine, CuI, $Pd(PPh_3)_4$, NEt_3, DMF, 25 °C		[176] see also [177]

Table 5.11. continued.

Entry	Starting resin	Conditions	Product	Ref.
10		PhI (0.08 mol/L, 1.5 eq), PdCl$_2$(PPh$_3$)$_2$ (0.05 eq), CuI (0.05 eq), NEt$_3$/DCM 1:6, 20 °C, 18 h		[178]
11		Pd$_2$(dba)$_3$ (2.5 mmol/L), CuI (4 mmol/L), PPh$_3$ (0.02 mol/L), NEt$_3$, 65 °C, 12 h	(Ar: 3-(Me$_3$SiC≡C)-5-(NC)C$_6$H$_3$)	[179] see also [180–182]
12		[Ph$_2$I][BF$_4$] (1.5 eq), CuI (0.1 eq), Pd(PPh$_3$)$_4$ (0.05 eq), NaHCO$_3$ (2 eq), DMF, 40 °C, 24 h		[142]
13		1-octyne, CuCl, HONH$_3$Cl, EtOH, PrNH$_2$, 20 °C, 2 d		[183]

Terminal alkynes can be deprotonated with butyllithium and then alkylated. Lezn-off and co-workers used this methodology in the solid-phase synthesis of insect phero-mones [52,68]. More amenable to automation are aminoalkylations of alkynes, which can be performed with either the alkyne, the aldehyde, or the amine linked to the sup-port (Table 5.11). The arylation of alkynes can be realized under mild reaction condi-tions using the Sonogashira procedure [169] (Table 5.11). This reaction, in which an aryl or heteroaryl iodide is coupled with a terminal alkyne in the presence of cop-per(I), palladium(0), and an amine, tolerates a number of functional groups such as alcohols, phenols, sulfonamides RSO$_2$NHR, and Boc-protected primary amines [145]. Oligonucleotides containing 5-iodouridine have been alkynylated on CPG under Sonogashira conditions (Table 15.28). In addition to the examples given in Table 5.11, further examples have been reported [170,171]. Conjugated diacetylenes have been prepared on cross-linked polystyrene from terminal alkynes and 1-bromo- or 1-iodoalkynes under the standard conditions of the Cadiot–Chodkiewicz coupling (Entry **13**, Table 5.11).

Isolable aliphatic organopalladium compounds can also be coupled with alkynes in the presence of copper(I). One example of such a process is outlined in Figure 5.19.

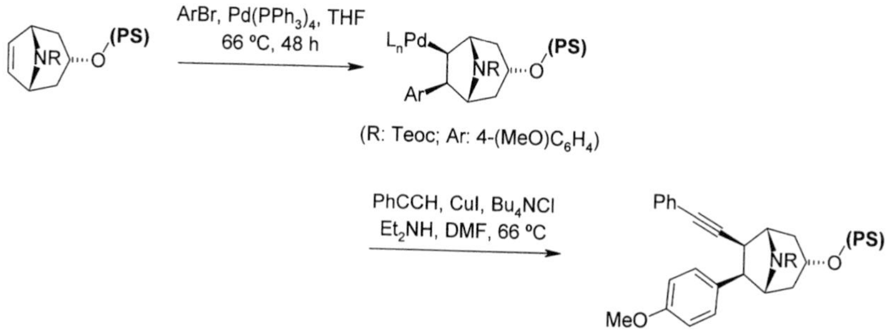

Figure 5.19. Coupling of alkylpalladium compounds with alkynes [38].

5.4 Preparation of Biaryls

The arylation of support-bound arenes has mainly been performed using the Suzuki and Stille coupling reactions. Both reactions proceed smoothly with arenes and heteroarenes such as furans, thiophenes, or pyridines. Examples of the arylation of heteroarenes are presented in Chapter 15.

In the Suzuki coupling, an aryl- or heteroarylboronic acid or ester is coupled with an arene bearing a good leaving group in the presence of palladium(0) and a base. Although examples based on both resin-bound boronic acids and resin-bound aryl halides have been reported, the latter variant is generally preferred. However, this might be simply because fewer bifunctional boronic acid derivatives are available than functionalized aryl halides.

The Suzuki coupling tolerates additional functional groups in the reacting arenes, such as ether, aldehyde [184], trifluoromethyl, and carbamate groups. Difficulties have been reported with arenes bearing hydroxymethyl, carboxyl, or amino groups [37]. Further problematic substrates are 2,5-dialkylphenylboronic acids, which react only slowly [185]. Such difficult couplings can occasionally be driven to completion by using higher concentrations of reagents and catalyst, by increasing the reaction temperature and time, by choosing a different catalyst or base, or by repeating the coupling several times.

The examples reported to date do not allow a clear ranking of palladium catalysts with regard to their ability to catalyze the Suzuki reaction. It should also be kept in mind that palladium(0) complexes are air-sensitive, and the quality of commercially available catalysts of this type can vary substantially. Some successful Suzuki couplings are listed in Table 5.12. Further examples have been reported [184,186–196].

Table 5.12. Preparation of biaryls by the Suzuki reaction on insoluble supports.

Entry	Starting resin	Conditions	Product	Ref.
1		PhB(OH)$_2$ (4 eq), K$_2$CO$_3$ (9 eq), Pd(OAc)$_2$ (0.1 eq), dioxane/H$_2$O 6:1, 100 °C, 6 h		[37]
2		PhB(OH)$_2$ (0.25 mol/L, 2 eq), Na$_2$CO$_3$ (2.5 eq), Pd(PPh$_3$)$_4$ (0.05 eq), DME/H$_2$O 7:1, 85 °C, overnight		[197]
3		PhB(OH)$_2$ (0.1 mol/L, 4 eq), PdCl$_2$(dppf) (0.2 eq), NEt$_3$/DMF 10:18, 65 °C, overnight		[139]
4		(HO)$_2$B—OCONEt$_2$ (3 eq), Pd(PPh$_3$)$_4$ (0.05 eq), Na$_2$CO$_3$ (8 eq), DME, 85 °C, 48 h		[185]
5		PhB(OH)$_2$, Pd$_2$(dba)$_3$ (0.1 eq), K$_2$CO$_3$ (2 eq), DMF, 20 °C, 18 h		[34]
6		(0.1 mol/L, 5 eq), Pd(PPh$_3$)$_4$ (0.02 eq), K$_3$PO$_4$ (5 eq), DMF/H$_2$O 20:1, 80 °C, 20 h		[198] see also [199]
7		[Ph$_2$I][BF$_4$] (0.05 mol/L, 1.5 eq), Pd(PPh$_3$)$_4$ (0.05 eq), Na$_2$CO$_3$ (3 eq), DMF, 20 °C, 20 h		[142] see also [34]

Experimental Procedure 5.2: Suzuki coupling with a polystyrene-bound aryl bromide [185]

To a suspension of polystyrene-bound aryl bromide (0.15 g, approx. 0.09 mmol) in DME (5 mL) under argon were added Pd(PPh$_3$)$_4$ (5.2 mg, 4.5 µmol, 0.05 equiv.) and a degassed aqueous solution of sodium carbonate (0.36 mL, 2 mol/L, 0.72 mmol, 8 equiv.). The boronate (86 mg, 0.27 mmol, 3 equiv.) was added and the mixture was heated to reflux under argon for 48 h. Extensive washing with DME, DME/water, aqueous hydrochloric acid (0.3 mol/L), water, DMF, ethyl acetate, and methanol, followed by drying under reduced pressure yielded the polystyrene-bound biaryl ready for cleavage.

The Stille coupling is closely related to the Suzuki coupling. The only difference is that a stannane is used instead of a boronic acid derivative, and that a base is not generally required. Additional functional groups tolerated in the reactants include hydroxyl, azide, nitro, oxirane, and carbamate groups, but neither primary aliphatic

Table 5.13. Preparation of biaryls by the Stille reaction on insoluble supports.

Entry	Starting resin	Conditions	Product	Ref.
1		PhSnBu$_3$ (3 eq), Pd(PPh$_3$)$_4$ (0.05 eq), DMF, 60 °C, 24 h		[153]
2		(3 eq), Pd(PPh$_3$)$_4$ (0.05 eq), DMF, 60 °C, 24 h		[153]
3		PhSnBu$_3$, CuI (0.1 eq), NaCl (2 eq), NMP, 100 °C, 24 h		[155]
4		PhSnBu$_3$ (3 eq), Pd(PPh$_3$)$_4$ (0.05 eq), DMF, 60 °C, 24 h		[153]
5		3-bromoaniline, NMP, Pd$_2$(OAc)$_2$[P(o-Tol)$_3$]$_2$ ('palladacycle'), LiCl, 90 °C, 18 h		[200]

amines [200] nor (2-alkoxyaryl)stannanes [37]. Most of the reported examples of solid-phase Stille reactions are couplings of a resin-bound aryl halide; few examples have been reported of the 'inverse' variant, in which the organostannane is linked to the support. Illustrative examples of Stille couplings on insoluble supports are listed in Table 5.13. Further examples have been reported [154,188,201,202].

Biaryls have also been prepared by coupling support-bound aryl halides with arylzinc compounds (Figure 5.20) or with aryl(fluoro)silanes [203]. As with Suzuki or Stille couplings, these reactions also require transition metal catalysis. An additional strategy for coupling arenes on solid phase is the oxidative dimerization of phenols (Figure 5.20).

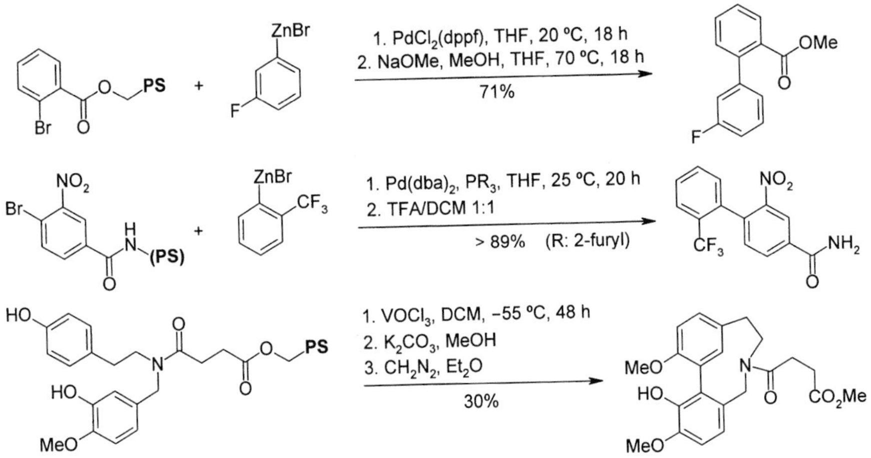

Figure 5.20. Biaryl formation from resin-bound aryl bromides and arylzinc compounds [32,204], and by oxidative coupling of phenols [205].

References for Chapter 5

[1]   O'Brien, R. A.; Chen, T.; Rieke, R. D. *J. Org. Chem.* **1992**, *57*, 2667–2677.
[2]   Lochmann, L.; Fréchet, J. M. J. *Macromolecules* **1996**, *29*, 1767–1771.
[3]   Itsuno, S.; Darling, G. D.; Stöver, H. D. H.; Fréchet, J. M. J. *J. Org. Chem.* **1987**, *52*, 4644–4645.
[4]   Colombo, R. *J. Chem. Soc., Chem. Commun.* **1981**, 1012–1013.
[5]   Schlatter, J. M.; Mazur, R. H. *Tetrahedron Lett.* **1977**, 2851–2852.
[6]   Lacombe, P.; Castagner, B.; Gareau, Y.; Ruel, R. *Tetrahedron Lett.* **1998**, *39*, 6785–6786.
[7]   Stranix, B. R.; Gao, J. P.; Barghi, R.; Salha, J.; Darling, G. D. *J. Org. Chem.* **1997**, *62*, 8987–8993.
[8]   Brummond, K. M.; Lu, J. *J. Org. Chem.* **1999**, *64*, 1723–1726.
[9]   Ojima, I.; Tsai, C. Y.; Zhang, Z. D. *Tetrahedron Lett.* **1994**, *35*, 5785–5788.
[10] Kobayashi, S.; Moriwaki, M. *Tetrahedron Lett.* **1997**, *38*, 4251–4254.
[11] Park, B. D.; Lee, H. I.; Ryoo, S. J.; Lee, Y. S. *Tetrahedron Lett.* **1997**, *38*, 591–594.
[12] Wang, S.; Merrifield, R. B. *J. Am. Chem. Soc.* **1969**, *91*, 6488–6491.
[13] Tomoi, M.; Kori, N.; Kakiuchi, H. *Reactive Polymers* **1985**, *5*, 341–349.
[14] Schiemann, K.; Showalter, H. D. H. *J. Org. Chem.* **1999**, *64*, 4972–4975.
[15] Wolters, E. T. M.; Tesser, G. I.; Nivard, R. J. F. *J. Org. Chem.* **1974**, *39*, 3388–3392.
[16] Scott, R. H.; Barnes, C.; Gerhard, U.; Balasubramanian, S. *Chem. Commun.* **1999**, 1331–1332.
[17] Gagnon, R.; Dory, Y. L.; Deslongchamps, P. *Tetrahedron Lett.* **2000**, *41*, 4751–4755.
[18] Soucy, P.; Dory, Y. L.; Deslongchamps, P. *Synlett* **2000**, 1123–1126.
[19] Ramaseshan, M.; Dory, Y. L.; Deslongchamps, P. *J. Comb. Chem.* **2000**, *2*, 615–623.

[20] Barrett, A. G. M.; de Miguel, Y. R. *Chem. Commun.* **1998**, 2079–2080.
[21] Ruel, G.; The, N. K.; Dumartin, G.; Delmond, B.; Pereyre, M. *J. Organomet. Chem.* **1993**, *444*, C18–C20.
[22] Li, Z.; Ganesan, A. *Synlett* **1998**, 405–406.
[23] Dygutsch, D. P.; Eilbracht, P. *Tetrahedron* **1996**, *52*, 5461–5468.
[24] Bonds, W. D.; Brubaker, C. H.; Chandrasekaran, E. S.; Gibbons, C.; Grubbs, R. H.; Kroll, L. C. *J. Am. Chem. Soc.* **1975**, *97*, 2128–2132.
[25] Bertini, V.; Lucchesini, F.; Pocci, M.; De Munno, A. *J. Org. Chem.* **2000**, *65*, 4839–4842.
[26] Jensen, A. E.; Dohle, W.; Knochel, P. *Tetrahedron* **2000**, *56*, 4197–4201.
[27] Boymond, L.; Rottländer, M.; Cahiez, G.; Knochel, P. *Angew. Chem. Int. Ed.* **1998**, *37*, 1701–1703.
[28] Kondo, Y.; Komine, T.; Fujinami, M.; Uchiyama, M.; Sakamoto, T. *J. Comb. Chem.* **1999**, *1*, 123–126.
[29] Hu, Y.; Porco, J. A.; Labadie, J. W.; Gooding, O. W.; Trost, B. M. *J. Org. Chem.* **1998**, *63*, 4518–4521.
[30] Doi, T.; Hijikuro, I.; Takahashi, T. *J. Am. Chem. Soc.* **1999**, *121*, 6749–6750.
[31] Hijikuro, I.; Doi, T.; Takahashi, T. *J. Am. Chem. Soc.* **2001**, *123*, 3716–3722.
[32] Rottländer, M.; Knochel, P. *Synlett* **1997**, 1084–1086.
[33] Vanier, C.; Wagner, A.; Mioskowski, C. *Tetrahedron Lett.* **1999**, *40*, 4335–4338.
[34] Guiles, J. W.; Johnson, S. G.; Murray, W. V. *J. Org. Chem.* **1996**, *61*, 5169–5171.
[35] Woolard, F. X.; Paetsch, J.; Ellman, J. A. *J. Org. Chem.* **1997**, *62*, 6102–6103.
[36] Backes, B. J.; Ellman, J. A. *J. Am. Chem. Soc.* **1994**, *116*, 11171–11172.
[37] Wendeborn, S.; Berteina, S.; Brill, W. K. D.; De Mesmaeker, A. *Synlett* **1998**, 671–675.
[38] Koh, J. S.; Ellman, J. A. *J. Org. Chem.* **1996**, *61*, 4494–4495.
[39] Zhu, X.; Ganesan, A. *J. Comb. Chem.* **1999**, *1*, 157–162.
[40] Bräse, S.; Enders, D.; Köbberling, J.; Avemaria, F. *Angew. Chem. Int. Ed.* **1998**, *37*, 3413–3415.
[41] Enholm, E. J.; Gallagher, M. E.; Jiang, S.; Batson, W. A. *Org. Lett.* **2000**, *2*, 3355–3357.
[42] Sibi, M. P.; Chandramouli, S. V. *Tetrahedron Lett.* **1997**, *38*, 8929–8932.
[43] Yim, A. M.; Vidal, Y.; Viallefont, P.; Martinez, J. *Tetrahedron Lett.* **1999**, *40*, 4535–4538.
[44] Nagashima, T.; Davies, H. M. L. *J. Am. Chem. Soc.* **2001**, *123*, 2695–2696.
[45] Tietze, L. F.; Steinmetz, A. *Angew. Chem. Int. Ed. Engl.* **1996**, *35*, 651–652.
[46] Vo, N. H.; Eyermann, C. J.; Hodge, C. N. *Tetrahedron Lett.* **1997**, *38*, 7951–7954.
[47] Cheng, W.-C.; Halm, C.; Evarts, J. B.; Olmstead, M. M.; Kurth, M. J. *J. Org. Chem.* **1999**, *64*, 8557–8562.
[48] Brown, R. C. D.; Keily, J.; Karim, R. *Tetrahedron Lett.* **2000**, *41*, 3247–3251.
[49] Nieuwstad, T. J.; Kieboom, A. P. G.; Breijer, A. J.; van der Linden, J.; van Bekkum, H. *Recl. Trav. Chim. Pays-Bas* **1976**, *95*, 225–254.
[50] Burkett, B. A.; Chai, C. L. L. *Tetrahedron Lett.* **1999**, *40*, 7035–7038.
[51] Cheng, W.-C.; Olmstead, M. M.; Kurth, M. J. *J. Org. Chem.* **2001**, *66*, 5528–5533.
[52] Fyles, T. M.; Leznoff, C. C.; Weatherston, J. *Can. J. Chem.* **1977**, *55*, 4135–4143.
[53] Burkett, B. A.; Chai, C. L. L. *Tetrahedron Lett.* **2000**, *41*, 6661–6664.
[54] Gosselin, F.; Di Renzo, M.; Ellis, T. H.; Lubell, W. D. *J. Org. Chem.* **1996**, *61*, 7980–7981.
[55] Melean, L. G.; Haase, W.-C.; Seeberger, P. H. *Tetrahedron Lett.* **2000**, *41*, 4329–4333.
[56] Kurth, M. J.; Randall, L. A. A.; Takenouchi, K. *J. Org. Chem.* **1996**, *61*, 8755–8761.
[57] Stranix, B. R.; Liu, H. Q.; Darling, G. D. *J. Org. Chem.* **1997**, *62*, 6183–6186.
[58] Stranix, B. R.; Darling, G. D. *J. Org. Chem.* **1997**, *62*, 9001–9004.
[59] Neumann, W. P.; Peterseim, M. *Reactive Polymers* **1993**, *20*, 189–205.
[60] Gerigk, U.; Gerlach, M.; Neumann, W. P.; Vieler, R.; Weintritt, V. *Synthesis* **1990**, 448–452.
[61] Moberg, C.; Rákos, L. *Reactive Polymers* **1991**, *15*, 25–35.
[62] Fréchet, J. M. J.; Eichler, E. *Polym. Bull.* **1982**, *7*, 345–351.
[63] Sylvain, C.; Wagner, A.; Mioskowski, C. *Tetrahedron Lett.* **1998**, *39*, 9679–9680.
[64] Leznoff, C. C.; Greenberg, S. *Can. J. Chem.* **1976**, *54*, 3824–3829.
[65] Ren, Q.; Huang, W.; Ho, P. *Reactive Polymers* **1989**, *11*, 237–244.
[66] Wong, J. Y.; Manning, C.; Leznoff, C. C. *Angew. Chem. Int. Ed. Engl.* **1974**, *13*, 666–667.
[67] Adams, J. H.; Cook, R. M.; Hudson, D.; Jammalamadaka, V.; Lyttle, M. H.; Songster, M. F. *J. Org. Chem.* **1998**, *63*, 3706–3716.
[68] Leznoff, C. C.; Fyles, T. M.; Weatherston, J. *Can. J. Chem.* **1977**, *55*, 1143–1153.
[69] Leznoff, C. C.; Sywanyk, W. *J. Org. Chem.* **1977**, *42*, 3203–3205.
[70] Chan, T.; Huang, W. *J. Chem. Soc., Chem. Commun.* **1985**, 909–911.
[71] Vagner, J.; Krchnák, V.; Lebl, M.; Barany, G. *Collect. Czech. Chem. Commun.* **1996**, *61*, 1697–1702.
[72] Veerman, J. J. N.; van Maarseveen, J. H.; Visser, G. M.; Kruse, C. G.; Schoemaker, H. E.; Hiemstra, H.; Rutjes, F. P. J. T. *Eur. J. Org. Chem.* **1998**, 2583–2589.

[73] Chen, C.; Randall, L. A. A.; Miller, R. B.; Jones, A. D.; Kurth, M. J. *J. Am. Chem. Soc.* **1994**, *116*, 2661–2662.

[74] Chin, J.; Fell, B.; Shapiro, M. J.; Tomesch, J.; Wareing, J. R.; Bray, A. M. *J. Org. Chem.* **1997**, *62*, 538–539.

[75] Marzinzik, A. L.; Felder, E. R. *J. Org. Chem.* **1998**, *63*, 723–727.

[76] Lyngsø, L. O.; Nielsen, J. *Tetrahedron Lett.* **1998**, *39*, 5845–5848.

[77] Leznoff, C. C.; Wong, J. Y. *Can. J. Chem.* **1973**, *51*, 3756–3764.

[78] Chen, C.; Randall, L. A. A.; Miller, R. B.; Jones, A. D.; Kurth, M. J. *Tetrahedron* **1997**, *53*, 6595–6609.

[79] Rotella, D. P. *J. Am. Chem. Soc.* **1996**, *118*, 12246–12247.

[80] Nicolaou, K. C.; Winssinger, N.; Pastor, J.; Ninkovic, S.; Sarabia, F.; He, Y.; Vourloumis, D.; Yang, Z.; Li, T.; Giannakakou, P.; Hamel, E. *Nature* **1997**, *387*, 268–272.

[81] Fréchet, J. M.; Schuerch, C. *J. Am. Chem. Soc.* **1971**, *93*, 492–496.

[82] Piscopio, A. D.; Miller, J. F.; Koch, K. *Tetrahedron Lett.* **1997**, *38*, 7143–7146.

[83] Hall, B. J.; Sutherland, J. D. *Tetrahedron Lett.* **1998**, *39*, 6593–6596.

[84] Hird, N. W.; Irie, K.; Nagai, K. *Tetrahedron Lett.* **1997**, *38*, 7111–7114.

[85] Barco, A.; Benetti, S.; De Risi, C.; Marchetti, P.; Pollini, G. P.; Zanirato, V. *Tetrahedron Lett.* **1998**, *39*, 7591–7594.

[86] Paris, M.; Heitz, A.; Guerlavais, V.; Cristau, M.; Fehrentz, J. A.; Martinez, J. *Tetrahedron Lett.* **1998**, *39*, 7287–7290.

[87] Zaragoza, F.; Stephensen, H. *J. Org. Chem.* **1999**, *64*, 2555–2557.

[88] Johnson, C. R.; Zhang, B. R. *Tetrahedron Lett.* **1995**, *36*, 9253–9256.

[89] Blaskovich, M. A.; Kahn, M. *J. Org. Chem.* **1998**, *63*, 1119–1125.

[90] Sun, S.; Murray, W. V. *J. Org. Chem.* **1999**, *64*, 5941–5945.

[91] Boldi, A. M.; Johnson, C. R.; Eissa, H. O. *Tetrahedron Lett.* **1999**, *40*, 619–622.

[92] Burns, C. J.; Groneberg, R. D.; Salvino, J. M.; McGeehan, G.; Condon, S. M.; Morris, R.; Morrissette, M.; Mathew, R.; Darnbrough, S.; Neuenschwander, K.; Scotese, A.; Djuric, S. W.; Ullrich, J.; Labaudiniere, R. *Angew. Chem. Int. Ed.* **1998**, *37*, 2848–2850.

[93] Salvino, J. M.; Kiesow, T. J.; Darnbrough, S.; Labaudiniere, R. *J. Comb. Chem.* **1999**, *1*, 134–139.

[94] Wipf, P.; Henninger, T. C. *J. Org. Chem.* **1997**, *62*, 1586–1587.

[95] Kuster, G. J.; Scheeren, H. W. *Tetrahedron Lett.* **2000**, *41*, 515–519.

[96] Hamper, B. C.; Kolodziej, S. A.; Scates, A. M. *Tetrahedron Lett.* **1998**, *39*, 2047–2050.

[97] Hamper, B. C.; Snyderman, D. M.; Owen, T. J.; Scates, A. M.; Owsley, D. C.; Kesselring, A. S.; Chott, R. C. *J. Comb. Chem.* **1999**, *1*, 140–150.

[98] Hamper, B. C.; Gan, K. Z.; Owen, T. J. *Tetrahedron Lett.* **1999**, *40*, 4973–4976.

[99] Gordeev, M. F.; Patel, D. V.; Wu, J.; Gordon, E. M. *Tetrahedron Lett.* **1996**, *37*, 4643–4646.

[100] Tietze, L. F.; Hippe, T.; Steinmetz, A. *Synlett* **1996**, 1043–1044.

[101] Zaragoza, F. *Tetrahedron Lett.* **1995**, *36*, 8677–8678.

[102] Lee, C. L.; Sim, M. M. *Tetrahedron Lett.* **2000**, *41*, 5729–5732.

[103] Hollinshead, S. P. *Tetrahedron Lett.* **1996**, *37*, 9157–9160.

[104] Chiu, C.; Tang, Z.; Ellingboe, J. W. *J. Comb. Chem.* **1999**, *1*, 73–77.

[105] Katritzky, A. R.; Serdyuk, L.; Chassaing, C.; Toader, D.; Wang, X.; Forood, B.; Flatt, B.; Sun, C.; Vo, K. *J. Comb. Chem.* **2000**, *2*, 182–185.

[106] Nicolaou, K. C.; Pfefferkorn, J. A.; Roecker, A. J.; Cao, G.-Q.; Barluenga, S.; Mitchell, H. J. *J. Am. Chem. Soc.* **2000**, *122*, 9939–9953.

[107] Beebe, X.; Chiappari, C. L.; Olmstead, M. M.; Kurth, M. J.; Schore, N. E. *J. Org. Chem.* **1995**, *60*, 4204–4212.

[108] Zaragoza, F., unpublished results.

[109] Ball, C. P.; Barrett, A. G. M.; Commercon, A.; Compère, D.; Kuhn, C.; Roberts, R. S.; Smith, M. L.; Venier, O. *Chem. Commun.* **1998**, 2019–2020.

[110] Guthrie, E. J.; Macritchie, J.; Hartley, R. C. *Tetrahedron Lett.* **2000**, *41*, 4987–4990.

[111] Zaragoza, F. *Metal Carbenes in Organic Synthesis*; Wiley-VCH: Weinheim, New York, **1999**.

[112] Ivin, K. J.; Mol, J. C. *Olefin Metathesis and Metathesis Polymerization*; Academic Press: London, **1997**.

[113] Reichwein, J. F.; Liskamp, R. M. J. *Eur. J. Org. Chem.* **2000**, 2335–2344.

[114] Schuster, M.; Pernerstorfer, J.; Blechert, S. *Angew. Chem. Int. Ed. Engl.* **1996**, *35*, 1979–1980.

[115] Pernerstorfer, J.; Schuster, M.; Blechert, S. *Synthesis* **1999**, 138–144.

[116] Heerding, D. A.; Takata, D. T.; Kwon, C.; Huffman, W. F.; Samanen, J. *Tetrahedron Lett.* **1998**, *39*, 6815–6818.

[117] Varray, S.; Gauzy, C.; Lamaty, F.; Lazaro, R.; Martinez, J. *J. Org. Chem.* **2000**, *65*, 6787–6790.

[118] Brown, R. C. D.; Castro, J. L.; Moriggi, J.-D. *Tetrahedron Lett.* **2000**, *41*, 3681–3685.

[119] Miller, S. J.; Blackwell, H. E.; Grubbs, R. H. *J. Am. Chem. Soc.* **1996**, *118*, 9606–9614.
[120] Lee, D.; Sello, J. K.; Schreiber, S. L. *J. Am. Chem. Soc.* **1999**, *121*, 10648–10649.
[121] Lee, D.; Sello, J. K.; Schreiber, S. L. *Org. Lett.* **2000**, *2*, 709–712.
[122] Reichwein, J. F.; Wels, B.; Kruijtzer, J. A. W.; Versluis, C.; Liskamp, R. M. J. *Angew. Chem. Int. Ed.* **1999**, *38*, 3684–3687.
[123] Reichwein, J. F.; Versluis, C.; Liskamp, R. M. J. *J. Org. Chem.* **2000**, *65*, 6187–6195.
[124] Louie, J.; Grubbs, R. H. *Angew. Chem. Int. Ed.* **2001**, *40*, 247–249.
[125] Rainier, J. D.; Cox, J. M.; Allwein, S. P. *Tetrahedron Lett.* **2001**, *42*, 179–181.
[126] Huang, J.; Stevens, E. D.; Nolan, S. P.; Petersen, J. L. *J. Am. Chem. Soc.* **1999**, *121*, 2674–2678.
[127] Biagini, S. C. G.; Gibson, S. E.; Keen, S. P. *J. Chem. Soc., Perkin Trans. 1* **1998**, 2485–2499.
[128] Cuny, G. D.; Cao, J.; Hauske, J. R. *Tetrahedron Lett.* **1997**, *38*, 5237–5240.
[129] Schuster, M.; Lucas, N.; Blechert, S. *Chem. Commun.* **1997**, 823–824.
[130] Nicolaou, K. C.; Pastor, J.; Winssinger, N.; Murphy, F. *J. Am. Chem. Soc.* **1998**, *120*, 5132–5133.
[131] Schuster, M.; Blechert, S. *Tetrahedron Lett.* **1998**, *39*, 2295–2298.
[132] Schürer, S. C.; Blechert, S. *Synlett* **1998**, 166–168.
[133] Conde-Frieboes, K.; Andersen, S.; Breinholt, J. *Tetrahedron Lett.* **2000**, *41*, 9153–9156.
[134] Blackwell, H. E.; Clemons, P. A.; Schreiber, S. L. *Org. Lett.* **2001**, *3*, 1185–1188.
[135] Beaver, K. A.; Siegmund, A. C.; Spear, K. L. *Tetrahedron Lett.* **1996**, *37*, 1145–1148.
[136] Hiroshige, M.; Hauske, J. R.; Zhou, P. *J. Am. Chem. Soc.* **1995**, *117*, 11590–11591.
[137] Pop, I. E.; Dhalluin, C. F.; Déprez, B. P.; Melnyk, P. C.; Lippens, G. M.; Tartar, A. L. *Tetrahedron* **1996**, *52*, 12209–12222.
[138] Hanessian, S.; Xie, F. *Tetrahedron Lett.* **1998**, *39*, 737–740.
[139] Ruhland, B.; Bombrun, A.; Gallop, M. A. *J. Org. Chem.* **1997**, *62*, 7820–7826.
[140] Shaughnessy, K. H.; Kim, P.; Hartwig, J. F. *J. Am. Chem. Soc.* **1999**, *121*, 2123–2132.
[141] Yu, K. L.; Deshpande, M. S.; Vyas, D. M. *Tetrahedron Lett.* **1994**, *35*, 8919–8922.
[142] Kang, S. K.; Yoon, S. K.; Lim, K. H.; Son, H. J.; Baik, T. G. *Synth. Commun.* **1998**, *28*, 3645–3655.
[143] Blettner, C. G.; König, W. A.; Stenzel, W.; Schotten, T. *Tetrahedron Lett.* **1999**, *40*, 2101–2102.
[144] Hiroshige, M.; Hauske, J. R.; Zhou, P. *Tetrahedron Lett.* **1995**, *36*, 4567–4570.
[145] Berteina, S.; Wendeborn, S.; Brill, W. K. D.; De Mesmaeker, A. *Synlett* **1998**, 676–678.
[146] Ma, S.; Duan, D.; Shi, Z. *Org. Lett.* **2000**, *2*, 1419–1422.
[147] Brown, S. D.; Armstrong, R. W. *J. Am. Chem. Soc.* **1996**, *118*, 6331–6332.
[148] Brown, S. D.; Armstrong, R. W. *J. Org. Chem.* **1997**, *62*, 7076–7077.
[149] Thompson, L. A.; Moore, F. L.; Moon, Y. C.; Ellman, J. A. *J. Org. Chem.* **1998**, *63*, 2066–2067.
[150] Dragoli, D. R.; Thompson, L. A.; O'Brien, J.; Ellman, J. A. *J. Comb. Chem.* **1999**, *1*, 534–539.
[151] Lee, C. Y.; Hanson, R. N. *Tetrahedron* **2000**, *56*, 1623–1629.
[152] Xu, G.; Loftus, T. L.; Wargo, H.; Turpin, J. A.; Buckheit, R. W.; Cushman, M. *J. Org. Chem.* **2001**, *66*, 5958–5964.
[153] Chamoin, S.; Houldsworth, S.; Snieckus, V. *Tetrahedron Lett.* **1998**, *39*, 4175–4178.
[154] Wendeborn, S.; De Mesmaeker, A.; Brill, W. K. D. *Synlett* **1998**, 865–868.
[155] Kang, S. K.; Kim, J. S.; Yoon, S. K.; Lim, K. H.; Yoon, S. S. *Tetrahedron Lett.* **1998**, *39*, 3011–3012.
[156] Yli-Kaulhaluoma, J. *Tetrahedron* **2001**, *57*, 7053–7071.
[157] Yedidia, V.; Leznoff, C. C. *Can. J. Chem.* **1980**, *58*, 1144–1150.
[158] Winkler, J. D.; McCoull, W. *Tetrahedron Lett.* **1998**, *39*, 4935–4936.
[159] Sun, S.; Turchi, I. J.; Xu, D.; Murray, W. V. *J. Org. Chem.* **2000**, *65*, 2555–2559.
[160] Winkler, J. D.; Kwak, Y. S. *J. Org. Chem.* **1998**, *63*, 8634–8635.
[161] Schürer, S. C.; Blechert, S. *Synlett* **1999**, 1879–1882.
[162] Graven, A.; Meldal, M. *J. Chem. Soc., Perkin Trans. 1* **2001**, 3198–3203.
[163] Graven, A.; Hilaire, P. M. S.; Sanderson, S. J.; Mottram, J. C.; Coombs, G. H.; Meldal, M. *J. Comb. Chem.* **2001**, *3*, 441–452.
[164] Crawshaw, M.; Hird, N. W.; Irie, K.; Nagai, K. *Tetrahedron Lett.* **1997**, *38*, 7115–7118.
[165] Nestler, H. P.; Bartlett, P. A.; Still, W. C. *J. Org. Chem.* **1994**, *59*, 4723–4724.
[166] Baldwin, J. J.; Burbaum, J. J.; Henderson, I.; Ohlmeyer, M. H. J. *J. Am. Chem. Soc.* **1995**, *117*, 5588–5589.
[167] Burbaum, J. J.; Ohlmeyer, M. H. J.; Reader, J. C.; Henderson, I.; Dillard, L. W.; Li, G.; Randle, T. L.; Sigal, N. H.; Chelsky, D.; Baldwin, J. J. *Proc. Natl. Acad. Sci. USA* **1995**, *92*, 6027–6031.
[168] Cano, M.; Camps, F.; Joglar, J. *Tetrahedron Lett.* **1998**, *39*, 9819–9822.
[169] Sonogashira, K.; Tohda, Y.; Hagihara, N. *Tetrahedron Lett.* **1975**, 4467–4470.
[170] Tan, D. S.; Foley, M. A.; Shair, M. D.; Schreiber, S. L. *J. Am. Chem. Soc.* **1998**, *120*, 8565–8566.
[171] Collini, M. D.; Ellingboe, J. W. *Tetrahedron Lett.* **1997**, *38*, 7963–7966.
[172] Dyatkin, A. B.; Rivero, R. A. *Tetrahedron Lett.* **1998**, *39*, 3647–3650.
[173] Stieber, F.; Grether, U.; Waldmann, H. *Angew. Chem. Int. Ed.* **1999**, *38*, 1073–1077.

[174] McNally, J. J.; Youngman, M. A.; Dax, S. L. *Tetrahedron Lett.* **1998**, *39*, 967–970.
[175] Youngman, M. A.; Dax, S. L. *Tetrahedron Lett.* **1997**, *38*, 6347–6350.
[176] Khan, S. I.; Grinstaff, M. W. *Tetrahedron Lett.* **1998**, *39*, 8031–8034.
[177] Khan, S. I.; Grinstaff, M. W. *J. Am. Chem. Soc.* **1999**, *121*, 4704–4705.
[178] Bolton, G. L.; Hodges, J. C.; Rubin, J. R. *Tetrahedron* **1997**, *53*, 6611–6634.
[179] Nelson, J. C.; Young, J. K.; Moore, J. S. *J. Org. Chem.* **1996**, *61*, 8160–8168.
[180] Huang, S.; Tour, J. M. *J. Am. Chem. Soc.* **1999**, *121*, 4908–4909.
[181] Huang, S.; Tour, J. M. *J. Org. Chem.* **1999**, *64*, 8898–8906.
[182] Shortell, D. B.; Palmer, L. C.; Tour, J. M. *Tetrahedron* **2001**, *57*, 9055–9065.
[183] Montierth, J. M.; DeMario, D. R.; Kurth, M. J.; Schore, N. E. *Tetrahedron* **1998**, *54*, 11741–11748.
[184] Blettner, C. G.; König, W. A.; Rühter, G.; Stenzel, W.; Schotten, T. *Synlett* **1999**, 307–310.
[185] Chamoin, S.; Houldsworth, S.; Kruse, C. G.; Bakker, W. I.; Snieckus, V. *Tetrahedron Lett.* **1998**, *39*, 4179–4182.
[186] Chenera, B.; Finkelstein, J. A.; Veber, D. F. *J. Am. Chem. Soc.* **1995**, *117*, 11999–12000.
[187] Han, Y. X.; Walker, S. D.; Young, R. N. *Tetrahedron Lett.* **1996**, *37*, 2703–2706.
[188] Larhed, M.; Lindeberg, G.; Hallberg, A. *Tetrahedron Lett.* **1996**, *37*, 8219–8222.
[189] Yoo, S. E.; Seo, J. S.; Yi, K. Y.; Gong, Y. D. *Tetrahedron Lett.* **1997**, *38*, 1203–1206.
[190] Raju, B.; Kogan, T. P. *Tetrahedron Lett.* **1997**, *38*, 3373–3376.
[191] Lago, M. A.; Nguyen, T. T.; Bhatnagar, P. *Tetrahedron Lett.* **1998**, *39*, 3885–3888.
[192] Huwe, C. M.; Künzer, H. *Tetrahedron Lett.* **1999**, *40*, 683–686.
[193] Gosselin, F.; Van Betsbrugge, J.; Hatam, M.; Lubell, W. D. *J. Org. Chem.* **1999**, *64*, 2486–2493.
[194] Blettner, C. G.; König, W. A.; Stenzel, W.; Schotten, T. *J. Org. Chem.* **1999**, *64*, 3885–3890.
[195] Xiong, Y.; Klopp, J.; Chapman, K. T. *Tetrahedron Lett.* **2001**, *42*, 8423–8427.
[196] Gravel, M.; Bérubé, C. D.; Hall, D. G. *J. Comb. Chem.* **2000**, *2*, 228–231.
[197] Frenette, R.; Friesen, R. W. *Tetrahedron Lett.* **1994**, *35*, 9177–9180.
[198] Piettre, S. R.; Baltzer, S. *Tetrahedron Lett.* **1997**, *38*, 1197–1200.
[199] Tempest, P. A.; Armstrong, R. W. *J. Am. Chem. Soc.* **1997**, *119*, 7607–7608.
[200] Brody, M. S.; Finn, M. G. *Tetrahedron Lett.* **1999**, *40*, 415–418.
[201] Forman, F. W.; Sucholeiki, I. *J. Org. Chem.* **1995**, *60*, 523–528.
[202] Tan, D. S.; Foley, M. A.; Stockwell, B. R.; Shair, M. D.; Schreiber, S. L. *J. Am. Chem. Soc.* **1999**, *121*, 9073–9087.
[203] Homsi, F.; Nozaki, K.; Hiyama, T. *Tetrahedron Lett.* **2000**, *41*, 5869–5872.
[204] Marquais, S.; Arlt, M. *Tetrahedron Lett.* **1996**, *37*, 5491–5494.
[205] ApSimon, J. W.; Dixit, D. M. *Can. J. Chem.* **1982**, *60*, 368–370.

6 Preparation of Alkyl and Aryl Halides

The introduction of halogens into organic substrates may or may not be accompanied by a formal oxidation of the substrate. Oxidative halogenations include additions to C=C double bonds, aromatic electrophilic substitutions, halomethylations, and the reaction of organometallic compounds with halogens or synthetic equivalents thereof. Because polystyrene is also susceptible to oxidative halogenation, only reactive substrates will undergo clean reaction with halogenating agents before significant halogenation of the support. Acid-labile linkers, which are usually based on electron-rich arenes (e.g. alkoxybenzyl alcohols or benzhydryl derivatives), are generally incompatible with strong oxidizing agents. Halogenation can also be realized by non-oxidative processes, such as aliphatic or aromatic nucleophilic substitutions. Strategies for the generation of alkyl, vinyl, and aryl halides upon cleavage of intermediates from insoluble supports are discussed in Section 3.13.

6.1 Preparation of Alkyl Halides

The support originally used for solid-phase synthesis was partially chloromethylated cross-linked polystyrene, which was prepared by chloromethylation of cross-linked polystyrene with chloromethyl methyl ether and tin(IV) chloride [1–3] or zinc chloride [4] (Figure 6.1). Haloalkylations of this type are usually only used for the functionalization of supports, and not for selective transformation of support-bound intermediates. Because of the mutagenicity of α-haloethers, other methods have been developed for the preparation of chloromethyl polystyrene. These include the chlorination of methoxymethyl polystyrene (Figure 6.1 [5]), the use of a mixture of dimethoxymethane, sulfuryl chloride, and chlorosulfonic acid instead of chloromethyl methyl ether [6], the chlorination of hydroxymethyl polystyrene [7], and the chlorination of cross-linked 4-methylstyrene–styrene copolymer with sodium hypochlorite [8], sulfuryl chloride [8], or cobalt(III) acetate/lithium chloride [9] (Figure 6.1, Table 6.1).

Ketones can be α-brominated on solid phase by treatment with synthetic equivalents of bromine, such as pyridinium tribromide (Entry **2**, Table 6.1) or phenyltrimethylammonium tribromide (DCM, 20 °C, 3 h [10]). Resin-bound organometallic compounds, such as vinylstannanes [11] or organozinc derivatives [12], react cleanly with iodine to yield the corresponding vinyl or alkyl iodides (see also Section 3.13). Additions of halogens or their synthetic equivalents to C=C double bonds on cross-

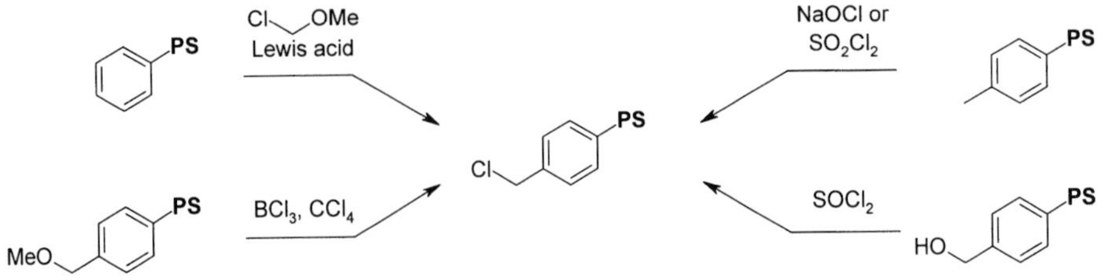

Figure 6.1. Methods for the preparation of chloromethyl polystyrene.

linked polystyrene have been mainly limited to reactive substrates, such as styrenes [13,14] or enol ethers (Table 6.1). Iodolactonizations with simultaneous cleavage from an insoluble support are discussed in Section 3.5.2.

Table 6.1. Oxidative halogenation of alkenes and alkanes on polystyrene.

Entry	Starting resin	Conditions	Product	Ref.
1	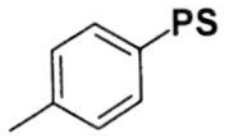	LiCl, AcOH/PhH 1:1, Co(OAc)$_3$, 50 °C, 2 h, or aq NaOCl, DCE, BnEt$_3$NCl, 25 °C, 16 h	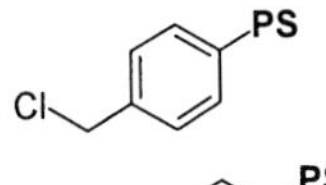	[8,9]
2	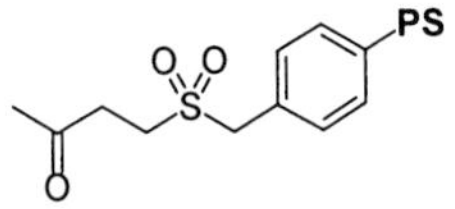	PyHBrBr$_2$, AcOH, 60 °C, 16 h	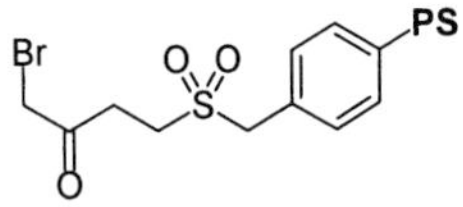	[15] see also [10]
3	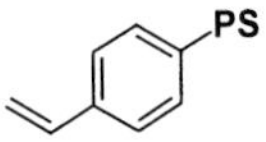	LiBr, Me$_3$SiCl, PhMe, H$_2$O, AIBN, 70 °C, 24 h	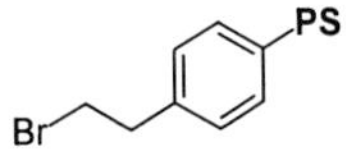	[16]
4	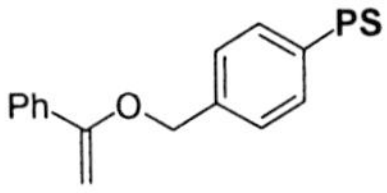	Br$_2$ (1 mol/L, 2 eq), DCM, 25 °C, 1 h	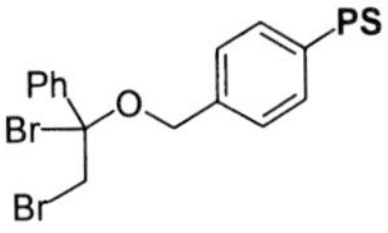	[17]
5	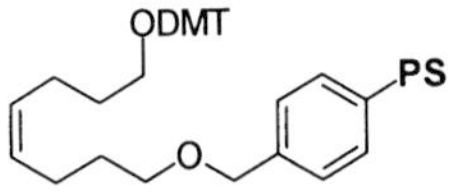	CuBr$_2$ (10 eq), LiBr (20 eq), THF, MeCN, 48 h	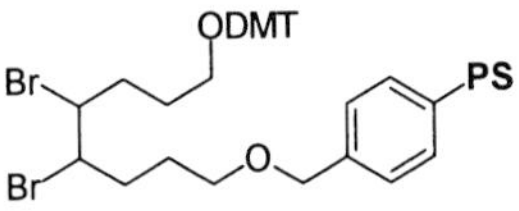	[18]
6	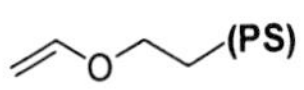	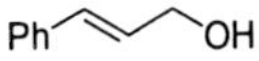 Ph⌒⌒OH (30 eq), NBS (5 eq), DCM, 0 °C, 6 d	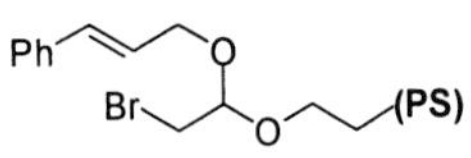	[19]
7	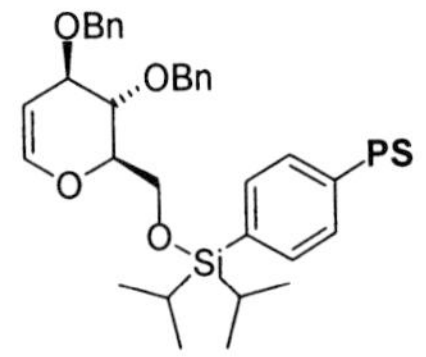	I(coll)$_2$ClO$_4$, PhSO$_2$NH$_2$, DCM, 0 °C	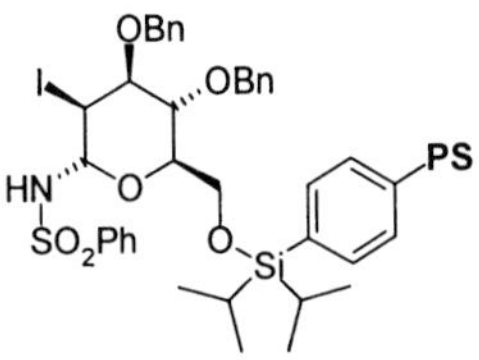	[20] see also [21]
8	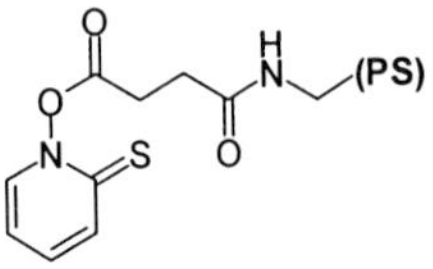	CBrCl$_3$ (50 eq), hv, DMF, 0.3 h	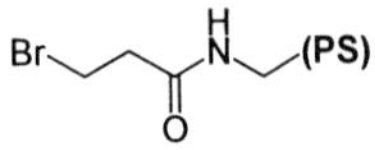	[22]

Nucleophilic substitutions offer a more versatile means of access to alkyl halides than oxidative halogenations. The most common starting materials for this purpose are resin-bound alcohols, which can be converted into halides either directly or via the intermediate formation of sulfonates [23]. Polystyrene-bound halides can also be

Table 6.2. Conversion of resin-bound benzylic and allylic alcohols into halides.

Entry	Starting resin	Conditions	Product	Ref.
1		HBr, DCM, 0 °C, 4 h		[36]
2		CBr₄, PPh₃, DCM, 20 °C, 16 h		[37]
3	Wang resin	MsCl (0.32 mol/L, 3 eq), DIPEA (4 eq), DMF, 25 °C, 3 d		[38]
4	Sasrin	NBS (0.47 mol/L, 50 eq), Me₂S (55 eq), DCM, 0 °C, 5 h (these conditions are not suitable for trityl or benzhydryl alcohols)		[39]
5		(3 eq), DCM, 20 °C, 5 h		[40]
6		CBr₄ (0.1 mol/L, 2 eq), PPh₃ (2 eq), THF, 16 h		[41]
7		1% HCl in THF/DCM; or PPh₃ (0.18 mol/L, 5.5 eq), C₂Cl₆ (5.5 eq), THF, 20 °C, 6 h		[42,43] see also [44]
8	Ar: 3,4-dialkoxyphenyl	Ph₃PBr₂ (0.12 mol/L, 1.5 eq), DCM, 20 °C, 1 h		[45] see also [46]
9		Ph₃PBr₂ (0.27 mol/L, 3 eq), DCM, 20 °C, 17 h		[47]
10		PBu₃ (0.12 mol/L, 10 eq), CCl₄, 0 °C, 0.3 h, 20 °C, 4 h		[48]

prepared from other halides by nucleophilic substitution. Iodomethyl polystyrene, for instance, has been prepared from Merrifield resin by treatment with sodium iodide in acetone (57 °C, 2 d [24]).

Polystyrene-bound allylic or benzylic alcohols react smoothly with hydrogen chloride or hydrogen bromide to yield the corresponding halides. The more stable the intermediate carbocation, the more easily the solvolysis will proceed. Alternatively, thionyl chloride can be used to convert benzyl alcohols into chlorides [7,25,26]. A milder alternative for preparing bromides or iodides, which is also suitable for non-benzylic alcohols, is the treatment of alcohols with phosphines and halogens or the preformed adducts thereof (Table 6.2, Experimental Procedure 6.1 [27–31]). Benzhydryl and trityl alcohols bound to cross-linked or non-cross-linked polystyrene are particularly prone to solvolysis, and can be converted into the corresponding chlorides by treatment with acetyl chloride in toluene or similar solvents (Table 6.2 [32–35]).

Experimental Procedure 6.1: Conversion of Wang resin into *p*-benzyloxybenzyl bromide resin [49]

A mixture of Wang resin (50.0 g, 36.5 mmol), DMF (500 mL), triphenylphosphine (47.9 g, 183 mmol, 5 equiv.), and carbon tetrabromide (60.5 g, 182 mmol, 5 equiv.) was shaken at room temperature for 2.5 h. The resin was filtered and washed sequentially with DMF, DCM, *i*PrOH, DMF, DCM, and *i*PrOH (2 × 300 mL of each solvent).

Aliphatic alcohols do not undergo solvolysis as readily as benzylic alcohols, and are generally converted into halides under basic reaction conditions via an intermediate sulfonate. Because of the hydrophobicity of polystyrene, however, nucleophilic substitutions with halides on this support do not always proceed as readily as in solution (Table 6.3). Alternatively, phosphorus-based reagents can also be used to convert aliphatic alcohols into halides.

Table 6.3. Conversion of non-benzylic alcohols into halides.

Entry	Starting resin	Conditions	Product	Ref.
1		SOCl$_2$ (3.3 eq), pyridine (0.16 eq), CCl$_4$, 76 °C, 16 h		[50] see also [51]
2		1. TsCl (1.3 eq), (*i*Pr)$_2$NH (2 eq), CCl$_4$, 76 °C, 6 h 2. MgBr$_2$, Et$_2$O, 35 °C, 17 h		[50]
3		NaI (2 eq), Me$_3$SiCl (2 eq), MeCN, 81 °C, 8 h		[50]
4		1. MsCl (12 eq), pyridine, 20 °C, 48 h 2. NaI (0.2 mol/L), HMPA, 70 °C, 48 h		[52]
5		I$_2$, PPh$_3$, imidazole (4 eq of each), DCM, 0 °C, 4 h		[28] see also [53]
6		NH$_4$Cl, DMSO, 80 °C, 24 h		[54]

6.2 Preparation of Aryl and Heteroaryl Halides

Cross-linked polystyrene can be directly brominated in carbon tetrachloride using bromine in the presence of Lewis acids (Experimental Procedure 6.2 [55–58]). Thallium(III) acetate is a particularly suitable catalyst for this reaction [59]. Harsher bromination conditions should be avoided, because these can lead to decomposition of the polymer. Considering that isopropylbenzene is dealkylated when treated with bromine to yield hexabromobenzene [60], the expected products of the extensive bromination of cross-linked polystyrene would be soluble poly(vinyl bromide) and hexabromobenzene. In fact, if the bromination of cross-linked polystyrene is attempted using bromine in acetic acid, the polymer dissolves and apparently depolymerizes [61].

The iodination of cross-linked polystyrene has been achieved using iodine under strongly acidic reaction conditions [55] or in the presence of thallium(III) acetate [61], but this reaction does not proceed as smoothly as the bromination. More electron-rich arenes, such as thiophenes [45,62–64], furans [46], purines [65], indoles [66], or phenols [67,68] are readily halogenated, even in the presence of oxidant-labile linkers (Figure 6.2). Polystyrene-bound thiophenes have also been iodinated by lithiation with LDA followed by treatment with iodine [64].

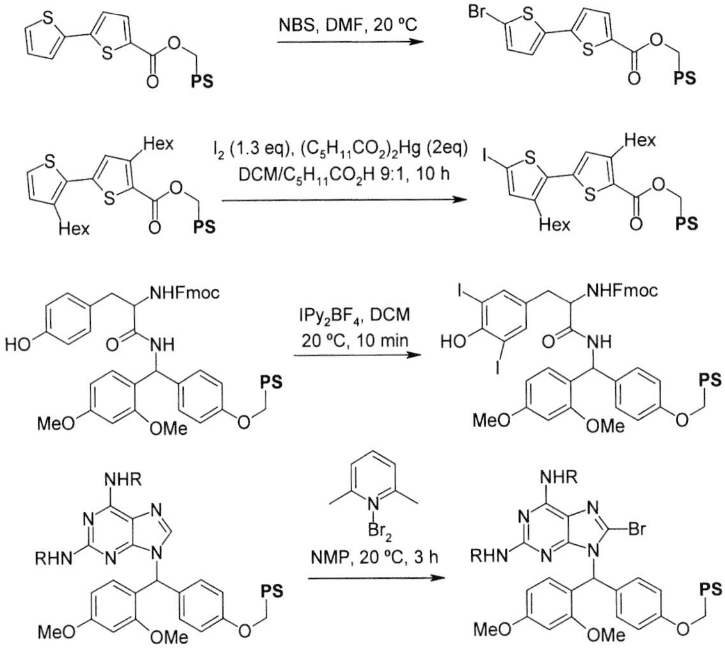

Figure 6.2. Halogenations of polystyrene-bound arenes [62,63,65,67,69].

Experimental Procedure 6.2: Bromination of cross-linked polystyrene [61]

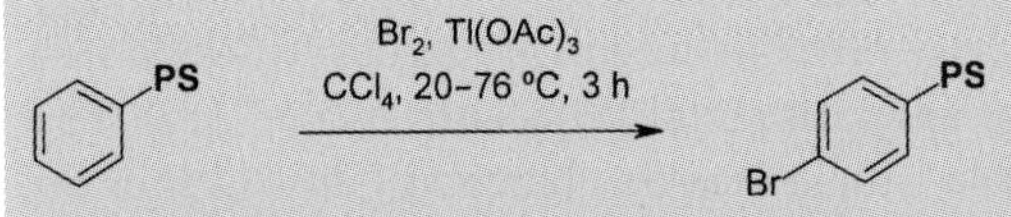

Thallium(III) acetate (1.18 g, 3.09 mmol) was added to a suspension of polystyrene (20 g, 1% cross-linked) in carbon tetrachloride (300 mL), and the mixture was stirred in the dark for 0.5 h. A solution of bromine (13.6 g, 85.1 mmol) in carbon tetrachloride (20 mL) was then added, and the mixture was stirred at room temperature in the dark for 1 h and at 76 °C for 1.5 h. Thereafter, the mixture was filtered and the resin was washed with carbon tetrachloride, acetone, acetone/water (2:1), acetone, benzene*, and methanol. Drying of the resin under reduced pressure yielded 26.3 g of brominated polystyrene with a bromine content of 3.1 mmol/g, which corresponds to a bromination of 43% of all the available phenyl groups.
* Benzene should be replaced by a less toxic solvent, e.g. toluene.

References for Chapter 6

[1]   Merrifield, R. B. *J. Am. Chem. Soc.* **1963**, *85*, 2149–2154.
[2]   Pepper, K. W.; Paisley, H. M.; Young, M. A. *J. Chem. Soc.* **1953**, 4097–4105.
[3]   Pinnell, R. P.; Khune, G. D.; Khatri, N. A.; Manatt, S. L. *Tetrahedron Lett.* **1984**, *25*, 3511–3514.
[4]   Ford, W. T.; Yacoub, S. A. *J. Org. Chem.* **1981**, *46*, 819–821.
[5]   Arshady, R.; Kenner, G. W.; Ledwith, A. *Makromol. Chem.* **1976**, *177*, 2911–2918.
[6]   Neumann, W. P.; Peterseim, M. *Reactive Polymers* **1993**, *20*, 189–205.
[7]   Bui, C. T.; Maeji, N. J.; Rasoul, F.; Bray, A. M. *Tetrahedron Lett.* **1999**, *40*, 5383–5386.
[8]   Mohanraj, S.; Ford, W. T. *Macromolecules* **1986**, *19*, 2470–2472.
[9]   Sheng, Q.; Stöver, H. D. H. *Macromolecules* **1997**, *30*, 6712–6714.
[10] Attanasi, O. A.; Filippone, P.; Guidi, B.; Hippe, T.; Mantellini, F.; Tietze, L. F. *Tetrahedron Lett.* **1999**, *40*, 9277–9280.
[11] Nicolaou, K. C.; Winssinger, N.; Pastor, J.; Murphy, F. *Angew. Chem. Int. Ed.* **1998**, *37*, 2534–2537.
[12] Karoyan, P.; Triolo, A.; Nannicini, R.; Giannotti, D.; Altamura, M.; Chassaing, G.; Perrotta, E. *Tetrahedron Lett.* **1999**, *40*, 71–74.
[13] Gerigk, U.; Gerlach, M.; Neumann, W. P.; Vieler, R.; Weintritt, V. *Synthesis* **1990**, 448–452.
[14] Tortolani, D. R.; Biller, S. A. *Tetrahedron Lett.* **1996**, *37*, 5687–5690.
[15] Barco, A.; Benetti, S.; De Risi, C.; Marchetti, P.; Pollini, G. P.; Zanirato, V. *Tetrahedron Lett.* **1998**, *39*, 7591–7594.
[16] Stranix, B. R.; Gao, J. P.; Barghi, R.; Salha, J.; Darling, G. D. *J. Org. Chem.* **1997**, *62*, 8987–8993.
[17] Ball, C. P.; Barrett, A. G. M.; Commercon, A.; Compère, D.; Kuhn, C.; Roberts, R. S.; Smith, M. L.; Venier, O. *Chem. Commun.* **1998**, 2019–2020.
[18] Melean, L. G.; Haase, W.-C.; Seeberger, P. H. *Tetrahedron Lett.* **2000**, *41*, 4329–4333.
[19] Watanabe, Y.; Ishikawa, S.; Takao, G.; Toru, T. *Tetrahedron Lett.* **1999**, *40*, 3411–3414.
[20] Zheng, C. S.; Seeberger, P. H.; Danishefsky, S. J. *Angew. Chem. Int. Ed.* **1998**, *37*, 786–789.
[21] Savin, K. A.; Woo, J. C. G.; Danishefsky, S. J. *J. Org. Chem.* **1999**, *64*, 4183–4186.
[22] Attardi, M. E.; Taddei, M. *Tetrahedron Lett.* **2001**, *42*, 3519–3522.
[23] Guibé, F.; Dangles, O.; Balavoine, G.; Loffet, A. *Tetrahedron Lett.* **1989**, *30*, 2641–2644.
[24] Majewski, M.; Ulaczyk, A.; Wang, F. *Tetrahedron Lett.* **1999**, *40*, 8755–8758.
[25] Wong, J. Y.; Manning, C.; Leznoff, C. C. *Angew. Chem. Int. Ed. Engl.* **1974**, *13*, 666–667.
[26] Bui, C. T.; Maeji, N. J.; Bray, A. M. *Biotech. Bioeng. (Comb. Chem.)* **2000**, *71*, 91–93.
[27] Bhandari, A.; Jones, D. G.; Schullek, J. R.; Vo, K.; Schunk, C. A.; Tamanaha, L. L.; Chen, D.; Yuan, Z. Y.; Needels, M. C.; Gallop, M. A. *Bioorg. Med. Chem. Lett.* **1998**, *8*, 2303–2308.
[28] Nicolaou, K. C.; Winssinger, N.; Vourloumis, D.; Ohshima, T.; Kim, S.; Pfefferkorn, J.; Xu, J. Y.; Li, T. *J. Am. Chem. Soc.* **1998**, *120*, 10814–10826.
[29] van Maarseveen, J. H.; den Hartog, J. A. J.; Engelen, V.; Finner, E.; Visser, G.; Kruse, C. G. *Tetrahedron Lett.* **1996**, *37*, 8249–8252.
[30] Ngu, K.; Patel, D. V. *Tetrahedron Lett.* **1997**, *38*, 973–976.
[31] Raju, B.; Kogan, T. P. *Tetrahedron Lett.* **1997**, *38*, 4965–4968.
[32] Brown, D. S.; Revill, J. M.; Shute, R. E. *Tetrahedron Lett.* **1998**, *39*, 8533–8536.
[33] Fréchet, J. M. J.; Haque, K. E. *Tetrahedron Lett.* **1975**, 3055–3056.
[34] Hayatsu, H.; Khorana, H. G. *J. Am. Chem. Soc.* **1966**, *88*, 3182–3183.
[35] Hayatsu, H.; Khorana, H. G. *J. Am. Chem. Soc.* **1967**, *89*, 3880–3887.
[36] Ajayaghosh, A.; Pillai, V. N. R. *Tetrahedron* **1988**, *44*, 6661–6666.
[37] Hall, B. J.; Sutherland, J. D. *Tetrahedron Lett.* **1998**, *39*, 6593–6596.
[38] Nugiel, D. A.; Wacker, D. A.; Nemeth, G. A. *Tetrahedron Lett.* **1997**, *38*, 5789–5790.
[39] Zoller, T.; Ducep, J.-B.; Hibert, M. *Tetrahedron Lett.* **2000**, *41*, 9985–9988.
[40] Berteina, S.; Wendeborn, S.; De Mesmaeker, A. *Synlett* **1998**, 1231–1233.
[41] Newlander, K. A.; Chenera, B.; Veber, D. F.; Yim, N. C. F.; Moore, M. L. *J. Org. Chem.* **1997**, *62*, 6726–6732.
[42] Mellor, S. L.; Chan, W. C. *Chem. Commun.* **1997**, 2005–2006.
[43] Garigipati, R. S. *Tetrahedron Lett.* **1997**, *38*, 6807–6810.
[44] Atkinson, G. E.; Fischer, P. M.; Chan, W. C. *J. Org. Chem.* **2000**, *65*, 5048–5056.
[45] Han, Y.; Giroux, A.; Lépine, C.; Laliberté, F.; Huang, Z.; Perrier, H.; Bayly, C. I.; Young, R. N. *Tetrahedron* **1999**, *55*, 11669–11685.
[46] Han, Y.; Roy, A.; Giroux, A. *Tetrahedron Lett.* **2000**, *41*, 5447–5451.
[47] Veerman, J. J. N.; van Maarseveen, J. H.; Visser, G. M.; Kruse, C. G.; Schoemaker, H. E.; Hiemstra, H.; Rutjes, F. P. J. T. *Eur. J. Org. Chem.* **1998**, 2583–2589.
[48] Ramaseshan, M.; Dory, Y. L.; Deslongchamps, P. *J. Comb. Chem.* **2000**, *2*, 615–623.

[49]   Morales, G. A.; Corbett, J. W.; DeGrado, W. F. *J. Org. Chem.* **1998**, *63*, 1172–1177.
[50]   Darling, G. D.; Fréchet, J. M. J. *J. Org. Chem.* **1986**, *51*, 2270–2276.
[51]   Mansour, A.; Portnoy, M. *J. Chem. Soc., Perkin Trans. 1* **2001**, 952–954.
[52]   Leznoff, C. C.; Fyles, T. M.; Weatherston, J. *Can. J. Chem.* **1977**, *55*, 1143–1153.
[53]   Nicolaou, K. C.; Winssinger, N.; Pastor, J.; Ninkovic, S.; Sarabia, F.; He, Y.; Vourloumis, D.; Yang, Z.; Li, T.; Giannakakou, P.; Hamel, E. *Nature* **1997**, *387*, 268–272.
[54]   Oh, H. S.; Hahn, H.-G.; Cheon, S. H.; Ha, D.-C. *Tetrahedron Lett.* **2000**, *41*, 5069–5072.
[55]   Heitz, W.; Michels, R. *Makromol. Chem.* **1971**, *148*, 9–17.
[56]   Bernard, M.; Ford, W. T. *J. Org. Chem.* **1983**, *48*, 326–332.
[57]   Weinshenker, N. M.; Crosby, G. A.; Wong, J. Y. *J. Org. Chem.* **1975**, *40*, 1966–1971.
[58]   Crosby, G. A.; Weinshenker, N. M.; Uh, H. S. *J. Am. Chem. Soc.* **1975**, *97*, 2232–2235.
[59]   Camps, F.; Castells, J.; Ferrando, M. J.; Font, J. *Tetrahedron Lett.* **1971**, 1713–1714.
[60]   Hennion, G. F.; Anderson, J. G. *J. Am. Chem. Soc.* **1946**, *68*, 424–426.
[61]   Farrall, M. J.; Fréchet, J. M. J. *J. Org. Chem.* **1976**, *41*, 3877–3882.
[62]   Malenfant, P. R. L.; Fréchet, J. M. J. *Chem. Commun.* **1998**, 2657–2658.
[63]   Kirschbaum, T.; Briehn, C. A.; Bäuerle, P. *J. Chem. Soc., Perkin Trans. 1* **2000**, 1211–1216.
[64]   Briehn, C. A.; Kirschbaum, T.; Bäuerle, P. *J. Org. Chem.* **2000**, *65*, 352–359.
[65]   Brill, W. K.-D.; Riva-Toniolo, C. *Tetrahedron Lett.* **2001**, *42*, 6515–6518.
[66]   Zhang, H.-C.; Ye, H.; White, K. B.; Maryanoff, B. E. *Tetrahedron Lett.* **2001**, *42*, 4751–4754.
[67]   Arsequell, G.; Espuña, G.; Valencia, G.; Barluenga, J.; Carlón, R. P.; González, J. M. *Tetrahedron Lett.* **1998**, *39*, 7393–7396.
[68]   Deleuze, H.; Sherrington, D. C. *J. Chem. Soc., Perkin Trans. 2* **1995**, 2217–2221.
[69]   Arsequell, G.; Espuña, G.; Valencia, G.; Barluenga, J.; Carlón, R. P.; González, J. M. *Tetrahedron Lett.* **1999**, *40*, 7279–7282.

7 Preparation of Alcohols and Ethers

7.1 Preparation of Alcohols

For many years, most efforts directed towards the development of solid-phase preparations of alcohols had been limited to the synthesis of biopolymers, such as oligonucleotides and oligosaccharides. Interest in the preparation and chemical transformation of all types of alcohol on insoluble supports only began to grow rapidly in the early 1990s, when chemists realized the potential of parallel solid-phase synthesis for high-throughput compound production.

Alcohols are generally prepared on insoluble supports by reduction of carbonyl compounds or by addition of carbon nucleophiles to carbonyl compounds, although other strategies have also been used (Figure 7.1).

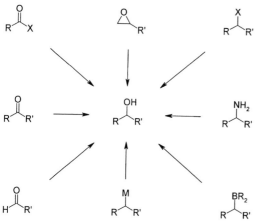

Figure 7.1. Strategies for the preparation of alcohols on solid phase. M: metal, X: leaving group.

7.1.1 By Reduction of Carbonyl Compounds

Support-bound aldehydes or ketones can be reduced to alcohols under mild reaction conditions that are compatible with most supports and linkers. Typical reducing agents are sodium borohydride or diisobutylaluminum hydride, which can penetrate cross-linked polystyrene provided that a solvent with sufficient swelling ability is

Table 7.1. Preparation of alcohols by reduction of carbonyl compounds.

Entry	Starting resin	Conditions	Product	Ref.
1		NaBH$_4$, (0.33 mol/L, 4 eq), THF/NMM/EtOH 16:7:7, 20 °C, 24 h		[1]
2		LiBH$_4$, THF, 60 °C		[2]
3		NaBH$_4$ (0.03 mol/L), THF/EtOH 1:1, 4 h		[3] see also [4]
4		L-Selectride (0.5 mol/L), THF, −75 °C to −65 °C, 20 h	(87:13)	[5] see also [6,7]
5		(HCHO)$_n$, KOH, DMF, 15 °C, 48 h		[8]
6		DIBAH (3 eq), THF, 20 °C, 4 h		[9] see also [10,11]
7		LiAlH$_4$ (2 eq), Et$_2$O, 30 °C, 6 h		[12]
8		LiAlH$_4$, Et$_2$O, 6 h		[13]
9		LiBH$_4$, THF, 20 °C, 12 h		[14]
10		1. *i*BuOCOCl, NMM, THF 2. NaBH$_4$, H$_2$O, THF		[15] see also [16]
11		1. EtOCOCl, NEt$_3$, THF, 0 °C 2. Bu$_4$NBH$_4$, MeOH		[17]

chosen. Reductions on polystyrene using sodium borohydride can be conducted in DMF or THF, or in mixtures of these solvents with alcohols (Table 7.1). Diastereoselective reductions have been performed on insoluble supports at low temperatures using Selectride (Entry **4**, Table 7.1). Formaldehyde can be used as an alternative to hydrides in reducing support-bound ketones (Cannizzaro reaction; Entry **5**, Table 7.1).

Polystyrene-bound carboxylic esters have been reduced with diisobutylaluminum hydride or lithium aluminum hydride. Use of the latter reagent can, however, lead to the formation of insoluble precipitates, which could readily cause problems if reactions are performed in fritted reactors. An alternative procedure for reducing carboxylic esters to alcohols involves saponification, followed by activation (e.g. as the mixed anhydride) and reduction with sodium borohydride (Entries **10** and **11**, Table 7.1). Care must be taken in the activation of carboxylic acids with alkyl chloroformates, because the resulting anhydrides decompose at room temperature to yield carboxylic esters.

7.1.2 By Addition of Carbon Nucleophiles to C=O Double Bonds

Support-bound carbonyl compounds can be converted into alcohols by treatment with suitable carbon nucleophiles. Aldehydes react readily with ketones or other C,H-acidic compounds under acid- or base-catalysis to yield the products of aldol addition (Table 7.2). Some types of C,H-acidic compound, such as 1,3-dicarbonyl compounds, can give the products of aldol condensation directly (Section 5.2.2.2).

Catalyzed aldol additions do not generally proceed with high diastereoselectivity at ambient temperature. Improved stereoselectivity can be achieved by using preformed, diastereomerically pure enolates at low temperatures (Entry **5**, Table 7.2). This strategy enables the solid-phase preparation of stereochemically defined polyketides. On cross-linked polystyrene, the observed diastereoselectivity in the addition of boron enolates to aldehydes is the same as that in the homogeneous phase reaction [14,18].

Support-bound carbonyl compounds have been converted into alcohols by treatment with Grignard reagents [26–31], organolithium compounds, allylindium compounds, allylsilanes, or organomanganese compounds (Table 7.3). These reactions, of course, require supports and linkers that do not react with the organometallic reagent. Furthermore, the support needs to show sufficient swelling in the chosen solvent. Because cross-linked polystyrene swells substantially in THF and is chemically stable towards strong nucleophiles, polystyrene is well suited for reactions involving Grignard or similar reagents. Careful drying is, however, often necessary to obtain good results. If the support-bound carbonyl compound can enolize, small amounts of water or alcohols can lead to substantial amounts of enolate on the support, which will not react with the organometallic reagent even if the latter is used in excess. Substrates with a high tendency to enolize are, for example, 1,3-dicarbonyl compounds and benzyl ketones. Enolate formation can be avoided by using less basic organometallic reagents, such as organotitanium, organozirconium, or organoindium reagents.

Table 7.2. Preparation of alcohols from support-bound carbonyl compounds by aldol addition.

Entry	Starting resin	Conditions	Product	Ref.
1		MeNO$_2$/EtOH/NEt$_3$/ THF 4:4:1:12, 20 °C, overnight		[19,20]
2		MeNO$_2$/NEt$_3$ 20:1, 20 °C, overnight (cross-linked PEG)		[21]
3		PhCOMe, K$_2$CO$_3$, THF, 67 °C, 48 h		[22] see also [23]
4		LDA, ZnCl$_2$, THF, −78 °C to −40 °C, 2 h		[24]
5		(C$_6$H$_{11}$)$_2$BO Et$_2$O, −78 °C, 1 h, 0 °C, 16 h	(> 97% ds)	[18] see also [14,25]

If enolate formation results in low conversions, the reaction may be repeated several times, which usually leads to improved results.

PEG-grafted polystyrene is also well suited for reactions with highly reactive organometallic reagents, provided that the support has been dried. PEG-containing polymers are generally more difficult to dry than pure polystyrene. Cross-linked PEG is stable towards Lewis acids, and can be used for SnCl$_4$-mediated allylations of aldehydes with allyl silanes [21].

The automated parallel synthesis of compounds in arrays of fritted reactors using organolithium or Grignard reagents is a special challenge. The transfer of these reagents to the reactor can readily lead to clogging of needles or tubing, and during the reaction or upon quenching precipitates might form and clog frits. All these potential problems should be carefully considered before starting library production.

Alcohols can also be prepared from support-bound carbon nucleophiles and carbonyl compounds (Table 7.4). Few examples have been reported of the α-alkylation of resin-bound esters with aldehydes or ketones. This reaction is complicated by the thermal instability of some ester enolates, which can undergo elimination of alkoxide to yield ketenes. Traces of water or alcohols can, furthermore, lead to saponification or transesterification and release of the substrate into solution. Less prone to base-induced cleavage are support-bound imides (Entry 2, Table 7.4; see also Entry 3, Table 13.8 [42]). Alternatively, support-bound thiol esters can be converted into stable silyl ketene acetals, which react with aldehydes under Lewis-acid catalysis (Entries 3 and 4, Table 7.4).

Table 7.3. Preparation of alcohols from support-bound carbonyl compounds and organometallic reagents.

Entry	Starting resin	Conditions	Product	Ref.
1		allyl bromide (14 eq), In (9 eq), THF/H$_2$O 1:1, ultrasound, 25 °C, 4 h (allyl boronates can also be used)		[32]
2		Me$_3$Si \\ SnCl$_4$, DCM, 2 h (cross-linked PEG)		[21]
3		Ph \\ BuLi (3 eq), THF, −78 °C, 4 h		[33]
4		PhMgBr, THF, −78 °C to −10 °C, 3 h (non-cross-linked PS)		[34,35]
5		Me$_3$Si≡Li (5 eq), THF, 0–23 °C, 2 h		[36]
6		PhMgBr, CeCl$_3$, THF, 3 h		[37]
7		PhMnI (0.25 mol/L), C$_6$H$_6$, 20 °C, 12 h		[38]
8		PhMgBr (large excess)		[39] see also [10,40]
9		allyl alcohol (0.7 mol/L, 7 eq), SnCl$_2$ (21 eq), PdCl$_2$(PhCN) (0.06 eq), THF, 20 °C, 5 h		[41]

Substituted allyl alcohols can be prepared on insoluble supports under mild conditions using the Baylis–Hillman reaction (Figure 7.2). In this reaction, an acrylate is treated with a nucleophilic tertiary amine (typically DABCO) or a phosphine in the presence of an aldehyde. Reversible Michael addition of the amine to the acrylate leads to an ester enolate, which then reacts with the aldehyde. The resulting allyl alcohols are valuable intermediates for the preparation of substituted carboxylic acids [43,44].

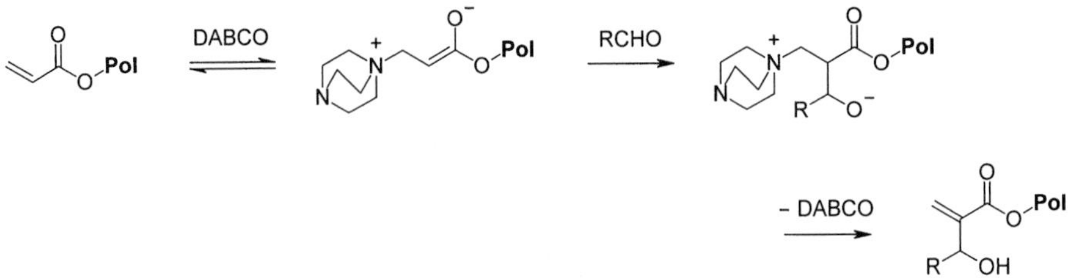

Figure 7.2. Mechanism of the Baylis–Hillman reaction.

Table 7.4. Preparation of alcohols from support-bound ester enolates and related carbon nucleophiles.

Entry	Starting resin	Conditions	Product	Ref.
1	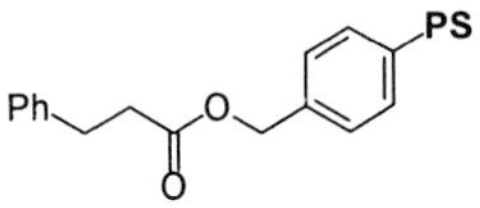	1. LDA, THF, −78 °C 2. ZnCl$_2$, 0 °C 3. 4-(MeO)C$_6$H$_4$CHO	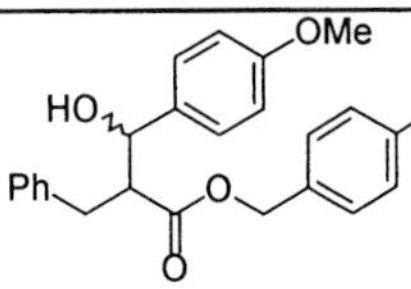	[45]
2	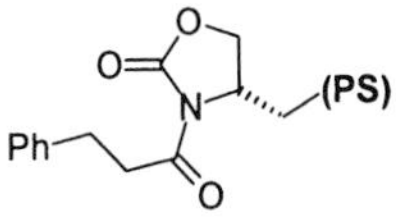	1. Bu$_2$BOTf (2 × 2 eq), DIPEA, DCM, −20 °C, 2 × 45 min 2. (iPr)CH$_2$CHO	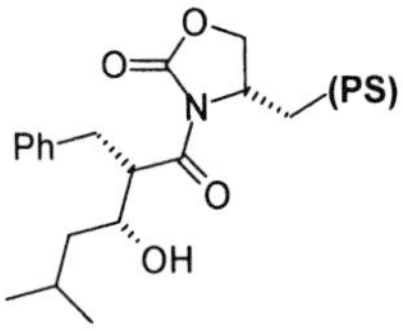	[46] see also [47]
3	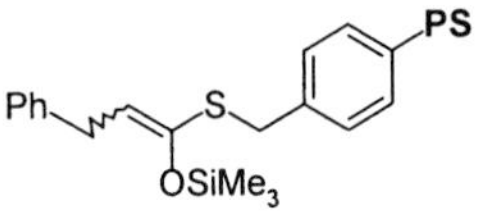	4-(MeO)C$_6$H$_4$CHO, Sc(OTf)$_3$, DCM, −78 °C, 20 h	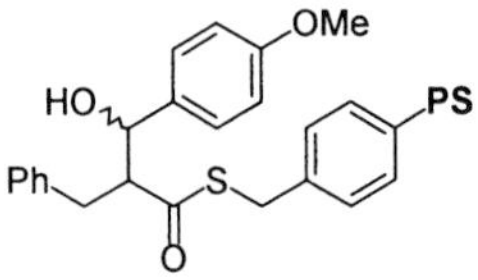	[48]
4	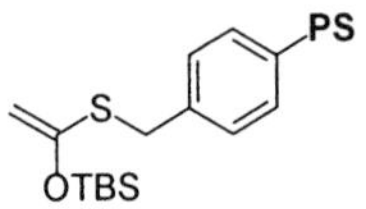	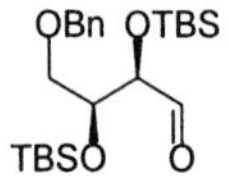 BF$_3$OEt$_2$, DCM, −78 °C, 20 h	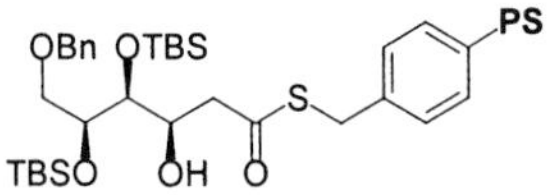 (> 98:2)	[49]
5	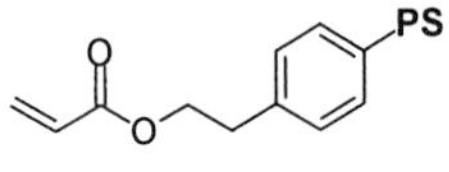	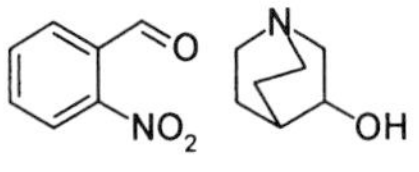 DMF, 22 °C, 18 h	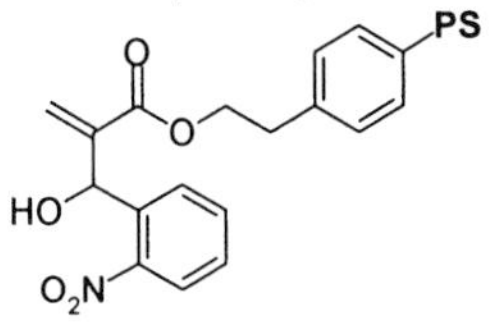	[43]
6	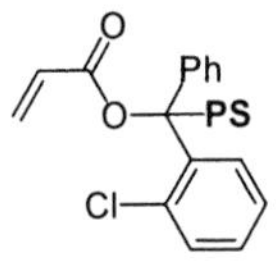	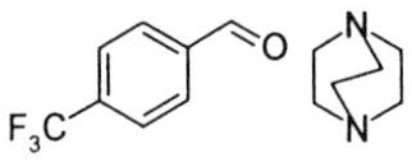 DMSO/CHCl$_3$ 1:1, 20 °C, 2 d	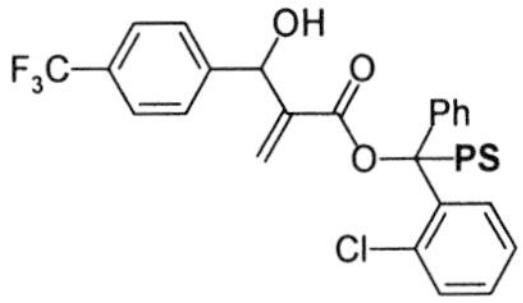	[44] see also [50]

More reactive carbon nucleophiles than enolates can also be prepared on insoluble supports (see Chapter 4) and are used to convert aldehydes or ketones into alcohols. Organolithium compounds have been generated on cross-linked polystyrene by deprotonation of formamidines and by metallation of aryl iodides (Table 7.5). Similarly, support-bound organomagnesium compounds can be prepared by metallation of aryl and vinyl iodides with Grignard reagents. The resulting organometallic compounds react with aldehydes or ketones to yield the expected alcohols (Table 7.5).

Table 7.5. Preparation of alcohols from support-bound organolithium compounds, organomagnesium compounds, and stannanes.

Entry	Starting resin	Conditions	Product	Ref.
1		*tert*-BuLi, THF, PhCHO, −78 °C to 20 °C, overnight		[51]
2		BuLi, THF, −78 °C, 15 min, then		[52]
3		*i*PrMgBr (0.19 mol/L, 7.3 eq), THF, −35 °C, 0.5 h, then PhCHO		[53] see also [54]
4		1. *i*PrMgBr (10 eq), THF/NMP 40:1, −40 °C, 1.5 h 2. PhCHO (15 eq), −40 °C, 2 h		[55]
5		BuLi, THF, −30 °C, 4 h; add TolCHO, −30 °C to 20 °C, 2 h		[56] see also [57,58]
6		C$_6$H$_{11}$CHO (20 eq), InBr$_3$ (1.2 eq), AcOEt, −78 °C to 20 °C, 5 h		[59] see also [60]

7.1.3 Miscellaneous Preparations of Alcohols

Hydroxymethyl polystyrene has been prepared from chloromethyl polystyrene, either by conversion into the acetate by nucleophilic substitution followed by saponification, or directly by treatment with a mixture of potassium acetate and tetrabutyl-ammonium hydroxide in 1,2-dichlorobenzene/water (85 °C, 2 d [61]).

A suitable means of access to functionalized alcohols that is also applicable to solid-phase synthesis is the ring-opening of epoxides with nucleophiles. Hydride (Entry **1**, Table 7.6), organolithium compounds (Entry **2**, Table 7.6), amines (Table 10.1), azide (Table 10.18), alcohols (Table 7.9), and thiols (Table 8.3) have been successfully used to convert (mostly support-bound) oxiranes into the corresponding alcohols. Acid-catalyzed ring-opening of oxiranes has also been used to prepare resin-bound alcohols: treatment of oxiranes with polystyrene-bound benzenesulfonic acid leads to 2-hydroxy-1-ethyl sulfonic esters [62]. The reaction of resin-bound organolithium compounds with epichlorohydrin has been used to prepare cyclobutanols (Figure 5.5 [63]). The oxidation of polystyrene-bound boranes or organolithium compounds has been used to prepare aliphatic alcohols and phenols (Table 7.6). Further oxidative protocols applicable to the solid-phase synthesis of alcohols include the α-hydroxylation of ester enolates with *N*-sulfonyloxaziridines, the oxidation of hydrazines [64], the ozonolysis of alkenes followed by reduction of the resulting ozonides, and the dihydroxylation of alkenes (Table 7.6). Attempts to perform asymmetric dihydroxylations of alkenes on either cross-linked polystyrene or PEG-grafted polystyrene resulted in lower enantioselectivities than those achieved in solution (Entries **6** and **7**, Table 7.6).

Table 7.6. Miscellaneous preparations of alcohols and phenols.

Entry	Starting resin	Conditions	Product	Ref.
1		LiBH$_4$, THF, 20 °C, 6 h		[67]
2		(190 eq), THF, −50 °C to 20 °C, 18 h		[68]
3		KN(SiMe$_3$)$_2$, Ph–N(SO$_2$Ph) oxaziridine		[11]
4		H$_2$O$_2$, NaOH, H$_2$O, THF, 20 °C, 4 h		[69,70]
5		1. O$_3$, DCM, −78 °C, 10 min 2. NaBH$_4$, *i*PrOH, ultrasound, 20 °C, overnight		[71]
6		K$_3$Fe(CN)$_6$, K$_2$CO$_3$, K$_2$OsO$_4$, (DHQD)$_2$PHAL (AD-mix-β), THF/H$_2$O 1:1, 20 °C, 12 h	(73% yield, 3% ee)	[72] see also [73]
7		K$_3$Fe(CN)$_6$, K$_2$CO$_3$, K$_2$OsO$_4$, (DHQD)$_2$PHAL (AD-mix-β), THF/H$_2$O 1:1, 20 °C, 18 h	(41% yield, 97% ee)	[72]

Table 7.6. continued.

Entry	Starting resin	Conditions	Product	Ref.
8	H₂N related structure with PS	NaNO₂, HCl (30% in H₂O), KCl, ultrasound, 3 h (PS-grafted crowns)	HO related structure with PS	[74]
9	structure with PS	1. BuLi, C₆H₁₂, 65 °C, 4 h 2. O₂, C₆H₁₂, 25 °C, 2 h	HO structure with PS	[75] see also [61]
10	X, O₂N structure with PS	BnMe₃N⁺OH⁻, H₂O, dioxane, 90 °C, 8 h (X: F, Cl, OMe)	HO, O₂N structure with PS	[76]
11	HO structure with (PS)	110 °C, PhMe	OH structure with (PS)	[52]
12	structure with PS	NaOH, EtOH, 78 °C, 1 h	Ph structure OH	[77]

Phenols have been prepared on solid phase by aromatic nucleophilic substitution with hydroxide, by thermal rearrangement of vinylcyclobutenones, by oxidative coupling of phenols (Figure 5.20 [65]), by cyclocondensation reactions with simultaneous release of the phenols into solution (Entry **12**, Table 7.6), and by Claisen rearrangement [66].

7.1.4 Protective Groups for Alcohols

Some of the strategies used for the protection of alcohols in solution can also be used on insoluble supports. The choice of protective groups is mainly limited by the type of linker chosen. The following section covers those protective groups for alcohols that can be either introduced or removed on insoluble supports. Orthogonal protection strategies for alcohols have mainly been developed for the solid-phase synthesis of oligonucleotides and oligosaccharides (Chapter 16).

7.1.4.1 Protective Groups Cleavable by Acids

Benzyl ethers bearing electron-donating groups can be cleaved by treatment with acids or oxidants. The rate of solvolysis increases with the number of electron-donating groups, and in the series benzyl < benzhydryl < trityl. For the solid-phase synthesis of oligonucleotides, the 5′-hydroxyl group is usually protected as 4,4′-dimethoxytrityl ether (DMT–OR; Entry **1**, Table 7.7), which can readily be cleaved by weak acids such as dichloroacetic acid (3% in DCM, 2 min [78–81]). Monomethoxytrityl ethers (MMT–OR) are only slightly more resistant towards acidolysis, and can, for example, be cleaved with 3% trichloroacetic acid in DCM [82] or 1% TsOH in DCM [5]. *tert*-Butyl ethers or *tert*-butyl carbamates (Boc-protected amines) are usually stable under these mildly acidic conditions.

As side-chain protection of serine or threonine in solid-phase peptide synthesis, unsubstituted benzyl ethers (Boc methodology), *tert*-butyl ethers [83], or trityl ethers (Fmoc methodology) can be used. Tyrosine is usually protected as its *tert*-butyl ether, benzyl ether, 2,6-dichlorobenzyl ether, or 2-bromobenzyl carbonate. Deprotection takes place during cleavage of the peptide from the support [84,85]. Benzyl ethers of CPG-bound carbohydrates have been cleaved by treatment with $NaBrO_3/Na_2S_2O_4$ in $AcOEt/H_2O$ [86]. 1-(Cyclopropyl)ethyl ethers have been used as acid-labile protective groups in the synthesis of oligosaccharides on PEG [87].

As an alternative to protection as ethers, alcohols can also be protected as acetals, the most common being tetrahydropyranyl ethers (THP–OR) and 1-ethoxyethyl ethers (EE–OR) (Table 7.7). Support-bound secondary aliphatic alcohols have been

Table 7.7. Acid-labile protective groups for alcohols.

Entry	Starting resin	Conditions	Product	Ref.
1		3% Cl_3CCO_2H in DCM, 8 s–50 min; or 80% AcOH, 0.5 h; or $ZnBr_2$, $MeOH/MeNO_2$		[90–93]
2		TsOH (0.03 mol/L), DCM/MeOH 97:3, 20 °C, 2 h		[94,95] see also [96]
3		AcOH, H_2O, THF		[88]
4		MeOH, dioxane, PPTS		[97]
5		TsOH (2% in DCM), 5 × 5 min (TsOH was dissolved in MeOH and diluted with DCM)		[5]

converted into THP ethers by treatment with dihydropyran and PPTS or TsOH [88,89]. Both THP and EE ethers can be cleaved under mildly acidic conditions (e.g. with dilute acetic acid [88]), under which no cleavage of *tert*-butyl ethers, *tert*-butyl carbamates, glycosides, or esters of Wang resin occurs.

7.1.4.2 Protective Groups Cleavable by Nucleophiles or Other Reagents

The solid-phase synthesis of oligosaccharides is usually performed using acid-resistant linkers and protective groups, because of the slightly acidic reaction conditions required for glycosylations (Section 16.3). Hydroxyl group protection is conveniently achieved by conversion into carboxylic esters, such as acetates, benzoates, or nitrobenzoates. Support-bound esters of primary or secondary aliphatic alcohols can be cleaved by treatment with alcoholates [97–99] (Table 7.8), with DBU in methanol, with hydrazine in DMF [100] or dioxane [101], or with ethylenediamine [102], provided that a linker resistant towards nucleophiles has been chosen.

Esters of 4-oxo carboxylic acids, such as levulinic acid (Entry **5**, Table 7.8) or 3-benzoylpropionic acid [103], can be cleaved under essentially neutral reaction conditions

Table 7.8. Protective groups for alcohols cleavable by nucleophiles and oxidants.

Entry	Starting resin	Conditions	Product	Ref.
1		NaOEt (0.1 mol/L), DMF/pyridine/EtOH 5:4:1, 20 °C, 1 h (macroporous PS)		[113]
2		NH₃ (30% in H₂O)/ THF/MeOH 2:6:2, 20 °C, 16 h		[114] see also [115]
3		K₂CO₃ (1.3 eq), MeOH/DCM 6:20, overnight		[116]
4		guanidine (0.1 mol/L, 2 eq), DMF, 20 °C, 2 × 16 h		[117] see also [106]
5		N₂H₄·AcOH, EtOH, 20 °C, 1 d		[118]

Table 7.8. continued.

Entry	Starting resin	Conditions	Product	Ref.
6	tBu(Ph)$_2$SiO, EtO, OMe (PS)	TBAF (10 eq), THF, 20 °C, 16 h	HO, EtO, OMe (PS)	[97] see also [119]
7	Et$_3$SiO, OAc (PS)	HF·pyridine, THF, 23 °C	HO, OAc (PS)	[106]
8	TBSO (PS), OR	TBAF (0.3 mol/L), AcOH (0.3 mol/L), 20 °C, overnight R: alkyl	HO (PS), OR	[115]
9	NO$_2$, F, PS, O, F, F, O (allyl)	Pd(PPh$_3$)$_4$, TolSO$_2$Na, MeOH, THF	NO$_2$, F, PS, O, F, F, OH	[120] see also [110,121]
10	MeO (CPG), O, O, (RO)$_2$OPO	CAN (0.1 mol/L), MeCN/H$_2$O 4:1, 25 °C, 5 min	(CPG), HO, O, (RO)$_2$OPO	[82]

using hydrazinium acetate [104]. The cleavage proceeds via hydrazone formation and ring-closure to yield the alcohol and a 2,3,4,5-tetrahydro-3-pyridazinone. Acetates may, however, also be hydrolyzed under these conditions [103].

Silyl ethers that have been used in solid-phase synthesis include TES, TIPS, TBS, and TBDPS ethers. Polystyrene-bound phenols can be converted into TIPS ethers by treatment with TIPS–OTf/imidazole in DMF for 5 min [105]. These silyl ethers are stable towards bases and weak acids, but can be selectively removed by treatment with TBAF (Entries **6** and **8**, Table 7.8) or pyridinium hydrofluoride (THF, 25 °C, 15 h [24,75,106]).

Allyl carbonates can be cleaved by nucleophiles under palladium(0) catalysis. Allyl carbonates have been proposed for side-chain protection of serine and threonine, and their stability under conditions of *N*-Fmoc or *N*-Boc deprotection has been demonstrated [107]. Prolonged treatment with nucleophiles (e.g., 20% piperidine in DMF, 24 h) can, however, lead to deprotection of Alloc-protected phenols [108,109]. Carbohydrates [110], tyrosine derivatives [107], and other phenols have been protected as allyl ethers, and deprotection could be achieved by palladium-mediated allylic substitution (Entry **9**, Table 7.8). 9-Fluorenyl carbonates have been used as protected intermediates for the solid-phase synthesis of oligosaccharides [111]. Deprotection was achieved by treatment with NEt$_3$/DCM (8:2) at room temperature.

4-Methoxyphenyl ethers can be cleaved by mild oxidants (Entry **10**, Table 7.8). Because many acid-labile linkers are also readily oxidized, care must be taken when applying this deprotection strategy. Benzyl ethers have been removed from Tentagel- or PEGA-bound carbohydrates by catalytic hydrogenation using palladium nanoparticles [112].

7.2 Preparation of Ethers

Ethers are widely used in solid-phase synthesis, either as linkers for alcohols or as target compounds. Almost all reported solid-phase syntheses of ethers are O-alkylations or O-arylations of alcohols, which differ only in the type of alkylating/arylating agent used and in the precise reaction conditions. This section covers mainly syntheses of acyclic ethers. Preparations of cyclic ethers are considered in Chapter 15.

7.2.1 Preparation of Dialkyl Ethers

Dialkyl ethers can be prepared either from support-bound electrophiles or from support-bound alcohols. Table 7.9 lists illustrative examples of the preparation of dialkyl ethers from support-bound electrophiles.

Cross-linked chloromethyl polystyrene (Merrifield resin) has frequently been used to O-alkylate aliphatic alcoholates, mainly for attaching linkers, ligands, or various synthetic auxiliaries to the support (Table 7.9). Examples other than those listed in Table 7.9 have been reported [24,119,122–126]. Strong bases, such as alkali metal hydrides, are required to deprotonate aliphatic alcohols. Suitable solvents for the conversion of Merrifield and related resins into alkyl benzyl ethers are THF, DMF, or diglyme, and high temperatures and long reaction times are generally necessary to achieve complete conversions. Support-bound halides bearing β-hydrogens are generally difficult to convert into dialkyl ethers, because the strongly basic reaction conditions required will usually lead to the formation of alkenes by dehydrohalogenation. Resin-bound benzhydryl or trityl halides react with alcohols according to an S_N1 mechanism, and, accordingly, do not require strong bases to O-alkylate aliphatic alcohols (Entry **7**, Table 7.9). Wang resin can also be etherified with aliphatic alcohols under acidic reaction conditions, by conversion of the polystyrene-bound alcohol into a trichloroacetimidate or a thiocarbonate (Entries **5** and **6**, Table 7.9). Support-bound aryldiazomethanes, which can be prepared from hydrazones or tosyl hydrazones, also react with alcohols under Lewis acid catalysis to yield benzyl ethers (Entry **8**, Table 7.9). Other support-bound electrophiles that react with aliphatic alcohols to yield ethers are epoxides, nitrostyrenes, and rhodium carbenoids generated from diazocarbonyl compounds and rhodium(II) acetate (Entries **10–12**, Table 7.9). The latter strategy enables the preparation of dialkyl ethers under essentially neutral reaction conditions, and thus the etherification of base- or acid-labile alcohols. Methyl ethers have been prepared by C-allylation of support-bound aldehydes or dimethylacetals with allylsilanes and TMS-methanolate (Entry **13**, Table 7.9). An elegant method for the preparation of monoethers of Wang resin with diols is the reduction of cyclic acetals (Entries **14** and **15**, Table 7.9). A more detailed exploration of the scope of this reaction, and the extension of this methodology to mono-hydroxy compounds remains to be performed. Because Wang resin derived ethers are stable towards bases or nucleophiles, but can readily be cleaved with dilute TFA, etherification represents a convenient method for linking alcohols to insoluble supports [10] (Section 3.11.1).

Polystyrene-bound aliphatic alcohols can be etherified with alkyl halides under strongly basic conditions (Table 7.10). Competing elimination is usually no concern, because a large excess of halide can be used and alkenes can be readily removed by filtration and washing of the support. Alternatively, the addition of resin-bound alco-

Table 7.9. Solid-phase synthesis of dialkyl ethers from support-bound electrophiles.

Entry	Starting resin	Conditions	Product	Ref.
1		KH, 18-crown-6, DMF, 80 °C, 5 d		[127] see also [128–130]
2		(3 eq), KOtBu (1 mol/L, 3 eq), THF, 20 °C, 3.5 d		[131]
3		NaH (0.1 mol/L), DMF, 0 °C, 48 h		[132]
4		NaH, Bu₄NI, 18-crown-6, THF, 45 °C, 2 h		[5]
5		DCM/C₆H₁₂ 1:1, BF₃OEt₂ (13 mmol/L), 20 °C, 10 min		[133] see also [9,134]
6		AgOTf, DCM, 5 h		[135] see also [11]
7		*N*-Fmoc-prolinol (0.05 mol/L, 3 eq), DIPEA (1.6 eq), DCE/THF 3:1, 48 h		[136]

Table 7.9. continued.

Entry	Starting resin	Conditions	Product	Ref.
8		MeO_2C...OH, NHFmoc (2 eq), BF_3OEt_2 (0.1 eq), DCM, 20 °C, 2 h		[137]
9		1. TFA/DCM 7:3, 20 °C, 5 h 2. $ArCH_2CH_2OH$ (0.5 mol/L), DCM, 1 h Ar: 2-naphthyl		[138]
10		dodecanol (0.1 mol/L, 8 eq), KOtBu (0.3 mol/L), diglyme, 100 °C, 30 h		[139]
11		HO...R NaH, THF, 20 °C, 20 h		[20]
12		iPrOH, $Rh_2(OAc)_4$, DCM, 20 °C, 20 h		[140]
13		MeO_2C, $PhMe_2Si$, Me_3SiOMe, Me_3SiOTf, DCM, −78 °C to −55 °C, 72 h		[141]
14		DIBAH (3 eq), DCM, −70 °C to −20 °C, 4 h		[142]
15		DIBAH (3 eq), DCM, −70 °C to −20 °C, 4 h		[142]

hols to acceptor-substituted alkenes can be used to prepare dialkyl ethers on insoluble supports (Entry **2**, Table 7.10). Treatment of polystyrene-bound alcohols with styrenes and N-iodosuccinimide under acidic conditions leads to the formation of 2-iodoalkyl ethers (Entry **3**, Table 7.10). These intermediates have been used to N-alkylate imidazole in the preparation of analogs of the antifungal agent miconazole [143]. Support-bound allylsilanes react with acetals under acidic conditions to yield homoallyl ethers (Entry **4**, Table 7.10).

Table 7.10. Solid-phase synthesis of dialkyl ethers from support-bound nucleophiles.

Entry	Starting resin	Conditions	Product	Ref.
1		NaH, THF/MeCN 5:1, 15-crown-5, ArCH$_2$Cl (0.3 mol/L), 67 °C, overnight Ar: 3,5-(MeO)$_2$C$_6$H$_3$		[51]
2		(0.7 mol/L), DBU (0.2 mol/L), DCM, 20 °C, overnight		[144]
3		(0.9 mol/L), NIS (0.9 mol/L), DME, TfOH (0.018 mol/L), 20 °C, 2 × 16 h		[143]
4		PhCH(OMe)$_2$, Me$_3$SiOTf, DCM, −78 °C to −55 °C, 72 h	(*syn/anti* 20:1)	[141]
5		1. NaH (3.3 eq), 18-crown-6 (0.6 eq), DMF, 23 °C, 18 h 2. add BnBr (8.9 eq), Bu$_4$NI (0.6 eq), 23 °C, 3 × 8 h		[145]

7.2.2 Preparation of Alkyl Aryl Ethers

Phenols attached to insoluble supports can be etherified either by treatment with alkyl halides and a base (Williamson ether synthesis) or by treatment with primary or secondary aliphatic alcohols, a phosphine, and an oxidant (typically DEAD; Mitsunobu reaction). The second methodology is generally preferred, because more alcohols than alkyl halides are commercially available, and because Mitsunobu etherifications proceed quickly at room temperature with high chemoselectivity, as illustrated by Entry 3 in Table 7.11. Thus, neither amines nor C,H-acidic compounds are usually alkylated under Mitsunobu conditions as efficiently as phenols. The reaction proceeds smoothly with both electron-rich and electron-poor phenols. Both primary and secondary aliphatic alcohols can be used to O-alkylate phenols, but variable results have been reported with 2-(Boc-amino)ethanols [146,147].

The Mitsunobu etherification of polystyrene-bound phenols is usually conducted in THF or NMP, simply by adding the alcohol, the phosphine, and DEAD. Some authors claim that the addition of tertiary amines is beneficial [148], but this seems not always to be the case [146]. It is, of course, important that neither the support nor any of the reagents is contaminated with traces of methanol (e.g. from previous washings of the resin) or ethanol (from the decomposition of DEAD), which would lead to the formation of product mixtures. It has been suggested [146] that the heat evolved upon mixing a phosphine with DEAD can contribute to the decomposition of the latter, and lead to the formation of ethyl ethers. This can be avoided by using another oxidant (e.g. DIAD or TMAD) or a preformed betaine (e.g. Entry **4**, Table 7.11).

The etherification of support-bound phenols with alkyl halides is usually performed in dipolar aprotic solvents (DMF, NMP, DMSO) in the presence of bases such as DBU, KN(SiMe$_3$)$_2$, phosphazenes [149], or cesium carbonate (Entries **6** and **7**, Table 7.11).

Polystyrene-bound alkylating agents can react with phenols under basic conditions to yield aryl ethers. The reaction conditions must be carefully adjusted for halides bearing β-hydrogens so as to prevent dehydrohalogenation. This is, of course, no prob-

Table 7.11. Preparation of alkyl aryl ethers from support-bound phenols.

Entry	Starting resin	Conditions	Product	Ref.
1		1,3-propanediol, PPh$_3$, DEAD, THF, 1 h		[146] see also [150]
2		*i*PrOH, PPh$_3$, DEAD, THF, 1 h		[146]
3		ArCH$_2$OH (0.1 mol/L, 5 eq), TMAD (5 eq), PBu$_3$ (5 eq), THF/DCM 1:1, 1 h		[151] see also [152]
4		2-(4-fluorophenyl)-ethanol, DCM/PhMe 1:1, 3 d		[115] see also [147,153]
5		EtOH, DEAD, PPh$_3$, THF, −15 °C, 1 h, then 20 °C, 24 h		[120]
6		BuI (0.25 mol/L), DBU (0.25 mol/L), DMSO/NMP 1:1, 20 h		[154] see also [149,155]
7		Cs$_2$CO$_3$, DMF, 25 °C, 30 h		[75]

lem when using benzylic halides, which are in fact the most commonly used support-bound alkylating agents for the O-alkylation of phenols (Table 7.12). These reactions are usually conducted in DMF or DMA, and bases such as sodium hydride [77,123,153,156,157], alkali metal carbonates (Entries **2**, **3**, and **7**; Table 7.12), sodium methanolate [29,158], or sodium hydroxide [159,160] can be used. Strongly acidic phenols (nitrophenol, pentafluorophenol) can even undergo alkylation in the presence of DIPEA [161], but the outcome of such etherifications using weak bases tends to be variable. Functionalized phenols, such as hydroxybenzaldehydes [160], hydroxybenzamides [159], or phenols containing keto [29] or hydroxyalkyl groups do not usually undergo side reactions under the conditions of aryl ether formation, and can be selec-

Table 7.12. Preparation of alkyl aryl ethers from support-bound alkylating agents.

Entry	Starting resin	Conditions	Product	Ref.
1		OHC—⟨⟩—OH, MeO, KOtBu, DMA, 90–50 °C, 14 h	('Sasrin-aldehyde')	[1,165] see also [27,166]
2		3-(HO)C₆H₄COMe, Cs₂CO₃, NaI, DMF		[167] see also [168]
3		2-cyanophenol, Cs₂CO₃, NMP, 70 °C, overnight		[169]
4		O₂N—⟨⟩—OH (0.5 mol/L), NMP/2,6-lutidine 4:1, 20 h		[170] see also [138]
5		Br—⟨⟩—CHO with OH (0.22 mol/L, 5 eq), DIPEA (5 eq), DMAP, DCM, 25 °C, 18 h		[171]
6		2-iodophenol (0.13 mol/L, 6 eq), BEMP (6 eq), dioxane, 100 °C, 90 h		[172]
7		K₂CO₃, KI, DMF, 80 °C, 16 h		[173] see also [174]

tively O-alkylated with support-bound benzyl halides (Table 7.12). PEG-bound mesylates and benzyl halides have been used to etherify phenols under conditions similar to those used in solution [162–164].

Alternatively, alkyl aryl ethers can be prepared from support-bound aliphatic alcohols by Mitsunobu etherification with phenols (Table 7.13). In this variant of the Mitsunobu reaction, the presence of residual methanol or ethanol is less critical than in the etherification of support-bound phenols, because no dialkyl ethers can be generated by the Mitsunobu reaction. For this reason, good results will also be obtained if the reaction mixture is allowed to warm upon mixing DEAD and the phosphine. Both triphenyl- and tributylphosphine can be used as the phosphine component. Tributylphosphine is a liquid and generally does not give rise to insoluble precipitates. This reagent must, however, be handled with care because it readily ignites in air when absorbed on paper.

Table 7.13. Preparation of alkyl aryl ethers from support-bound aliphatic alcohols.

Entry	Starting resin	Conditions	Product	Ref.
1		PPh₃, DEAD, THF, 20 °C, 2 × 24 h		[177]
2		Boc-Tyr-OMe (0.3 mol/L, 3 eq), PPh₃ (3 eq), DEAD (3 eq), NMM, 25 °C, 16 h		[175,178] see also [179]
3		phenol, PPh₃, DEAD, NMM		[15]
4		2-fluorophenol, PPh₃, DIAD, NMM, 20 °C, 12 h		[180]
5		4-(HO)C₆H₄CHO, DCM		[181]
6		TBAF, DMF, 25 °C, 17 h (partial epimerization occurs)		[182] see also [183]

It has been reported that tertiary amines (as additives or as the solvent) lead to increased yields when etherifying tyrosine derivatives with polystyrene-bound benzyl alcohols [175]. Nevertheless, other phenols react smoothly without the addition of a base [47,176]. When only a slight excess of phenol is used for the etherification of support-bound alcohols, *N,N'*-bis(ethoxycarbonyl)hydrazine (the by-product of the Mitsunobu reaction) can compete with the phenol to a significant extent and become attached to the support. This reaction can be suppressed by the use of a greater excess of phenol [168].

The most common resin-bound substrates for Mitsunobu etherification are primary benzylic alcohols, but a few non-benzylic alcohols have also been converted into aryl ethers (Table 7.13). Support-bound secondary alcohols are less suitable alkylating agents because elimination often predominates.

Experimental Procedure 7.1: Etherification of Fmoc-tyrosine methyl ester with a dialkoxybenzyl alcohol linker [179]

A solution of Fmoc-Tyr-OMe (96 mg, 0.23 mmol, 5.2 equiv.) and PPh$_3$ (72 mg, 0.28 mmol, 6.2 equiv.) in DCM (2 mL) was added to the DCM-swollen support (100 mg, 0.044 mmol). To this suspension, a solution of DEAD (0.042 mL, 0.27 mmol, 6.1 equiv.) in DCM (0.2 mL) was added and the resulting mixture was shaken at room temperature overnight. Filtration, washing with DCM, DMF, MeOH, and Et$_2$O, and drying under reduced pressure yielded a resin with a loading of 0.17 mmol/g. The residual hydroxyl groups were acetylated by treatment of the support with Ac$_2$O (6 equiv.) and DIPEA (6 equiv.) in DCM (0.5 mL) for 20 min.

7.2.3 Preparation of Diaryl Ethers

Acceptor-substituted haloarenes have been successfully used to O-arylate phenols by aromatic nucleophilic substitution (Table 7.14). The most common arylating agents are 2-fluoro-1-nitroarenes, 2-halopyridines, 2-halopyrimidines, and 2-halotriazines. When sufficiently reactive haloarenes are used, the reaction proceeds smoothly with either the arylating agent or the phenol linked to the support. The thallium(III) nitrate catalyzed arylation of phenols with aryl iodides has been used for macrocyclizations on solid phase [184]. Burgess and co-workers have developed a solid-phase synthesis of β-turn mimetics based on ring-closure by aromatic nucleophilic substitution (Entry **4**, Table 7.14; see also Table 10.5). Phenols, alkylamines, and thiols have been successfully used as nucleophiles for this type of macrocyclization [185].

Table 7.14. Preparation of diaryl ethers.

Entry	Starting resin	Conditions	Product	Ref.
1		salicylaldehyde, K$_2$CO$_3$, DMF, 40 °C, 16 h		[186]
2		1. NaOH, H$_2$O 2. cyanuric chloride, Me$_2$CO, 0 °C, 8 h, then 20 °C, 2 d		[101]
3		(0.16 mol/L), NaH, DMF, 2 h		[187]
4		1. TBAF, THF 2. K$_2$CO$_3$, DMF (cyclization also occurs without K$_2$CO$_3$ treatment)		[105] see also [185,188, 189]

References for Chapter 7

[1] Katritzky, A. R.; Toader, D.; Watson, K.; Kiely, J. S. *Tetrahedron Lett.* **1997**, *38*, 7849–7850.
[2] Brown, D. S.; Revill, J. M.; Shute, R. E. *Tetrahedron Lett.* **1998**, *39*, 8533–8536.
[3] Purandare, A. V.; Poss, M. A. *Tetrahedron Lett.* **1998**, *39*, 935–938.
[4] Chambers, S. L.; Ronald, R.; Hanesworth, J. M.; Kinder, D. H.; Harding, J. W. *Peptides* **1997**, *18*, 505–512.
[5] Lee, C. E.; Kick, E. K.; Ellman, J. A. *J. Am. Chem. Soc.* **1998**, *120*, 9735–9747.
[6] Thompson, L. A.; Moore, F. L.; Moon, Y. C.; Ellman, J. A. *J. Org. Chem.* **1998**, *63*, 2066–2067.
[7] Chen, S.; Janda, K. D. *Tetrahedron Lett.* **1998**, *39*, 3943–3946.
[8] Ren, Q.; Huang, W.; Ho, P. *Reactive Polymers* **1989**, *11*, 237–244.
[9] Furman, B.; Thürmer, R.; Kaluza, Z.; Lysek, R.; Voelter, W.; Chmielewski, M. *Angew. Chem. Int. Ed.* **1999**, *38*, 1121–1123.
[10] Hanessian, S.; Xie, F. *Tetrahedron Lett.* **1998**, *39*, 737–740.
[11] Hanessian, S.; Ma, J.; Wang, W. *Tetrahedron Lett.* **1999**, *40*, 4631–4634.
[12] Chandrasekhar, S.; Padmaja, M. B. *Synth. Commun.* **1998**, *28*, 3715–3720.
[13] Wang, S. *J. Am. Chem. Soc.* **1973**, *95*, 1328–1333.
[14] Reggelin, M.; Brenig, V.; Welcker, R. *Tetrahedron Lett.* **1998**, *39*, 4801–4804.
[15] Gayo, L. M.; Suto, M. J. *Tetrahedron Lett.* **1997**, *38*, 211–214.
[16] Bhandari, A.; Jones, D. G.; Schullek, J. R.; Vo, K.; Schunk, C. A.; Tamanaha, L. L.; Chen, D.; Yuan, Z. Y.; Needels, M. C.; Gallop, M. A. *Bioorg. Med. Chem. Lett.* **1998**, *8*, 2303–2308.
[17] Moran, E. J.; Wilson, T. E.; Cho, C. Y.; Cherry, S. R.; Schultz, P. G. *Biopolymers* **1995**, *37*, 213–219.
[18] Gennari, C.; Ceccarelli, S.; Piarulli, U.; Aboutayab, K.; Donghi, M.; Paterson, I. *Tetrahedron* **1998**, *54*, 14999–15016.
[19] Beebe, X.; Schore, N. E.; Kurth, M. J. *J. Org. Chem.* **1995**, *60*, 4196–4203.

[20] Beebe, X.; Chiappari, C. L.; Olmstead, M. M.; Kurth, M. J.; Schore, N. E. *J. Org. Chem.* **1995**, *60*, 4204–4212.
[21] Rademann, J.; Meldal, M.; Bock, K. *Chem. Eur. J.* **1999**, *5*, 1218–1225.
[22] Ruhland, T.; Künzer, H. *Tetrahedron Lett.* **1996**, *37*, 2757–2760.
[23] Graven, A.; Grøtli, M.; Meldal, M. *J. Chem. Soc., Perkin Trans. 1* **1999**, 955–962.
[24] Nicolaou, K. C.; Winssinger, N.; Pastor, J.; Ninkovic, S.; Sarabia, F.; He, Y.; Vourloumis, D.; Yang, Z.; Li, T.; Giannakakou, P.; Hamel, E. *Nature* **1997**, *387*, 268–272.
[25] Reggelin, M.; Brenig, V. *Tetrahedron Lett.* **1996**, *37*, 6851–6852.
[26] Leznoff, C. C.; Wong, J. Y. *Can. J. Chem.* **1973**, *51*, 3756–3764.
[27] Routledge, A.; Stock, H. T.; Flitsch, S. L.; Turner, N. J. *Tetrahedron Lett.* **1997**, *38*, 8287–8290.
[28] Xu, Z. H.; McArthur, C. R.; Leznoff, C. C. *Can. J. Chem.* **1983**, *61*, 1405–1409.
[29] Léger, R.; Yen, R.; She, M. W.; Lee, V. J.; Hecker, S. J. *Tetrahedron Lett.* **1998**, *39*, 4171–4174.
[30] Plater, M. J.; Murdoch, A. M.; Morphy, J. R.; Rankovic, Z.; Rees, D. C. *J. Comb. Chem.* **2000**, *2*, 508–512.
[31] Saha, A. K.; Liu, L.; Simoneaux, R. L.; Kukla, M. J.; Marichal, P.; Odds, F. *Bioorg. Med. Chem. Lett.* **2000**, *10*, 2175–2178.
[32] Cavallaro, C. L.; Herpin, T.; McGuinness, B. F.; Shimshock, Y. C.; Dolle, R. E. *Tetrahedron Lett.* **1999**, *40*, 2711–2714.
[33] Routledge, A.; Abell, C.; Balasubramanian, S. *Tetrahedron Lett.* **1997**, *38*, 1227–1230.
[34] Gosselin, F.; Van Betsbrugge, J.; Hatam, M.; Lubell, W. D. *J. Org. Chem.* **1999**, *64*, 2486–2493.
[35] Gosselin, F.; Van Betsbrugge, J.; Hatam, M.; Lubell, W. D. *J. Org. Chem.* **1999**, *64*, 2486–2493.
[36] Fraley, M. E.; Rubino, R. S. *Tetrahedron Lett.* **1997**, *38*, 3365–3368.
[37] Chen, C.; Munoz, B. *Tetrahedron Lett.* **1998**, *39*, 3401–3404.
[38] Leznoff, C. C.; Yedidia, V. *Can. J. Chem.* **1980**, *58*, 287–290.
[39] Liu, G. C.; Ellman, J. A. *J. Org. Chem.* **1995**, *60*, 7712–7713.
[40] Cossy, J.; Tresnard, L.; Pardo, D. G. *Synlett* **2000**, 409–411.
[41] Carde, L.; Llebaria, A.; Delgado, A. *Tetrahedron Lett.* **2001**, *42*, 3299–3302.
[42] Backes, B. J.; Ellman, J. A. *J. Am. Chem. Soc.* **1994**, *116*, 11171–11172.
[43] Prien, O.; Rölfing, K.; Thiel, M.; Künzer, H. *Synlett* **1997**, 325–326.
[44] Richter, H.; Walk, T.; Höltzel, A.; Jung, G. *J. Org. Chem.* **1999**, *64*, 1362–1365.
[45] Kurth, M. L.; Randall, L. A. A.; Chen, C.; Melander, C.; Miller, R. B.; McAlister, K.; Reitz, G.; Kang, R.; Nakatsu, T.; Green, C. *J. Org. Chem.* **1994**, *59*, 5862–5864.
[46] Purandare, A. V.; Natarajan, S. *Tetrahedron Lett.* **1997**, *38*, 8777–8780.
[47] Phoon, C. W.; Abell, C. *Tetrahedron Lett.* **1998**, *39*, 2655–2658.
[48] Kobayashi, S.; Hachiya, I.; Yasuda, M. *Tetrahedron Lett.* **1996**, *37*, 5569–5572.
[49] Kobayashi, S.; Wakabayashi, T.; Yasuda, M. *J. Org. Chem.* **1998**, *63*, 4868–4869.
[50] Kulkarni, B. A.; Ganesan, A. *J. Comb. Chem.* **1999**, *1*, 373–378.
[51] Furth, P. S.; Reitman, M. S.; Gentles, R.; Cook, A. F. *Tetrahedron Lett.* **1997**, *38*, 6643–6646.
[52] Tempest, P. A.; Armstrong, R. W. *J. Am. Chem. Soc.* **1997**, *119*, 7607–7608.
[53] Boymond, L.; Rottländer, M.; Cahiez, G.; Knochel, P. *Angew. Chem. Int. Ed.* **1998**, *37*, 1701–1703.
[54] Gelens, E.; Koot, W. J.; Menge, W. M. P. B.; Ottenheijm, H. C. J.; Timmerman, H. *Bioorg. Med. Chem. Lett.* **2000**, *10*, 1935–1938.
[55] Rottländer, M.; Knochel, P. *J. Comb. Chem.* **1999**, *1*, 181–183.
[56] Li, Z.; Ganesan, A. *Synlett* **1998**, 405–406.
[57] Garibay, P.; Vedsø, P.; Begtrup, M.; Hoeg-Jensen, T. *J. Comb. Chem.* **2001**, *3*, 332–340.
[58] Garibay, P.; Toy, P. H.; Hoeg-Jensen, T.; Janda, K. D. *Synlett* **1999**, 1438–1440.
[59] Cossy, J.; Rasamison, C.; Pardo, D. G.; Marshall, J. A. *Synlett* **2001**, 629–633.
[60] Cossy, J.; Rasamison, C.; Pardo, D. G. *J. Org. Chem.* **2001**, *66*, 7195–7198.
[61] Fréchet, J. M. J.; de Smet, M. D.; Farrall, M. J. *Polymer* **1979**, *20*, 675–680.
[62] Nicolaou, K. C.; Baran, P. S.; Zhong, Y.-L. *J. Am. Chem. Soc.* **2000**, *122*, 10246–10248.
[63] Cheng, W.-C.; Halm, C.; Evarts, J. B.; Olmstead, M. M.; Kurth, M. J. *J. Org. Chem.* **1999**, *64*, 8557–8562.
[64] Bonnet, D.; Rommens, C.; Gras-Masse, H.; Melnyk, O. *Tetrahedron Lett.* **1999**, *40*, 7315–7318.
[65] ApSimon, J. W.; Dixit, D. M. *Can. J. Chem.* **1982**, *60*, 368–370.
[66] Kumar, H. M. S.; Anjaneyulu, S.; Reddy, B. V. S.; Yadav, J. S. *Synlett* **2000**, 1129–1130.
[67] Rotella, D. P. *J. Am. Chem. Soc.* **1996**, *118*, 12246–12247.
[68] Darling, G. D.; Fréchet, J. M. J. *J. Org. Chem.* **1986**, *51*, 2270–2276.
[69] Stranix, B. R.; Gao, J. P.; Barghi, R.; Salha, J.; Darling, G. D. *J. Org. Chem.* **1997**, *62*, 8987–8993.
[70] Sylvain, C.; Wagner, A.; Mioskowski, C. *Tetrahedron Lett.* **1998**, *39*, 9679–9680.
[71] Sylvain, C.; Wagner, A.; Mioskowski, C. *Tetrahedron Lett.* **1997**, *38*, 1043–1044.
[72] Riedl, R.; Tappe, R.; Berkessel, A. *J. Am. Chem. Soc.* **1998**, *120*, 8994–9000.

[73] Xia, Y.; Yang, Z.-Y.; Brossi, A.; Lee, K.-H. *Org. Lett.* **1999**, *1*, 2113–2115.
[74] Bui, C. T.; Maeji, N. J.; Rasoul, F.; Bray, A. M. *Tetrahedron Lett.* **1999**, *40*, 5383–5386.
[75] Nicolaou, K. C.; Winssinger, N.; Pastor, J.; DeRoose, F. *J. Am. Chem. Soc.* **1997**, *119*, 449–450.
[76] Cohen, B. J.; Karoly-Hafeli, H.; Patchornik, A. *J. Org. Chem.* **1984**, *49*, 922–924.
[77] Katritzky, A. R.; Belyakov, S. A.; Fang, Y. F.; Kiely, J. S. *Tetrahedron Lett.* **1998**, *39*, 8051–8054.
[78] Yang, X. B.; Sierzchala, A.; Misiura, K.; Niewiarowski, W.; Sochacki, M.; Stec, W. J.; Wieczorek, M. W. *J. Org. Chem.* **1998**, *63*, 7097–7100.
[79] Wright, P.; Lloyd, D.; Rapp, W.; Andrus, A. *Tetrahedron Lett.* **1993**, *34*, 3373–3376.
[80] Adinolfi, M.; Barone, G.; De Napoli, L.; Iadonisi, A.; Piccialli, G. *Tetrahedron Lett.* **1998**, *39*, 1953–1956.
[81] Swayze, E. E. *Tetrahedron Lett.* **1997**, *38*, 8643–8646.
[82] Peyman, A.; Weiser, C.; Uhlmann, E. *Bioorg. Med. Chem. Lett.* **1995**, *5*, 2469–2472.
[83] Chang, C. D.; Waki, M.; Ahmad, M.; Meienhofer, J.; Lundell, E. O.; Haug, J. D. *Int. J. Pept. Prot. Res.* **1980**, *15*, 59–66.
[84] Fields, G. B.; Noble, R. L. *Int. J. Pept. Prot. Res.* **1990**, *35*, 161–214.
[85] Novabiochem Catalog and Peptide Synthesis Handbook, Läufelfingen, CH, **1999**.
[86] Adinolfi, M.; Barone, G.; Iadonisi, A.; Schiattarella, M. *Tetrahedron Lett.* **2001**, *42*, 5971–5972.
[87] Eichler, E.; Yan, F.; Sealy, J.; Whitfield, D. M. *Tetrahedron* **2001**, *57*, 6679–6693.
[88] Zhu, T.; Boons, G. J. *Angew. Chem. Int. Ed.* **1998**, *37*, 1898–1900.
[89] Gouault, N.; Cupif, J.-F.; Sauleau, A.; David, M. *Tetrahedron Lett.* **2000**, *41*, 7293–7297.
[90] Davis, P. W.; Vickers, T. A.; Wilson-Lingardo, L.; Wyatt, J. R.; Guinosso, C. J.; Sanghvi, Y. S.; De Baets, E. A.; Acevedo, O. L.; Cook, P. D.; Ecker, D. J. *J. Med. Chem.* **1995**, *38*, 4363–4366.
[91] Bayer, E.; Bleicher, K.; Maier, M. *Z. Naturforsch. Sect. B* **1995**, *50*, 1096–1100.
[92] Adams, S. P.; Kavka, K. S.; Wykes, E. J.; Holder, S. B.; Galluppi, G. R. *J. Am. Chem. Soc.* **1983**, *105*, 661–663.
[93] Eckstein, F. *Oligonucleotides and Analogues; A Practical Approach*; Oxford University Press: Oxford, **1991**.
[94] Kuisle, O.; Quiñoá, E.; Riguera, R. *Tetrahedron Lett.* **1999**, *40*, 1203–1206.
[95] Kuisle, O.; Quiñoá, E.; Riguera, R. *J. Org. Chem.* **1999**, *64*, 8063–8075.
[96] Gagnon, R.; Dory, Y. L.; Deslongchamps, P. *Tetrahedron Lett.* **2000**, *41*, 4751–4755.
[97] Wunberg, T.; Kallus, C.; Opatz, T.; Henke, S.; Schmidt, W.; Kunz, H. *Angew. Chem. Int. Ed.* **1998**, *37*, 2503–2505.
[98] Zehavi, U.; Patchornik, A. *J. Am. Chem. Soc.* **1973**, *95*, 5673–5677.
[99] Martin, G. E.; Shambhu, M. B.; Shakhshir, S. R.; Digenis, G. A. *J. Org. Chem.* **1978**, *43*, 4571–4574.
[100] Liang, R.; Yan, L.; Loebach, J.; Ge, M.; Uozumi, Y.; Sekanina, K.; Horan, N.; Gildersleeve, J.; Thompson, C.; Smith, A.; Biswas, K.; Still, W. C.; Kahne, D. *Science* **1996**, *274*, 1520–1522.
[101] Deleuze, H.; Sherrington, D. C. *J. Chem. Soc., Perkin Trans. 2* **1995**, 2217–2221.
[102] Rodebaugh, R.; Joshi, S.; Fraser-Reid, B.; Geysen, H. M. *J. Org. Chem.* **1997**, *62*, 5660–5661.
[103] Belorizky, N.; Excoffier, G.; Gagnaire, D.; Utille, J. P.; Vignon, M.; Vottero, P. *Bull. Soc. Chim. Fr.* **1972**, 4749–4753.
[104] Leikauf, E.; Barnekow, F.; Koster, H. *Tetrahedron* **1996**, *52*, 6913–6930.
[105] Burgess, K.; Lim, D.; Bois-Choussy, M.; Zhu, J. P. *Tetrahedron Lett.* **1997**, *38*, 3345–3348.
[106] Hunt, J. A.; Roush, W. R. *J. Am. Chem. Soc.* **1996**, *118*, 9998–9999.
[107] Loffet, A.; Zhang, H. X. *Int. J. Pept. Prot. Res.* **1993**, *42*, 346–351.
[108] Morley, A. D. *Tetrahedron Lett.* **2000**, *41*, 7401–7404.
[109] Morley, A. D. *Tetrahedron Lett.* **2000**, *41*, 7405–7408.
[110] Opatz, T.; Kunz, H. *Tetrahedron Lett.* **2000**, *41*, 10185–10188.
[111] Roussel, F.; Knerr, L.; Grathwohl, M.; Schmidt, R. R. *Org. Lett.* **2000**, *2*, 3043–3046.
[112] Kanie, O.; Grotenbreg, G.; Wong, C.-H. *Angew. Chem. Int. Ed.* **2000**, *39*, 4545–4547.
[113] Köster, H.; Cramer, F. *Liebigs Ann. Chem.* **1974**, 946–958.
[114] Fancelli, D.; Fagnola, M. C.; Severino, D.; Bedeschi, A. *Tetrahedron Lett.* **1997**, *38*, 2311–2314.
[115] Pavia, M. R.; Cohen, M. P.; Dilley, G. J.; Dubuc, G. R.; Durgin, T. L.; Forman, F. W.; Hediger, M. E.; Milot, G.; Powers, T. S.; Sucholeiki, I.; Zhou, S.; Hangauer, D. G. *Bioorg. Med. Chem.* **1996**, *4*, 659–666.
[116] Singh, R.; Nuss, J. M. *Tetrahedron Lett.* **1999**, *40*, 1249–1252.
[117] Heckel, A.; Mross, E.; Jung, K. H.; Rademann, J.; Schmidt, R. R. *Synlett* **1998**, 171–173.
[118] Shimizu, H.; Ito, Y.; Kanie, O.; Ogawa, T. *Bioorg. Med. Chem. Lett.* **1996**, *6*, 2841–2846.
[119] Nicolaou, K. C.; Winssinger, N.; Vourloumis, D.; Ohshima, T.; Kim, S.; Pfefferkorn, J.; Xu, J. Y.; Li, T. *J. Am. Chem. Soc.* **1998**, *120*, 10814–10826.
[120] Barber, A. M.; Hardcastle, I. R.; Rowlands, M. G.; Nutley, B. P.; Marriott, J. H.; Jarman, M. *Bioorg. Med. Chem. Lett.* **1999**, *9*, 623–626.

[121] Pelish, H. E.; Westwood, N. J.; Feng, Y.; Kirchhausen, T.; Shair, M. D. *J. Am. Chem. Soc.* **2001**, *123*, 6740–6741.
[122] Moberg, C.; Rákos, L. *Reactive Polymers* **1991**, *15*, 25–35.
[123] Furth, P. S.; Reitman, M. S.; Cook, A. F. *Tetrahedron Lett.* **1997**, *38*, 5403–5406.
[124] Worster, P. M.; McArthur, C. R.; Leznoff, C. C. *Angew. Chem. Int. Ed. Engl.* **1979**, *18*, 221–222.
[125] McArthur, C. R.; Worster, P. M.; Jiang, J.; Leznoff, C. C. *Can. J. Chem.* **1982**, *60*, 1836–1841.
[126] Rademann, J.; Schmidt, R. R. *J. Org. Chem.* **1997**, *62*, 3650–3653.
[127] Allin, S. M.; Shuttleworth, S. J. *Tetrahedron Lett.* **1996**, *37*, 8023–8026.
[128] Tietze, L. F.; Steinmetz, A. *Angew. Chem. Int. Ed. Engl.* **1996**, *35*, 651–652.
[129] Burgess, K.; Lim, D. *Chem. Commun.* **1997**, 785–786.
[130] Colwell, A. R.; Duckwall, L. R.; Brooks, R.; McManus, S. P. *J. Org. Chem.* **1981**, *46*, 3097–3102.
[131] Hernández, A. S.; Hodges, J. C. *J. Org. Chem.* **1997**, *62*, 3153–3157.
[132] Vidal-Ferran, A.; Bampos, N.; Moyano, A.; Pericàs, M. A.; Riera, A.; Sanders, J. K. M. *J. Org. Chem.* **1998**, *63*, 6309–6318.
[133] Hanessian, S.; Xie, F. *Tetrahedron Lett.* **1998**, *39*, 733–736.
[134] Craig, D.; Robson, M. J.; Shaw, S. J. *Synlett* **1998**, 1381–1383.
[135] Hanessian, S.; Huynh, H. K. *Tetrahedron Lett.* **1999**, *40*, 671–674.
[136] Atkinson, G. E.; Fischer, P. M.; Chan, W. C. *J. Org. Chem.* **2000**, *65*, 5048–5056.
[137] Mergler, M.; Dick, F.; Gosteli, J.; Nyfeler, R. *Tetrahedron Lett.* **1999**, *40*, 4663–4664.
[138] Tommasi, R. A.; Nantermet, P. G.; Shapiro, M. J.; Chin, J.; Brill, W. K. D.; Ang, K. *Tetrahedron Lett.* **1998**, *39*, 5477–5480.
[139] Brill, W. K. D.; De Mesmaeker, A.; Wendeborn, S. *Synlett* **1998**, 1085–1090.
[140] Zaragoza, F.; Petersen, S. V. *Tetrahedron* **1996**, *52*, 5999–6002.
[141] Panek, J. S.; Zhu, B. *J. Am. Chem. Soc.* **1997**, *119*, 12022–12023.
[142] Furman, B.; Thürmer, R.; Kaluza, Z.; Voelter, W.; Chmielewski, M. *Tetrahedron Lett.* **1999**, *40*, 5909–5912.
[143] Tortolani, D. R.; Biller, S. A. *Tetrahedron Lett.* **1996**, *37*, 5687–5690.
[144] Heinonen, P.; Lönnberg, H. *Tetrahedron Lett.* **1997**, *38*, 8569–8572.
[145] Phoon, C. W.; Abell, C. *J. Comb. Chem.* **1999**, *1*, 485–492.
[146] Krchnák, V.; Flegelová, Z.; Weichsel, A. S.; Lebl, M. *Tetrahedron Lett.* **1995**, *36*, 6193–6196.
[147] Johnson, M. G.; Bronson, D. D.; Gillespie, J. E.; Gifford-Moore, D. S.; Kalter, K.; Lynch, M. P.; McCowan, J. R.; Redick, C. C.; Sall, D. J.; Smith, G. F.; Foglesong, R. J. *Tetrahedron* **1999**, *55*, 11641–11652.
[148] Valerio, R. M.; Bray, A. M.; Patsiouras, H. *Tetrahedron Lett.* **1996**, *37*, 3019–3022.
[149] Du, X.; Armstrong, R. W. *J. Org. Chem.* **1997**, *62*, 5678–5679.
[150] Krchnák, V.; Weichsel, A. S.; Lebl, M.; Felder, S. *Bioorg. Med. Chem. Lett.* **1997**, *7*, 1013–1016.
[151] Rano, T. A.; Chapman, K. T. *Tetrahedron Lett.* **1995**, *36*, 3789–3792.
[152] Newlander, K. A.; Chenera, B.; Veber, D. F.; Yim, N. C. F.; Moore, M. L. *J. Org. Chem.* **1997**, *62*, 6726–6732.
[153] Brummond, K. M.; Lu, J. *J. Org. Chem.* **1999**, *64*, 1723–1726.
[154] Dankwardt, S. M.; Phan, T. M.; Krstenansky, J. L. *Mol. Diversity* **1995**, *1*, 113–120.
[155] Leznoff, C. C.; Dixit, D. M. *Can. J. Chem.* **1977**, *55*, 3351–3355.
[156] van Maarseveen, J. H.; den Hartog, J. A. J.; Engelen, V.; Finner, E.; Visser, G.; Kruse, C. G. *Tetrahedron Lett.* **1996**, *37*, 8249–8252.
[157] Bräse, S.; Köbberling, J.; Enders, D.; Lazny, R.; Wang, M.; Brandtner, S. *Tetrahedron Lett.* **1999**, *40*, 2105–2108.
[158] Kohlbau, H. J.; Tschakert, J.; Al-Qawasmeh, R. A.; Nizami, T. A.; Malik, A.; Voelter, W. *Z. Naturforsch. Sect. B* **1998**, *53*, 753–764.
[159] Kobayashi, S.; Aoki, Y. *Tetrahedron Lett.* **1998**, *39*, 7345–7348.
[160] Kobayashi, S.; Akiyama, R. *Tetrahedron Lett.* **1998**, *39*, 9211–9214.
[161] Lee, A.; Huang, L.; Ellman, J. A. *J. Am. Chem. Soc.* **1999**, *121*, 9907–9914.
[162] Zhao, X. Y.; Janda, K. D. *Tetrahedron Lett.* **1997**, *38*, 5437–5440.
[163] Chen, S.; Janda, K. D. *J. Am. Chem. Soc.* **1997**, *119*, 8724–8725.
[164] Benaglia, M.; Annunziata, R.; Cinquini, M.; Cozzi, F.; Ressel, S. *J. Org. Chem.* **1998**, *63*, 8628–8629.
[165] Sarantakis, D.; Bicksler, J. J. *Tetrahedron Lett.* **1997**, *38*, 7325–7328.
[166] Zhong, H. M.; Greco, M. N.; Maryanoff, B. E. *J. Org. Chem.* **1997**, *62*, 9326–9330.
[167] Hollinshead, S. P. *Tetrahedron Lett.* **1996**, *37*, 9157–9160.
[168] Heerding, D. A.; Takata, D. T.; Kwon, C.; Huffman, W. F.; Samanen, J. *Tetrahedron Lett.* **1998**, *39*, 6815–6818.
[169] Goff, D.; Fernandez, J. *Tetrahedron Lett.* **1999**, *40*, 423–426.

[170] Brill, W. K. D.; Schmidt, E.; Tommasi, R. A. *Synlett* **1998**, 906–908.
[171] Haap, W. J.; Kaiser, D.; Walk, T. B.; Jung, G. *Tetrahedron* **1998**, *54*, 3705–3724.
[172] Berteina, S.; De Mesmaeker, A.; Wendeborn, S. *Synlett* **1999**, 1121–1123.
[173] Rölfing, K.; Thiel, M.; Künzer, H. *Synlett* **1996**, 1036–1038.
[174] Matthews, D. P.; Green, J. E.; Shuker, A. J. *J. Comb. Chem.* **2000**, *2*, 19–23.
[175] Richter, L. S.; Gadek, T. R. *Tetrahedron Lett.* **1994**, *35*, 4705–4706.
[176] McNally, J. J.; Youngman, M. A.; Dax, S. L. *Tetrahedron Lett.* **1998**, *39*, 967–970.
[177] Devraj, R.; Cushman, M. *J. Org. Chem.* **1996**, *61*, 9368–9373.
[178] Winkler, J. D.; McCoull, W. *Tetrahedron Lett.* **1998**, *39*, 4935–4936.
[179] Cabrele, C.; Langer, M.; Beck-Sickinger, A. G. *J. Org. Chem.* **1999**, *64*, 4353–4361.
[180] Ruhland, T.; Andersen, K.; Pedersen, H. *J. Org. Chem.* **1998**, *63*, 9204–9211.
[181] Swayze, E. E. *Tetrahedron Lett.* **1997**, *38*, 8465–8468.
[182] Feng, Y.; Burgess, K. *Chem. Eur. J.* **1999**, *5*, 3261–3272.
[183] Goldberg, M.; Smith, L.; Tamayo, N.; Kiselyov, A. S. *Tetrahedron* **1999**, *55*, 13887–13898.
[184] Nakamura, K.; Nishiya, H.; Nishiyama, S. *Tetrahedron Lett.* **2001**, *42*, 6311–6313.
[185] Feng, Y.; Wang, Z.; Jin, S.; Burgess, K. *J. Am. Chem. Soc.* **1998**, *120*, 10768–10769.
[186] Wijkmans, J. C. H. M.; Culshaw, A. J.; Baxter, A. D. *Mol. Diversity* **1998**, *3*, 117–120.
[187] Mohan, R.; Yun, W. Y.; Buckman, B. O.; Liang, A.; Trinh, L.; Morrissey, M. M. *Bioorg. Med. Chem. Lett.* **1998**, *8*, 1877–1882.
[188] Kiselyov, A. S.; Eisenberg, S.; Luo, Y. *Tetrahedron Lett.* **1999**, *40*, 2465–2468.
[189] Kiselyov, A. S.; Smith, L.; Tempest, P. A. *Tetrahedron* **1999**, *55*, 14813–14822.

8 Preparation of Sulfur Compounds

8.1 Preparation of Thiols

Thiols bound irreversibly to supports have mainly been used as linkers for carboxylic acids (Sections 3.1.2 and 3.3.3), for pyrimidines and triazines (Section 3.8), or for other thiols (Section 3.12). When working with these resins, special care must be taken to prevent atmospheric oxidation of the thiols to the corresponding disulfides.

Some strategies used for the preparation of support-bound thiols are listed in Table 8.1. Oxidative thiolation of lithiated polystyrene has been used to prepare polymeric thiophenol (Entry 1, Table 8.1). Polystyrene functionalized with 2-mercaptoethyl groups has been prepared by radical addition of thioacetic acid to cross-linked vinylpolystyrene followed by hydrolysis of the intermediate thiol ester (Entry 2, Table 8.1). A more controllable introduction of thiol groups, suitable also for the selective transformation of support-bound substrates, is based on nucleophilic substitution with thiourea or potassium thioacetate. The resulting isothiouronium salts and thiol acetates can be saponified, preferably under reductive conditions, to yield thiols (Table 8.1). Thiol acetates have been saponified on insoluble supports with mercaptoethanol [1], propylamine [2], lithium aluminum hydride [3], sodium or lithium borohydride, alcoholates, or hydrochloric acid (Table 8.1).

Other functional groups that can be transformed into thiols on insoluble supports are methoxytrityl and trityl thioethers and mixed disulfides (Table 8.2). These functionalities can serve as protective groups for support-bound thiols that are cleavable without simultaneous cleavage of the thiol from the support. Most other protective groups for thiols, which have mainly been developed as side-chain protection of cysteine in solid-phase peptide synthesis, are removed either during or after cleavage of the peptide from the support. Conventional side-chain protective groups for cysteine include trityl, tert-butyl, and acetamidomethyl thioethers (AcNH–CH$_2$–SR; Acm–SR), as well as unsymmetrical disulfides. Deprotection is effected by treatment with TFA, mercury(II) salts, or silver(I) salts [17–20]. Alkoxycarbonyl groups are not suitable as protective groups for thiols because of their lability towards nucleophiles. For instance, allyloxycarbonyl thiols are completely deprotected within 3 h upon treatment with 30% piperidine in DMF [21].

Table 8.1. Preparation of thiols.

Entry	Starting resin	Conditions	Product	Ref.
1		1. S$_8$, THF, 20 °C, 1 h 2. LiAlH$_4$, THF		[4,5]
2		1. AcSH, AIBN, PhMe, 70 °C, 2 d 2. HCl (37% in H$_2$O)/ dioxane 1:2, 70 °C, 2 d		[6]
3		1. AcSH, PPh$_3$, DIAD, THF, 0 °C 2. NaBH$_4$, THF, MeOH, 65 °C		[7]
4		1. (H$_2$N)$_2$CS (0.9 mol/L), DMA, 85 °C, 20 h 2. pyrrolidine/dioxane 1:4, 110 °C, 2 h		[8] see also [9,10]
5		1. AcSK, DMF 2. LiBH$_4$, Et$_2$O (saponification under basic conditions failed)		[11,12] see also [2,13,14]
6		1. AcSH (5 eq), BF$_3$OEt$_2$ (0.8 eq), DME, 23 °C, 16 h 2. LiOBn, THF, 23 °C, 2 h		[15]
7		NaOMe (0.2 mol/L), THF/MeOH 3:1, 1 h		[16]

Table 8.2. Deprotection of support-bound, protected thiols.

Entry	Starting resin	Conditions	Product	Ref.
1		TFA/(*i*Bu)₃SiH/DCM 5:5:90		[22] see also [23–25]
2		3–5% TFA in DCM, 45 min		[26,27] see also [28]
3		PBu₃, DMF, *i*PrOH, H₂O		[29]
4		2-mercaptoethanol, DIPEA, DMF		[30]

8.2 Preparation of Thioethers

Thioethers are usually prepared on solid phase by S-alkylation or arylation of thiols. Thiols are powerful nucleophiles and, because of their high reactivity and ready availability, represent the most suitable starting materials for the solid-phase synthesis of thioethers. Numerous examples have been reported of both reactions of support-bound electrophiles with thiols and reactions of support-bound thiols with alkylating or arylating agents. Table 8.3 lists illustrative examples of the preparation of thioethers from support-bound alkylating agents. Alkyl halides and trichloroacetimidates have been used as alkylating agents for aromatic or aliphatic thiols. Additional suitable support-bound electrophiles are fluoroarenes, α,β-unsaturated carbonyl compounds, maleimides [31], and epoxides. Halides, thiolates, and even nitro groups can act as leaving groups in aromatic nucleophilic substitutions with thiols [32]. Nucleophilic substitutions with thiols do not require the use of strong bases and generally proceed smoothly and cleanly at room temperature to yield thioethers. Some S-alkylations and S-arylations can even be performed at room temperature in dilute acetic acid. Because thiols are sensitive to oxidation by air (disulfide formation), excess thiol should be used to ensure complete conversion of the substrate. As in other reactions of support-bound reagents with bifunctional compounds, the reaction of Merrifield resin with α,ω-dithiols can lead to significant cross-linking and thereby to a severe reduction of the swelling capacity of the support. By careful optimization of the reaction conditions, however, almost exclusive monoalkylation of α,ω-dithiols can be achieved (Entry **2**, Table 8.3).

Table 8.3. Preparation of thioethers from support-bound electrophiles.

Entry	Starting resin	Conditions	Product	Ref.
1		4-mercaptophenol (0.6 mol/L), KOH (0.4 mol/L), DMF, 90 °C, 16 h		[33] see also [34]
2		HS⌒⌒SH (1 mol/L), DBU (0.3 mol/L), PhMe, 20 °C, 24 h		[26] see also [35,36]
3	(1.35 mmol/g)	cysteine (0.07 mol/L, 2 eq), DMF/H$_2$O/DBU 200:1:4, overnight	(0.38 mmol/g)	[37] see also [38]
4		3,4-dichlorothiophenol, DIPEA, DMSO		[39] see also [40–42]
5		BnSH (0.65 mol/L, 10 eq), DIPEA (8.6 eq), NMP, 20 °C, 22 h		[43]
6		BnSH (5 eq), BF$_3$OEt$_2$ (0.7 eq), DCM, 23 °C, 16 h		[15]
7		MeO$_2$C⌒⌒SH NMM, DMA, 20 °C, 16 h		[44] see also [32,45]
8		PhSH (1.25 mol/L), NaOMe, THF, 20 °C, 3 d		[46] see also [47–49]
9		ArSH (0.5 mol/L, 10 eq), DBU (1 eq), DMF, 65 °C, 12 h Ar: 3-fluorophenyl		[50]
10		PhSNa, DMF, 20 °C, 12 h		[51]

Table 8.4. Preparation of thioethers from support-bound thiols.

Entry	Starting resin	Conditions	Product	Ref.
1		TMG, DMF, 23 °C		[7] see also [22,55]
2		(0.13 mol/L), 3% pyridine in DMF, 20 °C, 24 h		[10] see also [56]
3		AcOH, EtOH, 20 °C, 3 h		[12]
4		(0.13 mol/L), NEt$_3$ (0.13 mol/L), DMF, 1 h		[57]
5		(2 eq), 15-crown-5, THF, 20 °C, 16 h		[58] see also [59]
6		(10 eq), KOtBu (12 eq), THF/DMSO 2:1, 0 °C to 20 °C, 18 h		[60]
7		1. Hg(O$_2$CCF$_3$)$_2$, BnNH$_2$, THF, 20 °C 2. BnBr, THF, 20 °C		[61]

Support-bound thiols can be alkylated under mild reaction conditions with alkyl halides, α,β-unsaturated carbonyl compounds, or reactive aryl halides (Table 8.4). These reactions should be conducted under an inert atmosphere or in the presence of a reducing additive to prevent disulfide formation. PEG-bound thiophenol has been S-alkylated under conditions similar to those used in solution (DMF, Cs$_2$CO$_3$, 20 °C, 15 h [52–54]).

Because thioether formation is a rapid, irreversible reaction, it has been widely used for the macrocyclization of peptides or peptoids on supports (Entries **1–4**, Table 8.5 [62]). For this purpose, either S-protected cysteine or S-protected ω-aminomercaptans are linked to a support and then elongated by standard solid-phase peptide/peptoid chemistry. When a suitable length has been reached, the terminal amine is acy-

Table 8.5. Miscellaneous preparations of thioethers.

Entry	Starting resin	Conditions	Product	Ref.
1		TMG (2.5% in THF), 17 h		[29] see also [16,23, 66,67]
2		5% DBU in DMF, 20 °C, 24 h		[27] see also [63,68– 70]
3		1. TFA/TES/DCM 1:2:97, 2 × 10 min, 1 × 20 min 2. DIPEA, NMP		[71]
4		TBAF (4 eq), DCM, 20 °C, 3 h		[72]

lated with an acid bearing an electrophilic functionality. After deprotection of the thiol, cyclization is induced by treatment with a base. Although this strategy gives good results, even in the preparation of strained rings, it has certain limitations. As in most other macrocyclizations, low yields are mainly due to competing intermolecular bond formation, and for macrocyclization to occur efficiently some preorganization of the precursor is usually required [63]. A similar strategy for preparing macrocyclic peptides is based on the intramolecular reaction of N-terminal cysteine with an aldehyde to yield a thiazolidine [64].

Thioethers have also been prepared on cross-linked polystyrene by radical addition of thiols to support-bound alkenes and by reaction of support-bound carbon radicals (generated by addition of carbon radicals to resin-bound acrylates) with esters of 1-hydroxy-1,2-dihydro-2-pyridinethione ('Barton esters'; Entry 6, Table 8.5). Additional methods include the reaction of metallated supports with symmetric disulfides (Entries 7–9, Table 8.5) and the alkylation of polystyrene-bound, α-lithiated thioanisole [65].

Table 8.5. continued.

Entry	Starting resin	Conditions	Product	Ref.
5		1. H-Cys-OEt•HCl, AIBN, EtOH, PhMe, PEG, 70 °C, 24 h		[6]
6		(10 eq), hv, DCM, 0 °C, 4 h		[73]
7		MeSSMe, THF, 20 °C, 15 min, then 65 °C, 0.5 h		[4] see also [74,75]
8		*i*PrMgBr (0.19 mol/L, 7.3 eq), THF, −35 °C, 15 min, then CuCN•2 LiCl (10 eq), 15 min, then PhSSPh		[76]
9		1. BuLi (0.14 mol/L, 3 eq), THF, −60 °C, 20 min 2. MeSSMe (10 eq), −60 °C to 20 °C, 2 h		[77]

Thioglycosides have been prepared on solid phase by glycosylation of thiols with various types of glycosyl donor. Carbohydrate-derived thioethers have been used either to link carbohydrates to thiol-functionalized supports [9,26,78,79] or as glycosyl donors for the preparation of glycosides on solid phase (see Section 16.3.3).

8.3 Preparation of Sulfoxides and Sulfones

Sulfoxides and sulfones can be prepared on cross-linked polystyrene by oxidation of thioethers. The most commonly used reagent for this purpose is MCPBA in DCM [8,12,32,57,80–82] or dioxane [50,83] (Table 8.6), but other oxidants such as H_2O_2 in acetic acid [34], oxone (Entry **7**, Table 8.6), or oxaziridines [84] have also been used. PEG-bound thioethers have been converted into sulfones by oxidation with MCPBA in DCM [52,54] or with OsO_4/NMO [85]. The oxidation of thioethers to sulfoxides requires careful control of the reaction conditions to prevent the formation of sulfones. Sulfones have also been prepared by S-alkylation of polystyrene-bound sulfinates (Entries **8** and **9**, Table 8.6), by α-alkylation of sulfones (BuLi, THF, alkyl halide [86]), and by addition of sulfinyl radicals to resin-bound alkenes or alkynes (Entry **11**, Table 8.6).

Table 8.6. Preparation of sulfoxides and sulfones.

Entry	Starting resin	Conditions	Product	Ref.
1		MCPBA (0.05 mol/L, 1.4 eq), DCM, 0 °C, 20 h (Wang resin)	(94:6)	[87] see also [88,89]
2		MCPBA (0.12 mol/L, 1.2 eq), DCM, 20 °C, 18 h		[90]
3		MCPBA (0.2 mol/L, 5 eq), DCM, 20 °C, 96 h		[87]
4		MCPBA (1 mol/L, 3 eq), DCM, 20 °C, 15 h		[91]
5		MCPBA (0.15 mol/L, 3 eq), DCM, 25 °C, 1 h		[38]
6		MCPBA (0.6 mol/L, 10 eq), DCM, 20 °C, 16 h		[60]
7		Oxone, DMF, H$_2$O, 20 °C, 12 h		[36]
8		1. SO$_2$, THF 2. allyl bromide, THF, 67 °C, overnight		[86]
9		(5 eq), pyridine (5 eq), DMF, 80 °C, 48 h		[92] see also [93]
10		EtMgX, THF R: menthyl		[94]
11		TsBr (0.6 mol/L, 6 eq), AIBN (6 eq), PhMe, 70 °C, 22 h		[95]

8.4 Preparation of Sulfonamides and Sulfinamides

Sulfonamides have usually been prepared on solid supports by sulfonylation of aliphatic or aromatic amines with sulfonyl chlorides or by N-alkylation of other sulfonamides (Figure 8.1).

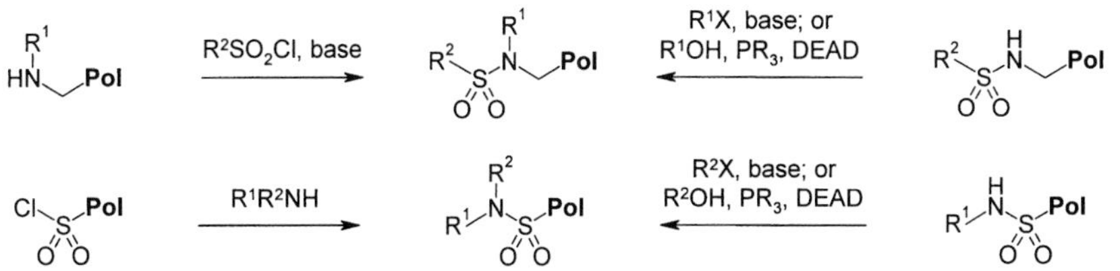

Figure 8.1. Strategies for the preparation of sulfonamides on solid phase.

The clean conversion of support-bound, primary amines into sulfonamides by treatment with sulfonyl chlorides is more difficult to perform than the acylation of amines with carboxylic acid derivatives, probably because of the oxidizing properties of sulfonyl chlorides and because primary amines can be doubly sulfonylated. Weak bases (pyridine, 2,6-lutidine, NMM, collidine), short reaction times, and only a slight excess of sulfonyl chloride should therefore be used to convert primary amines into sulfonamides (Table 8.7).

Sulfonyl chlorides having an α-hydrogen are unstable under basic reaction conditions and can give variable results [96,97]. For base-labile sulfonyl chlorides, the use of *O*-silyl ketene acetals as scavengers for HCl has been recommended [96]. Table 8.7 lists some illustrative procedures for the preparation of sulfonamides from primary amines on solid phase. Further examples have been reported [98–101].

Support-bound secondary amines are more readily sulfonylated than primary amines, as shown by the many examples reported in the literature (Table 8.8 [117–121]). Typically, the sulfonylation of polystyrene-bound amines is conducted in DCM, but THF [122], DMF [97,123], pyridine [124], or mixtures thereof can sometimes give better results [105]. NMM, DIPEA, and pyridine have mainly been used as bases.

Alternatively, sulfonamides can also be prepared by oxidation of sulfinamides with periodate (Entry **3**, Table 8.8) or with MCPBA [125]. Polystyrene-bound sulfonyl chlorides, which can be prepared from polystyrene-bound sulfonic acids by treatment with PCl$_5$, SOCl$_2$ [126–129], ClSO$_3$H [130], or SO$_2$Cl$_2$/PPh$_3$ [131], react smoothly with amines to yield the corresponding sulfonamides (Entry **4**, Table 8.8). Support-bound carbamates of primary aliphatic or aromatic amines can be N-sulfonylated in the presence of strong bases, and can therefore be used as backbone amide linkers for sulfonamides (Entries **5** and **6**, Table 8.8).

Sulfonamides of primary amines are readily deprotonated (pK_a 9–11) and can thus be N-alkylated or N-arylated. Because of their high nucleophilicity and low basicity, deprotonated sulfonamides also react smoothly with less reactive electrophiles, such as *n*-alkyl bromides [136] (Table 8.9). Sulfonamides can also be N-alkylated with aliphatic alcohols under Mitsunobu conditions. Suitable solvents for the N-alkylation of sulfonamides on polystyrene by Mitsunobu reaction are DCM, toluene, and THF.

Table 8.7. Preparation of sulfonamides from support-bound primary amines and sulfonyl chlorides.

Entry	Starting resin	Conditions	Product	Ref.
1		TsCl (0.15 mol/L, 2 eq), NEt$_3$ (3 eq), DCM, 7 h		[102]
2		ArSO$_2$Cl (3.5 eq), NEt$_3$ (3.5 eq), DCM Ar: 4-PrC$_6$H$_4$		[103] see also [104]
3		2-(O$_2$N)C$_6$H$_4$SO$_2$Cl (4 eq), DIPEA (6 eq), THF/DCM 2:1, 20 °C, 3 h		[105] see also [106–108]
4		TsCl (5 eq), pyridine/DCM 1:1, 20 °C, 15 h		[109]
5		BocHN (3 eq), NMM (6 eq), DCM, 2 × 2.5 h		[110] see also [96,111–113]
6		MsCl, pyridine, DCM, 28 h		[114]
7		PhSO$_2$Cl (3 eq), pyridine (1 eq), DMAP, DCM, 25 °C, 18 h		[115] see also [116]

Table 8.8. Miscellaneous preparations of sulfonamides from sulfonyl chlorides and from sulfinamides.

Entry	Starting resin	Conditions	Product	Ref.
1		(0.5 mol/L), DMF, pyridine, 2–18 h		[97]
2		TsCl (0.4 mol/L, 5 eq), DMAP, pyridine, 60 °C, 14 h		[132]
3		NMM, DCM, overnight; then NaIO₄, RuCl₃, DCM, MeCN, H₂O, 1 h		[133]
4		1. PCl₅, DMF, 23 °C, 4 h 2. 1,3-diaminopropane, DMF, pyridine		[129] see also [134]
5		LiN(SiMe₃)₂, THF, −78 °C, 45 min, then TsCl		[135]
6		NaH, DMA, 20 °C, 8 h, then add TsCl		[135]

Because many more alcohols than alkyl halides are commercially available, the Mitsunobu reaction enables the synthesis of larger and more diverse compound arrays than alkylation with alkyl halides. N-Acylsulfonamides are strongly acidic and can be alkylated with diazomethane (Entry 6, Table 8.9) or trimethylsilyldiazomethane [137]. Resin-bound sulfonamides have been N-acylated by treatment with acyl halides, and N-carbamoylated by treatment with isocyanates [138].

2-Nitro- or 2,4-dinitrobenzenesulfonamides of primary or secondary amines can be hydrolyzed under mildly basic conditions, and are increasingly being used for amine protection (see Section 10.1.10.7 [123,139,140]). N-(2-Nitrobenzenesulfonyl)amino acids can be used as an alternative to N-Fmoc amino acids for the solid-phase synthesis of peptides [141]. Deprotection is achieved by treatment of the polystyrene-bound sulfonamide with a solution of PhSH (0.5 mol/L) and K₂CO₃ (2 mol/L) in DMF for 10 min at room temperature [141], conditions that do not lead to cleavage of esters (e.g. of the Wang linker) or to racemization. The condensation of polystyrene-bound sulfinamides H₂N–SO–Pol with aldehydes yields N-sulfinylimines, which add

Grignard reagents in a highly diastereoselective manner to yield sulfinamides (Entry **10**, Table 8.9).

Sulfonamides can also be alkylated by support-bound electrophiles (Table 8.10). Polystyrene-bound allylic alcohols have been used to N-alkylate sulfonamides under the conditions of the Mitsunobu reaction. Oxidative iodosulfonylamidation of support-bound enol ethers (e.g. glycals; Entry **3**, Table 8.10) has been used to prepare *N*-sulfonyl aminals. Jung and co-workers have reported an interesting variant of the Baylis–Hillman reaction, in which tosylamide and an aromatic aldehyde were condensed with polystyrene-bound acrylic acid to yield 2-(sulfonamidomethyl)acrylates (Entry **4**, Table 8.10).

Table 8.9. N-Alkylation and N-arylation of support-bound sulfonamides and sulfinamides.

Entry	Starting resin	Conditions	Product	Ref.
1		KO*t*Bu (5 eq), BnBr (10 eq), THF, 50 °C, 2 h		[109]
2		phenacyl bromide (6 eq), DBU (6 eq), DMSO/ NMP 1:1, overnight; Ar: 4-(MeO)C$_6$H$_4$		[136]
3		(6 eq), DBU (6 eq), DMSO/NMP 1:1, overnight		[136]
4		propargyl bromide (0.2 mol/L, 4 eq), Cs$_2$CO$_3$ (2 eq), DMF, 20 h		[102]
5		iodoacetonitrile (1.2 mol/L), DIPEA, NMP, 24 h		[142,143]

Table 8.9. continued.

Entry	Starting resin	Conditions	Product	Ref.
6		CH_2N_2, Me_2CO, Et_2O		[144]
7		allyl-$O-CO_2Me$ (15 eq), $Pd_2(dba)_3$, PPh_3, THF, 2 h		[141]
8		EtOH (5 eq), PPh_3 (5 eq), DEAD (5 eq), DCM, 0.5 h		[123] see also [105,107, 136,145, 146]
9		Ph-OH PPh_3, DEAD, THF, 20 h		[147]
10		1. $PhCH_2CHO$ (0.25 mol/L, 10 eq), $Ti(OEt)_4$ (20 eq), THF, 12 h 2. EtMgBr (0.38 mol/L, 20 eq), DCM, -45 °C, 20 h, 20 °C, 4 h		[125]
11		$TolB(OH)_2$ (0.15 mol/L, 4 eq), $Cu(OAc)_2$ (2 eq), NEt_3 (2 eq), MS 4 Å, THF, 20 °C, 2 × 3 h		[148]

Table 8.10. N-Alkylation of sulfonamides with support-bound electrophiles.

Entry	Starting resin	Conditions	Product	Ref.
1	HO⌢⌢C6H4-PS	MeO2C⌢NHSO2Ar PPh3, DEAD, THF, 20 °C, 16 h Ar: 2,4-(O2N)2C6H3		[149] see also [139,150]
2	HO⌢⌢C6H4-PS	O2N⌢C6H3⌢NO2 BocHN-S(O2) PPh3, DIAD (3 eq of each), THF, 12 h		[151]
3	(glycal, OBn, OBn, O-Si-PS)	PhSO2NH2 (8 eq), I(coll)2ClO4 (5.5 eq), DCM, 0 °C, 6 h		[152]
4	(acrylate, O, Ph, Cl-C6H4, PS)	4-(F3C)C6H4CHO (16 eq), TsNH2 (20 eq), DABCO (1.6 eq), dioxane, 70 °C, 20 h		[153]

8.5 Preparation of Sulfonic Esters

Sulfonates of primary and secondary alcohols are strong alkylating agents, and are usually prepared on solid phase only as synthetic intermediates. Sulfonic esters of phenols are, however, sufficiently stable to serve as potential lead structures for various types of drug [154].

The most common synthesis of sulfonic esters, which can also be conducted on insoluble supports, is the sulfonylation of alcohols with sulfonyl chlorides under basic reaction conditions. Several examples of the sulfonylation of support-bound alcohols and of the reaction of support-bound sulfonyl chlorides with alcohols have been reported (Table 8.11). For the preparation of highly reactive sulfonates, bases of low nucleophilicity, such as DIPEA or 2,6-lutidine, should be used to prevent alkylation of the base by the newly formed sulfonate. This potential side reaction is, however, less likely to occur on cross-linked polystyrene than in solution, because quaternization on hydrophobic supports only proceeds sluggishly (see Section 10.2 and [155]).

Treatment of Wang resin with methanesulfonyl chloride in the presence of a weak base only yields the mesylate if the reaction is conducted at low temperature and for a short time (Entry 1, Table 8.11). Longer reaction times lead, in both DCM or DMF as solvents, to the formation of the corresponding benzyl chloride [132,156]. Both Wang resin and Sasrin can be converted to the corresponding tosylates or other sulfonates by treatment with the appropriate sulfonyl chlorides (Entry **1**, Table 10.16 [117]). The preparation of non-benzylic sulfonates of primary or secondary alcohols does not usually present problems (Table 8.11). Secondary alcohols do not react smoothly with aromatic sulfonyl chlorides but only with alkylsulfonyl chlorides. Tertiary alcohols are unreactive towards sulfonylation (Entry **7**, Table 8.11). Support-bound sulfonyl chlorides, which, as mentioned in Section 8.4, can be prepared from sulfonic acids by treatment with PCl_5 [129], $SOCl_2$ [126–128], $ClSO_3H$ [130], or SO_2Cl_2/PPh_3 [131], react with primary aliphatic alcohols or phenols to yield the expected sulfonates. Sulfonyl fluorides [157] show reactivity similar to that of the chlorides. Sulfonates have also been prepared by treating Merrifield resin with lithiated isopropyl methanesulfonate (Entry **12**, Table 8.11).

Table 8.11. Preparation of sulfonic esters.

Entry	Starting resin	Conditions	Product	Ref.
1		MsCl, DIPEA, DCM, 0–25 °C, 75 min		[158]
2		BuSO$_2$Cl (5 eq), DIPEA (10 eq), DMAP, DCM, 22 °C, overnight		[155]
3		MsCl (5 eq), pyridine, DCM, 0 °C, 24 h		[159] see also [160,161]
4		TsCl (20 eq), NEt$_3$ (20 eq), DCM, 20 °C, overnight		[2]
5		O_2N—⟨ ⟩—SO_2Cl pyridine, DCM, 20 °C, 9 h		[162]
6		Tf$_2$O (5 eq), 2,6-lutidine (7 eq), DCM, 0 °C, 6 h		[163]

Table 8.11. continued.

Entry	Starting resin	Conditions	Product	Ref.
7		MsCl, pyridine, DCM, 20 °C, 7 h		[164]
8		Tf$_2$O (8 eq), pyridine, DCM, 20 °C, 5–10 h		[165]
9		PhSO$_2$Cl (0.09 mol/L), DIPEA (0.09 mol/L), DCM, 16 h		[154]
10		(1.1 eq), NEt$_3$ (1.0 eq), DCM, 20 °C, overnight		[126] see also [127,166, 167]
11		tBuO$_2$C—⟨⟩—OH NEt$_3$, DCM		[128] see also [157]
12		(20 eq), THF, −78 °C to −25 °C, 20 h		[131]

8.6 Preparation of Sulfonium Salts

Sulfonium salts have been prepared on insoluble supports by S-alkylation of thioethers (Figure 8.2) only as synthetic intermediates. These compounds can be used to alkylate carboxylates [168] and halides [65], or as electrophiles for the Suzuki cross-coupling reaction (see Entry **7**, Table 3.48 [169]). Sulfonium salts are also C,H-acidic, and can be used as intermediates for the synthesis of epoxides (Entry **7**, Table 15.1 [170]).

The S-alkylation of polystyrene-bound thioethers requires the use of strong alkylating agents, such as methyl triflate [169] or Meerwein's salt (Figure 8.2). Attempted use of methyl iodide, even in large excess, did not lead to quantitative alkylation according to the second equation in Figure 8.2 [169]. This may have been due to the high nucleophilicity of iodide, which can react with sulfonium salts to yield thioethers and alkyl iodides [65].

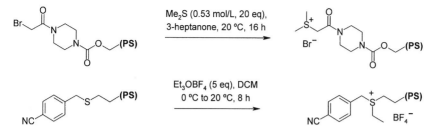

Figure 8.2. Preparation of sulfonium salts on cross-linked polystyrene [169,170].

References for Chapter 8

[1] Camarero, J. A.; Cotton, G. J.; Adeva, A.; Muir, T. W. *J. Pept. Res.* **1998**, *51*, 303–316.
[2] Bettinger, T.; Remy, J. S.; Erbacher, P.; Behr, J. P. *Bioconjugate Chem.* **1998**, *9*, 842–846.
[3] Bertini, V.; Lucchesini, F.; Pocci, M.; De Munno, A. *J. Org. Chem.* **2000**, *65*, 4839–4842.
[4] Farrall, M. J.; Fréchet, J. M. J. *J. Org. Chem.* **1976**, *41*, 3877–3882.
[5] Fréchet, J. M. J.; de Smet, M. D.; Farrall, M. J. *Polymer* **1979**, *20*, 675–680.
[6] Stranix, B. R.; Gao, J. P.; Barghi, R.; Salha, J.; Darling, G. D. *J. Org. Chem.* **1997**, *62*, 8987–8993.
[7] Moran, E. J.; Wilson, T. E.; Cho, C. Y.; Cherry, S. R.; Schultz, P. G. *Biopolymers* **1995**, *37*, 213–219.
[8] Masquelin, T.; Meunier, N.; Gerber, F.; Rossé, G. *Heterocycles* **1998**, *48*, 2489–2505.
[9] Chiu, S. H. L.; Anderson, L. *Carbohyd. Res.* **1976**, *50*, 227–238.
[10] Adamczyk, M.; Fishpaugh, J. R.; Mattingly, P. G. *Tetrahedron Lett.* **1999**, *40*, 463–466.
[11] Kobayashi, S.; Hachiya, I.; Suzuki, S.; Moriwaki, M. *Tetrahedron Lett.* **1996**, *37*, 2809–2812.
[12] Barco, A.; Benetti, S.; De Risi, C.; Marchetti, P.; Pollini, G. P.; Zanirato, V. *Tetrahedron Lett.* **1998**, *39*, 7591–7594.
[13] Bertini, V.; Lucchesini, F.; Pocci, M.; De Munno, A. *Tetrahedron Lett.* **1998**, *39*, 9263–9266.
[14] Kobayashi, S.; Wakabayashi, T.; Yasuda, M. *J. Org. Chem.* **1998**, *63*, 4868–4869.
[15] Phoon, C. W.; Oliver, S. F.; Abell, C. *Tetrahedron Lett.* **1998**, *39*, 7959–7962.
[16] Souers, A. J.; Virgilio, A. A.; Rosenquist, A.; Fenuik, W.; Ellman, J. A. *J. Am. Chem. Soc.* **1999**, *121*, 1817–1825.
[17] Yoshida, M.; Tatsumi, T.; Fujiwara, Y.; Iinuma, S.; Kimura, T.; Akaji, K.; Kiso, Y. *Chem. Pharm. Bull.* **1990**, *38*, 1551–1557.
[18] Houseman, B. T.; Mrksich, M. *J. Org. Chem.* **1998**, *63*, 7552–7555.
[19] Pearson, D. A.; Blanchette, M.; Baker, M. L.; Guindon, C. A. *Tetrahedron Lett.* **1989**, *30*, 2739–2742.
[20] Novabiochem Catalog and Peptide Synthesis Handbook, Läufelfingen, CH, **1999**.
[21] Loffet, A.; Zhang, H. X. *Int. J. Pept. Prot. Res.* **1993**, *42*, 346–351.
[22] Nefzi, A.; Giulianotti, M.; Houghten, R. A. *Tetrahedron Lett.* **1998**, *39*, 3671–3674.
[23] Cho, C. Y.; Youngquist, R. S.; Paikoff, S. J.; Beresini, M. H.; Hébert, A. R.; Berleau, L. T.; Liu, C. W.; Wemmer, D. E.; Keough, T.; Schultz, P. G. *J. Am. Chem. Soc.* **1998**, *120*, 7706–7718.
[24] Marsh, I. R.; Bradley, M. *Tetrahedron* **1997**, *53*, 17317–17334.
[25] Pátek, M.; Drake, B.; Lebl, M. *Tetrahedron Lett.* **1995**, *36*, 2227–2230.
[26] Rademann, J.; Schmidt, R. R. *J. Org. Chem.* **1997**, *62*, 3650–3653.
[27] Kiselyov, A. S.; Eisenberg, S.; Luo, Y. *Tetrahedron* **1998**, *54*, 10635–10640.
[28] Mourtas, S.; Gatos, D.; Kalaitzi, V.; Katakalou, C.; Barlos, K. *Tetrahedron Lett.* **2001**, *42*, 6965–6967.
[29] Virgilio, A. A.; Bray, A. A.; Zhang, W.; Trinh, L.; Snyder, M.; Morrissey, M. M.; Ellman, J. A. *Tetrahedron* **1997**, *53*, 6635–6644.
[30] Sucholeiki, I. *Tetrahedron Lett.* **1994**, *35*, 7307–7310.
[31] Hoveyda, H. R.; Hall, D. G. *Org. Lett.* **2001**, *3*, 3491–3494.
[32] Grimstrup, M.; Zaragoza, F. *Eur. J. Org. Chem.* **2001**, 3233–3246.
[33] Dressman, B. A.; Singh, U.; Kaldor, S. W. *Tetrahedron Lett.* **1998**, *39*, 3631–3634.
[34] Marshall, D. L.; Liener, I. E. *J. Org. Chem.* **1970**, *35*, 867–868.
[35] Farrall, M. J.; Fréchet, J. M. J. *J. Am. Chem. Soc.* **1978**, *100*, 7998–7999.
[36] Kroll, F. E. K.; Morphy, R.; Rees, D.; Gani, D. *Tetrahedron Lett.* **1997**, *38*, 8573–8576.

[37] Delaet, N. G. J.; Tsuchida, T. Lett. Pept. Sci. **1995**, 2, 325–331.

[38] Yamada, M.; Miyajima, T.; Horikawa, H. Tetrahedron Lett. **1998**, 39, 289–292.

[39] Bhandari, A.; Jones, D. G.; Schullek, J. R.; Vo, K.; Schunk, C. A.; Tamanaha, L. L.; Chen, D.; Yuan, Z. Y.; Needels, M. C.; Gallop, M. A. Bioorg. Med. Chem. Lett. **1998**, 8, 2303–2308.

[40] Tumelty, D.; Schwarz, M. K.; Needels, M. C. Tetrahedron Lett. **1998**, 39, 7467–7470.

[41] Karoyan, P.; Triolo, A.; Nannicini, R.; Giannotti, D.; Altamura, M.; Chassaing, G.; Perrotta, E. Tetrahedron Lett. **1999**, 40, 71–74.

[42] Barn, D. R.; Morphy, J. R.; Rees, D. C. Tetrahedron Lett. **1996**, 37, 3213–3216.

[43] Zaragoza, F.; Stephensen, H. Angew. Chem. Int. Ed. **2000**, 39, 554–556.

[44] Yan, B.; Kumaravel, G. Tetrahedron **1996**, 52, 843–848.

[45] Schwarz, M. K.; Tumelty, D.; Gallop, M. A. J. Org. Chem. **1999**, 64, 2219–2231.

[46] Chen, C.; Randall, L. A. A.; Miller, R. B.; Jones, A. D.; Kurth, M. J. Tetrahedron **1997**, 53, 6595–6609.

[47] Burns, C. J.; Groneberg, R. D.; Salvino, J. M.; McGeehan, G.; Condon, S. M.; Morris, R.; Morrissette, M.; Mathew, R.; Darnbrough, S.; Neuenschwander, K.; Scotese, A.; Djuric, S. W.; Ullrich, J.; Labaudiniere, R. Angew. Chem. Int. Ed. **1998**, 37, 2848–2850.

[48] Chen, C.; Randall, L. A. A.; Miller, R. B.; Jones, A. D.; Kurth, M. J. J. Am. Chem. Soc. **1994**, 116, 2661–2662.

[49] Salvino, J. M.; Mathew, R.; Kiesow, T.; Narensingh, R.; Mason, H. J.; Dodd, A.; Groneberg, R.; Burns, C. J.; McGeehan, G.; Kline, J.; Orton, E.; Tang, S.-Y.; Morrissette, M.; Labaudiniere, R. Bioorg. Med. Chem. Lett. **2000**, 10, 1637–1640.

[50] Salvino, J. M.; Patel, S.; Drew, M.; Krowlikowski, P.; Orton, E.; Kumar, N. V.; Caulfield, T.; Labaudiniere, R. J. Comb. Chem. **2001**, 3, 177–180.

[51] Le Hetet, C.; David, M.; Carreaux, F.; Carboni, B.; Sauleau, A. Tetrahedron Lett. **1997**, 38, 5153–5156.

[52] Zhao, X. Y.; Jung, K. W.; Janda, K. D. Tetrahedron Lett. **1997**, 38, 977–980.

[53] Jung, K. W.; Zhao, X. Y.; Janda, K. D. Tetrahedron **1997**, 53, 6645–6652.

[54] Zhao, X. Y.; Janda, K. D. Tetrahedron Lett. **1997**, 38, 5437–5440.

[55] Forman, F. W.; Sucholeiki, I. J. Org. Chem. **1995**, 60, 523–528.

[56] Katoh, M.; Sodeoka, M. Bioorg. Med. Chem. Lett. **1999**, 9, 881–884.

[57] Gayo, L. M.; Suto, M. J. Tetrahedron Lett. **1997**, 38, 211–214.

[58] Hummel, G.; Hindsgaul, O. Angew. Chem. Int. Ed. **1999**, 38, 1782–1784.

[59] Jobron, L.; Hummel, G. Org. Lett. **2000**, 2, 2265–2267.

[60] Kulkarni, B. A.; Ganesan, A. Tetrahedron Lett. **1999**, 40, 5633–5636.

[61] Girdwood, J. A.; Shute, R. E. Chem. Commun. **1997**, 2307–2308.

[62] Souers, A. J.; Ellman, J. A. Tetrahedron **2001**, 57, 7431–7448.

[63] Reyes, S.; Pattarawarapan, M.; Roy, S.; Burgess, K. Tetrahedron **2000**, 56, 9809–9818.

[64] Botti, P.; Pallin, T. D.; Tam, J. P. J. Am. Chem. Soc. **1996**, 118, 10018–10024.

[65] Crosby, G. A.; Kato, M. J. Am. Chem. Soc. **1977**, 99, 278–280.

[66] Roberts, K. D.; Lambert, J. N.; Ede, N. J.; Bray, A. M. Tetrahedron Lett. **1998**, 39, 8357–8360.

[67] Feng, Y.; Pattarawarapan, M.; Wang, Z.; Burgess, K. Org. Lett. **1999**, 1, 121–124.

[68] Feng, Y.; Wang, Z.; Jin, S.; Burgess, K. J. Am. Chem. Soc. **1998**, 120, 10768–10769.

[69] Fotsch, C.; Kumaravel, G.; Sharma, S. K.; Wu, A. D.; Gounarides, J. S.; Nirmala, N. R.; Petter, R. C. Bioorg. Med. Chem. Lett. **1999**, 9, 2125–2130.

[70] Feng, Y.; Burgess, K. Chem. Eur. J. **1999**, 5, 3261–3272.

[71] Sharma, S. K.; Wu, A. D.; Chandramouli, N. Tetrahedron Lett. **1996**, 37, 5665–5668.

[72] Ueki, M.; Ikeo, T.; Iwadate, M.; Asakura, T.; Williamson, M. P.; Slaninová, J. Bioorg. Med. Chem. Lett. **1999**, 9, 1767–1772.

[73] Zhu, X.; Ganesan, A. J. Comb. Chem. **1999**, 1, 157–162.

[74] Farrall, M. J.; Durst, T.; Fréchet, J. M. Tetrahedron Lett. **1979**, 203–206.

[75] Crosby, G. A.; Weinshenker, N. M.; Uh, H. S. J. Am. Chem. Soc. **1975**, 97, 2232–2235.

[76] Boymond, L.; Rottländer, M.; Cahiez, G.; Knochel, P. Angew. Chem. Int. Ed. **1998**, 37, 1701–1703.

[77] Havez, S.; Begtrup, M.; Vedsø, P.; Andersen, K.; Ruhland, T. J. Org. Chem. **1998**, 63, 7418–7420.

[78] Heckel, A.; Mross, E.; Jung, K. H.; Rademann, J.; Schmidt, R. R. Synlett **1998**, 171–173.

[79] Rademann, J.; Schmidt, R. R. Tetrahedron Lett. **1996**, 37, 3989–3990.

[80] Panek, J. S.; Zhu, B. Tetrahedron Lett. **1996**, 37, 8151–8154.

[81] Gosselin, F.; Di Renzo, M.; Ellis, T. H.; Lubell, W. D. J. Org. Chem. **1996**, 61, 7980–7981.

[82] García Echeverría, C. Tetrahedron Lett. **1997**, 38, 8933–8934.

[83] Flanigan, E.; Marshall, G. R. Tetrahedron Lett. **1970**, 2403–2406.

[84] Rolland, C.; Hanquet, G.; Ducep, J.-B.; Solladié, G. Tetrahedron Lett. **2001**, 42, 7563–7566.

[85] Han, H. S.; Janda, K. D. Tetrahedron Lett. **1997**, 38, 1527–1530.

[86] Halm, C.; Evarts, J.; Kurth, M. J. *Tetrahedron Lett.* **1997**, *38*, 7709–7712.

[87] Mata, E. G. *Tetrahedron Lett.* **1997**, *38*, 6335–6338.

[88] Burkett, B. A.; Chai, C. L. L. *Tetrahedron Lett.* **2000**, *41*, 6661–6664.

[89] Delpiccolo, C. M. L.; Mata, E. G. *Tetrahedron: Asymmetry* **1999**, *10*, 3893–3897.

[90] Masquelin, T.; Sprenger, D.; Baer, R.; Gerber, F.; Mercadal, Y. *Helv. Chim. Acta* **1998**, *81*, 646–660.

[91] Obrecht, D.; Abrecht, C.; Grieder, A.; Villalgordo, J. M. *Helv. Chim. Acta* **1997**, *80*, 65–72.

[92] Cheng, W.-C.; Olmstead, M. M.; Kurth, M. J. *J. Org. Chem.* **2001**, *66*, 5528–5533.

[93] Huang, W.; Cheng, S.; Sun, W. *Tetrahedron Lett.* **2001**, *42*, 1973–1974.

[94] Rolland, C.; Hanquet, G.; Ducep, J.-B.; Solladié, G. *Tetrahedron Lett.* **2001**, *42*, 9077–9080.

[95] Caddick, S.; Hamza, D.; Wadman, S. N. *Tetrahedron Lett.* **1999**, *40*, 7285–7288.

[96] Gude, M.; Piarulli, U.; Potenza, D.; Salom, B.; Gennari, C. *Tetrahedron Lett.* **1996**, *37*, 8589–8592.

[97] Floyd, C. D.; Lewis, C. N.; Patel, S. R.; Whittaker, M. *Tetrahedron Lett.* **1996**, *37*, 8045–8048.

[98] Tomasi, S.; Le Roch, M.; Renault, J.; Corbel, J. C.; Uriac, P.; Carboni, B.; Moncoq, D.; Martin, B.; Delcros, J. G. *Bioorg. Med. Chem. Lett.* **1998**, *8*, 635–640.

[99] Pop, I. E.; Déprez, B. P.; Tartar, A. L. *J. Org. Chem.* **1997**, *62*, 2594–2603.

[100] Gordeev, M. F.; Luehr, G. W.; Hui, H. C.; Gordon, E. M.; Patel, D. V. *Tetrahedron* **1998**, *54*, 15879–15890.

[101] Hone, N. D.; Payne, L. J. *Tetrahedron Lett.* **2000**, *41*, 6149–6152.

[102] Bolton, G. L.; Hodges, J. C.; Rubin, J. R. *Tetrahedron* **1997**, *53*, 6611–6634.

[103] Kim, S. W.; Hong, C. Y.; Koh, J. S.; Lee, E. J.; Lee, K. *Mol. Diversity* **1998**, *3*, 133–136.

[104] Kim, S. W.; Hong, C. Y.; Lee, K.; Lee, E. J.; Koh, J. S. *Bioorg. Med. Chem. Lett.* **1998**, *8*, 735–738.

[105] Yang, L.; Chiu, K. *Tetrahedron Lett.* **1997**, *38*, 7307–7310.

[106] Raman, P.; Stokes, S. S.; Angell, Y. M.; Flentke, G. R.; Rich, D. H. *J. Org. Chem.* **1998**, *63*, 5734–5735.

[107] Scicinski, J. J.; Barker, R. D.; Murray, P. J.; Jarvie, E. M. *Bioorg. Med. Chem. Lett.* **1998**, *8*, 3609–3614.

[108] Mohamed, N.; Bhatt, U.; Just, G. *Tetrahedron Lett.* **1998**, *39*, 8213–8216.

[109] Beaver, K. A.; Siegmund, A. C.; Spear, K. L. *Tetrahedron Lett.* **1996**, *37*, 1145–1148.

[110] de Bont, D. B. A.; Dijkstra, G. D. H.; den Hartog, J. A. J.; Liskamp, R. M. J. *Bioorg. Med. Chem. Lett.* **1996**, *6*, 3035–3040.

[111] Gennari, C.; Longari, C.; Ressel, S.; Salom, B.; Piarulli, U.; Ceccarelli, S.; Mielgo, A. *Eur. J. Org. Chem.* **1998**, 2437–2449.

[112] Brouwer, A. J.; Van der Linden, H. J.; Liskamp, R. M. J. *J. Org. Chem.* **2000**, *65*, 1750–1757.

[113] Monnee, M. C. F.; Marijne, M. F.; Brouwer, A. J.; Liskamp, R. M. J. *Tetrahedron Lett.* **2000**, *41*, 7991–7995.

[114] Zhang, H. C.; Brumfield, K. K.; Jaroskova, L.; Maryanoff, B. E. *Tetrahedron Lett.* **1998**, *39*, 4449–4452.

[115] Meyers, H. V.; Dilley, G. J.; Durgin, T. L.; Powers, T. S.; Winssinger, N. A.; Zhu, H.; Pavia, M. R. *Mol. Diversity* **1995**, *1*, 13–20.

[116] Wang, Y.; Huang, T. N. *Tetrahedron Lett.* **1998**, *39*, 9605–9608.

[117] Ngu, K.; Patel, D. V. *Tetrahedron Lett.* **1997**, *38*, 973–976.

[118] Fivush, A. M.; Willson, T. M. *Tetrahedron Lett.* **1997**, *38*, 7151–7154.

[119] Swayze, E. E. *Tetrahedron Lett.* **1997**, *38*, 8643–8646.

[120] Swayze, E. E. *Tetrahedron Lett.* **1997**, *38*, 8465–8468.

[121] Saha, A. K.; Liu, L.; Marichal, P.; Odds, F. *Bioorg. Med. Chem. Lett.* **2000**, *10*, 2735–2739.

[122] Dixit, D. M.; Leznoff, C. C. *Israel J. Chem.* **1978**, *17*, 248–252.

[123] Reichwein, J. F.; Liskamp, R. M. J. *Tetrahedron Lett.* **1998**, *39*, 1243–1246.

[124] Chenera, B.; Finkelstein, J. A.; Veber, D. F. *J. Am. Chem. Soc.* **1995**, *117*, 11999–12000.

[125] Dragoli, D. R.; Burdett, M. T.; Ellman, J. A. *J. Am. Chem. Soc.* **2001**, *123*, 10127–10128.

[126] ten Holte, P.; Thijs, L.; Zwanenburg, B. *Tetrahedron Lett.* **1998**, *39*, 7407–7410.

[127] Bicak, N.; Senkal, B. F. *Reactive and Functional Polymers* **1996**, *29*, 123–128.

[128] Jin, S.; Holub, D. P.; Wustrow, D. J. *Tetrahedron Lett.* **1998**, *39*, 3651–3654.

[129] Zhong, H. M.; Greco, M. N.; Maryanoff, B. E. *J. Org. Chem.* **1997**, *62*, 9326–9330.

[130] Hu, J.-B.; Zhao, G.; Ding, Z. *Angew. Chem. Int. Ed.* **2001**, *40*, 1109–1111.

[131] Hunt, J. A.; Roush, W. R. *J. Am. Chem. Soc.* **1996**, *118*, 9998–9999.

[132] Raju, B.; Kogan, T. P. *Tetrahedron Lett.* **1997**, *38*, 4965–4968.

[133] de Bont, D. B. A.; Moree, W. J.; Liskamp, R. M. J. *Bioorg. Med. Chem.* **1996**, *4*, 667–672.

[134] Zhang, H.-C.; Ye, H.; Moretto, A. F.; Brumfield, K. K.; Maryanoff, B. E. *Org. Lett.* **2000**, *2*, 89–92.

[135] Raju, B.; Kogan, T. P. *Tetrahedron Lett.* **1997**, *38*, 3373–3376.

[136] Dankwardt, S. M.; Smith, D. B.; Porco, J. A.; Nguyen, C. H. *Synlett* **1997**, 854–856.

[137] Quaderer, R.; Hilvert, D. *Org. Lett.* **2001**, *3*, 3181–3184.
[138] Xiong, Y.; Klopp, J.; Chapman, K. T. *Tetrahedron Lett.* **2001**, *42*, 8423–8427.
[139] Piscopio, A. D.; Miller, J. F.; Koch, K. *Tetrahedron Lett.* **1998**, *39*, 2667–2670.
[140] Wipf, P.; Henninger, T. C. *J. Org. Chem.* **1997**, *62*, 1586–1587.
[141] Miller, S. C.; Scanlan, T. S. *J. Am. Chem. Soc.* **1998**, *120*, 2690–2691.
[142] Backes, B. J.; Virgilio, A. A.; Ellman, J. A. *J. Am. Chem. Soc.* **1996**, *118*, 3055–3056.
[143] Backes, B. J.; Ellman, J. A. *J. Org. Chem.* **1999**, *64*, 2322–2330.
[144] Kenner, G. W.; McDermott, J. R.; Sheppard, R. C. *J. Chem. Soc., Chem. Commun.* **1971**, 636–637.
[145] Ngu, K.; Patel, D. V. *J. Org. Chem.* **1997**, *62*, 7088–7089.
[146] Boeijen, A.; Kruijtzer, J. A. W.; Liskamp, R. M. J. *Bioorg. Med. Chem. Lett.* **1998**, *8*, 2375–2380.
[147] Kay, C.; Murray, P. J.; Sandow, L.; Holmes, A. B. *Tetrahedron Lett.* **1997**, *38*, 6941–6944.
[148] Combs, A. P.; Rafalski, M. *J. Comb. Chem.* **2000**, *2*, 29–32.
[149] Piscopio, A. D.; Miller, J. F.; Koch, K. *Tetrahedron Lett.* **1997**, *38*, 7143–7146.
[150] Veerman, J. J. N.; van Maarseveen, J. H.; Visser, G. M.; Kruse, C. G.; Schoemaker, H. E.; Hiemstra, H.; Rutjes, F. P. J. T. *Eur. J. Org. Chem.* **1998**, 2583–2589.
[151] Piscopio, A. D.; Miller, J. F.; Koch, K. *Tetrahedron* **1999**, *55*, 8189–8198.
[152] Zheng, C. S.; Seeberger, P. H.; Danishefsky, S. J. *Angew. Chem. Int. Ed.* **1998**, *37*, 786–789.
[153] Richter, H.; Jung, G. *Tetrahedron Lett.* **1998**, *39*, 2729–2730.
[154] Dankwardt, S. M.; Phan, T. M.; Krstenansky, J. L. *Mol. Diversity* **1995**, *1*, 113–120.
[155] Hari, A.; Miller, B. L. *Org. Lett.* **1999**, *1*, 2109–2111.
[156] Nugiel, D. A.; Wacker, D. A.; Nemeth, G. A. *Tetrahedron Lett.* **1997**, *38*, 5789–5790.
[157] Pan, Y.; Holmes, C. P. *Org. Lett.* **2001**, *3*, 2769–2771.
[158] Richter, L. S.; Desai, M. C. *Tetrahedron Lett.* **1997**, *38*, 321–322.
[159] Rölfing, K.; Thiel, M.; Künzer, H. *Synlett* **1996**, 1036–1038.
[160] Fyles, T. M.; Leznoff, C. C.; Weatherston, J. *Can. J. Chem.* **1977**, *55*, 4135–4143.
[161] Leznoff, C. C.; Fyles, T. M.; Weatherston, J. *Can. J. Chem.* **1977**, *55*, 1143–1153.
[162] Lee, C. E.; Kick, E. K.; Ellman, J. A. *J. Am. Chem. Soc.* **1998**, *120*, 9735–9747.
[163] Uozumi, Y.; Danjo, H.; Hayashi, T. *J. Org. Chem.* **1999**, *64*, 3384–3388.
[164] Hanessian, S.; Huynh, H. K. *Synlett* **1999**, 102–104.
[165] Rottländer, M.; Knochel, P. *Synlett* **1997**, 1084–1086.
[166] Rueter, J. K.; Nortey, S. O.; Baxter, E. W.; Leo, G. C.; Reitz, A. B. *Tetrahedron Lett.* **1998**, *39*, 975–978.
[167] Hijikuro, I.; Doi, T.; Takahashi, T. *J. Am. Chem. Soc.* **2001**, *123*, 3716–3722.
[168] Dorman, L. C.; Love, J. *J. Org. Chem.* **1969**, *34*, 158–165.
[169] Vanier, C.; Lorgé, F.; Wagner, A.; Mioskowski, C. *Angew. Chem. Int. Ed.* **2000**, *39*, 1679–1683.
[170] Peschke, B.; Bundgaard, J. G.; Breinholt, J. *Tetrahedron Lett.* **2001**, *42*, 5127–5130.

9 Preparation of Organoselenium Compounds

Support-bound organoselenium compounds have mainly been used as synthetic intermediates and linkers. The use of organoselenium compounds as linkers for solid-phase synthesis has been investigated by several groups [1–4]. The selenium–carbon bond is stable under a broad variety of reaction conditions, but can be selectively cleaved by tin radicals or by oxidants (see Sections 3.15.1 and 3.16.4).

Polystyrene-bound arylselenium derivatives have been prepared from lithiated polystyrene and metallic selenium [2] or dimethyldiselenide [1] (Figure 9.1). These derivatives are suitable starting materials for the preparation of other selenium reagents, which enable the facile attachment of organic substrates to insoluble supports. Alkali metal selenides or selenoborate complexes are strong nucleophiles and react swiftly with organic halides (including DCM!) to yield selenides. Electrophilic selenium derivatives, such as the bromides BrSe–Pol, phthalimides PhtNSe–Pol, and sulfonates RSO$_2$Se–Pol [5], undergo addition to alkenes under mild reaction conditions. Alky-

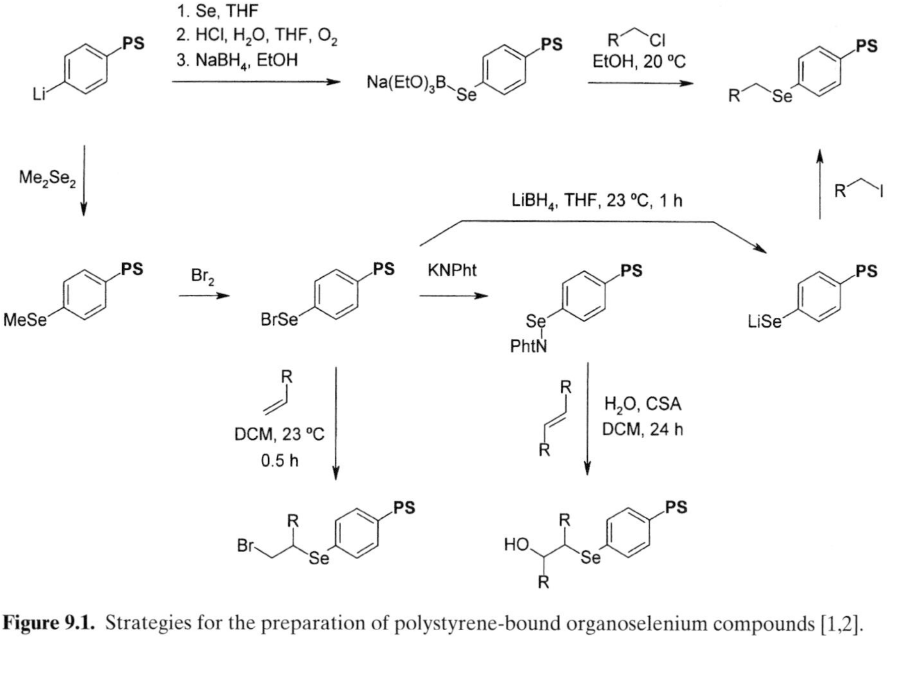

Figure 9.1. Strategies for the preparation of polystyrene-bound organoselenium compounds [1,2].

nylselenium compounds have been prepared by treating polystyrene-bound terminal alkynes with arylselenyl bromides (2 equiv.) in the presence of CuI (5 equiv.) and Bu₃N (4 equiv.) in DMF (60 °C, 24 h [6]). Some representative synthetic transformations of organoselenium compounds on insoluble supports are sketched in Figure 9.1.

The reaction of suitably functionalized alkenes with polystyrene-bound selenyl bromide can lead to cyclization of the alkene by C–C, C–O, or C–N bond formation with simultaneous attachment to the support. The cyclic product can then be released from

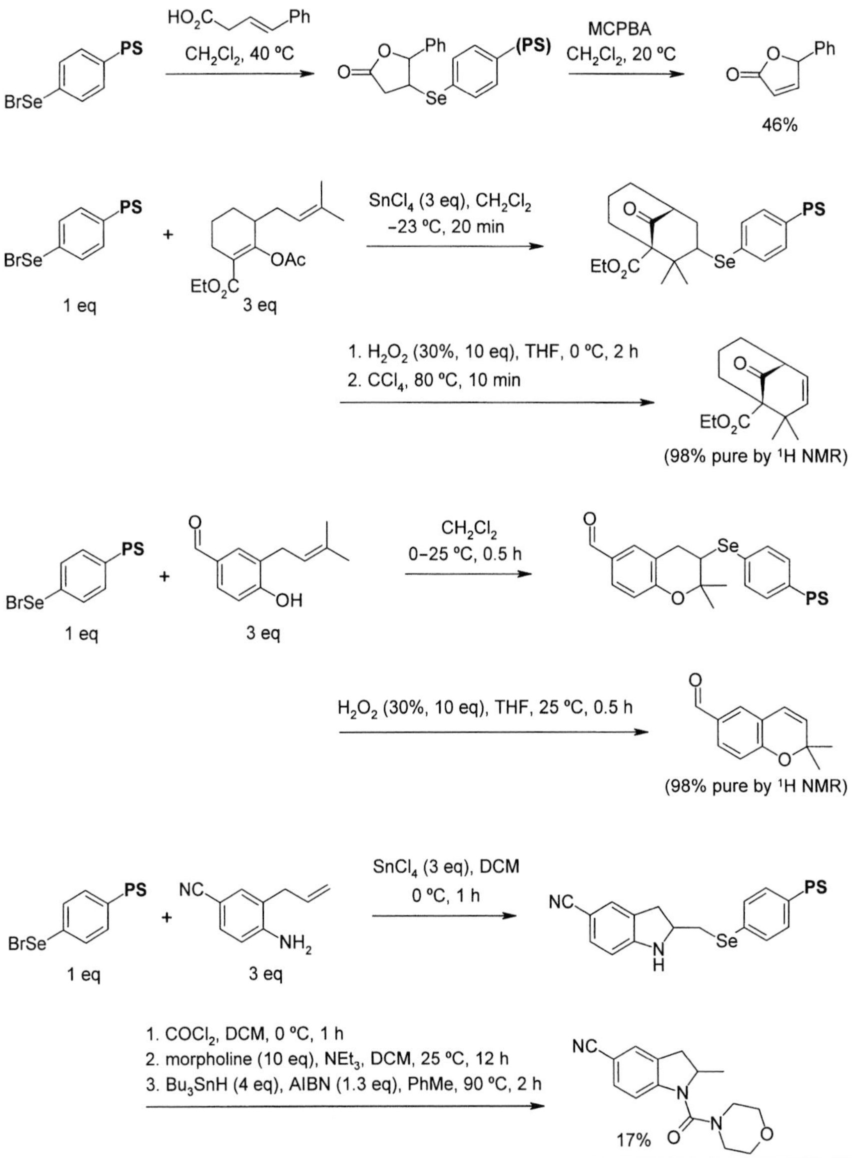

Figure 9.2. Oxidative cyclization of alkenes with polystyrene-bound selenyl bromide [8,9,12–14].

the support, either by treatment with an oxidant to yield a cyclic alkene, or by treatment with tin radicals to yield an alkane (see Sections 3.15.1 and 3.16.4). Illustrative examples of this technique are outlined in Figure 9.2. This strategy has been used for the preparation of arrays of diverse benzopyrans for screening [7–10]. Enantiomerically pure, support-bound selenyl bromides have been used for the preparation of enantiomerically enriched lactones, ethers, and acetals [11].

Polystyrene-bound selenides have also proven to be suitable intermediates for solid-phase carbohydrate chemistry. Figure 9.3 depicts a series of transformations of a resin-bound carbohydrate, which highlight the stability of the selenide linkage.

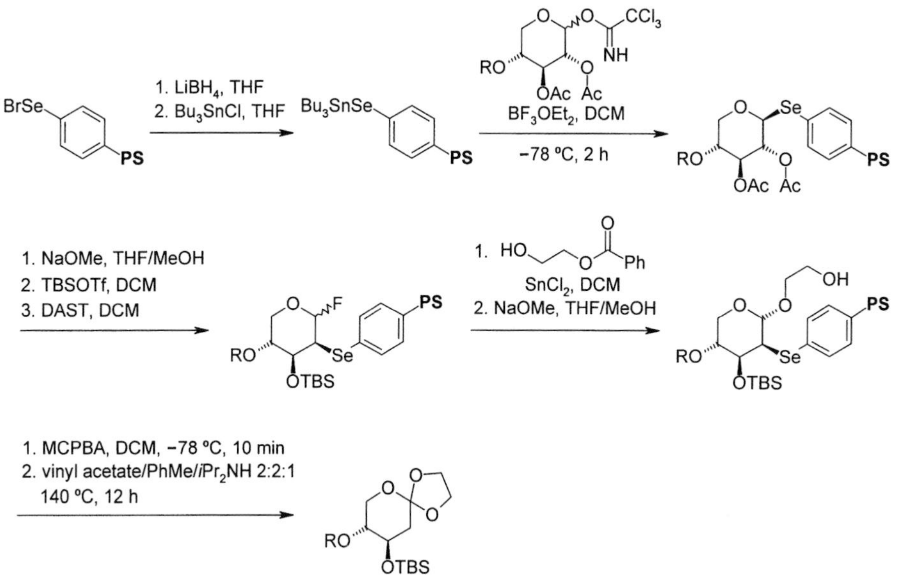

Figure 9.3. Transformations of a carbohydrate attached to cross-linked polystyrene by a selenium-based linkage [15].

References for Chapter 9

[1]   Nicolaou, K. C.; Pastor, J.; Barluenga, S.; Winssinger, N. *Chem. Commun.* **1998**, 1947–1948.
[2]   Ruhland, T.; Andersen, K.; Pedersen, H. *J. Org. Chem.* **1998**, *63*, 9204–9211.
[3]   Michels, R.; Kato, M.; Heitz, W. *Makromol. Chem.* **1976**, *177*, 2311–2320.
[4]   Zaragoza, F. *Angew. Chem. Int. Ed.* **2000**, *39*, 2077–2079.
[5]   Qian, H.; Huang, X. *Synlett* **2001**, 1913–1916.
[6]   Gendre, F.; Diaz, P. *Tetrahedron Lett.* **2000**, *41*, 5193–5197.
[7]   Nicolaou, K. C.; Pfefferkorn, J. A.; Schuler, F.; Roecker, A. J.; Cao, G.-Q.; Casida, J. E. *Chem. Biol.* **2000**, *7*, 979–992.
[8]   Nicolaou, K. C.; Pfefferkorn, J. A.; Cao, G.-Q. *Angew. Chem. Int. Ed.* **2000**, *39*, 734–739.
[9]   Nicolaou, K. C.; Cao, G.-Q.; Pfefferkorn, J. A. *Angew. Chem. Int. Ed.* **2000**, *39*, 739–743.
[10] Nicolaou, K. C.; Pfefferkorn, J. A.; Roecker, A. J.; Cao, G.-Q.; Barluenga, S.; Mitchell, H. J. *J. Am. Chem. Soc.* **2000**, *122*, 9939–9953.
[11] Uehlin, L.; Wirth, T. *Org. Lett.* **2001**, *3*, 2931–2933.
[12] Fujita, K.-I.; Watanabe, K.; Oishi, A.; Ikeda, Y.; Taguchi, Y. *Synlett* **1999**, 1760–1762.
[13] Nicolaou, K. C.; Pfefferkorn, J. A.; Cao, G.-Q.; Kim, S.; Kessabi, J. *Org. Lett.* **1999**, *1*, 807–810.

[14] Nicolaou, K. C.; Roecker, A. J.; Pfefferkorn, J. A.; Cao, G.-Q. *J. Am. Chem. Soc.* **2000**, *122*, 2966–2967.
[15] Nicolaou, K. C.; Mitchell, H. J.; Fylaktakidou, K. C.; Suzuki, H.; Rodríguez, R. M. *Angew. Chem. Int. Ed.* **2000**, *39*, 1089–1093.

10 Preparation of Nitrogen Compounds

10.1 Preparation of Amines

Amines are both valuable synthetic intermediates and interesting target molecules. A number of solid-phase syntheses have been developed, most of which are also suitable for the automated, parallel synthesis of this class of compound.

The most common preparations of amines on insoluble supports include nucleophilic aliphatic and aromatic substitutions, Michael-type additions, and the reduction of imines, amides, nitro groups, and azides (Figure 10.1). Further methods include the addition of carbon nucleophiles to imines (e.g. the Mannich reaction) and oxidative degradation of carboxylic acids or amides. Linkers for primary, secondary, and tertiary amines are discussed in Sections 3.6, 3.7, and 3.8.

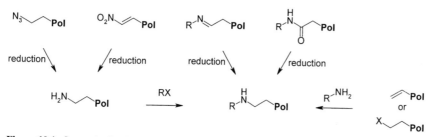

Figure 10.1. Strategies for the synthesis of amines on insoluble supports. X: leaving group; R: alkyl, aryl.

10.1.1 Preparation of Amines by Aliphatic Nucleophilic Substitution

10.1.1.1 With Support-Bound Alkylating Agents

Support-bound alkyl halides, sulfonates, or epoxides can be used to alkylate amines. Alkylating agents of low reactivity, such as non-benzylic alkyl halides or sulfonates, generally require high amine concentrations, dipolar aprotic solvents (e.g. DMSO), and high reaction temperatures to undergo substitution (Table 10.1). Forcing conditions can, however, also promote several other processes, such as elimination [1], cleavage of the linker [2], oxidation, or substitutions with competing nucleophiles (e.g. the solvent or water; see also Entry 6, Table 10.1). This is probably why few examples of N-alkylations with non-activated alkylating agents have been reported. The reactivity of the amine is also crucial for the success of N-alkylations, the best results usually being obtained with primary or secondary alkylamines. Particularly poor nucleophiles are amino acid esters and aromatic or heteroaromatic amines. Table 10.1 lists illustrative examples of the alkylation of amines with support-bound alkylating agents of low reactivity.

Table 10.1. Preparation of amines from support-bound weak alkylating agents.

Entry	Starting resin	Conditions	Product	Ref.
1	NosO, Ph, N₃ —(PS)	Ar⌢NH₂ (1 mol/L), NMP, 80 °C, 36 h Ar: 2,4-Cl₂C₆H₃	Ar⌢N, Ph, N₃ —(PS)	[3] see also [4,5]
2	Br⌢, O —(PS)	piperidine (0.7 mol/L), DMF, 18 h	N⌢, O —(PS)	[6]
3	MsO⌢S–S, N, PS	isobutylamine (1 mol/L), NMP, 50 °C, 16 h	N⌢S–S, N, PS	[7] see also [8,9]
4	MsO⌢, O —(PS)	benzylamine, MeCN, 81 °C, 20 h	Ph, N⌢, O —(PS)	[10] see also [1]
5	OEt, Br⌢O⌢TG	butylamine (2 mol/L), DMSO, 60 °C, 15 h	OEt, N⌢O⌢TG	[11]
6	Br⌢, N —(PS)	ArNH₂, DMSO Ar: halophenyl	Ar, HN⌢NH (PS) + N—(PS)	[12]
7	Ph, O —PS (epoxide)	piperidine, LiClO₄ (6 mol/L each), MeCN, 55 °C, 48 h	Ph, OH, N⌢O⌢PS (piperidine)	[13] see also [14–16]
8	Cl, O —(PS)	BnNH₂ (8 eq), dioxane, 80 °C, 68 h	Cl, Ph⌢N, OH, O —(PS)	[17] see also [18–21]

More reactive electrophiles, such as benzyl and allyl halides, as well as α- or β-halocarbonyl compounds, react smoothly with amines, often even at room temperature. Support-bound chloro- and bromoacetamides, for instance, react cleanly with a wide range of aliphatic and aromatic amines to yield glycine derivatives (Entries **1–4**, Table 10.2 [22–32]). This reaction is usually conducted in DMSO at room temperature (2–12 h), but for sensitive amines DMF or NMP might offer milder reaction conditions (Entry **3**, Table 10.2). Higher yields can often be obtained by increasing the reaction temperature and the concentration of the amine.

Table 10.2. Alkylation of amines with support-bound strong alkylating agents.

Entry	Starting resin	Conditions	Product	Ref.
1	Br—C(=O)—NH—(PS)	Ph$_2$CH—NH$_2$ (2.5 mol/L), DMSO, 20 °C, 2 h	Ph, Ph—CH—NH—C(=O)—NH—(PS)	[41]
2	Br—C(=O)—NH—(PS)	allylamine (0.5 mol/L), DMF, 20 °C, 12 h	allyl—NH—C(=O)—NH—(PS)	[42]
3	Br—C(=O)—NH—(PS)	sugar—O—CH$_2$CH$_2$—NH$_2$ (1 mol/L), DMF, DBU, 50 °C	sugar—O—CH$_2$CH$_2$—NH—CH$_2$—C(=O)—NH—(PS)	[43]
4	thienyl-CH$_2$—N(—(PS))—C(=O)—CH$_2$—Cl	BuNH$_2$ (2 mol/L), NMP, NaI, 80 °C, overnight	thienyl-CH$_2$—N(—(PS))—C(=O)—CH$_2$—NH—Bu	[44]
5	Cl—CH$_2$—C$_6$H$_4$—PS	BnNH$_2$, DMF, 60 °C, 48 h	Ph—CH$_2$—NH—CH$_2$—C$_6$H$_4$—PS	[45]
6	Cl-CH$_2$, I—C$_6$H$_3$—CH$_2$—O—C(=O)—O—(PS)	BuNH$_2$ (10 eq), Bu$_4$NBr (2 eq), NEt$_3$ (3 eq), 20 °C, 96 h	Bu-NH-CH$_2$, I—C$_6$H$_3$—CH$_2$—O—C(=O)—O—(PS)	[46]
7	Cl—CH$_2$—C$_6$H$_4$—C(=O)—NH—(PS)	allylamine (1 mol/L) DMF, NaI, 75 °C, overnight	allyl—NH—CH$_2$—C$_6$H$_4$—C(=O)—NH—(PS)	[47] see also [48]
8	Cl-CH$_2$, I—C$_6$H$_3$—CH$_2$—O—C(=O)—(PS)	PhNH$_2$ (10 eq), DIPEA (3 eq), Bu$_4$NI (2 eq), dioxane, 100 °C, 36 h	Ph-NH-CH$_2$, I—C$_6$H$_3$—CH$_2$—O—C(=O)—(PS)	[46]
9	Br-CH$_2$—C$_6$H$_4$—N(CH$_2$Ph)—C(=O)—CH=CH—C(=O)—O—CH$_2$—(PS)	ArCH$_2$NH$_2$ (0.2 mol/L), DMF, 20 °C, 10 h Ar: 4-methoxyphenyl	HN(CH$_2$Ar)-CH$_2$—C$_6$H$_4$—N(CH$_2$Ph)—C(=O)—CH=CH—C(=O)—O—CH$_2$—(PS)	[49] see also [50]
10	Cl—CH$_2$—C$_6$H$_4$—O—CH$_2$—PS	4-nitroaniline, proton sponge, NaI, DMF, 90 °C, 24 h	O$_2$N—C$_6$H$_4$—NH—CH$_2$—C$_6$H$_4$—O—CH$_2$—PS	[51]
11	Ar—C(=O)—O—CH$_2$—C$_6$H$_4$—O—CH$_2$—PS	piperazine (neat), 130 °C, 20 h Ar: 4-FC$_6$H$_4$	piperazine-N—CH$_2$—C$_6$H$_4$—O—CH$_2$—PS	[52]

Table 10.2. continued.

Entry	Starting resin	Conditions	Product	Ref.
12		(0.5 mol/L), NMP, 20 °C, 20 h Ar: 2-ClC$_6$H$_4$		[53]
13		(1.5 mol/L), THF, 20 °C, 48 h		[44]
14		1. TFA/DCM 7:3, 20 °C, 5 h 2. ArNH$_2$ (0.5 mol/L), DCM, 1 h Ar: 4-(PhO)C$_6$H$_4$		[54]
15		morpholine (20 eq), DEAD (10 eq), PPh$_3$ (10 eq), DCM, 0–25 °C, 48 h		[55]
16		2-iodoaniline (0.2 mol/L), DMF, DIPEA, 80 °C, 18 h		[56] see also [57,58]
17		Bn$_2$NH, Pd(dba)$_2$, PPh$_3$, THF, 24 h		[45]
18		cyclopropylamine, BEMP, DMF, 20 °C, 6 h		[59]
19		PrNH$_2$ (4 mol/L, 50 eq), DMF, 0.5 h		[60]

Merrifield resin [33–37] and other support-bound benzyl halides [38] have also been used to alkylate amines (Entries **5–10**, Table 10.2). Similarly, resin-bound allyl bromides react cleanly with aliphatic or aromatic amines (Entry **16**, Table 10.2). Entry **15** in Table 10.2 is a rare example of the N-alkylation of an amine under the conditions of the Mitsunobu reaction. This reaction cannot generally be used to alkylate amines, and only proceeds under carefully optimized conditions with benzyl or benzhydryl alcohols. Allylic esters or carbonates undergo nucleophilic substitution under palladium(0) catalysis (Entry **17**, Table 10.2). When using ammonia, for example to convert cross-linked chloromethyl polystyrene into aminomethyl polystyrene, good results are only obtained with anhydrous ammonia in DCM [39]. Other solvents or aqueous ammonia generally lead to extensive cross-linking. Consequently, the conversion of resin-bound halides into primary amines is more conveniently achieved via an intermediate azide. Aminomethyl polystyrene has also been prepared from Merrifield resin by treatment of the latter with potassium hexamethyldisilazide, $KN(SiMe_3)_2$ [40].

10.1.1.2 By Alkylation of Support-Bound Amines

Amines can be prepared on insoluble supports by treating support-bound primary or secondary amines with alkylating agents. The reaction is generally limited to strong alkylating agents, such as benzyl or allyl halides or α-halo carbonyl compounds. Because quaternization does not readily occur on hydrophobic supports, the alkylation of polystyrene-bound amines can be conducted using an excess of the alkylating agent to provide products of acceptable purity (Table 10.3). Primary amines will, however, usually be doubly alkylated under these conditions, unless alkylating agents are used that lead to a significant steric or electronic deactivation of the resulting secondary amine. Examples of the N-monoalkylation of resin-bound α-amino acid derivatives include the alkylation with α-halo ketones (Entry **6**, Table 10.3) or with 2-(bromomethyl)acrylates [61]. Diarylmethyl halides can be used to alkylate primary amines only once, because steric hindrance usually prevents twofold diarylmethylation (Entry **2**, Table 10.3). The resulting secondary benzhydrylamines can, however, be alkylated by sterically less demanding alkylating agents (Entries **4** and **5**, Table 10.3). Because alkoxy-substituted diarylmethyl groups can be cleaved from aliphatic amines by treatment with TFA (see Section 10.1.10.6), diarylmethylation can be used to protect primary amines from twofold alkylation by sterically less demanding alkylating agents.

The Mitsunobu reaction is usually only suitable for the alkylation of negatively charged nucleophiles rather than for the alkylation of amines, and only a few examples of such reactions (mainly intramolecular N-alkylations or N-benzylations) have been reported (Entry **15**, Table 10.2). Halides, however, are very efficiently alkylated under Mitsunobu conditions, and it has been found that the treatment of resin-bound ammonium iodides with benzylic alcohols, a phosphine, and an azodicarboxylate leads to clean benzylation of the amine (Entry **9**, Table 10.3). Unfortunately, alkylations with aliphatic alcohols do not proceed under these conditions. The latter can, however, also be used to alkylate resin-bound aliphatic amines when (cyanomethyl)-phosphonium iodides $[R_3P–CH_2CN^+][I^-]$ are used as coupling reagents [62]. These reagents convert aliphatic alcohols into alkyl iodides, which then alkylate the amine (Entry **10**, Table 10.3).

Polystyrene-bound secondary aliphatic amines and *N*-alkyl amino acids can be allylated by treatment with a diene and an aryl iodide or bromide in the presence of palladium(II) acetate (Entry 14, Table 10.3). As the diene, 1,3-, 1,4-, and 1,5-dienes can be used, and, besides aryl halides, heteroaryl bromides have also been successfully used [63]. This remarkable reaction is likely to proceed via the formation of an aryl palladium complex, with subsequent insertion of an alkene into the C–Pd bond. The resulting organopalladium compound does not undergo β-elimination (as in the Heck reaction), but isomerizes to an allyl palladium complex, which reacts with the amine to give the observed allyl amines.

Table 10.3. Alkylation of support-bound amines.

Entry	Starting resin	Conditions	Product	Ref.
1		Ph$_2$CHBr (0.7 mol/L), NMP, DIPEA, 80 °C, 18 h		[64] see also [65]
2		Ar$_2$CHCl (0.02 mol/L), DCM, DIPEA, 20 min Ar: 4-(MeO)C$_6$H$_4$		[66]
3		allyl bromide (4 eq), BEMP, dioxane, 20 °C, 70 h		[46] see also [59,67]
4		BnBr (3 eq), DIPEA, DMF/Me$_2$CO 1:1, 20 °C Ar: 4-fluorophenyl		[68]
5		phenacyl chloride (0.05 mol/L), DMF/ Me$_2$CO 1:1, 45 °C, 10 h		[68]
6		BocHN (2.5 eq), DIPEA, DMF, 3–6 h		[69] see also [70]
7	Ar: 2-nitrophenyl	PhtN (1 mol/L, 10 eq), DIPEA (10 eq), NMP, 80 °C, 45 h		[71]

Table 10.3. continued.

Entry	Starting resin	Conditions	Product	Ref.
8		(0.4 mol/L, 3 eq), NMP, 40 °C, 52 h		[72]
9		1. TsOH, NMP, 5 min 2. CsI, DMSO, 0.5 h 3. ArCH₂OH (0.7 mol/L, 22 eq), ADDP (19 eq), PBu₃ (14 eq), NMP, 20 °C, 23 h		[73]
10		2-phenylethanol (0.56 mol/L, 7 eq), Bu₃PCH₂CNI (6 eq), DIPEA (8 eq), MeCN, 81 °C, 23 h		[62]
11		HO–⧸⧹–OCO₂Me Pd(PPh₃)₄, AcOH, PhMe, 20 °C, 3 h		[74]
12		2-methoxybenzyl bromide (0.4 mol/L), DMSO, DBU, 3 h Ar: 2,5-(MeO)₂C₆H₃		[75]
13		ArCH₂Br (0.3 mol/L, 10 eq), DMF, DIPEA, 80 °C, 22 h Ar: 4-(F₃C)C₆H₄		[56]
14		(0.75 mol/L, 10 eq), PhI (2 eq), DIPEA (10 eq), LiCl (2 eq), Pd(OAc)₂ (0.1 eq), DMF, 100 °C, 2 d		[63]

10.1.2 Preparation of Amines by Aromatic Nucleophilic Substitution

10.1.2.1 With Support-Bound Arylating Agents

Aryl- and heteroaryl halides can undergo thermal or transition metal catalyzed substitution reactions with amines. These reactions proceed on insoluble supports under conditions similar to those used in solution. Not only halides, but also thiolates [76], nitro groups [76], sulfinates [77,78], and alcoholates [79] can serve as leaving groups for aromatic nucleophilic substitution.

Support-bound 4-fluoro-3-nitrobenzoic acid has become a widely used intermediate for the preparation of a variety of heterocycles (see Chapter 15). Aromatic nucleophilic substitution of fluoride in this reagent proceeds smoothly with a wide range of nucleophiles, including aliphatic and aromatic amines (Table 10.4). Aliphatic amines react at room temperature (Entry **1**, Table 10.4 [48,80]), whereas anilines generally require heating (Entry **3**, Table 10.4). Unprotected β-amino acids can be N-arylated directly using 4-fluoro-3-nitrobenzamides linked to PEG-grafted polystyrene (Entry **2**, Table 10.4). Less strongly activated fluorobenzenes have also been used as resin-bound arylating agents for amines, but higher reaction temperatures are often required (Entry **6**, Table 10.4).

Non-activated aryl bromides (but not fluorides) can be used as substrates for palladium(0)-catalyzed aromatic nucleophilic substitutions with aliphatic or aromatic amines. These reactions require sodium alcoholates or cesium carbonate as a base, and sterically demanding phosphines as ligands. Moreover, high reaction temperatures are often necessary to achieve complete conversion (Entries **7** and **8**, Table 10.4; Experimental Procedure 10.1). Unfortunately, the choice of substituents on the amine

Table 10.4. Preparation of aromatic amines from resin-bound arylating agents.

Entry	Starting resin	Conditions	Product	Ref.
1		EtO$_2$C~~NH$_2$ (0.22 mol/L, 10 eq), DMF, 20 °C, 2 h		[97]
2		HO$_2$C~~NH$_2$ NaHCO$_3$ (0.5 mol/L), H$_2$O/Me$_2$CO 1:1, 75 °C, 24 h		[98]
3		PhNH$_2$ (2 mol/L), DMSO, 50 °C, 12 h		[23]
4		piperazine, NMP, 110 °C, 4 h		[99]
5		BuNH$_2$ (5 mol/L), DMSO, 20 °C, 24 h		[100]
6		piperazine, NMP, 110 °C, 10 h		[52]
7		morpholine, dioxane, Pd$_2$(dba)$_3$, P(*o*-Tol)$_3$, NaO*t*Bu, 80 °C, 20 h		[82]

Table 10.4. continued.

Entry	Starting resin	Conditions	Product	Ref.
8		2,6-dimethylaniline, PhMe, Pd$_2$(dba)$_3$, P(o-Tol)$_3$, NaOtBu, 100 °C, 20 h		[81]
9		(1.9 mol/L), NMP, 24 h		[101]
10		RNH$_2$ (0.25 mol/L), DMSO, 70 °C, 12 h		[102] see also [103]
11		3-bromoaniline, HCl, iPrOH/DMF 1:1, 20 °C, overnight Ar: 3-bromophenyl		[104]

or aryl bromide is currently limited to non-protic functional groups [81]. As by-products of these N-arylations, the dehalogenated arenes are often observed [81,82].

N-Arylations of amines have also been realized with support-bound heteroaromatic halides (Entries **9–11**, Table 10.4). Several examples of the synthesis of substituted 1,3,5-triazines [83–85], purines [78,85–93], and pyrimidines [77,85,94–96] have been reported. The reactivity of these arylating agents depends strongly on their precise substitution pattern, and generally increases with decreasing electron density of the heteroarene. Illustrative examples are given in Table 10.4. The arylation of amines with simultaneous cleavage of the product from the support is discussed in Section 3.8.

Experimental Procedure 10.1: Arylation of benzylamine with a support-bound aryl bromide [82]

To the 3-bromobenzyl urea (bound to Tentagel via a Rink amide linker, 0.15 g) in dioxane (2 mL) under nitrogen were added benzylamine (15 equiv.) and a solution of Pd$_2$(dba)$_3$ (0.2 equiv.), BINAP (0.8 equiv.), and sodium *tert*-butoxide (18 equiv.) in dioxane (1 mL). The mixture was stirred at 80 °C for 16–20 h, filtered, and the support was washed sequentially with water, DMF, THF, DCM, iPrOH, DCM, and AcOH. Treatment of the resin with TFA/water (9:1) (2 × 20 min) and concentration of the filtrates yielded 3-(benzylamino)benzylurea (92% yield; 89% pure by HPLC, 220 nm). The product was found to contain 3% benzylurea.

Resin-bound arenes with more than one leaving group can be suitable starting materials for sequential aromatic nucleophilic substitutions. The synthesis outlined in Figure 10.2 enables the sequential attachment of three different nucleophilic reagents (a thiol and two aliphatic amines) at a resin-bound benzamide. Because no acylations are used in this synthesis, all three reagents can contain additional nucleophilic functionalities, without any need for protective groups. This feature makes this methodology suitable for the parallel synthesis of highly diverse arrays of compounds from cheap starting materials. Other suitable polyelectrophiles that can be used for sequential aromatic nucleophilic substitutions on solid phase are polyhalo triazines [85], pyrimidines [77,85], purines [85,86,88,93], and related heterocycles.

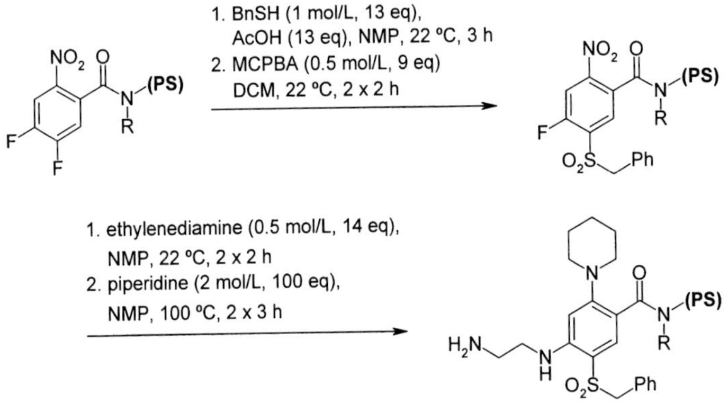

Figure 10.2. Sequential aromatic nucleophilic substitutions at a resin-bound polyelectrophilic arene [76].

10.1.2.2 By Arylation of Support-Bound Amines

Few examples have been reported of the arylation of support-bound amines. Unlike alkylations, the arylation of a primary amine usually leads to a strong reduction of the nucleophilicity of the nitrogen atom, and so twofold arylation of resin-bound primary amines cannot be expected. Illustrative examples are listed in Table 10.5.

Table 10.5. Preparation of aromatic amines by N-arylation of support-bound amines.

Entry	Starting resin	Conditions	Product	Ref.
1		(0.5 mol/L, 10 eq), DIPEA (5 eq), DMSO, 60 °C, 12 h		[105] see also [106]
2		(0.6 mol/L, 10 eq), NMP, DTBP, 60 °C, 11 h		[107]
3		2-fluoropyridine (neat), NBu₃, 110 °C, 30 h		[108]
4		2,4-dichloroquinazo- line (0.3 mol/L, 5 eq), NEt₃ (1.2 eq), DMF, 70 °C, 48 h		[109]
5		2-fluoro-6-chloropurine (0.17 mol/L, 5 eq), DIPEA (5 eq), THF, 60 °C, 16 h		[87] see also [88]
6		K₂CO₃, DMF, 25 °C, 30 h		[110] see also [111]

10.1.3 Preparation of Amines by Addition of Amines to C=C Double and C≡C Triple Bonds

Support-bound alkenes substituted with at least one electron-withdrawing group can react with primary or secondary amines (Table 10.6). If acrylates are used as Michael acceptors, the products (β-alanine derivatives) are generally stable and do not undergo β-elimination either upon N-acylation or upon treatment with TFA (for a longer synthetic sequence with a support-bound 2-(1-piperazinyl)propionate, see [42]). Suitable solvents for the addition of amines to electron-poor alkenes are THF, NMP, and DMSO. DMF is less suitable because it often contains dimethylamine, which also adds to Michael acceptors [112].

Acrylic acid esterified with cross-linked hydroxymethyl polystyrene or Wang resin reacts smoothly with primary or secondary aliphatic amines at room temperature (Entries **1** and **2**, Table 10.6). Only sterically demanding amines or amines of low nucleophilicity (anilines, α-amino acid esters) fail to add to polystyrene-bound acrylate. Support-bound acrylamides are less reactive than acrylic esters, and generally require heating to undergo addition with amines (Entries **4** and **5**, Table 10.6). α,β-Unsaturated esters with substituents in the 3-position (e.g. crotonates, Entry **3**, Table 10.6) react significantly more slowly with nucleophiles than do acrylates. The examples in Table 10.6 also show that polystyrene-bound esters are rather stable towards aminolysis, and provide for robust attachment even in the presence of high concentrations of amines. Entry **10** in Table 10.6 is an example of the alkylation of a resin-bound amine with an electron-poor alkene to yield a fluorinated peptide mimetic.

The addition of amines to electron-poor alkynes leads to the formation of enamines (Entry **11**, Table 10.6). These acceptor-substituted enamines are more stable towards hydrolysis than enamines derived from unsubstituted ketones. Alkynes devoid of electron-withdrawing groups do not react smoothly with amines, and usually require Hg(II) catalysis to be converted to resin-bound enamines (Entry **12**, Table 10.6).

Table 10.6. Addition of amines to alkenes and alkynes.

Entry	Starting resin	Conditions	Product	Ref.
1		tetrahydroisoquinoline (0.6 mol/L), DMF, 20 °C, 18 h		[64] see also [114,115]
2		Ph$_2$CHNH$_2$, (1.7 mol/L), DMSO, 20 °C, 48 h (no reaction with H-Phe-OMe or PhNH$_2$)	(55% conversion only)	[112]
3		EtNH$_2$ (0.9 mol/L), DMSO, 70 °C, 9 h		[116]
4		BnNH$_2$ (2.3 mol/L), DMSO, 50 °C, 24 h		[112]
5		(2.3 mol/L), DMSO, 50 °C, 72 h	(56% conversion only)	[112]
6		pyrrolidine (30 eq), DMF, 22 °C, 18 h		[117]

Table 10.6. continued.

Entry	Starting resin	Conditions	Product	Ref.
7		morpholine (10 eq), DMF, 20 °C, 16 h		[118] see also [119]
8		indoline (2 mol/L, 20 eq), DMF, 20 °C, overnight		[120]
9		4-chlorobenzylamine (0.7 mol/L, 10 eq), DBU (10 eq), NMP, 80 °C, 22 h		[121]
10		(3 eq), DCM, 20 °C, 3 d	3:1	[122]
11		iPrNH$_2$ (1.2 mol/L, 20 eq), 20 °C, 12 h		[123]
12		PhC≡CH (0.14 mol/L, 3.5 eq), Hg(OAc)$_2$ (3 eq), (iPr)$_2$NH (9 eq), THF/DCM 3:2, 20 °C, 72 h		[113]

10.1.4 Preparation of Amines by Reduction of Imines and Oximes

One of the most convenient preparations of alkylamines on insoluble supports is the reduction of imines, which can readily be prepared from either support-bound amines or support-bound carbonyl compounds (Figure 10.3).

Many aldehydes react spontaneously with primary aliphatic amines to yield the corresponding imines (Schiff bases). These are often sufficiently stable to enable filtration and washing of the support. Ketones are generally less readily converted into imines than are aldehydes, and the stoichiometric conversion of support-bound amines into ketone-derived imines is more conveniently achieved by treatment of the amine with an imine (Entry **10**, Table 10.7). The formation of imines can be promoted by acids [124–126], by dehydrating agents such as orthoformates [126–129], or by higher reaction temperatures. Highly electrophilic aldehydes (e.g. glyoxylic acid esters) are not always easy to convert into imines because dehydration of the intermediate hemiaminal does not proceed readily. One interesting strategy for the preparation of imines from resin-bound synthetic equivalents of glyoxylic acid is shown in Figure 10.4.

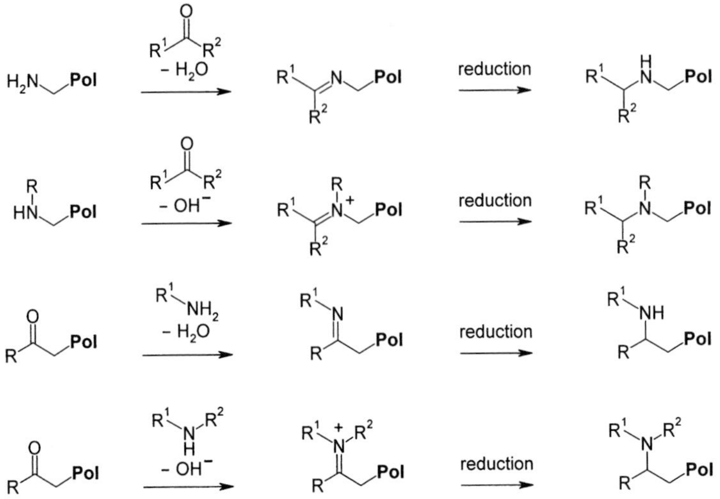

Figure 10.3. Reductive amination on solid phase.

Reduction of resin-bound imines can be performed with LiBH$_4$ [131] or NaBH$_4$ under basic reaction conditions, or with NaCNBH$_3$ in the presence of acetic acid. Other suitable reagents are given in Table 10.7. The two-step methodology (imine formation, washing, reduction) enables the clean monoalkylation of support-bound primary amines. Alternatively, amines can be alkylated on solid phase by one-pot imine formation and reduction. NaCNBH$_3$ is a particularly convenient reagent for this purpose because it does not usually reduce aldehydes, ketones, or (non-protonated) imines, but solely iminium salts. The main advantage of one-pot reductive aminations is that carbonyl compounds that do not readily form imines, e.g. ketones, can also be used. The most serious inconvenience of the one-pot method is that support-bound primary amines might be doubly alkylated.

Monoalkylation of support-bound amines can also be realized by a three-step protocol involving initial protection/activation of the amine (e.g. as 2-nitrobenzenesulfonamide [132,133], trifluoroacetamide [102], or 4,4′-dimethoxybenzhydrylamine [66]) followed by N-alkylation and deprotection. In particular, if monomethylations (see, e.g., Entry **7**, Table 10.7) or allylations are to be performed, this three-step strategy will generally give the best results.

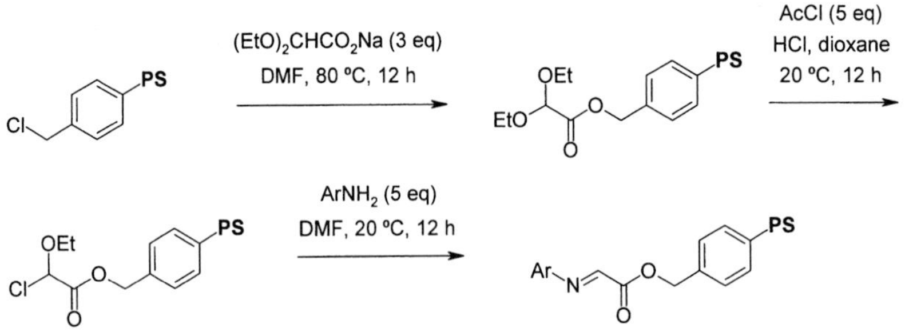

Figure 10.4. Preparation of glyoxylic acid derived imines [130].

Table 10.7. Reductive alkylation of support-bound amines.

Entry	Starting resin	Conditions	Product	Ref.
1		1. Me$_2$CHCHO (0.25 mol/L, 5 eq), 1% AcOH in DMF, 1.5 h 2. NaBH$_4$ (8 eq), DMF/MeOH 10:4, 2 h		[140] see also [131]
2		cyclohexanone (1.2 eq), THF/AcOH/ H$_2$O 90:5:5, NaCNBH$_3$, 23 °C, 3 h		[141] see also [68]
3		ArCHO (0.3 mol/L), 1% AcOH in DMA, NaCNBH$_3$ (1.7 mol/L), overnight		[75] see also [142]
4		C$_6$H$_{11}$CHO, NaBH(OAc)$_3$ (30 eq of each), DCM, 25 °C, 16 h		[143]
5		FmocHN—CHO (R^1) 0.5% AcOH in DMF, NaCNBH$_3$, 3 h		[144] see also [127]
6		Ph—CO$_2$Et (25 eq), 1% AcOH in DMF, NaCNBH$_3$ (40 eq)		[145]
7		HCHO, 1% AcOH in DMF, NaCNBH$_3$, 1 h Ar: 4-(MeO)C$_6$H$_4$		[66]
8		2-butanone, pyridine•BH$_3$ (10 eq of each), DMF/EtOH 3:1, 25 °C, 4 d (does not proceed with acetophenone)		[146] see also [147]
9		PhCHO (1.5 mol/L, 20 eq), DMF, AcOH (0.5 eq), NaBH(OAc)$_3$ (20 eq), 18 h		[115]
10		Ph—NH (Ph—CH(Me)NH) DCM, 20 °C, 16 h; then Ac$_2$O, DIPEA, DCM, 20 °C, 0.5 h; then NaCNBH$_3$, AcOH, DMA		[148] see also [149]

Illustrative examples of the reductive alkylation of support-bound amines are listed in Table 10.7. Further examples have been reported [134–138]. Closely related to the reduction of iminium salts is the reduction of *N*-alkylpyridinium salts, which has been successfully accomplished on Wang resin using NaBH$_4$ as the reducing agent (DMF/MeOH (1:1), 2 × 4 h; Entry **11**, Table 15.23 [139]).

Examples of the conversion of support-bound carbonyl compounds into amines are listed in Table 10.8. Additional examples have been reported [150,151]. The most important application of this reaction is the loading of backbone amide linkers (Section 3.3.1) with amines (Experimental Procedure 10.2). Because the reduction of imines with NaCNBH$_3$ or similar reducing agents generally requires slightly acidic reaction conditions, more acid than amine must always be used. Entry **7** in Table 10.8 is an interesting example of a preparation of MBHA resin, in which the initial product of reductive amination is a formamide (Leuckart reaction). This formamide can be hydrolyzed by treatment with aqueous hydrochloric acid, without cleavage of the benzylic C–N bond. As in the Leuckart reaction, reductive aminations with NaCNBH$_3$/AcOH under anhydrous conditions can lead to the formation of acetamides as side products [116]. This unwanted acetylation can be avoided by performing the reductive amination in the presence of small amounts of water (e.g. Entry **2**, Table 10.7).

Oximes generally require strong reducing agents to undergo complete conversion to primary amines (Entry **8**, Table 10.8); the use of weaker reducing reagents, such as borane, leads mainly to the formation of hydroxylamines (see Section 10.3).

Experimental Procedure 10.2: Racemization-free loading of amino acid esters on to a backbone amide linker [152]

The resin-bound aldehyde (0.50 g, approx. 0.3 mmol) was suspended in DMF (20 mL) containing 1% acetic acid, and NaBH(OAc)$_3$ (0.64 g, 3.00 mmol) was added. An α-amino acid methyl ester hydrochloride (3.00 mmol) was added to the suspension, and the mixture was stirred for 1 h. The resin was subsequently filtered off and washed sequentially with methanol, DMF, DCM, and further methanol, and dried under reduced pressure.

Table 10.8. Reductive amination of support-bound carbonyl compounds.

Entry	Starting resin	Conditions	Product	Ref.
1		Teoc–N, NH$_2$, DMTO–, pyridine•BH$_3$, 1% AcOH in MeOH/HC(OMe)$_3$	Teoc–N, DMTO–	[153] see also [154]
2	MeO	1. EtNH$_2$ (5 eq), THF, 18 h 2. filtration, washing 3. NaBH$_4$ (10 eq), THF/EtOH 3:1, 8 h	MeO	[155]
3		EtNH$_2$, Me$_4$NBH(OAc)$_3$, DCE, 16 h; then NaCNBH$_3$, MeOH, 6 h		[156]
4		BnNH$_2$ (15 eq), NaBH(OAc)$_3$ (5 eq), Na$_2$SO$_4$ (10 eq), 1% AcOH in DCM, ultrasound	Ph	[157] see also [158,159]
5		indoline (15 eq), NaBH(OAc)$_3$ (5 eq), Na$_2$SO$_4$ (10 eq), 1% AcOH in DCM, ultrasound		[157]
6		Ti(OiPr)$_4$ (0.6 mol/L), PhMe, BnNH$_2$ (0.4 mol/L, 2.5 eq), 2 h; then AcOH (10%), NaBH(OAc)$_3$ (2.4 mol/L, 10 eq), THF, 24 h	Ph	[160]
7		1. HCO$_2$NH$_4$, HCONH$_2$, HCO$_2$H/PhNO$_2$ 1:5, 165 °C, 22 h 2. aq HCl (12 mol/L)/EtOH 1:1, 78 °C, 1 h	NH$_2$	[161]
8	MeO, HO–N	LiAlH$_4$ (1 mol/L), THF, 20 °C, 24 h	MeO, H$_2$N	[162]

10.1.5 Preparation of Amines by Reaction of Carbon Nucleophiles with Imines or Aminals

Nucleophiles other than hydride can be added to support-bound imines to yield amines. These include C,H-acidic compounds, alkynes, electron-rich heterocycles, organometallic compounds, boronic acids, and ketene acetals (Table 10.9). When basic reaction conditions are used, stoichiometric amounts of the imine must be prepared on the support (Entries **1–3**, Table 10.9). Alternatively, if the carbon nucleophile is stable under acidic conditions, imines or iminium salts might be generated in situ, as, for instance, in the Mannich reaction. Few examples have been reported of Mannich reactions on insoluble supports, and most of these have been based on alkynes as C-nucleophiles.

Support-bound C-nucleophiles have also been successfully added to imines. Poly-styrene-bound thiol esters can be converted into ketene acetals by O-silylation, and then alkylated with imines in the presence of Lewis acids. Further examples include Mannich reactions of support-bound alkynes and indoles (Table 10.10). Some Man-nich-type products (e.g. 3-(aminomethyl)indoles, 2-(aminomethyl)phenols, β-amino ketones) are unstable and can decompose upon treatment with acids. 3-(Amino-

Table 10.9. Addition of C-nucleophiles to support-bound imines and hemiaminals.

Entry	Starting resin	Conditions	Product	Ref.
1	Tol–N (imine, PS resin, MeO, OMe)	BuLi, THF, −78 °C to 20 °C, 31 h	Tol–NH (PS resin, MeO, OMe)	[131]
2	Ph–N (allyl, O–PS, Ph Ph)	allyl–Li (10 eq), PhH, 20 °C, 1 h	Ph–N (O–PS, Ph Ph)	[163] see also [164]
3	N (pyridyl imine, PS, MeO, OMe)	Ph≡–MgBr, Et$_2$O, PhMe, 60 °C, 24 h	N (pyridyl, Ph, NH, PS, MeO, OMe)	[131]
4	Ph–N (O–PS)	OSiMe$_3$/OMe ketene acetal, Yb(OTf)$_3$, DCM/MeCN 1:1, 20 °C, 20 h	Ph–N (PS, CO$_2$Me)	[124] see also [130]
5	N=N benzotriazole, (PS), Ph	BrF$_2$CCO$_2$Et (0.12 mol/L, 10 eq), Zn (10 eq), Me$_3$SiCl (10 eq), THF, 67 °C, 2 h	F F (PS), EtO, Ph	[165]

Table 10.9. continued.

Entry	Starting resin	Conditions	Product	Ref.
6		MeO, PhCHO, CuCl, dioxane, 85 °C, 3 h		[166] see also [167]
7		Ph—≡ 1-phenylpiperazine, CuCl, dioxane, 75 °C		[166]
8		citric acid buffer (pH 4), 20 °C, 12 h		[168]
9		(2 eq), ArOK (1.3 eq), THF, 20 °C, overnight, then HCl, THF, 4 h		[169]
10		salicylaldehyde, 4-bromophenylboronic acid (each 1 mol/L, 26 eq), DMF/ DCE 2:3, 50 °C, 2 d		[170]
11		dibenzylamine (0.33 mol/L, 2.5 eq), 2-furylboronic acid (8 eq), DMF/DCE 2:3, 50 °C, 20 h		[170] see also [171]

methyl)indoles have, for instance, been used as acid-labile linkers for amines and amides (Tables 3.9 and 3.26). Hence, the choice of a suitable linker can be critical for Mannich reactions on insoluble supports.

Table 10.10. Addition of support-bound C-nucleophiles to imines.

Entry	Starting resin	Conditions	Product	Ref.
1		Ph-N-Ph / Sc(OTf)$_3$, DCM, 20 °C, 20 h		[172] see also [173]
2		Ph-N-C$_6$H$_{11}$ / Sc(OTf)$_3$, DCM, 20 °C, 20 h		[172]
3		1-naphthaldehyde, 1-cyclohexylpiperazine, CuCl, dioxane, 90 °C, 36 h R: cyclohexyl		[174] see also [175]
4		BnNH$_2$, HCHO, dioxane/AcOH 4:1, 23 °C, 1.5 h		[176] see also [177]

10.1.6 Preparation of Amines by Reduction of Amides and Carbamates

Polystyrene-bound amides, including peptides, can be reduced to the corresponding amines by treatment with borane in ethereal solvents. Other reagents, such as lithium aluminum hydride, are less convenient for reductions on insoluble supports, because insoluble precipitates can readily form and clog frits. Carbamates, *tert*-butyl ethers or thioethers, and trityl or benzhydryl amines remain unchanged upon treatment with borane, but carboxylic esters may undergo partial or complete reduction [178].

Carbamates can be reduced by lithium aluminum hydride (Entry **3**, Table 3.22 [179]) or by Red-Al (Entry **5**, Table 10.11). Illustrative examples of the reduction of amides and carbamates on solid phase are listed in Table 10.11.

The reduction of amides with borane leads to the formation of borane–amine adducts, which can be resistant towards acylating agents or hydrolysis. Such borane complexes can be cleaved either by treatment with a secondary amine (e.g. piperidine, 60 °C [180]), or oxidatively, by treatment with iodine (Entry **3**, Table 10.11 [181,182]).

Table 10.11. Reduction of support-bound amides and carbamates.

Entry	Starting resin	Conditions	Product	Ref.
1		BH₃ (1 mol/L), THF, 50 °C, 1 h; quenching with DBU (0.06 mol/L in NMP/MeOH 9:1)		[183] see also [184]
2		B(OH)₃, B(OMe)₃, BH₃ (1 mol/L), THF, 65 °C, 72 h		[185] see also [182,186, 187]
3		BH₃ (1 mol/L), THF, 65 °C, 24–72 h; then I₂ (3–5 eq, THF/DIPEA/AcOH 7:1:2, 20 °C, 4 h		[178]
4		BH₃•SMe₂ (0.26 mol/L), DME, 85 °C, 20 h		[124]
5		MeO—, MeO—AlH₂Na (70 eq), PhMe, 20 °C, overnight		[188]

10.1.7 Preparation of Amines by Reduction of Nitro Compounds

Anilines can readily be prepared on insoluble supports by reduction of nitroarenes (Table 10.12). The most commonly used reagent for this purpose is tin(II) chloride dihydrate, which is highly soluble in NMP or DMF and does not generally lead to insoluble precipitates in these solvents [189–193]. The reduction is usually performed with 1–5 M solutions of the reducing agent at 20–100 °C for 5–16 h. Solutions of tin(II) chloride can lead to the cleavage of some acid-labile linkers (e.g. trityl linkers or the Rink amide linker); this can be avoided by performing the reaction in the presence of a weak base such as NaOAc [194]. One additional disadvantage of the reduction of nitroarenes with tin(II) chloride is that the final products may contain small amounts of tin derivatives, which can be detrimental for some types of biological assay (e.g., for assays involving living cells [195]).

Other reagents that have been used to reduce support-bound aromatic nitro compounds include phenylhydrazine at high temperatures (Entry 5, Table 10.12), sodium borohydride in the presence of copper(II) acetylacetonate [100], chromium(II) chloride [196], Mn(0)/TMSCl/CrCl₂ [197], lithium aluminum hydride (Entry 3, Table

10.12), sodium dithionite (Entry **4**, Table 10.12), and indium [198]. The reduction of nitroarenes with tin(II) chloride proceeds via the intermediate formation of a hydroxylamine, which can be trapped inter- or intramolecularly with various electrophiles [199,200]. For instance, treatment of Wang resin bound benzonitriles with nitrobenzene and tin(II) chloride leads to resin-bound *N*-hydroxyamidines [116].

Table 10.12. Reduction of support-bound nitro compounds.

Entry	Starting resin	Conditions	Product	Ref.
1		SnCl$_2$·2 H$_2$O (1.6 mol/L), NMP, 20 °C, 16 h		[193]
2		SnCl$_2$·2 H$_2$O (8 eq), DMF, 25 °C, 16 h (does not proceed with Na$_2$S$_2$O$_4$ or FeSO$_4$·7 H$_2$O)		[143]
3		LiAlH$_4$, Et$_2$O, 35 °C, 48 h		[201]
4		Na$_2$S$_2$O$_4$, EtOH, 78 °C, 1.5 h		[202] see also [195]
5		PhNHNH$_2$ (neat), 200 °C, 1 h		[203]
6		LiAlH$_4$ (0.2 mol/L), AlCl$_3$ (0.2 mol/L), THF, 20 °C, overnight; (PS): Rink amide resin		[96]
7		SnCl$_2$·2 H$_2$O (2 mol/L, 10 eq), DMF, 20 °C, 4 h		[204]

10.1.8 Preparation of Amines by Reduction of Azides

The azido group is a further functional group that can serve as a precursor to an amine. The preparation of azides on solid phase is discussed in Section 10.4. Azides can be reduced to amines by several types of reagent, but few of these are compatible with insoluble supports. The most convenient reducing agents, which are also effective on cross-linked polystyrene, are phosphines in the presence of water, and thiols (Table 10.13). Phosphines react with azides to yield iminophosphoranes, which can be hydrolyzed to yield amines. The reduction can also be performed in the presence of an activated carboxylic acid derivative, whereby amides are formed directly. The latter strategy was used in Entry **5** in Table 10.13 because the intermediate amine would have undergone rapid intramolecular reaction with the ester linkage to yield a lactam. By performing the reduction in the presence of a HOBt ester, intramolecular nucleophilic cleavage was efficiently prevented. A further reducing agent for resin-bound azides is metallic indium [198]. α-Azido acids have been used for the solid-phase synthesis of peptides [205–207].

Table 10.13. Reduction of support-bound azides.

Entry	Starting resin	Conditions	Product	Ref.
1		dithiothreitol (2 mol/L), DIPEA (1 mol/L), DMF, 50 °C, 0.5 h		[208] see also [209]
2		SnCl$_2$ (0.2 mol/L), PhSH (0.8 mol/L), NEt$_3$ (1 mol/L), THF, 20 °C, 4 h		[3,4] see also [5,210, 211]
3		SnCl$_2$, PhSH, NEt$_3$, THF (acyl migration only in *cis* isomer)		[212]
4		HS⌒SH DIPEA 1:1, 24 h Ar: 9-anthracenyl		[213] see also [214]
5		PBu$_3$, EDC, PhCO$_2$H, HOBt, dioxane, 20 °C, 18 h	(no lactam formation)	[215] see also [216,217]
6		PPh$_3$ (10 eq), H$_2$O (30 eq), THF, 25 °C, 8 h		[218] see also [219– 221]

10.1.9 Miscellaneous Preparations of Amines

Amines have been prepared on insoluble supports by Hofmann degradation of amides [222] followed by hydrolysis of the intermediate isocyanates (Figure 10.5). One reagent suitable for this purpose is [bis(trifluoroacetoxy)iodo]benzene, which can be used both on cross-linked polystyrene [223] and on more hydrophilic supports such as polyacrylamides (Figure 10.6). Support-bound carboxylic acids can also be degraded via the acyl azides (Curtius degradation [224,225]) to yield isocyanates.

These can be converted into amines by direct hydrolysis, or into Fmoc-protected amines by reaction with 9-fluorenylmethanol [224].

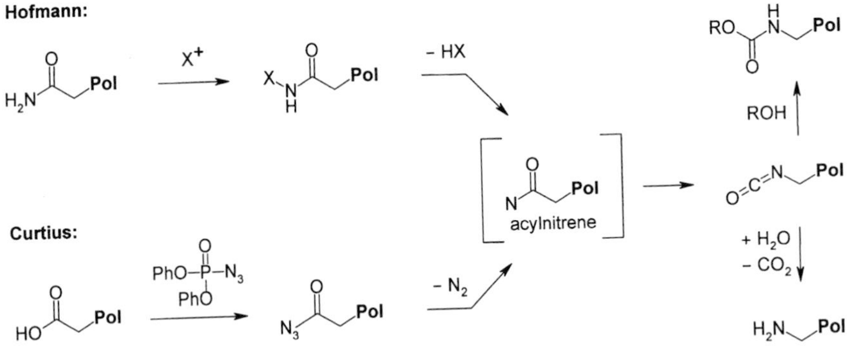

Hofmann:

Curtius:

Figure 10.5. Hofmann and Curtius degradation of support-bound amides and acids.

The degradation of support-bound α-amino acid amides has been used to prepare retro-inverso peptide mimetics ([226], second equation in Figure 10.6). These compounds are of interest because of their potentially improved metabolic stabilities, selectivities, and biological activities compared with peptides [226]. Although retro-inverso peptides are aminals susceptible to acid-catalyzed hydrolysis, N,N′-diacylated aminals can be sufficiently stable to withstand the conditions of Boc-group removal [227] or of acidolytic cleavage of peptides from Wang resin [226].

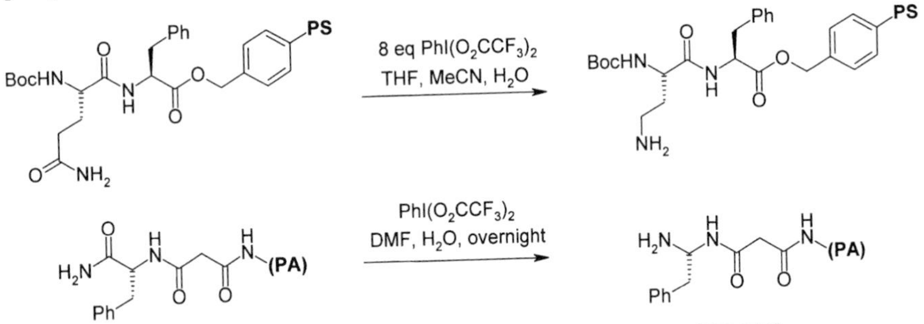

Figure 10.6. Conversion of amides into amines by oxidative degradation [222,228].

Substituted 2-aminonaphthalenes have been prepared on Wang resin by cyclocondensation of resin-bound 2-trifluoromethylphenyl acetate with arylacetonitriles (Figure 10.7). This reaction probably proceeds via an electrophilic o-quinone methide intermediate, formed by base-induced elimination of HF from the resin-bound ester [229].

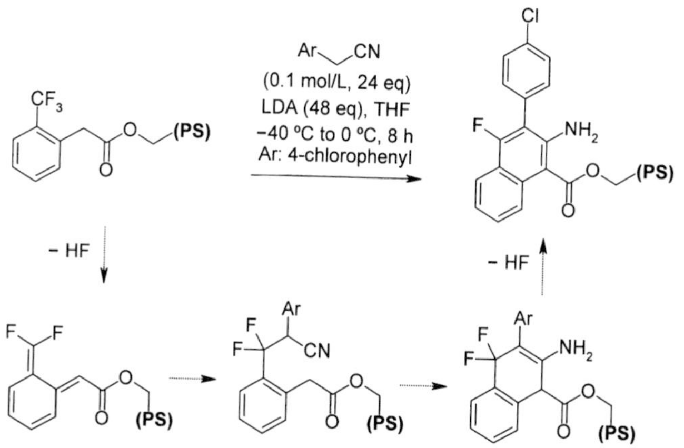

Figure 10.7. Formation of 2-aminonaphthalenes by cyclocondensation [229].

10.1.10 Protective Groups for Amines

Amines are important synthetic intermediates, and numerous protective groups have been developed for temporarily preventing amines from being acylated or alkylated. The following sections cover protective groups for amines that can be introduced or removed on insoluble supports. The development of such protective groups was mainly driven by solid-phase peptide synthesis. A more detailed collection of protective groups can be found in [230].

10.1.10.1 Carbamates

Carbamates are by far the most common type of amine protection used in solid-phase synthesis. Various types of carbamate have been developed that can be cleaved under mild reaction conditions on solid phase. Less well developed, however, are techniques that enable the protection of support-bound amines as carbamates. Protection of amino acids as carbamates (Boc or Fmoc) is usually performed in solution using aqueous base (Schotten–Baumann conditions). These conditions enable the selective protection of amines without simultaneous formation of imides or acylation of hydroxyl groups. Unfortunately, however, Schotten–Baumann conditions are not compatible with insoluble, hydrophobic supports. Other bases and solvents have to be used in order to prepare carbamates on, for example, cross-linked polystyrene, and more side reactions are generally observed than in aqueous solution.

Most carbamates used as protective groups for amines are either acid-labile or base-labile. Deprotection proceeds by the mechanisms outlined in Figure 10.8. During the deprotection of acid-labile carbamates, carbocations are formed, which can alkylate electron-rich structural elements in a given substrate (e.g. phenols, thiols, indoles, pyrroles) but not the deprotected (protonated) amine. On the other hand, some base-labile carbamates can lead to the formation of acceptor-substituted alkenes, which will alkylate the deprotected amine unless scavengers are added. As scavengers for such Michael acceptors, nucleophilic primary or secondary amines, such as pyrrolidine or piperidine, can be used.

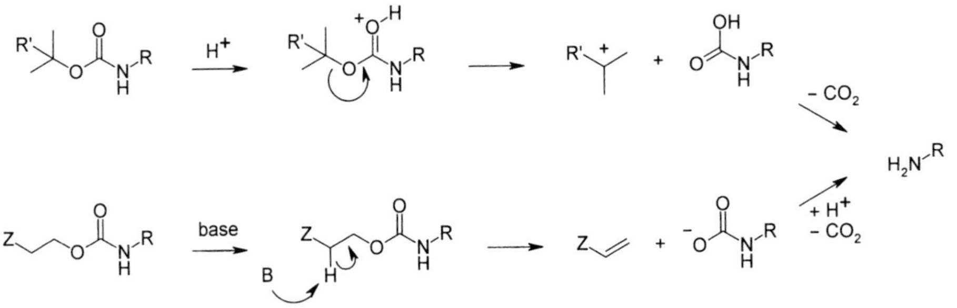

Figure 10.8. Mechanisms of deprotection of acid-labile and base-labile carbamates. R: alkyl, aryl; R': alkyl, aryl; Z: electron-withdrawing group; B: base.

tert-Butyl Carbamates (Boc Protection)

tert-Butyl carbamates (Boc amines) were developed as acid-labile protective groups for α-amino acids (see Section 16.1.2). Boc amines are stable towards bases, nucleophiles, weak acids, oxidants, and weak reducing agents, but can be selectively converted into amine trifluoroacetates by treatment with TFA/DCM (1:1).

The most commonly used reagent for converting amines into *tert*-butyl carbamates is di-*tert*-butyl dicarbonate ('Boc anhydride'). This reagent has almost completely replaced the more hazardous *tert*-butyloxycarbonyl azide, which had been the standard reagent for many years [231]. Few examples have been reported of the conversion of support-bound amines into *tert*-butyl carbamates (Table 10.14). Most substrates are secondary amines devoid of further acylable groups. It has been shown that polystyrene-bound anilides [232] or N-alkylamides in solution [233] can be N-alkoxycarbonylated with di-*tert*-butyl dicarbonate (NEt$_3$, DMAP, DCM, 6–72 h). These observations suggest that the selective Boc-protection of primary amines or amide-containing substrates on hydrophobic supports might require careful optimization.

tert-Butyl carbamates are stable towards bases and weak acids (half-life in 80% AcOH: 58 d [240]). Stronger acids, however, lead to dealkylation through the formation of a *tert*-butyl cation and decarboxylation of the intermediate carbamic acid. The removal of Boc groups from support-bound amines has been extensively studied. For the synthesis of peptides on cross-linked polystyrene using Boc methodology (Section 16.1.2), the standard deprotection protocol involves treatment of the resin with TFA/DCM (1:1) for 15–30 min at room temperature. Additional reagents useful for the acidolysis of Boc groups are 10% sulfuric acid in dioxane [241], aqueous hydrochloric acid (6 mol/L; only suitable for polyacrylamide-based supports [242]), hydrochloric acid (1.2 mol/L) in acetic acid [243], and hydrochloric acid (4–5 mol/L) in dioxane

Table 10.14. Conversion of support-bound amines into *tert*-butyl carbamates.

Entry	Starting resin	Conditions	Product	Ref.
1		Boc₂O, DIPEA		[234] see also [235,236]
2		Boc₂O, DIPEA, DCM, 20 °C, overnight		[30]
3		Boc₂O, THF, 25 °C, overnight		[237]
4		KF, NEt₃, DMF, 20 °C, 24 h		[238]
5		Boc₂O (4 eq), DMAP, THF, 20 °C, 16 h		[239]

[244] or DCM [245]. Mixtures of chlorotrimethylsilane (1 mol/L) and phenol (3 mol/L) in DCM (22 °C, 20 min) have been claimed to be superior to TFA/DCM [246]. Figure 10.9 depicts an interesting procedure, whereby Boc groups can be hydrolyzed from peptides bound to (TFA-labile) Rink amide resin [247]. The Wang linker is usually cleaved under the conditions of Boc-group removal, but it has been reported that the treatment of Wang resin bound, Boc-protected amines with sulfuric acid/dioxane (1:9) (10 °C, 2 h) leads to deprotection without significant cleavage of the linker [248].

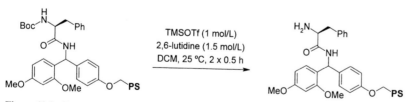

Figure 10.9. Boc group removal from Rink amide resin [247].

Other Acid-Labile Carbamates

Several other acid-labile carbamates have been developed, but most have found only limited application in solid-phase synthesis. A selection of these protective groups is listed in Table 10.15. Highly acid-labile carbamates, such as Bpoc, Mpc, or Ddz can be cleaved from substrates esterified with Wang resin without cleavage of the linker. Some of the protective groups listed in Table 10.15, e.g. Bpoc or Azoc, contain a chromophore, which enables photometric monitoring of the deprotection reaction. Such monitoring is important for automated solid-phase peptide synthesis, for quickly assessing the quality of a peptide (prior to cleavage from the support), and for localizing those positions in the peptide where peptide bond formation is difficult (see Section 16.1).

Benzyl carbamate protection (Cbz or Z group; see Table 10.15) was initially chosen by Merrifield for solid-phase peptide synthesis [255]. The strongly acidic conditions required for its solvolysis (30% HBr in AcOH, 25 °C, 5 h) demanded the use of an acid-resistant nitrobenzyl alcohol linker. Z-protection of the α-amino group in solid-phase peptide synthesis was, however, quickly abandoned and replaced by the more acid-labile Boc protection. Benzyl carbamates can be cleaved by strongly ionizing

Table 10.15. Selection of acid-labile carbamates.

Name	Structure	Cleavage conditions	Ref.
1-(4-biphenylyl)-1-methylethoxycarbonyl **Bpoc**		80% AcOH, 20 °C, 2.5 h; or 3% TFA in DCM, 5 min; or 0.5% TFA in DCM, 20 min; or Mg(ClO₄)₂ (10 eq), MeCN, 50 °C, 5 h	[240, 249–251]
1-(4-methylphenyl)-1-methylethoxycarbonyl **Mpc**		80% AcOH, 20 °C, 35 min; or 0.2% TFA in DCM, 1 min; or 5% Cl₂CHCO₂H in DCM, 5 min; or *B*-chlorocatecholborane (0.05 mol/L, 8 eq), DIPEA (4 eq), DCM, 2 min	[240, 249]
1-(3,5-dimethoxyphenyl)-1-methylethoxycarbonyl **Ddz**		AcOH/H₂O 8:2, 20 °C, 3 h; or 1% TFA in DCM, 20 °C, 1 h; or photolysis (280 nm) in THF	[252]
1-(1-adamantyl)-1-methylethoxycarbonyl **Adpoc**		3% TFA in DCM, 20 °C, 3 min	[253]
1-(4-phenylazophenyl)-1-methylethoxycarbonyl **Azoc**		1.5% TFA in DCM, 20 °C, 5 min; or AcOH/H₂O 8:2, 20 °C, 6 h	[254]
benzyloxycarbonyl **Cbz or Z**		30% HBr in AcOH, 25 °C, 5 h; or Me₃Sil (1.2 eq), CHCl₃, 25 °C, 10 min	[231, 255]

acids (HF, HBr, TfOH, HBF$_4$, etc.) or by hydrogenolysis, and today this group is only used in peptide synthesis for permanent side-chain protection of lysine. Deprotection is effected during cleavage of the peptide from the support or afterwards in solution. Secondary aliphatic amines bound to cross-linked polystyrene have been converted into benzyl carbamates by treatment with benzyl chloroformate (10 equiv.) and triethylamine (15 equiv.) in DCM (0 °C, 1 h, then 20 °C, 18 h) [256].

Fluorenylmethyl Carbamates (Fmoc Protection)

The 9-fluorenylmethoxycarbonyl group, developed by Carpino and co-workers in 1972 [257], has become one of the most widely used protective groups for aliphatic or aromatic amines in solid-phase synthesis. For solid-phase peptide synthesis in particular, this protective group plays an important role [258] (Section 16.1). The Fmoc group is not well suited for liquid-phase synthesis because non-volatile side products are formed during deprotection.

As in the case of Boc protection, the Fmoc group is not usually introduced on solid phase, but rather in solution, by the use of an activated Fmoc derivative (e.g. the chloroformate Fmoc-Cl or *O*-Fmoc-*N*-hydroxysuccinimide, Fmoc-OSu) and aqueous base (Experimental Procedure 10.3). *N*-Alkylamino acids bound to cross-linked polystyrene have been Fmoc-protected by treatment with Fmoc-Cl (4 equiv.) and DIPEA (6 equiv.) in DCM for 2 h [132,259]. Primary amines on insoluble supports can also be converted into Fmoc derivatives under these conditions [260].

The deprotection of Fmoc amines proceeds by base-induced β-elimination (Figure 10.10). The most commonly used reagent for this purpose is piperidine, which serves both as a base and as a scavenger of dibenzofulvene, which would otherwise react irreversibly with the deprotected amine.

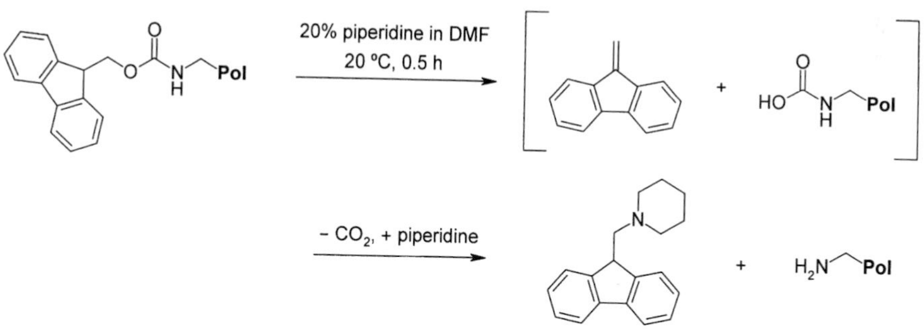

Figure 10.10. Mechanism of deprotection of Fmoc amines.

The stability of Fmoc amines towards various bases has been investigated in detail [258,261,262]. Ammonia and primary or secondary aliphatic amines in polar aprotic solvents lead to swift deprotection. Further reagents claimed to be useful for Fmoc deprotection are 2% DBU in DMF [263,264], 1% DBU + 1% HOBt in DMF (4 × 2 min; no cleavage of thiol esters occurs [265]), KF/NEt$_3$ in DMF [238], 40% Et$_2$NH in DCM (3 h [266]), and TBAF in DMF (20 °C, 2 min; this reagent also cleaves esters [267,268]). Tertiary amines do not deprotect Fmoc amines as quickly as secondary amines, and can therefore be used for Fmoc introduction. For instance, the half-life of Fmoc-Val-OH in DMF/DIPEA (1:1) is approximately 10 h [261]. No deprotection is generally observed upon treatment of Fmoc amines with pyridine, but in DMAP/ DMF (1:9), cleavage occurs with a half-life of about 1.5 h [261]. Deprotection of Fmoc amines has also been observed upon treatment with phosphines under the conditions of the Mitsunobu reaction [269].

Experimental Procedure 10.3: Fmoc protection of α-amino acids [270,271]

A solution of Fmoc-OSu (11.8 g, 35 mmol) in dioxane (100 mL) is added in one portion to an ice-cooled, stirred mixture of the α-amino acid (42 mmol), sodium carbonate (9.0 g, 85 mmol), and water (100 mL). The mixture is stirred at room temperature for 10 min, diluted with water (1.2 L), and washed with diethyl ether (80 mL) and ethyl acetate (2 × 100 mL). The aqueous layer is cooled and acidified to pH 2 with concentrated aqueous hydrochloric acid, whereupon a precipitate forms. The mixture is extracted with ethyl acetate (6 × 100 mL), and the combined extracts are washed with saturated aqueous sodium chloride solution (3 × 60 mL) and with water (2 × 60 mL), dried with sodium sulfate, and concentrated under reduced pressure. The Fmoc-amino acid precipitates upon addition of petroleum ether. This procedure is also suitable for the protection of proline (86% yield) and 4-hydroxyproline (78% yield).

Other Base-Labile Carbamates

Further carbamates susceptible to base-induced cleavage have been investigated, but none of these has been extensively used in solid-phase synthesis. 2,2-Bis(4-nitrophenyl)ethyl carbamates (Bnpeoc, Figure 10.11 [272]), (1,1-dioxobenzothiophen-2-yl)methyl carbamates (Bsmoc, Figure 10.11 [273]), and 2-(4-nitrophenyl)sulfonylethyl carbamates (Nsc [257,274]) show chemical stability similar to that of Fmoc amines. More stable towards bases than Fmoc amines are 2-(4-nitrophenyl)ethyl carbamates (Npeoc). This group has, for instance, been used in the solid-phase synthesis of peptidyl phosphonates [269,275]. The more base-labile Fmoc protection could not be used because deprotection occurred under the conditions of P–O bond formation (PAr₃, DIPEA, DIAD, THF) [269].

Alkyl carbamates are generally more stable towards nucleophiles than amides, and are therefore of limited utility as protective groups. Amines lacking other base-sensitive functionalities can, however, be protected as alkyl carbamates. An illustrative example of the use of ethyl carbamate as a protective group is sketched in Figure 10.11 [188].

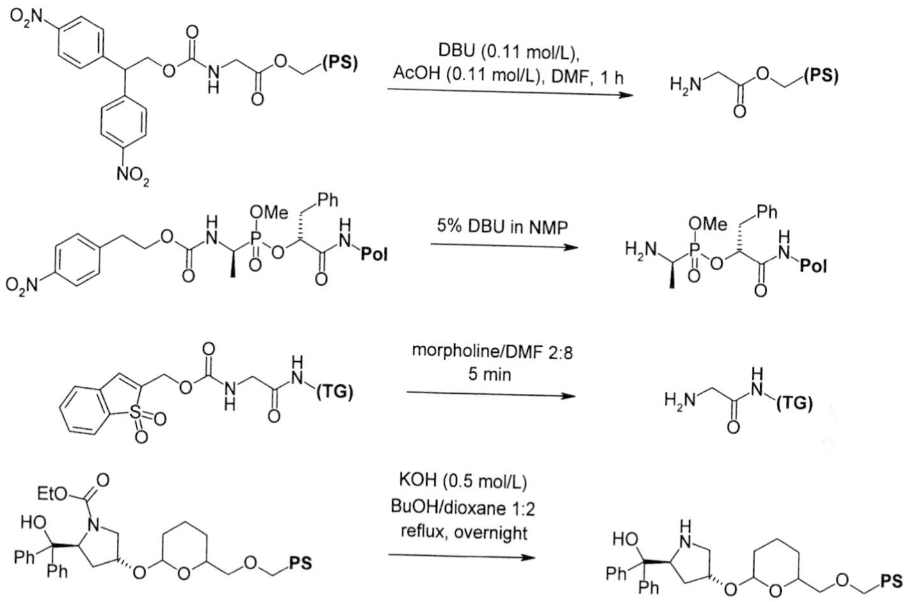

Figure 10.11. Base-induced cleavage of support-bound carbamates [188,269,272,273].

Miscellaneous Carbamates

Carbamate-based protective groups that can be cleaved by reagents other than acids or bases have also been developed. These supplement the protective groups presented above. Allyl carbamates (Alloc) have mainly been used for the side-chain protection of peptides (at lysine, arginine, or histidine) and can be selectively removed by soft nucleophiles in the presence of palladium(0) complexes [276–281]. Peptides have been prepared using Alloc N(α) protection [282,283]. Polystyrene-bound anilines have been converted into allyl carbamates by treatment with allyl chloroformate and NaHCO$_3$ in THF at 80 °C overnight [192]. Resin-bound α-amino acids and secondary aliphatic amines react cleanly with allyl chloroformate at room temperature in DCM in the presence of DIPEA; phenols remain unchanged under these conditions [279,284]. A typical deprotection, in which a stannane is used as a hydride source to reduce the intermediate allyl palladium complex, involves treatment of the allyl carbamate with a solution of PdCl$_2$(PPh$_3$)$_2$ (8 mmol/L), AcOH (0.5 mol/L), and Bu$_3$SnH (0.4 mol/L) in DCM for 10 min at room temperature [276]. Another example is shown in Figure 10.12.

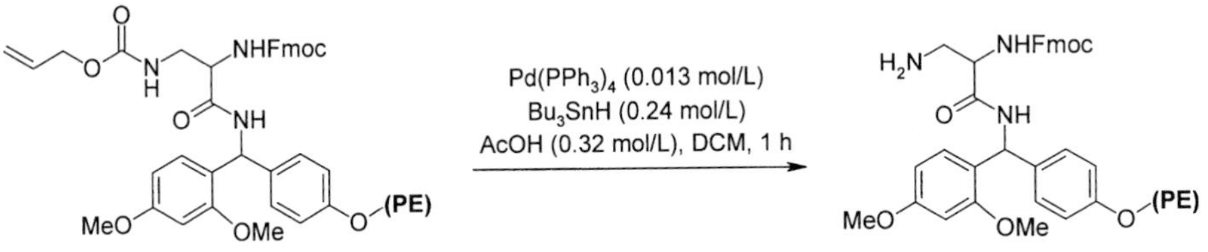

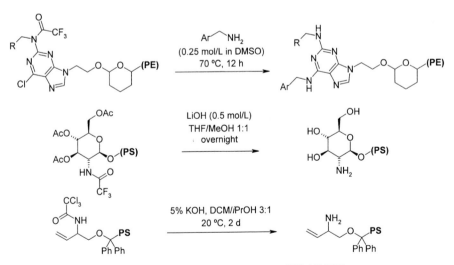

Figure 10.12. Deprotection of Alloc amines [285].

2-(Trimethylsilyl)ethyl carbamate (Teoc) can be selectively removed with fluoride (e.g. TBAF, THF, 50 °C, 5–20 h [147,286,287]), or by acidolysis with TFA [153,288]. 2-Nitrofluorenylmethyl carbamates can be cleaved photolytically or by treatment with bases [289].

10.1.10.2 Amides

Amides with electron-withdrawing substituents can be sufficiently labile towards nucleophilic attack to enable their use as protective groups. This is the case, for example, with trifluoro- [102,290] and trichloroacetamides [163], which are readily hydrolyzed under mild conditions (Figure 10.13). Suitable nucleophiles are hydrazine [291], aliphatic amines, and hydroxide, but if a hydrophobic support has been chosen, it must be borne in mind that the reactivity of alkali metal hydroxides will be reduced because of poor diffusion into the support. Amides of electron-poor amines (e.g. anilides) can also be readily cleaved by nucleophiles [292].

α-Amino acids can also be used to protect certain amines, because Edman degradation (treatment with an isothiocyanate followed by acid-catalyzed amide-bond clea-

Figure 10.13. Nucleophilic cleavage of support-bound amides [102,163,290].

vage and thiohydantoin formation) enables the selective removal of terminal amino acids. Sterically hindered *N*-alkylglycines are particularly useful, because these are not acylated by weak acylating agents (4-nitrophenyl esters, *N*-hydroxysuccinimidyl esters) and do not require further protection against acylation with such weak acylating agents [293]. One example of the use of small peptides as protective groups of amines, cleavable, not by a specific cleavage reagent but by a certain number of Edman degradations, is sketched in Figure 10.14.

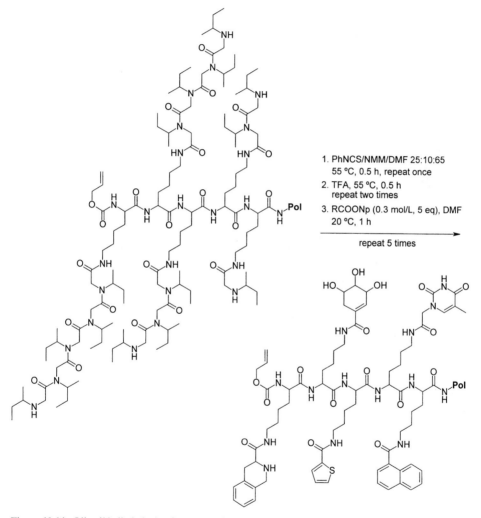

1. PhNCS/NMM/DMF 25:10:65
   55 °C, 0.5 h, repeat once
2. TFA, 55 °C, 0.5 h
   repeat two times
3. RCOONp (0.3 mol/L, 5 eq), DMF
   20 °C, 1 h

repeat 5 times

Figure 10.14. Oligo(*N*-alkyl glycines) as protective groups for amines [293].

10.1.10.3 Cyclic Imides

Phthalimide protection is stable towards acids and bases, but can be cleaved with strong nucleophiles, such as hydrazines or sulfides, or by reduction with sodium borohydride [230]. More sensitive towards nucleophilic attack than unsubstituted phthalimide is tetrachlorophthalimide [33]. This group has been successfully used as $N(\alpha)$ protection of amino acids in the solid-phase synthesis of peptides (deprotection: N_2H_4/DMF (15:85), 40 °C, 1 h; coupling: DIC/HOAt/amino acid (1:1:1), 3 equiv. of each, DMF, 25 °C, 4 h [294]). Typical conditions for the removal of phthaloyl protection on cross-linked polystyrene include treatment of the resin with hydrazine hydrate [295,296], with methyl hydrazine [297], or with primary aliphatic amines [298] in DMF, EtOH, or solvent mixtures for several hours at room temperature or above [296,299,300]. Illustrative examples are sketched in Figure 10.15. It has been claimed that the hydrazinolysis of polystyrene-bound phthalimides proceeds more readily in DCM or DCE than in DMF [301].

Polystyrene-bound amines can be converted into the corresponding phthalimides by heating with other phthalimides (BuOH, 85 °C, 18 h [33]). Support-bound phthalimides have also been prepared by N-alkylation of phthalimide derivatives [295] or by amidomethylation of cross-linked polystyrene with *N*-(hydroxymethyl) or *N*-(chloromethyl)phthalimide [296,299].

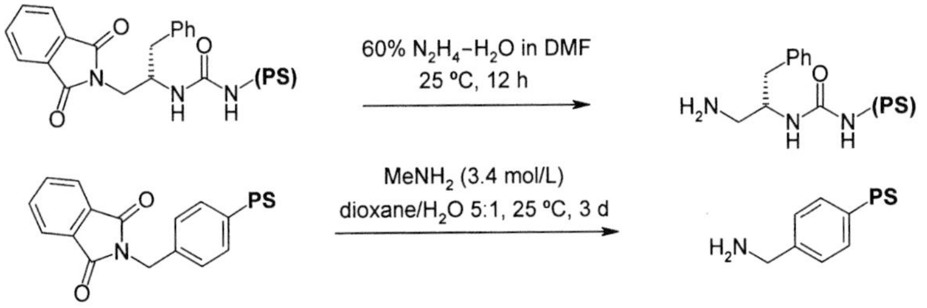

Figure 10.15. Examples of nucleophilic cleavage of phthaloyl groups [298,302].

Another cyclic imide that has been used as a protective group for primary amines is the dithiasuccinoyl group (Dts) [230,303,304]. This group is stable towards acids (e.g. during deprotection of Boc-protected amines), but can be cleaved with thiols under basic conditions (2-mercaptoethanol (0.2 mol/L), NEt₃ (0.5 mol/L), DCM, 25 °C, 5 min [303]).

10.1.10.4 Enamines

Enamines derived from simple ketones and aliphatic amines are too acid-labile and nucleophilic to be useful as protective group for amines. Triacylmethanes, however, form less basic enamines, which are sufficiently stable to be of use as amine protection.

Primary aliphatic and aromatic amines react reversibly with 2-acetyldimedone to yield non-nucleophilic enamines (1-[(4,4-dimethyl-2,6-dioxocyclohexylidene)ethyl] derivatives, 'Dde', Figure 10.16). Aliphatic amines react even at room temperature [266], whereas anilines generally require heating [305]. The resulting enamines are sufficiently stable towards TFA or piperidine to enable their use as efficient side-chain protection of lysine in solid-phase peptide synthesis [306,307]. The Dde group is also stable under the conditions of nucleophilic cleavage of 2,4-dinitrobenzenesulfon-amides (mercaptoacetic acid/DIPEA (1:1), 10 equiv. of each, DCM, 0.5 h [235]). A protective group similar to Dde, which is more stable under conditions of Fmoc group removal, is 2-(3-methylbutyryl)dimedone ('isovaleroyl' dimedone, ivDde [308]).

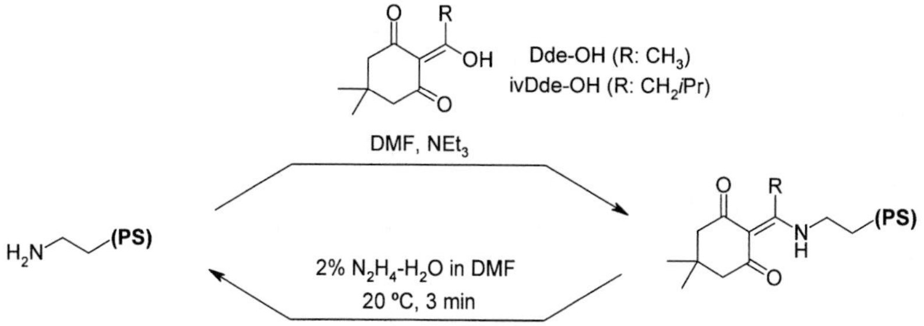

Figure 10.16. Protection of primary amines as enamines [234,306,309–311].

Secondary amines do not generally form enamines with 2-acetyldimedone, and this protective group can therefore be used for the selective protection of primary amines in the presence of secondary amines [234,309]. Support-bound 2-acetyldimedone has also been used as a linker for amines [312] (see Section 3.6).

10.1.10.5 Imines

Several types of imine have been used as protective groups for amines in solution [230]. Most are stable towards bases, but can be hydrolyzed by acids. Benzophenone-derived imines can be prepared by treating support-bound aliphatic primary amines with benzophenone imine [148,260], but usually not by treatment with benzophenone. Polystyrene-bound benzophenone imines of glycine are sufficiently C,H-acidic to enable C-alkylation with alkyl halides [260,313] or Michael acceptors [314], and have mainly been used for this purpose (see Section 13.4.4).

Deprotection of polystyrene-bound ketimines has been achieved by treatment with aqueous HCl or TFA in THF [260,313], or by benzophenone oxime formation with hydroxylamine hydrochloride [314,315] (Figure 10.17).

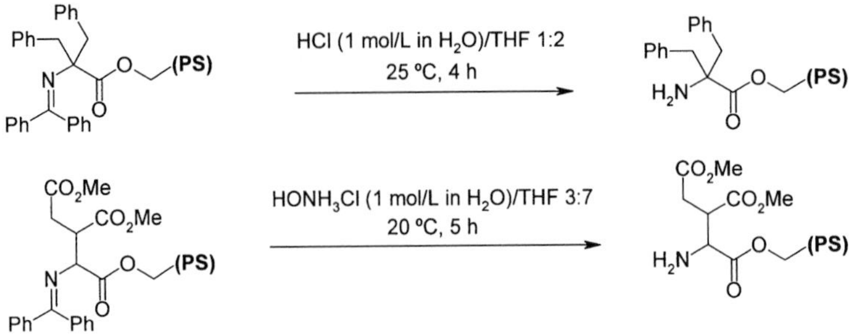

Figure 10.17. Deprotection of polystyrene-bound ketimines [313,314]. (PS): Wang resin.

10.1.10.6 *N*-Alkyl and *N*-Aryl Derivatives

Amines can also be protected from unwanted acylation or from twofold alkylation by alkylation with a sterically demanding group that is amenable to selective removal after completion of the synthesis. The most commonly used groups for this purpose are triphenylmethyl derivatives. Amines are readily tritylated using trityl chloride in the presence of a tertiary amine. Deprotection is achieved by treatment with dilute

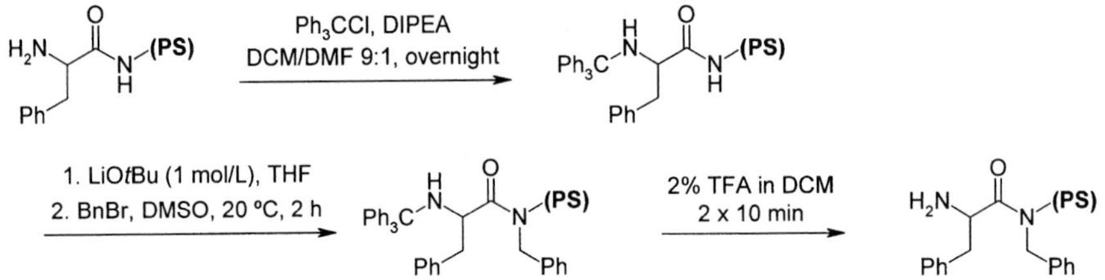

Figure 10.18. Protection of primary amines by tritylation [185,186]. (PS): MBHA polystyrene.

TFA [186,316] (Figure 10.18).

Barlos and co-workers investigated the synthesis of peptides on cross-linked polystyrene with TFA-labile linkers, using *N*-trityl amino acid HOBt esters as building blocks [317]. These authors found that *N*-trityl amino acids, esterified with Wang resin, could be selectively detritylated with 2% TFA in DCM (2 × 2 min), without significant cleavage of the Wang linker. More sensitive towards acidolytic cleavage than trityl amines are (4-methyltrityl)amines [310], which are cleavable in 2 min by 1% TFA in DCM, and (4-methoxytrityl)amines or (4,4′-dimethoxytrityl)amines [318], which can even be hydrolyzed with trichloroacetic acid (Figure 10.19 [319,320]). The 4,4′-dimethoxybenzhydryl group has been used to protect primary amines from twofold methylation during reductive alkylation with formaldehyde (Figure 10.19 [66]). Removal of this protective group requires 50% TFA and will therefore only be compatible with linkers that are more stable towards acids than the Wang linker.

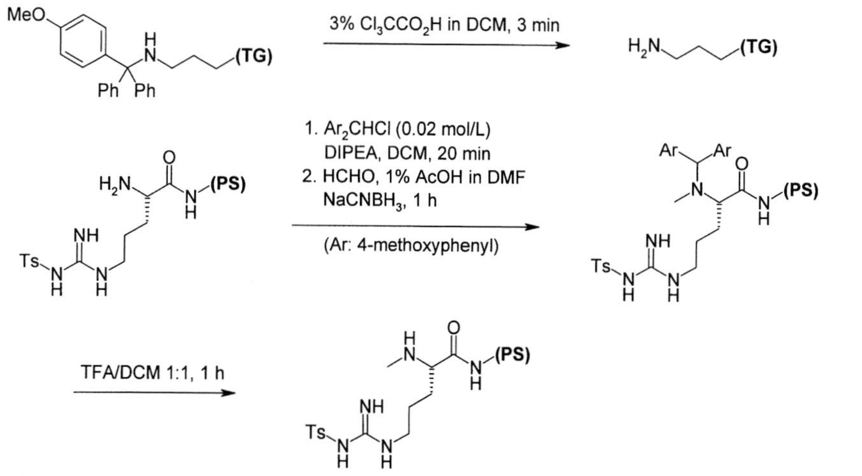

Figure 10.19. Methoxytrityl and 4,4'-dimethoxybenzhydryl protection of amines [66,319]. (PS): MBHA polystyrene.

10.1.10.7 *N*-Sulfenyl and *N*-Sulfonyl Derivatives

N-Sulfenylamines are prepared from sulfenyl chlorides ArSCl, and can be cleaved by acids, phosphines, or various nucleophiles [230,321]. The 2-nitrobenzenesulfenyl protective group (2-(O_2N)C_6H_4–S–NH–R), developed as amino group protection for α-amino acids [321,322], has been completely superseded by the Fmoc group.

Sulfonamides have found wider application as protective groups than sulfenamides (for the preparation of sulfonamides, see Section 8.4). Arenesulfonamides are generally very stable towards acidolytic or nucleophilic cleavage, unless substituted with electron-withdrawing groups. Thus, 2-nitrobenzene- and 2,4-dinitrobenzenesulfonamides derived from primary or secondary aliphatic amines can be selectively hydrolyzed under mild conditions with nucleophiles such as thiols or primary amines [323,324], even in the presence of carboxylic esters (Figure 10.20 [132,325–327]) or Dde-protected amines [235]. *N*-(2-Nitrobenzenesulfonyl)amino acids have been used

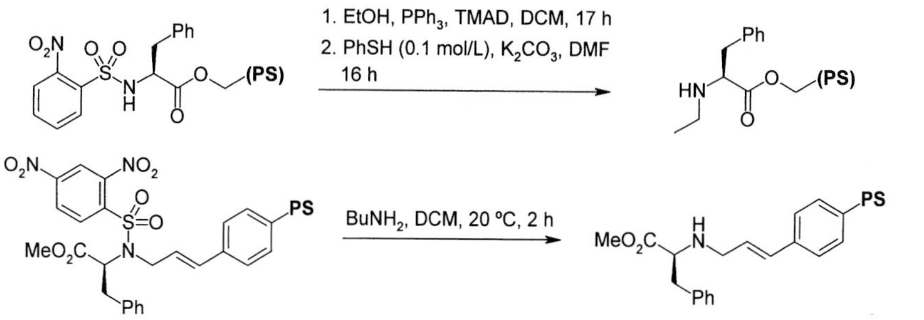

Figure 10.20. Nucleophilic cleavage of polystyrene-bound sulfonamides [133,323,334]. (PS): Wang resin.

to prepare *N*-alkylated peptides and other amino acid derivatives on insoluble supports [325,328–332]. 4-Nitrobenzenesulfonamides are not well suited as protective group for amines, because cleavage with thiols can lead to aromatic nucleophilic substitution of the nitro group and the formation of 4-thiobenzenesulfonamides, which are no longer easy to hydrolyze [333].

Sulfonamides prepared from 9-(chlorosulfonyl)anthracene and polystyrene-bound primary amines can be converted into amides by N-acylation of the sulfonamide (carboxylic acid anhydride, DMAP, pyridine, THF, 24 h) followed by nucleophilic desulfonylation with neat 1,3-propanedithiol/DIPEA [213] (Entry **4**, Table 10.13). An example of the use of sulfonamides as linkers for amines is given in Table 3.23.

10.2 Preparation of Quaternary Ammonium Salts

The quaternization of tertiary amines on cross-linked polystyrene has been investigated in detail. The most commonly used substrates in these studies have been *N,N*-dialkyl-β-alanine derivatives because, after their quaternization, pure tertiary amines can be released from the support by treatment with a base (see Section 3.7).

The formation of charged molecules within hydrophobic supports does not proceed as smoothly as in polar solvents. For instance, high reaction temperatures are required to quaternize polystyrene-bound phosphines [335,336], and the N-benzylation of pyridines with Merrifield resin also proceeds sluggishly [37]. Quaternization of tertiary

Table 10.16. Quaternization of support-bound tertiary amines.

Entry	Starting resin	Conditions	Product	Ref.
1		NEt$_3$ (0.7 mol/L), DMF, 23 °C, 20 h	< 12%	[338]
2		(0.29 mol/L), DMF, 20 °C, 18 h		[64] see also [118,119]
3		MeI (10 eq), DMF, 20 °C, 18 h		[64]
4		phenacyl bromide (2.4 mol/L), DMF, 20 °C, 18 h		[115]
5		BuI (2.6 mol/L, 20 eq), DMF, 58 °C, 60 h		[339]

amines on cross-linked polystyrene only proceeds well with reactive alkylating agents, such as methyl, allyl, propargyl, phenacyl, or benzyl halides or sulfonates (Table 10.16). Most reported examples were conducted at room temperature to avoid premature Hoffmann elimination and cleavage from the support [64]. Polystyrene-bound *O*-alkyl-*N,N*-dialkylhydroxylamines have been methylated at nitrogen by treatment with methyl triflate [236,337].

10.3 Preparation of Hydrazines and Hydroxylamines

Few examples of the preparation of hydrazines or hydroxylamines on insoluble supports have been reported (Table 10.17). Hydrazines have been prepared by the reduction of aromatic diazonium salts or *N*-nitroso amines (prepared from secondary amines by treatment with *tert*-butyl nitrite [340]), and by the N-amination of support-bound amines (Entry **3**, Table 10.17). The direct reduction of hydrazones with borane to yield hydrazines on solid phase has not been reported, and appears to be difficult because of the ease with which the N–N bond of hydrazines is cleaved by reducing agents [340].

The N-amination of amines does not proceed cleanly when using 3-aryloxaziridines as aminating reagents because benzaldehydes are formed as by-products and these undergo condensation with the support-bound amine to yield imines [341]. In the example reported, these were hydrolyzed by treatment with a hydrazine and the resulting amines were again treated with the oxaziridine to yield, after three repetitions, a sufficiently pure, Boc-protected hydrazine (Entry **3**, Table 10.17). Hydrazines have also been prepared by addition of organolithium compounds to polystyrene-bound hydrazones (Entry **4**, Table 10.17), and by the N-alkylation of free or protected hydrazines with resin-bound haloacetamides (Entry **5**, Table 10.17 [342]).

Support-bound hydroxylamines have been prepared by the reaction of hydroxylamines with support-bound alkylating agents (Entries **6** and **7**, Table 10.17). Derivatives of hydroxylamine protected at nitrogen (e.g. *N*-hydroxyphthalimide [29,343,344] or *N*-Fmoc hydroxylamine [65,345]) have been O-alkylated with Wang resin, with polystyrene-bound benzhydryl chloride [345], or with trityl chloride resins to yield supports suitable for the preparation of hydroxamic acids (Section 3.4). The deprotection of support-bound *N*-phthaloyl hydroxylamines can be accomplished under conditions similar to those used for the hydrolysis of phthalimides, e.g. by treatment with hydrazine, methylhydrazine [346], or methylamine. Polystyrene-bound hydroxylamines have also been prepared by the addition of alkyl radicals to support-bound *O*-alkyloximes (Entry **8**, Table 10.17), both inter- and intramolecularly (Entry **6**, Table 15.4). This reaction can proceed with high diastereoselectivity when using chiral auxiliaries [347]. The addition of alkyl radicals to sulfonyl oximes [ArSO$_2$–C(=NO–Pol)–CO$_2$Me] does not yield hydroxylamines but oximes [348]. Further methods for preparing hydroxylamines include the reduction of oximes with borane (Entries **9** and **10**, Table 10.17; this reduction does not proceed smoothly with sodium cyanoborohydride), the reductive alkylation of resin-bound *O*-alkylhydroxylamines (Entry **11**, Table 10.17), and the addition of Grignard reagents to resin-bound *N*-alkoxyaminals (Entry **12**, Table 10.17).

Table 10.17. Solid-phase synthesis of hydrazines and hydroxylamines.

Entry	Starting resin	Conditions	Product	Ref.
1		SnCl$_2$, conc HCl, 20 °C, 2 h, then 60 °C, 1 h		[349]
2		DIBAH (0.3 mol/L,10 eq), THF, DCM, 50 °C, 5 h		[340]
3		(1 eq), DCM, 20 °C, 0.5 h; 3 repetitions		[341, 350]
4		*t*BuLi (5 eq), THF, −78 °C to −20 °C		[340]
5		DIPEA, DMSO		[351, 352]
6		HONH$_3$Cl (0.23 mol/L, 10 eq), DIPEA (10 eq), DMF, 25 °C, 3 h		[353]
7		BnONH$_2$, 2,6-lutidine, DCM, 0 °C		[354]
8		*i*PrI (7.1 eq), Bu$_3$SnH (2.1 eq), Et$_3$B (1.1 eq), DCM, 20 °C, 1 h		[355] see also [347]
9		pyridine•BH$_3$ (10 eq), THF, 20 °C, 10 h		[356]
10		pyridine•BH$_3$ (15 eq), Cl$_2$HCCO$_2$H (22 eq), DCM, 20 °C, 18 h		[357] see also [358]
11		PhCHO (10 eq), NaBH(OAc)$_3$ (5 eq), THF, 20 °C, 16 h		[236]
12		4-FC$_6$H$_4$MgBr (10 eq), THF, 20 °C, 16 h		[337]

10.4 Preparation of Azides

Azides are convenient intermediates for the solid-phase preparation of primary amines (Section 10.1.8). Their preparation and handling on insoluble supports is significantly less hazardous than the handling of isolated organic azides. The most common strategy for preparing azides on insoluble supports is the nucleophilic substitution of halides or sulfonates (Table 10.18). Further suitable electrophiles are support-bound oxiranes, which can undergo ring-opening upon treatment with sodium azide to yield azido alcohols. This reaction can be catalyzed by enantiomerically pure chromium(III) complexes, and enantiomerically enriched azido alcohols have been prepared using this strategy (Entry **4**, Table 10.18). Resin-bound primary aliphatic alcohols can be directly converted into azides by treatment with DPPA, DEAD, and a phosphine (Entry **5**, Table 10.18).

Table 10.18. Solid-phase synthesis of azides.

Entry	Starting resin	Conditions	Product	Ref.
1		NaN₃ (1 mol/L), DMSO, 60 °C, overnight		[359] see also [211,360]
2		NaN₃ (1 mol/L), DMF, 45 °C, 24 h Ar: 4-(MeO)C₆H₄		[3] see also [221]
3		NaN₃ (0.2 mol/L), NH₄Cl (0.2 mol/L), DMF, 100 °C, 2 h		[361]
4		Me₃SiN₃ (20 eq), [(salen*)CrN₃] (0.2 eq), Et₂O, 24 h; salen*H₂: Ar:3,5-(*t*Bu)₂-2-(OH)C₆H₂	94% ee	[212]
5		PPh₃, DEAD, THF, (PhO)₂PON₃, overnight		[362] see also [218]

10.5 Preparation of Diazo Compounds

Diazocarbonyl compounds can be prepared on insoluble supports by diazo group transfer with sulfonyl azides or by diazotization of primary amines. Diazo group transfer from sulfonyl azides to 1,3-dicarbonyl compounds proceeds on cross-linked polystyrene as smoothly as in solution (Table 10.19). When 3-keto esters or amides are used as substrates for diazo group transfer, the resulting diazo ketones can readily be deacylated with secondary amines to yield diazoacetic esters or amides (Entry 3, Table 10.19). The direct diazotization of glycine is a less convenient method for the preparation of support-bound diazoacetic acid derivatives because this reaction generally gives good yields only when conducted in a biphasic solvent mixture (dilute aqueous HCl or H_2SO_4/DCM [363]). Aryldiazomethanes have been prepared on cross-linked polystyrene by heating tosylhydrazones with aqueous base or by oxidation of hydrazones (Table 10.19). These diazo compounds react quantitatively with carboxylic acids under mild conditions to yield the corresponding benzylic esters [364–366], and with alcohols in the presence of $BF_3 \cdot OEt_2$ to yield benzyl ethers [367].

Diazocarbonyl compounds can also be prepared by C-acylation of diazoalkanes with polystyrene-bound acyl halides (Entry 6, Table 10.19). As an alternative to diazomethane, the more stable α-(trimethylsilyl)diazomethane may be used, which is sufficiently nucleophilic to react with acyl halides. On heating, the resulting α-(trimethylsilyl)diazo ketones undergo Wolff rearrangement to yield ketenes, and have also been used as starting materials for the preparation of oxazoles [368].

Table 10.19. Solid-phase synthesis of diazo compounds.

Entry	Starting resin	Conditions	Product	Ref.
1		TsN$_3$ (0.5 mol/L), DIPEA (1.4 mol/L), DMF, 20 °C, 1 h		[369] see also [370–372]
2		NaNO$_2$ (4 eq), HCl, DCM, H$_2$O, 25 °C, 4 h		[373]
3		pyrrolidine (3 mol/L), DMF, 20 °C, 2 h		[369]
4		KOH (2 eq), MeOH/ THF 1:4, 90 °C, 7 min Ar: 2,4,6-(iPr)$_3$C$_6$H$_2$		[364]
5		AcOH, AcO$_2$H, TMG, I$_2$, DCM, 0 °C		[365,367]
6		Me$_3$SiCHN$_2$ (0.016 mol/L, 3 eq), THF/MeCN 1:1, 20 °C, 50 h		[368]

10.6 Preparation of Nitro Compounds

Cross-linked polystyrene can be directly nitrated with fuming nitric acid at low temperatures (–25 °C to 0 °C) [189,203,374]; polymers with up to one nitro group per arene result [203]. Partial nitration can be achieved with milder nitrating agents, such as acetyl nitrate [203]. Because direct nitrations are not compatible with most linkers (which are often acid- or oxidant-labile), nitro compounds are generally not prepared on supports but in solution, and are then linked to the support.

References for Chapter 10

[1]   Darling, G. D.; Fréchet, J. M. J. *J. Org. Chem.* **1986**, *51*, 2270–2276.
[2]   Karoyan, P.; Triolo, A.; Nannicini, R.; Giannotti, D.; Altamura, M.; Chassaing, G.; Perrotta, E. *Tetrahedron Lett.* **1999**, *40*, 71–74.
[3]   Lee, C. E.; Kick, E. K.; Ellman, J. A. *J. Am. Chem. Soc.* **1998**, *120*, 9735–9747.
[4]   Kick, E. K.; Ellman, J. A. *J. Med. Chem.* **1995**, *38*, 1427–1430.
[5]   Zhou, J.; Termin, A.; Wayland, M.; Tarby, C. M. *Tetrahedron Lett.* **1999**, *40*, 2729–2732.
[6]   Barn, D. R.; Morphy, J. R.; Rees, D. C. *Tetrahedron Lett.* **1996**, *37*, 3213–3216.
[7]   Virgilio, A. A.; Schürer, S. C.; Ellman, J. A. *Tetrahedron Lett.* **1996**, *37*, 6961–6964.
[8]   Kurokawa, K.; Kumihara, H.; Kondo, H. *Bioorg. Med. Chem. Lett.* **2000**, *10*, 1827–1830.
[9]   Renault, J.; Lebranchu, M.; Lecat, A.; Uriac, P. *Tetrahedron Lett.* **2001**, *42*, 6655–6658.
[10] Rölfing, K.; Thiel, M.; Künzer, H. *Synlett* **1996**, 1036–1038.
[11] Vojkovsky, T.; Weichsel, A.; Pátek, M. *J. Org. Chem.* **1998**, *63*, 3162–3163.
[12] Bhandari, A.; Jones, D. G.; Schullek, J. R.; Vo, K.; Schunk, C. A.; Tamanaha, L. L.; Chen, D.; Yuan, Z. Y.; Needels, M. C.; Gallop, M. A. *Bioorg. Med. Chem. Lett.* **1998**, *8*, 2303–2308.
[13] Vidal-Ferran, A.; Bampos, N.; Moyano, A.; Pericàs, M. A.; Riera, A.; Sanders, J. K. M. *J. Org. Chem.* **1998**, *63*, 6309–6318.
[14] Moberg, C.; Rákos, L. *Reactive Polymers* **1991**, *15*, 25–35.
[15] Peschke, B.; Bundgaard, J. G.; Breinholt, J. *Tetrahedron Lett.* **2001**, *42*, 5127–5130.
[16] Bryan, W. M.; Huffman, W. F.; Bhatnagar, P. K. *Tetrahedron Lett.* **2000**, *41*, 6997–7000.
[17] Wendeborn, S.; De Mesmaeker, A.; Brill, W. K. D. *Synlett* **1998**, 865–868.
[18] Brill, W. K. D.; De Mesmaeker, A.; Wendeborn, S. *Synlett* **1998**, 1085–1090.
[19] Fréchet, J. M. J.; Eichler, E. *Polym. Bull.* **1982**, *7*, 345–351.
[20] Berteina, S.; De Mesmaeker, A.; Wendeborn, S. *Synlett* **1999**, 1121–1123.
[21] Rossé, G.; Ouertani, F.; Schröder, H. *J. Comb. Chem.* **1999**, *1*, 397–401.
[22] Pei, Y. H.; Moos, W. H. *Tetrahedron Lett.* **1994**, *35*, 5825–5828.
[23] Tumelty, D.; Schwarz, M. K.; Needels, M. C. *Tetrahedron Lett.* **1998**, *39*, 7467–7470.
[24] Richter, L. S.; Zuckermann, R. N. *Bioorg. Med. Chem. Lett.* **1995**, *5*, 1159–1162.
[25] Goff, D. A.; Zuckermann, R. N. *J. Org. Chem.* **1995**, *60*, 5744–5745.
[26] Rano, T. A.; Chapman, K. T. *Tetrahedron Lett.* **1995**, *36*, 3789–3792.
[27] Goff, D. A.; Zuckermann, R. N. *J. Org. Chem.* **1995**, *60*, 5748–5749.
[28] Scott, B. O.; Siegmund, A. C.; Marlowe, C. K.; Pei, Y.; Spear, K. L. *Mol. Diversity* **1995**, *1*, 125–134.
[29] Floyd, C. D.; Lewis, C. N.; Patel, S. R.; Whittaker, M. *Tetrahedron Lett.* **1996**, *37*, 8045–8048.
[30] Byk, G.; Frederic, M.; Scherman, D. *Tetrahedron Lett.* **1997**, *38*, 3219–3222.
[31] Virgilio, A. A.; Bray, A. M.; Zhang, W.; Trinh, L.; Snyder, M.; Morrissey, M. M.; Ellman, J. A. *Tetrahedron* **1997**, *53*, 6635–6644.
[32] Révész, L.; Bonne, F.; Manning, U.; Zuber, J. F. *Bioorg. Med. Chem. Lett.* **1998**, *8*, 405–408.
[33] Stangier, P.; Hindsgaul, O. *Synlett* **1996**, 179–181.
[34] Hird, N. W.; Irie, K.; Nagai, K. *Tetrahedron Lett.* **1997**, *38*, 7111–7114.
[35] Conti, P.; Demont, D.; Cals, J.; Ottenheijm, H. C. J.; Leysen, D. *Tetrahedron Lett.* **1997**, *38*, 2915–2918.
[36] Adrian, F. M.; Altava, B.; Burguete, M. I.; Luis, S. V.; Salvador, R. V.; García-España, E. *Tetrahedron* **1998**, *54*, 3581–3588.

[37] Obika, S.; Nishiyama, T.; Tatematsu, S.; Nishimoto, M.; Miyashita, K.; Imanishi, T. *Heterocycles* **1998**, *49*, 261–267.
[38] Ngu, K.; Patel, D. V. *Tetrahedron Lett.* **1997**, *38*, 973–976.
[39] Rich, D. H.; Gurwara, S. K. *J. Am. Chem. Soc.* **1975**, *97*, 1575–1579.
[40] Sampson, D. F. J.; Simmonds, R. G.; Bradley, M. *Tetrahedron Lett.* **2001**, *42*, 5517–5519.
[41] Zuckermann, R. N.; Kerr, J. M.; Kent, S. B. H.; Moos, W. H. *J. Am. Chem. Soc.* **1992**, *114*, 10646–10647.
[42] Kiselyov, A. S.; Eisenberg, S.; Luo, Y. *Tetrahedron* **1998**, *54*, 10635–10640.
[43] Yuasa, H.; Kamata, Y.; Kurono, S.; Hashimoto, H. *Bioorg. Med. Chem. Lett.* **1998**, *8*, 2139–2144.
[44] Brown, D. S.; Revill, J. M.; Shute, R. E. *Tetrahedron Lett.* **1998**, *39*, 8533–8536.
[45] Bräse, S.; Enders, D.; Köbberling, J.; Avemaria, F. *Angew. Chem. Int. Ed.* **1998**, *37*, 3413–3415.
[46] Berteina, S.; Wendeborn, S.; De Mesmaeker, A. *Synlett* **1998**, 1231–1233.
[47] Marx, M. A.; Grillot, A. L.; Louer, C. T.; Beaver, K. A.; Bartlett, P. A. *J. Am. Chem. Soc.* **1997**, *119*, 6153–6167.
[48] Dankwardt, S. M.; Newman, S. R.; Krstenansky, J. L. *Tetrahedron Lett.* **1995**, *36*, 4923–4926.
[49] Bhalay, G.; Blaney, P.; Palmer, V. H.; Baxter, A. D. *Tetrahedron Lett.* **1997**, *38*, 8375–8378.
[50] Newlander, K. A.; Chenera, B.; Veber, D. F.; Yim, N. C. F.; Moore, M. L. *J. Org. Chem.* **1997**, *62*, 6726–6732.
[51] Raju, B.; Kogan, T. P. *Tetrahedron Lett.* **1997**, *38*, 4965–4968.
[52] Yamamoto, Y.; Ajito, K.; Ohtsuka, Y. *Chem. Lett.* **1998**, 379–380.
[53] Brill, W. K. D.; Schmidt, E.; Tommasi, R. A. *Synlett* **1998**, 906–908.
[54] Tommasi, R. A.; Nantermet, P. G.; Shapiro, M. J.; Chin, J.; Brill, W. K. D.; Ang, K. *Tetrahedron Lett.* **1998**, *39*, 5477–5480.
[55] Nicolaou, K. C.; Cao, G.-Q.; Pfefferkorn, J. A. *Angew. Chem. Int. Ed.* **2000**, *39*, 739–743.
[56] Zhang, H. C.; Maryanoff, B. E. *J. Org. Chem.* **1997**, *62*, 1804–1809.
[57] Goff, D. *Tetrahedron Lett.* **1998**, *39*, 1477–1480.
[58] van Maarseveen, J. H.; den Hartog, J. A. J.; Engelen, V.; Finner, E.; Visser, G.; Kruse, C. G. *Tetrahedron Lett.* **1996**, *37*, 8249–8252.
[59] Richter, H.; Walk, T.; Höltzel, A.; Jung, G. *J. Org. Chem.* **1999**, *64*, 1362–1365.
[60] Lee, A.; Huang, L.; Ellman, J. A. *J. Am. Chem. Soc.* **1999**, *121*, 9907–9914.
[61] Chai, W. Y.; Murray, W. V. *Tetrahedron Lett.* **1999**, *40*, 7185–7188.
[62] Zaragoza, F.; Stephensen, H. *J. Org. Chem.* **2001**, *66*, 2518–2521.
[63] Wang, Y.; Huang, T. N. *Tetrahedron Lett.* **1999**, *40*, 5837–5840.
[64] Brown, A. R.; Rees, D. C.; Rankovic, Z.; Morphy, J. R. *J. Am. Chem. Soc.* **1997**, *119*, 3288–3295.
[65] Mellor, S. L.; McGuire, C.; Chan, W. C. *Tetrahedron Lett.* **1997**, *38*, 3311–3314.
[66] Kaljuste, K.; Undén, A. *Int. J. Pept. Prot. Res.* **1993**, *42*, 118–124.
[67] Connoly, P. J.; Beers, K. N.; Wetter, S. K.; Murray, W. V. *Tetrahedron Lett.* **2000**, *41*, 5187–5191.
[68] Purandare, A. V.; Poss, M. A. *Tetrahedron Lett.* **1998**, *39*, 935–938.
[69] Chambers, S. L.; Ronald, R.; Hanesworth, J. M.; Kinder, D. H.; Harding, J. W. *Peptides* **1997**, *18*, 505–512.
[70] Alewood, P. F.; Brinkworth, R. I.; Dancer, R. J.; Garnham, B.; Jones, A.; Kent, S. B. H. *Tetrahedron Lett.* **1992**, *33*, 977–980.
[71] Manov, N.; Bienz, S. *Tetrahedron* **2001**, *57*, 7893–7898.
[72] Shreder, K.; Zhang, L.; Gleeson, J. P.; Ericsson, J. A.; Yalamoori, V. V.; Goodman, M. *J. Comb. Chem.* **1999**, *1*, 383–387.
[73] Zaragoza, F.; Stephensen, H. *Tetrahedron Lett.* **2000**, *41*, 1841–1844.
[74] Flegelová, Z.; Pátek, M. P. *J. Org. Chem.* **1996**, *61*, 6735–6738.
[75] Green, J. *J. Org. Chem.* **1995**, *60*, 4287–4290.
[76] Grimstrup, M.; Zaragoza, F. *Eur. J. Org. Chem.* **2001**, 3233–3246.
[77] Guillier, F.; Roussel, P.; Moser, H.; Kane, P.; Bradley, M. *Chem. Eur. J.* **1999**, *5*, 3450–3458.
[78] Ding, S.; Gray, N. S.; Ding, Q.; Schultz, P. G. *J. Org. Chem.* **2001**, *66*, 8273–8276.
[79] Scicinski, J. J.; Congreve, M. S.; Jamieson, C.; Ley, S. V.; Newman, E. S.; Vinader, V. M.; Carr, R. A. E. *J. Comb. Chem.* **2001**, *3*, 387–396.
[80] Pan, P. C.; Sun, C. M. *Bioorg. Med. Chem. Lett.* **1999**, *9*, 1537–1540.
[81] Ward, Y. D.; Farina, V. *Tetrahedron Lett.* **1996**, *37*, 6993–6996.
[82] Willoughby, C. A.; Chapman, K. T. *Tetrahedron Lett.* **1996**, *37*, 7181–7184.
[83] Scharn, D.; Wenschuh, H.; Reineke, U.; Schneider-Mergener, J.; Germeroth, L. *J. Comb. Chem.* **2000**, *2*, 361–369.
[84] Deleuze, H.; Sherrington, D. C. *J. Chem. Soc., Perkin Trans. 2* **1995**, 2217–2221.
[85] Scharn, D.; Germeroth, L.; Schneider-Mergener, J.; Wenschuh, H. *J. Org. Chem.* **2001**, *66*, 507–513.

[86] Brill, W. K.-D.; Riva-Toniolo, C. *Tetrahedron Lett.* **2001**, *42*, 6515–6518.
[87] Dorff, P. H.; Garigipati, R. S. *Tetrahedron Lett.* **2001**, *42*, 2771–2773.
[88] Brill, W. K.-D.; Riva-Toniolo, C.; Müller, S. *Synlett* **2001**, 1097–1100.
[89] Gray, N. S.; Kwon, S.; Schultz, P. G. *Tetrahedron Lett.* **1997**, *38*, 1161–1164.
[90] Brun, V.; Legraverend, M.; Grierson, D. S. *Tetrahedron Lett.* **2001**, *42*, 8169–8171.
[91] Brun, V.; Legraverend, M.; Grierson, D. S. *Tetrahedron Lett.* **2001**, *42*, 8165–8167.
[92] Brun, V.; Legraverend, M.; Grierson, D. S. *Tetrahedron Lett.* **2001**, *42*, 8161–8164.
[93] Kim, K.; Wang, B. *Chem. Commun.* **2001**, 2268–2269.
[94] Barillari, C.; Barlocco, D.; Raveglia, L. F. *Eur. J. Org. Chem.* **2001**, 4737–4741.
[95] Baxter, A. D.; Boyd, E. A.; Cox, P. B.; Loh, V.; Monteils, C.; Proud, A. *Tetrahedron Lett.* **2000**, *41*, 8177–8181.
[96] Di Lucrezia, R.; Gilbert, I. H.; Floyd, C. D. *J. Comb. Chem.* **2000**, *2*, 249–253.
[97] Shapiro, M. J.; Kumaravel, G.; Petter, R. C.; Beveridge, R. *Tetrahedron Lett.* **1996**, *37*, 4671–4674.
[98] Schwarz, M. K.; Tumelty, D.; Gallop, M. A. *Tetrahedron Lett.* **1998**, *39*, 8397–8400.
[99] MacDonald, A. A.; DeWitt, S. H.; Hogan, E. M.; Ramage, R. *Tetrahedron Lett.* **1996**, *37*, 4815–4818.
[100] Phillips, G. B.; Wei, G. P. *Tetrahedron Lett.* **1996**, *37*, 4887–4890.
[101] Mohan, R.; Yun, W. Y.; Buckman, B. O.; Liang, A.; Trinh, L.; Morrissey, M. M. *Bioorg. Med. Chem. Lett.* **1998**, *8*, 1877–1882.
[102] Norman, T. C.; Gray, N. S.; Koh, J. T.; Schultz, P. G. *J. Am. Chem. Soc.* **1996**, *118*, 7430–7431.
[103] Crimmins, M. T.; Zuercher, W. J. *Org. Lett.* **2000**, *2*, 1065–1067.
[104] Cobb, J. M.; Fiorini, M. T.; Goddard, C. R.; Theoclitou, M. E.; Abell, C. *Tetrahedron Lett.* **1999**, *40*, 1045–1048.
[105] Tumelty, D.; Schwarz, M. K.; Cao, K.; Needels, M. C. *Tetrahedron Lett.* **1999**, *40*, 6185–6188.
[106] Krchnák, V.; Smith, J.; Vágner, J. *Tetrahedron Lett.* **2001**, *42*, 2443–2446.
[107] Gordeev, M. F.; Luehr, G. W.; Hui, H. C.; Gordon, E. M.; Patel, D. V. *Tetrahedron* **1998**, *54*, 15879–15890.
[108] Brummond, K. M.; Lu, J. *J. Org. Chem.* **1999**, *64*, 1723–1726.
[109] Wu, Z.; Kim, J.; Soll, R. M.; Dhanoa, D. S. *Biotechnol. Bioeng. (Comb. Chem.)* **2000**, *71*, 87–90.
[110] Feng, Y.; Wang, Z.; Jin, S.; Burgess, K. *J. Am. Chem. Soc.* **1998**, *120*, 10768–10769.
[111] Fotsch, C.; Kumaravel, G.; Sharma, S. K.; Wu, A. D.; Gounarides, J. S.; Nirmala, N. R.; Petter, R. C. *Bioorg. Med. Chem. Lett.* **1999**, *9*, 2125–2130.
[112] Hamper, B. C.; Kolodziej, S. A.; Scates, A. M.; Smith, R. G.; Cortez, E. *J. Org. Chem.* **1998**, *63*, 708–718.
[113] Aznar, F.; Valdés, C.; Cabal, M.-P. *Tetrahedron Lett.* **2000**, *41*, 5683–5687.
[114] Kolodziej, S. A.; Hamper, B. C. *Tetrahedron Lett.* **1996**, *37*, 5277–5280.
[115] Ouyang, X. H.; Armstrong, R. W.; Murphy, M. M. *J. Org. Chem.* **1998**, *63*, 1027–1032.
[116] Zaragoza, F. unpublished results.
[117] Prien, O.; Rölfing, K.; Thiel, M.; Künzer, H. *Synlett* **1997**, 325–326.
[118] Kroll, F. E. K.; Morphy, R.; Rees, D.; Gani, D. *Tetrahedron Lett.* **1997**, *38*, 8573–8576.
[119] Heinonen, P.; Lönnberg, H. *Tetrahedron Lett.* **1997**, *38*, 8569–8572.
[120] Garibay, P.; Nielsen, J.; Høeg-Jensen, T. *Tetrahedron Lett.* **1998**, *39*, 2207–2210.
[121] Zaragoza, F.; Stephensen, H. *Angew. Chem. Int. Ed.* **2000**, *39*, 554–556.
[122] Volonterio, A.; Bravo, P.; Moussier, N.; Zanda, M. *Tetrahedron Lett.* **2000**, *41*, 6517–6521.
[123] Paulvannan, K.; Chen, T. *J. Org. Chem.* **2000**, *65*, 6160–6166.
[124] Kobayashi, S.; Aoki, Y. *Tetrahedron Lett.* **1998**, *39*, 7345–7348.
[125] Boyd, E. A.; Chan, W. C.; Loh, V. M. *Tetrahedron Lett.* **1996**, *37*, 1647–1650.
[126] Gordeev, M. F.; Gordon, E. M.; Patel, D. V. *J. Org. Chem.* **1997**, *62*, 8177–8181.
[127] Szardenings, A. K.; Burkoth, T. S.; Look, G. C.; Campbell, D. A. *J. Org. Chem.* **1996**, *61*, 6720–6722.
[128] Look, G. C.; Murphy, M. M.; Campbell, D. A.; Gallop, M. A. *Tetrahedron Lett.* **1995**, *36*, 2937–2940.
[129] Ruhland, B.; Bhandari, A.; Gordon, E. M.; Gallop, M. A. *J. Am. Chem. Soc.* **1996**, *118*, 253–254.
[130] Kobayashi, S.; Akiyama, R.; Kitagawa, H. *J. Comb. Chem.* **2000**, *2*, 438–440.
[131] Katritzky, A. R.; Xie, L.; Zhang, G.; Griffith, M.; Watson, K.; Kiely, J. S. *Tetrahedron Lett.* **1997**, *38*, 7011–7014.
[132] Yang, L.; Chiu, K. *Tetrahedron Lett.* **1997**, *38*, 7307–7310.
[133] Scicinski, J. J.; Barker, R. D.; Murray, P. J.; Jarvie, E. M. *Bioorg. Med. Chem. Lett.* **1998**, *8*, 3609–3614.
[134] Tourwé, D.; Piron, J.; Defreyn, P.; Van Binst, G. *Tetrahedron Lett.* **1993**, *34*, 5499–5502.

[135] Szardenings, A. K.; Burkoth, T. S.; Lu, H. H.; Tien, D. W.; Campbell, D. A. *Tetrahedron* **1997**, *53*, 6573–6593.
[136] Kim, S. W.; Ahn, S. Y.; Koh, J. S.; Lee, J. H.; Ro, S.; Cho, H. Y. *Tetrahedron Lett.* **1997**, *38*, 4603–4606.
[137] Matthews, J.; Rivero, R. A. *J. Org. Chem.* **1997**, *62*, 6090–6092.
[138] Matthews, J.; Rivero, R. A. *J. Org. Chem.* **1998**, *63*, 4808–4810.
[139] Eda, M.; Kurth, M. J. *Tetrahedron Lett.* **2001**, *42*, 2063–2068.
[140] Chan, W. C.; Mellor, S. L. *J. Chem. Soc., Chem. Commun.* **1995**, 1475–1477.
[141] Brown, E. G.; Nuss, J. M. *Tetrahedron Lett.* **1997**, *38*, 8457–8460.
[142] Devraj, R.; Cushman, M. *J. Org. Chem.* **1996**, *61*, 9368–9373.
[143] Arumugam, V.; Routledge, A.; Abell, C.; Balasubramanian, S. *Tetrahedron Lett.* **1997**, *38*, 6473–6476.
[144] Meyer, J. P.; Davis, P.; Lee, K. B.; Porreca, F.; Yamamura, H. I.; Hruby, V. J. *J. Med. Chem.* **1995**, *38*, 3462–3468.
[145] Blackburn, C.; Pingali, A.; Kehoe, T.; Herman, L. W.; Wang, H. Q.; Kates, S. A. *Bioorg. Med. Chem. Lett.* **1997**, *7*, 823–826.
[146] Nawaz, M. K.; Arumugam, V.; Balasubramanian, S. *Tetrahedron Lett.* **1996**, *37*, 4819–4822.
[147] Koh, J. S.; Ellman, J. A. *J. Org. Chem.* **1996**, *61*, 4494–4495.
[148] Lee, S. H.; Chung, S. H.; Lee, Y. S. *Tetrahedron Lett.* **1998**, *39*, 9469–9472.
[149] Steele, J.; Gordon, D. W. *Bioorg. Med. Chem. Lett.* **1995**, *5*, 47–50.
[150] Fivush, A. M.; Willson, T. M. *Tetrahedron Lett.* **1997**, *38*, 7151–7154.
[151] Hone, N. D.; Davies, S. G.; Devereux, N. J.; Taylor, S. L.; Baxter, A. D. *Tetrahedron Lett.* **1998**, *39*, 897–900.
[152] Boojamra, C. G.; Burow, K. M.; Thompson, L. A.; Ellman, J. A. *J. Org. Chem.* **1997**, *62*, 1240–1256.
[153] Swayze, E. E. *Tetrahedron Lett.* **1997**, *38*, 8643–8646.
[154] Swayze, E. E. *Tetrahedron Lett.* **1997**, *38*, 8465–8468.
[155] Sarantakis, D.; Bicksler, J. J. *Tetrahedron Lett.* **1997**, *38*, 7325–7328.
[156] Estep, K. G.; Neipp, C. E.; Stramiello, L. M. S.; Adam, M. D.; Allen, M. P.; Robinson, S.; Roskamp, E. J. *J. Org. Chem.* **1998**, *63*, 5300–5301.
[157] Ley, S. V.; Mynett, D. M.; Koot, W. J. *Synlett* **1995**, 1017–1020.
[158] Bray, A. M.; Chiefari, D. S.; Valerio, R. M.; Maeji, N. J. *Tetrahedron Lett.* **1995**, *36*, 5081–5084.
[159] Bui, C. T.; Bray, A. M.; Ercole, F.; Pham, Y.; Rasoul, F. A.; Maeji, N. J. *Tetrahedron Lett.* **1999**, *40*, 3471–3474.
[160] Breitenbucher, J. G.; Hui, H. C. *Tetrahedron Lett.* **1998**, *39*, 8207–8210.
[161] Matsueda, G. R.; Stewart, J. M. *Peptides* **1981**, *2*, 45–50.
[162] Larsen, S. D.; DiPaolo, B. A. *Org. Lett.* **2001**, *3*, 3341–3344.
[163] Schuster, M.; Pernerstorfer, J.; Blechert, S. *Angew. Chem. Int. Ed. Engl.* **1996**, *35*, 1979–1980.
[164] Chenera, B.; Finkelstein, J. A.; Veber, D. F. *J. Am. Chem. Soc.* **1995**, *117*, 11999–12000.
[165] Vidal, A.; Nefzi, A.; Houghten, R. A. *J. Org. Chem.* **2001**, *66*, 8268–8272.
[166] McNally, J. J.; Youngman, M. A.; Dax, S. L. *Tetrahedron Lett.* **1998**, *39*, 967–970.
[167] Youngman, M. A.; Dax, S. L. *J. Comb. Chem.* **2001**, *3*, 469–472.
[168] Jönsson, D.; Molin, H.; Undén, A. *Tetrahedron Lett.* **1998**, *39*, 1059–1062.
[169] O'Donnell, M. J.; Delgado, F.; Drew, M. D.; Pottorf, R. S.; Zhou, C. Y.; Scott, W. L. *Tetrahedron Lett.* **1999**, *40*, 5831–5835.
[170] Schlienger, N.; Bryce, M. R.; Hansen, T. K. *Tetrahedron* **2000**, *56*, 10023–10030.
[171] Klopfenstein, S. R.; Chen, J. J.; Golebiowski, A.; Li, M.; Peng, S. X.; Shao, X. *Tetrahedron Lett.* **2000**, *41*, 4835–4839.
[172] Kobayashi, S.; Hachiya, I.; Suzuki, S.; Moriwaki, M. *Tetrahedron Lett.* **1996**, *37*, 2809–2812.
[173] Kobayashi, S.; Moriwaki, M. *Tetrahedron Lett.* **1997**, *38*, 4251–4254.
[174] Dyatkin, A. B.; Rivero, R. A. *Tetrahedron Lett.* **1998**, *39*, 3647–3650.
[175] Youngman, M. A.; Dax, S. L. *Tetrahedron Lett.* **1997**, *38*, 6347–6350.
[176] Zhang, H. C.; Brumfield, K. K.; Jaroskova, L.; Maryanoff, B. E. *Tetrahedron Lett.* **1998**, *39*, 4449–4452.
[177] Kraxner, J.; Arlt, M.; Gmeiner, P. *Synlett* **2000**, 125–127.
[178] Manku, S.; Laplante, C.; Kopac, D.; Chan, T.; Hall, D. G. *J. Org. Chem.* **2001**, *66*, 874–885.
[179] Ho, C. Y.; Kukla, M. J. *Tetrahedron Lett.* **1997**, *38*, 2799–2802.
[180] Nefzi, A.; Giulianotti, M. A.; Ong, N. A.; Houghten, R. A. *Org. Lett.* **2000**, *2*, 3349–3350.
[181] Wang, F.; Manku, S.; Hall, D. G. *Org. Lett.* **2000**, *2*, 1581–1583.
[182] Hall, D. G.; Laplante, C.; Manku, S.; Nagendran, J. *J. Org. Chem.* **1999**, *64*, 698–699.
[183] Paikoff, S. J.; Wilson, T. E.; Cho, C. Y.; Schultz, P. G. *Tetrahedron Lett.* **1996**, *37*, 5653–5656.

[184] Karigiannis, G.; Mamos, P.; Balayiannis, G.; Katsoulis, I.; Papaioannou, D. *Tetrahedron Lett.* **1998**, *39*, 5117–5120.

[185] Nefzi, A.; Ostresh, J. M.; Houghten, R. A. *Tetrahedron* **1999**, *55*, 335–344.

[186] Nefzi, A.; Ostresh, J. M.; Meyer, J. P.; Houghten, R. A. *Tetrahedron Lett.* **1997**, *38*, 931–934.

[187] Ostresh, J. M.; Schoner, C. C.; Hamashin, V. T.; Nefzi, A.; Meyer, J. P.; Houghten, R. A. *J. Org. Chem.* **1998**, *63*, 8622–8623.

[188] Liu, G. C.; Ellman, J. A. *J. Org. Chem.* **1995**, *60*, 7712–7713.

[189] Dowling, L. M.; Stark, G. R. *Biochemistry* **1969**, *8*, 4728–4734.

[190] Meyers, H. V.; Dilley, G. J.; Durgin, T. L.; Powers, T. S.; Winssinger, N. A.; Zhu, H.; Pavia, M. R. *Mol. Diversity* **1995**, *1*, 13–20.

[191] Kiselyov, A. S.; Armstrong, R. W. *Tetrahedron Lett.* **1997**, *38*, 6163–6166.

[192] Morales, G. A.; Corbett, J. W.; DeGrado, W. F. *J. Org. Chem.* **1998**, *63*, 1172–1177.

[193] Zaragoza, F.; Stephensen, H. *J. Org. Chem.* **1999**, *64*, 2555–2557.

[194] Lee, C. L.; Chan, K. P.; Lam, Y.; Lee, S. Y. *Tetrahedron Lett.* **2001**, *42*, 1167–1169.

[195] Scheuerman, R. A.; Tumelty, D. *Tetrahedron Lett.* **2000**, *41*, 6531–6535.

[196] Hari, A.; Miller, B. L. *Tetrahedron Lett.* **1999**, *40*, 245–248.

[197] Hari, A.; Miller, B. L. *Angew. Chem. Int. Ed.* **1999**, *38*, 2777–2779.

[198] Kamal, A.; Reddy, G. S. K.; Reddy, K. L. *Tetrahedron Lett.* **2001**, *42*, 6969–6971.

[199] Stephensen, H.; Zaragoza, F. *Tetrahedron Lett.* **1999**, *40*, 5799–5802.

[200] Ruhland, T.; Künzer, H. *Tetrahedron Lett.* **1996**, *37*, 2757–2760.

[201] Beebe, X.; Chiappari, C. L.; Olmstead, M. M.; Kurth, M. J.; Schore, N. E. *J. Org. Chem.* **1995**, *60*, 4204–4212.

[202] Hughes, I. *Tetrahedron Lett.* **1996**, *37*, 7595–7598.

[203] Seliger, H. *Makromol. Chem.* **1973**, *169*, 83–93.

[204] Kuster, G. J.; Scheeren, H. W. *Tetrahedron Lett.* **2000**, *41*, 515–519.

[205] Tornøe, C. W.; Davis, P.; Porreca, F.; Meldal, M. *J. Pept. Sci.* **2000**, *6*, 594–602.

[206] Tornøe, C. W.; Sengeløv, H.; Meldal, M. *J. Pept. Sci.* **2000**, *6*, 314–320.

[207] Lundquist, J. T.; Pelletier, J. C. *Org. Lett.* **2001**, *3*, 781–783.

[208] Meldal, M.; Juliano, M. A.; Jansson, A. M. *Tetrahedron Lett.* **1997**, *38*, 2531–2534.

[209] Long, D. D.; Smith, M. D.; Marquess, D. G.; Claridge, T. D. W.; Fleet, G. W. J. *Tetrahedron Lett.* **1998**, *39*, 9293–9296.

[210] Kim, J. M.; Bi, Y. Z.; Paikoff, S. J.; Schultz, P. G. *Tetrahedron Lett.* **1996**, *37*, 5305–5308.

[211] Tortolani, D. R.; Biller, S. A. *Tetrahedron Lett.* **1996**, *37*, 5687–5690.

[212] Annis, D. A.; Helluin, O.; Jacobsen, E. N. *Angew. Chem. Int. Ed.* **1998**, *37*, 1907–1909.

[213] Savin, K. A.; Woo, J. C. G.; Danishefsky, S. J. *J. Org. Chem.* **1999**, *64*, 4183–4186.

[214] Oertel, K.; Zech, G.; Kunz, H. *Angew. Chem. Int. Ed.* **2000**, *39*, 1431–1433.

[215] Tang, Z. L.; Pelletier, J. C. *Tetrahedron Lett.* **1998**, *39*, 4773–4776.

[216] Tremblay, M. R.; Poirier, D. *Tetrahedron Lett.* **1999**, *40*, 1277–1280.

[217] Gouault, N.; Cupif, J.-F.; Sauleau, A.; David, M. *Tetrahedron Lett.* **2000**, *41*, 7293–7297.

[218] Nicolaou, K. C.; Winssinger, N.; Vourloumis, D.; Ohshima, T.; Kim, S.; Pfefferkorn, J.; Xu, J. Y.; Li, T. *J. Am. Chem. Soc.* **1998**, *120*, 10814–10826.

[219] Osborn, N. J.; Robinson, J. A. *Tetrahedron* **1993**, *49*, 2873–2884.

[220] Liang, R.; Yan, L.; Loebach, J.; Ge, M.; Uozumi, Y.; Sekanina, K.; Horan, N.; Gildersleeve, J.; Thompson, C.; Smith, A.; Biswas, K.; Still, W. C.; Kahne, D. *Science* **1996**, *274*, 1520–1522.

[221] Tremblay, M. R.; Poirier, D. *J. Comb. Chem.* **2000**, *2*, 48–65.

[222] Kowalski, J.; Lipton, M. A. *Tetrahedron Lett.* **1996**, *37*, 5839–5840.

[223] Blettner, C.; Bradley, M. *Tetrahedron Lett.* **1994**, *35*, 467–470.

[224] Richter, L. S.; Andersen, S. *Tetrahedron Lett.* **1998**, *39*, 8747–8750.

[225] Shao, H.; Colucci, M.; Tong, S.; Zhang, H.; Castelhano, A. L. *Tetrahedron Lett.* **1998**, *39*, 7235–7238.

[226] Fletcher, M. D.; Campbell, M. M. *Chem. Rev.* **1998**, *98*, 763–795.

[227] Rivier, J. E.; Jiang, G. C.; Koerber, S. C.; Porter, J.; Simon, L.; Craig, A. G.; Hoeger, C. A. *Proc. Natl. Acad. Sci. USA* **1996**, *93*, 2031–2036.

[228] Pessi, A.; Pinori, M.; Verdini, A. S.; Viscomi, G. C. *J. Chem. Soc., Chem. Commun.* **1983**, 195–197.

[229] Kiselyov, A. S. *Tetrahedron* **2001**, *57*, 5321–5326.

[230] Green, T. W.; Wuts, P. G. M. *Protective Groups in Organic Synthesis*; John Wiley & Sons: New York, **1991**.

[231] Jones, J. *The Chemical Synthesis of Peptides*; Oxford University Press: Oxford, **1994**.

[232] Hulme, C.; Peng, J.; Morton, G.; Salvino, J. M.; Herpin, T.; Labaudiniere, R. *Tetrahedron Lett.* **1998**, *39*, 7227–7230.

[233] Flynn, D. L.; Zelle, R. E.; Grieco, P. A. *J. Org. Chem.* **1983**, *48*, 2424–2426.

[234] Nash, I. A.; Bycroft, B. W.; Chan, W. C. *Tetrahedron Lett.* **1996**, *37*, 2625–2628.
[235] Hone, N. D.; Payne, L. J. *Tetrahedron Lett.* **2000**, *41*, 6149–6152.
[236] Blaney, P.; Grigg, R.; Rankovic, Z.; Thoroughgood, M. *Tetrahedron Lett.* **2000**, *41*, 6635–6638.
[237] Yamada, M.; Miyajima, T.; Horikawa, H. *Tetrahedron Lett.* **1998**, *39*, 289–292.
[238] Furlán, R. L. E.; Mata, E. G. *Tetrahedron Lett.* **1998**, *39*, 6421–6422.
[239] Panek, J. S.; Zhu, B. *Tetrahedron Lett.* **1996**, *37*, 8151–8154.
[240] Sieber, P.; Iselin, B. *Helv. Chim. Acta* **1968**, *51*, 614–622.
[241] Houghten, R. A.; Beckman, A.; Ostresh, J. M. *Int. J. Pept. Prot. Res.* **1986**, *27*, 653–658.
[242] Naharissoa, H.; Sarrade, V.; Follet, M.; Calas, B. *Pept. Res.* **1992**, *5*, 293–299.
[243] Bayer, E.; Dengler, M.; Hemmasi, B. *Int. J. Pept. Prot. Res.* **1985**, *25*, 178–186.
[244] Andreatta, R. H.; Rink, H. *Helv. Chim. Acta* **1973**, *56*, 1205–1218.
[245] Allin, S. M.; Shuttleworth, S. J. *Tetrahedron Lett.* **1996**, *37*, 8023–8026.
[246] Kaiser, E.; Picart, F.; Kubiak, T.; Tam, J. P.; Merrifield, R. B. *J. Org. Chem.* **1993**, *58*, 5167–5175.
[247] Zhang, A. J.; Russell, D. H.; Zhu, J. P.; Burgess, K. *Tetrahedron Lett.* **1998**, *39*, 7439–7442.
[248] Trivedi, H. S.; Anson, M.; Steel, P. G.; Worley, J. *Synlett* **2001**, 1932–1934.
[249] Plunkett, M. J.; Ellman, J. A. *J. Org. Chem.* **1997**, *62*, 2885–2893.
[250] Kuisle, O.; Quiñoá, E.; Riguera, R. *Tetrahedron Lett.* **1999**, *40*, 1203–1206.
[251] Wildemann, D.; Drewello, M.; Fischer, G.; Schutkowski, M. *J. Chem. Soc., Chem. Commun.* **1999**, 1809–1810.
[252] Birr, C.; Lochinger, W.; Stahnke, G.; Lang, P. *Liebigs Ann. Chem.* **1972**, *763*, 162–172.
[253] Kalbacher, H.; Voelter, W. *Angew. Chem. Int. Ed. Engl.* **1978**, *17*, 944–945.
[254] Tun-Kyi, A.; Schwyzer, R. *Helv. Chim. Acta* **1976**, *59*, 1642–1646.
[255] Merrifield, R. B. *J. Am. Chem. Soc.* **1963**, *85*, 2149–2154.
[256] Wen, J. J.; Spatola, A. F. *J. Pept. Res.* **1997**, *49*, 3–14.
[257] Carpino, L. A.; Han, G. Y. *J. Org. Chem.* **1972**, *37*, 3404–3409.
[258] Fields, G. B.; Noble, R. L. *Int. J. Pept. Prot. Res.* **1990**, *35*, 161–214.
[259] Ede, N. J.; Ang, K. H.; James, I. W.; Bray, A. M. *Tetrahedron Lett.* **1996**, *37*, 9097–9100.
[260] O'Donnell, M. J.; Zhou, C. Y.; Scott, W. L. *J. Am. Chem. Soc.* **1996**, *118*, 6070–6071.
[261] Atherton, E.; Logan, C. J.; Sheppard, R. C. *J. Chem. Soc., Perkin Trans. 1* **1981**, 538–546.
[262] Chang, C. D.; Waki, M.; Ahmad, M.; Meienhofer, J.; Lundell, E. O.; Haug, J. D. *Int. J. Pept. Prot. Res.* **1980**, *15*, 59–66.
[263] Wade, J. D.; Bedford, J.; Sheppard, R. C.; Tregear, G. W. *Pept. Res.* **1991**, *4*, 194–199.
[264] Tickler, A. K.; Barrow, C. J.; Wade, J. D. *J. Pept. Sci.* **2001**, *7*, 488–494.
[265] Clippingdale, A. B.; Barrow, C. J.; Wade, J. D. *J. Pept. Sci.* **2000**, *6*, 225–234.
[266] Page, P.; Bradley, M.; Walters, I.; Teague, S. *J. Org. Chem.* **1999**, *64*, 794–799.
[267] Ueki, M.; Kai, K.; Amemiya, M.; Horino, H.; Oyamada, H. *J. Chem. Soc., Chem. Commun.* **1988**, 414–415.
[268] Ueki, M.; Amemiya, M. *Tetrahedron Lett.* **1987**, *28*, 6617–6620.
[269] Campbell, D. A.; Bermak, J. C. *J. Am. Chem. Soc.* **1994**, *116*, 6039–6040.
[270] Lapatsanis, L.; Milias, G.; Froussios, K.; Kolovos, M. *Synthesis* **1983**, 671–673.
[271] Paquet, A. *Can. J. Chem.* **1982**, *60*, 976–980.
[272] Ramage, R.; Blake, A. J.; Florence, M. R.; Gray, T.; Raphy, G.; Roach, P. L. *Tetrahedron* **1991**, *47*, 8001–8024.
[273] Carpino, L. A.; Ismail, M.; Truran, G. A.; Mansour, E. M. E.; Iguchi, S.; Ionescu, D.; El-Faham, A.; Riemer, C.; Warrass, R. *J. Org. Chem.* **1999**, *64*, 4324–4338.
[274] Samukov, V. V.; Sabirov, A. N.; Pozdnyakov, P. I. *Tetrahedron Lett.* **1994**, *35*, 7821–7824.
[275] Campbell, D. A.; Bermak, J. C.; Burkoth, T. S.; Patel, D. V. *J. Am. Chem. Soc.* **1995**, *117*, 5381–5382.
[276] Loffet, A.; Zhang, H. X. *Int. J. Pept. Prot. Res.* **1993**, *42*, 346–351.
[277] Kunz, H.; Unverzagt, C. *Angew. Chem. Int. Ed. Engl.* **1984**, *23*, 436–437.
[278] Girdwood, J. A.; Shute, R. E. *Chem. Commun.* **1997**, 2307–2308.
[279] Pelish, H. E.; Westwood, N. J.; Feng, Y.; Kirchhausen, T.; Shair, M. D. *J. Am. Chem. Soc.* **2001**, *123*, 6740–6741.
[280] Zorn, C.; Gnad, F.; Salmen, S.; Herpin, T.; Reiser, O. *Tetrahedron Lett.* **2001**, *42*, 7049–7053.
[281] Fernández-Forner, D.; Casals, G.; Navarro, E.; Ryder, H.; Albericio, F. *Tetrahedron Lett.* **2001**, *42*, 4471–4474.
[282] Freund, E.; Vitali, F.; Linden, A.; Robinson, J. A. *Helv. Chim. Acta* **2000**, *83*, 2572–2579.
[283] Gomez-Martinez, P.; Dessolin, M.; Guibé, F.; Albericio, F. *J. Chem. Soc., Perkin Trans. 1* **1999**, 2871–2874.
[284] Caulfield, T. J.; Patel, S.; Salvino, J. M.; Liester, L.; Labaudiniere, R. *J. Comb. Chem.* **2000**, *2*, 600–603.

[285] Valerio, R. M.; Bray, A. M.; Stewart, K. M. *Int. J. Pept. Prot. Res.* **1996**, *47*, 414–418.
[286] Kim, S. W.; Hong, C. Y.; Lee, K.; Lee, E. J.; Koh, J. S. *Bioorg. Med. Chem. Lett.* **1998**, *8*, 735–738.
[287] Kim, S. W.; Hong, C. Y.; Koh, J. S.; Lee, E. J.; Lee, K. *Mol. Diversity* **1998**, *3*, 133–136.
[288] Davis, P. W.; Swayze, E. E. *Biotechnol. Bioeng. (Comb. Chem.)* **2000**, *71*, 19–27.
[289] Henkel, B.; Bayer, E. *J. Pept. Sci.* **2001**, *7*, 152–156.
[290] Silva, D. J.; Wang, H.; Allanson, N. M.; Jain, R. K.; Sofia, M. J. *J. Org. Chem.* **1999**, *64*, 5926–5929.
[291] Sofia, M. J.; Allanson, N.; Hatzenbuhler, N. T.; Jain, R.; Kakarla, R.; Kogan, N.; Liang, R.; Liu, D.; Silva, D. J.; Wang, H.; Gange, D.; Anderson, J.; Chen, A.; Chi, F.; Dulina, R.; Huang, B.; Kamau, M.; Wang, C.; Baizman, E.; Branstrom, A.; Bristol, N.; Goldman, R.; Han, K.; Longley, C.; Midha, S.; Axelrod, H. R. *J. Med. Chem.* **1999**, *42*, 3193–3198.
[292] Collini, M. D.; Ellingboe, J. W. *Tetrahedron Lett.* **1997**, *38*, 7963–7966.
[293] Miranda, L. P.; Meldal, M. *Angew. Chem. Int. Ed.* **2001**, *40*, 3655–3657.
[294] Cros, E.; Planas, M.; Mejías, X.; Bardají, E. *Tetrahedron Lett.* **2001**, *42*, 6105–6107.
[295] Aronov, A. M.; Gelb, M. H. *Tetrahedron Lett.* **1998**, *39*, 4947–4950.
[296] Zikos, C. C.; Ferderigos, N. G. *Tetrahedron Lett.* **1995**, *36*, 3741–3744.
[297] Wess, G.; Bock, K.; Kleine, H.; Kurz, M.; Guba, W.; Hemmerle, H.; Lopez-Calle, E.; Baringhaus, K. H.; Glombik, H.; Enhsen, A.; Kramer, W. *Angew. Chem. Int. Ed. Engl.* **1996**, *35*, 2222–2224.
[298] Adams, J. H.; Cook, R. M.; Hudson, D.; Jammalamadaka, V.; Lyttle, M. H.; Songster, M. F. *J. Org. Chem.* **1998**, *63*, 3706–3716.
[299] Mitchell, A. R.; Kent, S. B. H.; Erickson, B. W.; Merrifield, R. B. *Tetrahedron Lett.* **1976**, 3795–3798.
[300] Wells, N. J.; Basso, A.; Bradley, M. *Biopolymers* **1998**, *47*, 381–396.
[301] Nielsen, J.; Rasmussen, P. H. *Tetrahedron Lett.* **1996**, *37*, 3351–3354.
[302] Burgess, K.; Ibarzo, J.; Linthicum, D. S.; Russell, D. H.; Shin, H.; Shitangkoon, A.; Totani, R.; Zhang, A. J. *J. Am. Chem. Soc.* **1997**, *119*, 1556–1564.
[303] Barany, G.; Merrifield, R. B. *J. Am. Chem. Soc.* **1977**, *99*, 7363–7365.
[304] Planas, M.; Bardají, E.; Jensen, K. J.; Barany, G. *J. Org. Chem.* **1999**, *64*, 7281–7289.
[305] Chan, W. C.; Bycroft, B. W.; Evans, D. J.; White, P. D. *J. Chem. Soc., Chem. Commun.* **1995**, 2209–2210.
[306] Lelièvre, D.; Daguet, D.; Brack, A. *Tetrahedron Lett.* **1995**, *36*, 9317–9320.
[307] Bloomberg, G. B.; Askin, D.; Gargaro, A. R.; Tanner, M. J. A. *Tetrahedron Lett.* **1993**, *34*, 4709–4712.
[308] Chhabra, S. R.; Hothi, B.; Evans, D. J.; White, P. D.; Bycroft, B. W.; Chan, W. C. *Tetrahedron Lett.* **1998**, *39*, 1603–1606.
[309] Kellam, B.; Bycroft, B. W.; Chhabra, S. R. *Tetrahedron Lett.* **1997**, *38*, 4849–4852.
[310] Novabiochem Catalog and Peptide Synthesis Handbook, Läufelfingen, CH **1999**.
[311] Bycroft, B. W.; Chan, W. C.; Chhabra, S. R.; Hone, N. D. *J. Chem. Soc., Chem. Commun.* **1993**, 778–779.
[312] Chhabra, S. R.; Khan, A. N.; Bycroft, B. W. *Tetrahedron Lett.* **1998**, *39*, 3585–3588.
[313] Griffith, D. L.; O'Donnell, M. J.; Pottorf, R. S.; Scott, W. L.; Porco, J. A. *Tetrahedron Lett.* **1997**, *38*, 8821–8824.
[314] Domínguez, E.; O'Donnell, M. J.; Scott, W. L. *Tetrahedron Lett.* **1998**, *39*, 2167–2170.
[315] O'Donnell, M. J.; Lugar, C. W.; Pottorf, R. S.; Zhou, C. Y.; Scott, W. L.; Cwi, C. L. *Tetrahedron Lett.* **1997**, *38*, 7163–7166.
[316] Kaljuste, K.; Undén, A. *Tetrahedron Lett.* **1996**, *37*, 3031–3034.
[317] Barlos, K.; Gatos, D.; Kallitsis, J.; Papaioannou, D.; Sotiriu, P.; Schäfer, W. *Liebigs Ann. Chem.* **1987**, 1031–1035.
[318] Viirre, R. D.; Hudson, R. H. E. *Org. Lett.* **2001**, *3*, 3931–3934.
[319] Will, D. W.; Langner, D.; Knolle, J.; Uhlmann, E. *Tetrahedron* **1995**, *51*, 12069–12082.
[320] van der Laan, A. C.; Meeuwenoord, N. J.; Kuyl-Yeheskiely, E.; Oosting, R. S.; Brands, R.; van Boom, J. H. *Recl. Trav. Chim. Pays-Bas* **1995**, *114*, 295–297.
[321] Kessler, W.; Iselin, B. *Helv. Chim. Acta* **1966**, *49*, 1330–1344.
[322] Matsueda, R.; Maruyama, H.; Kitazawa, E.; Takahagi, H.; Mukaiyama, T. *J. Am. Chem. Soc.* **1975**, *97*, 2573–2575.
[323] Piscopio, A. D.; Miller, J. F.; Koch, K. *Tetrahedron* **1999**, *55*, 8189–8198.
[324] Combs, A. P.; Rafalski, M. *J. Comb. Chem.* **2000**, *2*, 29–32.
[325] Miller, S. C.; Scanlan, T. S. *J. Am. Chem. Soc.* **1998**, *120*, 2690–2691.
[326] Hidai, Y.; Kan, T.; Fukuyama, T. *Tetrahedron Lett.* **1999**, *40*, 4711–4714.
[327] Wipf, P.; Henninger, T. C. *J. Org. Chem.* **1997**, *62*, 1586–1587.
[328] Raman, P.; Stokes, S. S.; Angell, Y. M.; Flentke, G. R.; Rich, D. H. *J. Org. Chem.* **1998**, *63*, 5734–5735.

[329] Reichwein, J. F.; Wels, B.; Kruijtzer, J. A. W.; Versluis, C.; Liskamp, R. M. J. *Angew. Chem. Int. Ed.* **1999**, *38*, 3684–3687.
[330] Reichwein, J. F.; Versluis, C.; Liskamp, R. M. J. *J. Org. Chem.* **2000**, *65*, 6187–6195.
[331] Berst, F.; Holmes, A. B.; Ladlow, M.; Murray, P. J. *Tetrahedron Lett.* **2000**, *41*, 6649–6653.
[332] Lin, X.; Dorr, H.; Nuss, J. M. *Tetrahedron Lett.* **2000**, *41*, 3309–3313.
[333] Wuts, P. G. M.; Northuis, J. M. *Tetrahedron Lett.* **1998**, *39*, 3889–3890.
[334] Piscopio, A. D.; Miller, J. F.; Koch, K. *Tetrahedron Lett.* **1997**, *38*, 7143–7146.
[335] Leznoff, C. C.; Fyles, T. M.; Weatherston, J. *Can. J. Chem.* **1977**, *55*, 1143–1153.
[336] Nicolaou, K. C.; Winssinger, N.; Pastor, J.; Ninkovic, S.; Sarabia, F.; He, Y.; Vourloumis, D.; Yang, Z.; Li, T.; Giannakakou, P.; Hamel, E. *Nature* **1997**, *387*, 268–272.
[337] Blaney, P.; Grigg, R.; Rankovic, Z.; Thoroughgood, M. *Tetrahedron Lett.* **2000**, *41*, 6639–6642.
[338] Hari, A.; Miller, B. L. *Org. Lett.* **1999**, *1*, 2109–2111.
[339] Cai, J.; Wathey, B. *Tetrahedron Lett.* **2001**, *42*, 1383–1385.
[340] Kirchhoff, J. H.; Bräse, S.; Enders, D. *J. Comb. Chem.* **2001**, *3*, 71–77.
[341] Klinguer, C.; Melnyk, O.; Loing, E.; Gras-Masse, H. *Tetrahedron Lett.* **1996**, *37*, 7259–7262.
[342] Wilson, L. J.; Li, M.; Portlock, D. E. *Tetrahedron Lett.* **1998**, *39*, 5135–5138.
[343] Richter, L. S.; Desai, M. C. *Tetrahedron Lett.* **1997**, *38*, 321–322.
[344] Bauer, U.; Ho, W. B.; Koskinen, A. M. P. *Tetrahedron Lett.* **1997**, *38*, 7233–7236.
[345] Atkinson, G. E.; Fischer, P. M.; Chan, W. C. *J. Org. Chem.* **2000**, *65*, 5048–5056.
[346] Morvan, F.; Sanghvi, Y. S.; Perbost, M.; Vasseur, J. J.; Bellon, L. *J. Am. Chem. Soc.* **1996**, *118*, 255–256.
[347] Miyabe, H.; Konishi, C.; Naito, T. *Org. Lett.* **2000**, *2*, 1443–1445.
[348] Jeon, G.-H.; Yoon, J.-Y.; Kim, S.; Kim, S. S. *Synlett* **2000**, 128–130.
[349] Semenov, A. N.; Gordeev, K. Y. *Int. J. Pept. Prot. Res.* **1995**, *45*, 303–304.
[350] Bonnet, D.; Rommens, C.; Gras-Masse, H.; Melnyk, O. *Tetrahedron Lett.* **1999**, *40*, 7315–7318.
[351] Lohse, A.; Jensen, K. B.; Bols, M. *Tetrahedron Lett.* **1999**, *40*, 3033–3036.
[352] Lohse, A.; Jensen, K. B.; Lundgren, K.; Bols, M. *Bioorg. Med. Chem.* **1999**, *7*, 1965–1971.
[353] Haap, W. J.; Kaiser, D.; Walk, T. B.; Jung, G. *Tetrahedron* **1998**, *54*, 3705–3724.
[354] Hanessian, S.; Yang, R. Y. *Tetrahedron Lett.* **1996**, *37*, 5835–5838.
[355] Miyabe, H.; Fujishima, Y.; Naito, T. *J. Org. Chem.* **1999**, *64*, 2174–2175.
[356] Kobayashi, S.; Akiyama, R. *Tetrahedron Lett.* **1998**, *39*, 9211–9214.
[357] Robinson, D. E.; Holladay, M. W. *Org. Lett.* **2000**, *2*, 2777–2779.
[358] Tois, J.; Franzén, R.; Aitio, O.; Laakso, I.; Kylänlahti, I. *J. Comb. Chem.* **2001**, *3*, 542–545.
[359] Drewry, D. H.; Gerritz, S. W.; Linn, J. A. *Tetrahedron Lett.* **1997**, *38*, 3377–3380.
[360] Schneider, S. E.; Bishop, P. A.; Salazar, M. A.; Bishop, O. A.; Anslyn, E. V. *Tetrahedron* **1998**, *54*, 15063–15086.
[361] Le Hetet, C.; David, M.; Carreaux, F.; Carboni, B.; Sauleau, A. *Tetrahedron Lett.* **1997**, *38*, 5153–5156.
[362] Hanessian, S.; Xie, F. *Tetrahedron Lett.* **1998**, *39*, 737–740.
[363] Searle, N. E. *Org. Synth.* **1963**, *Coll. Vol. IV*, 424–426.
[364] Bhalay, G.; Dunstan, A. R. *Tetrahedron Lett.* **1998**, *39*, 7803–7806.
[365] Chapman, P. H.; Walker, D. *J. Chem. Soc., Chem. Commun.* **1975**, 690–691.
[366] Schlienger, N.; Bryce, M. R.; Hansen, T. K. *Tetrahedron Lett.* **2000**, *41*, 5147–5150.
[367] Mergler, M.; Dick, F.; Gosteli, J.; Nyfeler, R. *Tetrahedron Lett.* **1999**, *40*, 4663–4664.
[368] Iso, Y.; Shindo, H.; Hamana, H. *Tetrahedron* **2000**, *56*, 5353–5361.
[369] Zaragoza, F.; Petersen, S. V. *Tetrahedron* **1996**, *52*, 5999–6002.
[370] Gowravaram, M. R.; Gallop, M. A. *Tetrahedron Lett.* **1997**, *38*, 6973–6976.
[371] Whitehouse, D. L.; Nelson, K. H.; Savinov, S. N.; Löwe, R. S.; Austin, D. J. *Bioorg. Med. Chem.* **1998**, *6*, 1273–1282.
[372] Clapham, B.; Spanka, C.; Janda, K. D. *Org. Lett.* **2001**, *3*, 2173–2176.
[373] Cano, M.; Camps, F.; Joglar, J. *Tetrahedron Lett.* **1998**, *39*, 9819–9822.
[374] Kumari, K. A.; Sreekumar, K. *Polymer* **1996**, *37*, 171–176.

11 Preparation of Phosphorus Compounds

Compounds containing phosphorus can be both valuable synthetic intermediates and target compounds of solid-phase synthesis. Important synthetic intermediates include phosphonium salts and phosphorus ylides, which are key intermediates in carbonyl olefinations. Their preparation is discussed in Section 5.2.2.1. The preparation of oligonucleotides, these being the most important phosphorus-containing target molecules in solid-phase synthesis, is considered in Section 16.2. In this chapter, the preparation of phosphines, phosphonic acid derivatives, and phosphinic acid derivatives is discussed.

11.1 Preparation of Phosphines

Polymer-bound phosphines have mainly been used as ligands for the preparation of insoluble transition metal catalysts (see Section 4.4) or as starting materials for the preparation of resin-bound phosphonium salts. Phosphines have been prepared on solid phase by the reaction of metallated supports with chlorophosphines [1,2], and by treating resin-bound alkyl or aryl halides with lithiated phosphines [3,4]. Illustrative examples are sketched in Figure 11.1. Polymer-bound phosphines have also been prepared by reduction of resin-bound phosphine oxides with $HSiCl_3$ (N,N-dimethylaniline, dioxane, 100 °C [5]). α-Aminophosphines have been prepared by the addition of diphenylphosphine to polystyrene-bound imines [6]. Phosphines can be reversibly linked to cross-linked polystyrene through a P–N bond (Figure 11.1). This form of attachment is stable towards organometallic reagents, but can be cleaved by treatment with PCl_3, alcohols, or thiols.

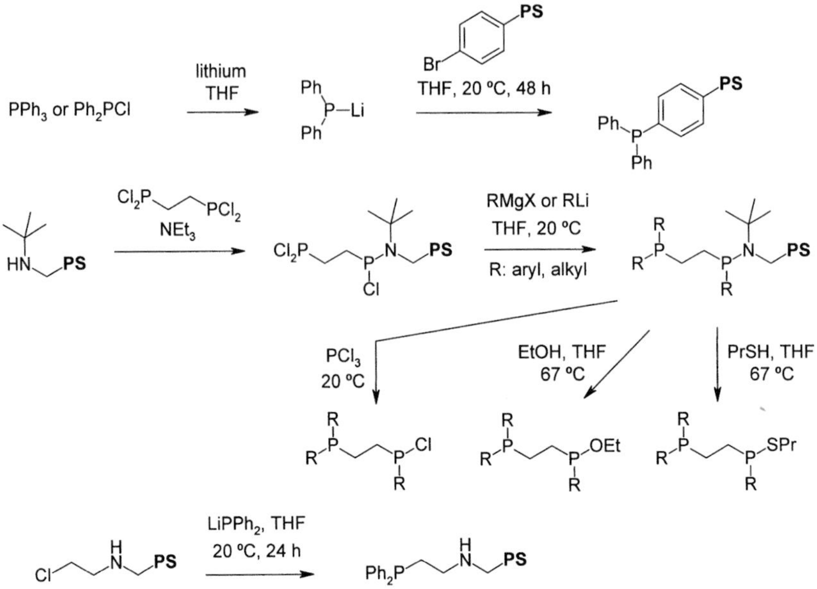

Figure 11.1. Preparation and transformations of resin-bound phosphines [4,7,8].

11.2 Preparation of Phosphonic Acid Derivatives

H-Phosphonates react with aldehydes or imines to yield α-hydroxy- or α-aminoalk-ylphosphonates, respectively. These reactions can also be conducted on cross-linked polystyrene. In Figure 11.2, a sequence is outlined in which polystyrene-bound *H*-phosphonates are treated with imines and aldehydes. The variant in which a support-bound imine is converted into an α-aminoalkylphosphonate has also been reported [9].

The formation of dialkyl phosphonates by coupling of monoalkyl phosphonates with support-bound alcohols is discussed in Section 16.2.3. Some additional examples, not related to the synthesis of oligonucleotides, are sketched in Figure 11.3.

Support-bound, enantiomerically pure alcohols can be converted into phospho-nates by Mitsunobu esterification, which results in complete inversion at the stereo-genic center. This strategy has been used to prepare peptidyl phosphonates on solid phase. These are interesting transition-state analogs with potential utility as peptidase inhibitors (Figure 11.3 [12,13]) or tyrosine phosphatase inhibitors [14]. Serine or threonine derivatives can be converted into phosphonates by direct phosphonylation with an activated monoalkyl phosphonate [15] or by treatment with phosphonamidites RP(OR)NR$_2$ in the presence of tetrazole followed by oxidation [16].

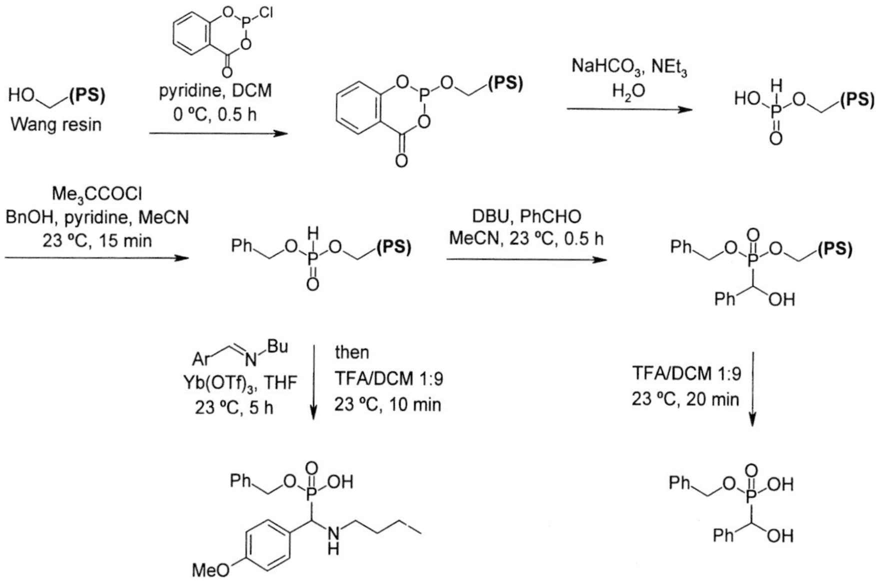

Figure 11.2. Preparation of α-hydroxy- and α-aminoalkylphosphonates on insoluble supports [10,11].

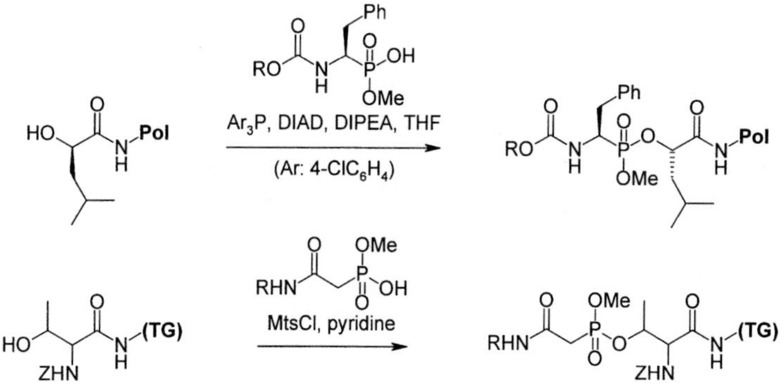

Figure 11.3. Preparation of phosphonates by phosphonylation of support-bound alcohols [12,13,15].

11.3 Preparation of Phosphinic Acid Derivatives

Peptidomimetics in which one amide bond is replaced by a phosphinic acid (R–P(OH)(=O)–R; 'phosphinic peptides') are of interest as potential protease inhibitors [17–19]. These compounds have been prepared either from orthogonally protected phosphorus-containing monomers [17,18,20], or by forming the phosphorus-containing fragments on solid phase, as sketched in Figure 11.4 [19,21]. Phosphinic acids have been prepared on solid phase mainly by reaction of carbon electrophiles with monoalkylphosphinates. As carbon electrophiles, acrylates, aldehydes, reactive alkyl halides, or α,β-unsaturated ketones can be used.

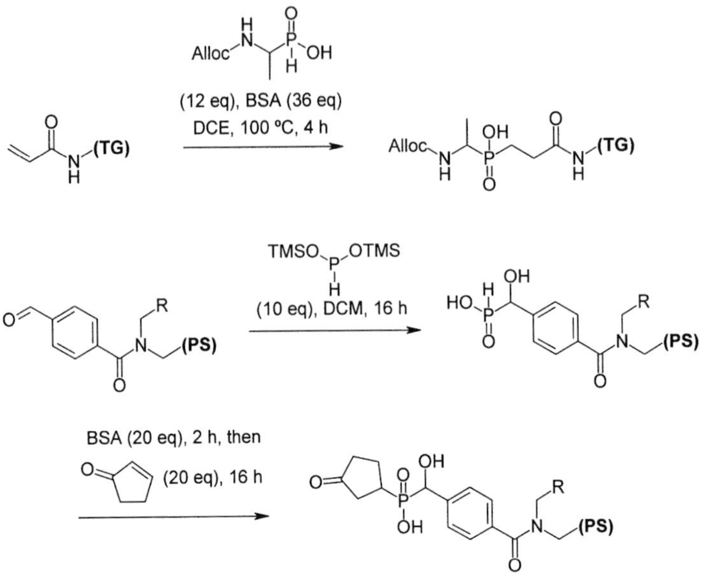

Figure 11.4. Preparation of phosphinic acids on solid phase [19,21,22].

References for Chapter 11

[1]   O'Brien, R. A.; Chen, T.; Rieke, R. D. *J. Org. Chem.* **1992**, *57*, 2667–2677.
[2]   Farrall, M. J.; Fréchet, J. M. J. *J. Org. Chem.* **1976**, *41*, 3877–3882.
[3]   Amos, R. A.; Emblidge, R. W.; Havens, N. *J. Org. Chem.* **1983**, *48*, 3598–3600.
[4]   Bernard, M.; Ford, W. T. *J. Org. Chem.* **1983**, *48*, 326–332.
[5]   Charette, A. B.; Boezio, A. A.; Janes, M. K. *Org. Lett.* **2000**, *2*, 3777–3779.
[6]   Ben-Aroya, B. B.-N.; Portnoy, M. *J. Comb. Chem.* **2001**, *3*, 524–527.
[7]   Li, G. Y.; Fagan, P. J.; Watson, P. L. *Angew. Chem. Int. Ed.* **2001**, *40*, 1106–1109.
[8]   Mansour, A.; Portnoy, M. *J. Chem. Soc., Perkin Trans. 1* **2001**, 952–954.
[9]   Boyd, E. A.; Chan, W. C.; Loh, V. M. *Tetrahedron Lett.* **1996**, *37*, 1647–1650.
[10] Cao, X. D.; Mjalli, A. M. M. *Tetrahedron Lett.* **1996**, *37*, 6073–6076.
[11] Zhang, C. Z.; Mjalli, A. M. M. *Tetrahedron Lett.* **1996**, *37*, 5457–5460.
[12] Campbell, D. A.; Bermak, J. C. *J. Am. Chem. Soc.* **1994**, *116*, 6039–6040.
[13] Campbell, D. A.; Bermak, J. C.; Burkoth, T. S.; Patel, D. V. *J. Am. Chem. Soc.* **1995**, *117*, 5381–5382.
[14] Hum, G.; Grzyb, J.; Taylor, S. D. *J. Comb. Chem.* **2000**, *2*, 234–242.
[15] Johnson, C. R.; Zhang, B. R. *Tetrahedron Lett.* **1995**, *36*, 9253–9256.
[16] Wijkmans, J. C. H. M.; Meeuwenoord, N. J.; Bloemhoff, W.; van der Marel, G. A.; van Boom, J. H. *Tetrahedron* **1996**, *52*, 2103–2112.
[17] Buchardt, J.; Ferreras, M.; Krog-Jensen, C.; Delaissé, J.-M.; Foged, N. T.; Meldal, M. *Chem. Eur. J.* **1999**, *5*, 2877–2884.
[18] Vassiliou, S.; Mucha, A.; Cuniasse, P.; Georgiadis, D.; Lucet-Levannier, K.; Beau, F.; Kannan, R.; Murphy, G.; Knäuper, V.; Rio, M.; Basset, P.; Yiotakis, A.; Dive, V. *J. Med. Chem.* **1999**, *42*, 2610–2620.
[19] Dorff, P. H.; Chiu, G.; Goldstein, S. W.; Morgan, B. P. *Tetrahedron Lett.* **1998**, *39*, 3375–3378.
[20] Yiotakis, A.; Vassiliou, S.; Jirácek, J.; Dive, V. *J. Org. Chem.* **1996**, *61*, 6601–6605.
[21] Buchardt, J.; Meldal, M. *J. Chem. Soc., Perkin Trans. 1* **2000**, 3306–3310.
[22] Cox, P. B.; Loh, V. M.; Monteils, C.; Baxter, A. D.; Boyd, E. A. *Tetrahedron Lett.* **2001**, *42*, 125–128.

12 Preparation of Aldehydes and Ketones

Aldehydes and ketones are usually prepared on insoluble supports by the acylation of arenes, C,H-acidic compounds, or organometallic reagents. Alcohols or other substrates can also be converted into carbonyl compounds by oxidation (Figure 12.1). Linkers that enable the generation of aldehydes and ketones upon cleavage from a support are considered in Section 3.14.

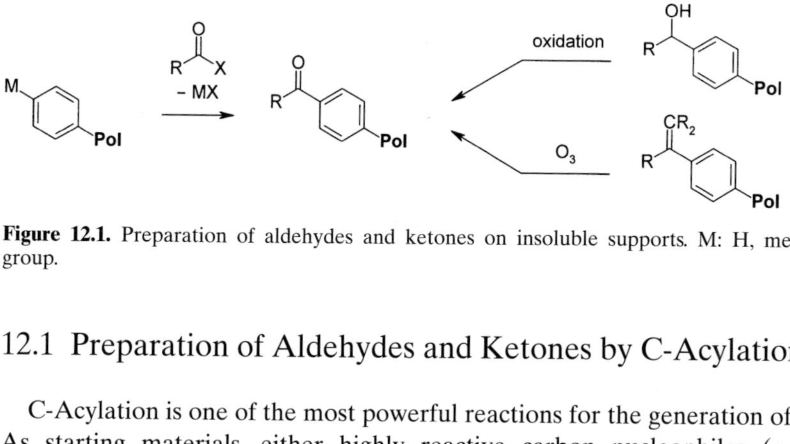

Figure 12.1. Preparation of aldehydes and ketones on insoluble supports. M: H, metal; X: leaving group.

12.1 Preparation of Aldehydes and Ketones by C-Acylation

C-Acylation is one of the most powerful reactions for the generation of C–C bonds. As starting materials, either highly reactive carbon nucleophiles (e.g. Grignard reagents) may be combined with unreactive acylating agents (e.g. Weinreb amides), or unreactive carbon nucleophiles (e.g. polystyrene) may be treated with strong acylating agents (e.g. acyl halides + AlCl$_3$). Most of these strategies have also been successfully used for the preparation of carbonyl compounds on insoluble supports.

Cross-linked polystyrene can be acylated with aliphatic and aromatic acyl halides in the presence of AlCl$_3$ (Friedel–Crafts acylation, Table 12.1). This reaction has mainly been used for the functionalization of polystyrene-based supports, and only rarely for the modification of support-bound substrates. Electron-rich arenes (Entry **3**, Table 12.1) or heteroarenes, such as indoles (Entry **5**, Table 15.7), undergo smooth Friedel–Crafts acylation without severe deterioration of the support. Suitable solvents for Friedel–Crafts acylations of cross-linked polystyrene are tetrachloroethene [1], DCE [2], CS$_2$ [3,4], nitrobenzene [5,6], and CCl$_4$ [7]. As in the bromination of polystyrene, Friedel–Crafts acylations at high temperatures (e.g. DCE, 83 °C, 15 min [2]) can lead to partial dealkylation of phenyl groups and yield a soluble polymer.

Table 12.1. Preparation of aldehydes and ketones by C-acylation of support-bound carbon nucleophiles.

Entry	Starting resin	Conditions	Product	Ref.
1		*i*PrCOCl, AlCl$_3$, CS$_2$, 46 °C, 4 h		[10] see also [2]
2		BrCHMeCOBr, AlCl$_3$, DCM, 0–20 °C, 21 h		[5]
3		PhCOCl (1 eq), SnCl$_4$ or FeCl$_3$ (5 eq), DCM, 20 °C, 6–12 h PS: aminomethyl-PS with PAM linker		[11]
4		ArCOCl, Pd$_2$(dba)$_3$, K$_2$CO$_3$, DIPEA, THF, 1 h Ar: 4-(MeO)C$_6$H$_4$		[12]
5		PhCO$_2$Et (1 mol/L, 30 eq), NaH, DMA, 90 °C, 1 h		[13] see also [14,15]
6		*t*BuOCH(NMe$_2$)$_2$, THF, 20 °C; then HCl (2 mol/L in H$_2$O)/THF 1:2 (Rink amide linker)		[16] see also [17]
7		PhCO$_2$H (10 eq), NEt$_3$ (20 eq), (EtO)$_2$(O)PCN (10 eq), DMF, 0–20 °C, 15 h		[18]
8		(PhCO)$_2$O, NEt$_3$, DMF, 20 °C, overnight		[19]
9		MgCl$_2$, NEt$_3$, PhMe, 20 °C, 12 h		[20]
10		BuLi (5 eq), THF, −30 °C, 4 h; then DMA (10 eq), 20 °C, 2 h		[21] see also [15]
11		BuLi (5 eq), THF, −30 °C, 4 h; then DMF (10 eq), 20 °C, 2 h		[21]

Aryl ketones can also be prepared by C-acylation of support-bound arylstannanes with acyl halides (Entry **4**, Table 12.1). The reaction conditions are mild and suitable for selective chemical transformations of polystyrene-bound intermediates. As an alternative, resin-bound arylstannanes can be converted into benzophenones by treatment with aryl halides and carbon monoxide ([(Pd(PPh$_3$)$_4$], DMSO, 80 °C, 18 h–3 d [8]).

C-Acylations of C,H-acidic compounds have also been realized on insoluble supports. The few examples that have been reported include the C-acylation of support-bound ester enolates with acyl halides [9], Claisen condensations of polystyrene-bound ketones with benzoic acid esters, the C-acylation of nitriles with acyl nitriles or anhydrides, and the C-acylation of phosphonates with acyl halides (Entries **5–9**, Table 12.1). The α-formylation of support-bound arylacetonitriles can be accomplished in two steps by α-aminomethylenation followed by hydrolysis of the intermediate enamine (Entry **6**, Table 12.1).

Polystyrene-bound Weinreb and related amides react with organolithium or Grignard reagents to yield ketones (Table 12.2). Amides have also been used for this purpose, without significant formation of tertiary alcohols. Aldehydes can be prepared on solid phase by reduction of *N,O*-dialkylhydroxamates with LiAlH$_4$ (Entry **1**, Table 12.2). The successful conversion of polystyrene-bound acyl chlorides into ketones by treatment with organocadmium compounds has also been reported (Entry **4**, Table 12.2).

Table 12.2. Preparation of aldehydes and ketones from support-bound acylating agents.

Entry	Starting resin	Conditions	Product	Ref.
1		LiAlH$_4$ (1.5 eq), THF, 0 °C, 25 min		[22] see also [23]
2		MeMgCl (5 eq), THF, overnight		[24]
3		BnMgBr (0.2 mol/L, 4.2 eq), THF, 4 °C, 8 h		[25] see also [26]
4		PhCdCl, C$_6$H$_6$, 20 °C, 24 h		[27]

12.2 Preparation of Aldehydes and Ketones by Oxidation

Substrates suitable for oxidative conversion into carbonyl compounds are alkenes, primary or secondary alcohols, and benzyl halides. Polystyrene-bound alkenes have been converted into aldehydes (with the loss of one carbon atom) by ozonolysis followed by reductive cleavage of the intermediate ozonide (Entry **1**, Table 12.3).

Alkenes can be transformed into ketones by Wacker oxidation (Entry **2**, Table 12.3), but this reaction does not seem to proceed cleanly on polymeric supports. Janda and co-workers were able to oxidize styrenes bound to macroporous polystyrene to the corresponding acetophenones, but reported that the reaction did not proceed on PEG

Table 12.3. Preparation of aldehydes and ketones by oxidation.

Entry	Starting resin	Conditions	Product	Ref.
1		O_3, DCM, −78 °C, then PPh$_3$, 20 °C, 16 h		[29] see also [30]
2		PdCl$_2$ (1 eq), CuCl$_2$ (3 eq), H$_2$O, air, overnight (macroporous PS)		[28]
3		DMSO, NaHCO$_3$, 155 °C, 6 h		[31–33] see also [7]
4		NMO (0.3 mol/L, 10 eq), TPAP (0.2 eq), 20 °C, 1.5 h		[34] see also [35]
5		SO$_3$·pyridine (0.56 mol/L), NEt$_3$/DMSO 1:1.8, 20 °C, 3 h		[36] see also [37]
6		(Dess–Martin periodinane; 1.5 eq), DCM, 25 °C, 6 h		[38]
7		Dess–Martin periodinane (10 eq), DCM, 20 °C, 18 h		[37]
8		(COCl)$_2$ (4 eq), DMSO (8 eq), NEt$_3$ (13 eq), −78 °C to −25 °C		[39] see also [40]
9		CrO$_2$Cl$_2$ (0.17 mol/L, 5 eq), tBuOH (10 eq), pyridine (15 eq), DCM, 20 °C, 28 h (formation of CrO$_2$(OtBu)$_2$) Ar: 4-(MeO)C$_6$H$_4$		[41] see also [42]
10		SO$_3$·pyridine (0.5 mol/L), NEt$_3$/DMSO 1:15, overnight, repeat once (oxidation did not proceed on pure PS)		[43] see also [44]

Table 12.3. continued.

Entry	Starting resin	Conditions	Product	Ref.
11	(resin structure with PS, Si, Bu Bu, MeO, HO)	Dess–Martin periodinane (3 eq), DCM, 40 °C, 2 h	(product structure with PS, Si, Bu Bu, MeO, O=)	[45] see also [46,47]
12	(pyrrolidine structure, HO···, N-Boc, (PA))	pyridinium dichromate (0.2 mol/L), DMF, 37 °C (Rink amide linker)	(product structure, N-Boc, O=, (PA))	[48]
13	(structure with HO, H₂N, (PEG))	NaIO$_4$ (0.23 mol/L, 10 eq), H$_2$O/NaH$_2$PO$_4$-buffer (pH 7), 3 h (cross-linked PEG)	(product structure, (PEG))	[49]
14	(structure with Tol, OH, OH, R, N, (PS))	Bu$_4$NIO$_4$ (0.28 mol/L, 8 eq), DCE, 20 °C, overnight	(product structure, R, N, (PS))	[50]
15	(indole structure, HN, Pr, (PS))	DDQ (2 eq), THF/H$_2$O 9:1, 20 °C, 0.3 h	(indole product structure, HN, Pr, (PS))	[51]

in homogeneous phase [28]. Polystyrene-bound benzaldehyde can be prepared directly from Merrifield resin by treatment of the latter with DMSO and a base at high temperatures (Entry **3**, Table 12.3). These reaction conditions are, however, not suitable for sensitive intermediates. The oxidation of alcohols to carbonyl compounds proceeds under mild conditions that are compatible with a number of linkers and additional functional groups. Suitable oxidants are NMO in the presence of catalytic amounts of TPAP, sulfur trioxide/pyridine, oxalyl chloride/DMSO (Swern oxidation), Dess–Martin periodinane, CrO$_2$(OtBu)$_2$, and pyridinium dichromate (Table 12.3). 1,2-Diols and aminoalcohols can be oxidatively cleaved with sodium periodate to yield aldehydes (Entries **13** and **14**, Table 12.3).

12.3 Miscellaneous Preparations of Aldehydes and Ketones

Other solid-phase preparations of carbonyl compounds include the hydrolysis of acetals (Table 12.4), inter- [52] and intramolecular Pauson–Khand reactions, the iso-merization of allyl alcohols, and the α-alkylation and α-arylation of other ketones. Tietze reported the generation of acetoacetyl dianions on cross-linked polystyrene and their selective alkylation at C-4 (Entry **6**, Table 12.4). The use of weaker bases resulted in single or twofold alkylation at C-2 [53].

Few examples of Mannich-type aminoalkylations of ketones on solid phase have been reported [54]. The resulting β-amino ketones are generally acid-sensitive and highly reactive, and mild cleavage conditions must be employed for the release of the

Table 12.4. Miscellaneous preparations of aldehydes and ketones.

Entry	Starting resin	Conditions	Product	Ref.
1		TsOH, NMP/Me$_2$CO 2:1, 50 °C, 3 × 17 h		[58]
2		MeOTf (10 eq), DCM, 20 °C, 20 min		[59] see also [60]
3		NMO (0.22 mol/L), DCM, 2 h		[61]
4		3-bromopyridine (10 eq), Pd$_2$(dba)$_3$ (0.33 eq), P(o-Tol)$_3$ (0.66 eq), NEt$_3$ (10 eq), DMF, 100 °C, 24 h		[62]
5		TBAF (1 mol/L, 10 eq), THF, 20 °C, 2 h, then EtI (44 eq), 20 °C, 2 h		[13]
6		1. LDA (6 eq), THF, 0 °C, 1 h 2. iPrI (5 eq), THF, 0–25 °C, 12 h		[63,64]
7		(3 eq), Pd(PPh$_3$)$_4$ (0.1 eq), THF, 20 °C, 1 h		[53]
8		2,4-pentanedione (0.7 mol/L, 17 eq), DMF/DBU 10:1, 20 °C, 16 h		[65]
9		(2-thienyl)Cu(CN)Li, THF, −20 °C, then 2-cyclohexenone		[66]
10		(5 eq), THF, −78 °C to 20 °C		[67] see also [68]
11		Ph⌒CN (0.8 mol/L, 32 eq), KN(SiMe$_3$)$_2$ (4 eq), air, DMF, 20 °C, 18 h		[65]

products from the polymeric support. Leznoff and co-workers reported the preparation of 1,2-diketones (in low yield) from polystyrene-bound benzaldehydes through benzoin condensation and simultaneous oxidation [55]. The 1,4-addition of support-bound organometallic compounds to enones and the oxidative degradation of benzyl cyanides have also been reported (Entries **9** and **11**, Table 12.4). Cyclohexenones have been prepared on solid phase by oxidative C-alkylation of phenols (see Entry **4**, Table 15.35 [56]). Cyclobutanones are accessible by [2 + 2] cycloaddition of ketenes to resin-bound alkenes (Figure 5.5 [57]).

References for Chapter 12

[1]   Chapman, P. H.; Walker, D. *J. Chem. Soc., Chem. Commun.* **1975**, 690–691.
[2]   Matsueda, G. R.; Stewart, J. M. *Peptides* **1981**, *2*, 45–50.
[3]   Fréchet, J. M. J.; Haque, K. E. *Tetrahedron Lett.* **1975**, 3055–3056.
[4]   Kobayashi, S.; Moriwaki, M. *Tetrahedron Lett.* **1997**, *38*, 4251–4254.
[5]   Mizoguchi, T.; Shigezane, K.; Takamura, N. *Chem. Pharm. Bull.* **1970**, *18*, 1465–1474.
[6]   Ajayaghosh, A.; Pillai, V. N. R. *Tetrahedron* **1988**, *44*, 6661–6666.
[7]   Kumari, K. A.; Sreekumar, K. *Polymer* **1996**, *37*, 171–176.
[8]   Yun, W.; Li, S.; Wang, B.; Chen, L. *Tetrahedron Lett.* **2001**, *42*, 175–177.
[9]   Patchornik, A.; Kraus, M. A. *J. Am. Chem. Soc.* **1970**, *92*, 7587–7589.
[10] Ren, Q.; Huang, W.; Ho, P. *Reactive Polymers* **1989**, *11*, 237–244.
[11] Bevacqua, F.; Basso, A.; Gitto, R.; Bradley, M.; Chimirri, A. *Tetrahedron Lett.* **2001**, *42*, 7683–7685.
[12] Plunkett, M. J.; Ellman, J. A. *J. Org. Chem.* **1997**, *62*, 2885–2893.
[13] Marzinzik, A. L.; Felder, E. R. *Tetrahedron Lett.* **1996**, *37*, 1003–1006.
[14] Stauffer, S. R.; Katzenellenbogen, J. A. *J. Comb. Chem.* **2000**, *2*, 318–329.
[15] Nicolaou, K. C.; Cao, G.-Q.; Pfefferkorn, J. A. *Angew. Chem. Int. Ed.* **2000**, *39*, 739–743.
[16] Wilson, R. D.; Watson, S. P.; Richards, S. A. *Tetrahedron Lett.* **1998**, *39*, 2827–2830.
[17] MacDonald, A. A.; DeWitt, S. H.; Hogan, E. M.; Ramage, R. *Tetrahedron Lett.* **1996**, *37*, 4815–4818.
[18] Sim, M. M.; Lee, C. L.; Ganesan, A. *Tetrahedron Lett.* **1998**, *39*, 2195–2198.
[19] Sim, M. M.; Lee, C. L.; Ganesan, A. *Tetrahedron Lett.* **1998**, *39*, 6399–6402.
[20] Kim, D. Y.; Suh, K. H. *Synth. Commun.* **1999**, *29*, 1271–1275.
[21] Li, Z.; Ganesan, A. *Synlett* **1998**, 405–406.
[22] Gosselin, F.; Van Betsbrugge, J.; Hatam, M.; Lubell, W. D. *J. Org. Chem.* **1999**, *64*, 2486–2493.
[23] Paris, M.; Douat, C.; Heitz, A; Gibbons, W.; Martinez, J.; Fehrentz, J. A. *Tetrahedron Lett.* **1999**, *40*, 5179–5182.
[24] Kim, S. W.; Bauer, S. M.; Armstrong, R. W. *Tetrahedron Lett.* **1998**, *39*, 6993–6996.
[25] Lee, C. E.; Kick, E. K.; Ellman, J. A. *J. Am. Chem. Soc.* **1998**, *120*, 9735–9747.
[26] Wallace, O. B. *Tetrahedron Lett.* **1997**, *38*, 4939–4942.
[27] Leznoff, C. C.; Yedidia, V. *Can. J. Chem.* **1980**, *58*, 287–290.
[28] Hori, M.; Gravert, D. J.; Wentworth, P.; Janda, K. D. *Bioorg. Med. Chem. Lett.* **1998**, *8*, 2363–2368.
[29] Gennari, C.; Ceccarelli, S.; Piarulli, U.; Aboutayab, K.; Donghi, M.; Paterson, I. *Tetrahedron* **1998**, *54*, 14999–15016.
[30] Sylvain, C.; Wagner, A.; Mioskowski, C. *Tetrahedron Lett.* **1997**, *38*, 1043–1044.
[31] Fréchet, J. M.; Schuerch, C. *J. Am. Chem. Soc.* **1971**, *93*, 492–496.
[32] Sheng, Q.; Stöver, H. D. H. *Macromolecules* **1997**, *30*, 6712–6714.
[33] Beebe, X.; Schore, N. E.; Kurth, M. J. *J. Org. Chem.* **1995**, *60*, 4196–4203.
[34] Yan, B.; Sun, Q.; Wareing, J. R.; Jewell, C. F. *J. Org. Chem.* **1996**, *61*, 8765–8770.
[35] Li, W.; Yan, B. *J. Org. Chem.* **1998**, *63*, 4092–4097.
[36] Chen, C.; Randall, L. A. A.; Miller, R. B.; Jones, A. D.; Kurth, M. J. *Tetrahedron* **1997**, *53*, 6595–6609.
[37] Reggelin, M.; Brenig, V.; Welcker, R. *Tetrahedron Lett.* **1998**, *39*, 4801–4804.
[38] Nicolaou, K. C.; Pastor, J.; Winssinger, N.; Murphy, F. *J. Am. Chem. Soc.* **1998**, *120*, 5132–5133.

[39] Nicolaou, K. C.; Winssinger, N.; Pastor, J.; Ninkovic, S.; Sarabia, F.; He, Y.; Vourloumis, D.; Yang, Z.; Li, T.; Giannakakou, P.; Hamel, E. *Nature* **1997**, *387*, 268–272.
[40] Marx, M. A.; Grillot, A. L.; Louer, C. T.; Beaver, K. A.; Bartlett, P. A. *J. Am. Chem. Soc.* **1997**, *119*, 6153–6167.
[41] Miller, P. C.; Owen, T. J.; Molyneaux, J. M.; Curtis, J. M.; Jones, C. R. *J. Comb. Chem.* **1999**, *1*, 223–234.
[42] Leznoff, C. C.; Fyles, T. M.; Weatherston, J. *Can. J. Chem.* **1977**, *55*, 1143–1153.
[43] Page, P.; Bradley, M.; Walters, I.; Teague, S. *J. Org. Chem.* **1999**, *64*, 794–799.
[44] Furth, P. S.; Reitman, M. S.; Cook, A. F. *Tetrahedron Lett.* **1997**, *38*, 5403–5406.
[45] Thompson, L. A.; Moore, F. L.; Moon, Y. C.; Ellman, J. A. *J. Org. Chem.* **1998**, *63*, 2066–2067.
[46] Nicolaou, K. C.; Baran, P. S.; Zhong, Y.-L. *J. Am. Chem. Soc.* **2000**, *122*, 10246–10248.
[47] Dragoli, D. R.; Thompson, L. A.; O'Brien, J.; Ellman, J. A. *J. Comb. Chem.* **1999**, *1*, 534–539.
[48] Bray, A. M.; Chiefari, D. S.; Valerio, R. M.; Maeji, N. J. *Tetrahedron Lett.* **1995**, *36*, 5081–5084.
[49] Rademann, J.; Meldal, M.; Bock, K. *Chem. Eur. J.* **1999**, *5*, 1218–1225.
[50] Schlienger, N.; Bryce, M. R.; Hansen, T. K. *Tetrahedron Lett.* **2000**, *41*, 5147–5150.
[51] Nishida, A.; Fuwa, M.; Naruto, S.; Sugano, Y.; Saito, H.; Nakagawa, M. *Tetrahedron Lett.* **2000**, *41*, 4791–4794.
[52] Spitzer, J. L.; Kurth, M. J.; Schore, N. E. *Tetrahedron* **1997**, *53*, 6791–6808.
[53] Tietze, L. F.; Hippe, T.; Steinmetz, A. *Chem. Commun.* **1998**, 793–794.
[54] Schunk, S.; Enders, D. *Org. Lett.* **2001**, *3*, 3177–3180.
[55] Leznoff, C. C.; Wong, J. Y. *Can. J. Chem.* **1973**, *51*, 3756–3764.
[56] Pelish, H. E.; Westwood, N. J.; Feng, Y.; Kirchhausen, T.; Shair, M. D. *J. Am. Chem. Soc.* **2001**, *123*, 6740–6741.
[57] Brown, R. C. D.; Keily, J.; Karim, R. *Tetrahedron Lett.* **2000**, *41*, 3247–3251.
[58] Veerman, J. J. N.; van Maarseveen, J. H.; Visser, G. M.; Kruse, C. G.; Schoemaker, H. E.; Hiemstra, H.; Rutjes, F. P. J. T. *Eur. J. Org. Chem.* **1998**, 2583–2589.
[59] Lee, H. B.; Balasubramanian, S. *J. Org. Chem.* **1999**, *64*, 3454–3460.
[60] Routledge, A.; Abell, C.; Balasubramanian, S. *Tetrahedron Lett.* **1997**, *38*, 1227–1230.
[61] Bolton, G. L.; Hodges, J. C.; Rubin, J. R. *Tetrahedron* **1997**, *53*, 6611–6634.
[62] Kulkarni, B. A.; Ganesan, A. *J. Comb. Chem.* **1999**, *1*, 373–378.
[63] Tietze, L. F.; Steinmetz, A. *Synlett* **1996**, 667–668.
[64] Tietze, L. F.; Evers, H.; Hippe, T.; Steinmetz, A.; Töpken, E. *Eur. J. Org. Chem.* **2001**, 1631–1634.
[65] Stephensen, H.; Zaragoza, F. *Tetrahedron Lett.* **1999**, *40*, 5799–5802.
[66] Kondo, Y.; Komine, T.; Fujinami, M.; Uchiyama, M.; Sakamoto, T. *J. Comb. Chem.* **1999**, *1*, 123–126.
[67] Ley, S. V.; Mynett, D. M.; Koot, W. J. *Synlett* **1995**, 1017–1020.
[68] Gutke, H.-J.; Spitzner, D. *Tetrahedron* **1999**, *55*, 3931–3936.

13 Preparation of Carboxylic Acid Derivatives

13.1 Preparation of Amides

The acylation of amines on insoluble supports is one of the most thoroughly investigated reactions in solid-phase synthesis. The continuous optimization of peptide bond formation in recent decades has led to protocols that enable racemization-free, quantitative acylations of support-bound peptides with protected amino acids. In recent years, the range of amides available by solid-phase synthesis has expanded significantly to include, for example, N-alkylated peptides and anilides. New strategies for the preparation of amides, such as C-carbamoylations and the Ugi reaction, have also been successfully realized on insoluble supports.

13.1.1 Acylation of Amines with Isolated Acylating Agents

Amides are usually prepared by treating an amine with an acylating agent (Figure 13.1). Carboxylic acids are not suitable acylating agents for this purpose, because amines are sufficiently basic to form stable salts with acids. These salts can, however, be converted into amides by treatment with a dehydrating reagent (e.g. a carbodiimide) or by strong heating [1]. Alternatively, carboxylic acids can be converted into acylating agents such as HOBt esters or symmetric anhydrides ('activation'), and then treated with an amine (Figure 13.1). Acylating agents such as symmetric anhydrides and acyl halides are sufficiently reactive to acylate ammonium salts, and can therefore be used to acylate support-bound amines in the presence of excess carboxylic acid.

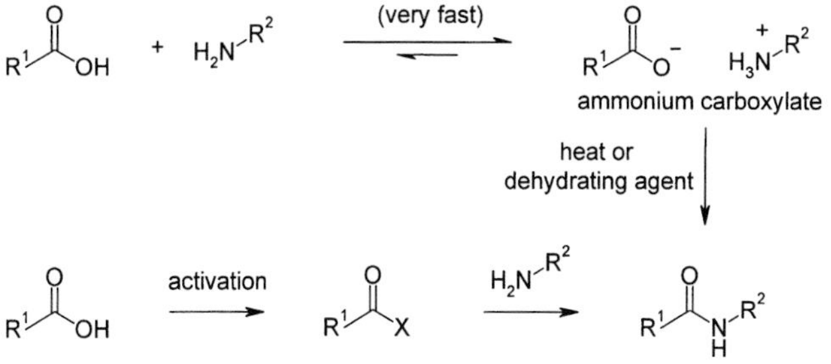

Figure 13.1. Acylation of amines with carboxylic acids. X: see Figure 13.2.

Many different acylating agents have been evaluated for their utility in solid-phase peptide synthesis. Ideally, the activated amino acid derivatives should be chemically and stereochemically stable, highly soluble in DCM, and lead to fast and complete acylation of all types of amine. Optimization of these parameters has led to the activated acid derivatives that are used today in solid-phase synthesis. Figure 13.2 shows a selection of the most common types of acylating agent used for the solid-phase synthesis of amides; these are ordered according to their approximate reactivity towards amines. It should be noted, however, that the rate of N-acylation depends on the solvent, and can be effectively increased by various catalysts (e.g. DMAP, HOBt [2], or HOAt [3,4]). For this reason, the ranking given in Figure 13.2 should be considered only as a rough guide. Comparative studies of various acylating agents used in solid-phase peptide synthesis have been reported [2,5,6].

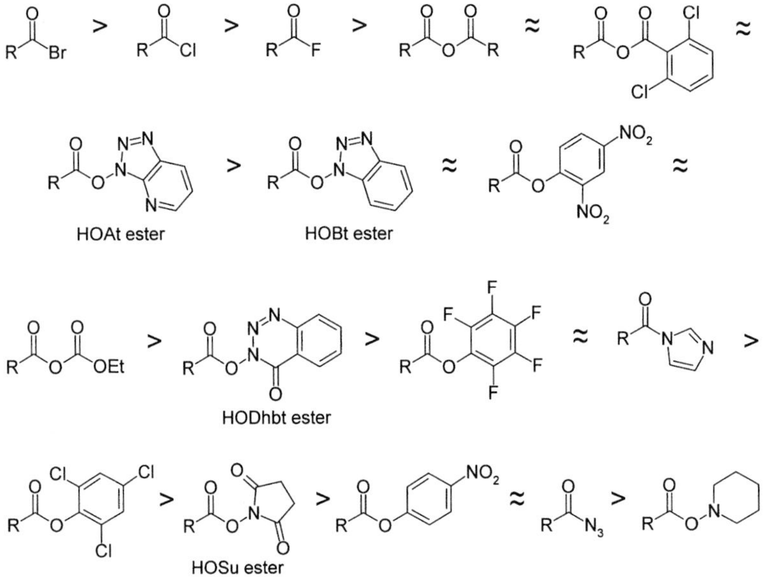

Figure 13.2. Representative acylating agents for the solid-phase synthesis of amides, ranked according to their approximate reactivity towards amines [3,6–12].

The reagents sketched in Figure 13.2 are stable and can be prepared either in solution or on insoluble supports. Activated Boc- or Fmoc-protected amino acid derivatives that are sufficiently stable to be isolated, some of which are commercially available, include acyl chlorides [9,13], fluorides [10,14,15], symmetric anhydrides [16], pentafluorophenyl esters, *N*-hydroxysuccinimidyl esters, and 4-nitrophenyl esters [17,18].

Difficult N-acylations, such as those of *N*-alkylanilines or α-alkylamino acid derivatives, are most conveniently performed with acyl halides in non-nucleophilic solvents (e.g. DCM, DCP) in the presence of pyridine or DIPEA [13]. Acyl halides can be prepared on insoluble supports under conditions similar to those used in solution. Typical reagents for the preparation of acyl chlorides include oxalyl chloride [19–21], thionyl chloride [22,23], and triphosgene [13]. Anhydrous solvents must be used for all washing of the polymer to prevent hydrolysis [24–26]. The acylation of support-bound amines with acyl halides is generally limited to substrates devoid of other acylable functionalities because of the high reactivity and low selectivity of these acylating agents.

Activated α-amino acids and other enantiomerically pure acids with a center of chirality at the α-position can racemize upon treatment with a base. The main mechanisms of racemization of such acylating agents are elimination with concomitant formation of ketenes, and enolate formation. For instance, acyl halides bearing an α-hydrogen will usually undergo dehydrohalogenation to yield ketenes if a tertiary amine is used as the base. In the case of α-acylamino acids, racemization is facilitated by the formation of oxazolones (see Figure 16.2). The danger of racemization of such activated acid derivatives generally increases with their reactivity. Hence, if amines are to be acylated with peptide fragments or other sensitive, enantiomerically pure acids, only derivatives of low acylating power, such as acyl azides, should be used. Moreover, only small amounts of weak bases should be added to the reaction mixture, in order to keep racemization minimal. Several additives (e.g. Cu(OBt)$_2$ [27]) have been reported to suppress racemization during solid-phase peptide synthesis. Examples of acylations with α-amino acid derivatives are given below. For additional notes

Table 13.1. Preparation of amides using isolated acyl halides.

Entry	Starting resin	Conditions	Product	Ref.
1		MeCOCl, NEt$_3$, DCM		[28] see also [29,30]
2		PhCOCl/DCM 1:1, pyridine (1 eq), 20 °C, 16 h		[31]
3		⌇COCl (0.2 mol/L, 2 eq), NEt$_3$ (3 eq), DCM, 20 °C, 2 h		[32]
4	(Ar: 4-(MeO)C$_6$H$_4$)	FmocHN⌇COF (0.2 mol/L, 10 eq), DTBMP (10 eq), DCM, 24 h, 20 °C		[33] see also [13]
5		isoquinoline (0.37 mol/L, 5 eq), DCM, 20 °C, 20 min, then Me$_3$SiCN (4.5 eq), 48 h		[22]

concerning the problem of racemization, see Section 13.4.1.1. Illustrative examples of the preparation of amides with isolated acyl halides on insoluble supports are listed in Table 13.1.

Less reactive than acyl halides, but still suitable for difficult couplings, are symmetric or mixed anhydrides (e.g. with pivalic or 2,6-dichlorobenzoic acid) and HOAt-derived active esters. HOBt esters smoothly acylate primary or secondary aliphatic amines, including amino acid esters or amides, without concomitant esterification of alcohols or phenols [34]. HOBt esters are the most commonly used type of activated esters in automated solid-phase peptide synthesis. For reasons not yet fully understood, acylations with HOBt esters or halophenyl esters can be effectively catalyzed by HOBt and HOAt [3], and mixtures of BOP (in situ formation of HOBt esters) and HOBt are among the most efficient coupling agents for solid-phase peptide synthesis [2]. In acylations with activated amino acid derivatives, the addition of HOBt or HOAt also retards racemization [4,12,35].

The synthesis of large peptides on insoluble supports requires rapid coupling with coupling yields > 99.95% [8]. Acylating agents of lower reactivity than HOBt esters are generally not suitable for solid-phase peptide synthesis (unless additives are used), because reaction rates are too low and quantitative acylations can no longer be attained. The lower reactivity of these reagents, however, also implies greater stability towards hydrolysis, a

Table 13.2. Preparation of amides by acylation of amines with isolated anhydrides and esters.

Entry	Starting resin	Conditions	Product	Ref.
1		$(ClCH_2CO)_2O$ (0.85 mol/L), DMAP (0.03 mol/L), DCE, 20 °C, 24 h		[36] see also [37]
2		HCO_2H, Ac_2O, DCM		[38]
3		(1.5 eq), HOBt, DIPEA, DMF, 20 °C, 5 h		[39] see also [40,41]
4		DCM, −15 °C to 20 °C, 2.5 h		[42]
5		(0.18 mol/L, 6 eq), DMF/pyridine 9:1, 20 °C, 16 h		[43]

Table 13.2. continued.

Entry	Starting resin	Conditions	Product	Ref.
6		H-His(Tr)-OLi, DMF		[44]
7		BnNH$_2$ (0.53 mol/L), DMF, 20 °C, 18 h		[45] see also [46]
8		Ph $\sim$ NH$_2$ (0.43 mol/L), DMF, 20 °C, 20 h		[47,48]
9		BnNH$_2$, AlMe$_3$, DCM, 50 °C, 3 h		[49] see also [50]
10		BnNH$_2$ (1 mol/L), NMP, 20 °C		[51]

reduced tendency to racemize (in the case of activated α-amino acids), and increased chemoselectivity. These properties turn out to be particularly profitable for *support-bound* acylating agents, and pentafluorophenyl esters, imidazolides, and acyl azides have proven suitable not only for the cyclization of peptides on insoluble supports, but also as resin-bound acylating agents for intermolecular acylations (Table 13.2). Simple alkyl esters (excluding formates) do not generally react with amines. The aminolysis of esters can, however, be efficiently catalyzed by trialkylalanes (Entry 9, Table 13.2).

13.1.2 Acylation of Amines with Acylating Reagents Formed In Situ

Acylating agents such as those shown in Figure 13.2 can either be prepared in pure form and then used to acylate amines, or, alternatively, may be generated in the presence of an amine ('in situ activation'). Because activated acid derivatives cannot be stored indefinitely, in situ activation is generally the preferred strategy for preparing amides on solid phase. Several coupling reagents have been developed for this purpose; these react only slowly or not at all with amines, but efficiently convert free acids or carboxylates into acylating agents. The most important reagents of this type are carbodiimides and phosphonium and uronium salts.

13.1.2.1 Activation of Acids with Carbodiimides

Carbodiimides were introduced as coupling reagents for peptide synthesis by Sheehan and Hess in 1955 [55]. Dicyclohexylcarbodiimide (DCC) has been the standard reagent for many years, but this compound is gradually being replaced by diisopropylcarbodiimide (DIC) and the water-soluble *N*-ethyl-*N'*-(3-dimethylaminopropyl)carbodiimide (EDC) (Figure 13.3). DIC is an easy-to-handle liquid, whereas DCC (mp 34 °C) is generally sold as a solidified mass that is difficult to remove from its container. Moreover, the by-product diisopropylurea is more soluble than dicyclohexylurea in most solvents, and can be washed away more easily during solid-phase synthesis.

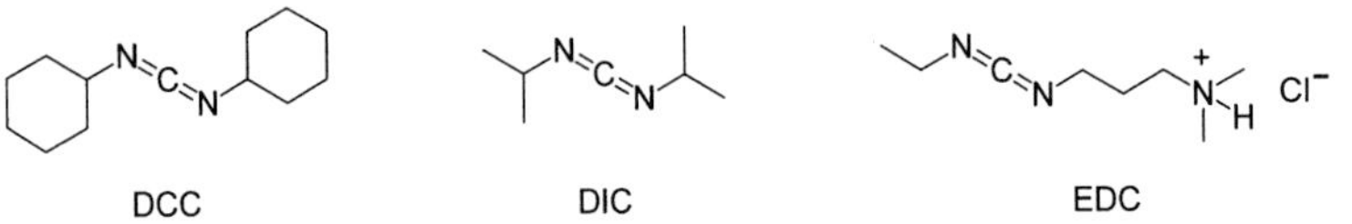

DCC   DIC   EDC

Figure 13.3. Representative carbodiimides used in solid-phase synthesis.

The mechanism of carbodiimide-mediated amide formation is outlined in Figure 13.4. The first intermediate formed during the reaction of a carbodiimide with a carboxylic acid (or an ammonium carboxylate [56]) is an *O*-acylisourea [1]. Under acidic conditions [57,58], this intermediate reacts with acids to yield anhydrides, and with alcohols, phenols, HOBt, or other compounds containing hydroxyl groups, to yield the corresponding esters. The carbodiimide is thereby transformed into a urea.

Under basic reaction conditions, however, reaction of the *O*-acylisourea with a carboxylate is slow [58], and its rearrangement to an *N*-acylurea can efficiently compete with anhydride formation [59]. This can, for instance, occur in the presence of excess carbodiimide [60], a tertiary amine [56,61], or pyridine [59]. *N*-Acylureas are stable products, and will generally not acylate amines. Generation of *O*-acylisoureas in the presence of primary or secondary amines can lead to the formation of amides, but *N*-acylureas can readily be formed as by-products, in particular if a large excess of amine is used. The formation of *N*-acylureas can be suppressed by using excess carboxylic acid or by adding an acidic, hydroxyl group containing compound (e.g. HOBt) to the reaction mixture. Under these conditions, the *O*-acylisourea is mainly converted into a symmetric anhydride or an active ester, which then acylates the amine. Amines react only slowly with carbodiimides (to yield guanidines), and this side reaction does not usually interfere significantly with amide formation.

For the solid-phase synthesis of amides, it makes a significant difference whether the amine or the acid is linked to the support. Resin-bound amines are readily acylated by adding first a carboxylic acid and then a carbodiimide (Table 13.3). The acid/carbodiimide ratio is not critical, because both the *O*-acylisourea (ratio 1:1) and the symmetric anhydride (ratio 2:1) will lead to N-acylation. It should, however, be borne in mind that the half-lives of *O*-acylisoureas are shorter than those of anhydrides, and for difficult couplings it might be advantageous to acylate with a symmetric anhydride (two equivalents of acid and one of carbodiimide).

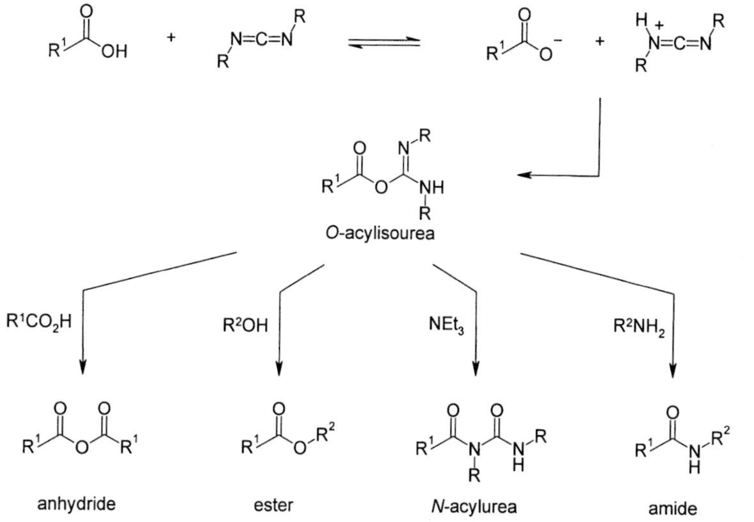

Figure 13.4. Activation of carboxylic acids with carbodiimides.

Illustrative examples of the acylation of support-bound amines with carbodiimides as coupling agents are listed in Table 13.3. Difficulties are usually encountered in the acylation of α-alkylamino acid derivatives (which are significantly less nucleophilic than simple secondary amines; Entries **3** and **4**) and N-alkyl (Entries **5** and **6**) or N-aryl anilines. Acylations with haloacetic or related acids containing a leaving group prone to nucleophilic displacement should not be performed with the aid of HOBt and bases because O-alkylation of HOBt by the product occurs readily.

Experimental Procedure 13.1: Typical acylation of support-bound amines with anhydrides formed in situ [71]

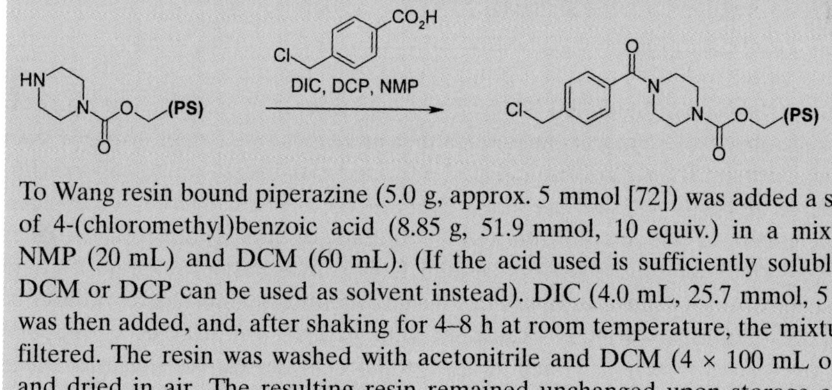

To Wang resin bound piperazine (5.0 g, approx. 5 mmol [72]) was added a solution of 4-(chloromethyl)benzoic acid (8.85 g, 51.9 mmol, 10 equiv.) in a mixture of NMP (20 mL) and DCM (60 mL). (If the acid used is sufficiently soluble, pure DCM or DCP can be used as solvent instead). DIC (4.0 mL, 25.7 mmol, 5 equiv.) was then added, and, after shaking for 4–8 h at room temperature, the mixture was filtered. The resin was washed with acetonitrile and DCM (4 × 100 mL of each) and dried in air. The resulting resin remained unchanged upon storage at room temperature for several months.

Table 13.3. Acylation of support-bound amines with carboxylic acids and carbodiimides.

Entry	Starting resin	Conditions	Product	Ref.
1		Boc-Gly-OH (4 eq), DCC (4 eq), DCM, 0.5 h		[62] see also [63,64]
2		RCO$_2$H (0.5 eq), C$_6$F$_5$OH (2.5 eq), DIC (2.5 eq), DMF, 24 h		[65]
3		(20 eq), DIC (10 eq), DCM/DMF 4:1, 2 × 2 h X: H or COR		[66] see also [67]
4		EDC (0.44 mol/L), anthranilic acid (0.37 mol/L), NMP, 12 h		[68]
5		BrCH$_2$CO$_2$H (0.6 mol/L, 12 eq), DIC (13 eq), DMF, 20 °C, 2 × 0.5 h		[69]
6		(3 eq), HOAt (3 eq), DIC (3 eq), DMF, 20 °C, 8 h		[70]

Because carbodiimide-mediated N-acylations lead mainly to acid-derived by-products, the conversion of support-bound acids into amides is less straightforward than the opposite variant. If a support-bound acid is treated with excess amine and a carbodiimide, large amounts of *N*-acylureas will generally be formed. This can be avoided by adding HOBt to the mixture, thereby generating an intermediate HOBt ester (Table 13.4). Yields of amides are not as high as for the acylation of support-bound amines, however, which makes this strategy unsuitable for the solid-phase synthesis of peptides. An additional problem associated with peptide synthesis using support-bound active esters is that activated *N*-acyl amino acid derivatives racemize more readily than activated *N*-alkoxycarbonyl amino acids (see, e.g., [5] and Figure 16.2). These problems are less critical for the solid-phase synthesis of small non-peptides, and several successful conversions of support-bound acids into amides have been reported (Table 13.4). As discussed above, support-bound acylating agents of low reactivity will often lead to cleaner products than highly activated acid derivatives.

Table 13.4. Acylation of amines with support-bound carboxylic acids and carbodiimides.

Entry	Starting resin	Conditions	Product	Ref.
1		BnNH$_2$, HOBt, EDC (5 eq of each), DMF, 20 °C, 16 h		[44] see also [73]
2		EDC, HOBt, NEt$_3$ (3.5 eq of each), DMF, 12 h		[74] see also [75]
3		HOBt (0.63 mol/L, 4 eq), N-(2-hydroxy-ethyl)aniline (4 eq), DIC (4 eq), DMF, 20 °C, 12–120 h		[76] see also [77]
4		HOBt, DCC (5 eq of each, 0.12 mol/L), DMF		[78] see also [79]
5		Ph⌒⌒NH$_2$ (0.17 mol/L, 5 eq), HOBt (4 eq), DIPEA (5 eq), EDC (4 eq), DMF/DCM 1:1, 20 °C, 2 × 6 h		[80]

Experimental Procedure 13.2: Synthesis of unsymmetric malonamides [71,77]

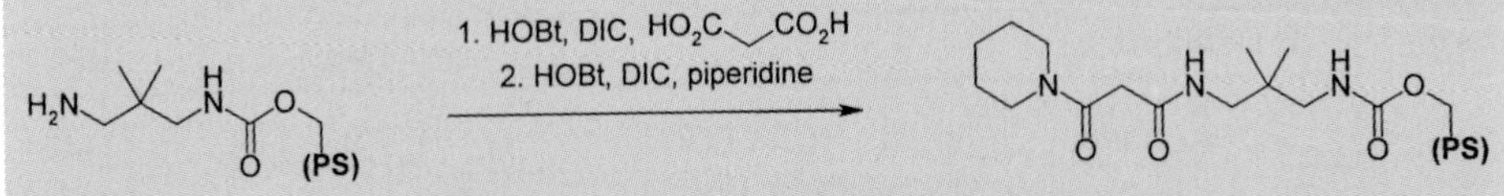

To Wang resin bound 1,3-diamino-2,2-dimethylpropane (40 mg, approx. 0.04 mmol [72,81]) was added a solution of HOBt (130 mg, 0.96 mmol, 24 equiv.) and malonic acid (48 mg, 0.46 mmol, 12 equiv.) in a mixture of NMP (1 mL) and DCP (1 mL). DIC (0.18 mL, 1.16 mmol, 29 equiv.) was added, and the resulting mixture was shaken at room temperature for 3–20 h. After filtration, the resin was washed with NMP (3 × 2 mL). A solution of HOBt (148 mg, 1.10 mmol, 27 equiv.) in a mixture of NMP (1 mL) and DCP (1 mL) was added, followed by DIC (0.16 mL, 1.03 mmol, 26 equiv.) and piperidine (0.12 mL, 1.21 mmol, 30 equiv.). The resulting mixture was shaken at room temperature for 6–20 h, then filtered, and the resin was washed with NMP and DCM.

13.1.2.2 Activation of Acids with Phosphonium Salts

Triaminophosphonium salts, unlike carbodiimides, only react with carboxylic acids under basic conditions, thereby yielding acyloxyphosphonium salts (Figure 13.5). Depending on the counterion X^- and on the precise reaction conditions, the acyloxyphosphonium salt can either acylate the amine directly or be transformed into the symmetric anhydride or an active ester [2,82], which then acylates the amine. At low

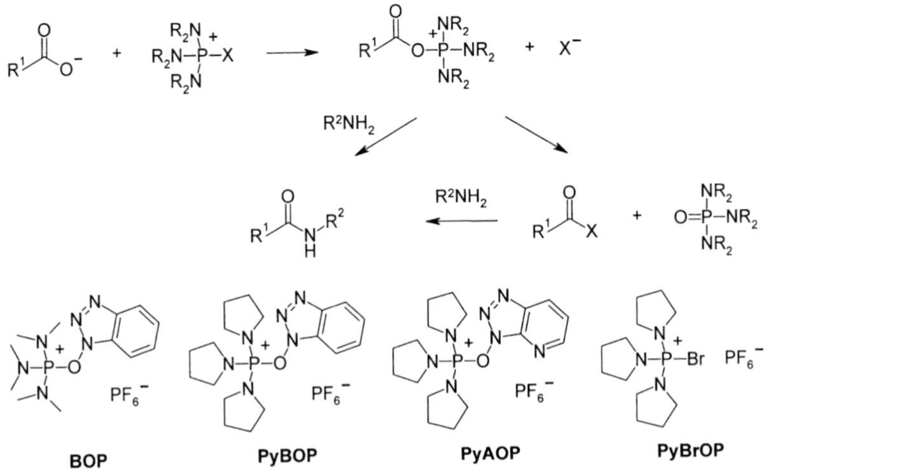

Figure 13.5. Activation of carboxylates by triaminophosphonium salts [17,82,85–88]. X: leaving group.

temperatures, direct acylation by the acyloxyphosphonium intermediate seems to be the prevalent mechanism [83].

A selection of commercially available phosphonium salts suitable for the activation of carboxylic acids in the presence of amines is sketched in Figure 13.5. Phosphonium salts such as those shown in Figure 13.5 do not react with amines, and are well suited for preparing amides on insoluble supports, with either the amine or the acid linked to the support (Table 13.5). Solutions of these reagents in dry DMF are quite stable and can be used even after standing at room temperature for several days [84]. BOP, one of the first phosphonium salts used for peptide synthesis [85], leads to the formation of mutagenic HMPA, and should therefore be replaced by the less hazardous PyBOP [86].

Table 13.5. Preparation of amides using phosphonium salts as coupling agents.

Entry	Starting resin	Conditions	Product	Ref.
1		quinaldic acid, PyBrOP, DIPEA (10 eq of each), NMP, 24 h		[89]
2		PyBrOP (each 0.5 mol/L), DMSO/NMM 1:1, 2 × 0.5 h Ar: 4-(MeO)C$_6$H$_4$		[90]
3		PyBOP, DIPEA, DMF		[91] see also [92]
4		PyBOP, NMM, 2-aminophenol (5 eq of each, 0.38 mol/L), DMF, 20 °C, 17 h		[93]
5		iPr$_2$NH, PyBrOP, DCM (low yield)		[94]

13.1.2.3 Activation of Acids with Uronium Salts

Uronium salts can be used to convert carboxylates into HOBt or related esters [5], into pentafluorophenyl esters [95], or into acyl halides [10] in the presence of amines (Figure 13.6). A highly reactive O-acylisouronium salt is formed initially. This intermediate cannot undergo intramolecular rearrangements and acts only as an acylating agent.

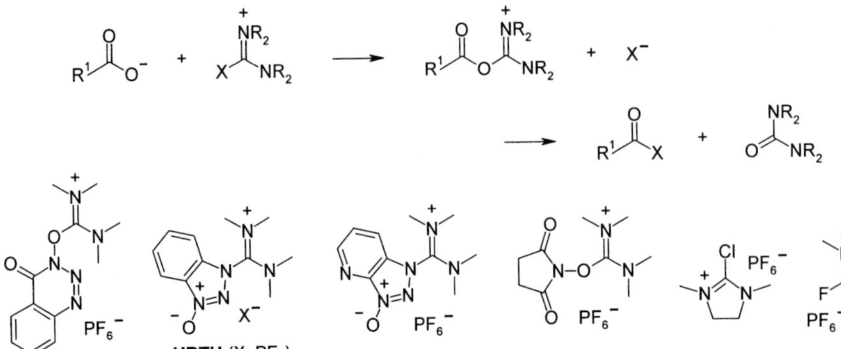

Figure 13.6. Activation of carboxylates by uronium salts [5,10,17,18,84,96]. X: leaving group.

Table 13.6. Preparation of amides using uronium salts as coupling agents.

Entry	Starting resin	Conditions	Product	Ref.
1	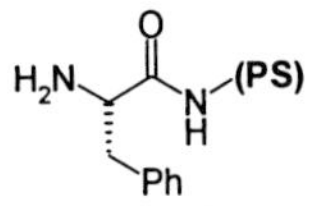	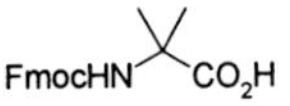 FmocHN—CO$_2$H (5 eq), TFFH (5 eq), DIPEA (10 eq), DMF, 0.5 h	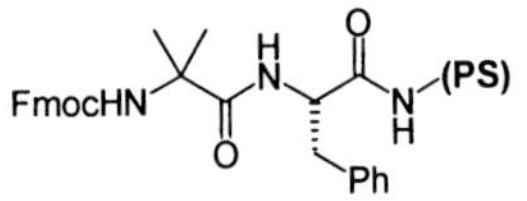	[10]
2	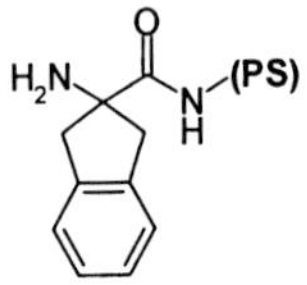	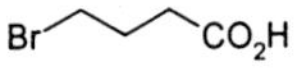 Br—CO$_2$H, HATU, DIPEA, DMF	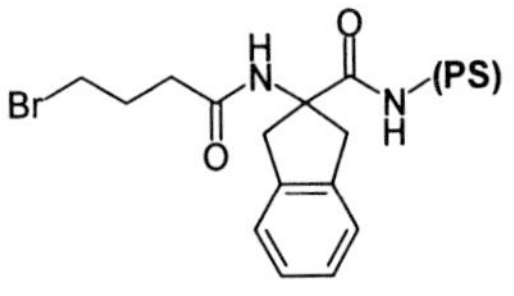	[97]
3	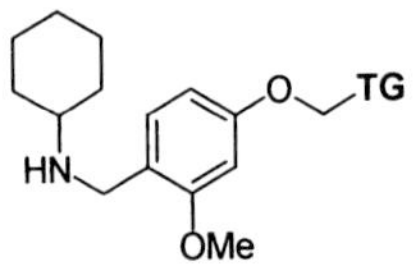	Boc-Phe-OH (5 eq), HATU (5 eq), DIPEA (10 eq), DCM/NMP 1:1, 12 h	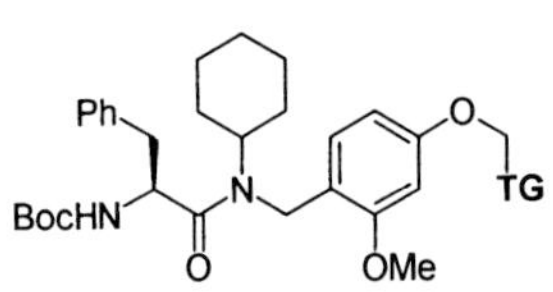	[98]
4	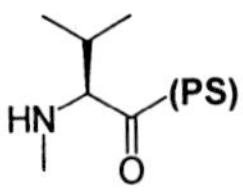	CIP (2.8 eq), HOAt (1.4 eq), Fmoc-Val-OH (2.8 eq), DIPEA (0.4 mol/L, 8.3 eq), DMF, 1.5 h	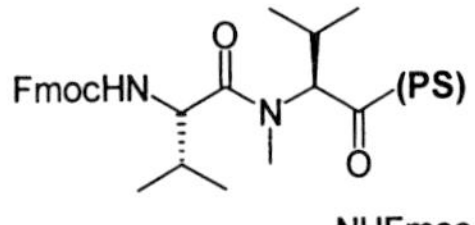	[99]
5	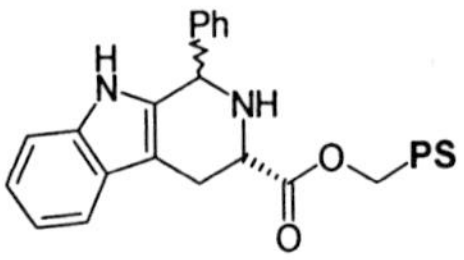	Fmoc-Ala-OH (3 eq), CIP (3 eq), DIPEA (6 eq), NMP, 20 °C, 2 × 16 h	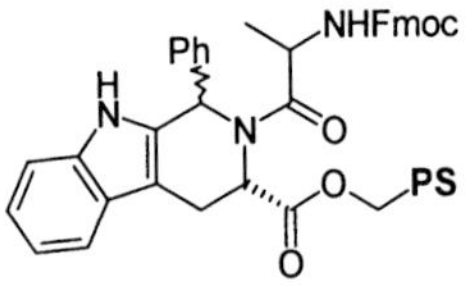	[100] see also [101,102]
6	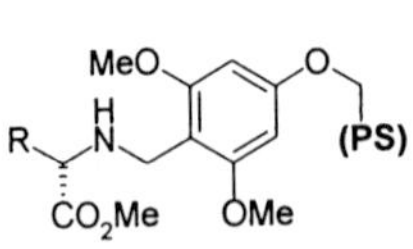	Fmoc-amino acid (10 eq), HATU (10 eq), DIPEA (20 eq), DCM/DMF 9:1, 25 °C, 2 × 2 h	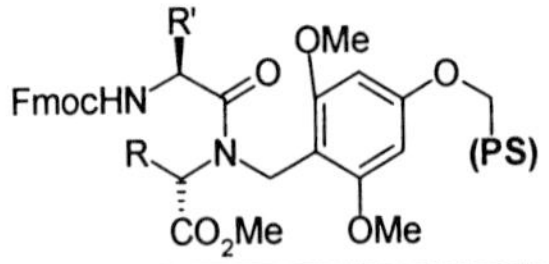	[103]

Table 13.6. continued.

Entry	Starting resin	Conditions	Product	Ref.
7		(0.67 mol/L), HBTU (0.63 mol/L), DMF/ DIPEA 2:1, 45 min		[104]
8		phenylacetic acid (5 eq, 0.1 mol/L), HBTU (5 eq), DIPEA (10 eq), DMF, 20 °C, 3 h		[105]
9		(0.07 mol/L), TBTU, HOBt, DIPEA (each 0.12 mol/L), DMF, 25 °C, 16 h		[106]
10		1. TBTU (5 eq), NMM (5 eq), DMF, 0.5 h, then wash resin 2. H-Phe-OMe, 4 h		[107]

The reactivity of uronium- (and phosphonium-) based coupling reagents is mainly determined by the type of activated acid derivative formed during activation (see Figure 13.2). Unlike phosphonium salts, uronium salts can react with amines to yield guanidines [84]. This side reaction can interfere with amide formation if more uronium salt than carboxylic acid is used. Illustrative examples of the use of uronium salts are listed in Table 13.6.

13.1.2.4 Activation of Acids with Miscellaneous Coupling Reagents

Figure 13.7 shows a selection of azolium-, ammonium-, and immonium-based coupling reagents, which are suitable for the activation of carboxylic acids in the presence of amines. The main requirement for the use of these reagents is that the reaction with carboxylate proceeds significantly more rapidly than that with amines. This can generally be achieved by using a positively charged electrophile. The reagents sketched in Figure 13.7 are well suited for the preparation of peptides, and can even be used for difficult couplings such as those required for the preparation of N-alkylated peptides [108,109].

Other less commonly used coupling reagents include EEDQ (formation of mixed carboxylic carbonic anhydrides), Bop-Cl (formation of mixed carboxylic phosphinic anhydrides [52,53]), DPPA (formation of acyl azides), DECP (formation of acyl cyanides), MSNT (formation of mixed carboxylic sulfonic anhydrides), and benzisoxazolium salts (generation of phenyl esters [54]).

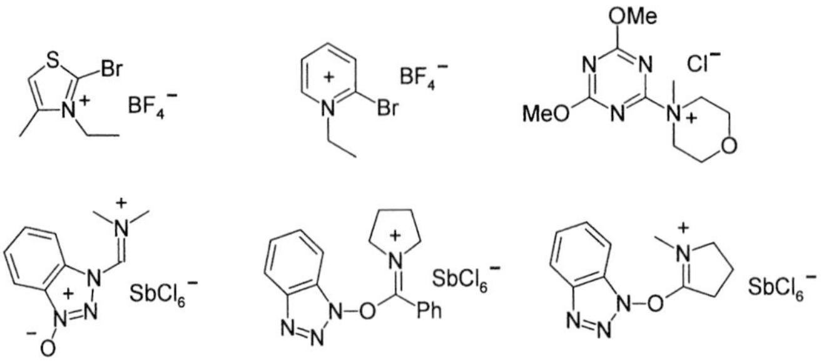

Figure 13.7. Azolium, ammonium, and immonium salts suitable as coupling reagents [108–110].

13.1.3 Miscellaneous Preparations of Amides

In addition to the reagents presented above, other acylating agents can be used to acylate support-bound amines (Table 13.7). Ketenes are highly reactive acylating agents that can be generated in situ by Wolff rearrangement of diazo ketones (Entry **1**, Table 13.7), by thermolysis of 2-acyl Meldrum acids or substituted 1,3-dioxin-4-ones (Entry **2**, Table 13.7), or by photolysis of chromium(II) carbene complexes (Entry **3**, Table 13.7). Thermolysis of 1,3-dioxin-4-ones in the presence of amines is a convenient method for preparing 3-oxo amides, which are valuable intermediates for the synthesis of several types of heterocycle. 3-Oxo amides are difficult to prepare by conventional acylation with an acid because most 3-oxo carboxylic acids are unstable and undergo decarboxylation even at room temperature.

A further strategy used to prepare amides on insoluble supports is based on the Ugi reaction (Figure 13.8). Simple mixing of an amine, an aldehyde, an acid, and an isonitrile can lead to the formation of α-amino acid amides. The mechanism of this remarkable reaction is outlined in Figure 13.8. Sometimes, the amine is first condensed with the aldehyde to form an imine, which is then combined with the acid and the isonitrile.

Although the Ugi reaction can, in principle, be performed with any of the four components linked to the support, in most of the reported examples, support-bound amines and acids have been used (Entries **4–8**, Table 13.7).

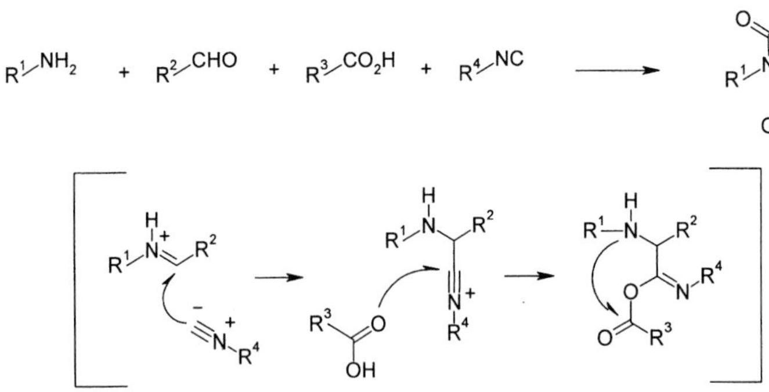

Figure 13.8. Mechanism of the Ugi reaction.

Support-bound C,H-acidic compounds, such as acetoacetamides, react with isocyanates under basic conditions to yield amides through C-carbamoylation [71]. Similarly, polystyrene-bound aryllithium compounds can be converted into benzamides by treatment with isocyanates [111]. These reactions are closely related to C-thiocarbamoylation, which has been used for the solid-phase synthesis of thioamides (see Section 13.9). Amides have also been prepared by C-alkylation of resin-bound N-acylaminals with allyltrimethylsilane or diethylzinc (Entry **11**, Table 13.7).

Table 13.7. Miscellaneous preparations of amides.

Entry	Starting resin	Conditions	Product	Ref.
1		FmocHN group, (4 eq), DMF/THF 1:1, PhCO$_2$Ag (0.35 eq), NEt$_3$, 0 °C, 4 h		[112]
2		(6 eq), PhMe, 65 °C, 6 h		[113] see also [114,115]
3		Cr(CO)$_5$, hv, CO, THF, 20 °C, 30 h	(89% de)	[116] see also [117]
4		(0.67 mol/L each), DCM, 8 h		[118]
5		(0.3 mol/L each), DCM/ MeOH 3:4, 1–8 h		[118]

Table 13.7. continued.

Entry	Starting resin	Conditions	Product	Ref.
6	HO₂C, N, S, PS; Ph	1. BnNH₂, iPrCHO (5 eq of each), dioxane/ MeOH 4:1, MS 4 Å, 20 °C, 3 h 2. add BuNC (5 eq) and resin (1 eq), 55 °C, 48 h		[119] see also [120,121]
7	HO, O, (PS)	BnNH₂, EtCHO, DCM/ MeOH 1:1, 23 °C		[122]
8	NH₂; MeO, OMe, PS	PhCHO (0.5 mol/L), DCM, 25 °C, 1 h; then ArCO₂H, BuNC, 72 h		[123] see also [124–126]
9	CN, (PS)	histamine, ArCHO (each 0.17 mol/L, 10 eq), THF/ MeOH 1:1, 20 °C, 3 d		[38] see also [127,128]
10	H₂N, N, S, PS; NC, N	KOH, EtOH, 80 °C		[129]
11	Ph, H, N, O, PS; MeO, O	allyltrimethylsilane (0.9 mol/L, 10 eq), BF₃OEt₂ (3 eq), DCM, 0 °C, 1 h, 20 °C, 12 h		[130]

13.1.4 Preparation of Amides by C-Alkylation of Other Amides

Support-bound amides bearing an α-hydrogen can be deprotonated and C-alkylated using soft electrophiles. The base required will depend on the acidity of the amide. Amides derived from simple alkanoic acids can only be deprotonated stoichiometrically using strong bases such as LDA (Table 13.8). Glycine esters and amides have been C-alkylated on cross-linked polystyrene after conversion into the corresponding benzophenone imines. This strategy enables the solid-phase synthesis of non-natural α-amino acid derivatives.

Various chiral auxiliaries have been linked to insoluble supports and used to prepare enantiomerically enriched carboxylic acid derivatives (Entries **2** and **3**, Table 13.8). Although the recovery of support-bound chiral auxiliaries is certainly easier than in solution reactions, there have been reports of variable results regarding the stereoselectivity of C-alkylations on insoluble supports [131]. It seems that the optimization of these diastereoselective alkylations is less straightforward on solid phase than in solution, and it remains to be determined if support-bound chiral auxiliaries of this type really represent a viable alternative to non-polymeric auxiliaries.

Table 13.8. Preparation of amides by C-alkylation of other amides.

Entry	Starting resin	Conditions	Product	Ref.
1		BnBr (2 eq), BEMP (2 eq), NMP, 20 °C, overnight		[132] see also [89,133–136]
2		LDA (2 eq), THF, 0 °C, 0.5 h, then allyl iodide (3 eq), 0 °C to 20 °C, 24 h	(87% de)	[137]
3		1. TiCl$_3$(O*i*Pr) (5 eq), DIPEA (6 eq), DCM, 0 °C, 3 h; wash 2. acrylonitrile (10 eq), DCM, 0 °C, 24 h	(78% de)	[138] see also [131, 139]
4		LDA (15 eq), *i*PrI (50 eq), THF, 0–20 °C, 10 h Ar: 4-BrC$_6$H$_4$		[140]

13.1.5 Preparation of Amides by N-Alkylation of Other Amides

Acylated primary amines bound to insoluble supports can be deprotonated and N-alkylated. In solution, this reaction generally only leads to acceptable results when reactive, soft alkylating agents devoid of a β-hydrogen are used, because otherwise competing elimination and O-alkylation can occur. Lactams and cyclic imides are often more readily N-alkylated than acyclic amides. Even under optimal conditions, however, the N-alkylation of amides in solution does not always proceed smoothly. Elimination is a less severe problem in the alkylation of support-bound amides because a large excess of the alkylating agent can be used. Examples of successful N-alkylations of amides are listed in Table 13.9. Strongly acidic amides, such as trifluoroacetanilides (and sulfonamides, see Section 8.4) can be N-alkylated with primary aliphatic alcohols under essentially neutral conditions by means of the Mitsunobu reaction (Entry **4**, Table 13.9).

Table 13.9. Preparation of amides by N-alkylation of other amides.

Entry	Starting resin	Conditions	Product	Ref.
1		1. LiO*t*Bu (1 mol/L), THF 2. MeI, DMSO, 20 °C, 2 h		[141] see also [142]
2		LiO*t*Bu (10 eq), THF, 20 °C, 1.5 h, then drain and add OMs (5 eq), DMSO, 2 × 5 h		[73]
3		PhN(Li)Ac, BuI, DMF, 20 °C, 4 h Ar: 2-(N$_3$)-5-BrC$_6$H$_3$		[143]
4		EtOH (0.2 mol/L), PPh$_3$ (0.4 mol/L), DEAD (0.2 mol/L), THF, −10 °C to 20 °C, 6 h		[144]
5		[Bu$_4$N][HSO$_4$], BuLi, THF, −70 °C to 20 °C, overnight		[145]

13.2 Preparation of Hydroxamic Acids and Hydrazides

Hydroxylamines and hydrazines can be acylated on insoluble supports using the same type of acylating agent as is used for the acylation of amines [146–149]. Because of their higher nucleophilicity, hydroxylamines or hydrazines can be acylated more readily than amines, and unreactive acylating agents such as carboxylic esters can sometimes be successfully employed (Table 13.10). Polystyrene-bound *O*-alkyl hydroxamic acids can be N-alkylated by treatment with reactive alkyl halides and bases such as DBU (Entry **5**, Table 13.10).

Table 13.10. Preparation of hydrazides and hydroxamic acid derivatives.

Entry	Starting resin	Conditions	Product	Ref.
1		$N_2H_4 \cdot H_2O/DMI$ 2:1, 90 °C, 6 h		[150]
2		$N_2H_4/BuOH$ 2:8, 20 °C, 6.5 h		[151]
3		$NH_2OH \cdot HCl$, KOH, MeOH, EtOH, DCM, 6 h, then AcOH		[152]
4		H_2NOBn (8 eq), DIC (4 eq), HOBt (4 eq), DMF, overnight		[153]
5		$ArCH_2Br$ (0.4 mol/L, 6 eq), DBU (6 eq), PhMe, 20 °C, 3 d Ar: $4\text{-}BrC_6H_4$		[146]
6		CO, K_2CO_3, Pd(OAc)$_2$, PPh$_3$, PhMe, 100 °C, 16 h		[154]

13.3 Preparation of Carboxylic Acids

Polymer-bound carboxylic acids have often been used as supports for alcohols or as intermediates for the preparation of other supports. Because supports with hydroxyl or amino groups are readily available, the most commonly used strategy for preparing supports with free carboxyl groups is the acylation of polymeric alcohols or amines with succinic or glutaric anhydride. Carboxyl groups can, however, also be prepared by chemical transformation of several other functional groups on insoluble supports. These include the carboxylation of organometallic reagents, the oxidation of benzyl halides and alkenes, and the saponification of carboxylic acid derivatives (Figure 13.9).

One convenient means of access to carboxyl-functionalized, cross-linked polystyrene involves metallation of the support, followed by treatment with carbon dioxide [150,155–158]. Alternatively, better control of the loading and homogeneity of carboxylated polystyrene can be achieved by converting chloromethyl polystyrene into

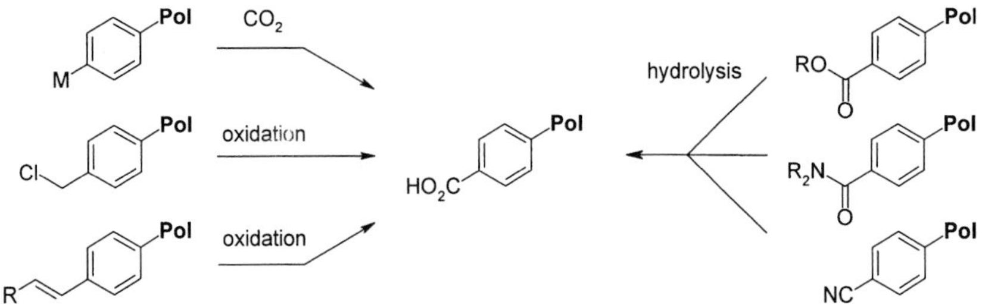

Figure 13.9. Methods for the preparation of support-bound benzoic acids. M: metal.

the corresponding polymeric benzoic acid. This can be accomplished by oxidation of Merrifield resin to the corresponding benzaldehyde by heating with DMSO, followed by exhaustive oxidation to the acid with MCPBA (Entry **1**, Table 13.11). Polystyrene-bound alkenes can be oxidatively degraded to carboxylic acids by ozonolysis and oxidation of the intermediate ozonide with oxygen (Entry **2**, Table 13.11). Another route to carboxyl-functionalized polystyrene is the acid-mediated hydrolysis of nitriles,

Table 13.11. Preparation of carboxylic acids.

Entry	Starting resin	Conditions	Product	Ref.
1	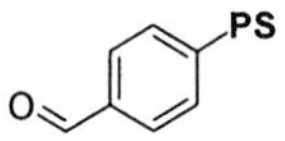	MCPBA, DME, 40 °C, 15 h	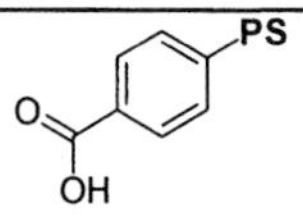	[161]
2	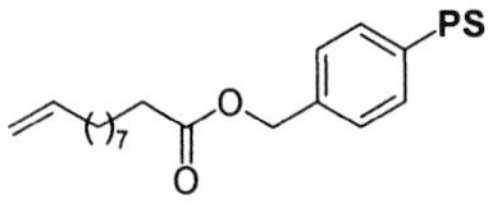	O_3, DCM/AcOH 7:1, −78 °C, 10 min, then O_2, 20 °C, overnight	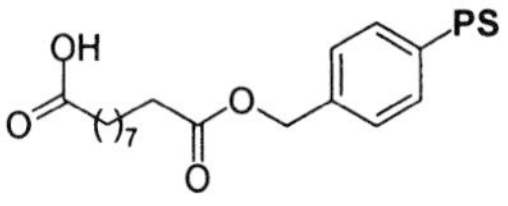	[162]
3	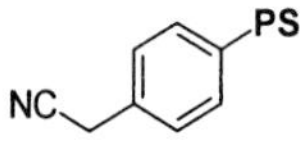	H_2SO_4/AcOH/H_2O 1:1:1, 120 °C, 10 h	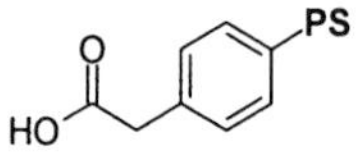	[163]
4	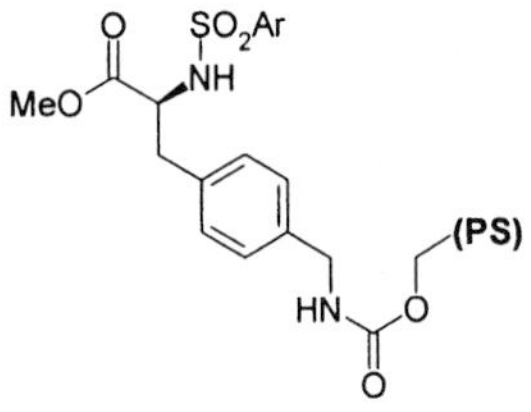	LiOH (5 eq), THF/H_2O 5:1, 20 °C, 1 h Ar: 2-naphthyl	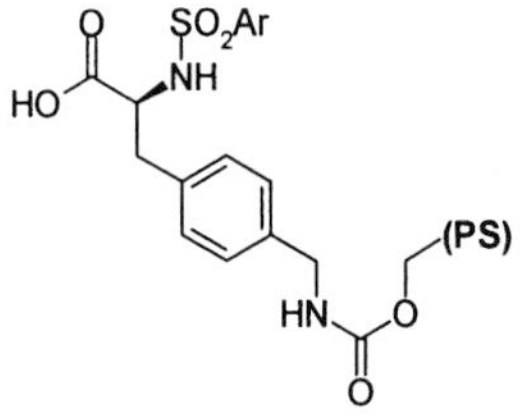	[164]
5	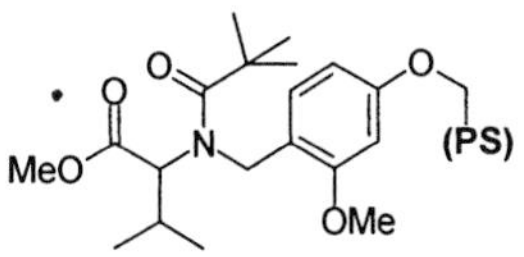	NaOH, MeOH, THF, 60 °C, 24 h	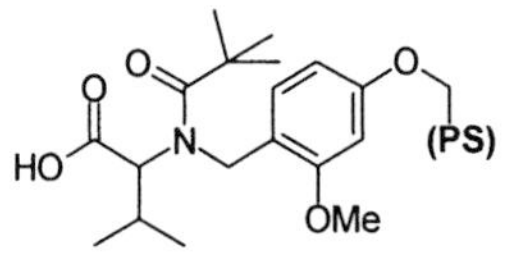	[165]
6	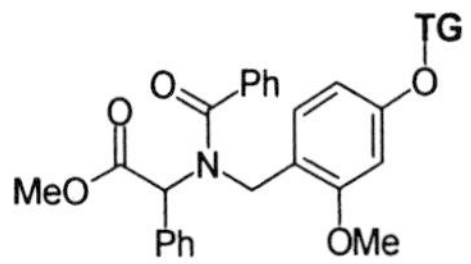	KOH (0.6 mol/L), dioxane/H_2O 3:1, 20 °C, 12 h	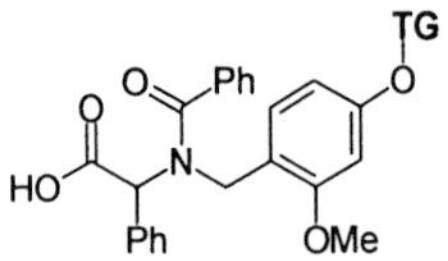	[29]

Table 13.11. continued.

Entry	Starting resin	Conditions	Product	Ref.
7		1. Boc$_2$O, NEt$_3$, DMAP, DCM, 23 °C, 18 h 2. LiOH (0.8 mol/L), H$_2$O/H$_2$O$_2$/THF 19:1:5		[121] see also [166]
8		KOSiMe$_3$, THF, 20 °C		[75,167]
9		Pd(PPh$_3$)$_4$, Me$_3$SiN$_3$, DCE, 20 °C		[75,167] see also [160,168]
10		Pd(PPh$_3$)$_4$ (3 eq), CHCl$_3$/AcOH/NMM 37:2:1, 20 °C, 14 h		[169]
11		HCl, THF, 23 °C		[122]
12		LiOH, H$_2$O$_2$, THF, H$_2$O, 0 °C, 4 h		[170]

which are available from chloromethyl polystyrene by nucleophilic substitution with cyanide (Entry **3**, Table 13.11). Polyethylene has been partially carboxylated by treatment with Cr$_2$O$_3$/H$_2$SO$_4$/H$_2$O [159].

Most of these procedures are incompatible with common linkers, and are therefore unsuitable for the transformation of support-bound substrates into carboxylic acids. A more versatile approach for this purpose is the saponification of carboxylic esters. Saponifications with KOH or NaOH usually proceed smoothly on hydrophilic supports, such as Tentagel [19] or polyacrylamides, but not on cross-linked polystyrene. Esters linked to hydrophobic supports are more conveniently saponified with LiOH [45] or KOSiMe$_3$ in THF or dioxane (Table 13.11). Alternatively, palladium(0)-mediated saponification of allyl esters [94] can be used to prepare acids on cross-linked polystyrene (Entries **9** and **10**, Table 13.11). Fmoc-protected amines are not deprotected under these conditions [160].

13.4 Preparation of Carboxylic Esters

The goal of most studies of esterification reactions on insoluble supports has been the attachment of N-protected amino acids to polystyrene-derived alcohols. In recent years, however, the scope of esterifications has been expanded to other types of alcohol and acid. In the following section, the most important strategies for the preparation of esters on solid phase are discussed. These have been organized according to type of functionality that is initially linked to the support.

13.4.1 Preparation of Esters from Support-Bound Alcohols

Support-bound primary or secondary aliphatic alcohols can be acylated under conditions similar to those used in solution, provided that these conditions are compatible with the chosen linker. For instance, acids can be activated with a carbodiimide either as symmetric anhydrides or as *O*-acylisoureas, which quickly react with alcohols in the presence of a catalyst, such as DMAP or another base, to yield esters (Table 13.12). Further acid derivatives suitable for esterification reactions on solid phase include acyl halides and imidazolides. HOBt esters react only slowly with alcohols, but enable the selective acylation of primary alcohols in the presence of secondary alcohols (Entry **5**, Table 13.12).

Esterification of resin-bound alcohols with 3-oxo carboxylic acids (which readily undergo decarboxylation) or with malonic acid is best performed using the corresponding ketenes, which can be generated in situ by thermolysis of dioxinones or other precursors (Entries **6–9**, Table 13.12). PEG can also be acetoacetylated with acetyl ketene generated by thermolysis [171].

Transesterification under strongly basic reaction conditions has been used to acylate support-bound alcohols with alkyl esters (Entry **10**, Table 13.12). For sensitive acids, the Mitsunobu reaction is a particularly mild method of esterification. This reaction gives high yields with support-bound primary aliphatic alcohols and proceeds under essentially neutral reaction conditions (Experimental Procedure 13.4). Mitsunobu esterification of PEG with *N*-Fmoc amino acids has also been reported [172].

Support-bound phenols [173–176] or other compounds with an acidic OH group (oximes [177–179], hydroxamic acids, *N*-hydroxy benzotriazoles, etc.) undergo rapid esterification with carbodiimides as coupling agents (Entries **13** and **14**, Table 13.12). The resulting products are susceptible to nucleophilic attack, and can be used as insoluble acylating agents.

Table 13.12. Acylation of support-bound alcohols.

Entry	Starting resin	Conditions	Product	Ref.
1		$O_2N\frown CO_2H$ (0.67 mol/L, 4.5 eq), DCC (4.5 eq), THF, 0 °C, 3 h, then 20 °C, ultrasound, overnight		[180]
2		TsO⁻ / ArOC(Me)$_2$CO$_2$H, DIPEA, DCM, 20 °C, 4 h		[181]
3		EtO$_2$CCOCl (0.45 mol/L, 8 eq), NEt$_3$ (1 eq), DMAP (0.15 eq), DCM, 0–20 °C, overnight		[182]
4		phthalic anhydride, (0.44 mol/L, 6.7 eq), NEt$_3$ (6.7 eq), DMAP (1.3 eq), DMF, 20 °C, 18 h		[183]
5		BOP, HOBt (2 eq of each), DIPEA (3 eq), DCM, 20 °C, 24 h		[184]
6	PS–allyl alcohol copolymer	EtO$_2$C$\frown$CO$_2$Et (neat), microwaves (400 W), 5 min		[185] see also [186]
7		THF, 65 °C		[21] see also [77]
8		(10 eq), PhMe, 100 °C, 3 h		[187]
9		Ar$\frown$OEt DMAP, PhMe, 110 °C, 18 h Ar: 2,4,5-F$_3$C$_6$H$_2$		[188]

Table 13.12. continued.

Entry	Starting resin	Conditions	Product	Ref.
10		LDA (1.1 eq), THF, −78 °C, 5 min, then 2 eq 20 °C, overnight		[189]
11		NHAc CO₂H PPh₃, DEAD, THF, 24 h		[190] see also [191]
12		(0.6 mol/L, 0.9 eq), pyridine (1.6 eq), DCM, 20 °C, 72 h		[192]
13		ZHN CO₂H (0.15 mol/L, 3 eq), EDC (3.3 eq), DMF/CHCl₃ 1:1, 20 °C, overnight		[193] see also [194]
14		R CO₂H (0.03 mol/L, 1 eq), DCC (1 eq), DCM, 0 °C, 20 min		[195] see also [196–198]

Experimental Procedure 13.3: Esterification of Wang resin with acryloyl chloride [32]

Acryloyl chloride (143 mL, 1.76 mol) was added dropwise to a stirred mixture of Wang resin (1.0 kg, 0.88 mol), DCM (7.0 L), and triethylamine (0.36 L, 2.6 mol) over a period of 20 min, keeping the temperature below 30 °C. After stirring for 2 h at room temperature, the resin was filtered (inverse filtration) and washed three times with DCM. The acylation was repeated using the same amounts of the reagents and DCM (4 L). After 2 h, the resin was removed by filtration, washed with DCM, methanol, and DMF, and capped by reaction with acetic anhydride (166 mL, 1.76 mol) and triethylamine (244 mL, 1.76 mol) in DMF (2 L) for 1 h at room temperature. Washing with DMF, methanol, DCM, and diethyl ether, followed by drying under reduced pressure, yielded 1.09 kg of resin-bound acrylate.

Experimental Procedure 13.4: Esterification of Wang resin under Mitsunobu conditions [199]

A solution of DEAD (0.27 mL, 1.73 mmol) in THF (2 mL) was added dropwise to a mixture of Wang resin (0.5 g, 0.35 mmol), THF (8 mL), bromoacetic acid (0.24 g, 1.73 mmol), and triphenylphosphine (0.45 g, 1.72 mmol). The resulting mixture was shaken at room temperature for 24 h, then filtered, and the resin was washed with THF and DCM.

13.4.1.1 Esterification of Support-Bound Alcohols with N-Protected α-Amino Acids

Standard solid-phase peptide synthesis requires the first (C-terminal) amino acid to be esterified with a polymeric alcohol. Partial racemization can occur during the esterification of N-protected amino acids with Wang resin or hydroxymethyl polystyrene [200,201]. *N*-Fmoc amino acids are particularly problematic because the bases required to catalyze the acylation of alcohols can also lead to deprotection. A comparative study of various esterification methods for the attachment of Fmoc amino acids to Wang resin [202] showed that the highest loadings with minimal racemization can be achieved under Mitsunobu conditions or by activation with 2,6-dichlorobenzoyl chloride (Experimental Procedure 13.5). *N*-Fmoc amino acid fluorides in the presence of DMAP also proved suitable for the racemization-free esterification of Wang resin (Entry **1**, Table 13.13). The most extensive racemization was observed when DMF or THF was used as solvent, whereas little or no racemization occurred in toluene or DCM [203].

The esterification of *O*-glycosyl-*N*-Fmoc serine with Wang resin leads to extensive racemization of the amino acid if acylation is performed with HOBt, HOAt, or pentafluorophenyl esters in the presence of DMAP [204]. Racemization is minimal, however, when MSNT is used as activating agent (in situ formation of a mixed carboxylic sulfonic anhydride) and 1-methylimidazole is used as base (Entry **2**, Table 13.13). For acylations with Fmoc-Ser(*t*Bu)-OH, which also undergoes extensive racemization when activated with, for example, HATU/HOAt in the presence of NMM [35], 2,4,6-trimethylpyridine (collidine) has been recommended as a base. Tertiary alcohols can usually only be esterified with highly reactive acylating agents, such as acyl chlorides, and few examples of such acylations have been reported (Entry **7**, Table 13.13).

Table 13.13. Acylation of support-bound aliphatic alcohols with α-amino acid derivatives.

Entry	Starting resin	Conditions	Product	Ref.
1		 (0.15 mol/L, 3 eq), DMAP (2 eq), PhMe, 10 min		[203]
2		 (3 eq), MSNT (0.1 mol/L, 3 eq), 1-methylimidazole (2.3 eq), DCM, 0.5 h (R: tetraacetyl-β-galactosyl)	 (+ 6% *R*-serine derivative)	[204] see also [201]
3		Boc-Val-OH (0.16 mol/L), CDI (0.14 eq), DCM, 20 °C, 8 h		[63]
4		Fmoc-Leu-OH (0.6 mol/L, 1.5 eq), PPh₃ (3 eq), DEAD (3 eq), THF, 20 °C, 0.5 h		[205]
5		 (3 eq), DCC (1.5 eq), pyridine, DCM, 20 °C, 24 h		[206] see also [207]
6		 DIC (3 eq), DMAP (0.1 eq), THF, 2 h		[208]
7		 (5 eq), pyridine/DCM 4:6, 25 °C, 10 h		[209] see also [210]
8		 (0.5 mol/L, 2 eq), DCC (2 eq), DMF, 0 °C, 1 h, 20 °C, 5 h		[211]
9		Boc-Ala-OH (0.2 mol/L, 2 eq), DCC (1 eq), DCM/pyridine 7:1, 16 h		[152]

Experimental Procedure 13.5: Esterification of Wang resin with Fmoc amino acids [202]

To a mixture of Wang resin (10 g, 7.4 mmol), Fmoc-Phe-OH (5.7 g, 14.7 mmol, 2 equiv.) and DMF (50 mL) were added pyridine (2 mL, 24.7 mmol, 3.3 equiv.) and 2,6-dichlorobenzoyl chloride (2.1 mL, 14.7 mmol, 2 equiv.). The suspension was shaken at room temperature for 15–20 h. After washing, the remaining hydroxyl groups were benzoylated by treatment of the resin with pyridine (3 mL) and benzoyl chloride (3 mL) in DCE (80 mL) for 2 h. The amount of epimeric amino acid ester remained < 1% for the following *N*-Fmoc amino acids [202]: Asp(O*t*Bu), Arg(Mtr), Glu(O*t*Bu), Ile, Leu, Lys(Boc), Met, Phe, Ser(*t*Bu), Thr(*t*Bu), Tyr(*t*Bu).

13.4.2 Preparation of Esters from Support-Bound Alkylating Agents

As an alternative to the acylation of support-bound alcohols, esters can be prepared by O-alkylation of carboxylates using resin-bound alkylating agents (Table 13.14). This strategy has often been used to link *N*-Boc amino acids to Merrifield resin. Because the tendency of salts to diffuse into cross-linked polystyrene is a function of their lipophilicity and polarizability, the yields of O-alkylations of carboxylates are highly dependent on the chosen cation. In a comparative study, Gisin [212] found that highest substitution rates were observed when cesium salts of Boc amino acids were treated with chloromethyl polystyrene (1% cross-linked), whereas rubidium, potassium, or sodium carboxylates [213,214] were less reactive. Surprisingly, alkylammonium and lithium carboxylates reacted about ten times slower with Merrifield resin than did the cesium salts. Benzyl halides bearing electron-donating groups or benzyl bromides are more reactive than Merrifield resin, and can O-alkylate carboxylates under mild conditions (Experimental Procedure 13.6). As an alternative to benzyl halides, dialkyl(benzyl)sulfonium salts have also been examined as alkylating agents for carboxylates [215].

Polystyrene-bound benzhydryl- or trityl halides react much more rapidly with carboxylates than chloromethyl polystyrene, and the base used to form the carboxylate no longer plays a decisive role in these reactions (see Experimental Procedure 13.7). Support-bound phenyldiazomethanes have been used to prepare esters directly from carboxylic acids under mild reaction conditions. Unfortunately, the diazomethanes required are not easy to prepare, and have not yet found widespread application.

Table 13.14. Preparation of esters from support-bound alkylating agents.

Entry	Starting resin	Conditions	Product	Ref.
1		BocHN—CO₂Cs (R) (0.06 mol/L, 1 eq), DMF, 50 °C, overnight		[212] see also [216]
2		CO₂Cs / OH DMF, 80 °C, 8 h		[217]
3		CO₂H (0.15 mol/L, 5 eq), NaH (5 eq), Bu₄NBr (0.2 eq), THF, 67 °C, 3 d		[218]
4		BocHN—CO₂Cs (3 eq), DMF, 40 °C, overnight		[219]
5		FmocHN—CO₂H DIPEA, DCE, 20 °C, 18–36 h		[220]
6		Fmoc-Ser-OH (1.2 eq), DCM, 20 °C, 5 min		[221]
7		Boc-Gly-OH, CHCl₃, 20 °C, < 1 h		[222] see also [223]

Experimental Procedure 13.6: Esterification by nucleophilic displacement [224]

p-Benzyloxybenzyl bromide resin (50.0 g, 36.5 mmol; prepared from Wang resin; see Experimental Procedure 6.1) was suspended in DMF and treated with 4-fluoro-3-nitrobenzoic acid (13.5 g, 72.9 mmol, 2 equiv.), cesium iodide (19.0 g, 73.1 mmol, 2 equiv.), and DIPEA (9.43 g, 73.0 mmol, 2 equiv.) at room temperature overnight. Filtration, washing with water and then with DMF, DCM, *i*PrOH, DMF, DCM, and *i*PrOH (2 × 300 mL of each), followed by drying under reduced pressure, yielded the polystyrene-bound ester.

Experimental Procedure 13.7: General procedure for the attachment of Fmoc amino acids to a 2-chlorotrityl linker [225]

A solution of DIPEA (0.44 mL, 2.5 mmol) in DCE (5 mL) was added dropwise to a mixture of 2-chlorotrityl chloride resin (1.0 g, 1.6 mmol), DCE (10 mL), and the Fmoc amino acid (1.0 mmol). The resulting mixture was stirred for 20 min, and then methanol (1 mL) was added. After stirring for a further 10 min, the resin was removed by filtration, washed with DCE, DMF, *i*PrOH, methanol, and diethyl ether, and dried under reduced pressure.

13.4.3 Preparation of Esters from Support-Bound Acylating Reagents

The esterification of support-bound carboxylic acids has not been investigated as thoroughly as the esterification of support-bound alcohols. Resin-bound activated acid derivatives that are well suited to the preparation of esters include *O*-acylisoureas (formed from acids and carbodiimides), acyl halides [23,226–228], and mixed anhydrides (Table 13.15). *N*-Acylurea formation does not compete with esterifications as efficiently as it does with the formation of amides from support-bound acids. Esters can also be prepared from carboxylic acids on insoluble supports by acid-catalyzed esterification [152,229]. Alternatively, support-bound carboxylic acids can be esterified by *O*-alkylation, either with primary or secondary aliphatic alcohols under Mitsunobu conditions or with reactive alkyl halides or sulfonates (Table 13.15).

Table 13.15. Preparation of esters from support-bound acids or acylating agents.

Entry	Starting resin	Conditions	Product	Ref.
1		CF₃CH₂OH, DIC, DMAP		[21] see also [218]
2		(5 eq), DIC (5 eq), pyridine, DMAP, DMF, 25 °C, 24 h		[230] see also [218]
3		DIC, DMAP, pyridine, DMF		[231]
4		NEt₃, DMAP, DCM		[20]
5		RR'CHCOH (0.2 eq), NEt₃ (2 eq), DMAP (0.1 eq), DCM, 25 °C, 8 h		[232]
6		Bu₄NI, DIPEA, PhMe, 80 °C		[165]
7		THF		[44]
8		PPh₃, DIAD		[233]
9		iPrOH, AcCl, 55 °C		[122]
10		BnOH (5 mol/L, 100 eq), KCN (0.1 eq), THF, 60 h		[234]

13.4.4 Preparation of Esters by Chemical Modification of Other Esters

Esters linked to insoluble supports can be converted into ester enolates and then C-alkylated under conditions similar to those used in solution. Illustrative examples of such alkylations are listed in Table 13.16. Strong bases such as LDA, KN(SiMe$_3$)$_2$, or BEMP are required to achieve quantitative deprotonation. Benzophenone imines of glycine esterified with Wang resin have been C-alkylated with the aid of phosphazene bases (Entries 3 and 4, Table 13.16). Alkylations of this type can be conducted enantioselectively (up to 89% *ee*) by adding enantiomerically pure, quinine-derived quaternary ammonium salts to a mixture of the base and the alkylating agent [235,236]. An additional method for C-alkylating polystyrene-bound benzophenone imines of glycine involves their oxidation with Pb(OAc)$_4$ to yield an α-acetoxyimine, followed by alkylation with a borane under basic reaction conditions (Entry 5, Table 13.16).

Table 13.16. Preparation of esters by C-alkylation of support-bound esters.

Entry	Starting resin	Conditions	Product	Ref.
1		KN(SiMe$_3$)$_2$, BnBr, THF, −78 °C to 0 °C		[50] see also [239,240]
2		KN(SiMe$_3$)$_2$, THF, −78 °C, allyl iodide		[241]
3		Ph∽∽Br (0.7 mol/L, 10 eq), BEMP (10 eq), NMP, 20 °C, 24 h		[134] see also [89,133, 235,242]
4		(5 eq), BEMP (3 eq), NMP, 20 °C, 16 h		[135] see also [236]
5		1. Pb(OAc)$_4$ (0.06 mol/L, 1.5 eq), DCM, 20 °C, overnight 2. 9-*t*Bu-9-BBN (2 eq), ArOK (1.3 eq), THF, 20 °C, overnight		[243]
6		THF, 50 °C, 5 h		[244]
7		Me$_2$CuLi, Me$_3$SiCl (repeat once)		[241]

To avoid the formation of ketenes by alkoxide elimination, ester enolates are often prepared at low temperatures. If unreactive alkyl halides are used, the addition of Bu$_4$NI to the reaction mixture can be beneficial [134]. Examples of the radical-mediated α-alkylation of support-bound α-haloesters are given in Table 5.4. Further methods for C-alkylating esters on insoluble supports include the Ireland–Claisen rearrangement of *O*-allyl ketene acetals (Entry **6**, Table 13.16). Malonic esters and similar strongly C,H-acidic compounds have been C-alkylated with Merrifield resin [237,238].

13.5 Preparation of Thiol Esters

Thiol esters RC(O)SR are stronger acylating agents than simple alkyl esters, and have been prepared on solid phase mainly as synthetic intermediates. The preparation of thiol esters as intermediates for the synthesis of support-bound thiols is discussed in Section 8.1. Further examples of the preparation of thiol esters on insoluble supports include the aldol addition of ketene thioacetals to polystyrene-bound aldehydes

Table 13.17. Preparation of thiol esters.

Entry	Starting resin	Conditions	Product	Ref.
1		tBuS image BF$_3$OEt$_2$, DCM, Et$_2$O, −78 °C to −5 °C, 1.5 h L: menthylmethyl	 (88% ee)	[218]
2		Boc-Gly-OH, DCC, DMAP (4 eq of each), DCM, 24 h		[245]
3		BnO image Cl NEt$_3$, DCM, 20 °C		[246] see also [247,248]
4		BnSH (30 eq), DCC, DMAP, THF, 20 °C, 18 h		[170]
5		morpholine thioamide image NaI (10 eq), aq DMF, 100 °C, 24 h		[249]
6		AllocHN image SH (0.3 mol/L, 5 eq) DIAD (5 eq), PAr$_3$ (5 eq), THF, 20 °C, 1.5 h Ar: 4-chlorophenyl		[250]

(Entry **1**, Table 13.17), the acylation of support-bound thiols (Entries **2** and **3**), and the acylation of thiols with support-bound carboxylic acids (Entry **4**, Table 13.17). Thiol esters are stable towards acids (e.g. 50% TFA in DCM), but are readily cleaved by nucleophiles and are reduced to alcohols by treatment with LiBH$_4$ [170].

13.6 Preparation of Amidines and Imino Ethers

Amidines are strongly basic compounds that often show interesting biological activity. Few preparations of amidines on insoluble supports have, however, been reported (Table 13.18). Linkers for amidines are considered in Section 3.9.

Amidines have been prepared from other amidines on insoluble supports by nucleophilic displacement of one amine by another (Entry **1**, Table 13.18). Polystyrene-bound thioamides can be converted into amidines by treatment with an aliphatic or aromatic amine in the presence of EDC (Entry **2**, Table 13.18). Alternatively, resin-

Table 13.18. Preparation of amidines and imino ethers.

Entry	Starting resin	Conditions	Product	Ref.
1		PhMe, (NH$_4$)$_2$SO$_4$, 110 °C, 18 h		[254] see also [255]
2		ArCH$_2$NH$_2$ (0.6 mol/L), EDC (1 mol/L), DMF, 20 °C, 24 h		[256]
3		piperidine/DCM 1:4, 20 min		[252]
4		20 °C, 18 h		[257]
5		POCl$_3$ (0.09 mol/L, 10 eq), dioxane, 110 °C, 2.5 h		[105] see also [258]
6		NaN(SiMe$_3$)$_2$ (0.9 eq), THF, 20 °C, 1 h, then Cl$_3$CCN (2 eq), 0–20 °C, 16 h, or (Wang resin) Cl$_3$CCN (1.5 mol/L), DBU/DCM 1:100, 0 °C, 40 min		[259,260] see also [253]
7		NEt$_3$, DCM, MeOH, 20 °C, 4 h		[261]

bound thioamides can be S-alkylated with MeI and then converted into amidines by treatment with ammonia [251] or amines. The starting material used in Entry **3** in Table 13.18 was prepared by dehydration of the corresponding *N*-acylanthranilamide (PPh$_3$, I$_2$, DIPEA, DCM, 20 °C, 14.5 h [252]).

Trichloroacetimidates are the only type of imino ethers to have found some application in solid-phase synthesis. Trichloroacetimidates can readily be prepared from support-bound alcohols by treatment with trichloroacetonitrile and a base (Entry **6**, Table 13.18). Because trichloroacetimidates are good alkylating agents, this reaction offers a convenient alternative for converting support-bound aliphatic alcohols into alkylating agents. Trichloroacetimidates prepared from Wang resin or from hydroxymethyl polystyrene are quite stable and can be stored for several months without decomposition [253].

13.7 Preparation of Nitriles and Isonitriles

Few preparations of nitriles have been performed on insoluble supports (Table 13.19). Aromatic and heteroaromatic nitriles have been prepared on solid phase from the corresponding iodoarenes by metallation followed by reaction with tosyl cyanide (Entry **1**, Table 13.19). Moreover, the reaction of chloromethyl polystyrene with NaCN has been used to prepare support-bound benzyl cyanide (Entry **2**, Table 13.19). Cleavage with simultaneous formation of nitriles can be achieved by treating polystyrene-bound sulfonylhydrazones with KCN (Entry **3**, Table 13.19) or by cleaving amides from a Rink or Sieber linker with TFA anhydride (Entry **10**, Table 3.38 [262]). Support-bound benzaldehydes have been converted into 3-aryl-2-propenenitriles by means of a Horner–Emmons reaction with (EtO)$_2$P(O)CH$_2$CN [263].

Nitrile formation from unprotected asparagine or glutamine is an unwanted side reaction in solid-phase peptide synthesis [9,17]. This dehydration occurs during the activation of asparagine or glutamine with carbodiimides (Figure 13.10), but can be prevented either by performing the coupling in the presence of HOBt, or by using asparagine or glutamine derivatives with a protected amido group (e.g. *N*-trityl, *N*-4,4′-dimethoxybenzhydryl, *N*-2,4,6-trimethoxybenzyl, or *N*-9-xanthenyl amides [17]). These derivatives are also more soluble than non-amide-protected asparagine or glutamine.

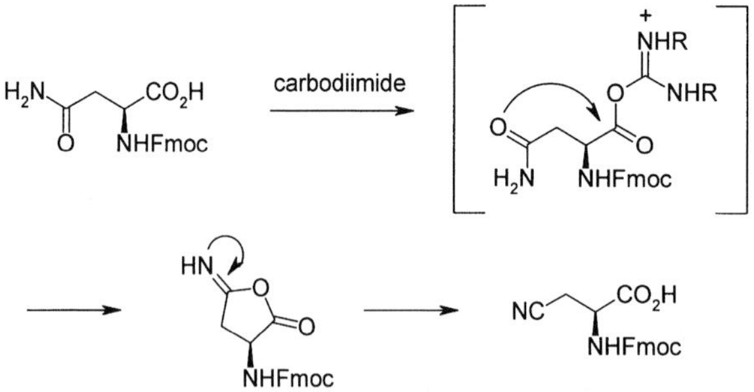

Figure 13.10. Mechanism of dehydration of asparagine during activation [9].

Isonitriles are valuable synthetic intermediates, and can be used to prepare peptoids and various heterocycles. Because of their toxicity and strong, unpleasant smell, however, chemists tend to avoid working with isonitriles. The handling of isolated isonitriles can be elegantly circumvented by generating them on insoluble supports. Despite the obvious advantages of using immobilized isonitriles, few examples have been reported. In most of these, the isonitrile functionality was formed by dehydration of formamides using phosphorus-derived reagents (Entries **4–7**, Table 13.19).

Table 13.19. Preparation of nitriles and isonitriles.

Entry	Starting resin	Conditions	Product	Ref.
1		*i*PrMgBr (0.19 mol/L, 7.3 eq), THF, −35 °C, 0.5 h, then TsCN		[264]
2		NaCN (1.13 mol/L), DMF/H$_2$O 10:2, 120 °C, 19.5 h		[163] see also [237]
3		KCN (0.2 mol/L, 3 eq), MeOH, 65 °C, 72 h		[265]
4		PPh$_3$, CCl$_4$, NEt$_3$, DCM		[266]
5		PPh$_3$ (4 eq), CCl$_4$ (4 eq), NEt$_3$ (10 eq), DCM, 20 °C, 16 h		[267]
6		PPh$_3$, CCl$_4$, NEt$_3$ (5 eq of each), DCM		[38] see also [268]
7		POCl$_3$ (5 eq), DIPEA (15 eq), DCM, 0 °C, 5 h, 20 °C, 1 h		[269]

13.8 Preparation of Imides

The intermolecular N-acylation of amides generally requires treatment of the latter with strong acylating agents, such as ketenes, acyl halides, or anhydrides, in the presence of a base. Few examples of such N-acylations have been reported (Entries **1–3**, Table 13.20; see also saftey-catch linkers, Section 3.1.2.3).

During the solid-phase synthesis of peptides containing *n*-alkyl esters of aspartic acid, under either acidic or basic reaction conditions, cyclic imides might form (Entry **5**, Table 13.20). Hydrolysis of the resulting succinimides can lead to the formation of peptides containing a β-amide bond (β-aspartyl peptides, Figure 13.11). This unwanted imide formation can be suppressed by protecting aspartic acid as a *tert*-alkyl ester (Fmoc methodology) or as a cyclohexyl ester (Boc methodology) [17,270]. Pep-

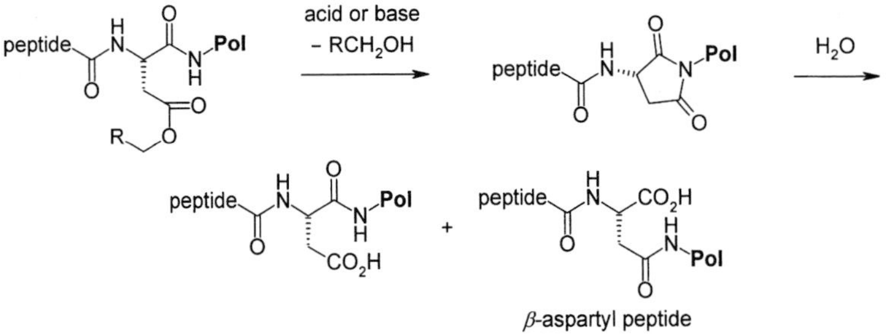

Figure 13.11. Formation and hydrolysis of succinimides derived from aspartic acid.

Table 13.20. Preparation of imides and related compounds.

Entry	Starting resin	Conditions	Product	Ref.
1	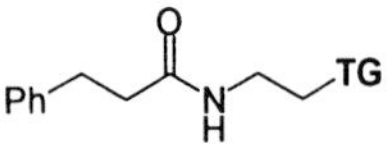 Ph—CO—N(H)—TG	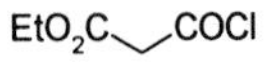 EtO$_2$C—COCl (3.9 mol/L), C$_6$H$_6$, 65 °C, 1.5 h	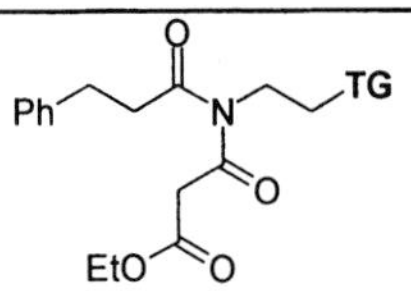 Ph—CO—N—TG, EtO—CO	[271–273]
2	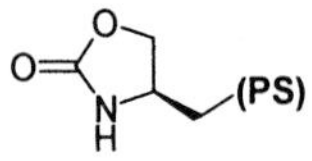 oxazolidinone (PS)	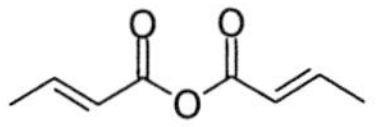 crotonic anhydride, DMAP, NEt$_3$, THF, 70 °C, 72 h	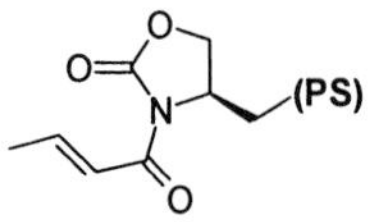 (PS) N-acyl oxazolidinone	[274] see also [139, 275]
3	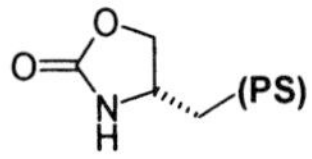 oxazolidinone (PS)	LiN(SiMe$_3$)$_2$ (2 eq), THF, −20 °C, then 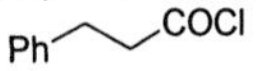 Ph—CO—COCl	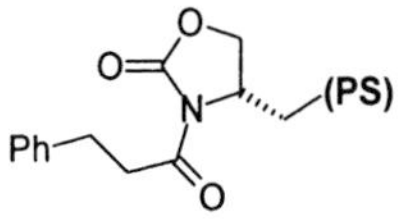 Ph—CO—N oxazolidinone (PS)	[276]
4	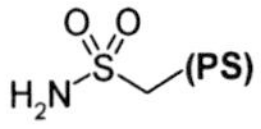 H$_2$N—S(=O)$_2$—(PS)	Fmoc-Phe-OH (0.27 mol/L, 3 eq), DIPEA (5 eq), PyBOP (3 eq), CHCl$_3$, −20 °C, 8 h	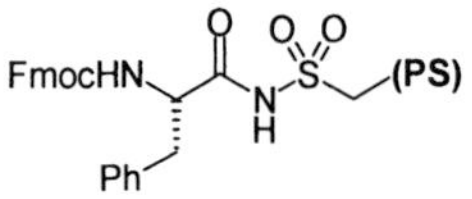 FmocHN—CO—NH—S(=O)$_2$—(PS), Ph	[83] see also [277]
5	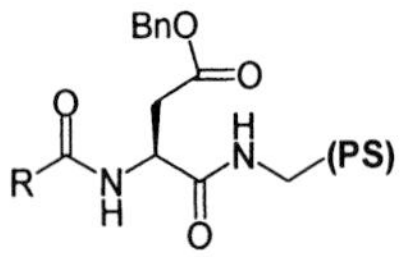 R—CO—NH—CH(CH$_2$OBn CO)—NH—(PS)	NEt$_3$, DCM, 25 °C, 24 h	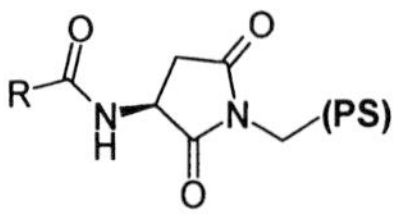 R—CO—NH—succinimide—(PS)	[270] see also [278]
6	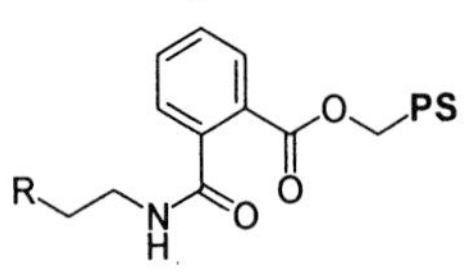 phthalic diamide-ester PS	AcOH/PhMe 25:75, 110 °C, 24 h	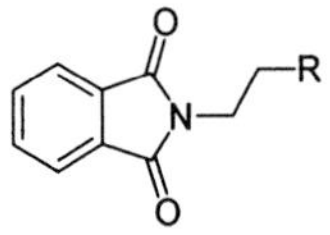 phthalimide—R	[183]
7	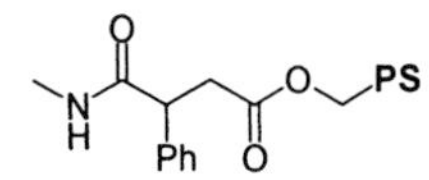 Me—NH—CO—CH(Ph)—CH$_2$—CO—O—PS	AcOH/DMF 1:20, 130 °C, 18 h	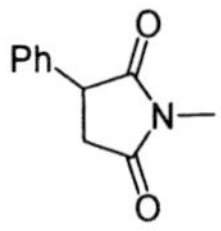 Ph—succinimide—N—Me	[183]
8	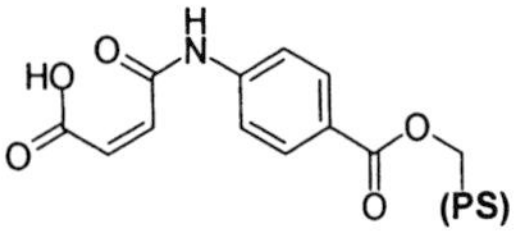 HO—CO—CH=CH—CO—NH—C$_6$H$_4$—CO—O—(PS)	Ac$_2$O (10 eq), NaOAc (1 eq), DMF, 20 °C, 4 h	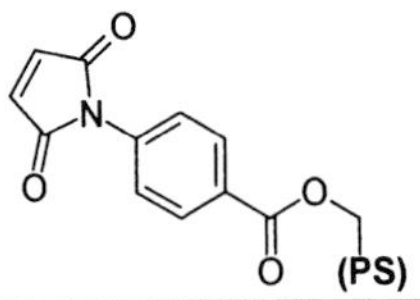 maleimide—C$_6$H$_4$—CO—O—(PS)	[279]

tides containing glutamic acid esters do not undergo imide formation as readily as those containing aspartic acid esters. Five-membered cyclic imides are also accessible by intramolecular nucleophilic cleavage of support-bound phthalic or succinic acid amides (Entries **6** and **7**, Table 13.20).

13.9 Preparation of Thioamides

Thioamides are suitable intermediates for the preparation of amidines, thiazoles, and thiophenes. Thioamides have mainly been prepared on insoluble supports by C-acylation of enamines or C,H-acidic compounds with isothiocyanates (Entries **1–3**,

Table 13.21. Preparation of thioamides.

Entry	Starting resin	Conditions	Product	Ref.
1		ArNCS, THF, 67 °C Ar: 4-(F$_3$CO)C$_6$H$_4$		[280]
2		PhNCS (0.7 mol/L), DBU/DMF 1:3.6, 20 °C, 18 h		[281] see also [256]
3		(1 mol/L), DBU/DMF 1:7, 20 °C, 15 h		[81]
4		Lawesson's reagent (0.16 mol/L), pyridine, 85 °C, 2 × 24 h		[282]
5		(1 mol/L), THF/H$_2$O 4:1, 70 °C, overnight		[283] see also [251]
6		1. BuLi (10 eq), THF, −78 °C to 0 °C, 2 h 2. ArNCS (10 eq), THF, −78 °C to 25 °C, 1 h Ar: 3,4-dimethoxyphenyl		[284]
7		FmocHN (0.34 mol/L, 3 eq), DCM, 1.5 h		[285]

Table 13.21) or from amides by treatment with Lawesson's reagent. Thioamides can also be prepared on cross-linked polystyrene by the addition of H₂S to nitriles (Entry **5**, Table 13.21), by thiocarbamoylation of resin-bound organolithium compounds (Entry **6**), or by the acylation of amines with reactive thio acid derivatives (Entry **7**, Table 13.21).

References for Chapter 13

[1]   March, J. *Advanced Organic Chemistry*; John Wiley & Sons: New York, **1992**.
[2]   Hudson, D. *J. Org. Chem.* **1988**, *53*, 617–624.
[3]   Klose, J.; El-Faham, A.; Henklein, P.; Carpino, L. A.; Bienert, M. *Tetrahedron Lett.* **1999**, *40*, 2045–2048.
[4]   Carpino, L. A.; El-Faham, A. *Tetrahedron* **1999**, *55*, 6813–6830.
[5]   Knorr, R.; Trzeciak, A.; Bannwarth, W.; Gillessen, D. *Tetrahedron Lett.* **1989**, *30*, 1927–1930.
[6]   Hudson, D. *Pept. Res.* **1990**, *3*, 51–55.
[7]   Jones, J. *The Chemical Synthesis of Peptides*; Oxford University Press: Oxford, **1994**.
[8]   Merrifield, B. in *Peptides; Synthesis, Structures, and Applications*; Gutte, B. Ed.; Academic Press: London, **1995**.
[9]   Fields, G. B.; Noble, R. L. *Int. J. Pept. Prot. Res.* **1990**, *35*, 161–214.
[10] Carpino, L. A.; El-Faham, A. *J. Am. Chem. Soc.* **1995**, *117*, 5401–5402.
[11] Chan, T. Y.; Chen, A.; Allanson, N.; Chen, R.; Liu, D. S.; Sofia, M. J. *Tetrahedron Lett.* **1996**, *37*, 8097–8100.
[12] Carpino, L. A. *J. Am. Chem. Soc.* **1993**, *115*, 4397–4398.
[13] Falb, E.; Yechezkel, T.; Salitra, Y.; Gilon, C. *J. Pept. Res.* **1999**, *53*, 507–517.
[14] Wenschuh, H.; Beyermann, M.; Krause, E.; Brudel, M.; Winter, R.; Schumann, M.; Carpino, L. A.; Bienert, M. *J. Org. Chem.* **1994**, *59*, 3275–3280.
[15] Carpino, L. A.; Sadat-Aalaee, D.; Chao, H. G.; DeSelms, R. H. *J. Am. Chem. Soc.* **1990**, *112*, 9651–9652.
[16] Heimer, E. P.; Chang, C. D.; Lambros, T.; Meienhofer, J. *Int. J. Pept. Prot. Res.* **1981**, *18*, 237–241.
[17] Novabiochem Catalog and Peptide Synthesis Handbook, Läufelfingen, CH, **1999**.
[18] Advanced Chemtech Handbook of Combinatorial and Solid-Phase Organic Chemistry, Louisville, USA, **1998**.
[19] Gayo, L. M.; Suto, M. J. *Tetrahedron Lett.* **1997**, *38*, 211–214.
[20] Panek, J. S.; Zhu, B. *J. Am. Chem. Soc.* **1997**, *119*, 12022–12023.
[21] Hamper, B. C.; Kolodziej, S. A.; Scates, A. M. *Tetrahedron Lett.* **1998**, *39*, 2047–2050.
[22] Lorsbach, B. A.; Bagdanoff, J. T.; Miller, R. B.; Kurth, M. J. *J. Org. Chem.* **1998**, *63*, 2244–2250.
[23] Leznoff, C. C.; Dixit, D. M. *Can. J. Chem.* **1977**, *55*, 3351–3355.
[24] Yan, B.; Kumaravel, G.; Anjaria, H.; Wu, A. Y.; Petter, R. C.; Jewell, C. F.; Wareing, J. R. *J. Org. Chem.* **1995**, *60*, 5736–5738.
[25] Goldwasser, J. M.; Leznoff, C. C. *Can. J. Chem.* **1978**, *56*, 1562–1568.
[26] Leznoff, C. C.; Goldwasser, J. M. *Tetrahedron Lett.* **1977**, 1875–1878.
[27] van den Nest, W.; Yuval, S.; Albericio, F. *J. Pept. Sci.* **2001**, *7*, 115–120.
[28] Zhu, Z. M.; McKittrick, B. *Tetrahedron Lett.* **1998**, *39*, 7479–7482.
[29] Bilodeau, M. T.; Cunningham, A. M. *J. Org. Chem.* **1998**, *63*, 2800–2801.
[30] Murphy, M. M.; Schullek, J. R.; Gordon, E. M.; Gallop, M. A. *J. Am. Chem. Soc.* **1995**, *117*, 7029–7030.
[31] Bicknell, A. J.; Hird, N. W. *Bioorg. Med. Chem. Lett.* **1996**, *6*, 2441–2444.
[32] Hamper, B. C.; Kolodziej, S. A.; Scates, A. M.; Smith, R. G.; Cortez, E. *J. Org. Chem.* **1998**, *63*, 708–718.
[33] Plunkett, M. J.; Ellman, J. A. *J. Org. Chem.* **1997**, *62*, 2885–2893.
[34] Krchnák, V.; Flegelová, Z.; Weichsel, A. S.; Lebl, M. *Tetrahedron Lett.* **1995**, *36*, 6193–6196.
[35] Di Fenza, A.; Tancredi, M.; Galoppini, C.; Rovero, P. *Tetrahedron Lett.* **1998**, *39*, 8529–8532.
[36] Zaragoza, F.; Stephensen, H. *J. Org. Chem.* **1999**, *64*, 2555–2557.
[37] Tumelty, D.; Schwarz, M. K.; Needels, M. C. *Tetrahedron Lett.* **1998**, *39*, 7467–7470.
[38] Hulme, C.; Peng, J.; Morton, G.; Salvino, J. M.; Herpin, T.; Labaudiniere, R. *Tetrahedron Lett.* **1998**, *39*, 7227–7230.

[39] Chao, H. G.; Bernatowicz, M. S.; Reiss, P. D.; Klimas, C. E.; Matsueda, G. R. *J. Am. Chem. Soc.* **1994**, *116*, 1746–1752.

[40] Greenberg, M. M.; Gilmore, J. L. *J. Org. Chem.* **1994**, *59*, 746–753.

[41] Albericio, F.; Barany, G. *Int. J. Pept. Prot. Res.* **1985**, *26*, 92–97.

[42] Trautwein, A. W.; Jung, G. *Tetrahedron Lett.* **1998**, *39*, 8263–8266.

[43] Gordeev, M. F.; Luehr, G. W.; Hui, H. C.; Gordon, E. M.; Patel, D. V. *Tetrahedron* **1998**, *54*, 15879–15890.

[44] Léger, R.; Yen, R.; She, M. W.; Lee, V. J.; Hecker, S. J. *Tetrahedron Lett.* **1998**, *39*, 4171–4174.

[45] Linn, J. A.; Gerritz, S. W.; Handlon, A. L.; Hyman, C. E.; Heyer, D. *Tetrahedron Lett.* **1999**, *40*, 2227–2230.

[46] Gordeev, M. F.; Gordon, E. M.; Patel, D. V. *J. Org. Chem.* **1997**, *62*, 8177–8181.

[47] Li, W.; Yan, B. *J. Org. Chem.* **1998**, *63*, 4092–4097.

[48] Marti, R. E.; Yan, B.; Jarosinski, M. A. *J. Org. Chem.* **1997**, *62*, 5615–5618.

[49] Hanessian, S.; Huynh, H. K. *Synlett* **1999**, 102–104.

[50] Hanessian, S.; Xie, F. *Tetrahedron Lett.* **1998**, *39*, 737–740.

[51] Thompson, L. A.; Moore, F. L.; Moon, Y. C.; Ellman, J. A. *J. Org. Chem.* **1998**, *63*, 2066–2067.

[52] Tung, R. D.; Rich, D. H. *J. Am. Chem. Soc.* **1985**, *107*, 4342–4343.

[53] Van der Auwera, C.; Anteunis, M. J. O. *Int. J. Pept. Prot. Res.* **1987**, *29*, 574–588.

[54] Kemp, D. S.; Wrobel, S. J.; Wang, S.-W.; Bernstein, Z.; Rebek, J. *Tetrahedron* **1974**, *30*, 3969–3980.

[55] Sheehan, J. C.; Hess, G. P. *J. Am. Chem. Soc.* **1955**, *77*, 1067–1068.

[56] DeTar, D. F.; Silverstein, R.; Rogers, F. F. *J. Am. Chem. Soc.* **1966**, *88*, 1024–1030.

[57] Doleschall, G.; Lempert, K. *Tetrahedron Lett.* **1963**, 1195–1199.

[58] Smith, M.; Moffatt, J. G.; Khorana, H. G. *J. Am. Chem. Soc.* **1958**, *80*, 6204–6212.

[59] Kurzer, F.; Douraghi-Zadeh, K. *Chem. Rev.* **1967**, *67*, 107–152.

[60] Balcom, B. J.; Petersen, N. O. *J. Org. Chem.* **1989**, *54*, 1922–1927.

[61] DeTar, D. F.; Silverstein, R. *J. Am. Chem. Soc.* **1966**, *88*, 1013–1019.

[62] Sarin, V. K.; Kent, S. B. H.; Mitchell, A. R.; Merrifield, R. B. *J. Am. Chem. Soc.* **1984**, *106*, 7845–7850.

[63] Mitchell, A. R.; Erickson, B. W.; Ryabtsev, M. N.; Hodges, R. S.; Merrifield, R. B. *J. Am. Chem. Soc.* **1976**, *98*, 7357–7362.

[64] Matthews, J.; Rivero, R. A. *J. Org. Chem.* **1998**, *63*, 4808–4810.

[65] Weigelt, D.; Magnusson, G. *Tetrahedron Lett.* **1998**, *39*, 2839–2842.

[66] Simmonds, R. G. *Int. J. Pept. Prot. Res.* **1996**, *47*, 36–41.

[67] Quibell, M.; Packman, L. C.; Johnson, T. *J. Am. Chem. Soc.* **1995**, *117*, 11656–11668.

[68] Boojamra, C. G.; Burow, K. M.; Thompson, L. A.; Ellman, J. A. *J. Org. Chem.* **1997**, *62*, 1240–1256.

[69] Zuckermann, R. N.; Kerr, J. M.; Kent, S. B. H.; Moos, W. H. *J. Am. Chem. Soc.* **1992**, *114*, 10646–10647.

[70] Ouyang, X.; Tamayo, N.; Kiselyov, A. S. *Tetrahedron* **1999**, *55*, 2827–2834.

[71] Zaragoza, F. unpublished results.

[72] Zaragoza, F.; Petersen, S. V. *Tetrahedron* **1996**, *52*, 5999–6002.

[73] Heerding, D. A.; Takata, D. T.; Kwon, C.; Huffman, W. F.; Samanen, J. *Tetrahedron Lett.* **1998**, *39*, 6815–6818.

[74] Kim, S. W.; Hong, C. Y.; Koh, J. S.; Lee, E. J.; Lee, K. *Mol. Diversity* **1998**, *3*, 133–136.

[75] Hoekstra, W. J.; Greco, M. N.; Yabut, S. C.; Hulshizer, B. L.; Maryanoff, B. E. *Tetrahedron Lett.* **1997**, *38*, 2629–2632.

[76] Miller, P. C.; Owen, T. J.; Molyneaux, J. M.; Curtis, J. M.; Jones, C. R. *J. Comb. Chem.* **1999**, *1*, 223–234.

[77] Hamper, B. C.; Snyderman, D. M.; Owen, T. J.; Scates, A. M.; Owsley, D. C.; Kesselring, A. S.; Chott, R. C. *J. Comb. Chem.* **1999**, *1*, 140–150.

[78] Zhang, H. C.; Maryanoff, B. E. *J. Org. Chem.* **1997**, *62*, 1804–1809.

[79] Wess, G.; Bock, K.; Kleine, H.; Kurz, M.; Guba, W.; Hemmerle, H.; Lopez-Calle, E.; Baringhaus, K. H.; Glombik, H.; Enhsen, A.; Kramer, W. *Angew. Chem. Int. Ed. Engl.* **1996**, *35*, 2222–2224.

[80] Devraj, R.; Cushman, M. *J. Org. Chem.* **1996**, *61*, 9368–9373.

[81] Stephensen, H.; Zaragoza, F. *J. Org. Chem.* **1997**, *62*, 6096–6097.

[82] Coste, J.; Dufour, M.-N.; Pantaloni, A.; Castro, B. *Tetrahedron Lett.* **1990**, *31*, 669–672.

[83] Backes, B. J.; Ellman, J. A. *J. Org. Chem.* **1999**, *64*, 2322–2330.

[84] Albericio, F.; Bofill, J. M.; El-Faham, A.; Kates, S. A. *J. Org. Chem.* **1998**, *63*, 9678–9683.

[85] Castro, B.; Dormoy, J. R.; Evin, G.; Selve, C. *Tetrahedron Lett.* **1975**, 1219–1222.

[86] Coste, J.; Le-Nguyen, D.; Castro, B. *Tetrahedron Lett.* **1990**, *31*, 205–208.

[87] Kiso, Y.; Yajima, H. in *Peptides; Synthesis, Structures, and Applications*; Gutte, B. Ed.; Academic Press: London, **1995**.
[88] Ehrlich, A.; Heyne, H.-U.; Winter, R.; Beyermann, M.; Haber, H.; Carpino, L. A.; Bienert, M. *J. Org. Chem.* **1996**, *61*, 8831–8838.
[89] Griffith, D. L.; O'Donnell, M. J.; Pottorf, R. S.; Scott, W. L.; Porco, J. A. *Tetrahedron Lett.* **1997**, *38*, 8821–8824.
[90] Richter, L. S.; Zuckermann, R. N. *Bioorg. Med. Chem. Lett.* **1995**, *5*, 1159–1162.
[91] Zhu, T.; Boons, G. J. *Angew. Chem. Int. Ed.* **1998**, *37*, 1898–1900.
[92] Tan, D. S.; Foley, M. A.; Shair, M. D.; Schreiber, S. L. *J. Am. Chem. Soc.* **1998**, *120*, 8565–8566.
[93] Wang, F.; Hauske, J. R. *Tetrahedron Lett.* **1997**, *38*, 6529–6532.
[94] Roussel, P.; Bradley, M.; Matthews, I.; Kane, P. *Tetrahedron Lett.* **1997**, *38*, 4861–4864.
[95] Habermann, J.; Kunz, H. *Tetrahedron Lett.* **1998**, *39*, 265–268.
[96] Carpino, L. A.; Henklein, P.; Foxman, B. M.; Abdelmoty, I.; Costisella, B.; Wray, V.; Domke, T.; El-Faham, A.; Mügge, C. *J. Org. Chem.* **2001**, *66*, 5245–5247.
[97] Bhandari, A.; Jones, D. G.; Schullek, J. R.; Vo, K.; Schunk, C. A.; Tamanaha, L. L.; Chen, D.; Yuan, Z. Y.; Needels, M. C.; Gallop, M. A. *Bioorg. Med. Chem. Lett.* **1998**, *8*, 2303–2308.
[98] Swayze, E. E. *Tetrahedron Lett.* **1997**, *38*, 8465–8468.
[99] Akaji, K.; Hayashi, Y.; Kiso, Y.; Kuriyama, N. *J. Org. Chem.* **1999**, *64*, 405–411.
[100] van Loevezijn, A.; van Maarseveen, J. H.; Stegman, K.; Visser, G. M.; Koomen, G. J. *Tetrahedron Lett.* **1998**, *39*, 4737–4740.
[101] Fantauzzi, P. P.; Yager, K. M. *Tetrahedron Lett.* **1998**, *39*, 1291–1294.
[102] Mohan, R.; Chou, Y. L.; Morrissey, M. M. *Tetrahedron Lett.* **1996**, *37*, 3963–3966.
[103] del Fresno, M.; Alsina, J.; Royo, M.; Barany, G.; Albericio, F. *Tetrahedron Lett.* **1998**, *39*, 2639–2642.
[104] Baird, E. E.; Dervan, P. B. *J. Am. Chem. Soc.* **1996**, *118*, 6141–6146.
[105] Acharya, A. N.; Ostresh, J. M.; Houghten, R. A. *J. Org. Chem.* **2001**, *66*, 8637–8676.
[106] Böhm, G.; Dowden, J.; Rice, D. C.; Burgess, I.; Pilard, J. F.; Guilbert, B.; Haxton, A.; Hunter, R. C.; Turner, N. J.; Flitsch, S. L. *Tetrahedron Lett.* **1998**, *39*, 3819–3822.
[107] Bleicher, K. H.; Wareing, J. R. *Tetrahedron Lett.* **1998**, *39*, 4591–4594.
[108] Li, P.; Xu, J.-C. *J. Org. Chem.* **2000**, *65*, 2951–2958.
[109] Li, P.; Xu, J.-C. *Tetrahedron* **2000**, *56*, 8119–8131.
[110] Falchi, A.; Giacomelli, G.; Porcheddu, A.; Taddei, M. *Synlett* **2000**, 275–277.
[111] Nicolaou, K. C.; Cao, G.-Q.; Pfefferkorn, J. A. *Angew. Chem. Int. Ed.* **2000**, *39*, 739–743.
[112] Marti, R. E.; Bleicher, K. H.; Bair, K. W. *Tetrahedron Lett.* **1997**, *38*, 6145–6148.
[113] Romoff, T. T.; Ma, L.; Wang, Y. W.; Campbell, D. A. *Synlett* **1998**, 1341–1342.
[114] Weber, L.; Iaiza, P.; Biringer, G.; Barbier, P. *Synlett* **1998**, 1156–1158.
[115] Trautwein, A. W.; Süssmuth, R. D.; Jung, G. *Bioorg. Med. Chem. Lett.* **1998**, *8*, 2381–2384.
[116] Zhu, J.; Hegedus, L. S. *J. Org. Chem.* **1995**, *60*, 5831–5837.
[117] Pulley, S. R.; Hegedus, L. S. *J. Am. Chem. Soc.* **1993**, *115*, 9037–9047.
[118] Szardenings, A. K.; Burkoth, T. S.; Lu, H. H.; Tien, D. W.; Campbell, D. A. *Tetrahedron* **1997**, *53*, 6573–6593.
[119] Obrecht, D.; Abrecht, C.; Grieder, A.; Villalgordo, J. M. *Helv. Chim. Acta* **1997**, *80*, 65–72.
[120] Tempest, P. A.; Brown, S. D.; Armstrong, R. W. *Angew. Chem. Int. Ed. Engl.* **1996**, *35*, 640–642.
[121] Mjalli, A. M. M.; Sarshar, S.; Baiga, T. J. *Tetrahedron Lett.* **1996**, *37*, 2943–2946.
[122] Strocker, A. M.; Keating, T. A.; Tempest, P. A.; Armstrong, R. W. *Tetrahedron Lett.* **1996**, *37*, 1149–1152.
[123] Cao, X. D.; Moran, E. J.; Siev, D.; Lio, A.; Ohashi, C.; Mjalli, A. M. M. *Bioorg. Med. Chem. Lett.* **1995**, *5*, 2953–2958.
[124] Paulvannan, K. *Tetrahedron Lett.* **1999**, *40*, 1851–1854.
[125] Hoel, A. M. L.; Nielsen, J. *Tetrahedron Lett.* **1999**, *40*, 3941–3944.
[126] Piscopio, A. D.; Miller, J. F.; Koch, K. *Tetrahedron* **1999**, *55*, 8189–8198.
[127] Short, K. M.; Ching, B. W.; Mjalli, A. M. M. *Tetrahedron* **1997**, *53*, 6653–6679.
[128] Hulme, C.; Ma, L.; Cherrier, M.-P.; Romano, J. J.; Morton, G.; Duquenne, C.; Salvino, J.; Labaudiniere, R. *Tetrahedron Lett.* **2000**, *41*, 1883–1887.
[129] Srivastava, S. K.; Haq, W.; Chauhan, P. M. S. *Bioorg. Med. Chem. Lett.* **1999**, *9*, 965–966.
[130] Vanier, C.; Wagner, A.; Mioskowski, C. *Chem. Eur. J.* **2001**, *7*, 2318–2323.
[131] Burgess, K.; Lim, D. *Chem. Commun.* **1997**, 785–786.
[132] O'Donnell, M. J.; Zhou, C. Y.; Scott, W. L. *J. Am. Chem. Soc.* **1996**, *118*, 6070–6071.
[133] Scott, W. L.; Zhou, C. Y.; Fang, Z. Q.; O'Donnell, M. J. *Tetrahedron Lett.* **1997**, *38*, 3695–3698.

[134] O'Donnell, M. J.; Lugar, C. W.; Pottorf, R. S.; Zhou, C. Y.; Scott, W. L.; Cwi, C. L. *Tetrahedron Lett.* **1997**, *38*, 7163–7166.

[135] Domínguez, E.; O'Donnell, M. J.; Scott, W. L. *Tetrahedron Lett.* **1998**, *39*, 2167–2170.

[136] O'Donnell, M. J.; Drew, M. D.; Pottorf, R. S.; Scott, W. L. *J. Comb. Chem.* **2000**, *2*, 172–181.

[137] Moon, H.; Schore, N. E.; Kurth, M. J. *Tetrahedron Lett.* **1994**, *35*, 8915–8918.

[138] Phoon, C. W.; Abell, C. *Tetrahedron Lett.* **1998**, *39*, 2655–2658.

[139] Allin, S. M.; Shuttleworth, S. J. *Tetrahedron Lett.* **1996**, *37*, 8023–8026.

[140] Backes, B. J.; Ellman, J. A. *J. Am. Chem. Soc.* **1994**, *116*, 11171–11172.

[141] Nefzi, A.; Ostresh, J. M.; Houghten, R. A. *Tetrahedron* **1999**, *55*, 335–344.

[142] Theoclitou, M.-E.; Ostresh, J. M.; Hamashin, V.; Houghten, R. A. *Tetrahedron Lett.* **2000**, *41*, 2051–2054.

[143] Woolard, F. X.; Paetsch, J.; Ellman, J. A. *J. Org. Chem.* **1997**, *62*, 6102–6103.

[144] Norman, T. C.; Gray, N. S.; Koh, J. T.; Schultz, P. G. *J. Am. Chem. Soc.* **1996**, *118*, 7430–7431.

[145] Furman, B.; Thürmer, R.; Kaluza, Z.; Lysek, R.; Voelter, W.; Chmielewski, M. *Angew. Chem. Int. Ed.* **1999**, *38*, 1121–1123.

[146] Salvino, J. M.; Mervic, M.; Mason, H. J.; Kiesow, T.; Teager, D.; Airey, J.; Labaudiniere, R. *J. Org. Chem.* **1999**, *64*, 1823–1830.

[147] Mellor, S. L.; Chan, W. C. *Chem. Commun.* **1997**, 2005–2006.

[148] Chang, J. K.; Shimizu, M.; Wang, S. S. *J. Org. Chem.* **1976**, *41*, 3255–3258.

[149] Wang, S.; Merrifield, R. B. *J. Am. Chem. Soc.* **1969**, *91*, 6488–6491.

[150] Kobayashi, S.; Furuta, T.; Sugita, K.; Okitsu, O.; Oyamada, H. *Tetrahedron Lett.* **1999**, *40*, 1341–1344.

[151] Wilson, M. W.; Hernández, A. S.; Calvet, A. P.; Hodges, J. C. *Mol. Diversity* **1998**, *3*, 95–112.

[152] Sophiamma, P. N.; Sreekumar, K. *Indian J. Chem. Sect. B - Org. Chem.* **1997**, *36*, 995–999.

[153] Chen, J. J.; Spatola, A. F. *Tetrahedron Lett.* **1997**, *38*, 1511–1514.

[154] Grigg, R.; Major, J. P.; Martin, F. M.; Whittaker, M. *Tetrahedron Lett.* **1999**, *40*, 7709–7711.

[155] Fyles, T. M.; Leznoff, C. C. *Can. J. Chem.* **1976**, *54*, 935–942.

[156] Itsuno, S.; Darling, G. D.; Stöver, H. D. H.; Fréchet, J. M. J. *J. Org. Chem.* **1987**, *52*, 4644–4645.

[157] O'Brien, R. A.; Chen, T.; Rieke, R. D. *J. Org. Chem.* **1992**, *57*, 2667–2677.

[158] Farrall, M. J.; Fréchet, J. M. J. *J. Org. Chem.* **1976**, *41*, 3877–3882.

[159] Luo, K. X.; Zhou, P.; Lodish, H. F. *Proc. Natl. Acad. Sci. USA* **1995**, *92*, 11761–11765.

[160] Annis, D. A.; Helluin, O.; Jacobsen, E. N. *Angew. Chem. Int. Ed.* **1998**, *37*, 1907–1909.

[161] Beebe, X.; Chiappari, C. L.; Kurth, M. J.; Schore, N. E. *J. Org. Chem.* **1993**, *58*, 7320–7321.

[162] Sylvain, C.; Wagner, A.; Mioskowski, C. *Tetrahedron Lett.* **1997**, *38*, 1043–1044.

[163] Kusama, T.; Hayatsu, H. *Chem. Pharm. Bull.* **1970**, *18*, 319–327.

[164] Kim, S. W.; Hong, C. Y.; Lee, K.; Lee, E. J.; Koh, J. S. *Bioorg. Med. Chem. Lett.* **1998**, *8*, 735–738.

[165] Yu, K. L.; Civiello, R.; Roberts, D. G. M.; Seiler, S. M.; Meanwell, N. A. *Bioorg. Med. Chem. Lett.* **1999**, *9*, 663–666.

[166] Flynn, D. L.; Zelle, R. E.; Grieco, P. A. *J. Org. Chem.* **1983**, *48*, 2424–2426.

[167] Hoekstra, W. J.; Maryanoff, B. E.; Andrade-Gordon, P.; Cohen, J. H.; Costanzo, M. J.; Damiano, B. P.; Haertlein, B. J.; Harris, B. D.; Kauffman, J. A.; Keane, P. M.; McComsey, D. F.; Villani, F. J.; Yabut, S. C. *Bioorg. Med. Chem. Lett.* **1996**, *6*, 2371–2376.

[168] Iso, Y.; Shindo, H.; Hamana, H. *Tetrahedron* **2000**, *56*, 5353–5361.

[169] Bourne, G. T.; Meutermans, W. D. F.; Alewood, P. F.; McGeary, R. P.; Scanlon, M.; Watson, A. A.; Smythe, M. L. *J. Org. Chem.* **1999**, *64*, 3095–3101.

[170] Reggelin, M.; Brenig, V.; Welcker, R. *Tetrahedron Lett.* **1998**, *39*, 4801–4804.

[171] Far, A. R.; Tidwell, T. T. *J. Org. Chem.* **1998**, *63*, 8636–8637.

[172] Blaskovich, M. A.; Kahn, M. *J. Org. Chem.* **1998**, *63*, 1119–1125.

[173] Parlow, J. J.; Normansell, J. E. *Mol. Diversity* **1996**, *1*, 266–269.

[174] Cohen, B. J.; Karoly-Hafeli, H.; Patchornik, A. *J. Org. Chem.* **1984**, *49*, 922–924.

[175] Marshall, D. L.; Liener, I. E. *J. Org. Chem.* **1970**, *35*, 867–868.

[176] Parlow, J. J.; Mischke, D. A.; Woodard, S. S. *J. Org. Chem.* **1997**, *62*, 5908–5919.

[177] Scarr, R. B.; Findeis, M. A. *Pept. Res.* **1990**, *3*, 238–241.

[178] DeGrado, W. F.; Kaiser, E. T. *J. Org. Chem.* **1982**, *47*, 3258–3261.

[179] Smith, R. A.; Bobko, M. A.; Lee, W. *Bioorg. Med. Chem. Lett.* **1998**, *8*, 2369–2374.

[180] Sylvain, C.; Wagner, A.; Mioskowski, C. *Tetrahedron Lett.* **1999**, *40*, 875–878.

[181] Brown, P. J.; Hurley, K. P.; Stuart, L. W.; Willson, T. M. *Synthesis* **1997**, 778–782.

[182] Cobb, J. M.; Fiorini, M. T.; Goddard, C. R.; Theoclitou, M. E.; Abell, C. *Tetrahedron Lett.* **1999**, *40*, 1045–1048.

[183] Barn, D. R.; Morphy, J. R. *J. Comb. Chem.* **1999**, *1*, 151–156.

[184] Whitehouse, D. L.; Savinov, S. N.; Austin, D. J. *Tetrahedron Lett.* **1997**, *38*, 7851–7852.
[185] Eynde, J. J. V.; Rutot, D. *Tetrahedron* **1999**, *55*, 2687–2694.
[186] Tietze, L. F.; Steinmetz, A. *Angew. Chem. Int. Ed. Engl.* **1996**, *35*, 651–652.
[187] Tietze, L. F.; Steinmetz, A. *Synlett* **1996**, 667–668.
[188] MacDonald, A. A.; DeWitt, S. H.; Hogan, E. M.; Ramage, R. *Tetrahedron Lett.* **1996**, *37*, 4815–4818.
[189] Karoyan, P.; Triolo, A.; Nannicini, R.; Giannotti, D.; Altamura, M.; Chassaing, G.; Perrotta, E. *Tetrahedron Lett.* **1999**, *40*, 71–74.
[190] Barbaste, M.; Rolland-Fulcrand, V.; Roumestant, M. L.; Viallefont, P.; Martinez, J. *Tetrahedron Lett.* **1998**, *39*, 6287–6290.
[191] Fancelli, D.; Fagnola, M. C.; Severino, D.; Bedeschi, A. *Tetrahedron Lett.* **1997**, *38*, 2311–2314.
[192] Hahn, H. G.; Chang, K. H.; Nam, K. D.; Bae, S. Y.; Mah, H. *Heterocycles* **1998**, *48*, 2253–2261.
[193] Adamczyk, M.; Fishpaugh, J. R.; Mattingly, P. G. *Tetrahedron Lett.* **1999**, *40*, 463–466.
[194] Adamczyk, M.; Fishpaugh, J. R.; Mattingly, P. G. *Bioorg. Med. Chem. Lett.* **1999**, *9*, 217–220.
[195] Huang, W.; Kalivretenos, A. G. *Tetrahedron Lett.* **1995**, *36*, 9113–9116.
[196] Kalir, R.; Warshawsky, A.; Fridkin, M.; Patchornik, A. *Eur. J. Biochem.* **1975**, *59*, 55–61.
[197] Dendrinos, K. G.; Kalivretenos, A. G. *Tetrahedron Lett.* **1998**, *39*, 1321–1324.
[198] Pop, I. E.; Déprez, B. P.; Tartar, A. L. *J. Org. Chem.* **1997**, *62*, 2594–2603.
[199] Nouvet, A.; Lamaty, F.; Lazaro, R. *Tetrahedron Lett.* **1998**, *39*, 3469–3470.
[200] Mergler, M.; Tanner, R.; Gosteli, J.; Grogg, P. *Tetrahedron Lett.* **1988**, *29*, 4005–4008.
[201] Blankemeyer-Menge, B.; Nimtz, M.; Frank, R. *Tetrahedron Lett.* **1990**, *31*, 1701–1704.
[202] Sieber, P. *Tetrahedron Lett.* **1987**, *28*, 6147–6150.
[203] Granitza, D.; Beyermann, M.; Wenschuh, H.; Haber, H.; Carpino, L. A.; Truran, G. A.; Bienert, M. *J. Chem. Soc., Chem. Commun.* **1995**, 2223–2224.
[204] Harth-Fritschy, E.; Cantacuzène, D. *J. Pept. Res.* **1997**, *50*, 415–420.
[205] Barlos, K.; Gatos, D.; Kallitsis, J.; Papaioannou, D.; Sotiriu, P.; Schäfer, W. *Liebigs Ann. Chem.* **1987**, 1031–1035.
[206] Ajayaghosh, A.; Pillai, V. N. R. *J. Org. Chem.* **1987**, *52*, 5714–5717.
[207] Routledge, A.; Stock, H. T.; Flitsch, S. L.; Turner, N. J. *Tetrahedron Lett.* **1997**, *38*, 8287–8290.
[208] Kuisle, O.; Quiñoá, E.; Riguera, R. *Tetrahedron Lett.* **1999**, *40*, 1203–1206.
[209] Akaji, K.; Kiso, Y.; Carpino, L. A. *J. Chem. Soc., Chem. Commun.* **1990**, 584–586.
[210] Blackburn, C.; Pingali, A.; Kehoe, T.; Herman, L. W.; Wang, H. Q.; Kates, S. A. *Bioorg. Med. Chem. Lett.* **1997**, *7*, 823–826.
[211] Kalir, R.; Fridkin, M.; Patchornik, A. *Eur. J. Biochem.* **1974**, *42*, 151–156.
[212] Gisin, B. F. *Helv. Chim. Acta* **1973**, *56*, 1476–1482.
[213] Yoo, S. E.; Seo, J. S.; Yi, K. Y.; Gong, Y. D. *Tetrahedron Lett.* **1997**, *38*, 1203–1206.
[214] Kurth, M. L.; Randall, L. A. A.; Chen, C.; Melander, C.; Miller, R. B.; McAlister, K.; Reitz, G.; Kang, R.; Nakatsu, T.; Green, C. *J. Org. Chem.* **1994**, *59*, 5862–5864.
[215] Dorman, L. C.; Love, J. *J. Org. Chem.* **1969**, *34*, 158–165.
[216] Frenette, R.; Friesen, R. W. *Tetrahedron Lett.* **1994**, *35*, 9177–9180.
[217] Hanessian, S.; Yang, R. Y. *Tetrahedron Lett.* **1996**, *37*, 5835–5838.
[218] Gennari, C.; Ceccarelli, S.; Piarulli, U.; Aboutayab, K.; Donghi, M.; Paterson, I. *Tetrahedron* **1998**, *54*, 14999–15016.
[219] Nicolás, E.; Clemente, J.; Ferrer, T.; Albericio, F.; Giralt, E. *Tetrahedron* **1997**, *53*, 3179–3194.
[220] Garigipati, R. S. *Tetrahedron Lett.* **1997**, *38*, 6807–6810.
[221] Bhalay, G.; Dunstan, A. R. *Tetrahedron Lett.* **1998**, *39*, 7803–7806.
[222] Chapman, P. H.; Walker, D. *J. Chem. Soc., Chem. Commun.* **1975**, 690–691.
[223] Mergler, M.; Dick, F.; Gosteli, J.; Nyfeler, R. *Tetrahedron Lett.* **1999**, *40*, 4663–4664.
[224] Morales, G. A.; Corbett, J. W.; DeGrado, W. F. *J. Org. Chem.* **1998**, *63*, 1172–1177.
[225] Barlos, K.; Chatzi, O.; Gatos, D.; Stavropoulos, G. *Int. J. Pept. Prot. Res.* **1991**, *37*, 513–520.
[226] Wong, J. Y.; Leznoff, C. C. *Can. J. Chem.* **1973**, *51*, 2452–2456.
[227] Kantorowski, E. J.; Kurth, M. J. *J. Org. Chem.* **1997**, *62*, 6797–6803.
[228] Nizi, E.; Botta, M.; Corelli, F.; Manetti, F.; Messina, F.; Maga, G. *Tetrahedron Lett.* **1998**, *39*, 3307–3310.
[229] Fréchet, J. M.; Schuerch, C. *J. Am. Chem. Soc.* **1971**, *93*, 492–496.
[230] Meyers, H. V.; Dilley, G. J.; Durgin, T. L.; Powers, T. S.; Winssinger, N. A.; Zhu, H.; Pavia, M. R. *Mol. Diversity* **1995**, *1*, 13–20.
[231] Barber, A. M.; Hardcastle, I. R.; Rowlands, M. G.; Nutley, B. P.; Marriott, J. H.; Jarman, M. *Bioorg. Med. Chem. Lett.* **1999**, *9*, 623–626.
[232] Nicolaou, K. C.; Winssinger, N.; Vourloumis, D.; Ohshima, T.; Kim, S.; Pfefferkorn, J.; Xu, J. Y.; Li, T. *J. Am. Chem. Soc.* **1998**, *120*, 10814–10826.

[233] Krchnák, V.; Weichsel, A. S.; Lebl, M.; Felder, S. *Bioorg. Med. Chem. Lett.* **1997**, *7*, 1013–1016.
[234] de Blas, J.; Domínguez, E.; Ezquerra, J. *Tetrahedron Lett.* **2000**, *41*, 4567–4571.
[235] O'Donnell, M. J.; Delgado, F.; Pottorf, R. S. *Tetrahedron* **1999**, *55*, 6347–6362.
[236] O'Donnell, M. J.; Delgado, F.; Domínguez, E.; de Blas, J.; Scott, W. L. *Tetrahedron: Asymmetry* **2001**, *12*, 821–828.
[237] Fréchet, J. M. J.; de Smet, M. D.; Farrall, M. J. *J. Org. Chem.* **1979**, *44*, 1774–1779.
[238] Wolters, E. T. M.; Tesser, G. I.; Nivard, R. J. F. *J. Org. Chem.* **1974**, *39*, 3388–3392.
[239] Camps, F.; Castells, J.; Ferrando, M. J.; Font, J. *Tetrahedron Lett.* **1971**, 1713–1714.
[240] Kraus, M. A.; Patchornik, A. *Israel J. Chem.* **1971**, *9*, 269–271.
[241] Hanessian, S.; Ma, J.; Wang, W. *Tetrahedron Lett.* **1999**, *40*, 4631–4634.
[242] Sauvagnat, B.; Kulig, K.; Lamaty, F.; Lazaro, R.; Martinez, J. *J. Comb. Chem.* **2000**, *2*, 134–142.
[243] O'Donnell, M. J.; Delgado, F.; Drew, M. D.; Pottorf, R. S.; Zhou, C. Y.; Scott, W. L. *Tetrahedron Lett.* **1999**, *40*, 5831–5835.
[244] Hu, Y.; Porco, J. A. *Tetrahedron Lett.* **1999**, *40*, 3289–3292.
[245] Vlattas, I.; Dellureficio, J.; Dunn, R.; Sytwu, I. I.; Stanton, J. *Tetrahedron Lett.* **1997**, *38*, 7321–7324.
[246] Kobayashi, S.; Wakabayashi, T.; Yasuda, M. *J. Org. Chem.* **1998**, *63*, 4868–4869.
[247] Kobayashi, S.; Hachiya, I.; Suzuki, S.; Moriwaki, M. *Tetrahedron Lett.* **1996**, *37*, 2809–2812.
[248] Kobayashi, S.; Hachiya, I.; Yasuda, M. *Tetrahedron Lett.* **1996**, *37*, 5569–5572.
[249] May, P. J.; Bradley, M.; Harrowven, D. C.; Pallin, D. *Tetrahedron Lett.* **2000**, *41*, 1627–1630.
[250] Greenlee, M. L.; Laub, J. B.; Balkovec, J. M.; Hammond, M. L.; Hammond, G. G.; Pompliano, D. L.; Epstein-Toney, J. H. *Bioorg. Med. Chem. Lett.* **1999**, *9*, 2549–2554.
[251] Heinelt, U.; Herok, S.; Matter, H.; Wildgoose, P. *Bioorg. Med. Chem. Lett.* **2001**, *11*, 227–230.
[252] Wang, H.; Ganesan, A. *J. Comb. Chem.* **2000**, *2*, 186–194.
[253] Phoon, C. W.; Oliver, S. F.; Abell, C. *Tetrahedron Lett.* **1998**, *39*, 7959–7962.
[254] Furth, P. S.; Reitman, M. S.; Gentles, R.; Cook, A. F. *Tetrahedron Lett.* **1997**, *38*, 6643–6646.
[255] Furth, P. S.; Reitman, M. S.; Cook, A. F. *Tetrahedron Lett.* **1997**, *38*, 5403–5406.
[256] Zaragoza, F. *Tetrahedron Lett.* **1997**, *38*, 7291–7294.
[257] Gopalsamy, A.; Yang, H.; Ellingboe, J. W.; Kees, K. L.; Yoon, J.; Murrills, R. *Bioorg. Med. Chem. Lett.* **2000**, *10*, 1715–1718.
[258] Zhang, J.; Barker, J.; Lou, B.; Saneii, H. *Tetrahedron Lett.* **2001**, *42*, 8405–8408.
[259] Craig, D.; Robson, M. J.; Shaw, S. J. *Synlett* **1998**, 1381–1383.
[260] Hanessian, S.; Xie, F. *Tetrahedron Lett.* **1998**, *39*, 733–736.
[261] Martínez-Teipel, B.; Michelotti, E.; Kelly, M. J.; Weaver, D. G.; Acholla, F.; Beshah, K.; Teixidó, J. *Tetrahedron Lett.* **2001**, *42*, 6455–6457.
[262] Hone, N. D.; Payne, L. J.; Tice, C. M. *Tetrahedron Lett.* **2001**, *42*, 1115–1118.
[263] Lyngsø, L. O.; Nielsen, J. *Tetrahedron Lett.* **1998**, *39*, 5845–5848.
[264] Boymond, L.; Rottländer, M.; Cahiez, G.; Knochel, P. *Angew. Chem. Int. Ed.* **1998**, *37*, 1701–1703.
[265] Kamogawa, H.; Kanzawa, A.; Kadoya, M.; Naito, T.; Nanasawa, M. *Bull. Chem. Soc. Jpn.* **1983**, *56*, 762–765.
[266] Blackburn, C. *Tetrahedron Lett.* **1998**, *39*, 5469–5472.
[267] Kulkarni, B. A.; Ganesan, A. *Tetrahedron Lett.* **1999**, *40*, 5633–5636.
[268] Zhang, C. Z.; Moran, E. J.; Woiwode, T. F.; Short, K. M.; Mjalli, A. M. M. *Tetrahedron Lett.* **1996**, *37*, 751–754.
[269] Chen, J. J.; Golebiowski, A.; McClenaghan, J.; Klopfenstein, S. R.; West, L. *Tetrahedron Lett.* **2001**, *42*, 2269–2271.
[270] Tam, J. P.; Riemen, M. W.; Merrifield, R. B. *Pept. Res.* **1988**, *1*, 6–18.
[271] Gowravaram, M. R.; Gallop, M. A. *Tetrahedron Lett.* **1997**, *38*, 6973–6976.
[272] Whitehouse, D. L.; Nelson, K. H.; Savinov, S. N.; Austin, D. J. *Tetrahedron Lett.* **1997**, *38*, 7139–7142.
[273] Whitehouse, D. L.; Nelson, K. H.; Savinov, S. N.; Löwe, R. S.; Austin, D. J. *Bioorg. Med. Chem.* **1998**, *6*, 1273–1282.
[274] Winkler, J. D.; McCoull, W. *Tetrahedron Lett.* **1998**, *39*, 4935–4936.
[275] Faita, G.; Paio, A.; Quadrelli, P.; Rancati, F.; Seneci, P. *Tetrahedron Lett.* **2000**, *41*, 1265–1269.
[276] Purandare, A. V.; Natarajan, S. *Tetrahedron Lett.* **1997**, *38*, 8777–8780.
[277] Maclean, D.; Hale, R.; Chen, M. *Org. Lett.* **2001**, *3*, 2977–2980.
[278] Alvarez-Gutierrez, J. M.; Nefzi, A.; Houghten, R. A. *Tetrahedron Lett.* **2000**, *41*, 609–612.
[279] Hoveyda, H. R.; Hall, D. G. *Org. Lett.* **2001**, *3*, 3491–3494.
[280] Albert, R.; Knecht, H.; Andersen, E.; Hungerford, V.; Schreier, M. H.; Papageorgiou, C. *Bioorg. Med. Chem. Lett.* **1998**, *8*, 2203–2208.
[281] Zaragoza, F. *Tetrahedron Lett.* **1996**, *37*, 6213–6216.
[282] Larsen, S. D.; DiPaolo, B. A. *Org. Lett.* **2001**, *3*, 3341–3344.

[283] Goff, D.; Fernandez, J. *Tetrahedron Lett.* **1999**, *40*, 423–426.
[284] Nicolaou, K. C.; Pfefferkorn, J. A.; Schuler, F.; Roecker, A. J.; Cao, G.-Q.; Casida, J. E. *Chem. Biol.* **2000**, *7*, 979–992.
[285] Caba, J. M.; Rodriguez, I. M.; Manzanares, I.; Giralt, E.; Albericio, F. *J. Org. Chem.* **2001**, *66*, 7568–7574.

14 Preparation of Carbonic Acid Derivatives

14.1 Preparation of Carbodiimides

Carbodiimides have been prepared on insoluble supports mainly as polymeric dehydrating agents or as intermediates for the synthesis of guanidines and various heterocycles. Synthetic strategies that enable the preparation of carbodiimides on insoluble supports include the dehydration of ureas (Entry 1, Table 14.1), and the condensation of isocyanates or isothiocyanates with iminophosphoranes RN=PPh$_3$ (Entries 2–4, Table 14.1; Entry 12, Table 14.3). The latter are obtained either by treating azides with phosphines or by treatment of primary amines with phosphines in the presence of an oxidant such as DEAD. An additional important route to resin-bound carbodiimides is the desulfurization of thioureas, which can be accomplished with other carbodiimides, with HgO [1], HgCl$_2$ [2,3], or AgNO$_3$ [1], or with 2-chloro-1-methylpyridinium iodide (DCM, NEt$_3$, 45 °C, 3.5 h [4,5]). Carbodiimides are generally less reactive than other heterocumulenes, and can be stored for long periods in closed containers at room temperature.

Table 14.1. Preparation of carbodiimides.

Entry	Starting resin	Conditions	Product	Ref.
1		TsCl (0.15 mol/L), NEt$_3$ (0.57 mol/L), DCM, 40 °C, 50 h		[6]
2		1. PPh$_3$ (0.5 mol/L, 5 eq), THF, 20 °C, 6 h 2. PhNCO, THF, 20 °C, 8 h		[7] see also [8]
3		1. PPh$_3$ (0.4 mol/L, 5 eq), DEAD (5 eq), THF, 23 °C, 36 h 2. PhNCO (0.4 mol/L, 5 eq), PhMe, 23 °C, 4 h		[9]
4		BnNCS (10 eq), PPh$_3$ (35 eq), THF, 25 °C, 4 h		[10] see also [11]

14.2 Preparation of Isocyanates and Isothiocyanates

Isocyanates and isothiocyanates are highly reactive heterocumulenes, and are usually only prepared on solid phase as intermediates for the synthesis of ureas, carbamates, thioureas, etc., using methods similar to those used in solution (Table 14.2).

Support-bound primary amines can be converted into isocyanates by treatment with phosgene or with synthetic equivalents thereof (e.g. bis(trichloromethyl) carbonate 'triphosgene' or trichloromethyl chloroformate 'diphosgene'). Isocyanates can also be prepared on insoluble supports by oxidative degradation of amides (Hofmann degradation), by the thermolysis of acyl azides (Curtius degradation), and by cycloreversion of certain heterocycles ([12,13]; see Section 15.5). The Curtius degradation (Entry **3**, Table 14.2) enables the conversion of support-bound dicarboxylic acids into derivatives of non-natural amino acids, thereby offering interesting possibilities for the design of peptide mimetics (see also Entries **4** and **5**, Table 14.8). Isocyanates can

Table 14.2. Preparation of isocyanates and isothiocyanates.

Entry	Starting resin	Conditions	Product	Ref.
1		$COCl_2$ (0.36 mol/L, 5 eq), pyridine (5 eq), DCM, reflux, 1 h		[16] see also [17,18]
2		triphosgene (3.3 eq), DIPEA (10 eq), DCM, 0.5 h Ar: 4-(BnO)C$_6$H$_4$		[19]
3		DPPA (4 eq), NEt$_3$, PhMe, 20 °C, 2 h, then PhMe, 90 °C, 4 h		[20] see also [21]
4		MeSiCl$_3$ (0.45 mol/L, 10 eq), NEt$_3$ (20 eq), DCM, 20 °C, 10 h		[22]
5		(0.14 mol/L), DMF, 20 °C, 1 h		[23]
6		CS_2, NEt$_3$, DMF, 0–20 °C, 4 h, then ClCO$_2$Et, 20 °C, overnight		[24]
7		1. (COCl)$_2$ (2.5 eq), DCE/DMF 95:5, 8 h (repeat once; 14 h) 2. Bu$_4$NNCS (0.4 mol/L), DCE/THF 1:1, 2 × 4 h		[15]
8		Bu$_4$NSCN (0.4 mol/L, 3 eq), DCM, 20 °C, 50 h		[25]

also be prepared by treating polystyrene-bound Fmoc amines with methyltrichlorosilane and triethylamine (Entry **4**, Table 14.2). The mechanism of this reaction probably involves amine-induced cleavage of the Fmoc group to yield a carbamic acid, which is then trapped and dehydrated by the chlorosilane.

Isothiocyanates can be prepared from support-bound primary amines by treatment with thiophosgene [14] or synthetic analogs thereof (Entry **5**, Table 14.2). In an alternative two-step procedure, the amine is first treated with CS_2 and a tertiary amine to yield an ammonium dithiocarbamate, which is subsequently desulfurized with TsCl or a chloroformate (Entry **6**, Table 14.2; Experimental Procedure 14.1). Highly reactive acyl isothiocyanates have been prepared from support-bound acyl chlorides and tetrabutylammonium thiocyanate (Entry **7**, Table 14.2). These acyl isothiocyanates react with amines to give the corresponding *N*-acylthioureas, which can be used to prepare guanidines on insoluble supports (Entry **6**, Table 14.3).

Experimental Procedure 14.1: Conversion of polystyrene-bound primary aliphatic amines into isothiocyanates [26]

Wang resin bearing 1,3-diamino-2,2-dimethylpropane (0.60 g, approx. 0.6 mmol; for preparation, see Experimental Procedure 14.2) was swollen for 1 min in DCE (7.0 mL). The solvent was removed by filtration, and the resin was treated with DCE (5.2 mL), carbon disulfide (0.8 mL), and DIPEA (0.52 mL). After shaking for 45 min, a solution of tosyl chloride (1.32 g, 6.91 mmol, 12 equiv.) in DCE (1.5 mL) was added and shaking was continued for 15 h. The mixture was then filtered and the resin was washed with DCM (5 × 8.0 mL). An IR spectrum of the dried support (KBr pellet) showed a strong absorption at 2091 cm^{-1}.

14.3 Preparation of Guanidines

Most solid-phase strategies for the preparation and protection of guanidines were developed to enable the use of arginine in solid-phase peptide synthesis. In recent years, however, the biological activity of many guanidines has spurred a quest for more versatile solid-phase syntheses of this class of compound.

Several reagents can be used to convert support-bound amines into guanidines (Figure 14.1, Table 14.3). Aliphatic amines react with *N,N'*-di(alkoxycarbonyl)thioureas in the presence of carbodiimides as condensing agents to yield protected guanidines. Instead of the in situ activation of thioureas with carbodiimides, isolated *S*-alkyl- or *S*-arylisothioureas can also sometimes be used to effect this transformation. If faster conversion is desired, or if sterically demanding amines are to be transformed into guanidines, more reactive guanylating agents are required. These include

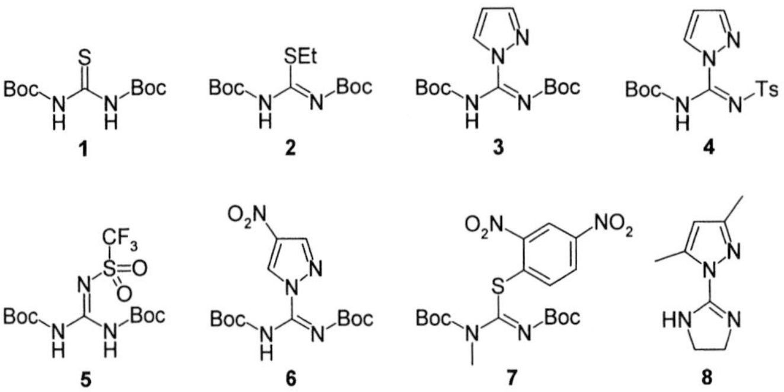

Figure 14.1. Reagents for the conversion of support-bound amines into protected or unprotected guanidines. **1** [29,30], **2** [29,31], **3** [29,32], **4** [33], **5** [34], **6** [27], **7** [10], **8** [31,35].

Table 14.3. Preparation of protected and unprotected guanidines.

Entry	Starting resin	Conditions	Product	Ref.
1		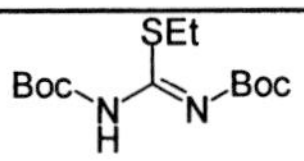 (2 eq), DIPEA (5 eq), DMF, 20 °C, 15 h	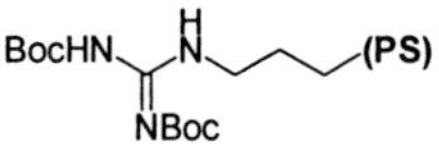	[31] see also [38,39]
2	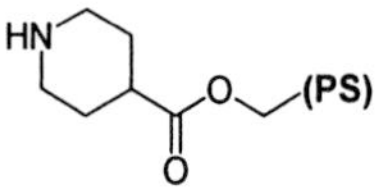	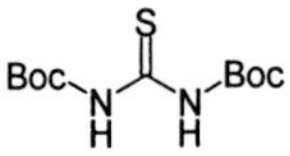 (1.5 eq), DIC (1.5 eq), DCE, 4 d	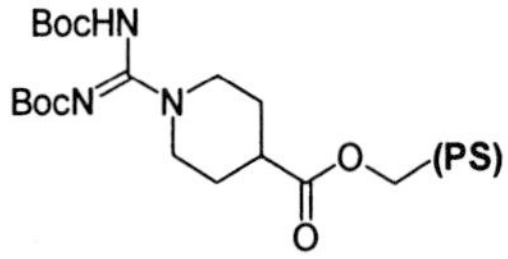	[30]
3	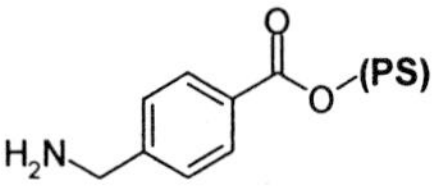	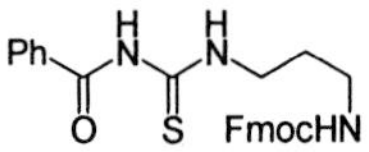 (3 eq), EDC (3 eq), DMF, 25 °C, 48 h	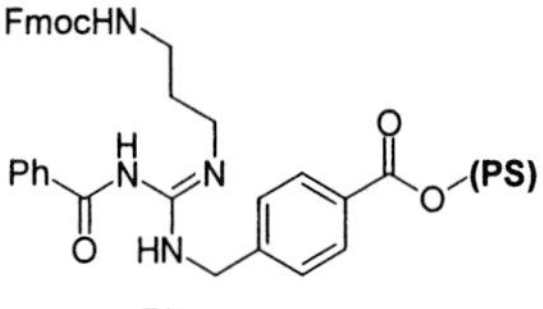	[10] see also [40]
4	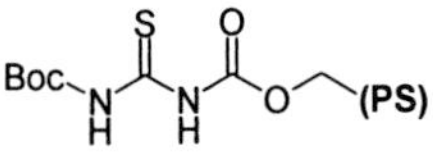	BnNHMe (3 eq), NMP	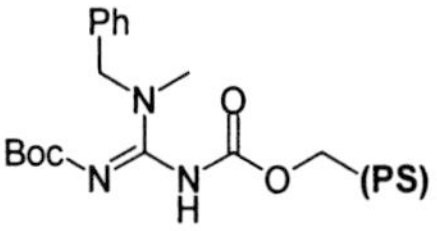	[36]
5	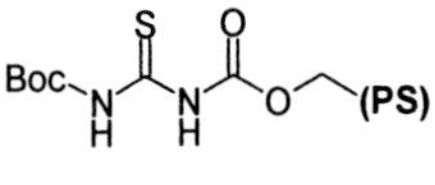	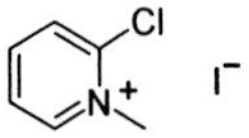 PhNH₂, NEt₃, NMP	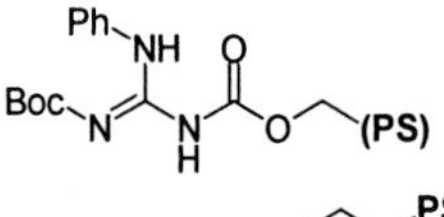	[36]
6	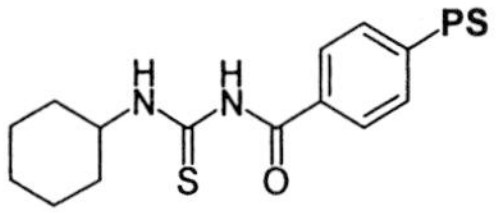	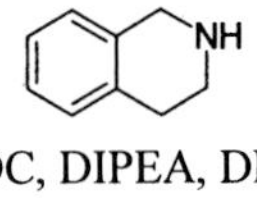 EDC, DIPEA, DMF	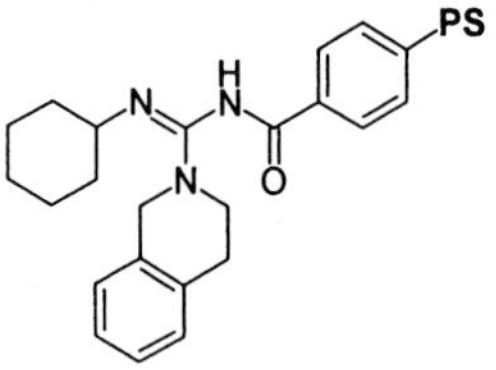	[15] see also [41–43]

Table 14.3. continued.

Entry	Starting resin	Conditions	Product	Ref.
7		morpholine (2 mol/L), DMSO, 80 °C, 12 h		[37]
8		BuNH$_2$ (0.56 mol/L, 10 eq), HgCl$_2$ (2 eq), NEt$_3$ (15 eq), DMF/DCM 5:1, 3 d		[3] see also [36,44]
9		4-hydroxyaniline (0.7 mol/L, 20 eq), MeCN, 38 °C, 2 h		[45] see also [46,47]
10		DCM, 16 h		[48] see also [49]
11		PrNH$_2$ (10 eq), DMSO, 25 °C, 18 h		[10] see also [4,5,9,11]
12		1. PhNCS (0.25 mol/L, 10 eq), PPh$_3$ (10 eq), THF, 23 °C, 2.5 h 2. Et$_2$NH (0.2 mol/L, 10 eq), NMP, 23 °C, 2 h		[50]

S-nitroaryl- and *S*-(1-methyl-2-pyridinium)isothioureas, as well as various pyrazole derivatives, of which the 4-nitropyrazole derivatives are among the most reactive. The latter reagents can even be used to convert aniline or diisopropylamine into guanidines (in solution; DMF, 25 °C [27]). Polystyrene-bound ethylenediamines can be converted into cyclic guanidines by treatment with cyanogen bromide (1.1 equiv., 0.02 mol/L, xylene, 20 °C, overnight [28]).

As an alternative to the guanylation of resin-bound amines, support-bound activated urea or guanidine derivatives can be prepared, reaction of which with amines yields guanidines. *N,N'*-Di(alkoxycarbonyl)thioureas can be converted into guanidines by treatment with aliphatic amines, without the need for other reagents (Entry **2**, Table 14.3). The addition of condensing agents, however, extends the scope of this reaction to include anilines (moderate yields only [36]). Activation of support-bound thioureas can be achieved by S-alkylation [37] or S-arylation [10]. Carbodiimides, which can be prepared from resin-bound thioureas by treatment with HgO [1], HgCl$_2$ [2,3], or AgNO$_3$ [1], have also been used as starting materials for the preparation of guanidines on cross-linked polystyrene (Entry **11**, Table 14.3).

14.4 Preparation of Ureas

The most commonly used strategies for preparing ureas on insoluble supports are outlined in Figure 14.2. The reaction conditions are similar to those used in solution as most of the reagents required are compatible with the widely used polymeric supports.

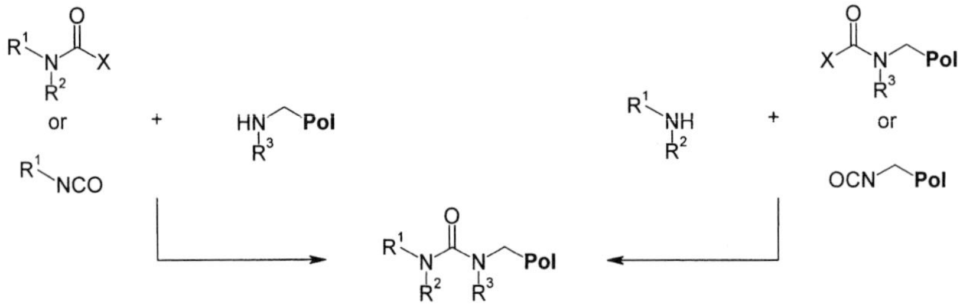

Figure 14.2. Strategies for the preparation of ureas on solid phase. X: leaving group.

Table 14.4. Preparation of ureas from support-bound amines.

Entry	Starting resin	Conditions	Product	Ref.
1	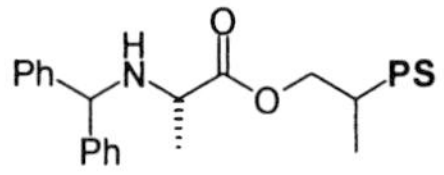	PhNCO (10 eq), DCE, 50 °C, 3 d	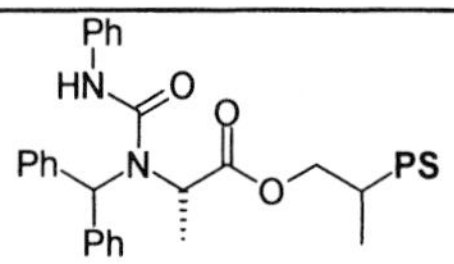	[52] see also [53–59]
2	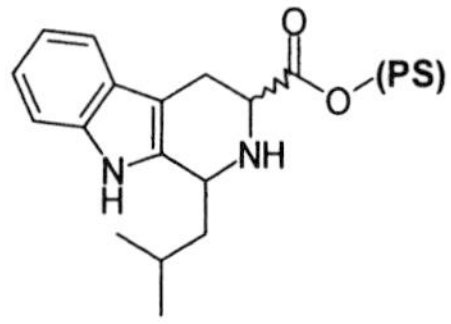	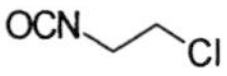 OCN ⌐ Cl (0.3 mol/L, 3 eq), DCM, 2 h	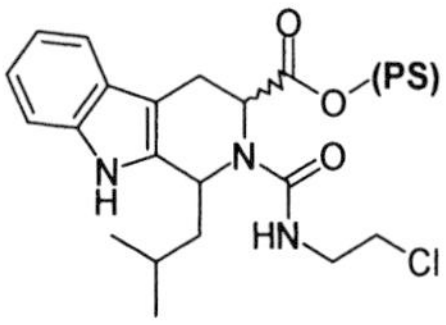	[60]
3	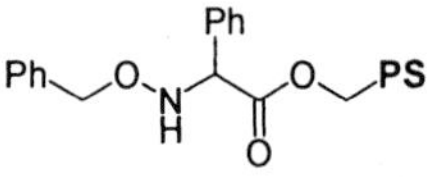	PhNCO, DCE, 83 °C, 24 h	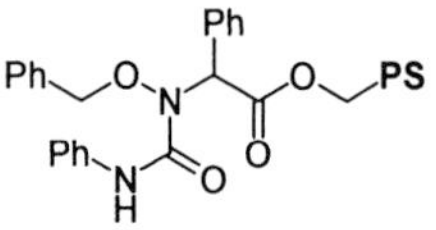	[61]
4	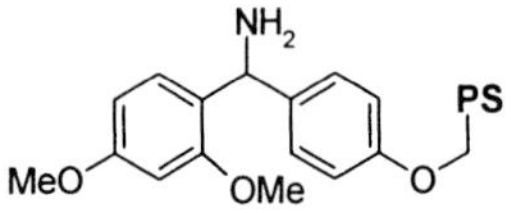	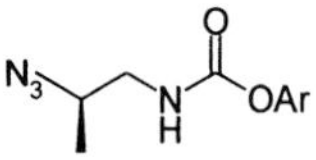 (5 eq, 0.07 mol/L), DIPEA (7 eq), DCM, 20 °C, 4 h Ar: 4-$(O_2N)C_6H_4$	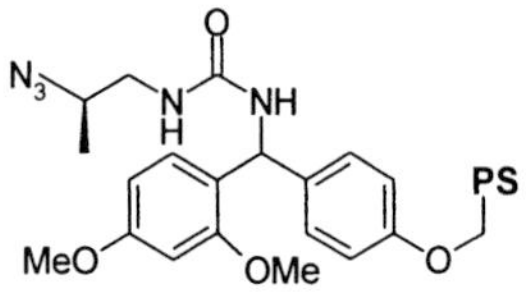	[62]
5	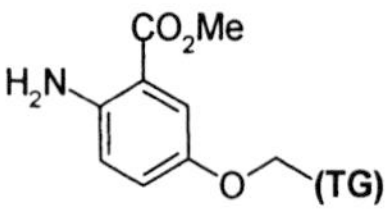	PhNCO (0.58 mol/L, 20 eq), DCM, 18 h	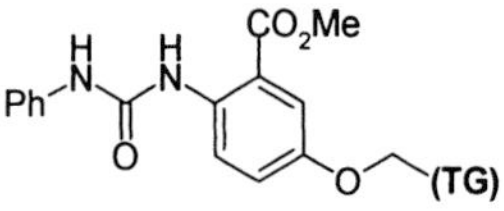	[63] see also [64]
6	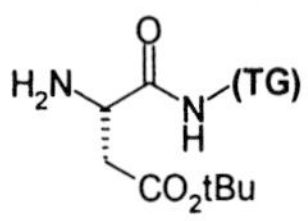	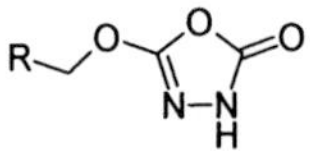 (0.13 mol/L, 3 eq), DCM, 20 °C, 1.5 h R: 9-fluorenyl	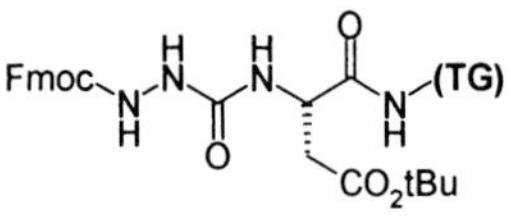	[65]

Isocyanates are the most efficient reagents for transforming support-bound amines into ureas. A broad range of amines, including sterically hindered aliphatic and aromatic amines, undergo clean aminocarbonylation with aliphatic and aromatic isocyanates (Table 14.4). The only disadvantage of isocyanates as reagents in solid-phase synthesis is their limited stability. Isocyanates can be generated immediately prior to use from carboxylic acids by conversion to the azide with DPPA followed by Curtius rearrangement [51]. Unfortunately, this reaction cannot be performed in the presence of a resin-bound amine because DPPA reacts with amines to yield phosphoric amides. Less reactive agents for aminocarbonylation include carbamoyl chlorides (which can be prepared from secondary amines and phosgene), N-(aminocarbonyl)benzotriazoles (Entry 2, Table 14.5), and aryl carbamates (Entry 4, Table 14.4). Aryl carbamates pre-

Table 14.5. Preparation of ureas from support-bound aminocarbonylating reagents.

Entry	Starting resin	Conditions	Product	Ref.
1		Ph⌒NH$_2$ (0.5 mol/L), THF, 20 °C, overnight		[67] see also [68]
2		H-Ser(*t*Bu)-OMe (0.13 mol/L, 3 eq), DIPEA (6 eq), DCM/DMF 1:1, 20 °C, overnight		[69] see also [70]
3		cyclopentylamine (20 eq), EtMgBr (20 eq), diglyme, 90 °C, 16–20 h (anilines can also be used)		[71]
4		DIPEA (0.5 mol/L each), THF/DCE 1:1, 20 °C, then HexNH$_2$ (0.35 mol/L), DIPEA, DMF, 20 °C, 12 h		[66] see also [72]
5		3-aminopyridine (0.5 mol/L), DCM, 12 h		[63] see also [73]
6		(0.56 mol/L, 10 eq), NEt$_3$ (10 eq), DMF, 24 h		[3]

Table 14.5. continued.

Entry	Starting resin	Conditions	Product	Ref.
7		1. COCl$_2$ (10 eq), DIPEA, DCM, 0–20 °C, 2 h 2. piperidin-4-ol (10 eq), pyridine (10 eq), DCM, 2 h		[74] see also [75]
8		1. Me$_3$SiCl (10 eq), NEt$_3$ (20 eq), DCM, 20 °C, 24 h, then drain 2. piperidine/DCM 1:9, 0.5 h		[22]
9		anthranilic acid, pyridine/DMF 1:9, 20 °C		[18]
10		(5 eq), DCM, 1 h Ar: 4-(BnO)C$_6$H$_4$		[19] see also [16]
11		DMA, 4 h		[20]

pared from secondary amines are significantly less reactive than aryl carbamates derived from primary amines and cannot usually be used for the preparation of ureas.

Ureas can also be prepared from support-bound aminocarbonylating reagents and amines. Suitable aminocarbonylating reagents that can also be prepared on insoluble supports are isocyanates (Section 14.2) and aryl carbamates, which can be obtained by exposing support-bound amines to phosgene and aryl chloroformates, respectively (Table 14.5). Aryl carbamates, and other aminocarbonylating agents derived from resin-bound *primary* amines, are generally much more reactive than those derived from secondary amines, probably due to the fact that in the case of primary amines, isocyanates can be formed as intermediates. Support-bound 4-nitrophenyl carbamates react with most aliphatic amines at room temperature or upon warming, but only slowly [18,66] or not at all with anilines (see, however, Entry **5**, Table 14.5). Aromatic amines can usually only be efficiently converted into *N*-arylureas by the use of support-bound isocyanates or carbamoyl chlorides (Entry **9**, Table 14.5). The reaction of immobilized aminocarbonylating reagents with amines to yield non-resin-bound ureas is considered in Section 3.3.3.

14.5 Preparation of Thioureas and Isothioureas

Thioureas can be prepared on solid phase in a similar manner as ureas, but using instead the corresponding thiocarbonyl derivatives. Hence, thioureas have been prepared on insoluble supports either by treating support-bound amines with isothiocyanates or by reaction of amines with support-bound aminothiocarbonylating reagents. Isothiocyanates are less reactive than isocyanates and even their reaction with primary aliphatic amines may require several days to reach completion (Table 14.6). A high concentration of isothiocyanate and high reaction temperatures can sometimes be helpful. More reactive than alkyl- or arylisothiocyanates are *N*-acyl- and *N*-alkoxycarbonylisothiocyanates (Entries **3**, **4**, and **6**; Table 14.6). Fmoc and Alloc isothiocya-

Table 14.6. Preparation of thioureas and isothioureas.

Entry	Starting resin	Conditions	Product	Ref.
1		(10 eq), DCM, 45 °C, 2–4 d		[77]
2		BnNCS (10 eq), DCM, 25 °C, 48 h		[10]
3		FmocNCS (0.2 mol/L), DCM, 20 min, then piperidine/DMF 2:8, 3 × 2.5 min		[78] see also [37]
4		PhMe, 1 h; then Pd(PPh$_3$)$_4$, Me$_2$NSiMe$_3$, CF$_3$CO$_2$SiMe$_3$, DCM, 6 h		[79]
5		(0.3 mol/L), DMF, overnight		[23]
6		2,6-dichloroaniline, DMF, 20 °C		[15]
7		1. (H$_2$N)$_2$CS (0.5 mol/L, 5 eq), DMF, 75 °C, 16 h 2. Boc$_2$O (0.26 mol/L, 6 eq), DIPEA (10 eq), DCM, 40 h 3. PhO(CH$_2$)$_2$OH (0.3 mol/L, 5 eq), PPh$_3$ (5 eq), DIAD (5 eq), THF, 14 h		[80] see also [81]
8		1. COCl$_2$, PhMe, 60 °C, 12 h 2. cyclohexylamine, PhMe, 60 °C, 12 h		[76]

nates have been used as synthetic equivalents of H–NCS to prepare monosubstituted thioureas from support-bound amines (Entries **3** and **4**, Table 14.6). These thioureas can be readily converted into thiazoles by treatment with α-halo ketones. Support-bound isothiocyanates (Section 14.2) can be converted into thioureas by treatment with amines, although few examples of this reaction have been reported (Entries **5** and **6**, Table 14.6).

Isothioureas can be prepared on insoluble supports by S-alkylation or S-arylation of thioureas (Entry **7**, Table 14.6). Further methods for the preparation of isothioureas on insoluble supports include the N-alkylation of polystyrene-bound, *N,N'*-di(alkoxycarbonyl)isothioureas with aliphatic alcohols by Mitsunobu reaction (Entry **7**, Table 14.6) and the addition of thiols to resin-bound carbodiimides [7]. Resin-bound dithiocarbamates, which can easily be prepared from Merrifield resin, carbon disulfide, and amines [76], react with phosgene to yield chlorothioformamidines, which can be converted into isothioureas by treatment with amines (Entry **8**, Table 14.6). The conversion of support-bound α-amino acids into thioureas can be accompanied by the release of thiohydantoins into solution (see Section 15.9). The rate of this cyclization depends, however, on the type of linker used and on the nucleophilicity of the intermediate thiourea.

14.6 Preparation of Carbamates

Carbamates have mainly been used in solid-phase synthesis as linkers and protective groups for amines (see Sections 3.6.2 and 10.1.10.1). Carbamates are generally prepared by treating amines with aryl carbonates or chloroformates, which can be prepared from alcohols and phosgene or synthetic equivalents thereof. The alternative route, in which carbamates, isocyanates, or carbamoyl chlorides are reacted with alcohols, is less widely used, but can also lead to satisfactory results on insoluble supports (Tables 14.7 and 14.8).

Support-bound aliphatic alcohols react smoothly with isocyanates in the presence of catalytic amounts of a base to yield carbamates (Table 14.7). In an interesting variant of this reaction, isocyanates were generated in situ by Curtius degradation of acyl azides (Entry **2**, Table 14.7).

Wang resin bound 4-nitrophenyl carbonate is a convenient intermediate for the attachment of amines to polystyrene as carbamates (see Experimental Procedure 14.2). Aliphatic amines [82–87], ammonia [88], and amino acids [89] react exothermically with this support, whereas anilines generally require catalysis and/or long reaction times (Entry **3**, Table 14.7). For the immobilization of anilines as carbamates, Wang resin derived chloroformate [90–92] generally leads to better results than resin-bound 4-nitrophenyl carbonates. Amidines also react with polystyrene-bound 4-nitrophenyl carbonates to yield *N*-alkoxycarbonyl amidines (Section 3.9 [93–95]). Support-bound alkoxycarbonyl hydrazines can be prepared by treating polystyrene-bound phenyl carbonates with hydrazine [96–98].

Experimental Procedure 14.2: Attachment of piperazine to Wang resin as carbamate [99]

A solution of 4-nitrophenyl chloroformate (43.0 g, 231 mmol, 5.5 equiv.) in DCM (200 mL) was added dropwise over a period of 0.5 h to a stirred suspension of Wang resin (45.0 g, 42.3 mmol) in DCM (600 mL) and pyridine (52.0 mL, 644 mmol, 15 equiv.). After completion of the addition, the mixture was stirred at room temperature for 3 h, and then filtered. The resin was washed with DCM (5 × 300 mL) and then added portionwise to a stirred solution of piperazine (38.2 g, 444 mmol, 11 equiv.) in DMF (600 mL), which led to an increase in the temperature of the mixture and a color change to yellow-orange. The resulting mixture was shaken at room temperature for 13 h, then filtered, and the resin was extensively washed with DMF, DCM, methanol, and finally with further DCM. After drying in air, about 45 g of resin-bound piperazine was obtained.

As alternatives to 4-nitrophenyl chloroformate, carbonyl diimidazole [100–102] or di-*N*-succinimidyl carbonate [103,104] can be used to convert polymeric alcohols into alkoxycarbonylating reagents suitable for the preparation of support-bound carbamates. Polystyrene-bound alkoxycarbonyl imidazole is less reactive than the corresponding 4-nitrophenyl carbonate, and sometimes requires heating to undergo reaction with amines. Additional activation of these imidazolides can be achieved by N-methylation (Entry **9**, Table 14.7).

Carbamates can also be prepared by treating support-bound amines with alkoxycarbonylating reagents such as chloroformates or aryl carbonates. Chloroformates or dicarbonates (e.g. Boc$_2$O) should not be used in large excess for the alkoxycarbonylation of primary amines because double derivatization can occur [119]. Less reactive reagents include 4-nitrophenyl carbonates and *N*-succinimidyl carbonates (Entries **2** and **3**, Table 14.8).

Support-bound isocyanates can be conveniently prepared from carboxylic acids by Curtius degradation. Because the reaction of the intermediate acyl azides with alcohols to yield esters is slow, Curtius degradation can be conducted in the presence of alcohols to yield carbamates directly (Entries **4** and **5**, Table 14.8).

Some carbamates can be cleanly N-alkylated on insoluble supports, either by treatment with strong bases and alkylating agents, or by reaction with aliphatic alcohols under Mitsunobu conditions (Table 14.9). Alkylation under Mitsunobu conditions

Table 14.7. Preparation of carbamates from support-bound alcohols or alkoxycarbonylating reagents.

Entry	Starting resin	Conditions	Product	Ref.
1		(0.83 mol/L, 6 eq), NEt$_3$, DCM, 6.5 h		[105] see also [106]
2		(0.36 mol/L, 3 eq), DPPA (5 eq), NEt$_3$ (10 eq), PhMe, 100 °C, 16 h		[107]
3		BSA, DMAP, DMF, 24 h		[108] see also [90,109]
4		(1.1 eq), KN(SiMe$_3$)$_2$ (1 eq), −78 °C to 0 °C, 0.5 h		[110]
5		N-Boc-guanidine, NEt$_3$, DMAP, DMF, 20 °C, 15 h		[47]
6		LiO$_2$C NH$_2$ (5 eq), DMF, 20 °C		[111] see also [35,89, 112,113]

only proceeds smoothly with acidic carbamates, such as aryl carbamates substituted with electron-withdrawing groups, and not with *N*-alkyl carbamates. The N-heteroarylation of polystyrene-bound carbamates with 2,4,6-trichloropyrimidine requires the use of strong bases such as KO*t*Bu (3 equiv., THF or DMF [123]).

Polystyrene-bound allylsilanes react with *N*-(alkoxycarbonyl)imines under Lewis acid catalysis to yield *N*-homoallylcarbamates (Entry **4**, Table 14.9). Similarly, Wang resin bound carbamates have been successfully N-alkylated with allylsilanes and aldehydes in a Mannich-type reaction (Entry **5**, Table 14.9). Resin-bound *N*-(alkoxycarbonyl)imines can be generated either from unsubstituted carbamates ROCONH$_2$ by

Table 14.7. continued.

Entry	Starting resin	Conditions	Product	Ref.
7		BnNH$_2$ (2 eq), DCM or DMF, 20 °C Ar: 4-(O$_2$N)C$_6$H$_4$		[114]
8		 (0.16 mol/L, 5 eq), NMM, THF/NMP 1:1, 60 °C, 4 h		[100]
9		MeOTf (0.11 mol/L, 1.7 eq), DCE, NEt$_3$ (5 eq), 10–20 °C, 15 min, then H-Leu-OMe (6 eq), 20 °C, 5.5 h		[115]
10		1. COCl$_2$, DCM, pyridine, 20 °C, 2 h 2. H$_2$NNHCO$_2$Me, NEt$_3$, DCM, 20 °C, 2 h	 (can be oxidized to the azo compound: NBS (1.1 eq), pyridine (1 eq), DCM, 20 °C, 1 h)	[116]
11		triphosgene (0.13 mol/L, 4 eq), 20 °C, overnight, then 2-aminopyridine (0.19 mol/L, 3.1 eq), DCM, 20 °C, overnight		[117] see also [91,118]

treatment with a carbonyl compound under acidic reaction conditions, or from α-sul-fonylcarbamates (Entry **6**, Table 14.9) by treatment with a base. Polystyrene-bound α-sulfonylcarbamates, which also undergo substitution of the sulfonyl group with carbon nucleophiles other than enolates, have been prepared by treatment of carbamates ROCONH$_2$ with aldehydes (6 equiv.) and sodium *p*-toluenesulfinate (3 equiv.) in the presence of TFA (6 equiv., DCM, 60 °C, 1 h [124]).

Carbamates can be prepared by O-alkylation of carbamic acids R$_2$N–CO$_2$H. Carbamic acids are usually susceptible to decarboxylation, but the corresponding salts can be prepared from amines and carbon dioxide under basic reaction conditions and

Table 14.8. Preparation of carbamates from support-bound amines or isocyanates.

Entry	Starting resin	Conditions	Product	Ref.
1	(structure, H_2N-O-benzyl-O-PS)	(allyl chloroformate) (1.1 eq), DIPEA (1.1 eq), DCM, 20 °C, 12 h	(structure)	[120]
2	(structure, HN-(PS), NHFmoc)	NHFmoc-O-OAr carbonate (5 eq), HOBt (10 eq), DIPEA (11 eq), THF, 50 °C, 2 × 5 h; Ar: 4-(O$_2$N)C$_6$H$_4$	(structure, NHFmoc)	[121] see also [122]
3	(structure, H_2N, Ph, R, -(PS))	(tetrahydrofuranyl succinimidyl carbonate) DIPEA, DCM	(structure, Ph, R, -(PS))	[75]
4	(structure, HO$_2$C-cyclobutane-(PS))	1. DPPA (0.83 mol/L, 10 eq), NEt$_3$ (15 eq), NMP, 20 °C, 1.5 h 2. 9-fluorenylmethanol (0.83 mol/L), *m*-xylene, 90 °C, 16 h	(structure, FmocHN, -(PS))	[21]
5	(structure, HO$_2$C-biphenyl-(PS))	same conditions as Entry **4**	(structure, FmocHN-biphenyl-(PS))	[21]
6	(structure, quinoline-O-(PS))	PhMgBr (5 eq), PhOCOCl (5 eq), THF, −40 °C to 10 °C	(structure, PhO, N, Ph, O-(PS))	[71]

then O-alkylated, for instance with Merrifield resin [125], to yield carbamates. Similarly, polystyrene-bound dithiocarbamates have been prepared by treating Merrifield resin with a mixture of an aliphatic or aromatic amine, carbon disulfide, DIPEA, and THF [126]. Carbamate salts are also the initial products formed upon removal of Fmoc protection, and it is possible to convert these salts into other carbamates by alkylation with alcohols under the conditions of the Mitsunobu reaction (Entry **8**, Table 14.9).

Table 14.9. Preparation of carbamates from other carbamates.

Entry	Starting resin	Conditions	Product	Ref.
1		BnBr (0.5 mol/L, 5 eq), DBU (5 eq), PhMe, 20 °C, 3 d		[120] see also [107]
2		(1.4 mol/L, 10 eq), LiI (1 eq), LiN(SiMe$_3$)$_2$ (1.1 eq), THF/NMP, 20 °C, 24 h		[105] see also [127]
3		EtOH, PBu$_3$, TMAD, THF, 60 °C, 15 h		[107]
4		(23 eq), BF$_3$OEt$_2$ (4 eq), DCM, 3 h		[128]
5		PhCHO, MeCN, BF$_3$OEt$_2$, −5 °C, 2 h		[88] see also [129,130]
6		PhC(OLi)=CH$_2$ (4 eq), THF, −78 °C, 2 h, then 20 °C, 0.5 h		[124]
7		1,3-diphenylpropanol (0.5 mol/L, 5 eq), PPh$_3$ (5 eq), DEAD (5 eq), THF, 12 h		[131]
8		2-phenylethanol (0.7 mol/L, 25 eq), ADDP (17 eq), PBu$_3$ (13 eq), DIPEA/NMP 17:83, 20 °C, 20 h		[132]

14.7 Preparation of Carbonates and Miscellaneous Carbonic Acid Derivatives

The preparation of carbonates is mechanistically closely related to the synthesis of carbamates, and similar reagents can be used for this purpose (Table 14.10). Resin-bound alcohols can be directly converted into carbonates by treatment with a chloroformate (see Experimental Procedure 14.2), in two steps by activation with phosgene or a synthetic equivalent thereof followed by reaction with an alcohol in the presence of a base, or by treatment of a resin-bound alcohol with carbon dioxide and an alkyl halide under basic reaction conditions [125]. Thiocarbonates can be prepared from

Wang resin and *S,S'*-di-2-pyridyl dithiocarbonate (Entry **6**, Table 14.10), and have been used as alkylating agents for the immobilization of aliphatic alcohols as benzyl ethers [133].

Table 14.10. Preparation of carbonates and thiocarbonates.

Entry	Starting resin	Conditions	Product	Ref.
1		O₂N— ... O–O–Cl, pyridine, 0 °C		[111] see also [112]
2		PhOCOCl (0.9 mol/L, 9 eq), pyridine (10 eq), DCM, 0 °C, overnight		[96]
3		(3 eq), DBU (3 eq), DCM, 20 °C, 15 min		[134] see also [135]
4		3-phenylpropanol (0.6 mol/L, 3 eq), DMAP (6 eq), 20 °C, 6 h		[136] see also [137]
5		benzyl alcohol (0.3 mol/L, 3 eq), Cs₂CO₃ (3 eq), Bu₄NI (3 eq), CO₂, DMF, 60 °C, overnight		[125]
6		(0.12 mol/L, 3 eq), NEt₃ (3 eq), DCM, 24 h		[133]

References for Chapter 14

[1] Dahmen, S.; Bräse, S. *Org. Lett.* **2000**, *2*, 3563–3565.
[2] Kojima, N.; Bruice, T. C. *Org. Lett.* **2000**, *2*, 81–84.
[3] Lin, P.; Ganesan, A. *Tetrahedron Lett.* **1998**, *39*, 9789–9792.
[4] Drewry, D. H.; Ghiron, C. *Tetrahedron Lett.* **2000**, *41*, 6989–6992.
[5] Chen, J.; Pattarawarapan, M.; Zhang, A. J.; Burgess, K. *J. Comb. Chem.* **2000**, *2*, 276–281.
[6] Weinshenker, N. M.; Shen, C. M.; Wong, J. Y. *Org. Synth.* **1988**, *Coll. Vol. VI*, 951–954.
[7] Villalgordo, J. M.; Obrecht, D.; Chucholowski, A. *Synlett* **1998**, 1405–1407.
[8] Gopalsamy, A.; Yang, H. *J. Comb. Chem.* **2000**, *2*, 378–381.
[9] Wang, F.; Hauske, J. R. *Tetrahedron Lett.* **1997**, *38*, 8651–8654.
[10] Schneider, S. E.; Bishop, P. A.; Salazar, M. A.; Bishop, O. A.; Anslyn, E. V. *Tetrahedron* **1998**, *54*, 15063–15086.
[11] Drewry, D. H.; Gerritz, S. W.; Linn, J. A. *Tetrahedron Lett.* **1997**, *38*, 3377–3380.
[12] Whitehouse, D. L.; Nelson, K. H.; Savinov, S. N.; Austin, D. J. *Tetrahedron Lett.* **1997**, *38*, 7139–7142.
[13] Gowravaram, M. R.; Gallop, M. A. *Tetrahedron Lett.* **1997**, *38*, 6973–6976.
[14] Zaragoza, F. unpublished results.
[15] Wilson, L. J.; Klopfenstein, S. R.; Li, M. *Tetrahedron Lett.* **1999**, *40*, 3999–4002.

[16] Matthews, J.; Rivero, R. A. *J. Org. Chem.* **1997**, *62*, 6090–6092.
[17] Annis, D. A.; Helluin, O.; Jacobsen, E. N. *Angew. Chem. Int. Ed.* **1998**, *37*, 1907–1909.
[18] Gordeev, M. F. *Biotechnology and Bioengineering* **1998**, *61*, 13–16.
[19] Limal, D.; Semetey, V.; Dalbon, P.; Jolivet, M.; Briand, J. P. *Tetrahedron Lett.* **1999**, *40*, 2749–2752.
[20] Shao, H.; Colucci, M.; Tong, S.; Zhang, H.; Castelhano, A. L. *Tetrahedron Lett.* **1998**, *39*, 7235–7238.
[21] Richter, L. S.; Andersen, S. *Tetrahedron Lett.* **1998**, *39*, 8747–8750.
[22] Chong, P. Y.; Petillo, P. A. *Tetrahedron Lett.* **1999**, *40*, 4501–4504.
[23] Wilson, M. W.; Hernández, A. S.; Calvet, A. P.; Hodges, J. C. *Mol. Diversity* **1998**, *3*, 95–112.
[24] Dowling, L. M.; Stark, G. R. *Biochemistry* **1969**, *8*, 4728–4734.
[25] Pirrung, M. C.; Pansare, S. V. *J. Comb. Chem.* **2001**, *3*, 90–96.
[26] Stephensen, H.; Zaragoza, F. *J. Org. Chem.* **1997**, *62*, 6096–6097.
[27] Yong, Y. F.; Kowalski, J. A.; Thoen, J. C.; Lipton, M. A. *Tetrahedron Lett.* **1999**, *40*, 53–56.
[28] Acharya, A. N.; Ostresh, J. M.; Houghten, R. A. *Tetrahedron* **2001**, *57*, 9911–9914.
[29] Shey, J. Y.; Sun, C. M. *Synlett* **1998**, 1423–1425.
[30] Robinson, S.; Roskamp, E. J. *Tetrahedron* **1997**, *53*, 6697–6705.
[31] Corbett, J. W.; Graciani, N. R.; Mousa, S. A.; DeGrado, W. F. *Bioorg. Med. Chem. Lett.* **1997**, *7*, 1371–1376.
[32] Sulyok, G. A. G.; Gibson, C.; Goodman, S. L.; Hölzemann, G.; Wiesner, M.; Kessler, H. *J. Med. Chem.* **2001**, *44*, 1938–1950.
[33] Zhang, Y.; Kennan, A. J. *Org. Lett.* **2001**, *3*, 2341–2344.
[34] Ho, K. C.; Sun, C. M. *Bioorg. Med. Chem. Lett.* **1999**, *9*, 1517–1520.
[35] Lee, Y.; Silverman, R. B. *Synthesis* **1999**, 1495–1499.
[36] Josey, J. A.; Tarlton, C. A.; Payne, C. E. *Tetrahedron Lett.* **1998**, *39*, 5899–5902.
[37] Kearney, P. C.; Fernandez, M.; Flygare, J. A. *Tetrahedron Lett.* **1998**, *39*, 2663–2666.
[38] Rockwell, A. L.; Rafalski, M.; Pitts, W. J.; Batt, D. G.; Petraitis, J. J.; DeGrado, W. F.; Mousa, S.; Jadhav, P. K. *Bioorg. Med. Chem. Lett.* **1999**, *9*, 937–942.
[39] Kowalski, J.; Lipton, M. A. *Tetrahedron Lett.* **1996**, *37*, 5839–5840.
[40] Mamai, A.; Madalengoitia, J. S. *Org. Lett.* **2001**, *3*, 561–564.
[41] Wilson, L. J. *Org. Lett.* **2001**, *3*, 585–588.
[42] Li, M.; Wilson, L. J.; Portlock, D. E. *Tetrahedron Lett.* **2001**, *42*, 2273–2275.
[43] Li, M.; Wilson, L. J. *Tetrahedron Lett.* **2001**, *42*, 1455–1458.
[44] Dodd, D. S.; Zhao, Y. *Tetrahedron Lett.* **2001**, *42*, 1259–1262.
[45] Pátek, M.; Smrcina, M.; Nakanishi, E.; Izawa, H. *J. Comb. Chem.* **2000**, *2*, 370–377.
[46] Wu, S.; Janusz, J. M. *Tetrahedron Lett.* **2000**, *41*, 1165–1169.
[47] Zapf, C. W.; Creighton, C. J.; Tomioka, M.; Goodman, M. *Org. Lett.* **2001**, *3*, 1133–1136.
[48] Ostresh, J. M.; Schoner, C. C.; Hamashin, V. T.; Nefzi, A.; Meyer, J. P.; Houghten, R. A. *J. Org. Chem.* **1998**, *63*, 8622–8623.
[49] Acharya, A. N.; Nefzi, A.; Ostresh, J. M.; Houghten, R. A. *J. Comb. Chem.* **2001**, *3*, 189–195.
[50] Jammalamadaka, V.; Berger, J. G. *Synth. Commun.* **2000**, 2077–2082.
[51] Migawa, M. T.; Swayze, E. E. *Org. Lett.* **2000**, *2*, 3309–3311.
[52] Lee, S. H.; Chung, S. H.; Lee, Y. S. *Tetrahedron Lett.* **1998**, *39*, 9469–9472.
[53] Boeijen, A.; Kruijtzer, J. A. W.; Liskamp, R. M. J. *Bioorg. Med. Chem. Lett.* **1998**, *8*, 2375–2380.
[54] Park, K. H.; Olmstead, M. M.; Kurth, M. J. *J. Org. Chem.* **1998**, *63*, 6579–6585.
[55] Swayze, E. E. *Tetrahedron Lett.* **1997**, *38*, 8465–8468.
[56] Fivush, A. M.; Willson, T. M. *Tetrahedron Lett.* **1997**, *38*, 7151–7154.
[57] Kim, S. W.; Ahn, S. Y.; Koh, J. S.; Lee, J. H.; Ro, S.; Cho, H. Y. *Tetrahedron Lett.* **1997**, *38*, 4603–4606.
[58] Kolodziej, S. A.; Hamper, B. C. *Tetrahedron Lett.* **1996**, *37*, 5277–5280.
[59] Burgess, K.; Ibarzo, J.; Linthicum, D. S.; Russell, D. H.; Shin, H.; Shitangkoon, A.; Totani, R.; Zhang, A. J. *J. Am. Chem. Soc.* **1997**, *119*, 1556–1564.
[60] Mohan, R.; Chou, Y. L.; Morrissey, M. M. *Tetrahedron Lett.* **1996**, *37*, 3963–3966.
[61] Hanessian, S.; Yang, R. Y. *Tetrahedron Lett.* **1996**, *37*, 5835–5838.
[62] Kim, J. M.; Bi, Y. Z.; Paikoff, S. J.; Schultz, P. G. *Tetrahedron Lett.* **1996**, *37*, 5305–5308.
[63] Buckman, B. O.; Mohan, R. *Tetrahedron Lett.* **1996**, *37*, 4439–4442.
[64] Wijkmans, J. C. H. M.; Culshaw, A. J.; Baxter, A. D. *Mol. Diversity* **1998**, *3*, 117–120.
[65] Gibson, C.; Goodman, S. L.; Hahn, D.; Hölzemann, G.; Kessler, H. *J. Org. Chem.* **1999**, *64*, 7388–7394.
[66] Wilson, L. J.; Li, M.; Portlock, D. E. *Tetrahedron Lett.* **1998**, *39*, 5135–5138.
[67] Xiao, X. Y.; Ngu, K.; Chao, C.; Patel, D. V. *J. Org. Chem.* **1997**, *62*, 6968–6973.
[68] Hutchins, S. M.; Chapman, K. T. *Tetrahedron Lett.* **1994**, *35*, 4055–4058.

[69] Nieuwenhuijzen, J. W.; Conti, P. G. M.; Ottenheijm, H. C. J.; Linders, J. T. M. *Tetrahedron Lett.* **1998**, *39*, 7811–7814.

[70] Bauser, M.; Winter, M.; Valenti, C. A.; Wiesmüller, K. H.; Jung, G. *Mol. Diversity* **1998**, *3*, 257–260.

[71] Wendeborn, S. *Synlett* **2000**, 45–48.

[72] Hutchins, S. M.; Chapman, K. T. *Tetrahedron Lett.* **1995**, *36*, 2583–2586.

[73] Gordeev, M. F.; Hui, H. C.; Gordon, E. M.; Patel, D. V. *Tetrahedron Lett.* **1997**, *38*, 1729–1732.

[74] Wang, G. T.; Chen, Y. W.; Wang, S. D.; Sciotti, R.; Sowin, T. *Tetrahedron Lett.* **1997**, *38*, 1895–1898.

[75] Kick, E. K.; Ellman, J. A. *J. Med. Chem.* **1995**, *38*, 1427–1430.

[76] Gomez, L.; Gellibert, F.; Wagner, A.; Mioskowski, C. *Chem. Eur. J.* **2000**, *6*, 4016–4020.

[77] Smith, J.; Liras, J. L.; Schneider, S. E.; Anslyn, E. V. *J. Org. Chem.* **1996**, *61*, 8811–8818.

[78] Kearney, P. C.; Fernandez, M.; Flygare, J. A. *J. Org. Chem.* **1998**, *63*, 196–200.

[79] Stadlwieser, J.; Ellmerer-Müller, E. P.; Takó, A.; Maslouh, N.; Bannwarth, W. *Angew. Chem. Int. Ed.* **1998**, *37*, 1402–1404.

[80] Dodd, D. S.; Wallace, O. B. *Tetrahedron Lett.* **1998**, *39*, 5701–5704.

[81] Lee, J.; Gauthier, D.; Rivero, R. A. *Tetrahedron Lett.* **1998**, *39*, 201–204.

[82] Dixit, D. M.; Leznoff, C. C. *J. Chem. Soc., Chem. Commun.* **1977**, 798–799.

[83] Marsh, I. R.; Smith, H.; Bradley, M. *Chem. Commun.* **1996**, 941–942.

[84] Ho, C. Y.; Kukla, M. J. *Tetrahedron Lett.* **1997**, *38*, 2799–2802.

[85] Kim, S. W.; Hong, C. Y.; Lee, K.; Lee, E. J.; Koh, J. S. *Bioorg. Med. Chem. Lett.* **1998**, *8*, 735–738.

[86] Brady, S. F.; Stauffer, K. J.; Lumma, W. C.; Smith, G. M.; Ramjit, H. G.; Lewis, S. D.; Lucas, B. J.; Gardell, S. J.; Lyle, E. A.; Appleby, S. D.; Cook, J. J.; Holahan, M. A.; Stranieri, M. T.; Lynch, J. J.; Lin, J. H.; Chen, I. W.; Vastag, K.; Naylor-Olsen, A. M.; Vacca, J. P. *J. Med. Chem.* **1998**, *41*, 401–406.

[87] Tomasi, S.; Le Roch, M.; Renault, J.; Corbel, J. C.; Uriac, P.; Carboni, B.; Moncoq, D.; Martin, B.; Delcros, J. G. *Bioorg. Med. Chem. Lett.* **1998**, *8*, 635–640.

[88] Meester, W. J. N.; Rutjes, F. P. J. T.; Hermkens, P. H. H.; Hiemstra, H. *Tetrahedron Lett.* **1999**, *40*, 1601–1604.

[89] Li, W.-R.; Lin, S. T.; Yang, J. H. *Synlett* **2000**, 1608–1612.

[90] Raju, B.; Kogan, T. P. *Tetrahedron Lett.* **1997**, *38*, 3373–3376.

[91] Burdick, D. J.; Struble, M. E.; Burnier, J. P. *Tetrahedron Lett.* **1993**, *34*, 2589–2592.

[92] Smith, A. L.; Thomson, C. G.; Leeson, P. D. *Bioorg. Med. Chem. Lett.* **1996**, *6*, 1483–1486.

[93] Mohan, R.; Yun, W. Y.; Buckman, B. O.; Liang, A.; Trinh, L.; Morrissey, M. M. *Bioorg. Med. Chem. Lett.* **1998**, *8*, 1877–1882.

[94] Kim, S. W.; Hong, C. Y.; Koh, J. S.; Lee, E. J.; Lee, K. *Mol. Diversity* **1998**, *3*, 133–136.

[95] Roussel, P.; Bradley, M.; Matthews, I.; Kane, P. *Tetrahedron Lett.* **1997**, *38*, 4861–4864.

[96] Wang, S. *J. Am. Chem. Soc.* **1973**, *95*, 1328–1333.

[97] Wang, S. *J. Org. Chem.* **1975**, *40*, 1235–1239.

[98] Wang, S.; Merrifield, R. B. *J. Am. Chem. Soc.* **1969**, *91*, 6488–6491.

[99] Zaragoza, F.; Petersen, S. V. *Tetrahedron* **1996**, *52*, 10823–10826.

[100] Hauske, J. R.; Dorff, P. *Tetrahedron Lett.* **1995**, *36*, 1589–1592.

[101] Rotella, D. P. *J. Am. Chem. Soc.* **1996**, *118*, 12246–12247.

[102] Munson, M. C.; Cook, A. W.; Josey, J. A.; Rao, C. *Tetrahedron Lett.* **1998**, *39*, 7223–7226.

[103] Alsina, J.; Rabanal, F.; Chiva, C.; Giralt, E.; Albericio, F. *Tetrahedron* **1998**, *54*, 10125–10152.

[104] Alsina, J.; Chiva, C.; Ortiz, M.; Rabanal, F.; Giralt, E.; Albericio, F. *Tetrahedron Lett.* **1997**, *38*, 883–886.

[105] Buchstaller, H. P. *Tetrahedron* **1998**, *54*, 3465–3470.

[106] Fitzpatrick, L. J.; Rivero, R. A. *Tetrahedron Lett.* **1997**, *38*, 7479–7482.

[107] Sunami, S.; Sagara, T.; Ohkubo, M.; Morishima, H. *Tetrahedron Lett.* **1999**, *40*, 1721–1724.

[108] Huang, W.; Scarborough, R. M. *Tetrahedron Lett.* **1999**, *40*, 2665–2668.

[109] Gouilleux, L.; Fehrentz, J. A.; Winternitz, F.; Martinez, J. *Tetrahedron Lett.* **1996**, *37*, 7031–7034.

[110] Smith, A. L.; Stevenson, G. I.; Lewis, S.; Patel, S.; Castro, J. L. *Bioorg. Med. Chem. Lett.* **2000**, *10*, 2693–2696.

[111] Léger, R.; Yen, R.; She, M. W.; Lee, V. J.; Hecker, S. J. *Tetrahedron Lett.* **1998**, *39*, 4171–4174.

[112] Singh, R.; Nuss, J. M. *Tetrahedron Lett.* **1999**, *40*, 1249–1252.

[113] Johansson, A.; Åkerblom, E.; Ersmark, K.; Lindeberg, G.; Hallberg, A. *J. Comb. Chem.* **2000**, *2*, 496–507.

[114] Dressman, B. A.; Singh, U.; Kaldor, S. W. *Tetrahedron Lett.* **1998**, *39*, 3631–3634.

[115] Hernández, A. S.; Hodges, J. C. *J. Org. Chem.* **1997**, *62*, 3153–3157.

[116] Arnold, L. D.; Assil, H. I.; Vederas, J. C. *J. Am. Chem. Soc.* **1989**, *111*, 3973–3976.

[117] Scialdone, M. A.; Shuey, S. W.; Soper, P.; Hamuro, Y.; Burns, D. M. *J. Org. Chem.* **1998**, *63*, 4802–4807.

[118] Scialdone, M. A. *Tetrahedron Lett.* **1996**, *37*, 8141–8144.

[119] Hulme, C.; Peng, J.; Morton, G.; Salvino, J. M.; Herpin, T.; Labaudiniere, R. *Tetrahedron Lett.* **1998**, *39*, 7227–7230.

[120] Salvino, J. M.; Mervic, M.; Mason, H. J.; Kiesow, T.; Teager, D.; Airey, J.; Labaudiniere, R. *J. Org. Chem.* **1999**, *64*, 1823–1830.

[121] Paikoff, S. J.; Wilson, T. E.; Cho, C. Y.; Schultz, P. G. *Tetrahedron Lett.* **1996**, *37*, 5653–5656.

[122] Cho, C. Y.; Youngquist, R. S.; Paikoff, S. J.; Beresini, M. H.; Hébert, A. R.; Berleau, L. T.; Liu, C. W.; Wemmer, D. E.; Keough, T.; Schultz, P. G. *J. Am. Chem. Soc.* **1998**, *120*, 7706–7718.

[123] Zucca, C.; Bravo, P.; Volonterio, A.; Zanda, M.; Wagner, A.; Mioskowski, C. *Tetrahedron Lett.* **2001**, *42*, 1033–1035.

[124] Schunk, S.; Enders, D. *Org. Lett.* **2001**, *3*, 3177–3180.

[125] Salvatore, R. N.; Flanders, V. L.; Ha, D.; Jung, K. W. *Org. Lett.* **2000**, *2*, 2797–2800.

[126] Gomez, L.; Gellibert, F.; Wagner, A.; Mioskowski, C. *J. Comb. Chem.* **2000**, *2*, 75–79.

[127] Srinivasan, T.; Gupta, P.; Kundu, B. *Tetrahedron Lett.* **2001**, *42*, 5993–5995.

[128] Brown, R. C. D.; Fisher, M. *Chem. Commun.* **1999**, 1547–1548.

[129] van Maarseveen, J. H.; Meester, W. J. N.; Veerman, J. J. N.; Kruse, C. G.; Hermkens, P. H. H.; Rutjes, F. P. J. T.; Hiemstra, H. *J. Chem. Soc., Perkin Trans. 1* **2001**, 994–1001.

[130] Vanier, C.; Wagner, A.; Mioskowski, C. *Chem. Eur. J.* **2001**, *7*, 2318–2323.

[131] Subramanyam, C. *Tetrahedron Lett.* **2000**, *41*, 6537–6540.

[132] Zaragoza, F.; Stephensen, H. *Tetrahedron Lett.* **2000**, *41*, 2015–2017.

[133] Hanessian, S.; Huynh, H. K. *Tetrahedron Lett.* **1999**, *40*, 671–674.

[134] Routledge, A.; Stock, H. T.; Flitsch, S. L.; Turner, N. J. *Tetrahedron Lett.* **1997**, *38*, 8287–8290.

[135] Routledge, A.; Abell, C.; Balasubramanian, S. *Tetrahedron Lett.* **1997**, *38*, 1227–1230.

[136] Li, W.-R.; Lin, Y.-S.; Yo, Y.-C. *Tetrahedron Lett.* **2000**, *41*, 6619–6622.

[137] Choo, H.; Chong, Y.; Chu, C. K. *Org. Lett.* **2001**, *3*, 1471–1473.

15 Preparation of Heterocycles

For more than two decades, most developments in solid-phase chemistry were focussed on the preparation of biopolymers. Only in recent years has interest in other synthetic targets, including heterocyclic compounds, begun to grow. Today, the field of solid-phase heterocyclic chemistry is rapidly expanding, and numerous preparations have been reported.

Interest in solid-phase heterocyclic chemistry originated mainly in the pharmaceutical industry. Heterocycles not only enable the spatial fixation of a set of structural elements relevant to reversible binding to proteins, but can also have a strong influence on the solubility and on other physicochemical properties of a compound. Because substituted heterocycles are often more easy to prepare than the corresponding carbocycles, heterocyclic chemistry has played, and continues to play, an important role in the development of new drugs (anti-inflammatory agents, antibiotics, antifungals, analgesics, etc.). Syntheses that enable the quick production of arrays of heterocycles, useful for the identification of new lead structures, are of critical importance to the pharmaceutical industry.

Several review articles have appeared covering the synthesis of heterocycles on insoluble supports for the production of compound libraries for drug discovery [1–4]. In this chapter, these syntheses have been organized according to the type of heterocycle prepared.

15.1 Preparation of Epoxides and Aziridines

Epoxides are reactive electrophiles, which enable the facile preparation of substituted alcohols by reaction with a broad range of nucleophiles. Epoxides can be prepared on insoluble supports either by epoxidation of alkenes or from aldehydes (Table 15.1).

Alkenes bound to cross-linked polystyrene can be epoxidized under conditions similar to those used in solution. The most commonly used reagent is m-chloroperbenzoic acid in DCM, but other reagents have also been used (Table 15.1). Because excess oxidant is usually required to furnish clean products, care must be taken with linkers or other functional groups prone to oxidation (ketones, amines, benzyl ethers, etc.).

The reaction of support-bound aldehydes with trimethylsulfonium halides in the presence of a strong base allows non-oxidative access to oxiranes (Entry **5**, Table

15.1). The same reaction has been performed with polystyrene-bound trialkylsulfo-nium salts, which react with aldehydes under basic reaction conditions to yield either non-support-bound [5] or resin-bound epoxides (Entry **7**, Table 15.1).

Aziridines have been prepared on insoluble supports by addition of primary amines to α-bromo acrylates and acrylamides (Entry **6**, Table 15.1). These aziridines are sufficiently stable to tolerate treatment with TFA [6].

Table 15.1. Preparation of epoxides and aziridines.

Entry	Starting resin	Conditions	Product	Ref.
1		MCPBA, NaHCO₃, DCM, 40 °C, 20 h	(82:18, main isomer shown)	[7] see also [8,9]
2		H₂O₂, K₂HPO₄, DCM, Cl₃CCN, H₂O, 40 °C, 3 h		[10]
3		(0.1 mol/L, 3 eq), Me₂CO, 20 °C, 1 h		[11]
4		(0.33 mol/L, 25 eq), DCM/ Me₂CO, 0 °C, 2 × 1 h		[12]
5		Me₃SCl, DCM, BnMe₃NCl, NaOH, H₂O, 20 °C, 3 h; or Me₃SI, KN(SiMe₃)₂, DMF, 20 °C, 10 h		[9,13]
6		BnNH₂, NMM, THF		[6]
7		3-chlorobenzaldehyde (0.4 mol/L, 15 eq), DBU (6 eq), DMSO, 20 °C, 16 h		[14]

15.2 Preparation of Azetidines and Thiazetidines

β-Lactams can be prepared from imines by reaction with ketenes (Table 15.2). The required imines are readily accessible by condensation of primary aliphatic or aromatic amines with carbonyl compounds, which, where necessary, may be performed in the presence of a dehydrating agent or catalytic amounts of an acid (see also Section 10.1.4). Ketenes can be generated by the dehydrohalogenation of acyl halides bearing an α-hydrogen using a tertiary amine. Alternatively, titanium ester enolates from 2-pyridinethiol esters or lithium ester enolates also undergo [2 + 2] cycloaddition to imines, and can be used to convert imines into β-lactams (Entry **3**, Table 15.2). Because ketenes are highly reactive intermediates, prone to numerous side reactions, support-bound ketenes are less suitable intermediates for the preparation of β-lactams.

Table 15.2. Preparation of azetidines and thiazetidines.

Entry	Starting resin	Conditions	Product	Ref.
1		PhO⌒COCl (0.8 mol/L), NEt$_3$ (1.1 mol/L), DCM, 0–25 °C, 16 h		[15] see also [16–19]
2		Ph⌒O⌒COCl (20 eq), NEt$_3$ (20 eq), DCM, 23 °C, 15 h		[20] see also [21]
3		(7 eq), N(octyl)$_3$, TiCl$_4$, DCM, −78 °C to 20 °C		[22] see also [23]
4		Hg(O$_2$CCF$_3$)$_2$, DCM, Me$_2$CO		[24]
5		MeO$_2$C⌒SO$_2$Cl (0.36 mol/L, 14 eq), pyridine (17 eq), THF, −78 °C, 3 h, then 20 °C, 24 h		[25]
6		1. LiN(SiMe$_3$)$_2$ (2.2 eq), THF, −78 °C, 1.5 h 2. PhN=CHPh (3 eq), −78 °C to 20 °C, 23 h		[26]
7		DEAD (5 eq), PPh$_3$ (10 eq), THF, 20 °C, 24 h		[27]

Sulfonyl chlorides possessing an α-hydrogen undergo elimination of hydrogen chloride when treated with bases to yield sulfenes. Like ketenes, sulfenes can undergo [2 + 2] cycloaddition to support-bound imines to yield β-sultams (Entry **5**, Table 15.2).

15.3 Preparation of Pyrroles and Pyrrolidines

Various approaches have been used to prepare pyrroles on insoluble supports (Figure 15.1). These include the condensation of α-halo ketones or nitroalkenes with enamines (Hantzsch pyrrole synthesis) and the decarboxylative condensation of *N*-acyl α-amino acids with alkynes (Table 15.3). The enamines required for the Hanztsch pyrrole synthesis are obtained by treating support-bound acetoacetamides with primary aliphatic amines. Unfortunately, 3-keto amides other than acetoacetamides are not readily accessible; this imposes some limitations on the range of substituents that may be incorporated into the products. Pyrroles have also been prepared by the treatment of polystyrene-bound vinylsulfones with isonitriles such as Tosmic [28] and by the reaction of resin-bound sulfonic esters of α-hydroxy ketones with enamines [29].

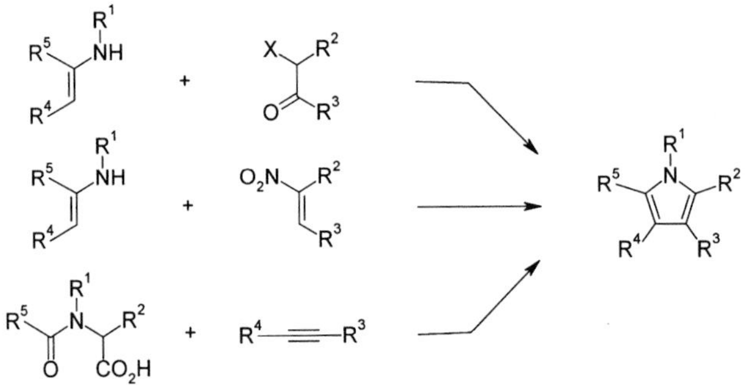

Figure 15.1. Strategies used for the preparation of pyrroles on insoluble supports.

2,5-Dihydropyrroles have recently become readily available by ring-closing metathesis. For this purpose, N-acylated or N-sulfonylated bis(allyl)amines are treated with catalytic amounts of a ruthenium carbene complex, whereupon cyclization to the dihydropyrrole occurs (Entries **6** and **7**, Table 15.3 [30,31]). Catalysis by carbene complexes is most efficient in aprotic, non-nucleophilic solvents, and can also be conducted on hydrophobic supports such as cross-linked polystyrene. Free amines or other soft nucleophiles might, however, compete with the alkene for electrophilic attack by the catalyst, and should therefore be avoided.

Pyrrolidines and pyrrolidinones can be synthesized on insoluble supports by intramolecular carbometallation of alkenes (Entry **1**, Table 15.4) and by Ugi reaction of support-bound isonitriles with amines and 4-oxo acids (Entry **4**, Table 15.4). In Entry **5** in Table 15.4, the pyrrolidinone ring is formed by intramolecular Diels–Alder reaction of a furan with a fumaric acid derivative. 2-Pyrrolidinones have also been prepared by treatment of resin-bound *O*-mesitylenesulfonyl cyclobutanone oximes with

Table 15.3. Preparation of pyrroles and 2,5-dihydropyrroles.

Entry	Starting resin	Conditions	Product	Ref.
1		ArCOCH$_2$Br, DTBP, DMF, 20 °C, 17 h Ar: 4-(Et$_2$N)C$_6$H$_4$		[32]
2		ArCHO, EtNO$_2$, HC(OMe)$_3$, piperidine, DMF/EtOH 1:1, 70 °C, 5 h Ar: 4-(F$_3$C)C$_6$H$_4$		[33]
3		O$_2$N— (cyclohexene) DMF/EtOH 1:1, 60 °C, 2 h		[33]
4		MeO$_2$C—≡—CO$_2$Me Ac$_2$O, 100 °C, 1–2 d		[34] see also [35]
5		PhHN—CO—CH$_2$—CO—Ph (10 eq), CuCl$_2$ (0.5 eq), THF, 20 °C, 3–5 h		[36]
6		Cl$_2$Ru=CHPh (PCy$_3$)$_2$ (0.15 eq), C$_6$H$_6$, 30 °C, 5 d		[37]
7		Cl$_2$Ru=CHPh (PCy$_3$)$_2$ (0.05 eq), C$_6$H$_6$, 75 °C, 18 h		[38]

KOSiMe$_3$ by a Beckmann-type rearrangement [39], and from glutamic acid derivatives by intramolecular acylation [40].

Numerous examples of the preparation of tetramic acids from N-acylated amino acid esters by a Dieckmann-type cyclocondensation have been reported (Entries **7–9**, Table 15.4). Deprotonated 1,3-dicarbonyl compounds and unactivated amide enolates can be used as carbon nucleophiles. In most of these examples, the ester that acts as electrophile also links the substrate to the support, so that cyclization and cleavage from the support occur simultaneously. The preparation of five-membered cyclic imides is discussed in Section 13.8.

A powerful means of access to pyrrolidines, which is also suitable for the solid-phase synthesis of this important class of heterocycles, is the cycloaddition of 2-azaallyl anions to alkenes (Figure 15.2). Various modifications of this reaction have been reported, the most commonly used being the addition of deprotonated, amino acid derived imines to acceptor-substituted alkenes (Entries **1–5**, Table 15.5).

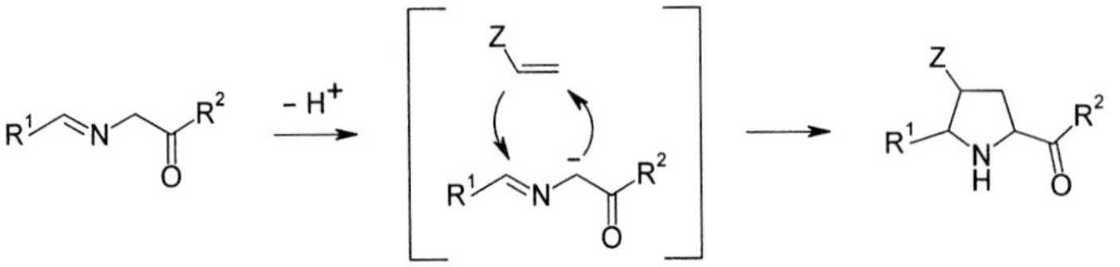

Figure 15.2. Formation of pyrrolidines by cycloaddition of 2-azaallyl anions to alkenes.

Table 15.4. Preparation of pyrrolidines and pyrrolidinones.

Entry	Starting resin	Conditions	Product	Ref.
1	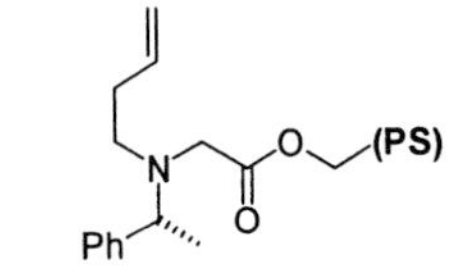	1. LDA (3 eq), THF, −78 °C to 0 °C 2. ZnBr$_2$ (5 eq) 3. H$_2$O, citric acid	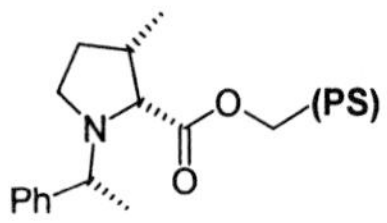	[41]
2	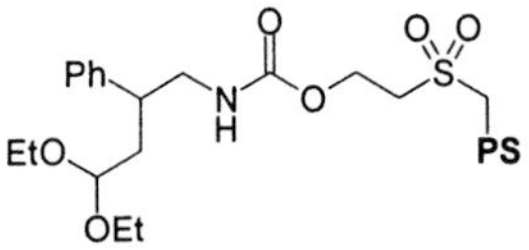	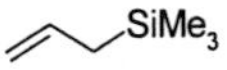 (10 eq), BF$_3$OEt$_2$ (3 eq), DCM, 20 °C, 17 h	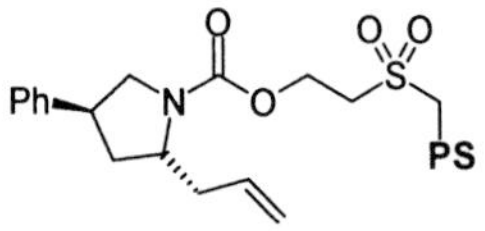	[42]
3	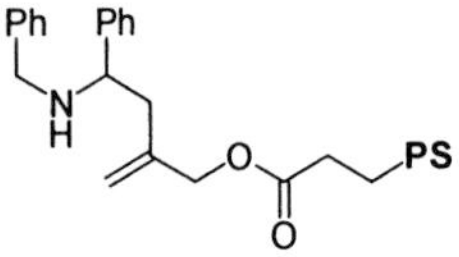	Pd(acac)$_2$ (0.1 eq), dppe (0.3 eq), THF, heat	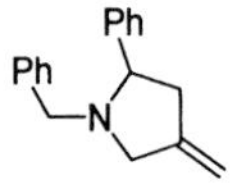 59–95% yield	[43]
4		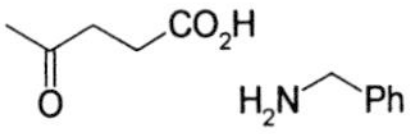 (0.3 mol/L, 5 eq of each), CHCl$_3$/MeOH 2:1, 20 °C, 48 h	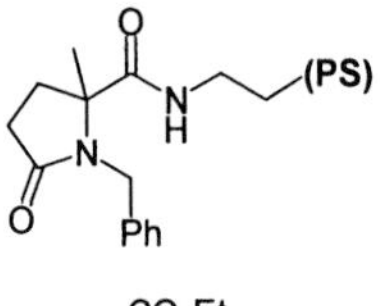	[44]
5	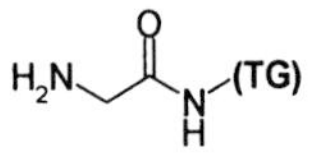	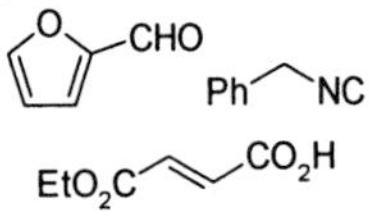 (0.2 mol/L, 10 eq of each), MeOH/DCM 2:1, 36 h	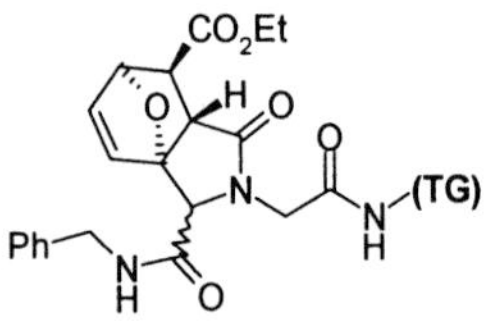	[45] see also [46,47]
6	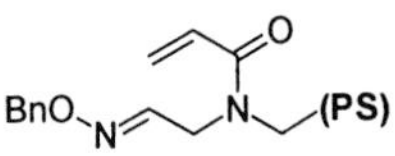	*i*PrI (30 eq), Et$_3$B, PhMe, 100 °C, 2 h	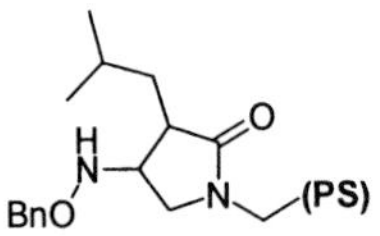	[48]

Table 15.4. continued.

Entry	Starting resin	Conditions	Product	Ref.
7		KOH (1 eq), DCM/MeOH 1:1, 0.5 h; or DIPEA/dioxane 3:7, 80 °C, 16 h		[49,50] see also [51]
8		NaOEt (0.1 mol/L), 85 °C, 24 h (Wang resin)		[52]
9		NaOEt (0.1 mol/L, 2 eq), 85 °C, 24 h (Wang resin)		[52]
10		ZnCl₂ (2 eq), LiN(SiMe₃)₂ (2 eq), THF, −78 °C to 0 °C, 2 h Ar: 4-(MeO)C₆H₄		[53]
11		KN(SiMe₃)₂		[54]

This cycloaddition can be performed with either the amino acid, the aldehyde, or the electron-poor alkene linked to the support. Intramolecular azaallyl anion cycloadditions have been used to prepare polycyclic systems on solid phase (Entries **5** and **7**, Table 15.5).

Less stabilized (and more reactive) azaallyl anions are formed upon transmetallation of α-stannylamine-derived imines with butyllithium. The resulting intermediates react with non-activated alkenes, such as stilbenes or vinyl sulfides, even at low temperature to yield N-metallated pyrrolidines (Entry **8**, Table 15.5).

Table 15.5. Preparation of pyrrolidines from 2-azaallyl anions.

Entry	Starting resin	Conditions	Product	Ref.
1		AgNO₃, NEt₃ (1 mol/L of each), MeCN, 8 h		[17] see also [55]
2		NEt₃ (0.05 mol/L, 3 eq of each), AcOH, DMF, 100 °C, 18 h		[56]
3		(0.16 mol/L, 10 eq), PhMe, 110 °C, 24 h; Ar: 4-(MeO)C₆H₄	(racemic)	[57] see also [55,58]
4		MeO₂C⌒N≔Ph LiBr, DBU, THF, 3 d Ar: 4-(MeO)C₆H₄		[59] see also [60,61]
5		AgOAc (0.2 mol/L, 3 eq), DIPEA (3 eq), MeCN, 20 °C, 48 h		[62]
6		(5 eq), NEt₃ (5 eq), THF, 20 °C, 1 h		[63] see also [64]
7		(0.11 mol/L, 1.7 eq), NEt₃ (1.7 eq), PhMe, 110 °C, 18 h		[65] see also [66]
8		1. BuLi, THF, PhSeCH=CH₂, –40 °C 2. MeI; 3. Bu₃SnH, AIBN, PhH, 80 °C		[67]

15.4 Preparation of Indoles and Indolines

The most important synthetic approaches to indoles that have been realized on insoluble supports are outlined in Figure 15.3. These include the Fischer indole synthesis, which has been successfully performed using polystyrene- or Tentagel-bound ketones and aldehydes (Entries **1** and **2**, Table 15.6). Indoles and indolines have also been prepared by palladium-mediated reaction of support-bound 2-iodo- or 2-bromoaniline derivatives with alkynes or 1,3-dienes (Table 15.6). These reactions, and the closely related cyclization of 2-alkynylanilines, probably proceed via an intermediate palladacycle, which undergoes reductive elimination to yield the observed indoles (Figure 15.3). Additional reactions suitable for synthesizing indoles on insoluble supports include the cyclocondensation of benzoquinones with enamines, and intramolecular Heck and Wittig reactions. The latter have been performed with phosphonium salts derived from polystyrene-bound triphenylphosphine, whereby indole formation and cleavage of the product from the support take place simultaneously (Entry **10**, Table 15.6). Indolines have also been prepared from 2-allylanilines using polystyrene-bound selenenyl bromide (Figure 9.2 [68]). Linkers suitable for the attachment of indoles to insoluble supports are discussed in Section 3.10.

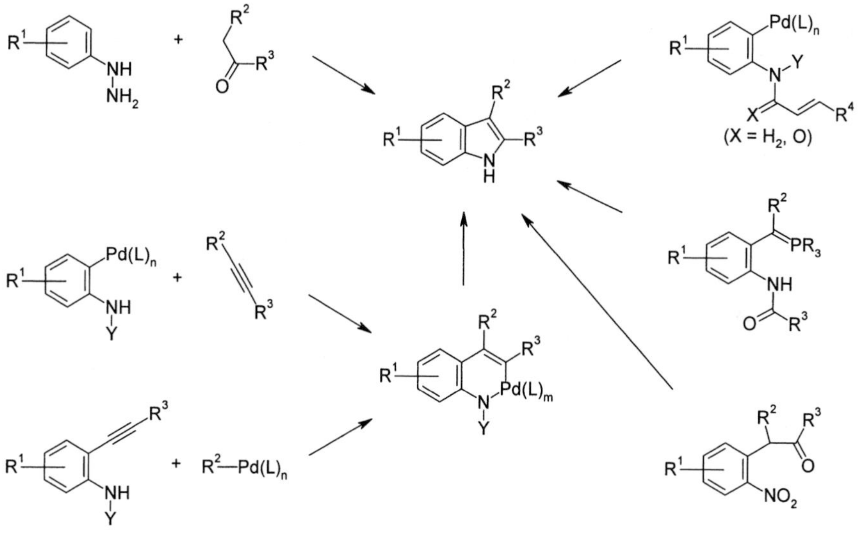

Figure 15.3. Strategies for the preparation of indoles on insoluble supports. Y: protective group.

Support-bound indoles can be modified in several ways. N-Alkylation of polystyrene-bound indoles has been achieved by treatment with reactive alkylating agents (MeI, BnBr, BrCH$_2$CO$_2$R) in conjunction with NaH or KOtBu as a base in DMF at room temperature (Entries **1** and **2**, Table 15.7). The aminomethylation of indole at C-3 proceeds smoothly on cross-linked polystyrene. The resulting (aminomethyl)indoles are thermally unstable and undergo substitution reactions with various carbon nucleophiles (e.g. cyanide or nitroacetates) at higher temperatures (Entry **4**, Table

Table 15.6. Preparation of indoles and indolines.

Entry	Starting resin	Conditions	Product	Ref.
1	Ph—(PS)	ZnCl₂ (each 0.5 mol/L), AcOH, 70 °C, 20 h	2-Ph indole (PS)	[69] see also [70]
2	piperidine aldehyde (TG)	(0.06 mol/L, 10 eq), TFA/DCM/ PhOMe 5:15:1, 40 °C, 17 h	O₂N spiro indoline (TG)	[71]
3	I, HN—(PS)	Pd(PPh₃)₂Cl₂ (0.2 eq), TMG (10 eq), DMF, 110 °C, 5 h (repeat once, 16 h)	indole (PS), SiMe₃	[72] see also [73–75]
4	acetamido iodo benzoate TG	PhS alkyne, Pd(PPh₃)₂Cl₂, CuI, TMG, dioxane, 90 °C, 18 h	PhS indole ester TG	[76] see also [77,78]
5	I benzamide (PS), HN—Tol sulfonyl	AcO (8 eq), Pd(OAc)₂ (0.1 eq), LiCl (2 eq), DIPEA (8 eq), DMF, 100 °C, 2 d	AcO indoline (PS), Tol sulfonyl	[79]
6	Ph, F₃C amide (PS)	cyclopentene—CO₂Me, OTf, K₂CO₃, DMF, Pd(PPh₃)₄, 20 °C, 24 h	Ph indole ester (PS), CO₂Me	[80]
7	Ph N, NH (PS), I	Pd(PPh₃)₂Cl₂ (1 mmol/L, 0.15 eq), Bu₄NCl (1.5 eq), NEt₃ (8 eq), DMF/H₂O 9:1, 80 °C, 24 h	indole NH (PS), Ph	[81]
8	N—(TG), Br	Pd(PPh₃)₄ (0.5 eq), PPh₃ (2 eq), NEt₃ (13 eq), DMA, 90 °C, 2 × 5 h	indole (TG)	[82]

Table 15.6. continued.

Entry	Starting resin	Conditions	Product	Ref.
9		Pd(OAc)₂ (7 mmol/L, 0.3 eq), Ag₂CO₃ (2 eq), PPh₃ (0.6 eq), DMF, 100 °C, 16 h		[83]
10		KO*t*Bu, PhMe, DMF, 110 °C, 45 min		[84]
11		SnCl₂•H₂O (2.4 mol/L, 53 eq), NMP, 20 °C, 10 h		[85]
12		SnCl₂•H₂O (2.4 mol/L, 53 eq), NMP, 20 °C, 10 h		[85]
13		(10 eq), ClRh(PPh₃)₃ (0.05 eq), EtOH/CHCl₃ 1:1, 80 °C, overnight		[86]
14		benzoquinone (5 eq), Me₂CO, 20 °C, 48 h		[87]

15.7). Resin-bound indoles can also be acylated at C-3, either with acyl halides (Entry 5, Table 15.7) or under the conditions of the Vilsmeier formylation (Entry 6, Table 15.7). C-Arylation of indoles can be accomplished by means of palladium-catalyzed coupling reactions, such as the Suzuki coupling (Entry 7, Table 15.7) or Stille coupling with resin-bound 2-bromoindoles [88] or 5-bromoindoles [75]. 2-Iodoindoles have been prepared on polystyrene by iododesilylation of 2-silylindoles with NIS (Entry 8, Table 15.7), and these can be C-arylated with arylboronic acids [73].

Table 15.7. Derivatization of indoles.

Entry	Starting resin	Conditions	Product	Ref.
1		NaH, DMF, 0.5 h, then BrCH$_2$CO$_2$Et, 20 °C, 4 h		[80]
2		BnBr (3 eq), KO*t*Bu (3 eq), DMF, 3 h		[78] see also [89]
3		pyrrolidine (8 eq), HCHO (8 eq), AcOH/dioxane 1:4, 23 °C, 1.5 h		[78] see also [90]
4		O$_2$N⌁CO$_2$Et (5 eq), PhMe, 100 °C, 9 h		[78]
5		C$_3$H$_5$COCl, AlCl$_3$, DCM, 12 h		[75]
6		POCl$_3$ (14 eq), Ph(Me)NCHO (14 eq), DCE, 20 °C, overnight		[91]
7		(9 eq), Na$_2$CO$_3$ (9 eq), Pd(PPh$_3$)$_4$ (0.04 eq), H$_2$O/EtOH/DME 3:2:8, microwaves (45 W), 3.8 min		[92]
8		NIS, DCM		[77]

15.5 Preparation of Furans and Tetrahydrofurans

Few solid-phase syntheses of furans have been reported. Most of these have been limited to furans with a specific substitution pattern and lack general applicability. Furans can be obtained by thermolysis of support-bound 7-oxa-2-azanorbornenes, which are prepared by 1,3-dipolar cycloaddition of electron-poor alkynes to iso-münchnones (Figure 15.4).

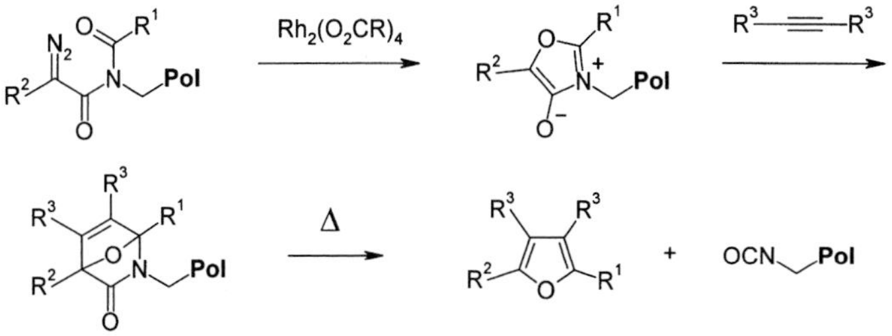

Figure 15.4. Generation of furans by thermolysis of support-bound 7-oxa-2-azanorbornenes [93–95].

Metallated furans can be generated on cross-linked polystyrene either by direct lithiation of furans with butyllithium [96] or by halogen–metal exchange with Grignard reagents [97]. The resulting furyllithium or -magnesium compounds are strongly nucleophilic and react smoothly with a number of electrophiles (Entries **3** and **4**, Table 15.8).

Vinylations and arylations of polystyrene-bound 2-bromofurans have been accomplished by treatment with stannanes [98] or boronic acids [99] in the presence of palladium complexes. Alternatively, 2-furylstannanes can be coupled with support-bound aryl iodides or bromides in the presence of palladium or copper complexes (Entries **5–7**, Table 15.8).

Tetrahydrofurans have been prepared on insoluble supports by treatment of 5-(3-butenyl)-4,5-dihydroisoxazole with electrophiles, whereby a remarkable C–C bond scission takes place to yield 2-(cyanomethyl)tetrahydrofurans (Figure 15.5; Entries **8** and **9**, Table 15.8).

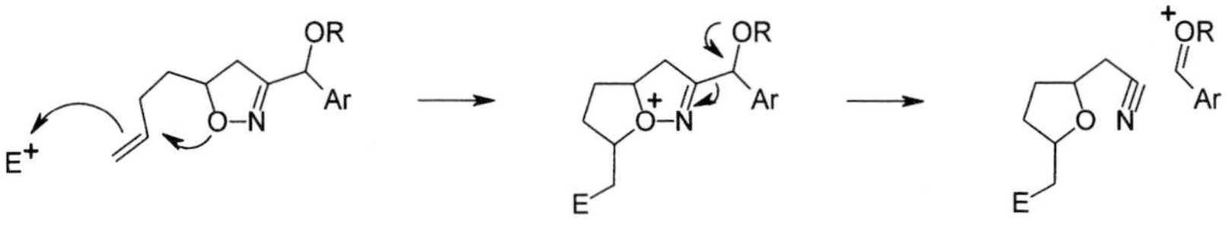

Figure 15.5. Mechanism of the formation of 2-cyanomethyltetrahydrofurans from butenylisoxazolines.

Tetrahydrofurans can also be prepared from support-bound haloalkenes by radical cyclization (Entries **10** and **11**, Table 15.8). Halogen abstraction by tin radicals leads to the formation of carbon-centered radicals, which undergo fast intramolecular addition to alkenes and alkynes. 2,5-Dihydrofurans have been prepared from 2-butene-1,4-diols etherified with Wang resin (Entry **12**, Table 15.8). TFA-mediated cleavage leads to simultaneous formation of the cyclic ethers. 1,3-Dihydroisobenzofurans can also be prepared using this strategy (Entry **11**, Table 15.9). The required alcohols were prepared by metallating support-bound vinyl iodides with iPrMgBr and then treating the resulting vinyl Grignard compound with aldehydes (see Entry **4**, Table 7.5).

Linkers that enable the preparation of γ-lactones by cleavage of hydroxy esters from insoluble supports are discussed in Section 3.5.2. Resin-bound γ-lactones have been prepared by Baeyer–Villiger oxidation of cyclobutanones [39], by intramolecular addition of alkyl radicals to oximes [48], by electrophilic addition of resin-bound selenenyl cyanide or bromide to β,γ-unsaturated acids (Figure 9.2 [100]), and by palladium-mediated coupling of resin-bound aryl iodides with allenyl carboxylic acids (Entry **10**, Table 5.7 [101]).

Table 15.8. Preparation of furans and tetrahydrofurans.

Entry	Starting resin	Conditions	Product	Ref.
1		C_6H_6, 70 °C, 0.5 h		[93]
2		C_6H_6, 80 °C, 1.5 h		[94]
3		iPrMgBr (0.19 mol/L, 7.3 eq), THF, −35 °C, 0.5 h, then add TsCN		[97]
4		BuLi (5 eq), THF, −30 °C, 4 h, then add DMA (10 eq), 20 °C, 2 h		[96]
5		NMP, CuI, NaCl, 100 °C, 24 h		[102]

Table 15.8. continued.

Entry	Starting resin	Conditions	Product	Ref.
6		 (3 eq), Pd(PPh₃)₄ (0.05 eq), DMF, 60 °C, 24 h		[98]
7		 (6 eq), Pd(PPh₃)₄ (0.2 eq), dioxane, 100 °C, 24 h		[11]
8		ICl (0.17 mol/L), DCM, −78 °C, 15 min		[103]
9		IBr (0.3 mol/L), DCM, −78 °C, 2.5 h	 (mixture of diastereomers)	[104]
10		Bu₃SnH (25 eq), AIBN (0.05 eq), PhMe, 70–80 °C, 2 h		[105] see also [106]
11		Bu₃SnH (0.1 mol/L, 5 eq), AIBN (0.5 eq), C₆H₆, 80 °C, 18 h		[107]
12		TFA/DCM 1:9, 20 °C, 10 min (Wang resin)		[108]

15.6 Preparation of Benzofurans and Dihydrobenzofurans

Benzofurans and dihydrobenzofurans have been prepared on polymeric supports by the palladium-mediated reaction of 2-iodophenols with dienes or alkynes (Entries **1** and **2**, Table 15.9). This reaction is closely related to the synthesis of indoles from 2-iodoanilines, and probably proceeds via an intermediate palladacycle (Figure 15.3). Benzofuran and isobenzofuran derivatives have also been prepared on cross-linked polystyrene by intramolecular addition of aryl radicals to C=C double bonds and by intramolecular Heck reaction.

2-(Allyloxy)iodoarenes react with SmI$_2$ to yield radical anions, which undergo thermal fragmentation. The resulting aryl radicals can cyclize to dihydrobenzofurans (Entries **8–10**, Table 15.9). The radical obtained after cyclization can be reduced and treated with a proton source such as water or an alcohol to yield alkanes, or with carbonyl compounds to yield alcohols (Entry **10**, Table 15.9).

Table 15.9. Preparation of benzofurans.

Entry	Starting resin	Conditions	Product	Ref.
1		Pd(OAc)$_2$, LiCl, DIPEA, DMF, 100 °C		[79]
2		Pd(PPh$_3$)$_2$Cl$_2$, CuI, DMF, 50 °C, 16 h		[109]
3		Bu$_3$SnH (3 eq), AIBN (0.6 eq), C$_6$H$_6$, 80 °C, 46 h		[110]
4		Bu$_3$SnH (3 eq), AIBN (0.6 eq), C$_6$H$_6$, 80 °C, 46 h		[110] see also [106]
5		Bu$_3$SnH (3 eq), AIBN (0.6 eq), C$_6$H$_6$, 80 °C, 48 h		[111]
6		Pd(PPh$_3$)$_2$Cl$_2$ (1 mmol/L, 0.15 eq), Bu$_4$NCl (1.5 eq), NEt$_3$ (8 eq), DMF/H$_2$O 9:1, 80 °C, 24 h		[81]

Table 15.9. continued.

Entry	Starting resin	Conditions	Product	Ref.
7	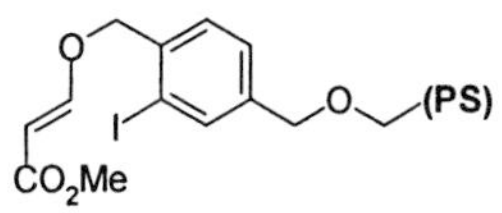	Pd(OAc)$_2$ (0.3 eq), PPh$_3$ (0.6 eq), Bu$_4$NCl (2 eq), K$_2$CO$_3$ (4 eq), DMA, 100 °C, 27 h	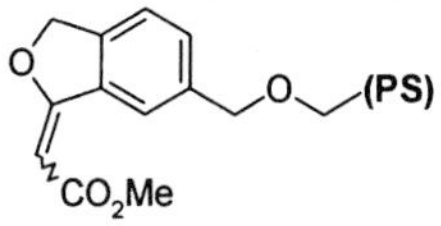	[111]
8	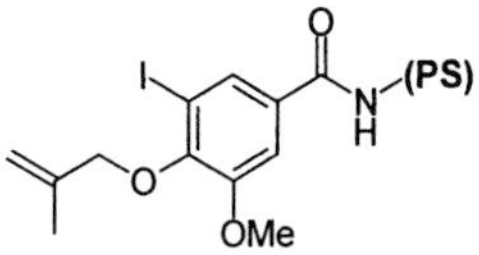	HMPA (40 eq), SmI$_2$ (10 eq), THF, 20 °C, 1 h	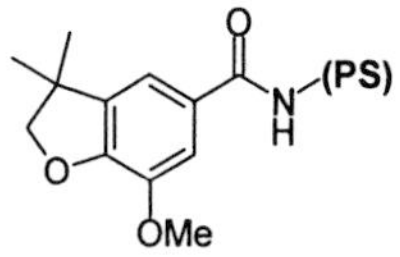	[112]
9	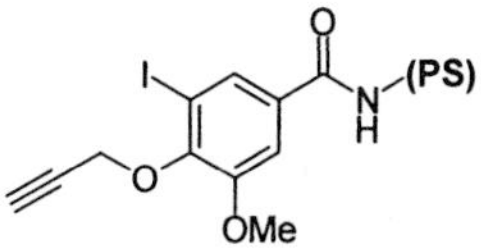	HMPA (40 eq), SmI$_2$ (10 eq), THF, 20 °C, 1 h	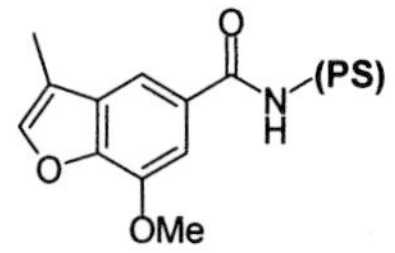	[112]
10	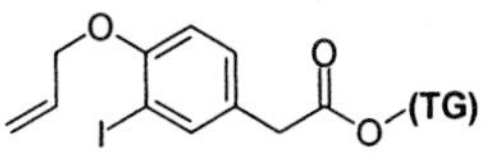	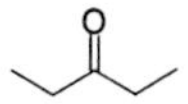 (20 eq), HMPA (40 eq), SmI$_2$ (10 eq), THF, 20 °C, 2 h	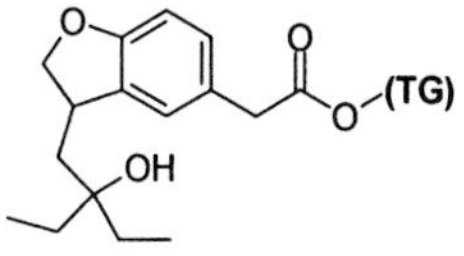	[113]
11	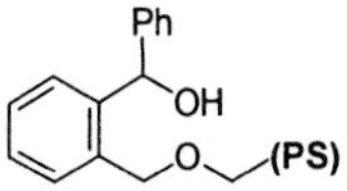	TFA/DCM 1:9, 20 °C, 10 min (Wang resin)	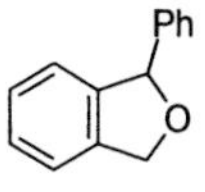	[108]
12	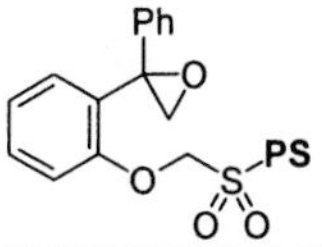	KOtBu (10 eq), THF, DMF, 0 °C to 25 °C, 15 min	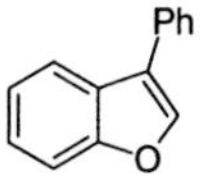	[114]

15.7 Preparation of Thiophenes

The few thiophene syntheses reported in which the formation of the heterocycle is realized on an insoluble support (Entries **1–3**, Table 15.10) are based on the intramolecular addition of C,H-acidic compounds to nitriles (Thorpe–Ziegler reaction), or on the Gewald thiophene synthesis. The mechanism of these cyclizations is outlined in Figure 15.6. In thiophene preparations performed on solid phase, the required α-(cya-

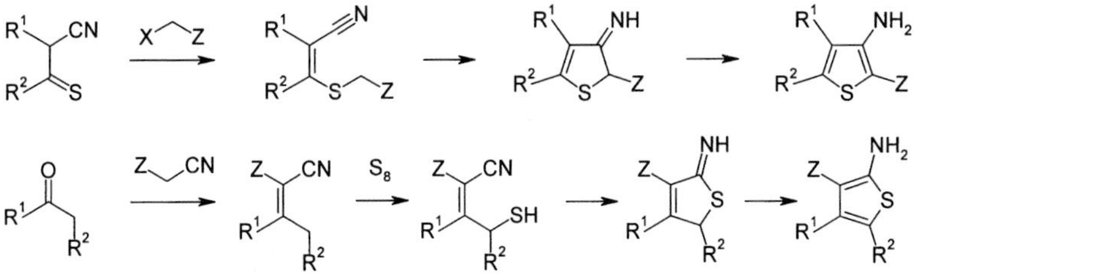

Figure 15.6. Mechanisms of thiophene formation by the Thorpe–Ziegler and Gewald reactions. X: leaving group; Z: electron-withdrawing group.

no)thiocarbonyl compounds have been thioamides, which can readily be prepared from nitriles and isothiocyanates. Either of these components can be linked to the support.

Thiophene is sufficiently acidic to be directly metallated upon treatment with *n*-BuLi (see Figure 4.1). This direct lithiation can also be realized with polystyrene-bound 3-(alkoxymethyl)thiophene [96]. The resulting organolithium compounds react as expected with several electrophiles, such as amides (to yield ketones), alkyl halides, aldehydes, and Me$_3$SiCl [96].

Table 15.10. Preparation of thiophenes.

Entry	Starting resin	Conditions	Product	Ref.
1		1. ArCOCH$_2$Br (1 mol/L, 10 eq), DMF/AcOH 20:1, 20 °C, 15 h 2. DMF/DBU 9:2, 20 °C, 15 h Ar: 4-PhC$_6$H$_4$		[115]
2		1. PhCOCH$_2$Cl (1 mol/L, 10 eq), DMF/AcOH 10:1, 20 °C, 20 h 2. DMF/DBU 2:1, 20 °C, 20 h		[116]
3		1-Boc-piperidin-4-one (0.75 mol/L, 27 eq), S$_8$ (27 eq), morpholine (27 eq), EtOH, 78 °C, 8 h		[117]
4		BuLi, THF, −30 °C, 4 h, then add allyl bromide (10 eq), −30 °C to 20 °C, 2 h		[96]
5		*i*PrMgBr (0.19 mol/L, 7.3 eq), THF, −35 °C, 0.5 h, then add TsCN		[97]
6		*i*PrMgBr (0.19 mol/L, 7.3 eq), THF, −35 °C, 15 min, then add CuCN• 2 LiCl (10 eq), 15 min, then add EtO$_2$C—Br		[97]
7		MeO—ZnBr Pd(dba)$_2$ (0.2 eq), P(2-furyl)$_3$ (0.4 eq), THF, 25 °C, 20 h		[118]
8		2-bromothiophene, Pd$_2$(dba)$_3$, NEt$_3$, P(*o*-Tol)$_3$, DMF, 100 °C, 20 h		[119]

Table 15.10. continued.

Entry	Starting resin	Conditions	Product	Ref.
9		ZnBr thiophene PdCl₂(dppf), THF, 20 °C, 18 h		[120]
10		OH thiophene B(OH)₂ (0.3 mol/L, 1.7 eq), Pd(PPh₃)₄ (0.03 eq), Na₂CO₃ (1.7 eq), DME/H₂O 6:1, 85 °C, overnight		[121]
11		thiophene B(OH)₂ Pd₂(dba)₃ (0.1 eq), K₂CO₃ (2 eq), DMF, 20 °C, 20 h		[122]
12		thiophene SnBu₃ (3 eq), Pd(PPh₃)₄ (0.05 eq), DMF, 60 °C, 24 h		[98]
13		thiophene SnBu₃ (3 eq), Pd(PPh₃)₄ (0.05 eq), DMF, 60 °C, 24 h		[98]
14		BF₄⁻ thiophene⁺–Ph (0.05 mol/L, 1.5 eq), Pd(PPh₃)₄ (0.05 eq), Na₂CO₃ (3 eq), DMF, 20 °C, 20 h (+ 20% biphenyl derivative)		[123] see also [122]
15		(HO)₂B thiophene (3 eq), Pd(PPh₃)₄ (0.05 eq), Na₂CO₃ (8 eq), DME, 85 °C, 48 h		[124] see also [125]
16		H₁₇C₈ thiophene SnMe₃ Pd(PPh₃)₂Cl₂, DMF, 80 °C		[126] see also [127]

Polystyrene-bound 2-bromothiophene can be metallated by treatment with Grignard reagents. The resulting thienylmagnesium compounds can be directly treated with carbon electrophiles to yield the corresponding derivatized thiophenes. For some types of electrophile, transmetallation with CuCN might be required in order to obtain clean products (Entry **6**, Table 15.10).

Thiophenes have been prepared on insoluble supports mainly by arylation or vinylation of halothiophenes and thienylstannanes (Table 15.10). Heck, Suzuki, and Stille couplings with thiophenes usually proceed as smoothly as those with substituted benzenes, and arylations or vinylations of thiophenes have often been used as examples to illustrate new conditions for the realization of these coupling reactions on solid phase.

15.8 Preparation of Imidazoles

Several types of cyclocondensation have been used to prepare imidazoles on insoluble supports (Figure 15.7). These include the condensation of 1,2-dicarbonyl compounds with aldehydes and amines, the 1,3-dipolar cycloaddition of *N*-sulfonylimines to münchnones, and the condensation of *N*-acylated α-amino ketones with ammonia (Table 15.11).

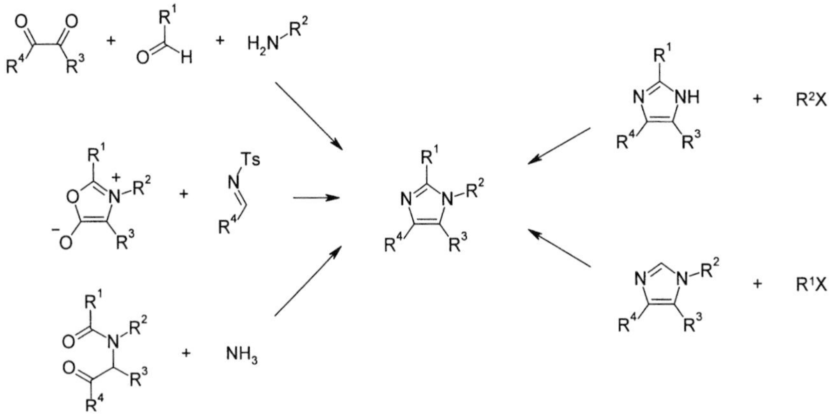

Figure 15.7. Strategies for the preparation of imidazoles on insoluble supports.

Support-bound 1,2-diamines can be readily converted into imidazolidinones by treatment with carbonyl diimidazole [128,129]. The required diamines have been prepared on cross-linked polystyrene by reduction of peptides bound to MBHA resin with borane. Similarly, bicyclic imidazolines have been prepared from triamines and thiocarbonyl diimidazole (Entry **10**, Table 14.3). Dehydration of polystyrene-bound monoacyl ethylene-1,2-diamines yields 4,5-dihydroimidazoles (cyclic amidines, Entry **5**, Table 13.18). Several groups have reported the synthesis of 2-aminoimidazol-4-ones from resin-bound amino acid derivatives (e.g., Entry **6**, Table 15.11). Most of these compounds are, however, unstable, and slowly decompose if dissolved in DMSO (Jesper Lau, private communication).

Imidazoles have been alkylated and arylated on cross-linked polystyrene at nitrogen and at C-2. Illustrative examples of these transformations are listed in Table 15.12. The reaction of polystyrene-bound 4-iodobenzoate with *N*-methylimidazole in the presence of Pd(0) (DMF, 120 °C, 16 h) leads to 4-arylimidazoles [141]. If CuI is

Table 15.11. Preparation of imidazoles and imidazolidinones.

Entry	Starting resin	Conditions	Product	Ref.
1		(0.41 mol/L, 10 eq), EDC (10 eq), DCM, 20 °C, 12 h		[130]
2		NH$_4$OAc (60 eq), AcOH, 100 °C, 20 h		[131] see also [132]
3		(1.2 mol/L, 20 eq), NH$_3$ (20 eq), NH$_4$OAc (1.4 eq), AcOH, 100 °C, 4 h		[133]
4		(1.2 mol/L each, 20 eq), NH$_4$OAc (40 eq), AcOH, 100 °C, 4 h Ar: 2-(HO)-4-BrC$_6$H$_3$		[133]
5		POCl$_3$ (15 eq, 0.1 mol/L), dioxane, 100 °C, 18 h		[134]
6		1. morpholine (5 eq), DIC (5 eq), DIPEA (5 eq), CHCl$_3$, 50 °C, 2 d 2. AcOH/DCM 1:9, 20 °C, 15 h (PS): Rink amide resin		[135] see also [136–138]
7		(0.05 mol/L, 5 eq), DCM, overnight		[139]
8		PhNCO (1 mol/L, 60 eq), NEt$_3$ (60 eq), DMF, 20–55 °C, overnight		[140]

added to the reaction mixture, however, 2-arylimidazoles are obtained as the main product [141]. 4-Acyl and 4-alkylimidazoles have been prepared from resin-bound 4-iodoimidazole by metallation and subsequent treatment with a suitable electrophile [142] (see Chapter 4).

Table 15.12. Alkylation and arylation of imidazoles and imidazolidinones.

Entry	Starting resin	Conditions	Product	Ref.
1		BuLi (3 eq), THF, −50 °C, 20 min, then add PhCHO (10 eq), −60 °C, 1 h, 20 °C, 2 h		[143]
2		Tol-B(OH)$_2$ (3 eq), Cu(OAc)$_2$ (5 eq), pyridine/NMP 1:1, microwaves, 5 × 30 s		[144]
3		Boc-His-OH (0.23 mol/L, 5 eq), DIPEA (5 eq), DMF, 2 d		[145]
4		FmocHN DIPEA, DCM, 20 °C, 2 h		[146]
5		(0.2 mol/L, 2 eq), DIPEA (2 eq), DCM, 1 h		[147]
6		imidazole (8 eq), dioxane, 80 °C, 68 h		[11]
7		(2.1 mol/L, 20 eq), AgOTf (1.3 eq), DMF, 85 °C, 2 × 16 h		[148]
8		imidazole (15 eq), DEAD (15 eq), PPh$_3$ (10 eq), THF, 25 °C, 48 h		[149]
9		imidazole (10 eq), MeCN, 100 °C, 20 h		[150]

15.9 Preparation of Hydantoins (2,4-Imidazolidinediones) and Thiohydantoins

One of the first examples of intramolecular nucleophilic cleavage of compounds from insoluble supports was the preparation of hydantoins from resin-bound α-amino acids. The latter, esterified with hydroxymethyl polystyrene or a similar support, were converted into ureas by treatment with an isocyanate, and cyclization to hydantoins with simultaneous cleavage from the support occurred upon heating or upon treatment with a base (Figure 15.8; Entries **1–4**, Table 15.13). Neat diisopropylamine seems to be particularly well suited for promoting this formation of hydantoins [151]. Thioureas can be cyclized in the same way to yield thiohydantoins. The formation of thiohydantoins by this method usually proceeds more readily than the formation of hydantoins [152]. This strategy for preparing hydantoins is hampered by the fact that the nucleophilicity of the intermediate urea depends on the type of isocyanate used (R^2 in the first equation in Figure 15.8), and so different yields might be obtained for each member of a hydantoin library. During the preparation of thiohydantoins by this method, thiourea formation and cyclization usually occur simultaneously, and if excess isothiocyanate is used, this will contaminate the crude thiohydantoins. Alternatively, hydantoins and thiohydantoins can be prepared on solid phase without simultaneous cleavage from the support. Two possible strategies are outlined in Figure 15.8 (Entries **11–14**, Table 15.13). Thiohydantoins have been prepared by the same strategy as in Entry **12** (Table 15.13), using thiocarbonyldiimidazole instead of phosgene [153].

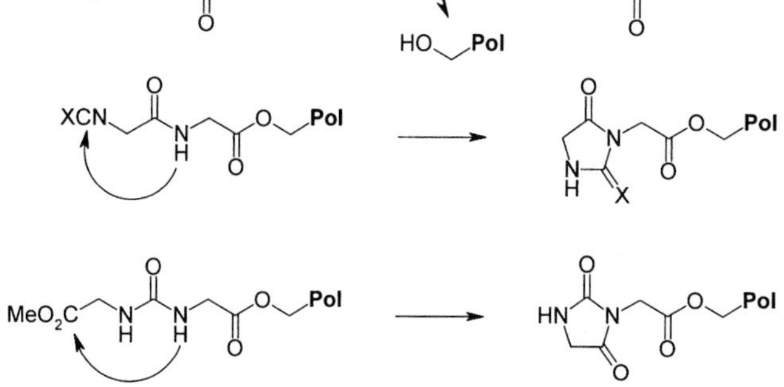

Figure 15.8. Strategies for the preparation of hydantoins and thiohydantoins on insoluble supports. X: O, S.

Table 15.13. Preparation of hydantoins and related heterocycles.

Entry	Starting resin	Conditions	Product	Ref.
1		iPrNH$_2$ (neat), 20 °C, 6 h		[155] see also [151, 156,157]
2		NEt$_3$ (0.82 mol/L), CHCl$_3$, 61 °C, 72 h (Wang resin)		[158] see also [151, 159,160]
3	(Wang resin)	1. triphosgene (3 eq), pyridine (10 eq), DCM, 3 h 2. pyridine (10 eq), (0.44 mol/L, 10 eq), DCM, 17 h		[161]
4		NEt$_3$/MeOH 1:9, 20 °C, 3 h		[162]
5		OSiMe$_3$ F$_3$C NSiMe$_3$ (0.1 eq), DCE, 80 °C, 8–24 h		[163] see also [164]
6		DIPEA (0.5 mol/L), DMF, 80 °C, 24 h		[165] see also [166]
7		1. PhNCO (3 eq), DCM, 20 °C, 24 h 2. DIPEA (1 mol/L, 10 eq), DMF, 100 °C, 36 h		[62]
8		PhNCS (0.05 mol/L, 0.7 eq), MeCN/CHCl$_3$ 1:1, reflux, 16 h (Wang resin)		[158]
9		KOtBu, EtOH, 20 °C Ar: 3-ClC$_6$H$_4$		[167]

Table 15.13. continued.

Entry	Starting resin	Conditions	Product	Ref.
10		NEt$_3$ (14 eq), MeOH, 90 °C, 48 h		[168]
11		MeSiCl$_3$ (0.6 mol/L, 20 eq), NEt$_3$ (40 eq), CHCl$_3$, 70 °C, 24 h		[169]
12		CDI (15 eq) or triphosgene (5 eq), DIPEA (15 eq), DCM		[170] see also [171]
13		NaOMe (10 eq), THF/MeOH 1:1, 3 h (complete racemization)		[154]
14		*i*Pr$_2$NH (neat), 20 °C, 1 h		[156]
15		(0.35 mol/L, 5 eq of each), KOCN (10 eq), pyridine•HCl (10 eq), CHCl$_3$/MeOH/H$_2$O 5:5:1, 20 °C, 24 h		[44]
16		allyl bromide (10 eq), BTPP (10 eq), THF, 12 h		[154]

A further method for preparing hydantoins is the N-alkylation of other hydantoins with reactive alkyl halides in the presence of strong bases [154]. Hydantoinimines have been synthesized from polystyrene-bound isonitriles by an Ugi-type multicomponent condensation (Entry **15**, Table 15.13).

15.10 Preparation of Benzimidazoles, Purines, and Other Fused Imidazoles

Several procedures have been reported for the preparation of benzimidazoles on insoluble supports (Table 15.14). The main synthetic strategies are the cyclocondensation of support-bound 1,2-diaminobenzenes with carboxylic acid derivatives and the oxidation of dihydrobenzimidazoles, which are readily formed from aldehydes and 1,2-diaminobenzenes (Figure 15.9). The approach based on carboxylic acid derivatives generally requires heating and/or strong acids (e.g. TFA; Entry **4**, Table 15.14), which limits the choice of supports and linkers. Preparations based on aldehydes, however, often proceed under significantly milder conditions. Oxidation of the cyclic aminals can often be achieved using weak oxidants, and sometimes occurs even in the absence of additional oxidants if excess aldehyde is used [13,172,173]. The most elegant strategy for preparing benzimidazoles on solid-phase is the direct reduction of 2-nitroanilines in the presence of aldehydes. Complete reduction of the nitro group is not required in this case because the intermediate hydroxylamine can undergo direct condensation with aldehydes [174]. This strategy cleanly yields benzimidazoles with a broad variety of aldehydes under mild reaction conditions [13]. When support-bound 1-amino-2-alkylaminobenzenes are treated with excess aldehyde, benzimidazolium salts (Figure 15.9) are occasionally formed as by-products [13,174]. Benzimidazolium salts have also been prepared by N-alkylation of resin-bound benzimidazoles with strong alkylating agents (e.g., benzyl bromides (50 equiv.), NMP, 70 °C, 18 h [175]).

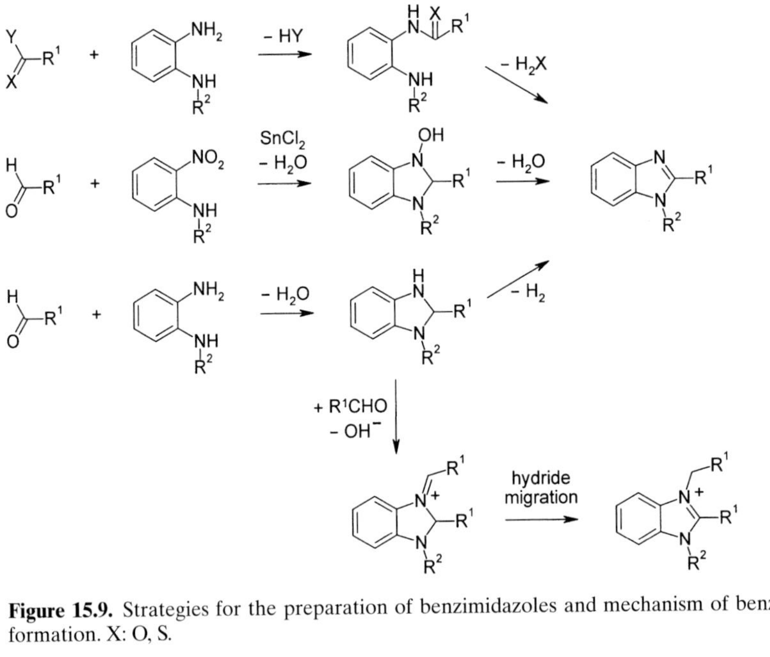

Figure 15.9. Strategies for the preparation of benzimidazoles and mechanism of benzimidazolium salt formation. X: O, S.

The 1,2-diaminobenzenes required for the synthesis of benzimidazoles are generally prepared on solid phase from 2-fluoronitrobenzenes. Aromatic nucleophilic substitution with an aliphatic or aromatic primary amine, followed by reduction of the nitro group (e.g. with $SnCl_2$), yields the desired intermediates. Because formic acid readily converts 1,2-diaminobenzenes into the corresponding benzimidazoles, the reduction of support-bound 2-aminonitrobenzenes should not be conducted in DMF but in NMP [176]. Resin-bound 1,2-diaminobenzenes can also be cyclized with carbonic acid derivatives, such as phosgene, thiophosgene, or synthetic equivalents thereof, to yield benzimidazolones or benzimidazolethiones (Entries **8** and **9**, Table 15.14), or with cyanogen bromide (Br–CN [177]) or isothiocyanates [178] to yield 2-aminobenzimidazoles. Purines and related benzannulated imidazoles can be prepared

Table 15.14. Preparation of benzimidazoles and related heterocycles.

Entry	Starting resin	Conditions	Product	Ref.
1		Ph~CHO (4 eq), DDQ (2 eq), DMF, 20 °C, 5 h		[184] see also [173]
2		(0.15 mol/L, 10 eq), NMP, 20 °C, 5 h, then 50 °C, 8 h		[172]
3		(0.25 mol/L, 30 eq), DMF/ BuOH 5:6, 90 °C, 24 h		[185]
4		1. $(BrCH_2CO)_2O$ (10 eq), DMF, 1 h; 2. pyrrolidine (2 mol/L), DMSO, 20 °C, 4 h; 3. TFA (neat), 16 h		[176]
5		AcOH, 90 °C, 16 h (PS): Wang resin		[186]
6		salicylaldehyde (0.7 mol/L, 14 eq), AcOH (26 eq), $SnCl \cdot H_2O$ (20 eq), NMP, 20 °C, 9 h R_2NH: morpholine		[13]

Table 15.14. continued.

Entry	Starting resin	Conditions	Product	Ref.
7		HC(OMe)₃/TFA/ DCM 1:1:2, 3 h		[187] see also [188]
8		(2 eq), THF/DCM/ DMF 8:2:1, 24 h		[189]
9		(10 eq), THF, 20 °C, overnight		[190] see also [191]
10		(0.6 mol/L, 10 eq), PhNO₂, 180 °C, 3 d		[192] see also [193]
11		(0.6 mol/L, 10 eq), PhNO₂, 180 °C, 3 d		[192]
12		AcOH		[194]
13		MeNCS (0.2 mol/L), C₆H₆, 80 °C, 15 min, then add DCC (0.2 mol/L), 80 °C, overnight		[195]
14		PhCHO (8 eq), 2-aminopyridine (8 eq), TsOH (1 eq), CHCl₃/MeOH/ HC(OMe)₃ 1:1:1, 20 °C, overnight (PS): Rink amide resin		[196] see also [183]

on solid phase by similar reactions to those used for the preparation of benzimidazoles.

Imidazo[1,2-*a*]pyridines are formed upon treatment of isonitriles with aldehydes and 2-aminopyridines (Entry **14**, Table 15.14). 2-Aminoazines other than 2-aminopyridine can also be used in this multicomponent reaction, yielding the corresponding fused imidazoles [179–183].

Support-bound benzimidazoles can be N-arylated by treatment with arylboronic acids in the presence of Cu(OAc)$_2$ (Entry **3**, Table 15.15). Benzimidazolones, which can be prepared from resin-bound 1,2-diaminobenzenes and di-*N*-succinimidyl carbonate (Entry **8**, Table 15.14), can also be N-alkylated on insoluble supports (Entry **4**, Table 15.15). Benzimidazole-2-thiones, on the other hand, are generally cleanly S-alkylated (Entry **5**, Table 15.15). Examples of amination by aromatic nucleophilic substitution at resin-bound halopurines are given in Section 10.1.2.

Table 15.15. Chemical modification of benzimidazoles and related heterocycles.

Entry	Starting resin	Conditions	Product	Ref.
1		2,6-dichloropurine (1.5 eq), Pd$_2$(dba)$_3$ (0.1 eq), PPh$_3$ (0.4 eq), 1,2,2,6,6-pentamethyl-piperidine (6 eq), THF/DMSO 1:1, 45 °C, 16 h		[197] see also [198]
2		PPh$_3$, DEAD, THF/DCM 1:1, 20 °C, 48 h		[199]
3		Tol-B(OH)$_2$ (3 eq), Cu(OAc)$_2$ (5 eq), pyridine/NMP 1:1, microwaves, 5 × 30 s	(two isomers, 1:1)	[144]
4		NaH (20 eq), DMF, 50 min; then *i*BuBr (40 eq), DMF, 24 h		[189]
5		BnBr, DIPEA, DMF, 20 °C, 16 h		[190] see also [191]

15.11 Preparation of Isoxazoles

Isoxazoles and their partially or fully saturated analogs have mainly been prepared, both in solution and on insoluble supports, by 1,3-dipolar cycloadditions of nitrile oxides or nitrones to alkenes or alkynes (Figure 15.10). Nitrile oxides can be generated in situ on insoluble supports by dehydration of nitroalkanes with isocyanates, or by conversion of aldehyde-derived oximes into α-chlorooximes and dehydrohalogenation of the latter. Nitrile oxides react smoothly with a wide variety of alkenes and alkynes to yield the corresponding isoxazoles. A less convergent approach to isoxazoles is the cyclocondensation of hydroxylamine with 1,3-dicarbonyl compounds or α,β-unsaturated ketones.

Dihydroisoxazoles with a substituent at nitrogen are most conveniently prepared by 1,3-dipolar cycloaddition of nitrones to alkenes or alkynes. Nitrones are usually prepared in situ from carbonyl compounds and *N*-(alkyl)hydroxylamines (Figure 15.10).

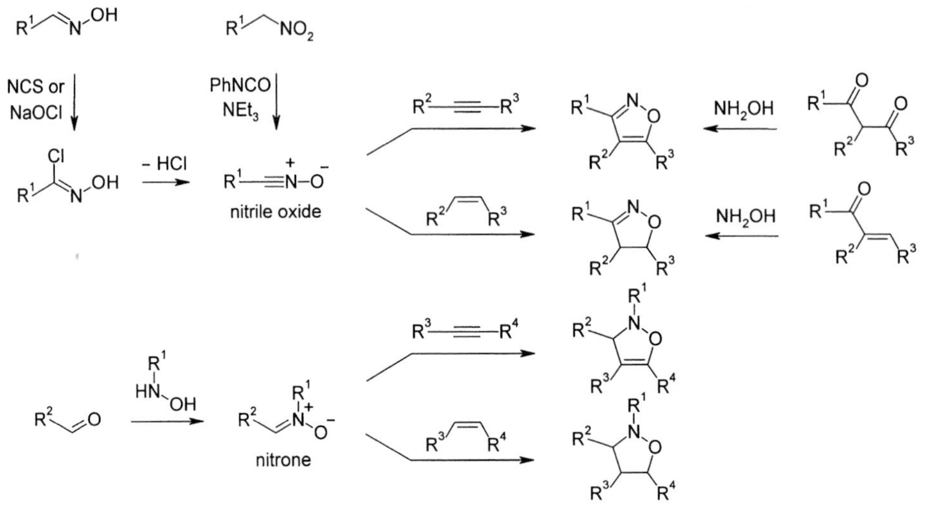

Figure 15.10. Synthetic strategies for the preparation of isoxazoles, isoxazolines, and isoxazolidines.

Most of the approaches outlined in Figure 15.10 have been successfully realized on insoluble supports, either with the alkene or alkyne linked to the support, or with support-bound 1,3-dipoles (Table 15.16). Nitrile oxides are highly reactive 1,3-dipoles and react smoothly with both electron-poor and electron-rich alkenes, including enol ethers [200]. The addition of resin-bound nitrile oxides to alkenes (Entries **5** and **6**, Table 15.16) has also been accomplished enantioselectively under catalysis by diisopropyl tartrate and EtMgBr [201]. The diastereoselectivity of the addition of nitrile oxides and nitrones to resin-bound chiral acrylates has been investigated [202]. Intramolecular 1,3-dipolar cycloadditions of nitrile oxides and nitrones to alkenes have been used to prepare polycyclic isoxazolidines on solid phase (Entries **7** and **9**, Table 15.16).

Strategies that lead to the formation of isoxazoles during cleavage from an insoluble support include the oxidative cleavage of *N*-(4-alkoxybenzyl)isoxazolidines with DDQ to yield isoxazolines (Entry **14**, Table 15.16), the nucleophilic cleavage of 2-acyl enamines with hydroxylamine (Entry **15**, Table 15.16), and the acidolysis of 2-cyanophenols etherified with an oxime resin (Entry **17**, Table 15.16). The required oxime ethers for the latter synthesis were prepared by reaction of the corresponding 2-fluorobenzonitriles with Kaiser oxime resin [203].

Table 15.16. Preparation of isoxazoles, isoxazolines, and isoxazolidines.

Entry	Starting resin	Conditions	Product	Ref.
1		BnNO$_2$ (0.05 mol/L), PhNCO (0.1 mol/L), NEt$_3$, THF, 60 °C, 20 h		[159] see also [204–206]
2		(5 eq), NaOCl (bleach; 0.4 mol/L, 10 eq), THF, 4 h		[207]
3		(5 eq), PhNCO (11 eq), NEt$_3$, C$_6$H$_6$, 80 °C, 30 h		[208] see also [206,209]
4		(5 eq), NEt$_3$ (5 eq), THF, 20 °C, 2 h		[63]
5		NaOCl (bleach; 0.4 mol/L, 10 eq), allyl alcohol (5 eq), THF, 4 h		[207] see also [201]
6		NCS (4 eq), DCM, 2 h; then 1-benzyl-2,5-dihydropyrrole (10 eq), NEt$_3$, DCM, overnight		[210]
7		PhNCO (1.2 mol/L), NEt$_3$ (0.12 mol/L), C$_6$H$_6$, 20 °C, 3 d		[104] see also [211]
8		NH$_2$OH (2.5 mol/L), DMA, 80 °C, 24 h		[212] see also [213]
9		HATU, DIPEA, DMAP, DCM, 20 °C		[214,215]

Table 15.16. continued.

Entry	Starting resin	Conditions	Product	Ref.
10		Ar²CHO (0.33 mol/L, 10 eq), HONHMe•HCl (10 eq), DIPEA (10 eq), PhMe, 80 °C, 5 h	 (mixture of diastereomers)	[216]
11		HONHMe•HCl (0.87 mol/L, 10 eq), DIPEA (10 eq), PhMe, 80 °C, 0.5 h, then ArCH=CH₂ (10 eq), 80 °C, 5 h Ar: 1-naphthyl		[216]
12		1. ArCHO (0.45 mol/L, 10 eq), PhMe, 80 °C, 0.5 h 2. PhSO₂CH=CH₂ (0.45 mol/L, 10 eq), PhMe, 80 °C, 5 h Ar: 4-(O₂N)C₆H₄		[216]
13		 Yb(OTf)₃ (0.2 eq), PhMe, 20 °C, 20 h		[217]
14		DDQ (3 eq), DCM/H₂O 10:1, 12 h		[217]
15		HONH₃Cl, NaHCO₃, EtOH/H₂O, 50 °C		[218]
16		SnCl₂•H₂O (1 mol/L, 31 eq), NMP, 20 °C, 18 h Ar: 4-(AcHN)C₆H₄		[85]
17		TFA/HCl (5 mol/L in H₂O) 4:1, 55 °C, 2 h		[203,219]

15.12 Preparation of Oxazoles and Oxazolidines

Few solid-phase syntheses of oxazoles have been reported (Table 15.17). The most general strategy is the dehydration of α-(acylamino) ketones (Entry **2**, Table 15.17) or 2-(acylamino)phenols (Entry **1**, Table 15.17). Oxazolidin-2-ones have been prepared by intramolecular nucleophilic cleavage of carbamates from insoluble supports (Entries **5** and **6**, Table 15.17). Resin-bound 2-aminoethanols, which are accessible by nucleophilic ring-opening of oxiranes with amines, undergo cyclocondensation with aldehydes to yield oxazolidines [220,221]. These compounds are unstable towards acids, and can be released from the support only under neutral or basic reaction conditions.

Table 15.17. Preparation of oxazoles and oxazolidines.

Entry	Starting resin	Conditions	Product	Ref.
1		PPh$_3$ (0.38 mol/L, 5 eq), DEAD (5 eq), THF, 20 °C, 17 h		[222]
2		Ph$_3$PCl$_2$ (2 eq), NEt$_3$ (6 eq), DCM, 16 h		[223] see also [224]
3		AcNH$_2$ (10 eq), BF$_3$OEt$_2$ (1 eq), xylenes, 140 °C, 48 h		[149]
4		PPh$_3$ (0.15 mol/L), Cl$_3$CCCl$_3$ (0.5 mol/L), NEt$_3$ (0.16 mol/L), MeCN/DCE 1:1, 60 °C, 2 × 12 h		[225]
5		DBN (3 eq), DCM, 20 °C, overnight		[226]
6		pyrrolidine (0.5 mol/L, 5 eq), LiClO$_4$ (5 eq), THF, 20 °C		[227]

15.13 Preparation of Thiazoles and Thiazolidines

Most reported solid-phase thiazole syntheses have been based on the cyclocondensation of thioamides or thioureas with α-halo ketones (Table 15.18). These reactions generally proceed smoothly under basic or acidic conditions, and are compatible with most common linkers.

Benzothiazoles have been prepared by oxidative condensation of aldehydes with resin-bound 4-mercapto-3-aminobenzoic acid derivatives (Entry **8**, Table 15.18), and by dehydration of 2-mercaptoanilides (Entry **9**, Table 15.18). Oxidation of 2,3-dihydrobenzothiazoles to benzothiazoles can occur spontaneously, without the need of any additional oxidant [231]. Polystyrene-bound 4-iodobenzoate reacts with thiazole in the presence of Pd(0) to yield a 4-(4-thiazolyl)benzoate (Entry **10**, Table 15.18). If CuI is added to the reaction mixture, however, the 4-(2-thiazolyl)benzoate results [141].

Thiazolidin-4-ones have been prepared by condensation of support-bound imines with α-mercapto carboxylic acids (Entry **11**, Table 15.18). These thioaminals are quite stable and tolerate, for example, treatment with TFA [228,229]. Thiazolidines, which can be prepared from resin-bound cysteine and aldehydes (Entry **12**, Table 15.18), are also remarkably stable towards acid-promoted hydrolysis. Libraries of thiazolidinones have been used to identify new cyclooxygenase-1 inhibitors [230].

Table 15.18. Preparation of thiazoles and thiazolidines.

Entry	Starting resin	Conditions	Product	Ref.
1		(0.5 mol/L), DMF, 70 °C, 4 h		[232]
2		(0.2 mol/L, 5 eq), dioxane, 3 × 1 h		[233]
3		dioxane, 60 °C, 16 h		[234]
4		4-FC$_6$H$_4$COCH$_2$Br (0.13 mol/L, 1.3 eq), DMF, 20 °C, 18 h		[235]
5		PhCOCH$_2$Cl (1 mol/L, 10 eq), DMF/AcOH 10:1, 20 °C, 20 h		[116]

Table 15.18. continued.

Entry	Starting resin	Conditions	Product	Ref.
6		thiourea (4 eq), MeOH, 65 °C, 12 h		[236] see also [29]
7		MeO$_2$C (5 eq), MeCN, 20 °C, 16 h		[237]
8		C$_6$H$_{11}$CHO (0.2 mol/L, 4 eq), DDQ (1 eq), DCM, 25 °C, 24 h		[238] see also [231]
9		1. TFA/DCM/TES 1:94:5, 7 × 2 min 2. DTT, DMF/MeOH 9:1, 20 °C, 3 h		[239]
10		thiazole (5 eq), Pd(OAc)$_2$ (0.1 eq), PPh$_3$ (0.2 eq), Cs$_2$CO$_3$ (1 eq), DMF, 120 °C, 16 h		[141]
11		HS–CO$_2$H (0.7 mol/L), THF, 70 °C, 3 h		[17] see also [228–230,240]
12		ArCHO, PhMe/MeCN/AcOH 45:45:10		[241]

15.14 Preparation of Pyrazoles

Most solid-phase syntheses of pyrazoles are based on the cyclocondensation of hydrazines with suitable 1,3-dielectrophiles. The reported examples include the reaction of hydrazines with support-bound α,β-unsaturated ketones, 1,3-diketones, 3-keto esters, α-(cyano)carbonyl compounds, and α,β-unsaturated nitriles (Table 15.19). Pyrazoles have also been prepared from polystyrene-bound 3-(hydrazino)esters, which are generated by the addition of ester enolates to hydrazones (Entry **7**, Table 15.19; see also Section 10.3). Benzopyrazoles can be prepared from support-bound hydrazones using the reaction sequence outlined in Figure 15.11. Oxidation of a polystyrene-bound benzophenone hydrazone yields an α-(acyloxy)azo compound. Upon treatment with a Lewis acid, this intermediate is converted into a 1,2-diazaallyl cation,

Table 15.19. Preparation of pyrazoles and pyrazolones.

Entry	Starting resin	Conditions	Product	Ref.
1		(0.5 mol/L, 10 eq), DMSO, 100 °C, 16 h		[243]
2		N_2H_4 (2.5 mol/L), DMA, 80 °C, 24 h		[212] see also [213, 244–246]
3		PhMe, 100 °C, 5 h		[247] see also [36,248]
4		BnN$_2$H$_3$, AcOH, EtOH, 70 °C		[249]
5		BnN$_2$H$_3$·HCl, AcOH/EtOH 1:9		[250]
6		PhN$_2$H$_3$ (0.14 mol/L, 10 eq), NaOEt (100 eq), EtOH, 78 °C, 24 h		[251]
7		NaOMe (0.42 mol/L), MeOH, 60 °C, 8 h		[252]
8		DMAD (0.4 mol/L, 4 eq), PhMe, 80 °C, 48 h		[253]
9		pyrazole (0.35 mol/L, 15 eq), K$_2$CO$_3$ (6 eq), MeCN, 60 °C, 2 d		[254]
10		Tol-B(OH)$_2$ (3 eq), Cu(OAc)$_2$ (5 eq), pyridine/NMP 1:1, microwaves, 5 × 30 s		[144]

which undergoes intramolecular aromatic electrophilic amination to yield the observed benzopyrazole.

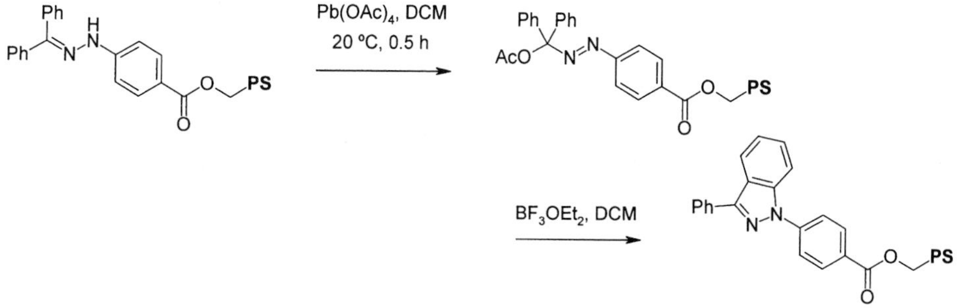

Figure 15.11. Synthesis of benzopyrazoles by intramolecular electrophilic amination [242].

Examples of the N-alkylation and N-arylation of pyrazoles on insoluble supports have also been reported. Unsubstituted pyrazole undergoes Michael addition to resin-bound α-acetamido acrylates to yield 1-pyrazolylalanine derivatives (Entry **9**, Table 15.19). N-Arylation of support-bound pyrazole has been accomplished using arylboronic acids in the presence of Cu(OAc)$_2$ under microwave irradiation (Entry **10**, Table 15.19).

15.15 Preparation of Triazoles, Tetrazoles, Oxadiazoles, and Thiadiazoles

1,2,3-Triazoles have been prepared on insoluble supports by 1,3-dipolar cycloaddition of alkyl azides to support-bound alkynes, and by diazo group transfer to 2-(acyl)-enamines (Entries **1** and **2**, Table 15.20). The second approach avoids the handling of hazardous reagents, and yields triazoles with complete control of the regioselectivity. 1-Hydroxybenzotriazoles can be prepared on cross-linked polystyrene from 2-chloro-nitroarenes, and their reduction to the corresponding benzotriazoles can be achieved by treatment with PCl$_3$ (Entries **3** and **6**, Table 15.20). 1,2,4-Triazoles have been prepared by condensation of amidines with polystyrene-bound hydrazides (Entry **5**, Table 15.20). The resulting triazoles can be N-alkylated by treatment with alkyl halides (0.25 mol/L, 30 equiv., DMF, NaOH), but mixtures of the 1-alkylated and 2-alkylated triazoles are obtained [255]. 1,2,4-Triazoles have also been prepared from N-amino-amidines (amidohydrazones; Entry **4**, Table 15.20), which were prepared from resin-bound thioamides by S-alkylation with methyl triflate followed by treatment with hydrazine [256]. 1,2,4-Triazoles undergo Michael addition to polystyrene-bound α-acetamido acrylates to yield triazole-derived α-amino acids (Entry **7**, Table 15.20). Benzotriazoles have been N-arylated on insoluble supports by treatment with aryl-boronic acids in the presence of catalytic amounts of copper salts (Entry **8**, Table 15.20). The multicomponent Ugi-type condensation of aldehydes, secondary amines, azide, and polystyrene-bound isonitriles has been reported to yield tetrazoles in acceptable yields (Entry **9**, Table 15.20). Tetrazoles have also been prepared from polystyrene-bound nitriles and Me_3SnN_3 (5 equiv., PhMe, 100 °C, 12 h [149]). This reaction yields *N*-trimethylstannyl tetrazoles, which can be destannylated by treatment with TFA. Resin-bound tetrazoles with a free NH group can be N-alkylated by treatment with alkyl halides in the presence of NEt_3 [149].

The reaction of Tentagel-bound carboxylic esters with amidooximes has been used to prepare oxadiazoles (Entry 11, Table 15.20). Thiadiazoles have been prepared from support-bound *N*-sulfonylhydrazones by treatment with thionyl chloride. Thiadiazole formation and cleavage from the support occurred simultaneously (Entry 12, Table 15.20). Perhydro-1-thia-2,5-diazole-2,2-dioxides (sulfahydantoins) have been prepared by aminosulfonylation of amino acids esterified with Wang resin, followed by ring-closure with simultaneous cleavage from the support [257].

Table 15.20. Preparation of triazoles, tetrazoles, oxadiazoles, and thiadiazoles.

Entry	Starting resin	Conditions	Product	Ref.
1		RCH_2N_3 (0.07 mol/L, 2 eq), PhMe, 110 °C, 12 h	(+ regioisomer)	[258] see also [259]
2		TsN_3 (0.5 mol/L, 6 eq), DMF/DIPEA 3:1, 20 °C, 24 h		[260]
3		1. N_2H_4 (0.1 mol/L), $EtOCH_2CH_2OH$, 114 °C, 20 h 2. dioxane/conc aq HCl 1:1, 100 °C, 20 h		[261]
4		1. $ClCH_2COCl$ (1 mol/L), DCM, 20 °C, 24 h 2. AcOH/DMF 1:1, 20 °C, 20 h		[256]
5		acetamidine (0.5 mol/L), MS 4 Å, 2-methoxyethanol, 100 °C, 7 d		[255]
6		PCl_3 (1.2 mol/L), $CHCl_3$, 61 °C, 20 h		[261]
7		1,2,4-triazole (0.14 mol/L, 6 eq), K_2CO_3 (3 eq), MeCN, 25 °C, 2 d		[254]
8		Tol-B(OH)$_2$ (3 eq), $Cu(OAc)_2$ (5 eq), pyridine/NMP 1:1, microwaves, 5 × 30 s	(two isomers, 1:1)	[144]

Table 15.20. continued.

Entry	Starting resin	Conditions	Product	Ref.
9	CN (PS)	PhCHO, piperidine (0.39 mol/L, 5 eq of each), NaN$_3$ (10 eq), pyridine•HCl (10 eq), CHCl$_3$/MeOH/H$_2$O 5:5:2, 20 °C, 4 d		[44]
10		DIC (3 mol/L, 5 eq), DMF, 100 °C, 18 h		[262]
11		NaOEt (10 eq of each), EtOH/DCM 1:1, 20 °C, 3 d		[263] see also [264,265]
12		SOCl$_2$ (4.2 mol/L, 20 eq), DCE, 60 °C, 5 h		[266]

15.16 Preparation of Pyridines and Dihydropyridines

Various approaches can be used to synthesize pyridines and partially saturated pyridines on insoluble supports. Dihydropyridines can be readily prepared by cyclocondensation of amines with ketones and aldehydes (Hantzsch pyridine synthesis, Figure 15.12). This synthesis proceeds particularly smoothly when using a β-keto carboxylic acid derivative as the ketone component. This cyclocondensation has been realized on

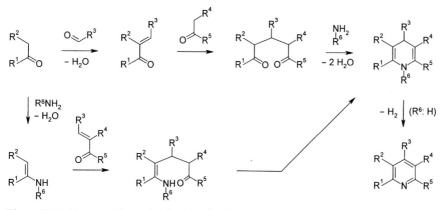

Figure 15.12. Two possible mechanisms for the formation of dihydropyridines and pyridines by cyclocondensation.

insoluble supports with either the amine or the β-keto carboxylic acid linked to the support (Entries **1** and **2**, Table 15.21). If ammonia is used as the amine, the initially formed dihydropyridines can be oxidized to the corresponding pyridines (Entries **3** and **4**, Table 15.21). Closely related to the Hantzsch pyridine synthesis is the cyclocon-

Table 15.21. Preparation of pyridines and dihydropyridines.

Entry	Starting resin	Conditions	Product	Ref.
1	(structure: MeS-, MeO₂C-, O-(PS) enone)	NH₂ reagent (MeO₂C-); DMF, MS 4 Å, 80 °C, 10 h	(dihydropyridine, MeO₂C, MeS, O-(PS))	[270] see also [271, 272]
2	(MeO-, O, N-(PS))	PhCHO, pentane-2,4-dione, pyridine, MS 4 Å, 45 °C, 24 h, then TFA/DCM (3:97), 45 min (Rink amide resin)	(MeO, O, NH, Ph dihydropyridine)	[273] see also [274]
3	(PS)-O, Ar-enone (Ar: 3,4-F₂C₆H₃)	OSiMe₃ furan enol ether (0.23 mol/L, 4 eq), CsF (1 eq), DMSO, 70 °C, 3 h; wash; NH₄OAc, AcOH/DMF 1:20, 100 °C, 18 h Ar: 3,4-F₂C₆H₃	(furyl pyridine, Ar, (PS)-O)	[275] see also [246, 276]
4	(dihydropyridine, MeO₂C, MeS, O-(PS))	CAN, DMA, 20 °C, 15 min	(pyridine, MeO₂C, MeS, O-(PS))	[270] see also [271]
5	Ph-N, O-PEG enamine ester	2,2,6-trimethyl-4H-1,3-dioxin-4-one; PhMe, 110 °C	Ph-N, O-PEG dihydropyridinone	[277]
6	Ph-N⁺, O-(PS) pyridinium	PhMgCl (4 eq), THF, 2 h	Ph-N, O-(PS) dihydropyridine	[278] see also [279]
7	Cl, PS benzyl chloride	Tol-S(O)-pyridine NLi (2 eq), THF, HMPA, 20 °C, 12 h	Tol-S(O), N, PS dihydropyridine	[280]
8	Cl, O-(PS) chloroacetate	pyridine (4.3 eq), DCM, 20 °C, 24 h	Cl⁻, pyridinium-N⁺, O-(PS)	[63] see also [281,282]
9	HO, Ph, N, O-PS piperidinol carbamate	TFA/DCM 2:1, O₂, 48 h	Ph-pyridine	[279]

densation of acetyl ketene (formed by thermolysis of Meldrum's acid) with PEG-bound enamines, whereby 1,4-dihydro-4-pyridinones are obtained (Entry 5, Table 15.21).

Dihydropyridines can be prepared on cross-linked polystyrene by the addition of organometallic reagents to pyridinium salts (Entry 6, Table 15.21). These reactions do not always give high yields because of several competing processes (e.g. cleavage of the linker, deacylation of the N-acylpyridinium salt).

Support-bound alkylating agents have been used to N-alkylate pyridines and dihydropyridines (Entries 7 and 8, Table 15.21). Similarly, resin-bound pyridines can be N-alkylated by treatment with α-halo ketones (DMF, 45 °C, 1 h [267]) or other alkylating agents [246]. Polystyrene-bound 1-[(alkoxycarbonyl)methyl]pyridinium salts can be prepared by N-alkylating pyridine with immobilized haloacetates (Entry 8, Table 15.21). These pyridinium salts react with acceptor-substituted alkenes to yield cyclopropanes (Section 5.1.3.6). Pyridinium salts have also been prepared by reaction of resin-bound primary amines with N-(2,4-dinitrophenyl)pyridinium salts [268,269].

Table 15.22. Preparation of pyridines by palladium-mediated vinylation and arylation of other pyridines.

Entry	Starting resin	Conditions	Product	Ref.
1		3-bromopyridine, Pd$_2$(dba)$_3$, NEt$_3$, DMF, P(o-Tol)$_3$, 100 °C, 20 h		[119]
2		Pd$_2$(dba)$_3$ (0.1 eq), K$_2$CO$_3$ (2 eq), DMF, 20 °C, 20 h		[122]
3		Pd$_2$(dba)$_3$, AsPh$_3$, dioxane, 50 °C, 24 h		[283]
4		(0.19 mol/L, 10 eq), Pd(PPh$_3$)$_4$ (0.05 eq), Na$_2$CO$_3$ (21 eq), PhMe/EtOH 9:1, 90 °C, 24 h		[281]
5		Pd(dba)$_2$ (0.2 eq), P(2-furyl)$_3$ (0.4 eq), THF, 25 °C, 20 h		[118]
6		PhSnBu$_3$, Pd(PPh$_3$)$_4$, DMF, 60 °C, 24 h		[98]

A convenient method for preparing vinylated or arylated pyridines on insoluble supports is palladium-mediated cross-coupling. The Heck, Suzuki, and Stille reactions have been successfully used for this purpose (Table 15.22). The conditions are essentially the same as those used for the related coupling of arenes (see Section 5.4).

15.17 Preparation of Tetrahydropyridines and Piperidines

Ring-closing metathesis is well suited for the preparation of five- or six-membered heterocycles, and has also been successfully used to prepare tetrahydropyridines on insoluble supports (Entries **1** and **2**, Table 15.23). Because metathesis catalysts (ruthenium or molybdenum carbene complexes) are electrophilic, reactions should be conducted with acylated amines to avoid poisoning of the catalyst.

Table 15.23. Preparation of tetrahydropyridines and piperidines.

Entry	Starting resin	Conditions	Product	Ref.
1		$Cl_2(PCy_3)_2Ru=Ph$ (0.13 eq), DCM, 40 °C, 12 h		[37] see also [10]
2		$Cl_2(PCy_3)_2Ru=Ph$ (0.05 eq), styrene (1 eq), PhMe, 50 °C, 18 h; Ar: 4-(MeO)C$_6$H$_4$		[284] see also [285]
3		Me$_3$SiO—OMe (0.9 mol/L, 5 eq), Yb(OTf)$_3$ (0.1 eq), THF, 60 °C, 3 h		[286] see also [287]
4		OHCCO$_2$Et (1.2 eq), isoprene (2.5 eq), Yb(OTf)$_3$ (0.1 eq), DCM, 48 h		[288]
5		(0.3 mol/L, 5 eq), BnNH$_2$ (5 eq), CHCl$_3$/MeOH 2:1, 20 °C, 48 h		[44]
6		BnNH$_2$, THF, 20 °C, 3 d		[289] see also [290]
7		acrylic acid anhydride (0.3 mol/L, 10 eq), DCM, 23 °C, 12 h		[291]

Table 15.23. continued.

Entry	Starting resin	Conditions	Product	Ref.
8	H₂N~~N(H)~PS (Ph Ph)	+ CO₂H coumarin + O=C(Ph) (5 eq + 10 eq), NMP/THF 1:1, MS 3 Å, 40 °C, 20 h	(bicyclic product, Ph, Ph, PS, NH)	[292]
9	O, OMe; O; MeO Ph O; (PS)	BnNH₂ (0.1 mol/L, 10 eq), AcOH (13 eq), MS 3 Å, PhMe, 70 °C, 60 h	(dihydropyridine product, Ph, MeO, N-Ph, (PS))	[293]
10	(tetrahydropyridinone, N-CO-O-PS)	ClMg~isobutyl; CuI, BF₃OEt₂, THF, 4 h	(piperidinone, N-CO-O-PS)	[294]
11	(t-Bu pyridinium, N⁺, O-CHPh₂-PS, RSO₃⁻)	NaBH₄ (0.3 mol/L, 10 eq), DMF/MeOH 1:1, 2 × 4 h	(t-Bu tetrahydropyridine, N, O-CHPh₂-PS)	[269]
12	Ph, Tol-C(=O)-N, dihydropyridine, O-CH₂-(PS)	(maleimide) N–Ph O=C...C=O (1 mol/L, 10 eq), THF, 65 °C, overnight	(polycyclic product, N-Ph, Ph, Tol-C(=O)-N, O-(PS))	[295]

One approach to tetrahydropyridinones is the Lewis acid mediated hetero-Diels–Alder reaction of electron-rich dienes with polystyrene-bound imines (Entries **3** and **4**, Table 15.23). The Ugi reaction of 5-oxo carboxylic acids and primary amines with support-bound isonitriles has been used to prepare piperidinones on insoluble supports (Entry **5**, Table 15.23). Entry **6** in Table 15.23 is an example of the preparation of a 4-piperidinone by amine-induced β-elimination of a resin-bound sulfinate followed by Michael addition of the amine to the newly generated divinyl ketone. The intramolecular Pauson–Khand reaction of propargyl(3-butenyl)amines, which yields cyclopenta[*c*]pyridin-6-ones, is depicted in Table 12.4.

Support-bound pyridines and partially saturated pyridines can be valuable synthetic intermediates, enabling various types of chemical transformation. Piperidinones can be prepared on cross-linked polystyrene by the addition of organometallic reagents to tetrahydropyridinones (Entry **10**, Table 15.23). 1,2-Dihydropyridines are electron-rich dienes that can undergo Diels–Alder reaction with electron-poor dienophiles. Diels–Alder cycloaddition of support-bound 1,2-dihydropyridines has been used to prepare nitrogen-containing polycyclic systems (Entry **12**, Table 15.23).

15.18 Preparation of Fused Pyridines

Tetrahydroisoquinolines can be readily prepared in solution and on insoluble supports by condensation of phenethylamines with aldehydes (Figure 15.13; Pictet–Spengler synthesis). The precise reaction conditions depend on the substitution pattern of the arene, but smooth cyclization generally only occurs with electron-rich arenes. Most of the reported examples have been performed under acidic reaction conditions, but phenols and imidazoles can also be alkylated intramolecularly by imines under basic or neutral conditions. Illustrative examples of Pictet–Spengler reactions on solid phase are listed in Table 15.24. Closely related to the Pictet–Spengler reaction is the Bischler–Napieralski synthesis of isoquinolines (Figure 15.13). In this reaction, an acylated phenethylamine is cyclized by treatment with a strong dehydrating agent. The resulting dihydroisoquinoline can either be reduced to the tetrahydroisoquinoline (e.g. with $NaBH_4$) or oxidized to yield an isoquinoline.

Histamine, tryptamine, and the corresponding amino acids histidine and tryptophan, can also be cyclized by treatment with aldehydes to yield annulated piperidines (Table 15.24). Numerous examples of the preparation of carbolines from support-bound tryptophan derivatives have been reported [60,296–301].

Table 15.24. Pictet–Spengler and Bischler–Napieralski reactions on insoluble supports.

Entry	Starting resin	Conditions	Product	Ref.
1		OHC (0.5 mol/L), pyridine, 100 °C, 14 h		[302]
2		PhCH₂CHO (10 eq), TFA/DCM 1:9, 20 °C, overnight		[303]
3		(4 eq), DCM/TFA 99:1, 20 °C, 48–72 h (Wang resin)		[304]
4		OHC (0.5 mol/L), pyridine, 100 °C, 14 h		[302]
5		1. POCl₃ (20 eq), PhMe, 90 °C, 18 h 2. NaBH₄ (20 eq), DCM, MeOH, 0–22 °C, 4 h		[305] see also [306]

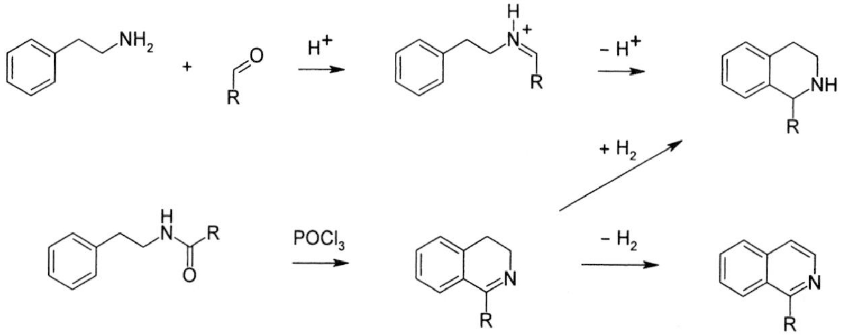

Figure 15.13. Pictet–Spengler and Bischler–Napieralski syntheses of tetrahydroisoquinolines and quinolines from 2-arylethylamines.

Table 15.25. Preparation of isoquinolines and indolizines.

Entry	Starting resin	Conditions	Product	Ref.
1	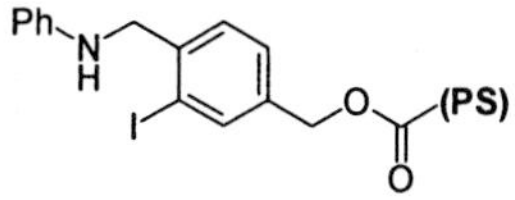	Bu₃SnH (16.2 eq), AIBN (13.5 eq), C₆H₆, 80 °C	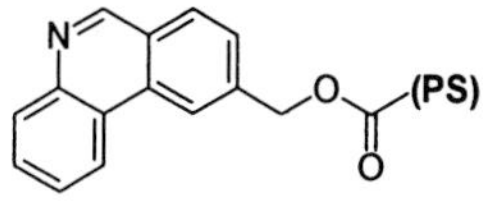	[111]
2	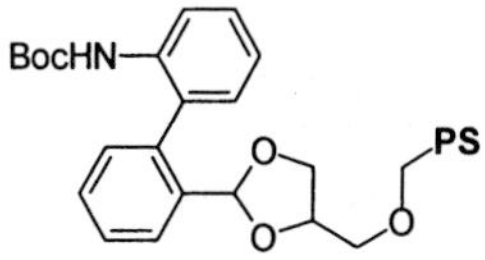	aq HCl (3 mol/L)/ dioxane 1:1, 80 °C, 48 h	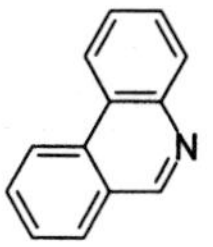	[124]
3	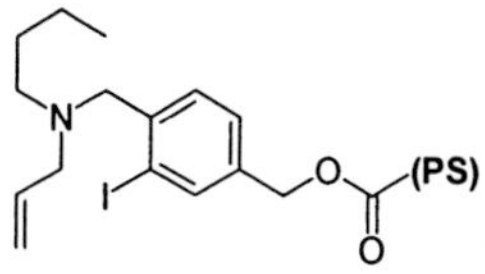	Pd(OAc)₂ (0.2 eq), PPh₃ (0.4 eq), Bu₄NCl (2 eq), K₂CO₃ (4 eq), DMA, 100 °C, 24 h	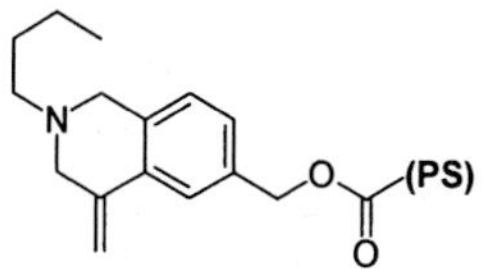	[111]
4	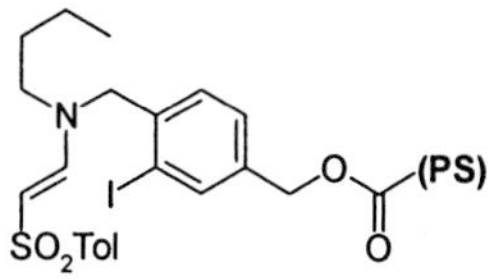	Pd(OAc)₂ (0.2 eq), PPh₃ (0.4 eq), Bu₄NCl (2 eq), K₂CO₃ (4 eq), DMA, 100 °C, 24 h	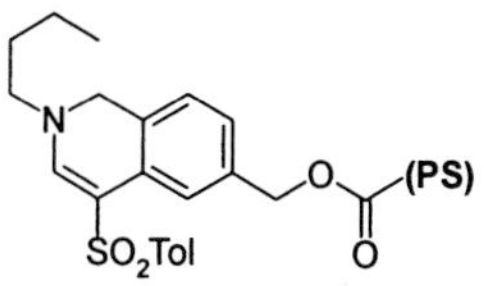	[111]
5	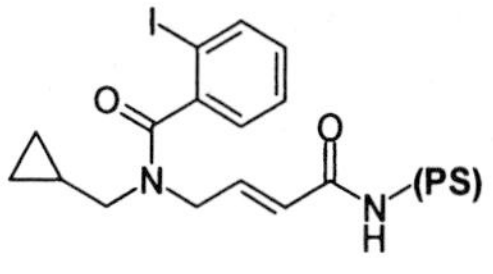	Pd(PPh₃)₄, NaOAc, PPh₃, DMA, 85 °C, 5 h	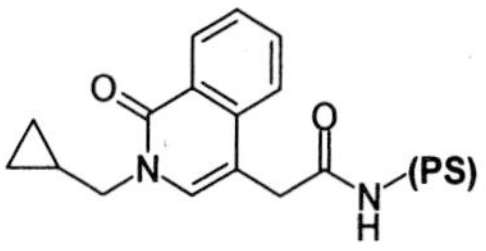	[307]
6	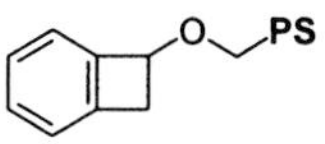	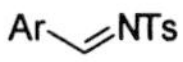 Ar⤳NTs (2 eq), EtCN, 108 °C, 14 h Ar: 4-(O₂N)C₆H₄	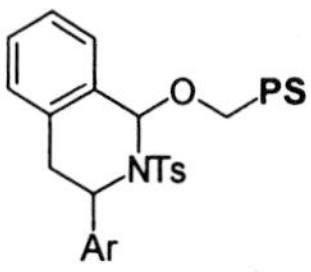	[308]
7	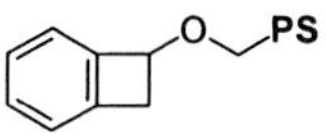	Cl₃CCN (2 eq), PhMe, 105 °C, 14 h	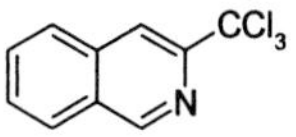	[308]

Table 15.25. continued.

Entry	Starting resin	Conditions	Product	Ref.
8		LDA, EtI, THF, −78 °C to 20 °C, 48 h		[309]
9		KOH (1 mol/L)/THF 1:1, 67 °C, 12 h		[309]
10		PhCOCl (0.11 mol/L, 3 eq), DCM, 0 °C, 15 min, then add indole, 0–20 °C, 2 h		[310]
11		O₂, DCE, 20 °C, 24 h		[111]
12		Ar (0.5 mol/L), [Co(Py)₄ (HCrO₄)₂], NEt₃ (0.5 mol/L), DMF, 80 °C, 2 h		[267]

Isoquinolines have been prepared on insoluble supports by radical-mediated cyclizations and by intramolecular Heck reaction (Table 15.25). Entry 1 in Table 15.25 is a rare example of the formation of a biaryl by intramolecular addition of an aryl radical to an arene. Oxidative aromatization was achieved by using a large excess of AIBN.

Polystyrene-bound *o*-quinodimethanes, which are formed upon thermolysis of benzocyclobutanes, can be converted into 1,2,3,4-tetrahydroisoquinoline derivatives by reaction with *N*-sulfonylimines. Reaction of *o*-quinodimethanes with electron-poor nitriles leads to the formation of 1,4-dihydroisoquinolines, which undergo elimination with simultaneous release of isoquinolines into solution (Entry 7, Table 15.25).

Reissert compounds (1-acyl-1,2-dihydro-2-quinolinecarbonitriles) have been prepared on cross-linked polystyrene and C-alkylated in the presence of strong bases (Entry 8, Table 15.25). Treatment of polystyrene-bound C-alkylated Reissert compounds with KOH leads to the release of isoquinolines into solution (Entry 9, Table 15.25). The reaction of support-bound quinoline- and isoquinoline *N*-oxides with acylating agents followed by treatment with electron-rich heteroarenes and enamines has been used to prepare alkylated and arylated derivatives of these heterocycles (Entry 10, Table 15.25; see also Table 15.26).

Various different types of cyclization have been used to prepare quinolines on insoluble supports. A reaction that enables the rapid preparation of substituted tetrahy-

Table 15.26. Preparation of quinolines.

Entry	Starting resin	Conditions	Product	Ref.
1	H₂N—⬡—(PS)	Yb(OTf)₃, MeCN/DCM 2:1, 24 h	(see structure)	[312] see also [313]
2	(see structure) —O—(PS)	PhNH₂ (3.3 eq), PhCHO (3.3 eq), MeCN/TFA 99:1, 12 h	(see structure)	[314]
3	(see structure) —(PS)	PhNH₂ (0.15 eq), cyclopentadiene (0.15 eq), MeCN/TFA 99:1, 24 h	(see structure)	[314] see also [315]
4	(see structure) (PS)	TMG, DCM, 55 °C, 18 h	(see structure) (PS)	[316] see also [317]
5	(see structure) —O—(PS)	PhMe, 80 °C, 24 h (Wang resin)	(see structure)	[318]
6	H₂N—⬡— (PS)	EtOCH=C(CO₂Et)₂, then 260 °C	(see structure) PS	[319]
7	(see structure) —(PS)	SnCl₂ (2 mol/L), DMF, 20 °C	(see structure) —(PS)	[18]
8	(see structure) —(PS)	(see structure) (0.11 mol/L, 10 eq of each), C₆H₆, 80 °C, 8 h; Ar: 4-(O₂N)C₆H₄	(see structure) —(PS)	[320]

Table 15.26. continued.

Entry	Starting resin	Conditions	Product	Ref.
9	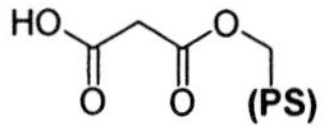	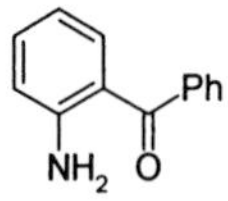 (10 eq), HOBt (10 eq), EDC (10 eq), DMF, 2 × 16 h, then add pyridine, piperidine, 16 h	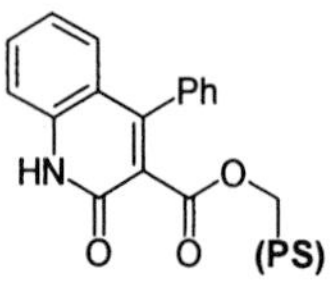	[321]
10	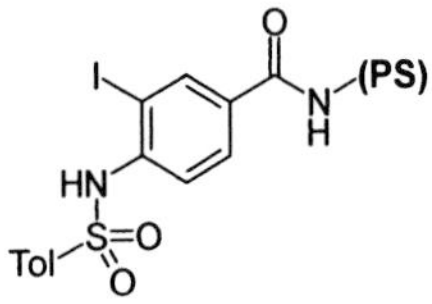	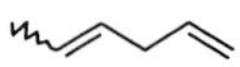 Pd(OAc)$_2$, LiCl, DIPEA, DMF, 100 °C	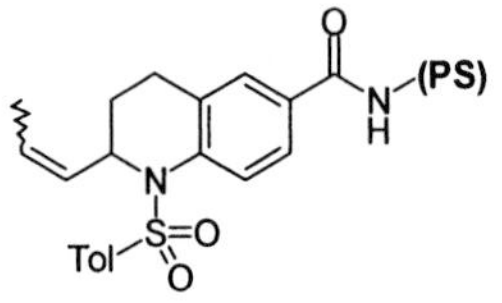	[79]
11	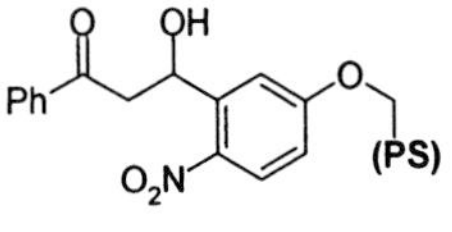	1. SnCl$_2$·2 H$_2$O, EtOH, 78 °C, 4 h (formation of quinoline-*N*-oxide) 2. TiCl$_3$, DCM, PhMe, 22 °C, overnight	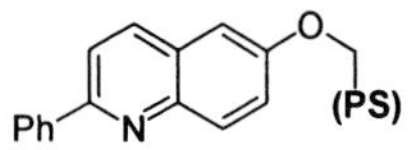	[322]
12	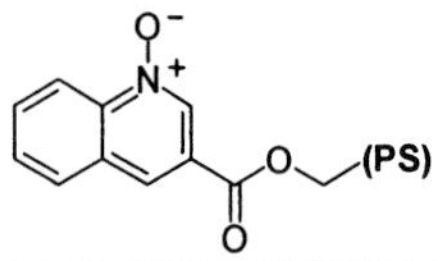	PhCOCl (0.11 mol/L, 3 eq), DCM, 0 °C, 15 min, then *N*-methylpyrrole, 0–20 °C, 2 h	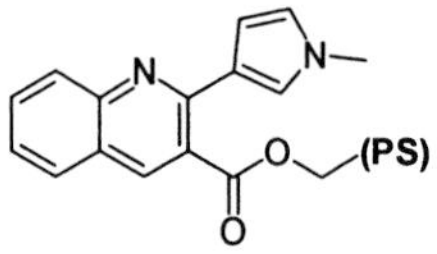	[310]

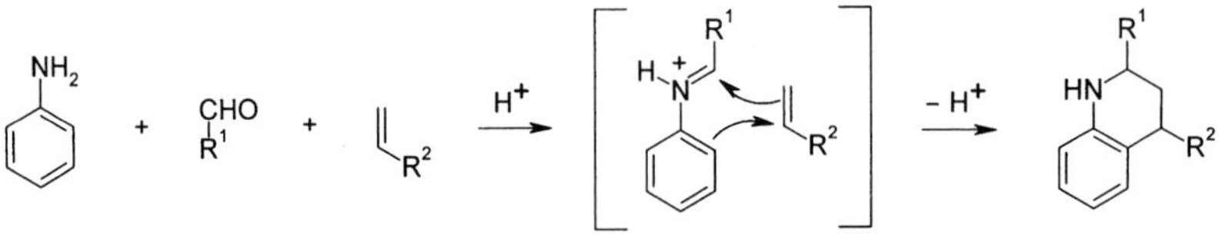

Figure 15.14. Preparation of tetrahydroquinolines from anilines, aldehydes, and alkenes.

droquinolines is the acid-catalyzed condensation of anilines with aldehydes and alkenes (Figure 15.14). Each of the required components can be linked to the support, whereby products with a large number of different substitution patterns become accessible (Entries **1–3**, Table 15.26).

Quinolines have also been prepared on insoluble supports by cyclocondensation reactions and by intramolecular aromatic nucleophilic substitution (Table 15.26). Entry **10** in Table 15.26 is an example of a remarkable palladium-mediated cycloaddition of support-bound 2-iodoanilines to 1,4-dienes. Reduction of the nitro group of polystyrene-bound 2-nitro-1-(3-oxoalkyl)benzenes with SnCl$_2$ (Entry **11**, Table 15.26) leads to the formation of quinoline *N*-oxides. These intermediates can be reduced to the quinolines on solid phase by treatment with TiCl$_3$. 4-Quinolones have been prepared by thermolysis of resin-bound 2-(arylamino)methylenemalonic esters [311].

15.19 Preparation of Pyridazines (1,2-Diazines)

Pyridazines and their partially saturated analogs have been prepared on insoluble supports by Diels–Alder reaction of electron-rich alkenes or alkynes with 1,2,4,5-tetrazines (Entries **1–3**, Table 15.27). The mechanism of this reaction is outlined in Figure 15.15. An additional approach, also based on the Diels–Alder reaction, is the cycloaddition of azo compounds to 1,3-dienes (Entries **4** and **5**, Table 15.27). The resulting tetrahydropyridazines (Entry **4**) have been used as constrained β-strand mimetics for the discovery of new protease inhibitors [323]. An example of the N-alkylation of hexahydropyridazines on solid phase is given in Section 10.3.

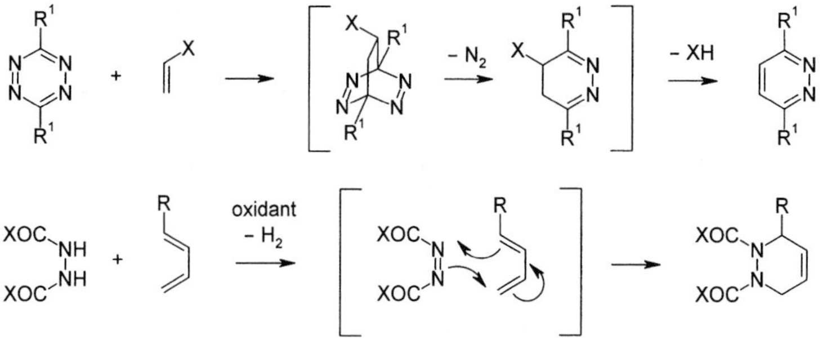

Figure 15.15. Preparation of 1,2-diazines by hetero-Diels–Alder reaction. X: OR, NR$_2$.

4-Halocinnolines can be prepared by treatment of polystyrene-bound (2-alkynyl-phenyl)triazenes with hydrogen chloride or hydrogen bromide (Entry **6**, Table 15.27). This reaction involves the initial release of 2-alkynylphenyldiazonium salts into solution, which then cyclize to the halocinnolines. The corresponding 4-hydroxycinnolines are obtained as by-products in variable amounts. Hydrogen fluoride or hydrogen iodide do not lead to the formation of cinnolines [324].

Phthalazine-1,4-diones have been prepared by nucleophilic cleavage of polystyrene-bound phthalimides (Entry **7**, Table 15.27). The reaction rate proved to be highly dependent on the solvent used, being low in EtOH but high in DCM. Using DMF as solvent, no phthalazinediones could be isolated [325]. Hexahydropyridazines have been prepared on cross-linked polystyrene in low yields by treatment of support-bound 1-acyloxy-1,3-dienes with diimide [326].

Table 15.27. Preparation of pyridazines and tetrahydropyridazines.

Entry	Starting resin	Conditions	Product	Ref.
1		(0.2 mol/L, 10 eq), 20 °C or 100 °C, 24 h		[327]
2		(0.2 mol/L, 10 eq), 100 °C, 16 h		[327]
3		(0.2 mol/L, 10 eq), 100 °C, 16 h		[327]
4		(4 eq), PhI(OAc)₂ (4 × 0.5 eq), dioxane, 2 × 4 h		[328] see also [323]
5		DEAD (1 mol/L, 10 eq), THF, 65 °C, overnight		[295]
6		HCl, H₂O, Me₂CO, 20 °C, 1 h		[324]
7		DCM, 20 °C		[325]
8		1. N₂H₄·H₂O (0.28 mol/L, 5 eq), DCM/MeOH 3:2, 20 °C, 4 h; 2. TsOH (5 eq), MeCN, 81 °C, 24 h (Wang resin)		[329]

15.20 Preparation of Pyrimidines (1,3-Diazines)

Ureas can be cyclized to hexahydro- and tetrahydro-2-pyrimidinones on insoluble supports either intra- or intermolecularly (Entries **1–3, 5**; Table 15.28). Because of the low nucleophilicity of ureas, harsh reaction conditions are generally required for these cyclizations. Similarly, polystyrene-bound guanidines can be condensed with 3-oxo esters to yield 2-amino-4-pyrimidinones (Entry **4**, Table 15.28). If a support-bound monoalkylguanidine is used as the starting material, the reaction with 3-oxo esters does not always proceed with high regioselectivity and mixtures of regioisomers (2-alkylamino-4-pyrimidinones and 2-amino-3-alkyl-4-pyrimidinones) can result [330,331]. The reverse strategy, in which amidines or guanidines undergo oxidative cyclocondensation with support-bound enones to yield pyrimidines, has also been reported (Entries **6–8**, Table 15.28).

Polystyrene-bound isothiuronium chloride, which is readily prepared from chloromethyl polystyrene and thiourea, has been used as a starting material for the preparation of various substituted pyrimidines (Entries **9–11**, Table 15.28). After oxidation to the corresponding sulfones, nucleophilic cleavage with amines proceeds smoothly to yield substituted 2-aminopyrimidine derivatives (see Section 3.8).

Palladium-mediated (Sonogashira) coupling [332] of 5-iodouracil with alkynes can be performed under mild reaction conditions, and is compatible with the supports, linkers, and protective groups used in solid-phase oligonucleotide synthesis (Entries **12** and **13**, Table 15.28). This coupling reaction has been used to prepare modified oligonucleotides on polystyrene and on CPG. The reaction of halopyrimidines with amines to yield aminopyrimidines is discussed in Section 10.1.2.

Table 15.28. Preparation of pyrimidines.

Entry	Starting resin	Conditions	Product	Ref.
1		PhMe saturated with HCl, sealed tube, 95 °C, 4 h (Wang resin)		[333] see also [166]
2		PhMe saturated with HCl, sealed tube, 95 °C, 4 h (Wang resin)		[333]
3		(0.13 mol/L, 4 eq of each), THF, HCl, 55 °C, 36 h		[334] see also [335, 336]
4		MeO$_2$C〜〜CO$_2$Me NaOMe (10 eq), MeOH, 65 °C, 16 h		[330, 331]
5		(0.5 mol/L, 10 eq), NaOEt (10 eq), DMA, 20 °C, 16 h		[243]
6		(0.5 mol/L, 10 eq), DMA, air, 100 °C, overnight		[243]
7		guanidine (0.5 mol/L, 10 eq), DMA, air, 100 °C, overnight		[243]
8		(10 eq), K$_2$CO$_3$, DMA, 70 °C, 8 h, then CAN (0.2 mol/L), DMA, 20 °C, 2 h		[337]
9		(0.4 mol/L, 1.2 eq), DIPEA (1.5 eq), DMF, 20 °C, 48 h		[338]

Table 15.28. continued.

Entry	Starting resin	Conditions	Product	Ref.
10		(0.17 mol/L, 2 eq), DIPEA (5 eq), DMF, 0–20 °C, 96 h		[339]
11		(0.19 mol/L, 1.8 eq), DIPEA (5 eq), DMF, 0–20 °C, 96 h		[339]
12		5-iodouridine, CuI, Pd(PPh₃)₄, NEt₃, DMF, 25 °C		[340]
13		Pd(PPh₃)₄, CuI, DMF/NEt₃ 7:3, 3 h		[341]

15.21 Preparation of Quinazolines

Numerous solid-phase preparations of quinazolinones have been reported. The main synthetic strategies used are summarized in Figure 15.16. Quinazolin-2,4-diones can be prepared from anthranilic acid derived ureas or from *N*-(alkoxycarbonyl)-anthranilamides. These reactions have been performed on insoluble supports either in such a way that the cyclized product remains linked to the support, or such that it is simultaneously cleaved from the support upon ring formation. Quinazolin-4-ones can be prepared by cyclocondensation of anthranilamides with aldehydes, orthoesters [342], or other carboxylic acid derivatives [343]. The selection of examples listed in Table 15.29 illustrates the variety of substitution patterns accessible by means of these cyclizations.

Support-bound quinazolin-2,4-diones can be N-alkylated, either with alkyl halides under basic conditions or with aliphatic alcohols by means of the Mitsunobu reaction (Entries **12–14**, Table 15.29). The methyl group of a 2-methylquinazolin-4-one is sufficiently acidic to undergo aldol condensations with aldehydes [343]. Aminations of chloroquinazolines are discussed in Section 10.1.2.

Table 15.29. Preparation of quinazolines.

Entry	Starting resin	Conditions	Product	Ref.
1		DMF, 125 °C, 16 h		[344]
2		NEt$_3$ (10 eq), MeOH, 60 °C, 24 h		[345]
3		TMG/NMP 5:95, 60 °C		[346] see also [347]
4		KOH (1 mol/L), EtOH, 1 h		[348] see also [349]
5		K$_2$CO$_3$, MeCN, 60 °C, 24 h		[350] see also [351,352]
6		O≈≈Ar (10 eq), AcOH/DMA 5:95, 100 °C, 24 h Ar: 2-(MeO)C$_6$H$_4$		[353]
7		Cl CONH$_2$ NH$_2$ (1.8 mol/L, 3.5 eq), CSA (1 eq), dioxane, 100 °C, 72 h		[354]
8		piperidine (0.4 mol/L, 5 eq), *m*-xylene, 23 °C, 2 h, 80 °C, 4 h		[355] see also [356]

Table 15.29. continued.

Entry	Starting resin	Conditions	Product	Ref.
9		HCO₂H, 20 °C		[357] see also [358]
10		MeCN/DCE 1:1, 81 °C, overnight R₂NH: piperidine		[359] see also [360]
11		TFA/H₂O 95:5, 80 °C, 2 × 16 h		[361]
12		Bu₄NI, TMG, DMSO, 20 °C, overnight		[346]
13		ArCH₂OH, DIAD, PPh₃, THF, 20 °C, overnight Ar: 4-(MeO)C₆H₄		[346]
14		(16 eq), THF, 20 °C, 1.5 h, then EtI (40 eq), DMF, 18 h		[348]
15		KMnO₄ (10 eq), Me₂CO, overnight		[353]

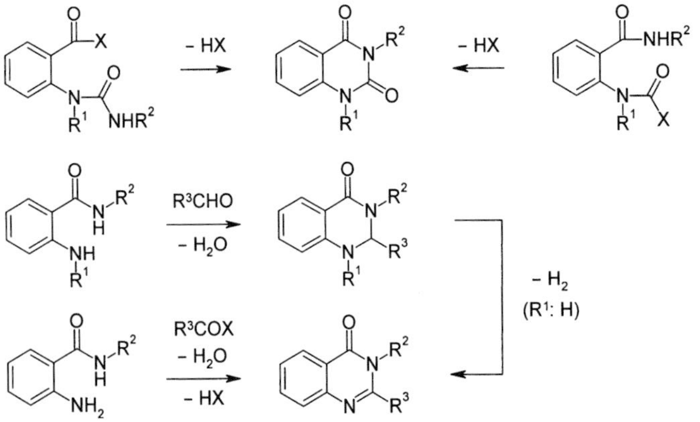

Figure 15.16. Strategies for the preparation of quinazolinones.

15.22 Preparation of Pyrazines and Piperazines (1,4-Diazines), and of their Fused Derivatives

15.22.1 Preparation of Diketopiperazines

An unwanted side reaction in solid-phase peptide synthesis is the formation of diketopiperazines during deprotection of the second amino acid (Figure 15.17). This reaction occurs particularly readily with dipeptides containing *N*-methylamino acids, proline [362], or glycine [363], and with dipeptides consisting of a D- and an L-amino acid. This reaction is one of the preferred strategies for preparing diketopiperazines. A second strategy involves linking an α-amino acid ester to a backbone amide linker through its amino group (see Section 3.3.1), and then acylating the resulting secondary amine with an N-protected α-amino acid. Deprotection will usually lead to spontaneous diketopiperazine formation, although without simultaneous cleavage from the support (Figure 15.17).

Diketopiperazine formation by intramolecular nucleophilic cleavage proceeds particularly smoothly if a linker prone to facile nucleophilic cleavage is used. In most of

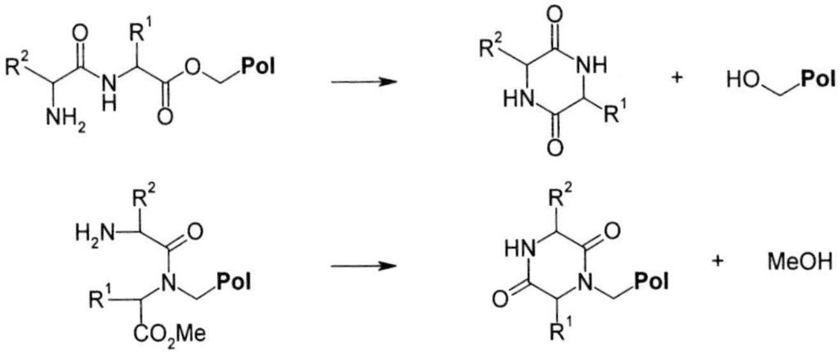

Figure 15.17. Strategies for the preparation of diketopiperazines from support-bound α-amino acid derivatives.

the reported examples, hydroxymethyl polystyrene or polystyrene with the PAM linker have been used. These linkers tolerate treatment with TFA, and are therefore compatible with the Boc methodology. Cleavage and simultaneous diketopiperazine formation can be performed under either basic or acidic conditions (Table 15.30).

Table 15.30. Preparation of diketopiperazines (piperazine-2,5-diones).

Entry	Starting resin	Conditions	Product	Ref.
1		DIPEA (2.2 eq), AcOH (5 eq), DCM, 16 h		[369] see also [370]
2		AcOH (1.25 mol/L), PhMe, 90 °C		[371]
3		AcOH (0.1 mol/L), DCM, 25 °C ($t_{1/2}$: 8.1 min)		[362]
4		AcOH/PhMe 1:99 or NEt$_3$/PhMe 4:96, 20 °C, 12 h (PAM resin)		[372]
5		isobutylamine (2 mol/L, 40 eq), DMSO, 20 °C, 21 h (Wang resin)		[373]
6		isobutylamine (2 mol/L, 20 eq), DMSO, 70 °C, 24 h, wash; TFA/H$_2$O 95:5, 1 h (Wang resin)		[373]
7		1. TFA, 3 h, concentrate 2. PhMe, 110 °C, 5 h (Wang resin; Ar: 4-(MeO)C$_6$H$_4$)		[374]
8		piperidine/THF 5:95, 20 °C, 16 h		[375] see also [376]

Table 15.30. continued.

Entry	Starting resin	Conditions	Product	Ref.
9		NH$_4$OAc, AcOH, PhMe, heat		[377]
10		NEt$_3$/PhMe 4:96, 20 °C, 12 h Ar: 4-(BnO)C$_6$H$_4$		[372]
11		piperidine/DMF 1:4, 25 °C, 3 × 1 min, 3 × 5 min		[367] see also [147,378, 379]
12		AcOH (2 mol/L), *i*PrOH, 50 °C, 18 h		[380]
13		K$_2$CO$_3$, DMF, 70 °C, 4 h R: 3-indolyl		[381]

Because cleavage from the support only occurs upon cyclization, most by-products formed during solid-phase synthesis of the dipeptide remain attached to the support, and the diketopiperazines obtained are usually very pure.

The Wang linker, although slightly more resistant towards nucleophilic cleavage than Merrifield or PAM resin, can also be used for the preparation of diketopiperazines. If spontaneous cyclization does not occur during TFA-mediated cleavage of a dipeptide from the support, heating of the crude product in toluene will usually bring about ring-closure (Entry **7**, Table 15.30). Diketopiperazines can also be prepared by the cyclization of *N*-alkylamino acid derivatives, e.g. those available from α-halo acids and primary amines (Entries **5** and **6**, Table 15.30). In this case, the linker used should be sufficiently stable towards aminolysis to avoid premature cleavage of the intermediates from the support. Moreover, a volatile amine should be used for the final substitution to facilitate removal of the excess amine from the product (Entry **5**, Table 15.30). 2,3-Diketopiperazines (cyclic oxalamides) have been prepared on solid phase by reaction of resin-bound ethylenediamines with oxalic acid diimidazolide [364–366].

Polystyrene-bound diketopiperazines, such as those prepared with the aid of a backbone amide linker, can be *N*-alkylated by treatment with alkyl halides in the presence of strong bases [367]. The conditions required for this N-alkylation are similar to those used for the N-alkylation of hydantoins.

As illustrated by the examples in Table 15.30, diketopiperazines are amenable to many structural variations. These heterocycles thus seem to be well suited for the preparation of libraries for lead discovery, and have, for instance, been used for the identification of new enzyme inhibitors [368]. Libraries of diketopiperazines will, however, be sparsely diverse because the diketopiperazine ring itself is 'pharmaco-phore-rich', and will contribute significantly to the overall properties of all the members of the library (see Section 1.4).

15.22.2 Preparation of Miscellaneous 1,4-Diazines and Quinoxalines

1,4-Diazines other than diketopiperazines can also be prepared on insoluble supports (Table 15.31; see also Figure 3.13 [382]). Most strategies are based on intramolecular nucleophilic substitutions or acylations. Several examples of the solid-phase preparation of quinoxalinones have been reported. In most cases, the compounds have been prepared from support-bound 2-fluoronitrobenzenes according to the strategies outlined in Figure 15.18. Alternatively, α-amino acid esters bound to polystyrene as *N*-benzyl derivatives can be N-arylated with 2-fluoronitrobenzene. Reduction of the resulting 2-nitroaniline leads to the formation of quinoxalinones [383]. 1,4-Diazines have been chemically modified by N- or C-alkylation on insoluble supports (Entries **9** and **10**, Table 15.31).

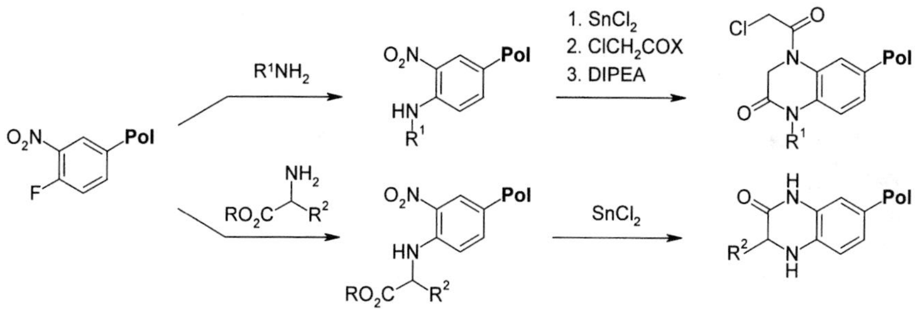

Figure 15.18. Strategies for the preparation of 1,2,3,4-tetrahydroquinoxalinones on insoluble supports.

Table 15.31. Preparation of 2-piperazinones and quinoxalines.

Entry	Starting resin	Conditions	Product	Ref.
1		DIPEA/NMP 1:9, 20 °C, 5 h		[384]
2		1. SnCl₂•2H₂O (2 mol/L), NMP, 20 °C, 2 h 2. DIPEA/NMP 1:99, 45 °C, 16 h		[385] see also [386]
3		SnCl₂ (2 mol/L, 20 eq), DMF, 80 °C, overnight	(no racemization)	[387] see also [388,389]
4		SnCl₂•H₂O (5 eq), DMF, 70 °C, 16 h		[390] see also [391]
5		H₂N–⬠ (2 mol/L), DMSO, 20 °C, 2 h		[392] see also [393]
6		PPh₃ (4 eq), DEAD (4 eq), MeCN/DMF		[394]

Table 15.31. continued.

Entry	Starting resin	Conditions	Product	Ref.
7		Br⌒⌒Br K₂CO₃, DMF, 60 °C Ar: 2-(O₂N)C₆H₄		[395]
8		TFA/H₂O 95:5, 3 h (PS): Wang resin Ar: 4-(*tert*-butoxy)phenyl		[396]
9		BnBr (1 mol/L, 25 eq), K₂CO₃ (25 eq), Me₂CO, 55 °C, 24 h		[388] see also [383]
10		BnBr (5 eq), LiN(SiMe₃)₂ (5 eq), THF, 20 °C, overnight		[397]
11		1,2-diaminobenzene (10 eq), THF, 20 °C, 15 h		[398]
12		1,2-diaminobenzene (10 eq), PPTS, PhMe, 110 °C, 16 h		[29]

15.23 Preparation of Triazines

Solid-phase syntheses of triazines (Table 15.32) include the cyclocondensation of polystyrene-bound isothiuronium salts with *N*-(cyano)iminodithiocarbonates and the cyclization with simultaneous cleavage from the support of derivatives of α-amino acid hydrazides (Table 15.32). (Benzylthio)triazines, such as those listed in Table 15.32 (Entries **1** and **2**), have been cleaved from a support by oxidation to sulfoxides or sulfones with *N*-benzenesulfonyl-3-phenyloxaziridine, followed by nucleophilic cleavage with primary or secondary aliphatic amines or with electron-rich anilines (see Section 3.8).

Table 15.32. Preparation of triazines.

Entry	Starting resin	Conditions	Product	Ref.
1		(0.2 mol/L, 1.2 eq), DIPEA (1.5 eq), DMA, 80 °C, 72 h		[399]
2		(0.37 mol/L, 2.5 eq), DIPEA (5 eq), DMA, 45 °C, 17 h		[399]
3		DIPEA (0.5 mol/L, 10 eq), DMF, 80 °C, 24 h		[165]
4		DIPEA (0.5 mol/L, 10 eq), DMF, 80 °C, 24 h		[165]
5		2-cyanoethylhydrazine, THF/EtOH 95:5, 50 °C, 24–48 h (products are unstable)		[400]
6		(0.22 mol/L, 4 eq), THF, 20 °C, 2 h		[401]

15.24 Preparation of Pyrans and Benzopyrans

The synthetic strategies used for the preparation of pyrans on insoluble supports
have mainly been hetero-Diels–Alder reactions of enones with enol ethers and ring-
closing olefin metathesis (Table 15.33). Benzopyrans have been prepared by hetero-
Diels–Alder reactions of polystyrene-bound *o*-quinodimethanes with aldehydes. The
required quinodimethanes were generated by thermolysis of benzocyclobutanes,
which were prepared in solution [308]. Other solid-phase procedures for the preparation of benzopyrans are the palladium-mediated reaction of support-bound 2-iodo-
phenols with 1,4-dienes (Entry **5**, Table 15.33) and the intramolecular Knoevenagel

condensation of malonic esters of salicylaldehyde (Entry **6**, Table 15.33). Chromones have been prepared from polystyrene-bound 2-acetylphenol by C-formylation followed by acidolytic cleavage from the support [402]. The preparation of benzopyrans

Table 15.33. Preparation of pyrans and benzopyrans.

Entry	Starting resin	Conditions	Product	Ref.
1		EtO (10 eq), DCM, 60 °C, 3 d		[403] see also [404]
2		AllO₂C Ph (3 eq), CuL (0.2 eq), MS 4 Å, THF, 0 °C L: enantiomerically pure bis(oxazolidinyl) ligand	> 94% ee	[405] see also [406]
3		(0.05 eq), DCM, 20 °C, 16 h		[285]
4		ArCHO (2 mol/L, 2 eq), PhMe, 108 °C, 14 h Ar: 4-(O₂N)C₆H₄		[308]
5		Pd(OAc)₂, LiCl, DIPEA, DMF, 100 °C		[79]
6		salicylaldehyde (20 eq), piperidine, pyridine, 20 °C, 2 × 16 h; then TFA/DCM 1:2, 1 h (Wang resin)		[407] see also [149]
7		(MeO)₂MeC(NMe₂) (0.3 mol/L), THF, 40 °C, 6 h		[408]
8		I₂ (10 eq), DMSO, 180 °C, 2 h Ar: 4-*t*BuC₆H₄		[149]
9		(10 eq), PhI(OAc)₂, DCM, THF, 25 °C, 2 h R: 2-bromobenzyl		[409]

by oxidative cyclization of 2-allylphenols with polystyrene-bound selenyl bromide is discussed in Chapter 9.

15.25 Preparation of Oxazines, Thiazines, and Thiadiazines

Examples of the preparation of oxazines and thiazines on insoluble supports are listed in Table 15.34. 3,1-Benzoxazin-4-ones can be prepared by intramolecular O-acylation of *N*-aminocarbonyl anthranilic acids (Entry 1, Table 15.34). The resulting benzoxazinones are sufficiently stable towards acids to enable TFA-mediated cleavage from a Wang linker [410]. 1,3-Oxazines have also been obtained by acidolytic cleavage of functionalized 3-amino-1-propanols from Wang resin (Entry 4, Table 3.30).

Morpholine-2,5-diones have been prepared by intramolecular, nucleophilic cleavage of polystyrene-bound amino acids acylated with α-hydroxy acids (Entry 2, Table 15.34). This preparation is analogous to the formation of diketopiperazines from dipeptide esters. In a similar approach, morpholine-2,5-diones were obtained by acidolytic cleavage of *N*-(bromoacetyl)amino acids from Wang resin (Entry 3, Table 15.34). In a remarkable cycloaddition, 1,2-oxazines have been prepared from polystyrene-bound acrylates by reaction with enol ethers and nitroalkenes under high-pressure (Entry 4, Table 15.34).

1,3-Thiazin-4-ones can be prepared either by one-pot condensation of 3-mercaptopropionic acids with primary amines and aldehydes (Entry 5, Table 15.34), or by condensation of 3-mercaptopropionamides with aldehydes (Entry 6, Table 15.34).

Table 15.34. Preparation of oxazines, thiazines, and thiadiazines.

Entry	Starting resin	Conditions	Product	Ref.
1		DIC, THF; or TsCl, pyridine; or Ac₂O, THF, 20 °C, overnight		[410]
2		NEt₃/DCM 5:95, 3 h		[372]
3		TFA/H₂O 95:5, 1 h (Wang resin)		[373]
4		(2 eq of each), DCM, 15 kbar, 20 °C, 48 h		[412] see also [413]

Table 15.34. continued.

Entry	Starting resin	Conditions	Product	Ref.
5		HS‿‿CO₂H (0.5 mol/L), PhCHO (0.25 mol/L), MS 3 Å, THF, 70 °C, 2 h		[228]
6		HCO₂H, 20 °C		[357] see also [414,415]
7		SnCl₂·H₂O, NaOAc, DMF, then PhMe, 110 °C, 12 h (PS): Rink amide resin		[416]
8		1. TFA/(*i*Bu)₃SiH/DCM 5:5:90 2. NMM, DMF		[417] see also [418]
9		HATU (5 eq), DIPEA (5 eq), DMF, overnight		[417]
10		1,3-dichloro-2-propanone (5 eq), NaCNBH₃ (5 eq), DMF/AcOH 99:1, 24 h		[238]
11		HCHO (20 eq), dioxane, 3 h, then add DIPEA (1 eq), asparagine, 6 h		[411]

1,3-Thiazin-4-ones are rather stable towards acids, and can be cleaved from Wang resin by treatment with TFA/DCM (1:1) without decomposition [228].

3-Thiomorpholinones have been prepared on cross-linked polystyrene by intramolecular thioether formation and by the lactamization of suitable amino acids (Entries **8** and **9**, Table 15.34). 1,3,5-Thiadiazine-2-thiones (Entry **11**, Table 15.34) are not stable towards TFA/DCM (1:1) and have therefore been prepared on Sasrin; cleavage from the support could be achieved using 3% TFA in DCM [411].

15.26 Preparation of Azepines and Larger Heterocycles with one Nitrogen Atom

Whereas five- and six-membered systems are usually easy to prepare and often form spontaneously from suitable intermediates, seven-membered heterocycles can be quite difficult to synthesize. Only if the precursor has little conformational flexibility and a conformation suitable for ring-closure can readily be attained, will the cyclization proceed smoothly (as, e.g., in the formation of benzodiazepines from 2-(aminoacetylamino)benzophenones). Otherwise, ring-closure must be forced by chemical activation or high reaction temperatures, and a low loading might be required to suppress polymerization.

Tetrahydroazepines can be prepared on insoluble supports by ring-closing olefin metathesis (Entries **1–3**, Table 15.35; see also Figure 5.14), either with or without simultaneous cleavage from the support. The yields of this cyclization are not as high as for the formation of five- or six-membered rings, but under optimal conditions pure products can be obtained. Larger nitrogen-containing heterocycles are also available by ring-closing metathesis on solid phase (Entry **5**, Table 15.35). If one of the two alkenes that undergo metathesis also acts as a linker, ring-closing metathesis and cleavage from the support will occur simultaneously. For this type of cyclization, yields are usually higher if an additional alkene (e.g. styrene or ethene, see Entry **2**, Table 15.35) is added to the reaction mixture. This alkene enables efficient regeneration of the catalyst by reaction with the support-bound carbene complex, and allows fast ring-closing metathesis to occur with only small amounts of catalyst (see Figure 3.39).

Benzazepinones can be prepared on cross-linked polystyrene by intramolecular Heck reaction (Entry **6**, Table 15.35). In the presence of sodium formate, the intramolecular Heck reaction of iodoarenes with alkynes yields methylene benzazepinones (Entry **7**, Table 15.35). Surprisingly, when this reaction was performed in solution, the main product (65% yield) was a dehalogenated, non-cyclized benzamide. In the synthesis on cross-linked polystyrene, however, this product was not observed [419].

Caprolactams can be prepared by intramolecular nucleophilic cleavage of support-bound 6-aminohexanoates (Entry **9**, Table 15.35). The yields of such reactions are usually not high, even if reactive esters are used for attachment to the support. High loading should be avoided to prevent the formation of oligomers by intermolecular nucleophilic cleavage.

Table 15.35. Preparation of azepines and azocines.

Entry	Starting resin	Conditions	Product	Ref.
1		 (0.05 eq), DCE, 80 °C, 16 h		[285] see also [420–422]
2		 (0.05 eq), styrene (1 eq), PhMe, 50 °C, 18 h		[284]
3		 (0.11 eq), DCM, 40 °C, overnight		[10]
4		PhI(OAc)₂, (CF₃)₂CHOH, DCM, 23 °C		[423]
5		 (0.08 eq), DCM, 40 °C, overnight		[10]
6		Pd(OAc)₂ (0.2 eq), PPh₃, Bu₄NCl, KOAc, DMF, 70 °C, 5 h		[419]
7		Pd(OAc)₂ (0.2 eq), PPh₃, Bu₄NCl, HCO₂Na, DMF, 70 °C		[419]
8	 (Rink amide resin)	 (0.27 mol/L, 5.4 eq), DCM/MeOH 1:1, 2 d		[424]
9		TFA, DCM, then 5% NEt₃ in DCM, 25 °C, 24 h		[425]

15.27 Preparation of Diazepines, Thiazepines, and Larger Heterocycles with more than one Heteroatom

Diazepanones have been prepared on insoluble supports by intramolecular nucleophilic cleavage, by intramolecular Mitsunobu reaction of sulfonamides with alcohols, and by intramolecular acylations (Table 15.36). As in the case of azepines, these reactions do not always proceed smoothly, and care must be taken to prevent potential side reactions from occurring. For instance, intramolecular acylations in peptides containing aspartic acid (Entry **2**, Table 15.36) will generally lead to the formation of succinimides (see Table 13.20) unless *N*-alkylamino acids are used.

The cyclization of *N*-alkenyl derivatives of resin-bound peptides by ruthenium-catalyzed ring-closing metathesis has been investigated [426–428]. These reactions do not proceed well on insoluble supports, and cyclizations in solution are generally more successful.

Table 15.36. Preparation of diazepanes and related heterocycles.

Entry	Starting resin	Conditions	Product	Ref.
1		DIPEA (2.2 eq), AcOH (5 eq), DCM, 16 h		[369]
2		DPPA, DIPEA, overnight; or HATU (3 eq), DIPEA (3 eq), DMF		[429,430] see also [431]
3		PPh₃, DEAD, THF, 0 °C, 2 h, 20 °C, 36 h		[432]
4		NEt₃, THF, 67 °C, 5 d		[433]

Benzodiazepines, on the other hand, are easier to prepare than non-fused diazepines, because the precursors are less flexible and suitably predisposed to facilitate ring-closure. Most benzodiazepines do not form spontaneously, however, and additional activation is usually required to promote the cyclization (Table 15.37). Even under forcing conditions, cyclizations can sometimes still be difficult. For instance, the cyclization shown in Entry **2** in Table 15.37 only proceeded when pure DIC was used as the coupling agent, whereupon the desired benzodiazepinone together with

Table 15.37. Preparation of benzodiazepines and related heterocycles.

Entry	Starting resin	Conditions	Product	Ref.
1		HOBt (0.38 mol/L, 4 eq), DIC (4 eq), DMF, 20 °C, overnight		[435] see also [436]
2		DIC, DCM/C$_6$H$_6$ 1:1, 20 °C, 6 h		[434] see also [437]
3		HBTU (3 eq), DIPEA (3 eq), DMF, overnight		[438]
4		Ph-N=C(OLi)... (0.22 mol/L, 20 eq), DMF/THF 1:1, 20 °C, 30 h, then allyl bromide (40 eq), 6 h		[439]
5		*p*-xylene, 130 °C, 7 h		[440] see also [441,442]
6		SnCl$_2$·2 H$_2$O (2 mol/L), DMF, overnight		[443]
7		AcOH/DMF 5:95, 60 °C, 12 h Ar: 4-(MeO)C$_6$H$_4$		[444]
8		Dess-Martin periodinane (0.2 mol/L, 5 eq), DCM, 20 °C, 2 h		[445]

Table 15.37. continued.

Entry	Starting resin	Conditions	Product	Ref.
9		(0.24 mol/L, 1.4 eq), pyridine, 115 °C, 48 h		[446]
10		NaO*t*Bu, THF, 60 °C, 24 h		[447]
11		NEt₃/DCM 1:1, 27 °C, 4 h		[296]
12		NaOMe (0.4 mol/L, 2 eq), THF/MeOH 4:1, 20 °C, 10 h (Wang resin)		[448]
13		1. SnCl₂ (0.07 mol/L, 3 eq), NEt₃ (8 eq), PhSH (11 eq), C₆H₆, 20 °C, 1 h 2. DMF/TFA/H₂O 7:2:1, 20 °C, 14 h		[449]
14		PhO NCO (0.04 mol/L, 5 eq), DMF, 60 °C, 24 h (PS)NH₂: MBHA resin		[450]
15		hydrazine (10 eq), EtOH, 78 °C, 24 h PS: aminomethyl-PS with PAM linker		[451]
16		DBU/DMF 5:95, 20 °C, 24 h		[452] see also [453,454]

Table 15.37. continued.

Entry	Starting resin	Conditions	Product	Ref.
17	(TG) Ar: 2-MeC₆H₄	hexyl iodide (0.25 mol/L), THF/ DMF 1:1, 20 °C, 1 h Ar: 2-MeC₆H₄	(TG)	[436]
18	(PS) Ph	BnBr (0.9 mol/L, 20 eq), K₂CO₃ (20 eq), Me₂CO, 55 °C, overnight	(PS) Ph	[435]

20–50% DIC-derived *N*-acylurea were obtained. Additives such as HOBt or DMAP, or other coupling reagents (EDC, HATU, DECP) did not lead to cyclization in this case [434].

Several examples have been reported in which anthranilamides of α-amino acid esters have been cyclized to yield benzodiazepinediones (Entries **4, 5,** and **10,** Table 15.37). Examples of cyclizations yielding benzodiazepines with simultaneous cleavage from the support have also been reported (Entries **9, 10** and **12**; Table 15.37). Unfortunately, in most of these examples the desired products are contaminated with non-volatile by-products, and purification of the products will probably be required for most applications.

Support-bound benzodiazepines can be chemically modified by N-alkylation (Entries **17** and **18,** Table 15.37). Strong bases and a large excess of reactive alkylating agents are often required to achieve complete conversion of the support-bound substrates.

References for Chapter 15

[1] Corbett, J. W. *Org. Prep. Proc. Int.* **1998**, *30*, 491–550.
[2] Nefzi, A.; Ostresh, J. M.; Houghten, R. A. *Chem. Rev.* **1997**, *97*, 449–472.
[3] Nuss, J. M.; Desai, M. C.; Zuckermann, R. N.; Singh, R.; Renhowe, P. A.; Goff, D. A.; Chinn, J. P.; Wang, L.; Dorr, H.; Brown, E. G.; Subramanian, S. *Pure Appl. Chem.* **1997**, *69*, 447–452.
[4] Franzén, R. G. *J. Comb. Chem.* **2000**, *2*, 195–214.
[5] Farrall, M. J.; Durst, T.; Fréchet, J. M. *Tetrahedron Lett.* **1979**, 203–206.
[6] Filigheddu, S. N.; Masala, S.; Taddei, M. *Tetrahedron Lett.* **1999**, *40*, 6503–6506.
[7] Rotella, D. P. *J. Am. Chem. Soc.* **1996**, *118*, 12246–12247.
[8] Le Hetet, C.; David, M.; Carreaux, F.; Carboni, B.; Sauleau, A. *Tetrahedron Lett.* **1997**, *38*, 5153–5156.
[9] Fréchet, J. M. J.; Eichler, E. *Polym. Bull.* **1982**, *7*, 345–351.
[10] Pernerstorfer, J.; Schuster, M.; Blechert, S. *Synthesis* **1999**, 138–144.
[11] Wendeborn, S.; De Mesmaeker, A.; Brill, W. K. D. *Synlett* **1998**, 865–868.
[12] Savin, K. A.; Woo, J. C. G.; Danishefsky, S. J. *J. Org. Chem.* **1999**, *64*, 4183–4186.
[13] Zaragoza, F. unpublished results.
[14] Peschke, B.; Bundgaard, J. G.; Breinholt, J. *Tetrahedron Lett.* **2001**, *42*, 5127–5130.
[15] Ruhland, B.; Bhandari, A.; Gordon, E. M.; Gallop, M. A. *J. Am. Chem. Soc.* **1996**, *118*, 253–254.

[16] Ruhland, B.; Bombrun, A.; Gallop, M. A. *J. Org. Chem.* **1997**, *62*, 7820–7826.
[17] Ni, Z. J.; Maclean, D.; Holmes, C. P.; Murphy, M. M.; Ruhland, B.; Jacobs, J. W.; Gordon, E. M.; Gallop, M. A. *J. Med. Chem.* **1996**, *39*, 1601–1608.
[18] Pei, Y. Z.; Houghten, R. A.; Kiely, J. S. *Tetrahedron Lett.* **1997**, *38*, 3349–3352.
[19] Gordon, K.; Bolger, M.; Khan, N.; Balasubramanian, S. *Tetrahedron Lett.* **2000**, *41*, 8621–8625.
[20] Molteni, V.; Annunziata, R.; Cinquini, M.; Cozzi, F.; Benaglia, M. *Tetrahedron Lett.* **1998**, *39*, 1257–1260.
[21] Singh, R.; Nuss, J. M. *Tetrahedron Lett.* **1999**, *40*, 1249–1252.
[22] Benaglia, M.; Cinquini, M.; Cozzi, F. *Tetrahedron Lett.* **1999**, *40*, 2019–2020.
[23] Annunziata, R.; Benaglia, M.; Cinquini, M.; Cozzi, F. *Chem. Eur. J.* **2000**, *6*, 133–138.
[24] Kobayashi, S.; Moriwaki, M.; Akiyama, R.; Suzuki, S.; Hachiya, I. *Tetrahedron Lett.* **1996**, *37*, 7783–7786.
[25] Gordeev, M. F.; Gordon, E. M.; Patel, D. V. *J. Org. Chem.* **1997**, *62*, 8177–8181.
[26] Schunk, S.; Enders, D. *Org. Lett.* **2000**, *2*, 907–910.
[27] Meloni, M. M.; Taddei, M. *Org. Lett.* **2001**, *3*, 337–340.
[28] Cheng, W.-C.; Olmstead, M. M.; Kurth, M. J. *J. Org. Chem.* **2001**, *66*, 5528–5533.
[29] Nicolaou, K. C.; Baran, P. S.; Zhong, Y.-L. *J. Am. Chem. Soc.* **2000**, *122*, 10246–10248.
[30] Zaragoza, F. *Metal Carbenes in Organic Synthesis*; Wiley-VCH: Weinheim, New York, **1999**.
[31] Ivin, K. J.; Mol, J. C. *Olefin Metathesis and Metathesis Polymerization*; Academic Press: London, **1997**.
[32] Trautwein, A. W.; Süssmuth, R. D.; Jung, G. *Bioorg. Med. Chem. Lett.* **1998**, *8*, 2381–2384.
[33] Trautwein, A. W.; Jung, G. *Tetrahedron Lett.* **1998**, *39*, 8263–8266.
[34] Mjalli, A. M. M.; Sarshar, S.; Baiga, T. J. *Tetrahedron Lett.* **1996**, *37*, 2943–2946.
[35] Strocker, A. M.; Keating, T. A.; Tempest, P. A.; Armstrong, R. W. *Tetrahedron Lett.* **1996**, *37*, 1149–1152.
[36] Attanasi, O. A.; De Crescentini, L.; Filippone, P.; Mantellini, F.; Tietze, L. F. *Tetrahedron* **2001**, *57*, 5855–5863.
[37] Schuster, M.; Pernerstorfer, J.; Blechert, S. *Angew. Chem. Int. Ed. Engl.* **1996**, *35*, 1979–1980.
[38] Heerding, D. A.; Takata, D. T.; Kwon, C.; Huffman, W. F.; Samanen, J. *Tetrahedron Lett.* **1998**, *39*, 6815–6818.
[39] Brown, R. C. D.; Keily, J.; Karim, R. *Tetrahedron Lett.* **2000**, *41*, 3247–3251.
[40] Alvarez-Gutierrez, J. M.; Nefzi, A.; Houghten, R. A. *Tetrahedron Lett.* **2000**, *41*, 851–854.
[41] Karoyan, P.; Triolo, A.; Nannicini, R.; Giannotti, D.; Altamura, M.; Chassaing, G.; Perrotta, E. *Tetrahedron Lett.* **1999**, *40*, 71–74.
[42] Veerman, J. J. N.; Rutjes, F. P. J. T.; van Maarseveen, J. H.; Hiemstra, H. *Tetrahedron Lett.* **1999**, *40*, 6079–6082.
[43] Brown, R. C. D.; Fisher, M. *Chem. Commun.* **1999**, 1547–1548.
[44] Short, K. M.; Ching, B. W.; Mjalli, A. M. M. *Tetrahedron* **1997**, *53*, 6653–6679.
[45] Paulvannan, K. *Tetrahedron Lett.* **1999**, *40*, 1851–1854.
[46] Sun, S.; Murray, W. V. *J. Org. Chem.* **1999**, *64*, 5941–5945.
[47] Paulvannan, K.; Chen, T.; Jacobs, J. W. *Synlett* **1999**, 1609–1611.
[48] Miyabe, H.; Fujii, K.; Tanaka, H.; Naito, T. *Chem. Commun.* **2001**, 831–832.
[49] Romoff, T. T.; Ma, L.; Wang, Y. W.; Campbell, D. A. *Synlett* **1998**, 1341–1342.
[50] Weber, L.; Iaiza, P.; Biringer, G.; Barbier, P. *Synlett* **1998**, 1156–1158.
[51] Kulkarni, B. A.; Ganesan, A. *Tetrahedron Lett.* **1998**, *39*, 4369–4372.
[52] Matthews, J.; Rivero, R. A. *J. Org. Chem.* **1998**, *63*, 4808–4810.
[53] Miller, P. C.; Owen, T. J.; Molyneaux, J. M.; Curtis, J. M.; Jones, C. R. *J. Comb. Chem.* **1999**, *1*, 223–234.
[54] Smith, A. B.; Liu, H.; Okumura, H.; Favor, D. A.; Hirschmann, R. *Org. Lett.* **2000**, *2*, 2041–2044.
[55] Murphy, M. M.; Schullek, J. R.; Gordon, E. M.; Gallop, M. A. *J. Am. Chem. Soc.* **1995**, *117*, 7029–7030.
[56] Hamper, B. C.; Dukesherer, D. R.; South, M. S. *Tetrahedron Lett.* **1996**, *37*, 3671–3674.
[57] Bicknell, A. J.; Hird, N. W. *Bioorg. Med. Chem. Lett.* **1996**, *6*, 2441–2444.
[58] Barrett, A. G. M.; Boffey, R. J.; Frederiksen, M. U.; Newton, C. G.; Roberts, R. S. *Tetrahedron Lett.* **2001**, *42*, 5579–5581.
[59] Hollinshead, S. P. *Tetrahedron Lett.* **1996**, *37*, 9157–9160.
[60] Dondas, H. A.; Grigg, R.; MacLachlan, W. S.; MacPherson, D. T.; Markandu, J.; Sridharan, V.; Suganthan, S. *Tetrahedron Lett.* **2000**, *41*, 967–970.
[61] Hoveyda, H. R.; Hall, D. G. *Org. Lett.* **2001**, *3*, 3491–3494.
[62] Gong, Y. D.; Najdi, S.; Olmstead, M. M.; Kurth, M. J. *J. Org. Chem.* **1998**, *63*, 3081–3086.
[63] Bicknell, A. J.; Hird, N. W.; Readshaw, S. A. *Tetrahedron Lett.* **1998**, *39*, 5869–5872.

[64] Brooking, P.; Crawshaw, M.; Hird, N. W.; Jones, C.; MacLachlan, W. S.; Readshaw, S. A.; Wilding, S. *Synthesis* **1999**, 1986–1992.

[65] Marx, M. A.; Grillot, A. L.; Louer, C. T.; Beaver, K. A.; Bartlett, P. A. *J. Am. Chem. Soc.* **1997**, *119*, 6153–6167.

[66] Peng, G.; Sohn, A.; Gallop, M. A. *J. Org. Chem.* **1999**, *64*, 8342–8349.

[67] Pearson, W. H.; Clark, R. B. *Tetrahedron Lett.* **1997**, *38*, 7669–7672.

[68] Nicolaou, K. C.; Roecker, A. J.; Pfefferkorn, J. A.; Cao, G.-Q. *J. Am. Chem. Soc.* **2000**, *122*, 2966–2967.

[69] Hutchins, S. M.; Chapman, K. T. *Tetrahedron Lett.* **1996**, *37*, 4869–4872.

[70] Kim, R. M.; Manna, M.; Hutchins, S. M.; Griffin, P. R.; Yates, N. A.; Bernick, A. M.; Chapman, K. T. *Proc. Natl. Acad. Sci. USA* **1996**, *93*, 10012–10017.

[71] Cheng, Y.; Chapman, K. T. *Tetrahedron Lett.* **1997**, *38*, 1497–1500.

[72] Smith, A. L.; Stevenson, G. I.; Swain, C. J.; Castro, J. L. *Tetrahedron Lett.* **1998**, *39*, 8317–8320.

[73] Zhang, H.-C.; Ye, H.; White, K. B.; Maryanoff, B. E. *Tetrahedron Lett.* **2001**, *42*, 4751–4754.

[74] Zhang, H.-C.; Ye, H.; Moretto, A. F.; Brumfield, K. K.; Maryanoff, B. E. *Org. Lett.* **2000**, *2*, 89–92.

[75] Wu, T. Y. H.; Ding, S.; Gray, N. S.; Schultz, P. G. *Org. Lett.* **2001**, *3*, 3827–3830.

[76] Fagnola, M. C.; Candiani, I.; Visentin, G.; Cabri, W.; Zarini, F.; Mongelli, N.; Bedeschi, A. *Tetrahedron Lett.* **1997**, *38*, 2307–2310.

[77] Zhang, H. C.; Brumfield, K. K.; Maryanoff, B. E. *Tetrahedron Lett.* **1997**, *38*, 2439–2442.

[78] Zhang, H. C.; Brumfield, K. K.; Jaroskova, L.; Maryanoff, B. E. *Tetrahedron Lett.* **1998**, *39*, 4449–4452.

[79] Wang, Y.; Huang, T. N. *Tetrahedron Lett.* **1998**, *39*, 9605–9608.

[80] Collini, M. D.; Ellingboe, J. W. *Tetrahedron Lett.* **1997**, *38*, 7963–7966.

[81] Zhang, H. C.; Maryanoff, B. E. *J. Org. Chem.* **1997**, *62*, 1804–1809.

[82] Yun, W. Y.; Mohan, R. *Tetrahedron Lett.* **1996**, *37*, 7189–7192.

[83] Arumugam, V.; Routledge, A.; Abell, C.; Balasubramanian, S. *Tetrahedron Lett.* **1997**, *38*, 6473–6476.

[84] Hughes, I. *Tetrahedron Lett.* **1996**, *37*, 7595–7598.

[85] Stephensen, H.; Zaragoza, F. *Tetrahedron Lett.* **1999**, *40*, 5799–5802.

[86] Sun, Q.; Zhou, X.; Islam, K.; Kyle, D. J. *Tetrahedron Lett.* **2001**, *42*, 6495–6497.

[87] Ketcha, D. M.; Wilson, L. J.; Portlock, D. E. *Tetrahedron Lett.* **2000**, *41*, 6253–6257.

[88] Smith, A. L.; Stevenson, G. I.; Lewis, S.; Patel, S.; Castro, J. L. *Bioorg. Med. Chem. Lett.* **2000**, *10*, 2693–2696.

[89] Heinelt, U.; Herok, S.; Matter, H.; Wildgoose, P. *Bioorg. Med. Chem. Lett.* **2001**, *11*, 227–230.

[90] Hübner, H.; Kraxner, J.; Gmeiner, P. *J. Med. Chem.* **2000**, *43*, 4563–4569.

[91] Tois, J.; Franzèn, R.; Aitio, O.; Laakso, I.; Kylänlahti, I. *J. Comb. Chem.* **2001**, *3*, 542–545.

[92] Larhed, M.; Lindeberg, G.; Hallberg, A. *Tetrahedron Lett.* **1996**, *37*, 8219–8222.

[93] Whitehouse, D. L.; Nelson, K. H.; Savinov, S. N.; Löwe, R. S.; Austin, D. J. *Bioorg. Med. Chem.* **1998**, *6*, 1273–1282.

[94] Gowravaram, M. R.; Gallop, M. A. *Tetrahedron Lett.* **1997**, *38*, 6973–6976.

[95] Whitehouse, D. L.; Nelson, K. H.; Savinov, S. N.; Austin, D. J. *Tetrahedron Lett.* **1997**, *38*, 7139–7142.

[96] Li, Z.; Ganesan, A. *Synlett* **1998**, 405–406.

[97] Boymond, L.; Rottländer, M.; Cahiez, G.; Knochel, P. *Angew. Chem. Int. Ed.* **1998**, *37*, 1701–1703.

[98] Chamoin, S.; Houldsworth, S.; Snieckus, V. *Tetrahedron Lett.* **1998**, *39*, 4175–4178.

[99] Han, Y.; Roy, A.; Giroux, A. *Tetrahedron Lett.* **2000**, *41*, 5447–5451.

[100] Fujita, K.-I.; Watanabe, K.; Oishi, A.; Ikeda, Y.; Taguchi, Y. *Synlett* **1999**, 1760–1762.

[101] Ma, S.; Duan, D.; Shi, Z. *Org. Lett.* **2000**, *2*, 1419–1422.

[102] Kang, S.; Kim, J. S.; Yoon, S. K.; Lim, K. H.; Yoon, S. S. *Tetrahedron Lett.* **1998**, *39*, 3011–3012.

[103] Beebe, X.; Schore, N. E.; Kurth, M. J. *J. Org. Chem.* **1995**, *60*, 4196–4203.

[104] Beebe, X.; Chiappari, C. L.; Olmstead, M. M.; Kurth, M. J.; Schore, N. E. *J. Org. Chem.* **1995**, *60*, 4204–4212.

[105] Routledge, A.; Abell, C.; Balasubramanian, S. *Synlett* **1997**, 61–62.

[106] Berteina, S.; De Mesmaeker, A.; Wendeborn, S. *Synlett* **1999**, 1121–1123.

[107] Watanabe, Y.; Ishikawa, S.; Takao, G.; Toru, T. *Tetrahedron Lett.* **1999**, *40*, 3411–3414.

[108] Rottländer, M.; Knochel, P. *J. Comb. Chem.* **1999**, *1*, 181–183.

[109] Fancelli, D.; Fagnola, M. C.; Severino, D.; Bedeschi, A. *Tetrahedron Lett.* **1997**, *38*, 2311–2314.

[110] Berteina, S.; De Mesmaeker, A. *Tetrahedron Lett.* **1998**, *39*, 5759–5762.

[111] Berteina, S.; Wendeborn, S.; De Mesmaeker, A. *Synlett* **1998**, 1231–1233.

[112] Du, X.; Armstrong, R. W. *J. Org. Chem.* **1997**, *62*, 5678–5679.

[113] Du, X.; Armstrong, R. W. *Tetrahedron Lett.* **1998**, *39*, 2281–2284.

[114] Nicolaou, K. C.; Snyder, S. A.; Bigot, A.; Pfefferkorn, J. A. *Angew. Chem. Int. Ed.* **2000**, *39*, 1093–1096.

[115] Stephensen, H.; Zaragoza, F. *J. Org. Chem.* **1997**, *62*, 6096–6097.

[116] Zaragoza, F. *Tetrahedron Lett.* **1996**, *37*, 6213–6216.

[117] Castanedo, G. M.; Sutherlin, D. P. *Tetrahedron Lett.* **2001**, *42*, 7181–7184.

[118] Rottländer, M.; Knochel, P. *Synlett* **1997**, 1084–1086.

[119] Yu, K. L.; Deshpande, M. S.; Vyas, D. M. *Tetrahedron Lett.* **1994**, *35*, 8919–8922.

[120] Marquais, S.; Arlt, M. *Tetrahedron Lett.* **1996**, *37*, 5491–5494.

[121] Han, Y.; Giroux, A.; Lépine, C.; Laliberté, F.; Huang, Z.; Perrier, H.; Bayly, C. I.; Young, R. N. *Tetrahedron* **1999**, *55*, 11669–11685.

[122] Guiles, J. W.; Johnson, S. G.; Murray, W. V. *J. Org. Chem.* **1996**, *61*, 5169–5171.

[123] Kang, S. K.; Yoon, S. K.; Lim, K. H.; Son, H. J.; Baik, T. G. *Synth. Commun.* **1998**, *28*, 3645–3655.

[124] Chamoin, S.; Houldsworth, S.; Kruse, C. G.; Bakker, W. I.; Snieckus, V. *Tetrahedron Lett.* **1998**, *39*, 4179–4182.

[125] Kirschbaum, T.; Briehn, C. A.; Bäuerle, P. *J. Chem. Soc., Perkin Trans. 1* **2000**, 1211–1216.

[126] Malenfant, P. R. L.; Fréchet, J. M. J. *Chem. Commun.* **1998**, 2657–2658.

[127] Briehn, C. A.; Kirschbaum, T.; Bäuerle, P. *J. Org. Chem.* **2000**, *65*, 352–359.

[128] Nefzi, A.; Ostresh, J. M.; Meyer, J. P.; Houghten, R. A. *Tetrahedron Lett.* **1997**, *38*, 931–934.

[129] Nefzi, A.; Ostresh, J. M.; Giulianotti, M.; Houghten, R. A. *J. Comb. Chem.* **1999**, *1*, 195–198.

[130] Bilodeau, M. T.; Cunningham, A. M. *J. Org. Chem.* **1998**, *63*, 2800–2801.

[131] Zhang, C. Z.; Moran, E. J.; Woiwode, T. F.; Short, K. M.; Mjalli, A. M. M. *Tetrahedron Lett.* **1996**, *37*, 751–754.

[132] Lee, H. B.; Balasubramanian, S. *Org. Lett.* **2000**, *2*, 323–326.

[133] Sarshar, S.; Siev, D.; Mjalli, A. M. M. *Tetrahedron Lett.* **1996**, *37*, 835–838.

[134] Yu, Y.; El Abdellaoui, H. M.; Ostresh, J. M.; Houghten, R. A. *Tetrahedron Lett.* **2001**, *42*, 623–625.

[135] Li, M.; Wilson, L. J. *Tetrahedron Lett.* **2001**, *42*, 1455–1458.

[136] Drewry, D. H.; Ghiron, C. *Tetrahedron Lett.* **2000**, *41*, 6989–6992.

[137] Fu, M.; Fernandez, M.; Smith, M. L.; Flygare, J. A. *Org. Lett.* **1999**, *1*, 1351–1353.

[138] Yu, Y.; Ostresh, J. M.; Houghten, R. A. *J. Comb. Chem.* **2001**, *3*, 521–523.

[139] Nefzi, A.; Giulianotti, M. A.; Ong, N. A.; Houghten, R. A. *Org. Lett.* **2000**, *2*, 3349–3350.

[140] Goff, D. *Tetrahedron Lett.* **1998**, *39*, 1477–1480.

[141] Kondo, Y.; Komine, T.; Sakamoto, T. *Org. Lett.* **2000**, *2*, 3111–3113.

[142] Gelens, E.; Koot, W. J.; Menge, W. M. P. B.; Ottenheijm, H. C. J.; Timmerman, H. *Bioorg. Med. Chem. Lett.* **2000**, *10*, 1935–1938.

[143] Havez, S.; Begtrup, M.; Vedsø, P.; Andersen, K.; Ruhland, T. *J. Org. Chem.* **1998**, *63*, 7418–7420.

[144] Combs, A. P.; Saubern, S.; Rafalski, M.; Lam, P. Y. S. *Tetrahedron Lett.* **1999**, *40*, 1623–1626.

[145] Stahl, G. L.; Walter, R.; Smith, C. W. *J. Am. Chem. Soc.* **1979**, *101*, 5383–5394.

[146] Eleftheriou, S.; Gatos, D.; Panagopoulos, A.; Stathopoulos, S.; Barlos, K. *Tetrahedron Lett.* **1999**, *40*, 2825–2828.

[147] Sabatino, G.; Chelli, M.; Mazzucco, S.; Ginanneschi, M.; Papini, A. M. *Tetrahedron Lett.* **1999**, *40*, 809–812.

[148] Tortolani, D. R.; Biller, S. A. *Tetrahedron Lett.* **1996**, *37*, 5687–5690.

[149] Nicolaou, K. C.; Pfefferkorn, J. A.; Roecker, A. J.; Cao, G.-Q.; Barluenga, S.; Mitchell, H. J. *J. Am. Chem. Soc.* **2000**, *122*, 9939–9953.

[150] Rueter, J. K.; Nortey, S. O.; Baxter, E. W.; Leo, G. C.; Reitz, A. B. *Tetrahedron Lett.* **1998**, *39*, 975–978.

[151] Kim, S. W.; Koh, J. S.; Lee, E. J.; Ro, S. *Mol. Diversity* **1998**, *3*, 129–132.

[152] Sim, M. M.; Ganesan, A. *J. Org. Chem.* **1997**, *62*, 3230–3235.

[153] Nefzi, A.; Giulianotti, M. A.; Houghten, R. A. *Tetrahedron Lett.* **2000**, *41*, 2283–2287.

[154] Bauser, M.; Winter, M.; Valenti, C. A.; Wiesmüller, K. H.; Jung, G. *Mol. Diversity* **1998**, *3*, 257–260.

[155] Lee, S. H.; Chung, S. H.; Lee, Y. S. *Tetrahedron Lett.* **1998**, *39*, 9469–9472.

[156] Kim, S. W.; Ahn, S. Y.; Koh, J. S.; Lee, J. H.; Ro, S.; Cho, H. Y. *Tetrahedron Lett.* **1997**, *38*, 4603–4606.

[157] Migawa, M. T.; Swayze, E. E. *Org. Lett.* **2000**, *2*, 3309–3311.

[158] Matthews, J.; Rivero, R. A. *J. Org. Chem.* **1997**, *62*, 6090–6092.

[159] Park, K. H.; Abbate, E.; Najdi, S.; Olmstead, M. M.; Kurth, M. J. *Chem. Commun.* **1998**, 1679–1680.

[160] Boeijen, A.; Liskamp, R. M. J. *Eur. J. Org. Chem.* **1999**, 2127–2135.

[161] Scicinski, J. J.; Barker, R. D.; Murray, P. J.; Jarvie, E. M. *Bioorg. Med. Chem. Lett.* **1998**, *8*, 3609–3614.

[162] Boeijen, A.; Kruijtzer, J. A. W.; Liskamp, R. M. J. *Bioorg. Med. Chem. Lett.* **1998**, *8*, 2375–2380.
[163] Wilson, L. J.; Li, M.; Portlock, D. E. *Tetrahedron Lett.* **1998**, *39*, 5135–5138.
[164] Yoon, J.; Cho, C. W.; Han, H.; Janda, K. D. *Chem. Commun.* **1998**, 2703–2704.
[165] Hamuro, Y.; Marshall, W. J.; Scialdone, M. A. *J. Comb. Chem.* **1999**, *1*, 163–172.
[166] Wu, S.; Janusz, J. M. *Tetrahedron Lett.* **2000**, *41*, 1165–1169.
[167] Hanessian, S.; Yang, R. Y. *Tetrahedron Lett.* **1996**, *37*, 5835–5838.
[168] Dressman, B. A.; Spangle, L. A.; Kaldor, S. W. *Tetrahedron Lett.* **1996**, *37*, 937–940.
[169] Chong, P. Y.; Petillo, P. A. *Tetrahedron Lett.* **1999**, *40*, 2493–2496.
[170] Nefzi, A.; Ostresh, J. M.; Giulianotti, M.; Houghten, R. A. *Tetrahedron Lett.* **1998**, *39*, 8199–8202.
[171] Xiao, X. Y.; Ngu, K.; Chao, C.; Patel, D. V. *J. Org. Chem.* **1997**, *62*, 6968–6973.
[172] Tumelty, D.; Schwarz, M. K.; Cao, K.; Needels, M. C. *Tetrahedron Lett.* **1999**, *40*, 6185–6188.
[173] Farrant, E.; Rahman, S. S. *Tetrahedron Lett.* **2000**, *41*, 5383–5386.
[174] Wu, Z.; Rea, P.; Wickham, G. *Tetrahedron Lett.* **2000**, *41*, 9871–9874.
[175] Tumelty, D.; Cao, K.; Holmes, C. P. *Org. Lett.* **2001**, *3*, 83–86.
[176] Tumelty, D.; Schwarz, M. K.; Needels, M. C. *Tetrahedron Lett.* **1998**, *39*, 7467–7470.
[177] Lee, J.; Doucette, A.; Wilson, N. S.; Lord, J. *Tetrahedron Lett.* **2001**, *42*, 2635–2638.
[178] Smith, J. M.; Gard, J.; Cummings, W.; Kanizsai, A.; Krchnák, V. *J. Comb. Chem.* **1999**, *1*, 368–370.
[179] Bienaymé, H.; Bouzid, K. *Angew. Chem. Int. Ed.* **1998**, *37*, 2234–2237.
[180] Blackburn, C.; Guan, B.; Fleming, P.; Shiosaki, K.; Tsai, S. *Tetrahedron Lett.* **1998**, *39*, 3635–3638.
[181] Blackburn, C. *Tetrahedron Lett.* **1998**, *39*, 5469–5472.
[182] Blackburn, C.; Guan, B. *Tetrahedron Lett.* **2000**, *41*, 1495–1500.
[183] Chen, J. J.; Golebiowski, A.; Klopfenstein, S. R.; McClenaghan, J.; Peng, S. X.; Portlock, D. E.; West, L. *Synlett* **2001**, 1263–1265.
[184] Mayer, J. P.; Lewis, G. S.; McGee, C.; Bankaitis-Davis, D. *Tetrahedron Lett.* **1998**, *39*, 6655–6658.
[185] Phillips, G. B.; Wei, G. P. *Tetrahedron Lett.* **1996**, *37*, 4887–4890.
[186] Kilburn, J. P.; Lau, J.; Jones, R. C. F. *Tetrahedron Lett.* **2000**, *41*, 5419–5421.
[187] Huang, W.; Scarborough, R. M. *Tetrahedron Lett.* **1999**, *40*, 2665–2668.
[188] Smith, J. M.; Krchnák, V. *Tetrahedron Lett.* **1999**, *40*, 7633–7636.
[189] Wei, G. P.; Phillips, G. B. *Tetrahedron Lett.* **1998**, *39*, 179–182.
[190] Lee, J.; Gauthier, D.; Rivero, R. A. *Tetrahedron Lett.* **1998**, *39*, 201–204.
[191] Yeh, C. M.; Sun, C. M. *Synlett* **1999**, 810–812.
[192] Blettner, C. G.; König, W. A.; Rühter, G.; Stenzel, W.; Schotten, T. *Synlett* **1999**, 307–310.
[193] Sun, Q.; Yan, B. *Bioorg. Med. Chem. Lett.* **1998**, *8*, 361–364.
[194] Heizmann, G.; Eberle, A. N. *Mol. Diversity* **1997**, *2*, 171–174.
[195] Di Lucrezia, R.; Gilbert, I. H.; Floyd, C. D. *J. Comb. Chem.* **2000**, *2*, 249–253.
[196] Chen, J. J.; Golebiowski, A.; McClenaghan, J.; Klopfenstein, S. R.; West, L. *Tetrahedron Lett.* **2001**, *42*, 2269–2271.
[197] Crimmins, M. T.; Zuercher, W. J. *Org. Lett.* **2000**, *2*, 1065–1067.
[198] Brill, W. K.-D.; Riva-Toniolo, C. *Tetrahedron Lett.* **2001**, *42*, 6515–6518.
[199] Gray, N. S.; Kwon, S.; Schultz, P. G. *Tetrahedron Lett.* **1997**, *38*, 1161–1164.
[200] Barrett, A. G. M.; Procopiou, P. A.; Voigtmann, U. *Org. Lett.* **2001**, *3*, 3165–3168.
[201] Zou, N.; Jiang, B. *J. Comb. Chem.* **2000**, *2*, 6–7.
[202] Faita, G.; Paio, A.; Quadrelli, P.; Rancati, F.; Seneci, P. *Tetrahedron Lett.* **2000**, *41*, 1265–1269.
[203] Lepore, S. D.; Wiley, M. R. *J. Org. Chem.* **1999**, *64*, 4547–4550.
[204] Park, K. H.; Olmstead, M. M.; Kurth, M. J. *J. Org. Chem.* **1998**, *63*, 6579–6585.
[205] Kurth, M. J.; Randall, L. A. A.; Takenouchi, K. *J. Org. Chem.* **1996**, *61*, 8755–8761.
[206] Pei, Y. H.; Moos, W. H. *Tetrahedron Lett.* **1994**, *35*, 5825–5828.
[207] Cheng, J. F.; Mjalli, A. M. M. *Tetrahedron Lett.* **1998**, *39*, 939–942.
[208] Kantorowski, E. J.; Kurth, M. J. *J. Org. Chem.* **1997**, *62*, 6797–6803.
[209] De Luca, L.; Giacomelli, G.; Riu, A. *J. Org. Chem.* **2001**, *66*, 6823–6825.
[210] Shankar, B. B.; Yang, D. Y.; Girton, S.; Ganguly, A. K. *Tetrahedron Lett.* **1998**, *39*, 2447–2448.
[211] Cereda, E.; Ezhaya, A.; Quai, M.; Barbaglia, W. *Tetrahedron Lett.* **2001**, *42*, 4951–4953.
[212] Marzinzik, A. L.; Felder, E. R. *Tetrahedron Lett.* **1996**, *37*, 1003–1006.
[213] Shen, D.-M.; Shu, M.; Chapman, K. T. *Org. Lett.* **2000**, *2*, 2789–2792.
[214] Tan, D. S.; Foley, M. A.; Shair, M. D.; Schreiber, S. L. *J. Am. Chem. Soc.* **1998**, *120*, 8565–8566.
[215] Tan, D. S.; Foley, M. A.; Stockwell, B. R.; Shair, M. D.; Schreiber, S. L. *J. Am. Chem. Soc.* **1999**, *121*, 9073–9087.
[216] Haap, W. J.; Kaiser, D.; Walk, T. B.; Jung, G. *Tetrahedron* **1998**, *54*, 3705–3724.
[217] Kobayashi, S.; Akiyama, R. *Tetrahedron Lett.* **1998**, *39*, 9211–9214.
[218] Albert, R.; Knecht, H.; Andersen, E.; Hungerford, V.; Schreier, M. H.; Papageorgiou, C. *Bioorg. Med. Chem. Lett.* **1998**, *8*, 2203–2208.

[219] Lepore, S. D.; Wiley, M. R. *J. Org. Chem.* **2000**, *65*, 2924–2932.
[220] Tremblay, M. R.; Wentworth, P.; Lee, G. E.; Janda, K. D. *J. Comb. Chem.* **2000**, *2*, 698–709.
[221] Oh, H. S.; Hahn, H.-G.; Cheon, S. H.; Ha, D.-C. *Tetrahedron Lett.* **2000**, *41*, 5069–5072.
[222] Wang, F.; Hauske, J. R. *Tetrahedron Lett.* **1997**, *38*, 6529–6532.
[223] Clapham, B.; Spanka, C.; Janda, K. D. *Org. Lett.* **2001**, *3*, 2173–2176.
[224] Nishida, A.; Fuwa, M.; Naruto, S.; Sugano, Y.; Saito, H.; Nakagawa, M. *Tetrahedron Lett.* **2000**, *41*, 4791–4794.
[225] Beebe, X.; Wodka, D.; Sowin, T. J. *J. Comb. Chem.* **2001**, *3*, 360–366.
[226] ten Holte, P.; Thijs, L.; Zwanenburg, B. *Tetrahedron Lett.* **1998**, *39*, 7407–7410.
[227] Buchstaller, H. P. *Tetrahedron* **1998**, *54*, 3465–3470.
[228] Holmes, C. P.; Chinn, J. P.; Look, G. C.; Gordon, E. M.; Gallop, M. A. *J. Org. Chem.* **1995**, *60*, 7328–7333.
[229] Munson, M. C.; Cook, A. W.; Josey, J. A.; Rao, C. *Tetrahedron Lett.* **1998**, *39*, 7223–7226.
[230] Look, G. C.; Schullek, J. R.; Holmes, C. P.; Chinn, J. P.; Gordon, E. M.; Gallop, M. A. *Bioorg. Med. Chem. Lett.* **1996**, *6*, 707–712.
[231] Lee, C. L.; Lam, Y.; Lee, S. Y. *Tetrahedron Lett.* **2001**, *42*, 109–111.
[232] Goff, D.; Fernandez, J. *Tetrahedron Lett.* **1999**, *40*, 423–426.
[233] Kearney, P. C.; Fernandez, M.; Flygare, J. A. *J. Org. Chem.* **1998**, *63*, 196–200.
[234] Stadlwieser, J.; Ellmerer-Müller, E. P.; Takó, A.; Maslouh, N.; Bannwarth, W. *Angew. Chem. Int. Ed.* **1998**, *37*, 1402–1404.
[235] Baer, R.; Masquelin, T. *J. Comb. Chem.* **2001**, *3*, 16–19.
[236] Ball, C. P.; Barrett, A. G. M.; Commercon, A.; Compère, D.; Kuhn, C.; Roberts, R. S.; Smith, M. L.; Venier, O. *Chem. Commun.* **1998**, 2019–2020.
[237] Pirrung, M. C.; Pansare, S. V. *J. Comb. Chem.* **2001**, *3*, 90–96.
[238] Yokum, T. S.; Alsina, J.; Barany, G. *J. Comb. Chem.* **2000**, *2*, 282–292.
[239] Mourtas, S.; Gatos, D.; Barlos, K. *Tetrahedron Lett.* **2001**, *42*, 2201–2204.
[240] Schlienger, N.; Bryce, M. R.; Hansen, T. K. *Tetrahedron Lett.* **2000**, *41*, 5147–5150.
[241] Pátek, M.; Drake, B.; Lebl, M. *Tetrahedron Lett.* **1995**, *36*, 2227–2230.
[242] Yan, B.; Gstach, H. *Tetrahedron Lett.* **1996**, *37*, 8325–8328.
[243] Marzinzik, A. L.; Felder, E. R. *J. Org. Chem.* **1998**, *63*, 723–727.
[244] Stauffer, S. R.; Katzenellenbogen, J. A. *J. Comb. Chem.* **2000**, *2*, 318–329.
[245] Spivey, A. C.; Diaper, C. M.; Adams, H.; Rudge, A. J. *J. Org. Chem.* **2000**, *65*, 5253–5263.
[246] Grosche, P.; Höltzel, A.; Walk, T. B.; Trautwein, A. W.; Jung, G. *Synthesis* **1999**, 1961–1970.
[247] Tietze, L. F.; Steinmetz, A. *Synlett* **1996**, 667–668.
[248] Tietze, L. F.; Evers, H.; Hippe, T.; Steinmetz, A.; Töpken, E. *Eur. J. Org. Chem.* **2001**, 1631–1634.
[249] Wilson, R. D.; Watson, S. P.; Richards, S. A. *Tetrahedron Lett.* **1998**, *39*, 2827–2830.
[250] Watson, S. P.; Wilson, R. D.; Judd, D. B.; Richards, S. A. *Tetrahedron Lett.* **1997**, *38*, 9065–9068.
[251] Lyngsø, L. O.; Nielsen, J. *Tetrahedron Lett.* **1998**, *39*, 5845–5848.
[252] Kobayashi, S.; Furuta, T.; Sugita, K.; Okitsu, O.; Oyamada, H. *Tetrahedron Lett.* **1999**, *40*, 1341–1344.
[253] Washizuka, K.-I.; Nagai, K.; Minakata, S.; Ryu, I.; Komatsu, M. *Tetrahedron Lett.* **2000**, *41*, 691–695.
[254] Barbaste, M.; Rolland-Fulcrand, V.; Roumestant, M. L.; Viallefont, P.; Martinez, J. *Tetrahedron Lett.* **1998**, *39*, 6287–6290.
[255] Katritzky, A. R.; Qi, M.; Feng, D.; Zhang, G.; Griffith, M. C.; Watson, K. *Org. Lett.* **1999**, *1*, 1189–1191.
[256] Larsen, S. D.; DiPaolo, B. A. *Org. Lett.* **2001**, *3*, 3341–3344.
[257] Albericio, F.; Garcia, J.; Michelotti, E. L.; Nicolás, E.; Tice, C. M. *Tetrahedron Lett.* **2000**, *41*, 3161–3163.
[258] Moore, M.; Norris, P. *Tetrahedron Lett.* **1998**, *39*, 7027–7030.
[259] Freeze, S.; Norris, P. *Heterocycles* **1999**, *51*, 1807–1817.
[260] Zaragoza, F.; Petersen, S. V. *Tetrahedron* **1996**, *52*, 10823–10826.
[261] Schiemann, K.; Showalter, H. D. H. *J. Org. Chem.* **1999**, *64*, 4972–4975.
[262] Brown, B. J.; Clemens, I. R.; Neesom, J. K. *Synlett* **2000**, 131–133.
[263] Liang, G.-B.; Qian, X. *Bioorg. Med. Chem. Lett.* **1999**, *9*, 2101–2104.
[264] Sams, C. K.; Lau, J. *Tetrahedron Lett.* **1999**, *40*, 9359–9362.
[265] Hébert, N.; Hannah, A. L.; Sutton, S. C. *Tetrahedron Lett.* **1999**, *40*, 8547–8550.
[266] Hu, Y. H.; Baudart, S.; Porco, J. A. *J. Org. Chem.* **1999**, *64*, 1049–1051.
[267] Goff, D. A. *Tetrahedron Lett.* **1999**, *40*, 8741–8745.
[268] Eda, M.; Kurth, M. J.; Nantz, M. H. *J. Org. Chem.* **2000**, *65*, 5131–5135.
[269] Eda, M.; Kurth, M. J. *Tetrahedron Lett.* **2001**, *42*, 2063–2068.

[270] Gordeev, M. F.; Patel, D. V.; Wu, J.; Gordon, E. M. *Tetrahedron Lett.* **1996**, *37*, 4643–4646.
[271] Tadesse, S.; Bhandari, A.; Gallop, M. A. *J. Comb. Chem.* **1999**, *1*, 184–187.
[272] Breitenbucher, J. G.; Figliozzi, G. *Tetrahedron Lett.* **2000**, *41*, 4311–4315.
[273] Gordeev, M. F.; Patel, D. V.; England, B. P.; Jonnalagadda, S.; Combs, J. D.; Gordon, E. M. *Bioorg. Med. Chem.* **1998**, *6*, 883–889.
[274] Gordeev, M. F.; Patel, D. V.; Gordon, E. M. *J. Org. Chem.* **1996**, *61*, 924–928.
[275] Chiu, C.; Tang, Z.; Ellingboe, J. W. *J. Comb. Chem.* **1999**, *1*, 73–77.
[276] Bhandari, A.; Li, B.; Gallop, M. A. *Synthesis* **1999**, 1951–1960.
[277] Far, A. R.; Tidwell, T. T. *J. Org. Chem.* **1998**, *63*, 8636–8637.
[278] Chen, C.; Munoz, B. *Tetrahedron Lett.* **1998**, *39*, 6781–6784.
[279] Chen, C.; Munoz, B. *Tetrahedron Lett.* **1998**, *39*, 3401–3404.
[280] Obika, S.; Nishiyama, T.; Tatematsu, S.; Nishimoto, M.; Miyashita, K.; Imanishi, T. *Heterocycles* **1998**, *49*, 261–267.
[281] Lago, M. A.; Nguyen, T. T.; Bhatnagar, P. *Tetrahedron Lett.* **1998**, *39*, 3885–3888.
[282] Vo, N. H.; Eyermann, C. J.; Hodge, C. N. *Tetrahedron Lett.* **1997**, *38*, 7951–7954.
[283] Wendeborn, S.; Berteina, S.; Brill, W. K. D.; De Mesmaeker, A. *Synlett* **1998**, 671–675.
[284] Veerman, J. J. N.; van Maarseveen, J. H.; Visser, G. M.; Kruse, C. G.; Schoemaker, H. E.; Hiemstra, H.; Rutjes, F. P. J. T. *Eur. J. Org. Chem.* **1998**, 2583–2589.
[285] Piscopio, A. D.; Miller, J. F.; Koch, K. *Tetrahedron Lett.* **1997**, *38*, 7143–7146.
[286] Wang, Y. H.; Wilson, S. R. *Tetrahedron Lett.* **1997**, *38*, 4021–4024.
[287] Creighton, C. J.; Zapf, C. W.; Bu, J. H.; Goodman, M. *Org. Lett.* **1999**, *1*, 1407–1409.
[288] Zhang, W.; Xie, W.; Fang, J.; Wang, P. G. *Tetrahedron Lett.* **1999**, *40*, 7929–7933.
[289] Barco, A.; Benetti, S.; De Risi, C.; Marchetti, P.; Pollini, G. P.; Zanirato, V. *Tetrahedron Lett.* **1998**, *39*, 7591–7594.
[290] Barco, A.; Benetti, S.; De Risi, C.; Marchetti, P.; Pollini, G. P.; Zanirato, V. *J. Comb. Chem.* **2000**, *2*, 337–340.
[291] Paulvannan, K.; Chen, T. *J. Org. Chem.* **2000**, *65*, 6160–6166.
[292] Jönsson, D.; Erlandsson, M.; Undén, A. *Tetrahedron Lett.* **2001**, *42*, 6953–6956.
[293] Wagman, A. S.; Wang, L.; Nuss, J. M. *J. Org. Chem.* **2000**, *65*, 9103–9113.
[294] Chen, C.; McDonald, I. A.; Munoz, B. *Tetrahedron Lett.* **1998**, *39*, 217–220.
[295] Chen, C.; Munoz, B. *Tetrahedron Lett.* **1999**, *40*, 3491–3494.
[296] Fantauzzi, P. P.; Yager, K. M. *Tetrahedron Lett.* **1998**, *39*, 1291–1294.
[297] Chou, Y. L.; Morrissey, M. M.; Mohan, R. *Tetrahedron Lett.* **1998**, *39*, 757–760.
[298] Mohan, R.; Chou, Y. L.; Morrissey, M. M. *Tetrahedron Lett.* **1996**, *37*, 3963–3966.
[299] Kaljuste, K.; Undén, A. *Tetrahedron Lett.* **1995**, *36*, 9211–9214.
[300] Li, X.; Zhang, L.; Zhang, W.; Hall, S. E.; Tam, J. P. *Org. Lett.* **2000**, *2*, 3075–3078.
[301] Tóth, G. K.; Kele, Z.; Fülöp, F. *Tetrahedron Lett.* **2000**, *41*, 10095–10098.
[302] Hutchins, S. M.; Chapman, K. T. *Tetrahedron Lett.* **1996**, *37*, 4865–4868.
[303] Yang, L. H.; Guo, L. Q. *Tetrahedron Lett.* **1996**, *37*, 5041–5044.
[304] Mayer, J. P.; Bankaitis-Davis, D.; Zhang, J. W.; Beaton, G.; Bjergarde, K.; Andersen, C. M.; Goodman, B. A.; Herrera, C. J. *Tetrahedron Lett.* **1996**, *37*, 5633–5636.
[305] Rölfing, K.; Thiel, M.; Künzer, H. *Synlett* **1996**, 1036–1038.
[306] Meutermans, W. D. F.; Alewood, P. F. *Tetrahedron Lett.* **1995**, *36*, 7709–7712.
[307] Goff, D. A.; Zuckermann, R. N. *J. Org. Chem.* **1995**, *60*, 5748–5749.
[308] Craig, D.; Robson, M. J.; Shaw, S. J. *Synlett* **1998**, 1381–1383.
[309] Lorsbach, B. A.; Bagdanoff, J. T.; Miller, R. B.; Kurth, M. J. *J. Org. Chem.* **1998**, *63*, 2244–2250.
[310] Hoemann, M. Z.; Melikian-Badalian, A.; Kumaravel, G.; Hauske, J. R. *Tetrahedron Lett.* **1998**, *39*, 4749–4752.
[311] Huang, X.; Liu, Z. *Tetrahedron Lett.* **2001**, *42*, 7655–7657.
[312] Kiselyov, A. S.; Smith, L.; Armstrong, R. W. *Tetrahedron* **1998**, *54*, 5089–5096.
[313] Kiselyov, A. S.; Armstrong, R. W. *Tetrahedron Lett.* **1997**, *38*, 6163–6166.
[314] Kiselyov, A. S.; Smith, L.; Virgilio, A.; Armstrong, R. W. *Tetrahedron* **1998**, *54*, 7987–7996.
[315] Hong, B.-C.; Chen, Z.-Y.; Chen, W.-H. *Org. Lett.* **2000**, *2*, 2647–2649.
[316] MacDonald, A. A.; DeWitt, S. H.; Hogan, E. M.; Ramage, R. *Tetrahedron Lett.* **1996**, *37*, 4815–4818.
[317] Hay, A. M.; Hobbs-DeWitt, S.; MacDonald, A. A.; Ramage, R. *Synthesis* **1999**, 1979–1985.
[318] Sim, M. M.; Lee, C. L.; Ganesan, A. *Tetrahedron Lett.* **1998**, *39*, 6399–6402.
[319] Srivastava, S. K.; Haq, W.; Murthy, P. K.; Chauhan, P. M. S. *Bioorg. Med. Chem. Lett.* **1999**, *9*, 1885–1888.
[320] Gopalsamy, A.; Pallai, P. V. *Tetrahedron Lett.* **1997**, *38*, 907–910.
[321] Watson, B. T.; Christiansen, G. E. *Tetrahedron Lett.* **1998**, *39*, 9839–9840.

[322] Ruhland, T.; Künzer, H. *Tetrahedron Lett.* **1996**, *37*, 2757–2760.
[323] Ogbu, C. O.; Qabar, M. N.; Boatman, P. D.; Urban, J.; Meara, J. P.; Ferguson, M. D.; Tulinsky, J.; Lum, C.; Babu, S.; Blaskovich, M. A.; Nakanishi, H.; Ruan, F. Q.; Cao, B. L.; Minarik, R.; Little, T.; Nelson, S.; Nguyen, M.; Gall, A.; Kahn, M. *Bioorg. Med. Chem. Lett.* **1998**, *8*, 2321–2326.
[324] Bräse, S.; Dahmen, S.; Heuts, J. *Tetrahedron Lett.* **1999**, *40*, 6201–6203.
[325] Nielsen, J.; Rasmussen, P. H. *Tetrahedron Lett.* **1996**, *37*, 3351–3354.
[326] Gaviña, F.; Gil, P.; Palazón, B. *Tetrahedron Lett.* **1979**, 1333–1336.
[327] Panek, J. S.; Zhu, B. *Tetrahedron Lett.* **1996**, *37*, 8151–8154.
[328] Boldi, A. M.; Johnson, C. R.; Eissa, H. O. *Tetrahedron Lett.* **1999**, *40*, 619–622.
[329] Steger, M.; Young, D. W. *Tetrahedron* **1999**, *55*, 7935–7956.
[330] Nizi, E.; Botta, M.; Corelli, F.; Manetti, F.; Messina, F.; Maga, G. *Tetrahedron Lett.* **1998**, *39*, 3307–3310.
[331] Botta, M.; Corelli, F.; Maga, G.; Manetti, F.; Renzulli, M.; Spadari, S. *Tetrahedron* **2001**, *57*, 8357–8367.
[332] Sonogashira, K.; Tohda, Y.; Hagihara, N. *Tetrahedron Lett.* **1975**, 4467–4470.
[333] Kolodziej, S. A.; Hamper, B. C. *Tetrahedron Lett.* **1996**, *37*, 5277–5280.
[334] Wipf, P.; Cunningham, A. *Tetrahedron Lett.* **1995**, *36*, 7819–7822.
[335] Wahhab, A.; Leban, J. *Tetrahedron Lett.* **2000**, *41*, 1487–1490.
[336] Kappe, C. O. *Bioorg. Med. Chem. Lett.* **2000**, *10*, 49–51.
[337] Hamper, B. C.; Gan, K. Z.; Owen, T. J. *Tetrahedron Lett.* **1999**, *40*, 4973–4976.
[338] Obrecht, D.; Abrecht, C.; Grieder, A.; Villalgordo, J. M. *Helv. Chim. Acta* **1997**, *80*, 65–72.
[339] Masquelin, T.; Sprenger, D.; Baer, R.; Gerber, F.; Mercadal, Y. *Helv. Chim. Acta* **1998**, *81*, 646–660.
[340] Khan, S. I.; Grinstaff, M. W. *Tetrahedron Lett.* **1998**, *39*, 8031–8034.
[341] Khan, S. I.; Grinstaff, M. W. *J. Am. Chem. Soc.* **1999**, *121*, 4704–4705.
[342] Makino, S.; Suzuki, N.; Nakanishi, E.; Tsuji, T. *Synlett* **2000**, 1670–1672.
[343] Theoclitou, M.-E.; Ostresh, J. M.; Hamashin, V.; Houghten, R. A. *Tetrahedron Lett.* **2000**, *41*, 2051–2054.
[344] Smith, A. L.; Thomson, C. G.; Leeson, P. D. *Bioorg. Med. Chem. Lett.* **1996**, *6*, 1483–1486.
[345] Gouilleux, L.; Fehrentz, J. A.; Winternitz, F.; Martinez, J. *Tetrahedron Lett.* **1996**, *37*, 7031–7034.
[346] Gordeev, M. F.; Hui, H. C.; Gordon, E. M.; Patel, D. V. *Tetrahedron Lett.* **1997**, *38*, 1729–1732.
[347] Gordeev, M. F.; Luehr, G. W.; Hui, H. C.; Gordon, E. M.; Patel, D. V. *Tetrahedron* **1998**, *54*, 15879–15890.
[348] Buckman, B. O.; Mohan, R. *Tetrahedron Lett.* **1996**, *37*, 4439–4442.
[349] Makino, S.; Suzuki, N.; Nakanishi, E.; Tsuji, T. *Tetrahedron Lett.* **2000**, *41*, 8333–8337.
[350] Shao, H.; Colucci, M.; Tong, S.; Zhang, H.; Castelhano, A. L. *Tetrahedron Lett.* **1998**, *39*, 7235–7238.
[351] Villalgordo, J. M.; Obrecht, D.; Chucholowski, A. *Synlett* **1998**, 1405–1407.
[352] Yang, R.-Y.; Kaplan, A. *Tetrahedron Lett.* **2000**, *41*, 7005–7008.
[353] Mayer, J. P.; Lewis, G. S.; Curtis, M. J.; Zhang, J. W. *Tetrahedron Lett.* **1997**, *38*, 8445–8448.
[354] Cobb, J. M.; Fiorini, M. T.; Goddard, C. R.; Theoclitou, M. E.; Abell, C. *Tetrahedron Lett.* **1999**, *40*, 1045–1048.
[355] Wang, F.; Hauske, J. R. *Tetrahedron Lett.* **1997**, *38*, 8651–8654.
[356] Gopalsamy, A.; Yang, H. *J. Comb. Chem.* **2000**, *2*, 378–381.
[357] Vojkovsky, T.; Weichsel, A.; Pátek, M. *J. Org. Chem.* **1998**, *63*, 3162–3163.
[358] Eguchi, M.; Lee, M. S.; Stasiak, M.; Kahn, M. *Tetrahedron Lett.* **2001**, *42*, 1237–1239.
[359] Wang, H.; Ganesan, A. *J. Comb. Chem.* **2000**, *2*, 186–194.
[360] Zhang, J.; Barker, J.; Lou, B.; Saneii, H. *Tetrahedron Lett.* **2001**, *42*, 8405–8408.
[361] Wilson, L. *J. Org. Lett.* **2001**, *3*, 585–588.
[362] Gisin, B. F.; Merrifield, R. B. *J. Am. Chem. Soc.* **1972**, *94*, 3102–3106.
[363] Fields, G. B.; Noble, R. L. *Int. J. Pept. Prot. Res.* **1990**, *35*, 161–214.
[364] Nefzi, A.; Giulianotti, M. A.; Houghten, R. A. *Tetrahedron Lett.* **1999**, *40*, 8539–8542.
[365] Nefzi, A.; Giulianotti, M. A.; Houghten, R. A. *Tetrahedron* **2000**, *56*, 3319–3326.
[366] Nefzi, A.; Giulianotti, M. A.; Houghten, R. A. *J. Comb. Chem.* **2001**, *3*, 68–70.
[367] del Fresno, M.; Alsina, J.; Royo, M.; Barany, G.; Albericio, F. *Tetrahedron Lett.* **1998**, *39*, 2639–2642.
[368] Szardenings, A. K.; Antonenko, V.; Campbell, D. A.; DeFrancisco, N.; Ida, S.; Shi, L.; Sharkov, N.; Tien, D.; Wang, Y.; Navre, M. *J. Med. Chem.* **1999**, *42*, 1348–1357.
[369] Smith, R. A.; Bobko, M. A.; Lee, W. *Bioorg. Med. Chem. Lett.* **1998**, *8*, 2369–2374.
[370] Flanigan, E.; Marshall, G. R. *Tetrahedron Lett.* **1970**, 2403–2406.
[371] Kowalski, J.: Lipton, M. A. *Tetrahedron Lett.* **1996**, *37*, 5839–5840.

[372] Szardenings, A. K.; Burkoth, T. S.; Lu, H. H.; Tien, D. W.; Campbell, D. A. *Tetrahedron* **1997**, *53*, 6573–6593.
[373] Scott, B. O.; Siegmund, A. C.; Marlowe, C. K.; Pei, Y.; Spear, K. L. *Mol. Diversity* **1995**, *1*, 125–134.
[374] Steele, J.; Gordon, D. W. *Bioorg. Med. Chem. Lett.* **1995**, *5*, 47–50.
[375] van Loevezijn, A.; van Maarseveen, J. H.; Stegman, K.; Visser, G. M.; Koomen, G. J. *Tetrahedron Lett.* **1998**, *39*, 4737–4740.
[376] Wang, H.; Ganesan, A. *Org. Lett.* **1999**, *1*, 1647–1649.
[377] Li, W. R.; Peng, S. Z. *Tetrahedron Lett.* **1998**, *39*, 7373–7376.
[378] Bianco, A.; Sonksen, C. P.; Roepstorff, P.; Briand, J.-P. *J. Org. Chem.* **2000**, *65*, 2179–2187.
[379] Bianco, A.; Furrer, J.; Limal, D.; Guichard, G.; Elbayed, K.; Raya, J.; Piotto, M.; Briand, J.-P. *J. Comb. Chem.* **2000**, *2*, 681–690.
[380] Golebiowski, A.; Klopfenstein, S. R.; Shao, X.; Chen, J. J.; Colson, A.-O.; Grieb, A. L.; Russell, A. F. *Org. Lett.* **2000**, *2*, 2615–2617.
[381] Perrotta, E.; Altamura, M.; Barani, T.; Bindi, S.; Giannotti, D.; Harmat, N. J. S.; Nannicini, R.; Maggi, C. A. *J. Comb. Chem.* **2001**, *3*, 453–460.
[382] Berst, F.; Holmes, A. B.; Ladlow, M.; Murray, P. J. *Tetrahedron Lett.* **2000**, *41*, 6649–6653.
[383] Krchnák, V.; Szabo, L.; Vágner, J. *Tetrahedron Lett.* **2000**, *41*, 2835–2838.
[384] Zaragoza, F.; Stephensen, H. *J. Org. Chem.* **1999**, *64*, 2555–2557.
[385] Krchnák, V.; Smith, J.; Vágner, J. *Tetrahedron Lett.* **2001**, *42*, 2443–2446.
[386] Wu, Z.; Ede, N. J. *Tetrahedron Lett.* **2001**, *42*, 8115–8118.
[387] Morales, G. A.; Corbett, J. W.; DeGrado, W. F. *J. Org. Chem.* **1998**, *63*, 1172–1177.
[388] Lee, J.; Murray, W. V.; Rivero, R. A. *J. Org. Chem.* **1997**, *62*, 3874–3879.
[389] Mazurov, A. *Tetrahedron Lett.* **2000**, *41*, 7–10.
[390] Baxter, A. D.; Boyd, E. A.; Cox, P. B.; Loh, V.; Monteils, C.; Proud, A. *Tetrahedron Lett.* **2000**, *41*, 8177–8181.
[391] Makara, G. M.; Ewing, W.; Ma, Y.; Wintner, E. *J. Org. Chem.* **2001**, *66*, 5783–5789.
[392] Goff, D. A. *Tetrahedron Lett.* **1998**, *39*, 1473–1476.
[393] Goff, D. A.; Zuckermann, R. N. *Tetrahedron Lett.* **1996**, *37*, 6247–6250.
[394] Kung, P. P.; Swayze, E. *Tetrahedron Lett.* **1999**, *40*, 5651–5654.
[395] Mohamed, N.; Bhatt, U.; Just, G. *Tetrahedron Lett.* **1998**, *39*, 8213–8216.
[396] Shreder, K.; Zhang, L.; Gleeson, J. P.; Ericsson, J. A.; Yalamoori, V. V.; Goodman, M. *J. Comb. Chem.* **1999**, *1*, 383–387.
[397] Zhu, Z. M.; McKittrick, B. *Tetrahedron Lett.* **1998**, *39*, 7479–7482.
[398] Attanasi, O. A.; De Crescentini, L.; Filippone, P.; Mantellini, F.; Santeusanio, S. *Helv. Chim. Acta* **2001**, *84*, 2379–2386.
[399] Masquelin, T.; Meunier, N.; Gerber, F.; Rossé, G. *Heterocycles* **1998**, *48*, 2489–2505.
[400] Martínez-Teipel, B.; Michelotti, E.; Kelly, M. J.; Weaver, D. G.; Acholla, F.; Beshah, K.; Teixidó, J. *Tetrahedron Lett.* **2001**, *42*, 6455–6457.
[401] Gopalsamy, A.; Yang, H. *J. Comb. Chem.* **2001**, *3*, 278–283.
[402] Borrell, J. I.; Teixidó, J.; Schuler, E.; Michelotti, E. *Tetrahedron Lett.* **2001**, *42*, 5331–5334.
[403] Tietze, L. F.; Hippe, T.; Steinmetz, A. *Synlett* **1996**, 1043–1044.
[404] Leconte, S.; Dujardin, G.; Brown, E. *Eur. J. Org. Chem.* **2000**, 639–643.
[405] Stavenger, R. A.; Schreiber, S. L. *Angew. Chem. Int. Ed.* **2001**, *40*, 3417–3421.
[406] Dujardin, G.; Leconte, S.; Coutable, L.; Brown, E. *Tetrahedron Lett.* **2001**, *42*, 8849–8852.
[407] Watson, B. T.; Christiansen, G. E. *Tetrahedron Lett.* **1998**, *39*, 6087–6090.
[408] Harikrishnan, L. S.; Showalter, H. D. H. *Tetrahedron* **2000**, *56*, 515–519.
[409] Lindsley, C. W.; Chan, L. K.; Goess, B. C.; Joseph, R.; Shair, M. D. *J. Am. Chem. Soc.* **2000**, *122*, 422–423.
[410] Gordeev, M. F. *Biotechnology and Bioengineering* **1998**, *61*, 13–16.
[411] Pérez, R.; Reyes, O.; Suarez, M.; Garay, H. E.; Cruz, L. J.; Rodríguez, H.; Molero-Vilchez, M. D.; Ochoa, C. *Tetrahedron Lett.* **2000**, *41*, 613–616.
[412] Kuster, G. J.; Scheeren, H. W. *Tetrahedron Lett.* **1998**, *39*, 3613–3616.
[413] Kuster, G. J.; Scheeren, H. W. *Tetrahedron Lett.* **2000**, *41*, 515–519.
[414] Eguchi, M.; Lee, M. S.; Nakanishi, H.; Stasiak, M.; Lovell, S.; Kahn, M. *J. Am. Chem. Soc.* **1999**, *121*, 12204–12205.
[415] Kohn, W. D.; Zhang, L. *Tetrahedron Lett.* **2001**, *42*, 4453–4457.
[416] Lee, C. L.; Chan, K. P.; Lam, Y.; Lee, S. Y. *Tetrahedron Lett.* **2001**, *42*, 1167–1169.
[417] Nefzi, A.; Giulianotti, M.; Houghten, R. A. *Tetrahedron Lett.* **1998**, *39*, 3671–3674.
[418] Mortezaei, R.; Ida, S.; Campbell, D. A. *Mol. Diversity* **1999**, *4*, 143–148.
[419] Bolton, G. L.; Hodges, J. C. *J. Comb. Chem.* **1999**, *1*, 130–133.

[420] Piscopio, A. D.; Miller, J. F.; Koch, K. *Tetrahedron Lett.* **1998**, *39*, 2667–2670.
[421] van Maarseveen, J. H.; den Hartog, J. A. J.; Engelen, V.; Finner, E.; Visser, G.; Kruse, C. G. *Tetrahedron Lett.* **1996**, *37*, 8249–8252.
[422] Brown, R. C. D.; Castro, J. L.; Moriggi, J.-D. *Tetrahedron Lett.* **2000**, *41*, 3681–3685.
[423] Pelish, H. E.; Westwood, N. J.; Feng, Y.; Kirchhausen, T.; Shair, M. D. *J. Am. Chem. Soc.* **2001**, *123*, 6740–6741.
[424] Gauzy, L.; Le Merrer, Y.; Depezay, J. C.; Clerc, F.; Mignani, S. *Tetrahedron Lett.* **1999**, *40*, 6005–6008.
[425] Huang, W.; Kalivretenos, A. G. *Tetrahedron Lett.* **1995**, *36*, 9113–9116.
[426] Reichwein, J. F.; Liskamp, R. M. J. *Eur. J. Org. Chem.* **2000**, 2335–2344.
[427] Reichwein, J. F.; Wels, B.; Kruijtzer, J. A. W.; Versluis, C.; Liskamp, R. M. J. *Angew. Chem. Int. Ed.* **1999**, *38*, 3684–3687.
[428] Reichwein, J. F.; Versluis, C.; Liskamp, R. M. J. *J. Org. Chem.* **2000**, *65*, 6187–6195.
[429] Krchnák, V.; Weichsel, A. S. *Tetrahedron Lett.* **1997**, *38*, 7299–7302.
[430] Nefzi, A.; Ostresh, J. M.; Houghten, R. A. *Tetrahedron Lett.* **1997**, *38*, 4943–4946.
[431] Giovannoni, J.; Subra, G.; Amblard, M.; Martinez, J. *Tetrahedron Lett.* **2001**, *42*, 5389–5292.
[432] Nouvet, A.; Binard, M.; Lamaty, F.; Martinez, J.; Lazaro, R. *Tetrahedron* **1999**, *55*, 4685–4698.
[433] de Bont, D. B. A.; Moree, W. J.; Liskamp, R. M. J. *Bioorg. Med. Chem.* **1996**, *4*, 667–672.
[434] Schwarz, M. K.; Tumelty, D.; Gallop, M. A. *J. Org. Chem.* **1999**, *64*, 2219–2231.
[435] Lee, J.; Gauthier, D.; Rivero, R. A. *J. Org. Chem.* **1999**, *64*, 3060–3065.
[436] Schwarz, M. K.; Tumelty, D.; Gallop, M. A. *Tetrahedron Lett.* **1998**, *39*, 8397–8400.
[437] Morton, G. C.; Salvino, J. M.; Labaudinière, R. F.; Herpin, T. F. *Tetrahedron Lett.* **2000**, *41*, 3029–3033.
[438] Nefzi, A.; Ong, N. A.; Giulianotti, M. A.; Ostresh, J. M.; Houghten, R. A. *Tetrahedron Lett.* **1999**, *40*, 4939–4942.
[439] Boojamra, C. G.; Burow, K. M.; Thompson, L. A.; Ellman, J. A. *J. Org. Chem.* **1997**, *62*, 1240–1256.
[440] Goff, D. A.; Zuckermann, R. N. *J. Org. Chem.* **1995**, *60*, 5744–5745.
[441] Kamal, A.; Reddy, G. S. K.; Raghavan, S. *Bioorg. Med. Chem. Lett.* **2001**, *11*, 387–389.
[442] Kamal, A.; Reddy, G. S. K.; Reddy, K. L. *Tetrahedron Lett.* **2001**, *42*, 6969–6971.
[443] Nefzi, A.; Ong, N. A.; Houghten, R. A. *Tetrahedron Lett.* **2001**, *42*, 5141–5143.
[444] Plunkett, M. J.; Ellman, J. A. *J. Org. Chem.* **1997**, *62*, 2885–2893.
[445] Berry, J. M.; Howard, P. W.; Thurston, D. E. *Tetrahedron Lett.* **2000**, *41*, 6171–6174.
[446] Camps, F.; Cartells, J.; Pi, J. *Anales de Química* **1974**, *70*, 848–849.
[447] Mayer, J. P.; Zhang, J. W.; Bjergarde, K.; Lenz, D. M.; Gaudino, J. J. *Tetrahedron Lett.* **1996**, *37*, 8081–8084.
[448] Bhalay, G.; Blaney, P.; Palmer, V. H.; Baxter, A. D. *Tetrahedron Lett.* **1997**, *38*, 8375–8378.
[449] Woolard, F. X.; Paetsch, J.; Ellman, J. A. *J. Org. Chem.* **1997**, *62*, 6102–6103.
[450] Yu, Y.; Ostresh, J. M.; Houghten, R. A. *Org. Lett.* **2001**, *3*, 2797–2799.
[451] Bevacqua, F.; Basso, A.; Gitto, R.; Bradley, M.; Chimirri, A. *Tetrahedron Lett.* **2001**, *42*, 7683–7685.
[452] Ouyang, X.; Kiselyov, A. S. *Tetrahedron* **1999**, *55*, 8295–8302.
[453] Ouyang, X.; Kiselyov, A. S. *Tetrahedron Lett.* **1999**, *40*, 5827–5830.
[454] Ouyang, X.; Chen, Z.; Liu, L.; Dominguez, C.; Kiselyov, A. S. *Tetrahedron* **2000**, *56*, 2369–2377.

16 Preparation of Oligomeric Compounds

In this chapter, the term 'oligomeric compound' refers to products prepared by repeatedly linking one or several types of monomer to a growing chain of monomers. Included are biopolymers, such as peptides or oligonucleotides, and non-natural products, such as peptide nucleic acids, peptoids, or other synthetic polyamides.

Biopolymers play a central role in all living systems. The functions that proteins and peptides can perform are unparalleled by any other class of substance, and, in terms of efficiency and specificity, generally far exceed any non-peptide-mediated process. These functions include catalysis (enzymes), molecular recognition (enzymes, receptors, hormones), signal transduction (receptors), conversion of chemical into mechanical energy (muscle tissue), transport and storage of other compounds, and so on. Many of these functions depend on the precise arrangement and the correct dynamic behavior of a set of functional groups within the protein (e.g. mobility of these groups, the ability to 'collapse' onto a ligand or substrate, the ability to bind a substrate but quickly release a product). These dynamic parameters, which are crucial for the function of proteins, are a result of the rigidity of the peptide backbone and the three-dimensional (tertiary) structure of the protein.

Although attempts to produce non-peptidic oligomers with properties similar to those of peptides or proteins have so far met with only limited success at most, the continuous development of more efficient strategies for parallel solid-phase synthesis and the screening of compound libraries has made it possible to prepare and screen millions of compounds rather quickly [1–4]. With the aid of these technologies, it might be possible in the near future to identify synthetic oligomeric compounds capable of performing functions similar to those of proteins. Several potential applications of such synthetic oligomers can be conceived, such as their use as synthetic enzymes or as synthetic receptors for small molecules [5–11]. The development of new synthetic sequences enabling the solid-phase synthesis of non-natural, functionalized oligomeric compounds will, therefore, remain an active and challenging area of research in years to come.

16.1 Peptides

16.1.1 Merrifield's Peptide Synthesis

The first successful solid-phase synthesis of a peptide (H-Leu-Ala-Gly-Val-OH) was reported by Merrifield in 1963 [12]. The strategy used is outlined in Figure 16.1. The support was chloromethylated, partially nitrated (or brominated), 2% cross-linked polystyrene. Nitration was necessary to suppress acidolytic cleavage of the benzyl ester attachment during Z-group removal. Acylations of deprotected, support-bound amino acids were performed with DCC, and at the end of the synthesis the peptide was cleaved from the support by saponification with sodium hydroxide.

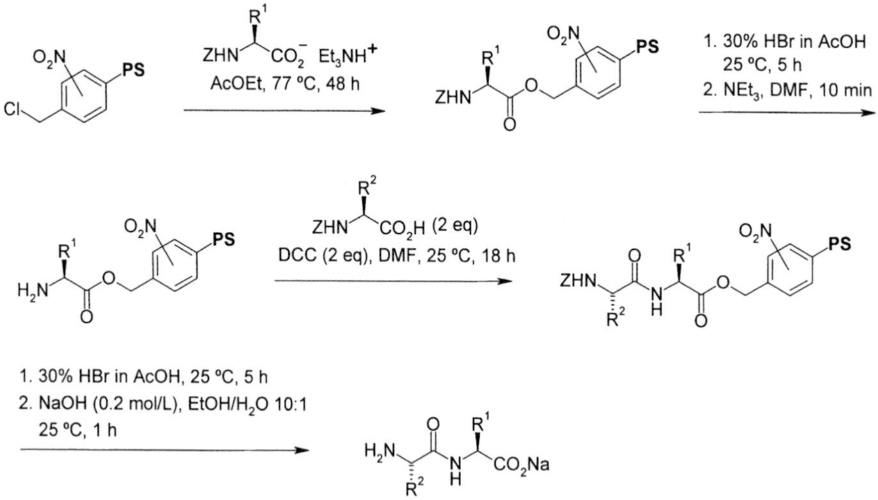

Figure 16.1. The first solid-phase peptide synthesis, developed by Merrifield [12].

This peptide synthesis, although hampered by some disadvantages, already featured most of the key elements of solid-phase peptide synthesis that have ultimately led to the development of the powerful synthetic protocols used today.

One critical issue is the direction in which the peptide is prepared. In Merrifield's strategy, the peptide is linked to the support via the carboxyl group, and coupling of monomers is effected by acylation of a support-bound amine. This strategy has two main advantages, which were critical for its success. Acylations of support-bound amines readily reach yields > 99.95% when a sufficiently activated acylating agent is used in excess. The opposite variant, i.e. the reaction of a support-bound acylating agent with an amine, cannot usually be performed with such high yields because strongly activated acylating agents will also undergo hydrolysis to some extent and less activated acylating agents will react only slowly and not to completion. Furthermore, acylating agents can be susceptible to intramolecular reactions leading to the formation of by-products (e.g. *N*-acylureas), which would accumulate on the support and reduce the yield and purity of the desired peptide. This small difference between yields might be of no consequence in the synthesis of short peptides or small non-pep-

tides, but for the preparation of large peptides quantitative coupling yields are abso-
lutely essential to keep the amount of deletion peptides (peptides missing one or
more amino acids) as low as possible.

A second advantage of preparing peptides by sequential acylation of support-
bound amines arises from the fact that activated *N*-acyl amino acids readily form oxazolones, which quickly racemize under basic conditions, such as in the presence of
excess amine. Hence, carboxyl group activation of support-bound peptides in the presence of an amine will readily lead to racemization (Figure 16.2).

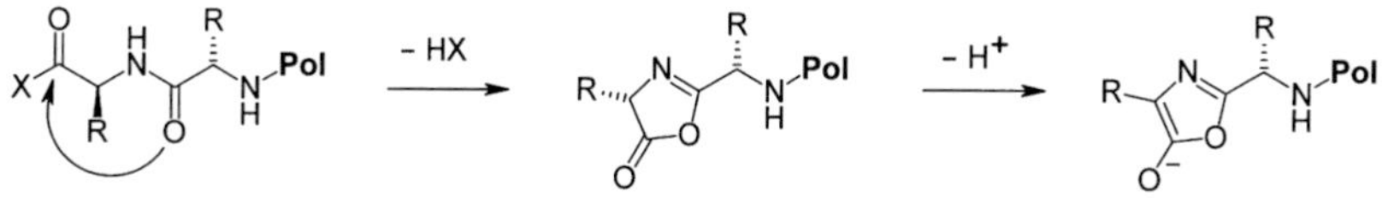

Figure 16.2. Racemization of activated *N*-acyl amino acids as a result of oxazolone formation. X: leaving group.

Activated *N*-alkoxycarbonyl amino acid derivatives, on the other hand, do not
cyclize as readily as *N*-acyl amino acids, and therefore racemize more slowly. Accordingly, solid-phase peptide synthesis is generally performed by acylation of support-
bound amines with activated *N*-alkoxycarbonyl amino acids. Examples of the preparation of peptides by the 'inverse' strategy (first amino acid linked to the support via
its amino group as carbamate; activation of support-bound *N*-acylamino acids) have,
nevertheless, been reported [13–16].

The first solid-phase peptide synthesis reported by Merrifield had some disadvantages, which were corrected in later versions of this synthesis. The main improvements
were the replacement of benzyloxycarbonyl protective groups by the TFA-labile Boc
protection, and the use of TFA-resistant linkers cleavable by HF.

16.1.2 The Boc Strategy

Figure 16.3 shows a representative example of the synthesis of peptides using the
Boc strategy. Hydroxymethyl polystyrene can be used as the support or, if large peptides are to be prepared, polystyrene with the more acid-resistant PAM linker. Polyacrylamides [17] with a benzyl alcohol linker are also compatible with the Boc strategy.

Attachment of the first amino acid is usually performed using a symmetric anhydride in the presence of DMAP, but mixed anhydrides or acid fluorides can also be
used (see Section 13.4.1). Alternatively, chloromethyl polystyrene can be treated with
the cesium salt of the first amino acid to yield the corresponding benzyl ester (Section
13.4.2). All proteinogenic Boc-protected α-amino acids linked to hydroxymethyl polystyrene or PAM resin are commercially available [18,19] and can be stored for long
periods without deterioration.

Boc-group removal usually requires treatment of the resin-bound carbamate with
50% TFA in DCM for 0.5 h (Section 10.1.10.1). Under these conditions, neither cleavage of the peptide from the support nor hydrolysis of side-chain protective groups

should occur. Loss of peptide by premature cleavage from the support can be minimized by the use of PAM resin, which is approximately 100 times more stable towards acidolysis than hydroxymethyl polystyrene [20]. *N*-Alkylated peptides are generally less stable towards acids than normal peptides, and the hydrolysis of *N*-alkyl peptides during Boc-group removal has been reported [21].

After deprotection, the amino group can be neutralized by washing with a tertiary amine. The main reason for this neutralization step is the removal of TFA, which could otherwise lead to the formation of trifluoroacetamides when a coupling reagent is added. However, it has been reported that the neutralization can be omitted if the coupling is performed in the presence of a base (e.g. RCO_2H (4 equiv.), TBTU (4 equiv.), DIPEA (2 equiv.), DMF, 20 min [22]). Surprisingly, the formation of trifluoroacetamides does not seem to occur under these conditions, despite the presence of TFA in the reaction mixture.

During neutralization of dipeptides linked to alcohol resins as esters, diketopiperazines can be formed by intramolecular nucleophilic cleavage (Section 15.22.1). Diketopiperazine formation during peptide synthesis can be avoided by acylating the first support-bound amino acid with a preformed N-protected dipeptide, or by using the 2-chlorotrityl linker, which cannot be readily cleaved by nucleophiles.

Standard coupling is generally conducted with active esters formed in situ, but for difficult couplings, other acylating agents, such as symmetric anhydrides or acid fluorides, longer reaction times, and twofold couplings might be required (Section 13.1).

Final cleavage of the peptide from the support requires HF or other acids of high ionizing power (TfOH, HBr/AcOH [18]). HF is a good solvent for amino acids and proteins, and does not usually lead to the cleavage of peptide bonds [23]. Because of its toxicity and capacity to dissolve glass, HF must be handled with great care in special HF-resistant containers.

Before cleavage with HF, those side-chain protective groups that will not be cleaved by HF must be removed (e.g. *N*-formyl groups from tryptophan, *N*-dinitrophenyl groups from histidine). The *N*-terminal Boc group is also usually removed before treatment with HF, to minimize the risk of alkylation of electron-rich function-

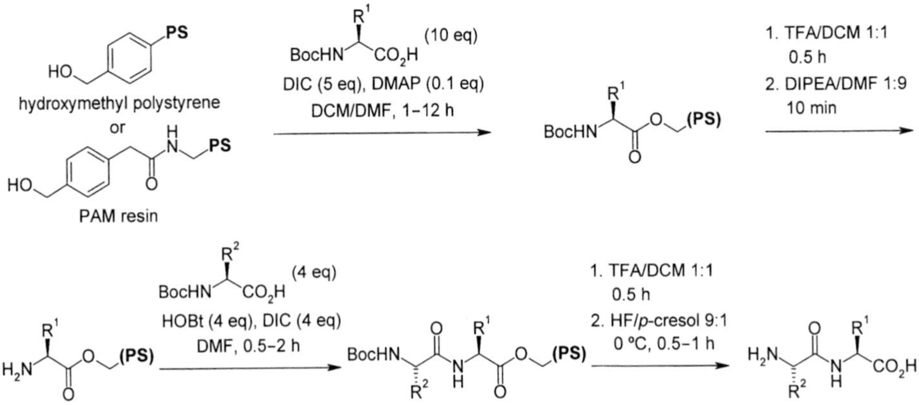

Figure 16.3. Solid-phase peptide synthesis using the Boc strategy [18,24]. All reactions, except the final cleavage, are performed at room temperature.

alities in the peptide by carbocations. Some of the side-chain protective groups will, however, be removed from the peptide during cleavage from the support. These include, for instance, benzyl esters of aspartic and glutamic acids, benzyl ethers of serine, threonine, and cysteine, and benzyloxycarbonyl groups from lysine, tyrosine, or histidine [18]. Because the resulting carbocations can readily and irreversibly alkylate the peptide at susceptible positions (e.g. at tyrosine or tryptophan), the use of scavengers is required to trap carbocations during cleavage from the support. Anisole, dimethyl sulfide, EDT, thiocresol, triethylsilane, or other compounds that react faster with alkylating agents than the peptide can be used as scavengers.

The Boc strategy has been successfully used for the preparation of peptides containing up to about 140 amino acids. Peptides of this size or beyond are, however, more conveniently prepared by the condensation of protected fragments in solution [25,26] or on solid phase [27,28]. One particular advantage of the Boc strategy, compared with the Fmoc strategy (see below), is that treatment with TFA after each coupling leads to the destruction of β-sheets, which can readily form during solid-phase peptide synthesis, and which often cause coupling difficulties [22]. One of the main disadvantages of the Boc strategy is that coupling efficiencies cannot be monitored as readily as in peptide synthesis with Fmoc protection. The repeated treatment with TFA also limits the choice of special amino acids that can be incorporated into the peptide. For instance, glycopeptides containing O-glycosylated serine or threonine would be difficult to prepare using the Boc strategy because of the acid-lability of glycosidic bonds. Most importantly, the Boc strategy requires the handling of HF, which implies the availability of trained staff and special equipment. Because such equipment is not always readily available, many chemists choose the Fmoc strategy rather than the Boc strategy.

16.1.3 The Fmoc Strategy

In the 1970s, milder methods for the solid-phase synthesis of peptides were developed. To avoid repetitive treatment of the growing peptide with strong acids, which could lead to premature release of the peptide from the support and to cleavage of sensitive amide bonds (e.g. in *N*-alkyl peptides), the use of base-labile N-protection was investigated. The Fmoc group, developed by Carpino in 1972 [29], proved particularly suitable because of the mild conditions required for its removal and because of the formation of a readily detectable by-product [9-(1-piperidinylmethyl)fluorene], which enables the automated monitoring of each deprotection step (for the mechanism of Fmoc deprotection, see Figure 10.10). A convenient solid-phase protocol for peptide synthesis based on Fmoc protection was developed by Atherton and Sheppard [30], and has become the most commonly used strategy for synthesizing peptides [31] (Figure 16.4).

Because no treatment with acid is required during peptide assembly, peptide synthesis with Fmoc amino acids can be conducted on acid-sensitive supports (e.g. Tentagel) and with acid-labile linkers. Wang resin is suitable for most purposes, but other supports, such as Sasrin or 2-chlorotrityl resin, can also be used. CPG, macroporous polystyrene, and polyacrylamides are also compatible with Fmoc peptide synthesis [32]. Premature cleavage of the peptide from Wang resin during Fmoc removal (nucleophilic cleavage with piperidine) is negligible.

As for Boc amino acids, all proteinogenic and several non-natural Fmoc amino acids esterified with Wang or similar resins are commercially available. Deprotection with 20% piperidine in DMF is usually complete within a few minutes, and the release of the fluorene derivative is easily monitored spectrophotometrically at 300–320 nm (see, e.g., [33]). The peak area can be used to determine the amount of chromophore released, which is proportional to the efficiency of the preceding coupling reaction. At the end of an automated peptide synthesis, inspection of the chromatogram of all Fmoc releases enables rapid assessment of the quality of the resin-bound peptide and quick location of positions in the peptide where coupling was unsuccessful or difficult.

During acylations with Fmoc-protected amino acids, addition of bases should be avoided as these could lead to partial deprotection and thence to multiple incorporation of the amino acid. Small amounts of DIPEA or pyridine, however, do not usually cause major problems (see Experimental Procedure 13.5).

The conditions used for final cleavage of the peptide from the support depend on the type of support and linker chosen. Because side-chain protective groups must also be removed, the use of scavengers is required during cleavage to avoid alkylation of the peptide.

Peptides with up to 166 amino acids (aglycon of erythropoietin [34]) have been prepared using the Fmoc strategy on cross-linked polystyrene. The preparation of peptides containing several N-methylamino acids (e.g. cyclosporine [35,36]) has also been successfully performed using Fmoc protection. However, such difficult peptides require careful planning of the synthesis and often the use of special coupling techniques.

The release of a readily detectable compound during each deprotection and the absence of corrosive reagents during peptide assembly makes the Fmoc strategy particularly well suited for automation. A typical programmable, fully automated peptide synthesizer can perform one complete coupling and deprotection cycle in approximately 1 h (20–40 min coupling, 10 min Fmoc removal, 10 min washing). Cleavage of

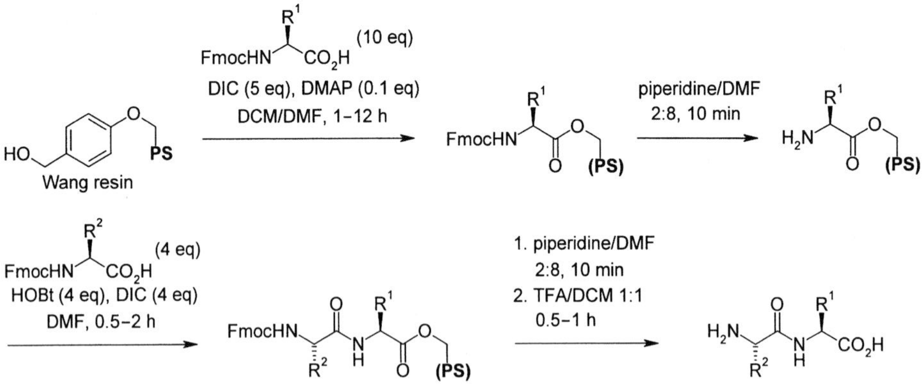

Figure 16.4. Solid-phase peptide synthesis using the Fmoc strategy.

the peptide from the support is generally not performed on the synthesizer but manually in a hood.

The mild reaction conditions during peptide synthesis using the Fmoc strategy also enable the use of sensitive amino acid derivatives, such as glycosylated serine, threonine, and asparagine (Figure 16.5), in the solid-phase synthesis of glycopeptides [37–43]. Although glycopeptides can be prepared by means of the Boc strategy [38], the harsh conditions used for Boc group removal can lead to the hydrolysis of glycosidic bonds, and the Fmoc strategy will usually be better suited for glycopeptide synthesis.

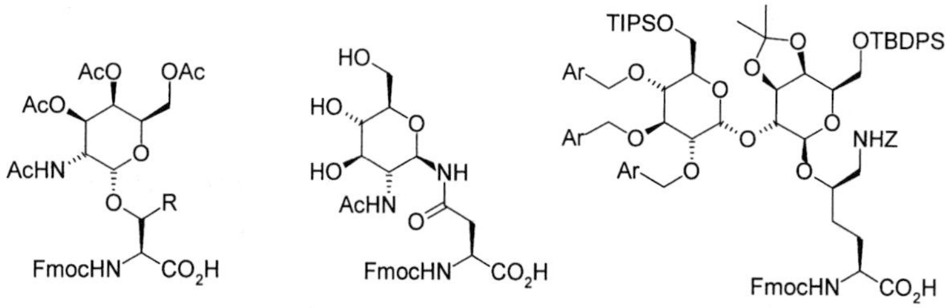

Figure 16.5. Typical protected glycosylated amino acid derivatives suitable for the solid-phase synthesis of glycopeptides [40,44–49]. R: H, Me; Ar: 4-methoxyphenyl.

16.1.4 Side-Chain Protection

Some amino acids contain functional groups that need to be masked during peptide synthesis. Protective groups suitable for the Boc strategy must be resistant towards TFA, whereas for the Fmoc strategy stability towards piperidine must be ensured.

Table 16.1. Recommended side-chain protective groups for solid-phase peptide synthesis [18].

Amino acid	Boc strategy	Fmoc strategy
Arg	Ts, Mts	Mtr, Pmc, Pbf
Asn	-	Tr
Asp	Bn, Cy	*t*-Bu
Cys	4-MeBn, Acm	Tr, Acm
Gln	-	Tr
Glu	Bn	*t*-Bu
His	Bom, Dnp	Tr
Lys	2-Cl-Z	Boc
Ser	Bn	*t*-Bu
Thr	Bn	*t*-Bu
Trp	formyl	Boc
Tyr	2-Br-Z	*t*-Bu

Choosing the right side-chain protection for solid-phase peptide synthesis can be a difficult task, in particular if large peptides are to be prepared. Many factors play a role, and careful planning of the synthesis and deprotection is usually required. Details concerning the compatibility of various protective groups and strategies for their removal can be found in more specialized literature [18,24,31,50]. Table 16.1 lists some of the most commonly used side-chain protective groups for solid-phase peptide synthesis, which are suitable for most applications.

16.1.5 Backbone Protection

It has often been observed that during solid-phase synthesis the growing peptide forms aggregates with the support, with other peptides, or with itself. These aggregates can, for instance, be β-sheets or other secondary structures stabilized by hydrogen bonds between amido groups RCONHR [18,51]. The formation of such aggregates occurs particularly readily in peptides containing an uninterrupted sequence of amino acids with short, lipophilic side chains (Ala, Val, Leu, etc.), and usually significantly reduces the rate of couplings and deprotections. The formation of such aggregates can be partly prevented by the use of dipolar aprotic solvents instead of DCM during N-acylation, by adding salts (LiBr, LiCl, NaClO$_4$, KSCN) or protic solvents (e.g. hexa-fluoroisopropanol), by the use of polar supports (e.g. polyacrylamide), or by perform-ing the synthesis using the Boc strategy [18,22]. Most elegantly, hydrogen-bond forma-tion may be precluded by introducing an *N*-alkylamino acid at certain positions in the peptide. This disrupts secondary structures and keeps the growing peptide well sol-vated and accessible to reagents. One type of backbone protection for peptides, suit-able for peptide synthesis using the Fmoc strategy, is Hmb protection (2-*hydroxy-4-methoxybenzyl*; Figure 16.6 [51–54]). The requisite N-alkylated amino acids are avail-able commercially [18], and their incorporation at every sixth to eighth position is usually sufficient to prevent aggregation of the peptide [55]. Alternatively, support-bound amino acids can be Hmb-protected by reductive alkylation [56]. Other related backbone-protecting groups have been described, such as the 2-hydroxy-4-methoxy-5-(methylthio)benzyl group [57].

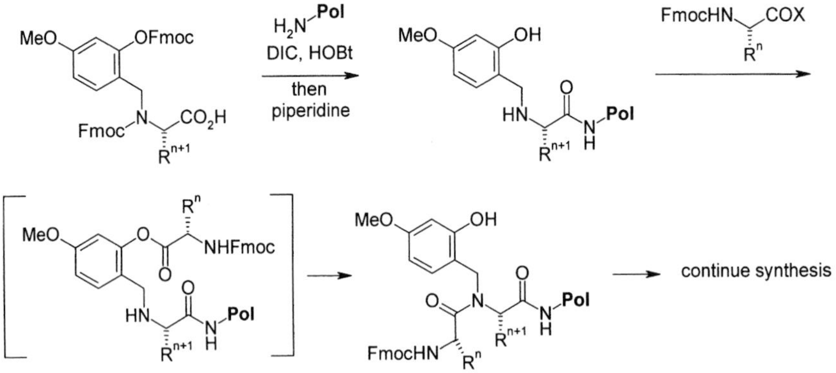

Figure 16.6. Incorporation and acylation of *N*-(Hmb) amino acids during solid-phase peptide synthesis.

The acylation of support-bound *N*-benzylamino acids is difficult and requires symmetric anhydrides or similarly activated amino acids [54]. (2-Hydroxybenzyl)amino acids are probably acylated by intramolecular *N,O*-acyl migration (Figure 16.6). After the introduction of Hmb protection, the free phenolic hydroxyl group is not acylated unless highly reactive acylating agents are used. The resulting esters are, however, unstable and will be cleaved during Fmoc-group removal. Deprotection of the peptide is achieved during cleavage from the support by treatment with TFA and a scavenger.

16.1.6 Cyclic Peptides

Cyclic peptides are of interest as turn mimetics or as conformationally constrained peptide analogs. The cyclization of peptides can either be performed after cleavage from the support in solution (see, e.g. [58]) or while still on the support. Besides cyclizations of the peptide backbone, peptides can be cyclized through their side chains or by means of a non-natural spacer. To cyclize the backbone of a peptide on an insoluble support, the peptide must be attached to the support either by a backbone amide linker [59] or via the side chain of an amino acid [60] (Figure 16.7).

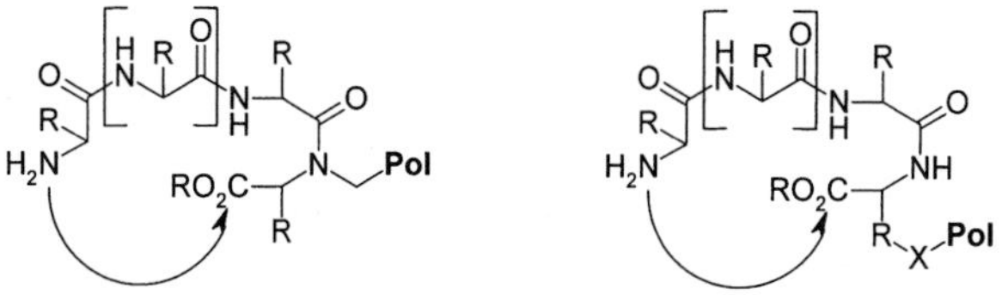

Figure 16.7. Strategies for the backbone cyclization of peptides on solid phase without simultaneous cleavage from the support.

Peptides with more than seven amino acids can usually be cyclized without problems, but smaller cyclopeptides can be difficult to prepare. Few examples of the preparation of cyclotetrapeptides have been reported [61,62]. The preparation of cyclopentapeptides proceeds satisfactorily in solution with HOAt-derived coupling reagents [63,64].

As shown in Figure 16.7, when using backbone attachment, the first amino acid is usually protected as an ester, which needs to be saponified before cyclization. The use of allyl esters enables palladium-mediated hydrolysis under mild reaction conditions [59]. During the cyclization, basic reaction conditions should be avoided to prevent racemization of the activated *N*-acylamino acid. Because such racemization can readily occur (see Figure 16.2), it is advisable to use coupling reagents with a low tendency to induce racemization. These include HOAt-derived reagents [63–66], DPPA, and PyBOP/HOBt [67].

Peptides can also be cyclized through their side-chain functionalities (e.g. using lysine, cysteine, glutamic acid, or aspartic acid [68–71]), or with the aid of synthetic spacers ([72]; see also Table 8.5). Examples have been reported of the preparation of cyclic peptides using safety-catch linkers (e.g. (4-alkylmercapto)phenol [62], sulfonamide [73], or hydrazide [74]), 2-nitrophenyl ester attachment [61], or oxime ester attachment [75–78], in which cyclization and release of the cyclopeptide by intramolecular nucleophilic cleavage occurred spontaneously after removal of the terminal amine protection.

Amino acids suitable for side-chain attachment to supports that have been used to prepare cyclic peptides on insoluble supports include tyrosine (etherified with Wang resin [60]), histidine (N-alkylated with trityl resin [79]), aspartic acid [66,67,80], asparagine [80], glutamic acid [80,81], lysine [65], and serine [65].

The main side reaction that can occur during the cyclization of peptides on insoluble supports is their oligomerization. This competing process can only be prevented by choosing a support with a lower loading and thereby reducing the concentration of the peptide. The use of *N*-alkylamino acids (proline, sarcosine, Hmb-protected amino acids [63]) will increase the flexibility of the peptide by reducing the energy barrier to amide bond rotation, and thereby facilitate cyclization.

Analogs of cyclic peptides have been prepared using a variety of spacers and reaction conditions to achieve the macrocyclization. These include olefin metathesis (Section 5.2.3), nucleophilic substitution (Sections 8.2 and 7.2.3), and the Heck reaction [82].

16.2 Oligonucleotides

The synthesis of oligonucleotides and analogs thereof on insoluble supports has become an extensive area of research, and detailed treatment would far exceed the scope of this book. In the following sections, only the most important features of oligonucleotide synthesis will be discussed. For further details, more specialized sources should be consulted [83]. Review articles have appeared, in which supports for the synthesis of oligonucleotides [84], and the uses of synthetic oligonucleotides in gene technology [85] and in drug discovery [86,87] are discussed.

16.2.1 Historical Overview

Unlike solid-phase peptide synthesis, the synthesis of oligonucleotides on supports was complicated by numerous severe problems, and the development of useful synthetic protocols took approximately 15 years. The first attempts to prepare oligodeoxynucleotides on insoluble supports were reported by Letsinger in 1965 [88] (Figure 16.8). He chose cross-linked polystyrene as the support, and immobilized the first nucleoside via its heterocyclic base. Phosphorylation was achieved by coupling cyanoethyl phosphate to the 5'-hydroxyl group of the resin-bound nucleoside with the aid of DCC. The resulting phosphodiester was then activated by treatment with mesitylenesulfonyl chloride and coupled with a second nucleoside. Cleavage from the support was realized by saponification, whereby simultaneous removal of the cyanoethyl protective group occurred (by β-elimination).

Several groups began to work on improving Letsinger's approach, not only by modifying the protective group strategy, but also by choosing other attachment sites and

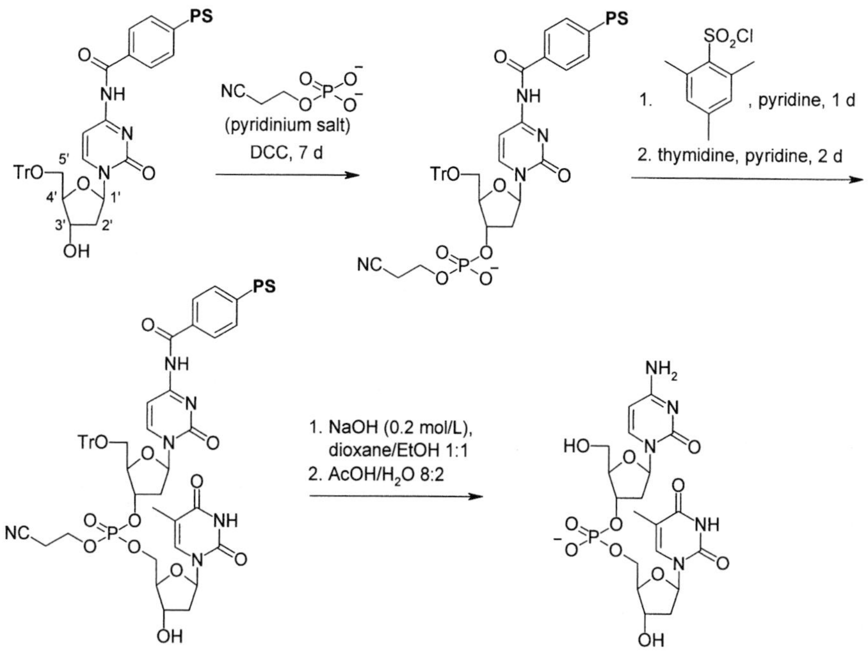

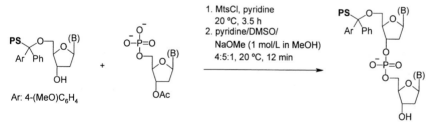

Figure 16.8. The first solid-phase synthesis of a dinucleotide [88].

by testing different supports. In 1966, two groups presented a synthetic strategy [89–91] in which the first nucleoside was attached to non-cross-linked (soluble) polystyrene by a methoxytrityl linker via its 5′-hydroxyl group. The resulting support-bound nucleoside was then coupled with 3′-O-acetylthymidine-5′-phosphate in pyridine with mesitylenesulfonyl chloride as the coupling agent, and, after saponification of the 3′-acetoxy group, the 3′-hydroxyl group became available for a new coupling (Figure 16.9).

Figure 16.9. Synthesis of oligonucleotides on non-cross-linked polystyrene using nucleoside 5′-phosphates [91].

This so-called phosphodiester method [92] was later also performed on cross-linked polystyrene [93], macroporous 25% cross-linked polystyrene [94], silica [95], poly(-vinyl alcohol) [96,97], proteins [98], PEG [99], and polyacrylamide [100]. This extensive evaluation of different supports stemmed from the observation that cross-linked polystyrene or other gel-type supports were not ideal for the preparation of phosphoric esters [93]. The highly polar phosphate backbone of DNA seemed to be incompatible with hydrophobic supports. Moreover, diffusion of activated phosphates into gel-type supports seemed to be too slow to enable the efficient and high-yielding coupling of nucleotides. Significantly higher reaction rates could be attained on macroporous supports and on silica, on which the growing oligonucleotide is only located on the surface of the support material. Such non-swellable supports are compatible with a broader choice of solvents than gel-type supports, and are also better suited to continuous-flow synthesis because the volume of the support material remains constant throughout the synthesis.

16.2.2 The Phosphotriester Method

In the 1970s, a significant modification of previous oligonucleotide syntheses was implemented by several groups [101–103]; this later became known as the phosphotriester method (Figure 16.10; for a review, see [104]). In this new strategy, not a simple nucleoside phosphate but a phosphodiester was coupled with the 5′-hydroxyl group of the support-bound nucleoside to yield a non-charged phosphotriester (as in Letsinger's approach, Figure 16.8). The lower polarity of the phosphotriester, as compared with a phosphodiester, made the triester method more appropriate for hydrophobic supports such as polystyrene [92], and enabled the preparation of larger oligonucleotides.

Suitable alcohols for phosphate protection include chlorophenols [102,103,105] and 2,2,2-trichloroethanol [101], which enable selective, nucleophilic deprotection by treatment with NaOH, without destruction of the oligonucleotide. Thiophenol has also been used as an (oxidant-labile) protective group for the phosphate backbone [106]. Supports used for the phosphotriester approach include cross-linked polystyrene [105], kieselguhr–polyamide [107], cellulose [107], polystyrene grafted on PTFE [108], and LCAA-CPG [109]. In most examples of the triester method, the first nucleoside was no longer attached to the support through its 5′-hydroxyl group but through its 3′-hydroxyl group as a hemisuccinate (Figure 16.10).

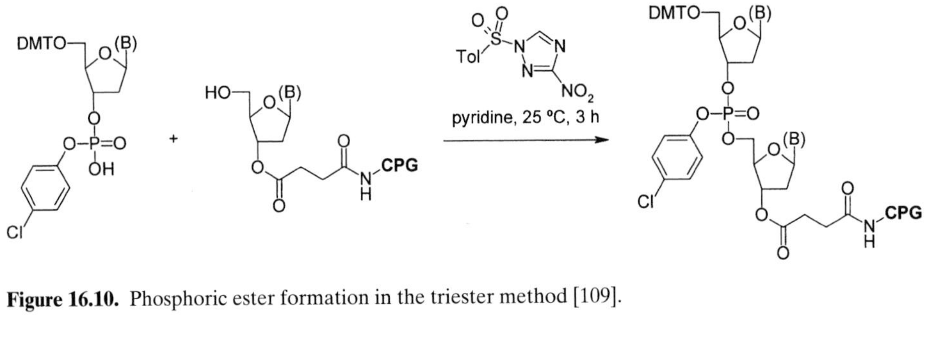

Figure 16.10. Phosphoric ester formation in the triester method [109].

16.2.3 The Phosphoramidite and *H*-Phosphonate Methods

In 1975, a further significant improvement in the synthesis of oligonucleotides was reported by Letsinger [110]. He found that the formation of phosphites from phosphinic chlorides (RO)PCl$_2$ or (RO)$_2$PCl proceeds much more rapidly than the formation of phosphates from chlorophosphates, and that the oxidation of phosphites to phosphates by iodine also proceeds quickly and under mild reaction conditions compatible with protected oligonucleotides. This new strategy, i.e. phosphite formation followed by oxidation, was later successfully used on silica for the preparation of oligoribonucleotides [111] and oligodeoxyribonucleotides (Figure 16.11). The required phosphinic chlorides were prepared shortly prior to use by treatment of 5′-DMT-protected nucleosides with methyl dichlorophosphite. These activated nucleosides react quickly with the support-bound nucleoside under mild conditions to yield a phosphite, which must be oxidized to the more stable trialkyl phosphate after each coupling. After capping with phenyl isocyanate (to derivatize all non-phosphorylated 5′-hydroxyl groups), the newly introduced nucleotide is detritylated by brief treatment with an

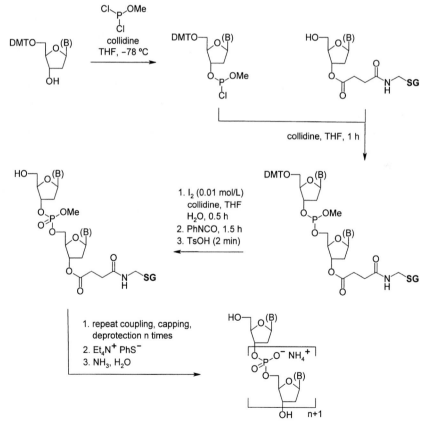

Figure 16.11. Early version of oligonucleotide synthesis involving the intermediate formation of phosphites [112,113].

acid. At the end of a synthesis, the methyl phosphates are demethylated by treatment with thiophenolate, and the oligonucleotide is finally cleaved from the support by treatment with ammonia.

It was later found that dialkyl(dialkylamino) phosphites (phosphoramidites, $(RO)_2PNR_2$ [114]), which are stable towards air and water and can be stored for longer, can readily be converted into the phosphinic chlorides $(RO)_2PCl$ by treatment with dimethylaniline hydrochloride, or into the corresponding tetrazolides by treatment with tetrazole. Tetrazolides $(RO)_2P-(1\text{-tetrazolyl})$ had proven excellent reagents for the phosphorylation of nucleosides [113], and the treatment of phosphoramidites with alcohols in the presence of tetrazole was found to be a satisfactory method for the rapid preparation of trialkyl phosphites [115,116].

A typical solid-phase synthesis of oligodeoxyribonucleotides from phosphoramidites, as usually performed on a modern synthesizer, is shown in Figure 16.12 [83,117]. The synthesis is conducted under continuous flow on non-swellable support materials, such as CPG, macroporous polystyrene [118], polymethacrylate [119], or polystyrene grafted on PTFE [120,121]. Larger amounts of oligonucleotide can be prepared on PEG–polystyrene graft supports [122–125] or on soluble supports [126–128]. Because of the high price and low loading of LCAA-CPG, reusable supports of this type are being developed [129].

The most remarkable aspect of the phosphoramidite approach is its speed: taking into account the time required for washing (30 s with acetonitrile after each step), it takes only approximately 5 min to couple each nucleotide onto the growing oligonucleotide. Cyanoethyl groups are usually used as protection of the phosphate backbone; these groups are removed with concentrated aqueous ammonia during or after cleavage of the oligonucleotide from the support. Some of the heterocyclic bases (adenine, guanine, cytosine) contain primary amino groups; these are generally also protected to avoid problems during solid-phase assembly of oligonucleotides. For this purpose, the amino groups in the monomers are acylated with benzoic or isobutyric acid. The resulting amides are cleaved by ammonolysis during final cleavage of the product from the support. It has recently been suggested [130] that base protection of nucleosides is not required if imidazolium triflate is used as a promoter for phosphite formation.

The synthesis sketched in Figure 16.12 can be used to prepare oligodeoxynucleotides of up to about 150 bases, and examples of the preparation of cyclic oligodeoxynucleotides [122] and oligoribonucleotides [131] using this technique have also been reported.

In a closely related strategy, nucleoside-3'-H-phosphonates $RO-P(OH)(H)=O$, which are tautomers of monoalkyl phosphites $RO-P(OH)_2$, are used as the monomers and are activated with coupling agents such as HOBt-derived phosphonium salts [132] or with pivaloyl or 1-adamantoyl chloride [125] (Figure 16.13; for the preparation of H-phosphonates, see Figure 16.28). The resulting H-phosphonate diesters are more stable than trialkyl phosphites, and their oxidation to the phosphates with iodine only needs to be performed once at the end of the synthesis. Because H-phosphonates are readily converted into phosphoric acid derivatives other than phosphates, they represent valuable intermediates for the preparation of modified DNA analogs, such as

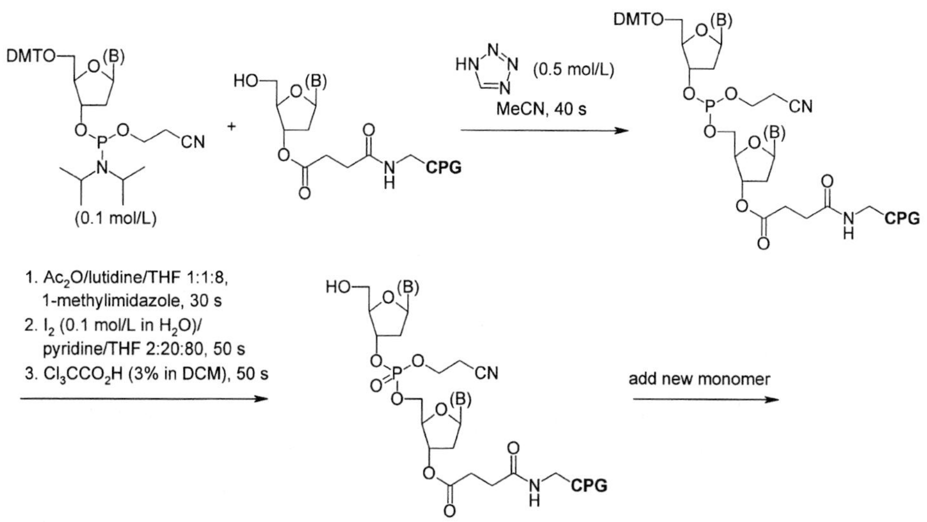

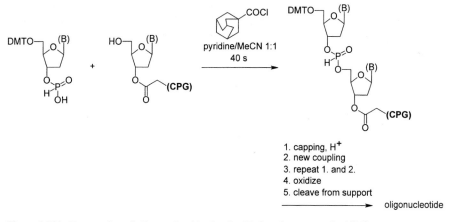

Figure 16.12. Solid-phase oligonucleotide synthesis using phosphoramidites [83].

phosphorothioates [83], phosphoramidates [133], and boranophosphates [134] (see Section 16.2.4). Furthermore, the phosphonate method is cheaper than the amidite method [125], and should be considered if large amounts of oligonucleotides are to be prepared. Coupling yields with activated *H*-phosphonates are, however, usually lower (approx. 95%) than those with phosphoramidites [83]. The *H*-phosphonate method has also been used to phosphorylate support-bound peptides containing serine or tyrosine [135].

Figure 16.13. Preparation of oligonucleotides by the *H*-phosphonate method [83].

Oligoribonucleotides are more difficult to handle than oligodeoxyribonucleotides, mainly because of their sensitivity towards nucleases. For the synthesis of oligoribonucleotides on insoluble supports, protection of the 2′-hydroxyl group is required, which must be stable towards oxidation with iodine and the acidic conditions of DMT deprotection, but which should be easy to remove under mild conditions during or after cleavage of the oligonucleotide from the support. Of the many protective groups evaluated, the TBS group seems to fulfil these requirements, and oligoribonucleotides with up to 30–40 bases can be prepared on supports by the use of TBS-protected ribonucleotides according to the phosphoramidite method [83,136]. Photosensitive protective groups (e.g. 2-nitrobenzyloxymethyl [137]) have also been evaluated.

16.2.4 Oligonucleotide Analogs

Oligonucleotides and analogs thereof can interact with proteins, DNA, and RNA in a highly specific manner, and might therefore be useful for the treatment of various human diseases [83,86,87,138,139]. Several different approaches have been described for the modification of oligonucleotides with the aim of modulating their biological activity and pharmacokinetic properties. One important modification of oligonucleotides is the replacement of phosphates by phosphorodithioates $(RO)_2P(S)S^-$ [83,140] and phosphorothioates $(RO)_2P(O)S^-$ (Figure 16.14 [141–145]), producing analogs with increased stability towards enzymatic and chemical degradation. Other examples of the backbone modification of oligonucleotides, compatible with solid-phase synthesis, are the replacement of phosphates by phosphonates [83,146], phosphoramidates [133,147,148], boranophosphates [134,149], guanidines [150,151], ureas [152], thioureas [153], and hydroxylamines [154] (Figure 16.14).

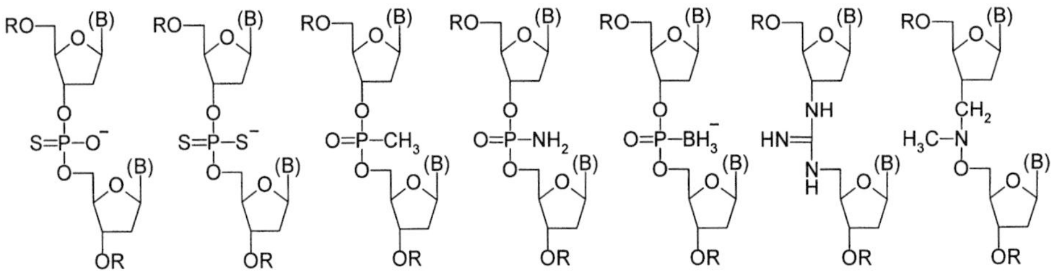

Figure 16.14. Backbone-modified DNA analogs that can be prepared on insoluble supports.

Other known oligonucleotide analogs include oligomers in which the heterocyclic bases have been modified [155] or replaced by other types of compound [156]. Solid-phase syntheses of hybrids of DNA with peptides [157–163], with carbohydrates [164,165], and with PNA (peptide nucleic acids, see Section 16.4.1.2 [166–168]) have also been reported, and several strategies have been developed that enable the preparation of oligonucleotides with a modified 5′- or 3′-terminus [169–174] or of cyclic oligonucleotides [122,175]. Various techniques for the parallel solid-phase synthesis of oligonucleotides have been developed for the preparation of compound libraries [176–181].

16.3 Oligosaccharides

The development of strategies for the solid-phase synthesis of oligosaccharides received little attention in the 1960s and 1970s, compared with the synthesis of peptides and oligonucleotides. One reason for this might have been the inherent difficulty of oligosaccharide synthesis, because, unlike peptide bond formation or phosphodiester formation, glycosylation can lead to mixtures of diastereomers. In recent years, however, powerful methods have been developed that enable the realization of highly stereoselective glycosidation on supports [38,182,183].

Two different strategies can be envisaged for the preparation of glycosides on insoluble supports (Figure 16.15). Support-bound, partially protected carbohydrates can be glycosylated by the addition of a suitable glycosyl donor and a promoter. Alternatively, a glycosyl donor can be generated on solid phase and coupled with a partially protected carbohydrate. Since most of the reported examples have been based on the first strategy, involving the glycosidation of support-bound alcohols, it would seem that this protocol is better suited to the preparation of oligosaccharides. The second strategy has mainly been realized by Danishefsky and co-workers, using glycals as latent support-bound glycosyl donors (see below).

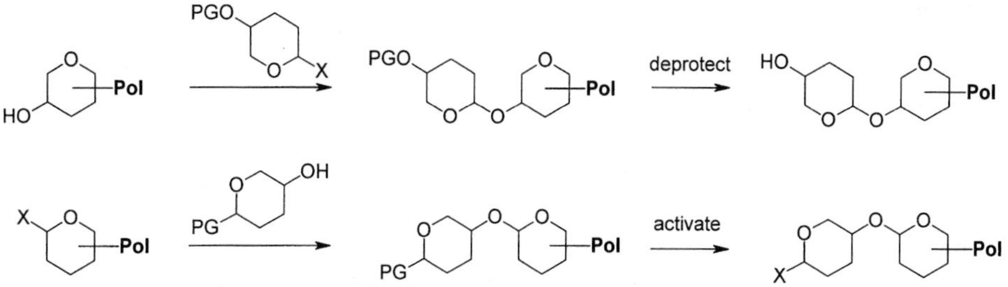

Figure 16.15. Strategies for the assembly of oligosaccharides on insoluble supports. PG: protective group; X: leaving group.

Oligosaccharides have been successfully synthesized on various types of support, including gel-type, hydrophobic or hydrophilic supports, and non-swellable hydrophobic or hydrophilic supports.

16.3.1 Glycosylation with Glycosyl Bromides

In early attempts directed towards the solid-phase synthesis of oligosaccharides, glycosidic bonds were formed by treating pyranosyl bromides with partially protected carbohydrates (Figure 16.16). These syntheses could be performed with either the pyranosyl bromide (glycosyl donor) [184,185] or the alcohol (glycosyl acceptor) [186–188] linked to the support.

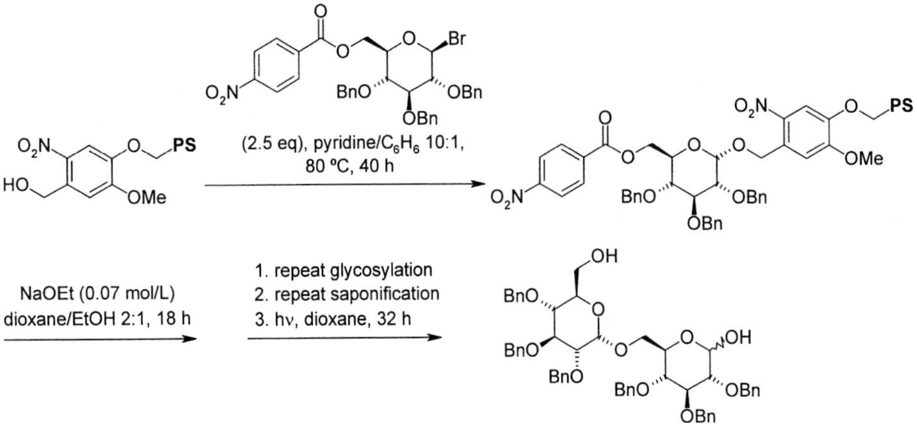

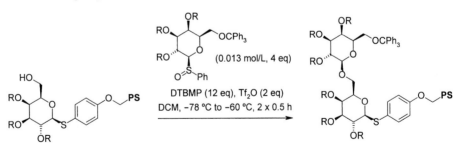

Figure 16.16. Early solid-phase synthesis of oligosaccharides using glycosyl bromides [187].

16.3.2 Glycosylation with Glycosyl Sulfoxides

More efficient methods for the preparation of glycosides were subsequently developed and were successfully performed on insoluble supports. One such method was based on the use of glycosyl sulfoxides, which react with alcohols upon treatment with Tf₂O at low temperatures to yield the corresponding glycosides with high stereoselectivity [189–191]. Using this strategy, di- and trisaccharides have been prepared on cross-linked polystyrene using a thioacetal-based linkage (Figure 16.17). Cleavage from the support was accomplished by treatment with Hg(O₂CCF₃)₂/DCM/H₂O [189]. This strategy has also been applied to the solid-phase synthesis of oligosaccharide libraries [192].

Figure 16.17. Glycosylation of polystyrene-bound alcohols with pyranosyl sulfoxides [189,193]. R: pivaloyl.

16.3.3 Glycosylation with Glycosyl Thioethers

Glycosyl thioethers can be converted into highly reactive glycosylating agents by S-methylation with MeOTf or [Me$_2$SSMe$^+$][OTf$^-$] [194,195]. Glycosylations with pyranosyl thioethers have been performed with either the glycosyl donor [196,197] or the alcohol bound to the support [194,198–200]. Illustrative examples of this type of glycosylation are sketched in Figures 16.18 and 16.20. The glycosylation of support-bound alcohols with glycosyl thioethers is one of the most efficient and flexible strategies for the preparation of oligosaccharides, and even branched heptasaccharides can be synthesized on polystyrene using this methodology [194].

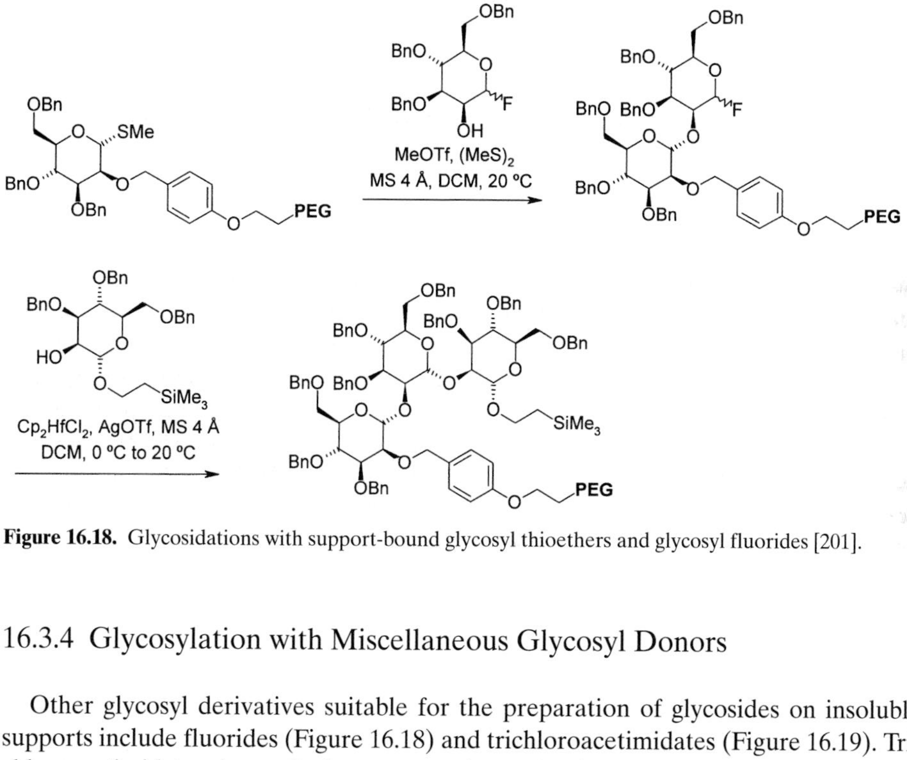

Figure 16.18. Glycosidations with support-bound glycosyl thioethers and glycosyl fluorides [201].

16.3.4 Glycosylation with Miscellaneous Glycosyl Donors

Other glycosyl derivatives suitable for the preparation of glycosides on insoluble supports include fluorides (Figure 16.18) and trichloroacetimidates (Figure 16.19). Trichloroacetimidates, in particular, seem to be convenient reagents for glycosidations on solid phase, and numerous examples of their use have been reported [202–205]. Glycosyl trichloroacetimidates also react with support-bound thiols under Lewis acid catalysis to yield glycosyl thioethers. Such thioethers can be used as linkers for carbohydrates. Oxidative cleavage with NBS in the presence of water leads to the release of the carbohydrate [206,207]. Cross-linked polystyrene [195,207,208], PEGA [209], CPG [164,206,210], polystyrene–PEG graft polymers [211], and PEG [212,213] have all been successfully used as supports for glycosylation with trichloroacetimidates.

The use of glycals as monomers for the solid-phase synthesis of oligosaccharides has been explored by Danishefsky (Figure 16.20; [183,215]). Glycals can be converted into glycosyl donors by epoxidation or haloamination, which usually proceed with

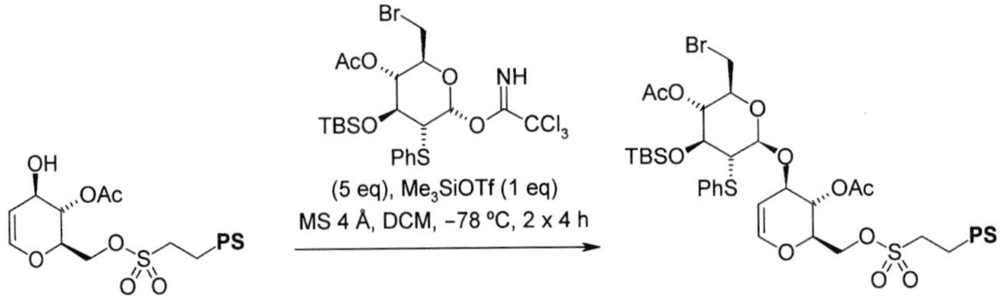

Figure 16.19. Glycosidation of support-bound alcohols with trichloroacetimidates [214].

high diastereoselectivity. The ensuing Lewis acid mediated glycosylations generally proceed in a stereochemically unambiguous manner.

Further glycosyl donors for the solid-phase synthesis of oligosaccharides include glycosyldisulfides [218], dinitrophenyl ethers [219], pent-4-en-1-yl glycosides [220] (activation by NIS), and phosphates [202,221]. Some types of support, such as CPG, PEGA, or soluble supports, enable the use of enzymes to promote various chemical transformations. Enzymes are also well suited to the regio- and stereoselective preparation of glycosides from unprotected carbohydrates, and several examples of enzyme-mediated preparations of glycosides on both insoluble [222–224] and soluble supports [225] have been reported.

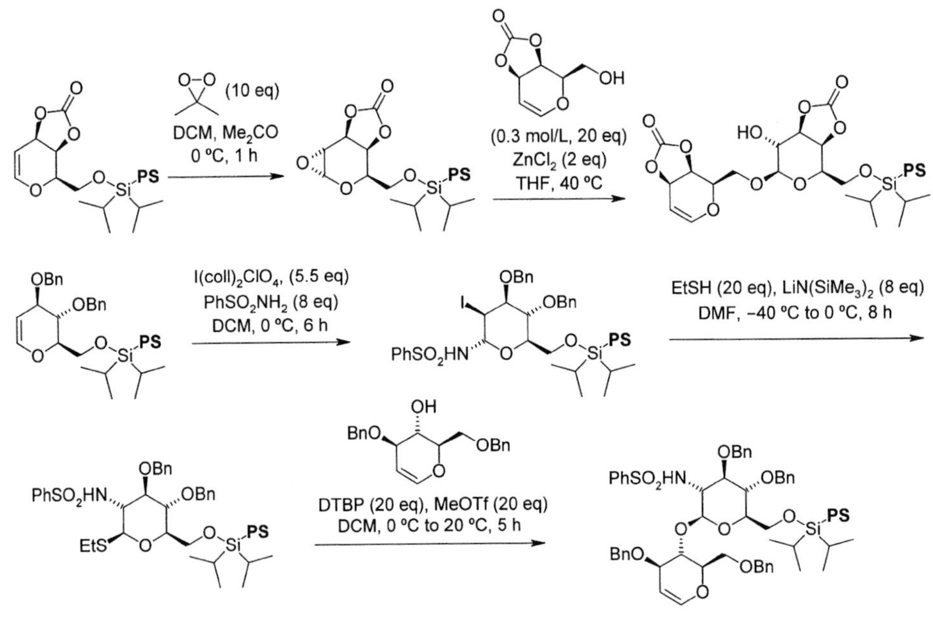

Figure 16.20. Solid-phase synthesis of oligosaccharides from glycals [216,217].

16.4 Miscellaneous Oligomeric Compounds

Several other types of compound have been prepared on insoluble supports by sequential addition of monomers to a growing chain [226,227]. Some of the most important oligomers of this kind are discussed in the following sections.

16.4.1 Oligoamides

Because amides can readily be prepared on solid phase, the synthesis of oligomeric amides other than peptides has received much attention in recent years, in particular with regard to the production of compound libraries. Several protected non-natural amino acids have been prepared and used for the synthesis of oligomers by means of the standard procedures of solid-phase peptide synthesis. A selection of such artificial amino acids is given in Table 16.2. Oligoamides have also been prepared by sequential coupling of diamines with diacids on solid phase [228]. Figure 10.14 depicts a strategy for the preparation of acylated oligo(lysine) derivatives.

Table 16.2. Synthetic amino acids or precursors thereof, suitable for the solid-phase preparation of oligoamides.

Entry	Monomer	Application	Reference
1	(Boc)FmocHN ... CO$_2$H	hybrids with peptides as sequence-specific DNA ligands	[229–232]
2	BocHN ... CO$_2$H	hybrids with peptides as sequence-specific DNA ligands	[231]
3	FmocHN, CO$_2$H R^{1}CO–N R^2 H	peptide mimetics	[233,234]
4	N$_3$ ... CO$_2$H AcO OAc		[235]
5	R FmocN ... CO$_2$H	peptide mimetics	[33]
6	OH HO ... CO$_2$H FmocHN O OBn	oligosaccharide mimetics	[236]
7	Boc N ... CO$_2$H OH	oligosaccharide mimetics	[237]

16.4.1.1 Oligoglycines

One type of oligoamide that can readily be prepared on supports without the need for any partially protected monomers (which are often tedious and expensive to synthesize) are N-substituted oligoglycines (Figure 16.21). These compounds are prepared by a sequence of acylation of a support-bound amine with bromoacetic acid, displacement of the bromide with a primary aliphatic or aromatic amine, and repeated acylation with bromoacetic acid. Because primary amines are cheap and available in large number, this approach enables the cost-efficient production of large, diverse compound libraries. Alternatively, protected N-substituted glycines can also be prepared in solution and then assembled on insoluble supports (Entry **5**, Table 16.2).

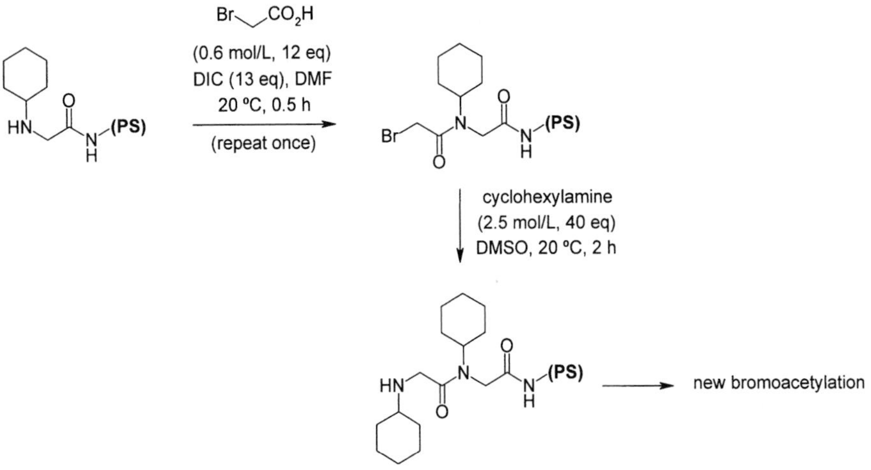

Figure 16.21. Preparation of N-substituted oligoglycines on cross-linked polystyrene [238].

The scope of the chemistry outlined in Figure 16.21 can be further expanded by including other acids susceptible to irreversible reaction with a primary amine. These can be other halocarboxylic acids, such as (chloromethyl)benzoic acids, or acrylic acid, which reacts with amines to yield N-substituted β-alanines. N-Substituted oligo(β-alanines) have been successfully prepared by sequentially acylating support-bound amines with acrylic acid and then performing a Michael addition with primary amines [239].

16.4.1.2 Peptide Nucleic Acids (PNA)

In 1991, Nielsen and co-workers [240] presented a new type of DNA mimetic, called peptide nucleic acid (PNA), in which the backbone of DNA had been replaced by oligo-[N-(2-aminoethyl)glycine] (Figure 16.22). Surprisingly, PNA forms very stable Watson–Crick duplexes with DNA. Because PNA is stable towards nucleases, PNA and hybrids of PNA and DNA are convenient DNA mimetics with a number of potentially interesting applications, for example as diagnostic tools or as antisense therapeutics.

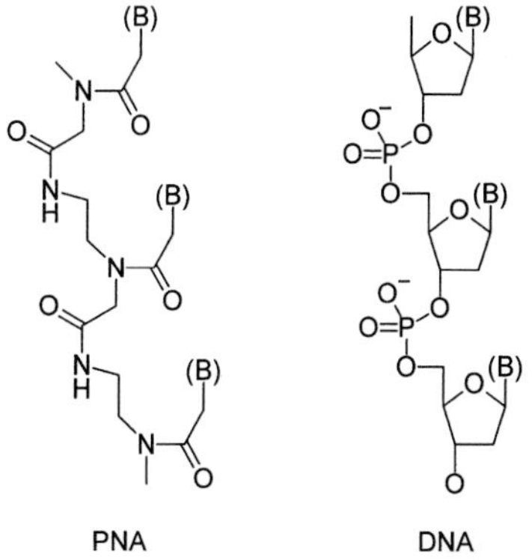

Figure 16.22. Structure of peptide nucleic acids and deoxyribonucleic acids.

PNA can be prepared on insoluble supports either from protected monomers [167,241–243] (e.g. the first equation in Figure 16.23), or by assembly of the required *N*-(2-aminoethyl)glycines [244,245] on solid phase, as outlined in Figure 16.23.

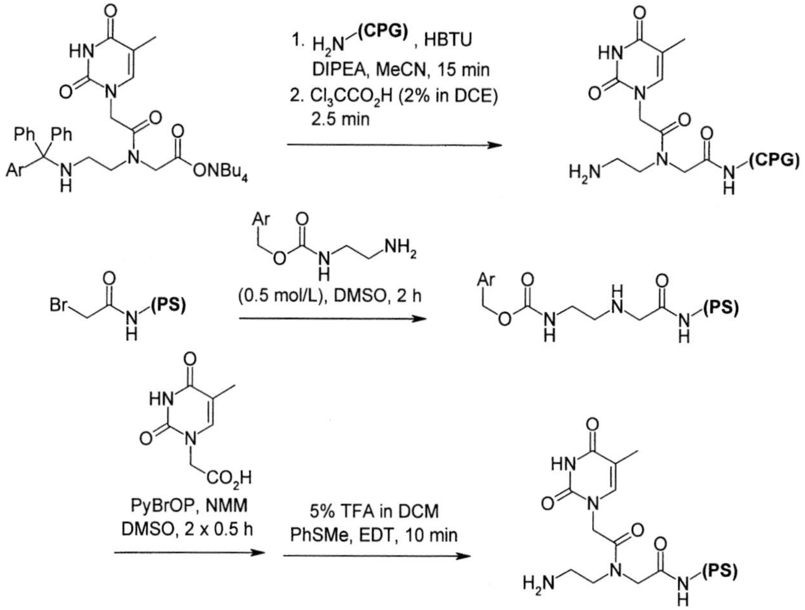

Figure 16.23. Preparation of PNA using protected monomers [166,246] or by preparation of the monomers on insoluble supports [247,248]. Ar: 4-(MeO)C₆H₄.

16.4.2 Oligoesters

Although oligomers of α-hydroxy acids (depsides) and mixed oligomers of α-hydroxy and α-amino acids (depsipeptides) are found in nature (antibiotics, ion-transporters [249,250]), little solid-phase methodology has been developed for their preparation. One recent strategy is based on sequential acylation with THP-protected α-hydroxy acids (DIC, DMAP, THF, 2 h; see Entry **6**, Table 13.13), followed by deprotection with TsOH/MeOH [249,251]. Under these conditions, no racemization was observed. A similar approach to the solid-phase synthesis of depsipeptides is outlined in Figure 16.24.

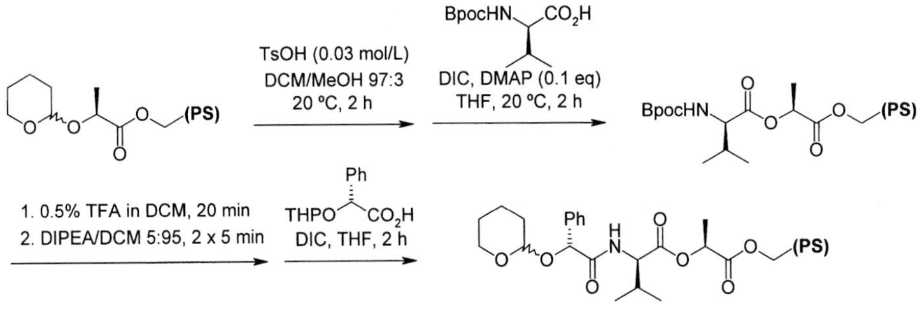

Figure 16.24. Solid-phase synthesis of depsipeptides [249,251].

16.4.3 Oligoureas and Oligothioureas

α-Amino acids can be converted into monomers suitable for the solid-phase synthesis of oligoureas, which are interesting peptide mimetics with greater resistance to enzymatic degradation. Two approaches for realizing this concept are outlined in Figure 16.25.

In the first approach, an amino acid derived 4-nitrophenyl carbamate is used as the carbamoylating reagent. 4-Nitrophenyl carbamates are generally not very reactive, and carbamoylations with these reagents only proceed smoothly with sufficiently nucleophilic aliphatic amines. 4-Nitrophenyl carbamates [252] and *O*-succinimidyl carbamates [253] in combination with *N*-Fmoc protection have also been used for the solid-phase synthesis of oligoureas.

The second approach uses the more reactive isocyanates, which can be prepared from the corresponding primary amines by treatment with phosgene [254]. Oligomers of hydrazine-derived ureas have been prepared as peptide mimetics on PEG in solution [255]. For this purpose, protected pentafluorophenyl carbazates have been used to convert support-bound hydrazines into ureas (Figure 16.25).

Oligothioureas have been prepared on cross-linked polystyrene from Boc-protected ω-aminoisothiocyanates, which were synthesized in solution from symmetric diamines [258]. The formation of thioureas from support-bound amines and isothiocyanates is a rather sluggish reaction (e.g. 3 d, 45 °C, mean coupling yield: 90% [258]), and only short oligomers can be prepared using this strategy.

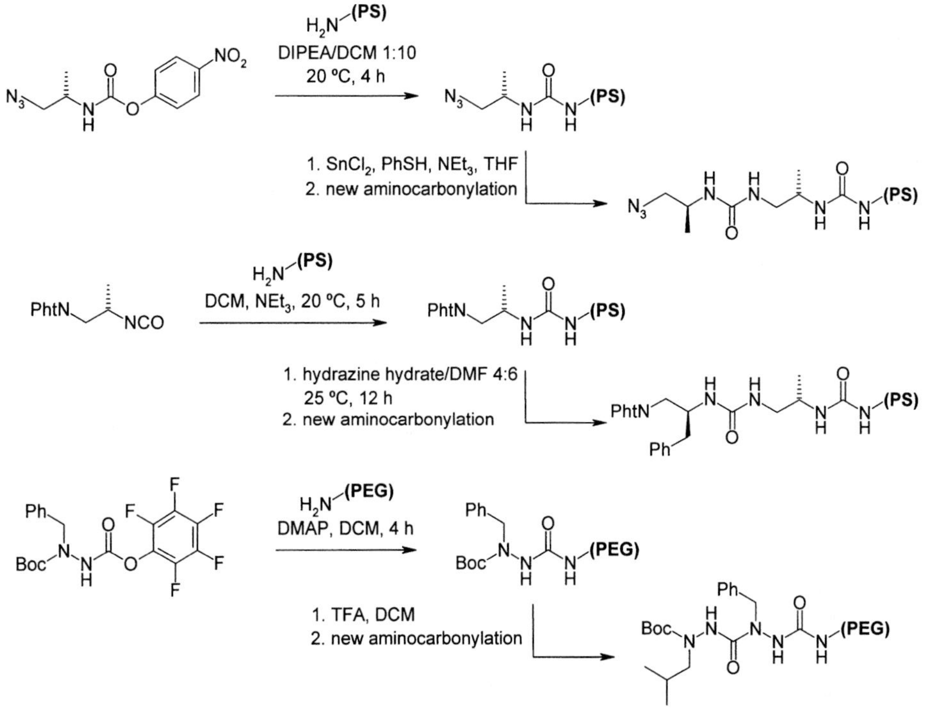

Figure 16.25. Strategies for the preparation of oligoureas [254–257].

16.4.4 Oligocarbamates

Enantiomerically pure aminoalcohols, which are readily available by reduction of α-amino acids, can be converted into alkoxycarbonylating reagents suitable for the solid-phase synthesis of oligocarbamates (Figure 16.26). Particularly convenient alkoxycarbonylating reagents are 4-nitrophenyl carbonates, which can be prepared from alcohols and 4-nitrophenyl chloroformate, and which react smoothly with aliphatic primary or secondary amines to yield the corresponding carbamates.

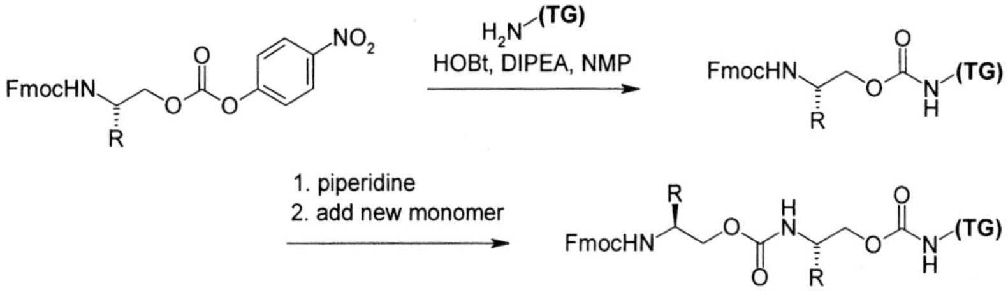

Figure 16.26. Preparation of oligocarbamates on insoluble supports [259,260].

16.4.5 Oligosulfonamides

Because sulfonamides are generally more stable towards enzymatic degradation than amides, the preparation of oligosulfonamides as peptide mimetics has been thoroughly investigated. The synthesis of α-aminosulfonamides has not been successful, probably because of the instability of this class of compound [261]. For the more stable β-amino sulfonamides and vinylogous α-aminosulfonamides (3-amino-1-propene-1-sulfonamides), however, suitable synthetic routes have been developed. Unfortunately, the preparations of the required monomers are not as easy as those of related amino acid derived monomers (Figure 16.27).

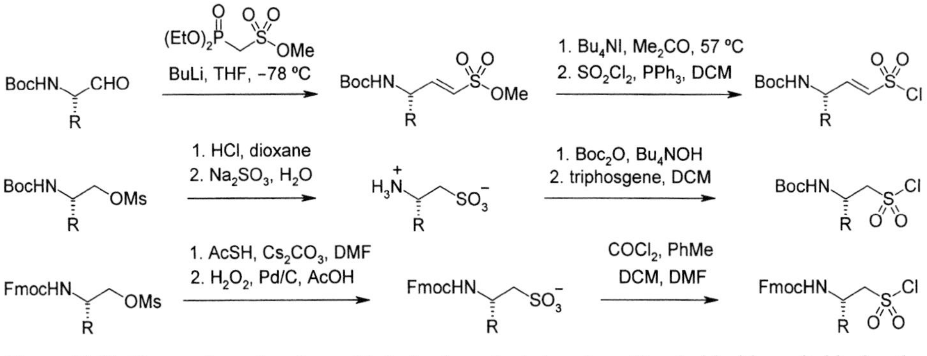

Figure 16.27. Preparation of amino acid derived, protected aminosulfonyl chlorides suitable for the solid-phase synthesis of oligosulfonamide peptidomimetics [261–264].

The sulfonyl chlorides shown in Figure 16.27 have been used to prepare peptide mimetics on solid phase, either alone or in combination with protected α-amino acids [263–267] (see also Section 8.4).

16.4.6 Oligomeric Phosphoric Acid Derivatives

The synthetic protocols used for the preparation of oligonucleotides on supports can also be used to prepare oligomers from diols other than nucleosides. Symmetric or unsymmetric diols, such as N-acylated 4-hydroxyprolinol [268] or cyclopentane-derived diols (carbocyclic deoxyribose analogs [269]), can be selectively mono-tritylated and then converted into a phosphoramidite that is suitable for the solid-phase synthesis of oligophosphates. An illustrative synthesis of protected *H*-phosphonates from diols, as well as their conversion on CPG into oligomeric phosphoramidates, are outlined in Figure 16.28.

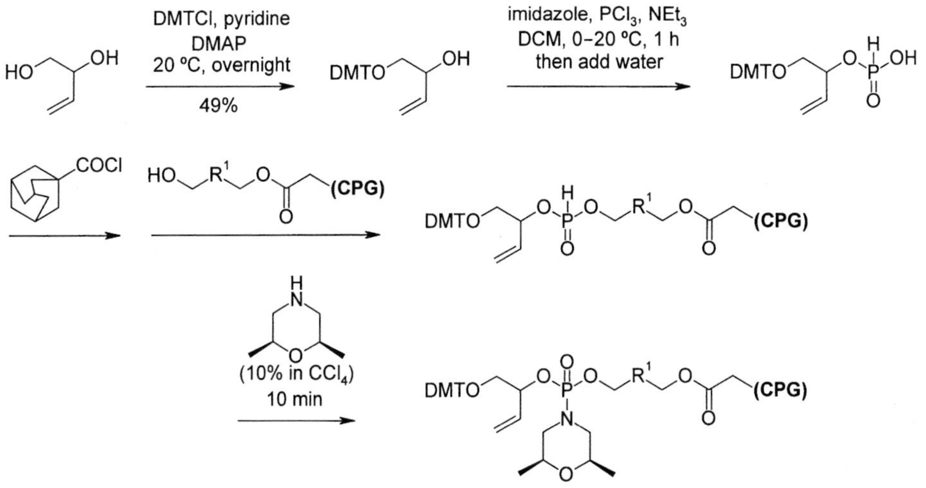

Figure 16.28. Conversion of diols into protected *H*-phosphonates and the preparation of oligomeric phosphonate diesters and phosphoramidates [133].

16.4.7 Peptide-Derived Oligomeric Compounds

Peptides can be chemically transformed on insoluble supports to yield other types of oligomer. These transformations include N-alkylation and reduction. N-Alkylation of polystyrene-bound peptides enables the preparation of peptide mimetics with improved enzymatic stability [270]. Polyamines have been prepared by exhaustive reduction of peptides with borane [271] (see Section 10.1.6). N-Alkylated polyamines can be prepared on solid phase by reduction of the corresponding N-alkylated peptides [272].

16.4.8 Oligomers Prepared by C–C Bond Formation

Few examples have been reported of the preparation of oligomers by C–C bond formation. Reggelin described a synthetic protocol that enables the sequential stereo-controlled assembly of polyketides on cross-linked polystyrene (Figure 16.29; see also [273]). However, this strategy involves many steps and only short polyketides have so far been prepared. Shorter and more efficient synthetic protocols will probably have to be developed to give access to larger and structurally more complex polyketides.

Another example of oligomer preparation by C–C bond formation is outlined in Figure 16.30. In this synthesis, nitroalkyl phenyl selenides are converted into nitrile oxides in the presence of support-bound terminal alkenes, forming isoxazolines. Oxidative elimination of the selenide yields a new alkene, which can then be subjected to further 1,3-dipolar cycloaddition with a new nitrile oxide. Although this synthesis is short and easy to perform, the cycloadditions proceed with low diastereoselectivity

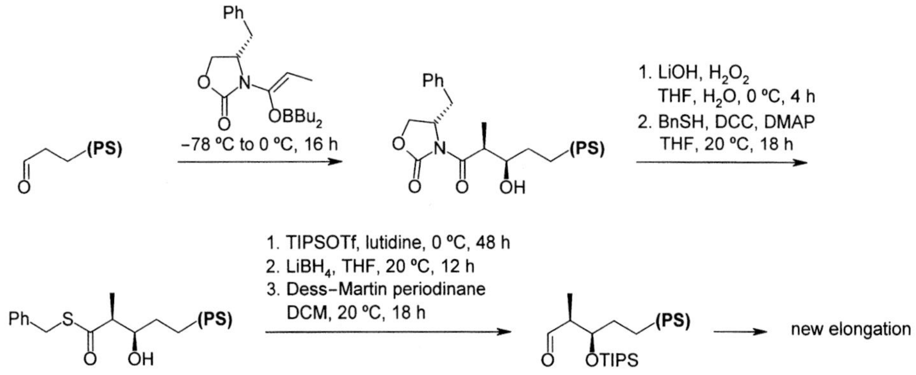

Figure 16.29. Iterative, stereocontrolled preparation of polyketides on cross-linked polystyrene [274].

and mixtures of stereoisomers are obtained. Furthermore, these oligomers cannot be readily derivatized with functionalized side chains.

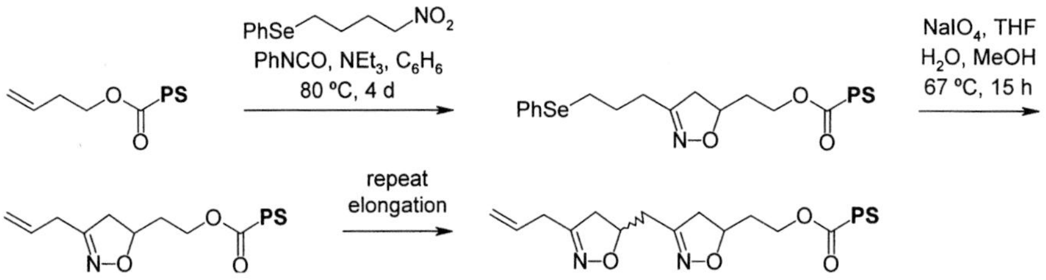

Figure 16.30. Solid-phase synthesis of oligoisoxazolines [275].

Oligomeric phenylalkynes, which are potentially interesting as liquid crystals [276] or for the construction of molecular electronic devices [277,278], have been prepared on insoluble supports by strategies such as that outlined in Figure 16.31. Substituted TMS-protected (ethynyl)iodobenzenes were used as monomers in this instance. Oligothiophenes have been prepared by a similar synthetic strategy [278].

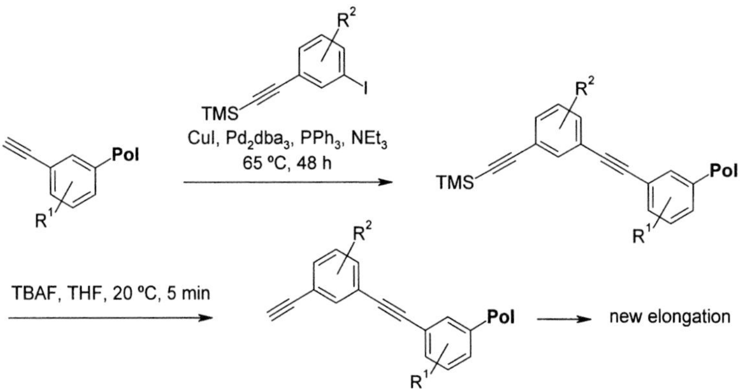

Figure 16.31. Solid-phase synthesis of oligomeric phenylalkynes [276,277,279,280].

16.4.9 Dendrimers

Dendrimers are branched, spherical oligomers, which can be prepared both in solution and on supports. Several interesting applications of dendrimers have been proposed including, for instance, their use as supports for organic synthesis [281–283], their use as carriers for peptides to induce immunogenic response in vivo (multiple antigen peptide system, MAP [18,284]), and as carriers for transporting DNA or other molecules into cells [284]. Dendrimers linked to insoluble supports have also been used to increase the loading of commercially available resins [285,286]. Some of the common strategies used to prepare dendrimers on supports are outlined in Figure 16.32. Most of the reported examples have been oligoamides of lysine, or oligomers prepared from alkanediamines and acrylic acid ('polyamidoamines'; Figure 16.32). Branched oligonucleotides [287] and branched oligoethers [288] have also been prepared on insoluble supports.

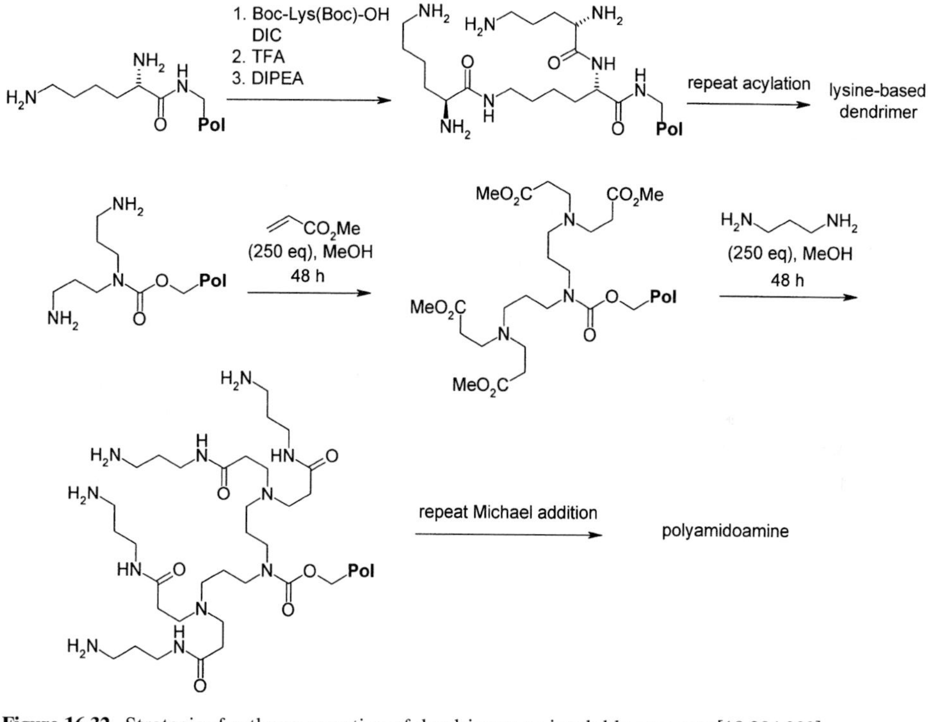

Figure 16.32. Strategies for the preparation of dendrimers on insoluble supports [18,284,289].

References for Chapter 16

[1]   Lam, K. S.; Lebl, M.; Krchnák, V. *Chem. Rev.* **1997**, *97*, 411–448.
[2]   Ganesan, A. *Angew. Chem. Int. Ed.* **1998**, *37*, 2828–2831.
[3]   Gordon, E. M.; Gallop, M. A.; Patel, D. V. *Acc. Chem. Res.* **1996**, *29*, 144–154.
[4]   Williard, X.; Pop, I.; Bourel, L.; Horváth, D.; Baudelle, R.; Melnyk, P.; Déprez, B.; Tartar, A. *Eur. J. Med. Chem.* **1996**, *31*, 87–98.
[5]   Shao, Y. F.; Still, W. C. *J. Org. Chem.* **1996**, *61*, 6086–6087.
[6]   Still, W. C. *Acc. Chem. Res.* **1996**, *29*, 155–163.
[7]   Cheng, Y. A.; Suenaga, T.; Still, W. C. *J. Am. Chem. Soc.* **1996**, *118*, 1813–1814.
[8]   Gennari, C.; Nestler, H. P.; Salom, B.; Still, W. C. *Angew. Chem. Int. Ed. Engl.* **1995**, *34*, 1765–1768.
[9]   Yoon, S. S.; Still, W. C. *Tetrahedron* **1995**, *51*, 567–578.
[10] Still, W. C.; Borchardt, A. *J. Am. Chem. Soc.* **1994**, *116*, 373–374.
[11] Fessmann, T.; Kilburn, J. D. *Angew. Chem. Int. Ed.* **1999**, *38*, 1993–1996.
[12] Merrifield, R. B. *J. Am. Chem. Soc.* **1963**, *85*, 2149–2154.
[13] Letsinger, R. L.; Kornet, M. J. *J. Am. Chem. Soc.* **1963**, *85*, 3045–3046.
[14] Matsueda, R.; Maruyama, H.; Kitazawa, E.; Takahagi, H.; Mukaiyama, T. *J. Am. Chem. Soc.* **1975**, *97*, 2573–2575.
[15] Johansson, A.; Åkerblom, E.; Ersmark, K.; Lindeberg, G.; Hallberg, A. *J. Comb. Chem.* **2000**, *2*, 496–507.
[16] Thieriet, N.; Guibé, F.; Albericio, F. *Org. Lett.* **2000**, *2*, 1815–1817.
[17] Arshady, R.; Atherton, E.; Clive, D. L. J.; Sheppard, R. C. *J. Chem. Soc., Perkin Trans. 1* **1981**, 529–537.
[18] Novabiochem Catalog and Peptide Synthesis Handbook, Läufelfingen, CH **1999**.
[19] Advanced Chemtech Handbook of Combinatorial and Solid-Phase Organic Chemistry, Louisville, USA **1998**.
[20] Mitchell, A. R.; Erickson, B. W.; Ryabtsev, M. N.; Hodges, R. S.; Merrifield, R. B. *J. Am. Chem. Soc.* **1976**, *98*, 7357–7362.
[21] Urban, J.; Vaisar, T.; Shen, R.; Lee, M. S. *Int. J. Pept. Prot. Res.* **1996**, *47*, 182–189.
[22] Beyermann, M.; Bienert, M. *Tetrahedron Lett.* **1992**, *33*, 3745–3748.
[23] Lenard, J. *Chem. Rev.* **1969**, *69*, 625–638.
[24] Jones, J. *The Chemical Synthesis of Peptides*; Oxford University Press: Oxford, **1994**.
[25] Lloyd-Williams, P.; Albericio, F.; Giralt, E. *Tetrahedron* **1993**, *49*, 11065–11133.
[26] Nishiuchi, Y.; Nishio, H.; Inui, T.; Bódi, J.; Kimura, T. *J. Pept. Sci.* **2000**, *6*, 84–93.
[27] Canne, L. E.; Botti, P.; Simon, R. J.; Chen, Y.; Dennis, E. A.; Kent, S. B. H. *J. Am. Chem. Soc.* **1999**, *121*, 8720–8727.
[28] Brik, A.; Keinan, E.; Dawson, P. E. *J. Org. Chem.* **2000**, *65*, 3829–3835.
[29] Carpino, L. A.; Han, G. Y. *J. Org. Chem.* **1972**, *37*, 3404–3409.
[30] Atherton, E.; Hübscher, W.; Sheppard, R. C.; Woolley, V. *Hoppe-Seyler's Z. Physiol. Chem.* **1981**, *362*, 833–839.
[31] Fields, G. B.; Noble, R. L. *Int. J. Pept. Prot. Res.* **1990**, *35*, 161–214.
[32] Albericio, F.; Pons, M.; Pedroso, E.; Giralt, E. *J. Org. Chem.* **1989**, *54*, 360–366.
[33] Kruijtzer, J. A. W.; Hofmeyer, L. J. F.; Heerma, W.; Versluis, C.; Liskamp, R. M. J. *Chem. Eur. J.* **1998**, *4*, 1570–1580.
[34] Robertson, N.; Ramage, R. *J. Chem. Soc., Perkin Trans. 1* **1999**, 1015–1021.
[35] Raman, P.; Stokes, S. S.; Angell, Y. M.; Flentke, G. R.; Rich, D. H. *J. Org. Chem.* **1998**, *63*, 5734–5735.
[36] Li, P.; Xu, J. C. *J. Org. Chem.* **2000**, *65*, 2951–2958.
[37] Kunz, H. *Pure Appl. Chem.* **1993**, *65*, 1223–1232.
[38] Osborn, H. M. I.; Khan, T. H. *Tetrahedron* **1999**, *55*, 1807–1850.
[39] Nakamura, K.; Ishii, A.; Ito, Y.; Nakahara, Y. *Tetrahedron* **1999**, *55*, 11253–11266.
[40] Broddefalk, J.; Forsgren, M.; Sethson, I.; Kihlberg, J. *J. Org. Chem.* **1999**, *64*, 8948–8953.
[41] Ishii, A.; Hojo, H.; Kobayashi, A.; Nakamura, K.; Nakahara, Y.; Ito, Y. *Tetrahedron* **2000**, *56*, 6235–6243.
[42] Ando, S.; Nakahara, Y.; Ito, Y.; Ogawa, T. *Carbohyd. Res.* **2000**, *329*, 773–780.
[43] Nakahara, Y.; Ando, S.; Itakura, M.; Kumabe, N.; Hojo, H.; Ito, Y. *Tetrahedron Lett.* **2000**, *41*, 6489–6493.
[44] Habermann, J.; Kunz, H. *Tetrahedron Lett.* **1998**, *39*, 265–268.

[45] Meldal, M.; Auzanneau, F. I.; Hindsgaul, O.; Palcic, M. M. *J. Chem. Soc., Chem. Commun.* **1994**, 1849–1850.
[46] Paulsen, H.; Merz, G.; Peters, S.; Weichert, U. *Liebigs Ann. Chem.* **1990**, 1165–1173.
[47] Christiansen-Brams, I.; Meldal, M.; Bock, K. *J. Chem. Soc., Perkin Trans. 1* **1993**, 1461–1471.
[48] Chadwick, R. J.; Thompson, J. S.; Tomalin, G. *Biochem. Soc. Trans.* **1991**, *19*, 406S.
[49] Hilaire, P. M. S.; Lowary, T. L.; Meldal, M.; Bock, K. *J. Am. Chem. Soc.* **1998**, *120*, 13312–13320.
[50] Atherton, E.; Sheppard, R. C. *Solid Phase Peptide Synthesis; A Practical Approach*; Oxford University Press: Oxford, **1989**.
[51] Simmonds, R. G. *Int. J. Pept. Prot. Res.* **1996**, *47*, 36–41.
[52] Hyde, C.; Johnson, T.; Owen, D.; Quibell, M.; Sheppard, R. C. *Int. J. Pept. Prot. Res.* **1994**, *43*, 431–440.
[53] Johnson, T.; Quibell, M.; Sheppard, R. C. *J. Pept. Sci.* **1995**, *1*, 11–25.
[54] Quibell, M.; Packman, L. C.; Johnson, T. *J. Am. Chem. Soc.* **1995**, *117*, 11656–11668.
[55] Quibell, M.; Turnell, W. G.; Johnson, T. *J. Chem. Soc., Perkin Trans. 1* **1995**, 2019–2024.
[56] Ede, N. J.; Ang, K. H.; James, I. W.; Bray, A. M. *Tetrahedron Lett.* **1996**, *37*, 9097–9100.
[57] Howe, J.; Quibell, M.; Johnson, T. *Tetrahedron Lett.* **2000**, *10*, 3997–4001.
[58] McMurray, J. S.; Lewis, C. A. *Tetrahedron Lett.* **1993**, *34*, 8059–8062.
[59] Bourne, G. T.; Meutermans, W. D. F.; Alewood, P. F.; McGeary, R. P.; Scanlon, M.; Watson, A. A.; Smythe, M. L. *J. Org. Chem.* **1999**, *64*, 3095–3101.
[60] Cabrele, C.; Langer, M.; Beck-Sickinger, A. G. *J. Org. Chem.* **1999**, *64*, 4353–4361.
[61] Fridkin, M.; Patchornik, A.; Katchalski, E. *J. Am. Chem. Soc.* **1965**, *87*, 4646–4648.
[62] Flanigan, E.; Marshall, G. R. *Tetrahedron Lett.* **1970**, 2403–2406.
[63] Ehrlich, A.; Heyne, H.-U.; Winter, R.; Beyermann, M.; Haber, H.; Carpino, L. A.; Bienert, M. *J. Org. Chem.* **1996**, *61*, 8831–8838.
[64] Klose, J.; El-Faham, A.; Henklein, P.; Carpino, L. A.; Bienert, M. *Tetrahedron Lett.* **1999**, *40*, 2045–2048.
[65] Alsina, J.; Rabanal, F.; Chiva, C.; Giralt, E.; Albericio, F. *Tetrahedron* **1998**, *54*, 10125–10152.
[66] Spatola, A. F.; Darlak, K.; Romanovskis, P. *Tetrahedron Lett.* **1996**, *37*, 591–594.
[67] Eichler, J.; Lucka, A. W.; Pinilla, C.; Houghten, R. A. *Mol. Diversity* **1996**, *1*, 233–240.
[68] Bloomberg, G. B.; Askin, D.; Gargaro, A. R.; Tanner, M. J. A. *Tetrahedron Lett.* **1993**, *34*, 4709–4712.
[69] Aletras, A.; Barlos, K.; Gatos, D.; Koutsogianni, S.; Mamos, P. *Int. J. Pept. Prot. Res.* **1995**, *45*, 488–496.
[70] Romanovskis, P.; Spatola, A. F. *J. Pept. Res.* **1998**, *52*, 356–374.
[71] Lanter, C. L.; Guiles, J. W.; Rivero, R. A. *Mol. Diversity* **1999**, *4*, 149–153.
[72] Annis, D. A.; Helluin, O.; Jacobsen, E. N. *Angew. Chem. Int. Ed.* **1998**, *37*, 1907–1909.
[73] Yang, L.; Morriello, G. *Tetrahedron Lett.* **1999**, *40*, 8197–8200.
[74] Rosenbaum, C.; Waldmann, H. *Tetrahedron Lett.* **2001**, *42*, 5677–5680.
[75] Nishino, N.; Xu, M.; Mihara, H.; Fujimoto, T.; Ohba, M.; Ueno, Y.; Kumagai, H. *J. Chem. Soc., Chem. Commun.* **1992**, 180–181.
[76] Ösapay, G.; Profit, A.; Taylor, J. W. *Tetrahedron Lett.* **1990**, *31*, 6121–6124.
[77] Ösapay, G.; Taylor, J. W. *J. Am. Chem. Soc.* **1990**, *112*, 6046–6051.
[78] Mihara, H.; Yamabe, S.; Niidome, T.; Aoyagi, H.; Kumagai, H. *Tetrahedron Lett.* **1995**, *36*, 4837–4840.
[79] Sabatino, G.; Chelli, M.; Mazzucco, S.; Ginanneschi, M.; Papini, A. M. *Tetrahedron Lett.* **1999**, *40*, 809–812.
[80] Crozet, Y.; Wen, J. J.; Loo, R. O.; Andrews, P. C.; Spatola, A. F. *Mol. Diversity* **1998**, *3*, 261–276.
[81] Eichler, J.; Lucka, A. W.; Houghten, R. A. *Pept. Res.* **1994**, *7*, 300–307.
[82] Akaji, K.; Teruya, K.; Akaji, M.; Aimoto, S. *Tetrahedron* **2001**, *57*, 2293–2303.
[83] Eckstein, F. *Oligonucleotides and Analogues; A Practical Approach*; Oxford University Press: Oxford, **1991**.
[84] Ghosh, P. K.; Kumar, P.; Gupta, K. C. *J. Indian Chem. Soc.* **1998**, *75*, 206–218.
[85] Engels, J. W.; Uhlmann, E. *Angew. Chem. Int. Ed. Engl.* **1989**, *28*, 716–734.
[86] Trotta, P. P.; Beutel, B. A.; Sherman, M. I. *Med. Res. Rev.* **1995**, *15*, 277–298.
[87] Uhlmann, E.; Peyman, A. *Chem. Rev.* **1990**, *90*, 543–584.
[88] Letsinger, R. L.; Mahadevan, V. *J. Am. Chem. Soc.* **1965**, *87*, 3526–3527.
[89] Hayatsu, H.; Khorana, H. G. *J. Am. Chem. Soc.* **1966**, *88*, 3182–3183.
[90] Cramer, F.; Helbig, R.; Hettler, H.; Scheit, K. H.; Seliger, H. *Angew. Chem.* **1966**, *78*, 640–641; *Angew. Chem. Int. Ed. Engl.* **1966**, *5*, 601–602.
[91] Hayatsu, H.; Khorana, H. G. *J. Am. Chem. Soc.* **1967**, *89*, 3880–3887.

[92] Köster, H.; Biernat, J.; McManus, J.; Wolter, A.; Stumpe, A.; Narang, C. K.; Sinha, N. D. *Tetrahedron* **1984**, *40*, 103–112.

[93] Kusama, T.; Hayatsu, H. *Chem. Pharm. Bull.* **1970**, *18*, 319–327.

[94] Köster, H.; Cramer, F. *Liebigs Ann. Chem.* **1974**, 946–958.

[95] Köster, H. *Tetrahedron Lett.* **1972**, 1527–1530.

[96] Seliger, H.; Aumann, G. *Tetrahedron Lett.* **1973**, 2911–2914.

[97] Schott, H.; Brandstetter, F.; Bayer, E. *Makromol. Chem.* **1973**, *173*, 247–251.

[98] Chapman, T. M.; Kleid, D. G. *J. Chem. Soc., Chem. Commun.* **1973**, 193–194.

[99] Brandstetter, F.; Schott, H.; Bayer, E. *Tetrahedron Lett.* **1973**, 2997–3000.

[100] Gait, M. J.; Sheppard, R. C. *Nucleic Acids Res.* **1977**, *4*, 1135–1158.

[101] Catlin, J. C.; Cramer, F. *J. Org. Chem.* **1973**, *38*, 245–250.

[102] Itakura, K.; Bahl, C. P.; Katagiri, N.; Michniewicz, J. J.; Wightman, R. H.; Narang, S. A. *Can. J. Chem.* **1973**, *51*, 3649–3651.

[103] Dembek, P.; Miyoshi, K.; Itakura, K. *J. Am. Chem. Soc.* **1981**, *103*, 706–708.

[104] Reese, C. B. *Tetrahedron* **1978**, *34*, 3143–3179.

[105] Ito, H.; Ike, Y.; Ikuta, S.; Itakura, K. *Nucleic Acids Res.* **1982**, *10*, 1755–1769.

[106] Sekine, M.; Hata, T. *Curr. Org. Chem.* **1999**, *3*, 25–66.

[107] Gait, M. J.; Matthes, H. W. D.; Singh, M.; Sproat, B. S.; Titmas, R. C. *Nucleic Acids Res.* **1982**, *10*, 6243–6254.

[108] Witkowski, W.; Birch-Hirschfeld, E.; Weiss, R.; Zarytova, V. F.; Gorn, V. V. *J. Prakt. Chem.* **1984**, *326*, 320–328.

[109] Gough, G. R.; Brunden, M. J.; Gilham, P. T. *Tetrahedron Lett.* **1981**, *22*, 4177–4180.

[110] Letsinger, R. L.; Finnan, J. L.; Heavner, G. A.; Lunsford, W. B. *J. Am. Chem. Soc.* **1975**, *97*, 3278–3279.

[111] Ogilvie, K. K.; Nemer, M. J. *Tetrahedron Lett.* **1980**, *21*, 4159–4162.

[112] Matteucci, M. D.; Caruthers, M. H. *Tetrahedron Lett.* **1980**, *21*, 719–722.

[113] Matteucci, M. D.; Caruthers, M. H. *J. Am. Chem. Soc.* **1981**, *103*, 3185–3191.

[114] Beaucage, S. L.; Caruthers, M. H. *Tetrahedron Lett.* **1981**, *22*, 1859–1862.

[115] Sinha, N. D.; Biernat, J.; McManus, J.; Köster, H. *Nucleic Acids Res.* **1984**, *12*, 4539–4557.

[116] Adams, S. P.; Kavka, K. S.; Wykes, E. J.; Holder, S. B.; Galluppi, G. R. *J. Am. Chem. Soc.* **1983**, *105*, 661–663.

[117] Beier, M.; Pfleiderer, W. *Helv. Chim. Acta* **1999**, *82*, 633–644.

[118] McCollum, C.; Andrus, A. *Tetrahedron Lett.* **1991**, *32*, 4069–4072.

[119] Reddy, M. P.; Michael, M. A.; Farooqui, F.; Girgis, S. *Tetrahedron Lett.* **1994**, *35*, 5771–5774.

[120] Birch-Hirschfeld, E.; Földes-Papp, Z.; Gührs, K. H.; Seliger, H. *Helv. Chim. Acta* **1996**, *79*, 137–150.

[121] Birch-Hirschfeld, E.; Eickhoff, H.; Stelzner, A.; Greulich, K. O.; Földes-Papp, Z.; Seliger, H.; Gührs, K. H. *Collect. Czech. Chem. Commun.* **1996**, *61*, S311–S314.

[122] Micura, R. *Chem. Eur. J.* **1999**, *5*, 2077–2082.

[123] Wright, P.; Lloyd, D.; Rapp, W.; Andrus, A. *Tetrahedron Lett.* **1993**, *34*, 3373–3376.

[124] Bayer, E.; Bleicher, K.; Maier, M. *Z. Naturforsch. Sect. B* **1995**, *50*, 1096–1100.

[125] Gao, H.; Gaffney, B. L.; Jones, R. A. *Tetrahedron Lett.* **1991**, *32*, 5477–5480.

[126] Bonora, G. M.; Biancotto, G.; Maffini, M.; Scremin, C. L. *Nucleic Acids Res.* **1993**, *21*, 1213–1217.

[127] Wörl, R.; Köster, H. *Tetrahedron* **1999**, *55*, 2941–2956.

[128] Bonora, G. M.; Baldan, A.; Schiavon, O.; Ferruti, P.; Veronese, F. M. *Tetrahedron Lett.* **1996**, *37*, 4761–4764.

[129] Pon, R. T.; Yu, S.; Guo, Z.; Deshmukh, R.; Sanghvi, Y. S. *J. Chem. Soc., Perkin Trans. 1* **2001**, 2638–2643.

[130] Hayakawa, Y.; Kataoka, M. *J. Am. Chem. Soc.* **1998**, *120*, 12395–12401.

[131] Frieden, M.; Grandas, A.; Pedroso, E. *Chem. Commun.* **1999**, 1593–1594.

[132] Wada, T.; Mochizuki, A.; Sato, Y.; Sekine, M. *Tetrahedron Lett.* **1998**, *39*, 5593–5596.

[133] Fathi, R.; Rudolph, M. J.; Gentles, R. G.; Patel, R.; MacMillan, E. W.; Reitman, M. S.; Pelham, D.; Cook, A. F. *J. Org. Chem.* **1996**, *61*, 5600–5609.

[134] Sergueev, D. S.; Shaw, B. R. *J. Am. Chem. Soc.* **1998**, *120*, 9417–9427.

[135] Kupihár, Z.; Kele, Z.; Tóth, G. K. *Org. Lett.* **2001**, *3*, 1033–1035.

[136] Pon, R. T.; Ogilvie, K. K. *Tetrahedron Lett.* **1984**, *25*, 713–716.

[137] Stutz, A.; Pitsch, S. *Synlett* **1999**, 930–934.

[138] Ecker, D. J.; Vickers, T. A.; Hanecak, R.; Driver, V.; Anderson, K. *Nucleic Acids Res.* **1993**, *21*, 1853–1856.

[139] Schneider, D. J.; Feigon, J.; Hostomsky, Z.; Gold, L. *Biochemistry* **1995**, *34*, 9599–9610.

[140] Seeberger, P. H.; Caruthers, M. H.; Bankaitis-Davis, D.; Beaton, G. *Tetrahedron* **1999**, *55*, 5759–5772.

[141] Yang, X. B.; Sierzchala, A.; Misiura, K.; Niewiarowski, W.; Sochacki, M.; Stec, W. J.; Wieczorek, M. W. *J. Org. Chem.* **1998**, *63*, 7097–7100.

[142] Zhang, Z.; Nichols, A.; Tang, J. X.; Han, Y.; Tang, J. Y. *Tetrahedron Lett.* **1999**, *40*, 2095–2098.

[143] Guo, M. J.; Yu, D.; Iyer, R. P.; Agrawal, S. *Bioorg. Med. Chem. Lett.* **1998**, *8*, 2539–2544.

[144] Ecker, D. J.; Wyatt, J. R.; Vickers, T.; Buckheit, R.; Roberson, J.; Imbach, J. L. *Nucleosides and Nucleotides* **1995**, *14*, 1117–1127.

[145] Ecker, D. J.; Wyatt, J. R.; Vickers, T.; Buckheit, R.; Roberson, J.; Imbach, J. L. *Nucleosides and Nucleotides* **1995**, *14*, 1117–1127.

[146] Le Bec, C.; Wickstrom, E. *J. Org. Chem.* **1996**, *61*, 510–513.

[147] Fathi, R.; Patel, R.; Cook, A. F. *Mol. Diversity* **1997**, *2*, 125–134.

[148] Robles, J.; Ibañez, V.; Grandas, A.; Pedroso, E. *Tetrahedron Lett.* **1999**, *40*, 7131–7134.

[149] Lin, J.; Shaw, B. R. *Chem. Commun.* **1999**, 1517–1518.

[150] Linkletter, B. A.; Szabo, I. E.; Bruice, T. C. *J. Am. Chem. Soc.* **1999**, *121*, 3888–3896.

[151] Kojima, N.; Bruice, T. C. *Org. Lett.* **2000**, *2*, 81–84.

[152] Linkletter, B. A.; Bruice, T. C. *Bioorg. Med. Chem.* **2000**, *8*, 1893–1901.

[153] Arya, D. P.; Bruice, T. C. *Bioorg. Med. Chem. Lett.* **2000**, *10*, 691–693.

[154] Morvan, F.; Sanghvi, Y. S.; Perbost, M.; Vasseur, J. J.; Bellon, L. *J. Am. Chem. Soc.* **1996**, *118*, 255–256.

[155] Khan, S. I.; Grinstaff, M. W. *J. Am. Chem. Soc.* **1999**, *121*, 4704–4705.

[156] Davis, P. W.; Vickers, T. A.; Wilson-Lingardo, L.; Wyatt, J. R.; Guinosso, C. J.; Sanghvi, Y. S.; De Baets, E. A.; Acevedo, O. L.; Cook, P. D.; Ecker, D. J. *J. Med. Chem.* **1995**, *38*, 4363–4366.

[157] Bergmann, F.; Bannwarth, W. *Tetrahedron Lett.* **1995**, *36*, 1839–1842.

[158] Basu, S.; Wickstrom, E. *Tetrahedron Lett.* **1995**, *36*, 4943–4946.

[159] de la Torre, B. G.; Aviñó, A.; Tarrason, G.; Piulats, J.; Albericio, F.; Eritja, R. *Tetrahedron Lett.* **1994**, *35*, 2733–2736.

[160] Juby, C. D.; Richardson, C. D.; Brousseau, R. *Tetrahedron Lett.* **1991**, *32*, 879–882.

[161] Robles, J.; Beltrán, M.; Marchán, V.; Pérez, Y.; Travesset, I.; Pedroso, E.; Grandas, A. *Tetrahedron* **1999**, *55*, 13251–13264.

[162] Sakakura, A.; Hayakawa, Y. *Tetrahedron* **2000**, *56*, 4427–4435.

[163] Antopolsky, M.; Azhayev, A. *Tetrahedron Lett.* **2000**, *41*, 9113–9117.

[164] Adinolfi, M.; Barone, G.; De Napoli, L.; Guariniello, L.; Iadonisi, A.; Piccialli, G. *Tetrahedron Lett.* **1999**, *40*, 2607–2610.

[165] Adinolfi, M.; De Napoli, L.; Di Fabio, G.; Guariniello, L.; Iadonisi, A.; Messere, A.; Montesarchio, D.; Piccialli, G. *Synlett* **2001**, 745–748.

[166] van der Laan, A. C.; Meeuwenoord, N. J.; Kuyl-Yeheskiely, E.; Oosting, R. S.; Brands, R.; van Boom, J. H. *Recl. Trav. Chim. Pays-Bas* **1995**, *114*, 295–297.

[167] Bergmann, F.; Bannwarth, W.; Tam, S. *Tetrahedron Lett.* **1995**, *36*, 6823–6826.

[168] Capasso, D.; De Napoli, L.; Di Fabio, G.; Messere, A.; Montesarchio, D.; Pedone, C.; Piccialli, G.; Saviano, M. *Tetrahedron* **2001**, *57*, 9481–9486.

[169] McMinn, D. L.; Hirsch, R.; Greenberg, M. M. *Tetrahedron Lett.* **1998**, *39*, 4155–4158.

[170] Peyman, A.; Weiser, C.; Uhlmann, E. *Bioorg. Med. Chem. Lett.* **1995**, *5*, 2469–2472.

[171] Yoo, D. J.; Greenberg, M. M. *J. Org. Chem.* **1995**, *60*, 3358–3364.

[172] McMinn, D. L.; Greenberg, M. M. *Tetrahedron* **1996**, *52*, 3827–3840.

[173] Hovinen, J.; Guzaev, A.; Azhayev, A.; Lönnberg, H. *Tetrahedron* **1994**, *50*, 7203–7218.

[174] Guzaev, A.; Lönnberg, H. *Tetrahedron* **1999**, *55*, 9101–9116.

[175] Bleczinski, C. F.; Richert, C. *Org. Lett.* **2000**, *2*, 1697–1700.

[176] Frank, R.; Meyerhans, A.; Schwellnus, K.; Blöcker, H. *Methods Enzymol.* **1987**, *154*, 221–249.

[177] Frank, R.; Heikens, W.; Heisterberg-Moutsis, G.; Blöcker, H. *Nucleic Acids Res.* **1983**, *11*, 4365–4377.

[178] Ott, J.; Eckstein, F. *Nucleic Acids Res.* **1984**, *12*, 9137–9142.

[179] Markiewicz, W. T.; Markiewicz, M.; Astriab, A.; Godzina, P. *Collect. Czech. Chem. Commun.* **1996**, *61*, S315–S318.

[180] Dittrich, F.; Tegge, W.; Frank, R. *Bioorg. Med. Chem. Lett.* **1998**, *8*, 2351–2356.

[181] Nanthakumar, A.; Pon, R. T.; Mazumder, A.; Yu, S.; Watson, A. *Bioconjugate Chem.* **2000**, *11*, 282–288.

[182] Ito, Y.; Manabe, S. *Curr. Opinion Chem. Biol.* **1998**, *2*, 701–708.

[183] Seeberger, P. H.; Danishefsky, S. J. *Acc. Chem. Res.* **1998**, *31*, 685–695.

[184] Guthrie, R. D.; Jenkins, A. D.; Roberts, G. A. F. *J. Chem. Soc., Perkin Trans. 1* **1973**, 2414–2417.

[185] Excoffier, G.; Gagnaire, D.; Utille, J. P.; Vignon, M. *Tetrahedron Lett.* **1972**, 5065–5068.

[186] Fréchet, J. M.; Schuerch, C. *J. Am. Chem. Soc.* **1971**, *93*, 492–496.
[187] Zehavi, U.; Patchornik, A. *J. Am. Chem. Soc.* **1973**, *95*, 5673–5677.
[188] Chiu, S. H. L.; Anderson, L. *Carbohyd. Res.* **1976**, *50*, 227–238.
[189] Yan, L.; Taylor, C. M.; Goodnow, R.; Kahne, D. *J. Am. Chem. Soc.* **1994**, *116*, 6953–6954.
[190] Silva, D. J.; Wang, H.; Allanson, N. M.; Jain, R. K.; Sofia, M. J. *J. Org. Chem.* **1999**, *64*, 5926–5929.
[191] Sofia, M. J.; Allanson, N.; Hatzenbuhler, N. T.; Jain, R.; Kakarla, R.; Kogan, N.; Liang, R.; Liu, D.; Silva, D. J.; Wang, H.; Gange, D.; Anderson, J.; Chen, A.; Chi, F.; Dulina, R.; Huang, B.; Kamau, M.; Wang, C.; Baizman, E.; Branstrom, A.; Bristol, N.; Goldman, R.; Han, K.; Longley, C.; Midha, S.; Axelrod, H. R. *J. Med. Chem.* **1999**, *42*, 3193–3198.
[192] Liang, R.; Yan, L.; Loebach, J.; Ge, M.; Uozumi, Y.; Sekanina, K.; Horan, N.; Gildersleeve, J.; Thompson, C.; Smith, A.; Biswas, K.; Still, W. C.; Kahne, D. *Science* **1996**, *274*, 1520–1522.
[193] Wang, Y.; Zhang, H.; Voelter, W. *Chem. Lett.* **1995**, 273–274.
[194] Nicolaou, K. C.; Winssinger, N.; Pastor, J.; DeRoose, F. *J. Am. Chem. Soc.* **1997**, *119*, 449–450.
[195] Doi, T.; Sugiki, M.; Yamada, H.; Takahashi, T.; Porco, J. A. *Tetrahedron Lett.* **1999**, *40*, 2141–2144.
[196] Zheng, C.; Seeberger, P. H.; Danishefsky, S. J. *J. Org. Chem.* **1998**, *63*, 1126–1130.
[197] Zhu, T.; Boons, G. J. *Angew. Chem. Int. Ed.* **1998**, *37*, 1898–1900.
[198] Zhu, T.; Boons, G. J. *J. Am. Chem. Soc.* **2000**, *122*, 10222–10223.
[199] Kanemitsu, T.; Kanie, O.; Wong, C.-H. *Angew. Chem. Int. Ed.* **1998**, *37*, 3415–3418.
[200] Takahashi, T.; Inoue, H.; Yamamura, Y.; Doi, T. *Angew. Chem. Int. Ed.* **2001**, *40*, 3230–3233.
[201] Ito, Y.; Kanie, O.; Ogawa, T. *Angew. Chem. Int. Ed. Engl.* **1996**, *35*, 2510–2512.
[202] Hewitt, M. C.; Seeberger, P. H. *Org. Lett.* **2001**, *3*, 3699–3702.
[203] Roussel, F.; Takhi, M.; Schmidt, R. R. *J. Org. Chem.* **2001**, *66*, 8540–8548.
[204] Roussel, F.; Knerr, L.; Schmidt, R. R. *Eur. J. Org. Chem.* **2001**, 2067–2073.
[205] Knerr, L.; Schmidt, R. R. *Eur. J. Org. Chem.* **2000**, 2803–2808.
[206] Heckel, A.; Mross, E.; Jung, K. H.; Rademann, J.; Schmidt, R. R. *Synlett* **1998**, 171–173.
[207] Rademann, J.; Schmidt, R. R. *J. Org. Chem.* **1997**, *62*, 3650–3653.
[208] Shimizu, H.; Ito, Y.; Kanie, O.; Ogawa, T. *Bioorg. Med. Chem. Lett.* **1996**, *6*, 2841–2846.
[209] Paulsen, H.; Schleyer, A.; Mathieux, N.; Meldal, M.; Bock, K. *J. Chem. Soc., Perkin Trans. 1* **1997**, 281–293.
[210] Adinolfi, M.; Barone, G.; De Napoli, L.; Iadonisi, A.; Piccialli, G. *Tetrahedron Lett.* **1998**, *39*, 1953–1956.
[211] Adinolfi, M.; Barone, G.; De Napoli, L.; Iadonisi, A.; Piccialli, G. *Tetrahedron Lett.* **1996**, *37*, 5007–5010.
[212] Douglas, S. P.; Whitfield, D. M.; Krepinsky, J. J. *J. Am. Chem. Soc.* **1995**, *117*, 2116–2117.
[213] Wang, Z. G.; Douglas, S. P.; Krepinsky, J. J. *Tetrahedron Lett.* **1996**, *37*, 6985–6988.
[214] Hunt, J. A.; Roush, W. R. *J. Am. Chem. Soc.* **1996**, *118*, 9998–9999.
[215] Savin, K. A.; Woo, J. C. G.; Danishefsky, S. J. *J. Org. Chem.* **1999**, *64*, 4183–4186.
[216] Danishefsky, S. J.; McClure, K. F.; Randolph, J. T.; Ruggeri, R. B. *Science* **1993**, *260*, 1307–1309.
[217] Zheng, C. S.; Seeberger, P. H.; Danishefsky, S. J. *Angew. Chem. Int. Ed.* **1998**, *37*, 786–789.
[218] Davis, B. G.; Ward, S. J.; Rendle, P. M. *Chem. Commun.* **2001**, 189–190.
[219] Petersen, L.; Jensen, K. J. *J. Chem. Soc., Perkin Trans. 1* **2001**, 2175–2182.
[220] Rodebaugh, R.; Joshi, S.; Fraser-Reid, B.; Geysen, H. M. *J. Org. Chem.* **1997**, *62*, 5660–5661.
[221] Andrade, R. B.; Plante, O. J.; Melean, L. G.; Seeberger, P. H. *Org. Lett.* **1999**, *1*, 1811–1814.
[222] Schuster, M.; Wang, P.; Paulson, J. C.; Wong, C. H. *J. Am. Chem. Soc.* **1994**, *116*, 1135–1136.
[223] Yamada, K.; Nishimura, S. I. *Tetrahedron Lett.* **1995**, *36*, 9493–9496.
[224] Blixt, O.; Norberg, T. *J. Org. Chem.* **1998**, *63*, 2705–2710.
[225] Huang, X.; Witte, K. L.; Bergbreiter, D. E.; Wong, C.-H. *Adv. Synth. Catal.* **2001**, *343*, 675–681.
[226] Moran, E. J.; Wilson, T. E.; Cho, C. Y.; Cherry, S. R.; Schultz, P. G. *Biopolymers* **1995**, *37*, 213–219.
[227] Liskamp, R. M. J. *Angew. Chem. Int. Ed. Engl.* **1994**, *33*, 633–636.
[228] Rose, K.; Vizzavona, J. *J. Am. Chem. Soc.* **1999**, *121*, 7034–7038.
[229] Vázquez, E.; Caamaño, A. M.; Castedo, L.; Mascareñas, J. L. *Tetrahedron Lett.* **1999**, *40*, 3621–3624.
[230] Herman, D. M.; Turner, J. M.; Baird, E. E.; Dervan, P. B. *J. Am. Chem. Soc.* **1999**, *121*, 1121–1129.
[231] Baird, E. E.; Dervan, P. B. *J. Am. Chem. Soc.* **1996**, *118*, 6141–6146.
[232] König, B.; Rödel, M. *Chem. Commun.* **1998**, 605–606.
[233] Rivier, J. E.; Jiang, G. C.; Koerber, S. C.; Porter, J.; Simon, L.; Craig, A. G.; Hoeger, C. A. *Proc. Natl. Acad. Sci. USA* **1996**, *93*, 2031–2036.
[234] Fletcher, M. D.; Campbell, M. M. *Chem. Rev.* **1998**, *98*, 763–795.
[235] Long, D. D.; Smith, M. D.; Marquess, D. G.; Claridge, T. D. W.; Fleet, G. W. J. *Tetrahedron Lett.* **1998**, *39*, 9293–9296.
[236] Müller, C.; Kitas, E.; Wessel, H. P. *J. Chem. Soc., Chem. Commun.* **1995**, 2425–2426.

[237] Byrgesen, E.; Nielsen, J.; Willert, M.; Bols, M. *Tetrahedron Lett.* **1997**, *38*, 5697–5700.

[238] Zuckermann, R. N.; Kerr, J. M.; Kent, S. B. H.; Moos, W. H. *J. Am. Chem. Soc.* **1992**, *114*, 10646–10647.

[239] Hamper, B. C.; Kolodziej, S. A.; Scates, A. M.; Smith, R. G.; Cortez, E. *J. Org. Chem.* **1998**, *63*, 708–718.

[240] Nielsen, P. E.; Egholm, M.; Berg, R. H.; Buchardt, O. *Science* **1991**, *254*, 1497–1500.

[241] Will, D. W.; Langner, D.; Knolle, J.; Uhlmann, E. *Tetrahedron* **1995**, *51*, 12069–12082.

[242] Seitz, O. *Tetrahedron Lett.* **1999**, *40*, 4161–4164.

[243] Egholm, M.; Nielsen, P. E.; Buchardt, O.; Berg, R. H. *J. Am. Chem. Soc.* **1992**, *114*, 9677–9678.

[244] Seitz, O.; Bergmann, F.; Heindl, D. *Angew. Chem. Int. Ed.* **1999**, *38*, 2203–2206.

[245] Viirre, R. D.; Hudson, R. H. E. *Org. Lett.* **2001**, *3*, 3931–3934.

[246] Stetsenko, D. A.; Lubyako, E. N.; Potapov, V. K.; Azhikina, T. L.; Sverdlov, E. D. *Tetrahedron Lett.* **1996**, *37*, 3571–3574.

[247] Richter, L. S.; Zuckermann, R. N. *Bioorg. Med. Chem. Lett.* **1995**, *5*, 1159–1162.

[248] Aldrian-Herrada, G.; Rabié, A.; Wintersteiger, R.; Brugidou, J. *J. Pept. Sci.* **1998**, *4*, 266–281.

[249] Kuisle, O.; Quiñoá, E.; Riguera, R. *Tetrahedron Lett.* **1999**, *40*, 1203–1206.

[250] López-Macià, A.; Jiménez, J. C.; Royo, M.; Giralt, E.; Albericio, F. *Tetrahedron Lett.* **2000**, *41*, 9765–9769.

[251] Kuisle, O.; Quiñoá, E.; Riguera, R. *J. Org. Chem.* **1999**, *64*, 8063–8075.

[252] Boeijen, A.; van Ameijde, J.; Liskamp, R. M. J. *J. Org. Chem.* **2001**, *66*, 8454–8462.

[253] Guichard, G.; Semetey, V.; Rodriguez, M.; Briand, J.-P. *Tetrahedron Lett.* **2000**, *41*, 1553–1557.

[254] Burgess, K.; Ibarzo, J.; Linthicum, D. S.; Russell, D. H.; Shin, H.; Shitangkoon, A.; Totani, R.; Zhang, A. J. *J. Am. Chem. Soc.* **1997**, *119*, 1556–1564.

[255] Han, H.; Janda, K. D. *J. Am. Chem. Soc.* **1996**, *118*, 2539–2544.

[256] Kim, J. M.; Bi, Y. Z.; Paikoff, S. J.; Schultz, P. G. *Tetrahedron Lett.* **1996**, *37*, 5305–5308.

[257] Boeijen, A.; Liskamp, R. M. J. *Eur. J. Org. Chem.* **1999**, 2127–2135.

[258] Smith, J.; Liras, J. L.; Schneider, S. E.; Anslyn, E. V. *J. Org. Chem.* **1996**, *61*, 8811–8818.

[259] Wendeborn, S.; De Mesmaeker, A.; Brill, W. K. D. *Synlett* **1998**, 865–868.

[260] Cho, C. Y.; Moran, E. J.; Cherry, S. R.; Stephans, J. C.; Fodor, S. P. A.; Adams, C. L.; Sundaram, A.; Jacobs, J. W.; Schultz, P. G. *Science* **1993**, *261*, 1303–1305.

[261] Gennari, C.; Salom, B.; Potenza, D.; Williams, A. *Angew. Chem. Int. Ed. Engl.* **1994**, *33*, 2067–2069.

[262] Gude, M.; Piarulli, U.; Potenza, D.; Salom, B.; Gennari, C. *Tetrahedron Lett.* **1996**, *37*, 8589–8592.

[263] de Bont, D. B. A.; Dijkstra, G. D. H.; den Hartog, J. A. J.; Liskamp, R. M. J. *Bioorg. Med. Chem. Lett.* **1996**, *6*, 3035–3040.

[264] Monnee, M. C. F.; Marijne, M. F.; Brouwer, A. J.; Liskamp, R. M. J. *Tetrahedron Lett.* **2000**, *41*, 7991–7995.

[265] Gennari, C.; Nestler, H. P.; Salom, B.; Still, W. C. *Angew. Chem. Int. Ed. Engl.* **1995**, *34*, 1763–1765.

[266] de Bont, D. B. A.; Moree, W. J.; Liskamp, R. M. J. *Bioorg. Med. Chem.* **1996**, *4*, 667–672.

[267] Gennari, C.; Longari, C.; Ressel, S.; Salom, B.; Piarulli, U.; Ceccarelli, S.; Mielgo, A. *Eur. J. Org. Chem.* **1998**, 2437–2449.

[268] Hébert, N.; Davis, P. W.; De Baets, E. L.; Acevedo, O. L. *Tetrahedron Lett.* **1994**, *35*, 9509–9512.

[269] Domínguez, B. M.; Cullis, P. M. *Tetrahedron Lett.* **1999**, *40*, 5783–5786.

[270] Dörner, B.; Husar, G. M.; Ostresh, J. M.; Houghten, R. A. *Bioorg. Med. Chem.* **1996**, *4*, 709–715.

[271] Ostresh, J. M.; Schoner, C. C.; Hamashin, V. T.; Nefzi, A.; Meyer, J. P.; Houghten, R. A. *J. Org. Chem.* **1998**, *63*, 8622–8623.

[272] Nefzi, A.; Ostresh, J. M.; Houghten, R. A. *Tetrahedron* **1999**, *55*, 335–344.

[273] Gennari, C.; Ceccarelli, S.; Piarulli, U.; Aboutayab, K.; Donghi, M.; Paterson, I. *Tetrahedron* **1998**, *54*, 14999–15016.

[274] Reggelin, M.; Brenig, V.; Welcker, R. *Tetrahedron Lett.* **1998**, *39*, 4801–4804.

[275] Kurth, M. J.; Randall, L. A. A.; Takenouchi, K. *J. Org. Chem.* **1996**, *61*, 8755–8761.

[276] Nelson, J. C.; Young, J. K.; Moore, J. S. *J. Org. Chem.* **1996**, *61*, 8160–8168.

[277] Huang, S.; Tour, J. M. *J. Am. Chem. Soc.* **1999**, *121*, 4908–4909.

[278] Huang, S.; Tour, J. M. *J. Org. Chem.* **1999**, *64*, 8898–8906.

[279] Chi, C.; Wu, J.; Wang, X.; Zhao, X.; Li, J.; Wang, F. *Macromolecules* **2001**, *34*, 3812–3814.

[280] Shortell, D. B.; Palmer, L. C.; Tour, J. M. *Tetrahedron* **2001**, *57*, 9055–9065.

[281] Zhang, J.; Aszodi, J.; Chartier, C.; L'hermite, N.; Weston, J. *Tetrahedron Lett.* **2001**, *42*, 6683–6686.

[282] Haag, R. *Chem. Eur. J.* **2001**, *7*, 327–335.

[283] Basso, A.; Evans, B.; Pegg, N.; Bradley, M. *Tetrahedron Lett.* **2000**, *41*, 3763–3767.

[284] Wells, N. J.; Basso, A.; Bradley, M. *Biopolymers* **1998**, *47*, 381–396.

[285] Mahajan, A.; Chhabra, S. R.; Chan, W. C. *Tetrahedron Lett.* **1999**, *40*, 4909–4912.
[286] Fromont, C.; Bradley, M. *Chem. Commun.* **2000**, 283–284.
[287] Hudson, R. H. E.; Robidoux, S.; Damha, M. J. *Tetrahedron Lett.* **1998**, *39*, 1299–1302.
[288] Basso, A.; Evans, B.; Pegg, N.; Bradley, M. *Chem. Commun.* **2001**, 697–698.
[289] Swali, V.; Wells, N. J.; Langley, G. J.; Bradley, M. *J. Org. Chem.* **1997**, *62*, 4902–4903.

17 Index

A

Acetals 107
 – hydrolysis of   222, 322, 433, 443
 – as linkers   107, 108, 117–121, 135, 433
 – oxidation of   83
 – reduction of   226, 227
Acetamidomethyl group   239
Acetic acid, as cleavage reagent   43, 44, 100,
   120, 121, 131
Acetoacetic esters, see 3-Oxo carboxylic acid
   derivatives
Acetoacetylation   328, 346
Acetoacetyl dianion   321, 322
2-Acetyldimedone
 – as linker   89, 90
 – as protective group   297
Acetylenes, see Alkynes
Acetyl ketene   346, 429
Acid-labile linkers, see Linkers
Acrylamides
 – polymers based on   27–29
 – reaction with amines   273–275
Acrylates
 – 1,4-addition of cuprates to   355
 – Baylis–Hillman reaction of   217, 218, 250,
   252
 – as dienophiles   193
 – Heck reaction of   187–190
 – as linkers for tertiary amines   92
 – polymers based on   32
 – preparation of   125
   – Experimental Procedure   348
 – reaction with amines   92, 273–275, 497
 – reaction with radicals   175, 404
Activation of carboxylic acids   214, 215,
   325–340, 346–351, 353, 354
C–H Activation of polyethylene   32
Active hydrogen compounds   130, 131
 – acylation of   318, 319, 395
 – alkylation of   135

 – arylation of   321, 322
 – condensations of   180, 181
 – conversion to diazo compounds   303, 304
Acylation
 – of C,H-acidic compounds   318, 319
 – of alcohols   79–81, 346–351, 353, 354
 – of amides   359–361
 – of amines   68–75, 325–340
   – kinetics   21
 – of arenes   317
 – of enamines   361
 – of organometallic compounds   121, 122,
   317–319
 – of thiols   356, 357
Acyl azides
 – as acylating reagents   326–329, 337
 – Curtius degradation of   285, 286, 370,
   379, 382
N-Acylcarbamates   81, 288
Acyl chlorides, see Acyl halides
Acyl cyanides   337
Acyl fluorides   326, 327, 349, 350
Acyl halides   326, 370
 – reaction of
   – with alcohols 346–351, 353, 354
   – with amides   359, 360
   – with amines   62, 326, 327
   – with arenes   317–319
   – with imines   391
   – with organometallic compounds   217,
   317–319
   – with thiols   356
O-Acyl hydroxamic acids, as linkers   72, 350
Acyl isothiocyanates   370, 371
O-Acylisoureas   330, 331
Acyloins, as linkers   52, 53, 113
Acyloxyphosphonium salts   334
N-Acylsulfonamides   50, 51, 73, 74, 251, 300
N-Acylureas   330–332
Adamantoyl chloride   483, 495
Addition to double bonds, see Alkenes